Encyclopedia of Complexity and Systems Science Series

Editor-in-Chief
Robert A. Meyers

The Encyclopedia of Complexity and Systems Science Series of topical volumes provides an authoritative source for understanding and applying the concepts of complexity theory together with the tools and measures for analyzing complex systems in all fields of science and engineering. Many phenomena at all scales in science and engineering have the characteristics of complex systems and can be fully understood only through the transdisciplinary perspectives, theories, and tools of self-organization, synergetics, dynamical systems, turbulence, catastrophes, instabilities, nonlinearity, stochastic processes, chaos, neural networks, cellular automata, adaptive systems, genetic algorithms, and so on. Examples of near-term problems and major unknowns that can be approached through complexity and systems science include: the structure, history, and future of the universe; the biological basis of consciousness; the integration of genomics, proteomics, and bioinformatics as systems biology; human longevity limits; the limits of computing; sustainability of human societies and life on earth; predictability, dynamics, and extent of earthquakes, hurricanes, tsunamis, and other natural disasters; the dynamics of turbulent flows; lasers or fluids in physics; microprocessor design; macromolecular assembly in chemistry and biophysics; brain functions in cognitive neuroscience; climate change; ecosystem management; traffic management; and business cycles. All these seemingly diverse kinds of phenomena and structure formation have a number of important features and underlying structures in common. These deep structural similarities can be exploited to transfer analytical methods and understanding from one field to another. This unique work will extend the influence of complexity and system science to a much wider audience than has been possible to date.

More information about this series at https://link.springer.com/bookseries/15581

Brian Dangerfield

Editor

System Dynamics

Theory and Applications

A Volume in the Encyclopedia of Complexity
and Systems Science, Second Edition

With 185 Figures and 21 Tables

Springer

Editor
Brian Dangerfield
School of Management
University of Bristol
Bristol, UK

ISBN 978-1-4939-8789-4   ISBN 978-1-4939-8790-0 (eBook)
ISBN 978-1-4939-8791-7 (print and electronic bundle)
https://doi.org/10.1007/978-1-4939-8790-0

This Springer imprint is published by the registered company Springer Science+Business Media,
LLC, part of Springer Nature.
The registered company address is: 233 Spring Street, New York, NY 10013, U.S.A.

This volume is dedicated to the many thousands of researchers and scholars worldwide who have embraced the feedback approach in their studies of organizational systems. May their expositions propel yet further growth in a field which has so much to offer for policy design.

A recent *Living Planet Report* observes: "Systems thinking can help us ask the right questions by examining complex problems layer by layer and then analysing the connections between these layers." The Report homes in on root causes of problems and how solving these in a complex world requires knowledge of the hierarchical relationships between events or symptoms, patterns or behaviours, systemic structures, and mental models. Further, it emphasizes the need to consider and analyze the interactions in a system in order to be better positioned to bring about positive and compelling change.

In addition, leaders of international organizations such as the UN, OECD, UNESCO, and WHO, as well as leaders of major business, public sector, and charitable and professional organizations, have all declared that "systems thinking is an essential leadership skill for managing the complexity of the economic, social and environmental issues that confront decision makers."

System Dynamics (SD) is a major systems science approach that has been adopted on a wide scale to address these issues. Initially developed by Professor Jay W Forrester at MIT in the USA in the late 1950s, SD has continued to grow and gain acceptance in all types of organizational structure the world over.

This book contains a comprehensive set of contemporary System Dynamics readings that contribute not only to methodological and theoretical developments in SD but also address a wide range of important application-specific issues documented by international experts in different social, economic, and environmental domains where the methodology has been applied.

This is an important collection of 23 "state-of-the-art" papers in System Dynamics which help define the current state and scope of the field just after its 60th anniversary. In 2009, 18 of these papers were originally published in the System Dynamics section of the *Encyclopedia of Complexity and Systems Science* (Robert A Meyers (Ed.), Springer, New York).

This revised collection, edited by Brian Dangerfield of the University of Bristol, UK, contains the original 18 chapters, the vast majority of which have been updated (some extensively), together with 5 new chapters. The complete set now includes:

- General overview and context to System Dynamics – core elements of SD, SD in the evolution of systems approaches, scenario-driven planning, engineering of strategy through a unified theory of performance

- Methodological contributions – structural dominance analysis, optimization methods, validation techniques and the engineering roots of model validation, group model building
- Social, economic, and behavioral issues – income distribution in a market economy, contributions to economics and economic modeling
- Business management applications – project management, delays and disruption in projects, operations management, diffusion of innovations, organizational learning, workforce planning, business policy and strategy
- Public policy applications – healthcare in the UK, Europe, and USA; defense strategic decision-making; environmental, energy, and climate change modeling; public policy per se

This volume is a major addition to the professional bookshelf and a vital component in the tool kit for SD modelers, policy analysts, consultants, students, and academics.

I recommend this book as essential reading for all those interested in the frontiers of systems thinking and system dynamics and the use of these tools for the betterment of our global society.

Victoria Business School   Robert Y. Cavana
Victoria University of Wellington
Wellington, New Zealand

The Encyclopedia of Complexity and System Science Series is a multivolume authoritative source for understanding and applying the basic tenets of complexity and systems theory as well as the tools and measures for analyzing complex systems in science, engineering, and many areas of social, financial, and business interactions. It is written for an audience of advanced university undergraduate and graduate students, professors, and professionals in a wide range of fields who must manage complexity on scales ranging from the atomic and molecular to the societal and global.

Complex systems are systems that comprise many interacting parts with the ability to generate a new quality of collective behavior through selforganization, e.g., the spontaneous formation of temporal, spatial, or functional structures. They are therefore adaptive as they evolve and may contain self-driving feedback loops. Thus, complex systems are much more than a sum of their parts. Complex systems are often characterized as having extreme sensitivity to initial conditions as well as emergent behavior that are not readily predictable or even completely deterministic. The conclusion is that a reductionist (bottom-up) approach is often an incomplete description of a phenomenon.

This recognition that the collective behavior of the whole system cannot be simply inferred from the understanding of the behavior of the individual components has led to many new concepts and sophisticated mathematical and modeling tools for application to many scientific, engineering, and societal issues that can be adequately described only in terms of complexity and complex systems.

Examples of Grand Scientific Challenges which can be approached through complexity and systems science include: the structure, history, and future of the universe; the biological basis of consciousness; the true complexity of the genetic makeup and molecular functioning of humans (genetics and epigenetics) and other life forms; human longevity limits; unification of the laws of physics; the dynamics and extent of climate change and the effects of climate change; extending the boundaries of and understanding the theoretical limits of computing; sustainability of life on the earth; workings of the interior of the earth; predictability, dynamics, and extent of earthquakes, tsunamis, and other natural disasters; dynamics of turbulent flows and the motion of granular materials; the structure of atoms as expressed in the Standard Model and the formulation of the Standard Model and gravity into a Unified Theory; the structure of water; control of global infectious diseases; and also evolution and quantification of (ultimately) human cooperative behavior in politics, economics, business systems, and social interactions. In fact, most of these issues have identified nonlinearities and are beginning to be addressed with nonlinear techniques, e.g., human longevity limits, the Standard Model, climate change, earthquake prediction, workings of the earth's interior, natural disaster prediction, etc.

The individual complex systems mathematical and modeling tools and scientific and engineering applications that comprised the *Encyclopedia of Complexity and Systems Science* are being completely updated and the majority will be published as individual books edited by experts in each field who are eminent university faculty members.

The topics are as follows:

Agent Based Modeling and Simulation
Applications of Physics and Mathematics to Social Science
Cellular Automata, Mathematical Basis of
Chaos and Complexity in Astrophysics
Climate Modeling, Global Warming, and Weather Prediction
Complex Networks and Graph Theory
Complexity and Nonlinearity in Autonomous Robotics
Complexity in Computational Chemistry
Complexity in Earthquakes, Tsunamis, and Volcanoes, and Forecasting and
    Early Warning of Their Hazards
Computational and Theoretical Nanoscience
Control and Dynamical Systems
Data Mining and Knowledge Discovery
Ecological Complexity
Ergodic Theory
Finance and Econometrics
Fractals and Multifractals
Game Theory
Granular Computing
Intelligent Systems
Nonlinear Ordinary Differential Equations and Dynamical Systems
Nonlinear Partial Differential Equations
Percolation
Perturbation Theory
Probability and Statistics in Complex Systems
Quantum Information Science
Social Network Analysis
Soft Computing
Solitons
Statistical and Nonlinear Physics
Synergetics
System Dynamics
Systems Biology

Each entry in each of the Series books was selected and peer reviews organized by one of our university-based book Editors with advice and consultation provided by our eminent Board Members and the Editor-in-Chief.

This level of coordination assures that the reader can have a level of confidence in the relevance and accuracy of the information far exceeding than that generally found on the World Wide Web. Accessibility is also a priority and for this reason each entry includes a glossary of important terms and a concise definition of the subject. In addition, we are pleased that the mathematical portions of our Encyclopedia have been selected by Math Reviews for indexing in MathSciNet. Also, ACM, the world's largest educational and scientific computing society, recognized our *Computational Complexity: Theory, Techniques, and Applications* book, which contains content taken exclusively from the *Encyclopedia of Complexity and Systems Science*, with an award as one of the notable Computer Science publications. Clearly, we have achieved prominence at a level beyond our expectations, but consistent with the high quality of the content!

Palm Desert, CA, USA   Robert A. Meyers
January 2020   Editor-in-Chief

The genesis of this volume was the System Dynamics section of the Encyclopedia of Complexity and Systems Science, a huge reference work which was published in its first edition in 2009. Many authors have taken the opportunity to augment and revise their earlier contributions, and along with another five new additional chapters, the whole now stands as a testimony to the status of achievements and developments in System Dynamics as we approach the third decade of the twenty-first century. It is hoped this volume will be a focus for many citations in the future scientific literature, for each chapter embraces the state of the art, along with extensive references, in respect of the various topics covered.

The book comprises a mix of methodological and applications-oriented contributions, all written by eminent specialists in that sphere of system dynamics. In particular, the 14 chapters devoted to applications of the methodology stand as evidence of the now extensive range of applications of the stock, flow, and information feedback concepts which lie at the heart of the principles of system dynamics. The book represents a unique contribution to the literature because such an all-encompassing volume has been absent from the growing list of works in the field; the totality is far more than would be found in a general textbook.

Meanwhile, the Encyclopedia of Complexity and Systems Science, in addition to the print editions of the topical volumes in the Second Edition, is available as an online reference work where authors of these chapters will be free to update their contributions once knowledge in particular specialist domains advances; the reference work has become a living entity.

Bristol, UK   Brian Dangerfield
January 2020   Volume Editor

Acknowledgments

I am indebted to the authors of the book chapters who have contributed their wisdom in various specialisms of the system dynamics landscape. I hope they feel that their efforts, often under stringent time pressure, have been adequately rewarded.

Also, I owe a debt of gratitude to the editorial staff at Springer who have been extremely patient as the various chapters have been assembled, reviewed, and proofed.

University of Bristol   Brian Dangerfield
UK

Contents

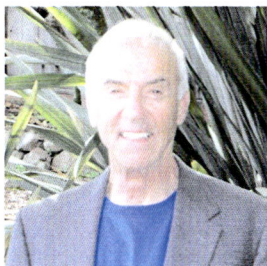

Dr. Robert A. Meyers
President: RAMTECH Limited
Manger, Chemical Process Technology, TRW Inc.
Post doctoral Fellow: California Institute of Technology
Ph.D. Chemistry, University of California at Los Angeles
B.A. Chemistry, California State University, San Diego

Biography

Dr. Meyers has worked with more than 20 Nobel laureates during his career and is the originator and serves as Editor-in-Chief of both the Springer Nature *Encyclopedia of Sustainability Science and Technology* and the related and supportive Springer Nature *Encyclopedia of Complexity and Systems Science*.

Education

Postdoctoral Fellow: California Institute of Technology
Ph.D. in Organic Chemistry, University of California at Los Angeles
B.A. Chemistry with minor in Mathematics, California State University, San Diego

Dr. Meyers holds more than 20 patents and is the author or Editor-in-Chief of 12 technical books including the *Handbook of Chemical Production Processes*, *Handbook of Synfuels Technology*, and *Handbook of Petroleum Refining Processes* now in 4th Edition, and the *Handbook of Petrochemical*

Production Processes, now in its second edition, (McGraw-Hill) and the *Handbook of Energy Technology and Economics*, published by John Wiley & Sons; *Coal Structure*, published by Academic Press; and *Coal Desulfurization* as well as the *Coal Handbook* published by Marcel Dekker. He served as Chairman of the Advisory Board for *A Guide to Nuclear Power Technology*, published by John Wiley & Sons, which won the Association of American Publishers Award as the best book in technology and engineering.

Brian Dangerfield is currently employed in the School of Management at the University of Bristol, joining in 2014. At the start of his academic career, and following a short spell in Industrial Operational Research, he was recruited by the University of Liverpool, UK, before moving on to the University of Salford where he was promoted to a Chair in Systems Modelling in 2000. Educated in the UK, Brian holds a bachelor's degree in Economics, Statistics, and Operational Research from Swansea University, a postgraduate diploma in Industrial Administration from Bradford University, and a Ph.D. (System Dynamics) from Salford University.

Brian's research interest is in the application of system dynamics (SD) simulation models to policy-level issues in economics, business, health, and public policy generally. Domains have included the epidemiology of HIV/AIDS, patient pathways, managing capacity enhancements in tertiary care, economies of scale and capacity management in the steel industry, macroeconomic modelling, and competitiveness in the UK construction industry. Numerous journal articles have resulted from this work, and he is a coeditor of the books *Discrete-Event Simulation and System Dynamics for Management Decision Making* (Wiley, 2014) and *Feedback Economics* (Springer, forthcoming 2020).

Brian was awarded the UK Operational Research Society's President's Medal in 1991 and their Goodeve Medal in 2005, both awards being for health-related papers applying the system dynamics methodology. For the 10 years to 2011 he was Executive Editor of the *System Dynamics Review*. In 2018, he was awarded a gold medal from the UK System Dynamics Society for an Outstanding Contribution to UK System Dynamics. Brian was a member of the National Council of the Operational Research Society from 1989 to 1991 and returned as a member of their General Council and

Education and Research Committee from 2012 to 2017. During 2009–2010, he was the President of both the Economics Chapter and the UK Chapter of the System Dynamics Society. In 2011, he acted as a Senior Visiting Professor at the Universiti Teknologi Malaysia.

In 2005, Brian completed a major economic modelling project funded by the State government of Sarawak in East Malaysia. From 2005 to 2008 he was Principal Investigator on a Research Council–financed project (joint with Reading and Loughborough universities) examining the systemic basis for sustained competitiveness in the UK construction industry. This project won the 2009 Chartered Institute of Building Innovation Research Paper Award.

Brian has supervised numerous Ph.D. and several M.Sc. research graduates on topics such as the effects of the introduction of a minimum wage, the future for health and social security funding given an aging population, the sustainability of mass tourism in island tourist economies, the rise and fall of superclubs in dance music culture, the supply of and demand for UK higher educated manpower, shifting the balance of care in health, the provision of a more vocationally skilled human resource base from schools in Sarawak, an early warning system for twin crises in sovereign and banking debts, childhood obesity, and modeling reverse logistics policies. He has held external examinership appointments for 17 Ph.D. vivas in the UK and abroad.

Brian's current interests include SD projects in health care, particularly childhood obesity and the management and planning for dementia at local authority level. He is also researching closed loop supply chains and how to employ social marketing campaigns to effect behavior change.

Fran Ackermann Curtin Business School, Curtin University, Perth, Western Australia, Australia

David F. Andersen Rockefeller College of Public Affairs and Policy, University at Albany, Albany, USA

Siôn Cave Decision Analysis Services Limited, Basingstoke, UK

Brian Dangerfield School of Management, University of Bristol, Bristol, UK

Jim Duggan School of Computer Science, NUI Galway, Galway, Ireland

Colin Eden Strathclyde Business School, University of Strathclyde, Glasgow, UK

Sondoss Elsawah School of Engineering and Information Technology, Capability Systems Centre, University of New South Wales, Canberra, Australian Defence Force Academy, Canberra, Australia

Andrew Ford School of the Environment, Washington State University, Pullman, WA, USA

David N. Ford Zachry Department of Civil Engineering, Texas A&M University, College Station, TX, USA

Nicholas C. Georgantzas Fordham University Business Schools, New York, USA

Andreas Größler Department of Operations Management, University of Stuttgart, Stuttgart, Germany

Stefan Groesser School of Engineering and Information Technology, Bern University of Applied Science, Bern, Switzerland

Gary B. Hirsch Creator of Learning Environments, Wayland, MA, USA

Jack Homer Homer Consulting, New York, USA

Susan Howick Strathclyde Business School, University of Strathclyde, Glasgow, UK

Christian Erik Kampmann Department of Innovation and Organizational Economics, Copenhagen Business School, Copenhagen, Denmark

James M. Lyneis System Design and Management Program, Massachusetts Institute of Technology, Cambridge, MA, USA

Kambiz Maani Massey Business School, Massey University, Albany, Auckland, New Zealand

Roderick Macdonald Initiative for System Dynamics in the Public Sector, University at Albany, Albany, USA

Frank H. Maier Geschäftsführender Gesellschafter, Simcon GmbH, Bensheim, Germany

Alan C. McLucas School of Engineering and Information Technology, Capability Systems Centre, University of New South Wales, Canberra, Australian Defence Force Academy, Canberra, Australia

Peter M. Milling Industrieseminar der Universität Mannheim, Mannheim University, Mannheim, Germany

Camilo Olaya Universidad de Los Andes, Bogotá, Colombia

Rogelio Oliva Mays Business School, Texas A&M University, College Station, TX, USA

Michael J. Radzicki Department of Social Science and Policy Studies, Worcester Polytechnic Institute, Worcester, MA, USA

Eliot Rich School of Business, University at Albany, Albany, USA

George P. Richardson Rockefeller College of Public Affairs and Policy, University at Albany, State University of New York, Albany, USA

Etiënne A. J. A. Rouwette Institute for Management Research, Radboud University, Nijmegen, The Netherlands

Khalid Saeed Worcester Polytechnic Institute, Worcester, USA

Markus Schwaninger Institute of Management and Strategy, University of St. Gallen, St. Gallen, Switzerland

Jac A. M. Vennix Institute for Management Research, Radboud University, Nijmegen, The Netherlands

Kim Warren Strategy Dynamics Limited, Princes Risborough, UK

Terence Williams University of Hull, Hull, UK

Graham Willis Robust Futures Limited, Winchester, UK

Eric Wolstenholme Symmetric Scenarios, Edinburgh, UK

Part I

Introduction

System Dynamics: Introduction

Brian Dangerfield
School of Management, University of Bristol, Bristol, UK

When the late Jay Wright Forrester published his first management-related paper in 1958, he subtitled it *"a major breakthrough for decision-makers."* At the time, some thought this rather an exaggeration if not pompous. Now that 60 years of system dynamics (SD) has elapsed, we can at least point to the achievements made and restate continuing progress in the pages of this volume. Was it a "major breakthrough"? It certainly has the potential to raise the standards in evidence-based policy making to warrant this description and some startlingly good examples of such work are set out here. After 60 years, perhaps the conclusion is that the use of SD as a methodology in academic papers has grown enormously whereas its influence in real-world policy making is less conspicuous – although the significant (and usually unpublished) efforts by the consulting industry are acknowledged.

The key to this disparity might be connected to the skills required to formulate good SD models – those which address a real-world problem with devastating simplicity and insight. It is deceptively easy to produce an SD model but there are subtleties involved in producing a really effective model for policy purposes. An uplift in modelling skills is something which a subset of the (now significant) amount of published material on SD is aimed at, and this volume will add to that corpus of work. In addition, it will illustrate the extent to which SD applications have spread from its origins in business and organizational issues to embrace health care, environmental, energy and climate issues, project management, some aspects of biological science and human physiology, government and public policy generally, economics, defense, the diffusion of innovations, workforce planning, and finally social and economic development. Other new applications are being encountered as the power of the methodology is becoming appreciated. It has long since justified the change of title from ***Industrial*** Dynamics (1958) to ***System*** Dynamics (1970 onwards).

What follows is divided into two sections: the first section covers theoretical and methodological aspects of SD, and this is supplemented by a final section detailing, over 14 chapters, a substantial range of applications of the methodology.

Richardson contributes an overview of the basics of SD modelling (see ▶ "Core of System Dynamics"). The underlying conceptual framework is that of the information feedback loop together with resource stocks and flows and an endogenous perspective on causation. The simplicity of the loop concept is apt to contribute to the apparent ease with which SD models can be created (along with the icon-based suites of SD software). But the novice reader should appreciate that it can take time to assimilate the modelling skills necessary to execute well an SD model-based application. Practice is essential and the references included will connect to further published material to assist the steep climb up the learning curve. So-called experts are still being confronted with the subtleties of SD modelling after years of involvement.

To place the SD methodology in context, the contribution by Schwaninger (see ▶ "System Dynamics in the Evolution of the Systems Approach") profiles it alongside various other systems-based approaches which have emerged in the management and social sciences. Those professing to become experts in SD need to know about the other range of approaches which co-exist in the field of systems science. All these other methodologies have their own enthusiasts, and this may even extend to the formation of societies with annual conferences. His Fig. 1 shows a diagram of the different systems approaches and their interrelationships. The text also elaborates on complementarities and potential synergies between certain approaches.

© Springer Science+Business Media, LLC, part of Springer Nature 2020
B. Dangerfield (ed.), *System Dynamics*,
https://doi.org/10.1007/978-1-4939-8790-0_538

Originally published in
R. A. Meyers (ed.), *Encyclopedia of Complexity and Systems Science*, © Springer Science+Business Media LLC 2019, https://doi.org/10.1007/978-3-642-27737-5_538-4

Running an SD model creates a time-path of output behavior covering all the variables it is deemed necessary to include in the model. The various runs of the model are, most frequently, addressed in comparative fashion rather than taken in isolation. They can therefore be described as computer-based scenarios each of which charts a possible but not assured future. Georgantzas (see ▸ "Scenario-Driven Planning with System Dynamics") describes environmental (traditional) scenario generation for which there is a considerable body of literature. But he emphasizes that successful strategy design involves the integration of three things: a knowledge of the business environment; the effects of unstated assumptions about change in the environment and strategy on performance; and finally the need to *compute* the effects on organizational performance. These three facets are accomplished by the process of SD modelling.

Warren's contribution considers how system dynamics provides a rigorous foundation for a key piece of management theory – how an enterprise actually functions and delivers performance (see ▸ "Engineering of Strategy: A General, Unified Theory of Performance and Strategic Management"). Existing theory in strategic management is fragmented and unreliable, and as a result is largely unused by professionals unlike, say, Finance theory. SD-based theory offers an entirely general and unified foundation for understanding and managing the development and performance of any organization.

The practice of SD when applied to real-world applications essentially involves managerial learning and will often involve an interaction with client teams rather than one individual. How best to organize such structured approaches to participative model building is described by Rouwette and Vennix (see ▸ "Group Model Building"). Client participation is a prerequisite for successful SD modelling in a real-world project.

The foundations of the SD methodology are characterized by Jay Forrester's engineering background. Therefore, *ipso facto*, the origins of SD can be traced back to engineering. Olaya's contribution (see ▸ "System Dynamics: Engineering Roots of Model Validation") highlights core engineering aspects of SD that inform model validation. Models can serve a wide range of purposes. A model's effectiveness for accomplishing a purpose, regardless of any type of justification (be it scientific, based on evidence or any other kind) is the crux of the matter and the main aspect to bear in mind.

In the theory and methodology section, this introductory roadmap has included, thus far, all the background for contextualizing SD and creating an SD model. This phase is concluded by three detailed expositions associated with *ex post* modelling activities. The following are covered: model validation, model optimization, and analytical methods to explain behavior and determine dominant loops.

Schwaninger and Groesser (see ▸ "System Dynamics Modeling: Validation for Quality Assurance") range over the various aspects of model validation, beginning with its epistemological foundations. In real-world modelling studies, testing and validation is a *sine qua non* of the process. SD is arguably unique in the management science discipline because of the range of tests that are available and the attention given to the task of model validation. Few other methodologies get near to the variety of tests which can be applied to an SD model. The authors consider the range of tests under three headings: model-related context, model structure, and model behavior.

Dangerfield and Duggan describe the methods for improving model performance (see ▸ "Optimization of System Dynamics Models"). The task can be categorized under two headings: calibration and policy optimization. The former relates to the determination of optimal parameter sets which deliver the best fit of the model to past time series data. Policy optimization on the other hand seeks to establish policies which deliver the "best" performance against a suitable metric, such as minimum cost or maximum revenue. Using such an approach can accelerate the learning which comes from repeated runs of the model. Sadly, in the existing SD literature, there are few accounts of its use in real-world studies; the number is growing, but only slowly. Against this, however, in both aspects of model optimization, various recent methodological advances have been put forward (from 2008), and these are outlined in the latter two sections of the chapter.

Kampmann and Oliva deal with the behavioral analysis issue (see ▶ "Analytical Methods for Structural Dominance Analysis in System Dynamics"). This activity tries to shed light on the model's dynamic behavior: why does it behave as it does? What loop structures are responsible for the dominant behavior – and indeed shifts in that behavior where it occurs? In other words, they explore the link between system structure and dynamic behavior. Early methods used eigenvalue analysis but, since then, more sophisticated approaches have been put forward. A major advance will occur when one or more of these is refined enough to be included in an SD software package; this is likely to take some time.

The methodology of SD exists for no other reason than to offer a quantum leap in the standards of policy analysis. Therefore, any portrayal of the field must include an exploration of the landscape which defines areas of application. There are 11 such areas covered in this section (some twice) and the choice has been made in the knowledge that there are others which also may have been included, together with some new application domains which are only just being exposed to the tools of SD modelling.

Business strategy was the starting point for SD applications and rightly takes pride of place. This is the domain which boasts the most numerous SD applications. Lyneis (see ▶ "Business Policy and Strategy, System Dynamics Applications to") concentrates on the process of how SD models are used in the task of strategy formulation. He goes on to consider the various drivers of business dynamics such as oscillations in supply chains and boom and bust life cycles. Detailed references are provided for a wide range of business application case studies.

Health care is consuming a higher share of GDP in many Western industrialized countries. This is due to the age profile of the population and advances in pharmacological and medical technologies. It is unsurprising that SD methods have been applied in tackling some of the most high-profile issues in health care and the relatively recent literature is testimony to the success of SD-based analysis. Indeed, it is arguable that some of the best modelling applications have surfaced in this sector. To do justice to the field of health care, two contributions were solicited, in part because of the different funding systems which exist on either side of the Atlantic: Wolstenholme surveys the work done by the UK and European authors (see ▶ "System Dynamics Applications to Health and Social Care in the United Kingdom and Europe"), while Hirsch and Homer concentrate on work published by the US authors (see ▶ "System Dynamics Applications to Health Care in the United States").

Wolstenholme describes work carried out in the UK and Continental Europe but gives particular emphasis to three areas where models have been deployed. He starts with the problem of delayed hospital discharge which generates hospital capacity problems. Epidemiology is also reviewed, in particular research on the epidemiology of HIV/AIDS. Finally, recent work on mental health reform in the UK is described.

Hirsch and Homer note that the system in the USA is comparatively difficult to manage because of its free-market approach and relative lack of regulation. They concentrate on three main areas: disease epidemiology including heart disease and diabetes; substance abuse; and health care capacity and delivery.

The application of SD to public policy generally is dealt with by Andersen, Rich, and MacDonald (see ▶ "System Dynamics Applications to Public Policy"). They emphasize how public policy issues are complex, cross organizational boundaries, involve stakeholders with widely different perspectives and evolve over time, such that longer term results may be wholly different from short-term outcomes. Detail is provided for one public policy case involving the Governor's Office of Regulatory Assistance in New York State. They conclude with coverage of studies in a range of public domains such as defense, health care, education, and the environment.

Operations management is concerned with inter- and intra-organizational value creation. This entails two elements: operations management in the narrow sense is about the efficiency of processes used to manufacture products or to provide services (doing things right); operations strategy deals with questions of effectiveness of these value creation processes (doing the right things). Größler (see ▶ "System Dynamics and Operations Management") describes how SD relates to operations management and reports on the various applications of the method in that domain.

One of the earliest (and still active) SD application domains is project management. First brought to prominence in 1980 via Cooper's paper on naval ship production, he then introduced the concept of the rework cycle in 1993. Building on earlier work, Ford and Lyneis (see ▶ "System Dynamics Applied to Project Management: A Survey, Assessment, and Directions for Future Research") review the various structures which underpin SD project modelling, together with the type of behavior modes encountered in project models. The authors conclude that there appears to be scope for enhancing management education in respect of project management. Few would contest that this is one area where SD has proven to be a great success; it has enhanced the traditional tools of project management significantly.

A particular area of SD application has brought the methodology into the legal arena. Disruption and delay in the execution of complex projects invariably finds two parties in dispute. Such disputes often center upon time delays and use of resources on projects – and what might have happened if things had been managed differently. SD models have been employed by parties to such disputes to attempt to justify the occurrence of these events. Howick, Ackermann, Eden, and Williams (see ▶ "Delay and Disruption in Complex Projects") report on how cognitive mapping, cause mapping, and SD can be fused into what they describe as a cascade model building process. The result is a rigorous process for explaining why a project behaved in a certain way.

Challenges facing governments, military commanders, and those managing defense resources are diverse and complex in their detail and dynamics. Taking one course of action precludes others.

Having once committed to a strategy or course of action and then seeing the consequences of that strategy being enacted, stopping, and changing direction may be difficult or practically impossible. The defense management challenges are exacerbated by limited opportunities ahead of making real decisions and enacting them to gain experience without unacceptable loss and unintended consequence. McLucas and El Sawah (see ▶ "System Dynamics Modeling to Inform Defense Strategic Decision-Making") describe qualitative and quantitative system dynamics research into defense problems and how research findings have been translated into advice and guidance to those responsible for formulating defense strategy and policy, defense capability, and the acquisition and management of capability assets.

Along with health care, the depletion of environmental resources and its effects has consumed many thousands of column inches in printed news media. SD has been employed in the search for stronger analyses in this sector and the efforts go back to the highly publicized *Limits to Growth* study in 1970–1972. Ford charts the most notable efforts which have emerged (see ▶ "System Dynamics Models of Environment, Energy, and Climate Change"). He ranges over environmental resource problems in the western USA together with models for greater understanding of climate change, the carbon cycle, and global warming. Mention is made of another "famous" SD model: C-ROADS. This award-winning model computes the changing average global temperature to 2100 under a range of scenarios. It is commonly employed as a basis for role-playing exercises and has been displayed in some prominent venues, such as The White House and at the various International Climate Change conferences. Ford concludes with studies in energy, specifically two applications to the electric power industry.

The field of economics is one where SD has received a mostly hostile reception. The statistical economic modelling tool of econometrics has an extensive history and as a preferred modelling methodology seems hard to dislodge. However, there are an increasing number of heterodox economists who are prepared to embrace SD concepts and Radzicki (see ▶ "System Dynamics and Its Contribution to Economics and Economic Modeling") describes the advances taking place. While some of the literature embodies the translation of existing economic models into an SD format (which is a laudable objective), he calls for more economic dynamics models to be built from scratch embodying the best practice in SD modelling. Economic policy is too important to be informed by a single, seemingly unassailable, modelling methodology, and it is to be hoped that in the future SD will become even more accepted as a viable tool for use in this field.

If the promotion of learning and understanding is the primary *raison d'etre* of SD, then achievement of this goal in an individual can be a significant accomplishment, especially if that person is the most senior in the client team. But there is a further goal to be pursued should the study fully reap the benefits of the SD methodology: how can we foster *organizational* learning? Maani tackles this head on (see ▶ "System Dynamics and Organizational Learning"). He defines the core capabilities of a learning organization and goes on to list the developing literature on organizational learning and, most importantly, how SD can aid the process through learning laboratories and microworlds.

Strategic workforce planning is concerned with aligning the workforce of an organization with its strategic goals and priorities. It has two core elements: determining the demand that will be placed on the workforce at some time in the future in terms of required effort; and determining how this demand can best be met through developing workforce plans. Cave and Willis (see ▶ "System Dynamics and Workforce Planning") describe how SD is uniquely placed to support the various elements of the strategic workforce planning process, where it has been successfully applied, and potential future developments.

New products and processes are emerging at an ever-increasing rate in modern times. We need to understand the myriad mechanisms which form the basis for their rate of adoption. Milling and Maier range over various SD models which have been created to understand and improve the management of the diffusion of innovations (see ▶ "System Dynamics Analysis of the Diffusion of Innovations"). From the often-cited Bass diffusion model (1969), the authors develop a series of additional features in a modular fashion. These features include competition, network externalities, dynamic pricing, and research and development. They conclude by stressing how it is not possible to offer general recommendations for strategies in dynamic and complex environments; such recommendations can only be given in the context of the specific case under scrutiny.

The final chapter is contributed by Saeed (see ▶ "Dynamics of Income Distribution in a Market Economy: Possibilities for Poverty Alleviation"). He takes an economic modelling perspective and describes an SD model which explains resource allocation, production, and entitlements in a market economy. Its purpose is to understand better how poverty might be reduced in the context of the redistribution of income. A comprehensive listing of the model is provided in an appendix.

There is an extensive coverage of SD in this volume. Its perspective is to portray the current standing of the field after 60+ years, rather than claim to be another introductory text on how to create SD models. Readers are encouraged to dip in and out of the many chapters (23 in all). It is hoped you enlighten your knowledge of the field as you do so.

Core of System Dynamics

George P. Richardson
Rockefeller College of Public Affairs and Policy,
University at Albany, State University of
New York, Albany, USA

Article Outline

Glossary

Endogenous Generated from within. Contrasting with "exogenous," meaning generated by forces external to a system or point of view.

Feedback Loop A closed path of causal influences and information, forming a circular-causal loop of information and action.

System Dynamics System dynamics is a computer-aided approach to theory building, policy analysis, and strategic decision support emerging from an endogenous point of view.

Definition of the Subject

System dynamics is a computer-aided approach to theory building, policy analysis, and strategic decision support emerging from an endogenous point of view (Nicholis and Prigogine 1977; Richardson 1991a). It applies to dynamic problems arising in complex social, managerial, economic, or ecological systems – literally any dynamic systems characterized by interdependence, mutual interaction, information feedback, and circular causality.

Introduction

The field of system dynamics developed initially from the work of Jay W. Forrester. His seminal book *Industrial Dynamics* (Forrester 1961) is still a significant statement of philosophy and methodology in the field. Within ten years of its publication, the span of applications grew from corporate and industrial problems to include the management of research and development, urban stagnation and decay, commodity cycles, and the dynamics of growth in a finite world. It is now applied in economics, public policy, environmental studies, defense, theory building in social science, and other areas, as well as in its home-field, management. The name industrial dynamics no longer does justice to the breadth of the field (for extensive examples, see Richardson (1996b), Sterman (2001)), so it has become generalized to system dynamics. The modern name suggests links to other system methodologies, but the links are weak and misleading. System dynamics emerges out of servomechanisms engineering, not general systems theory or cybernetics (Richardson 1991b).

The system dynamics approach is founded on the scientific method. The goal of a system dynamics project is sometimes to build theoretical understanding, sometimes to implement policies for improvement, and often both. To do so, system dynamics modelers seek to: include a broad model boundary that captures important feedbacks relevant to the problem to be addressed; represent important structures in the system including accumulations and state variables,

© Springer Science+Business Media, LLC, part of Springer Nature 2020
B. Dangerfield (ed.), *System Dynamics*,
https://doi.org/10.1007/978-1-4939-8790-0_536

Originally published in
R. A. Meyers (ed.), *Encyclopedia of Complexity and Systems Science*, © Springer Science+Business
Media LLC 2019, https://doi.org/10.1007/978-3-642-27737-5_536-4

delays, and nonlinearities; use behavioral decision rules for the actors and agents in the system that are grounded in first-hand study; and use the widest range of empirical data to formulate the model, estimate parameters, and build confidence in the conclusions.

Although the points below are presented as a list, modeling (and any scientific activity) is iterative – a continual process of formulating hypotheses, testing against data of all types, and revision of both formal and mental models.

The approach can be summarized as:

- **Beginning with a *problem*** to focus systems thinking and modeling, involving the stakeholders whose understanding and action is required to implement the change.
- **Defining problems *dynamically***, in terms of graphs over time (time series), employing actual data wherever possible.
- **Striving for an *endogenous, behavioral view* of the significant dynamics of a system**, a focus inward on the structures and decision rules in a system that themselves generate or exacerbate the perceived problem.
- **Thinking of all concepts in the real system as quantities interconnected in loops** of information feedback and circular causality, a consequence of the endogenous point of view.
- **Identifying the key variables** essential to address the problem and deciding on an appropriate level of aggregation for them. System dynamics models range from highly disaggregate representations such as individual items or agents to highly aggregated representations, and can be deterministic or stochastic, as needed to address the purpose of the study.
- **Formulating a richly explanatory behavioral model** capable of reproducing, by itself, the dynamic problem of concern, drawing on all relevant evidence, including qualitative and quantitative data. The model is usually a computer simulation model, but is occasionally left unquantified as a map capturing the important accumulations (stocks) in the system, the flows that alter them, and the causal feedback structure determining the flows.

- **Testing the structure and behavior of the model** against all relevant evidence to deepen understanding and to build confidence in it, including the model's ability to replicate historical data, ensuring the model is robust under extreme conditions, exploring the sensitivity of results to uncertainty in assumptions, and diagnosing the sources of unexpected model behavior.
- **Designing and testing policies to address the problem of concern**, testing these against data and comparing to real-world policies that have been tried in the system or similar settings.
- **Documenting the model and its supporting** sources so that it is as transparent as possible and enabling others to critique, use, and extend the work.
- **Working with stakeholders** and others to help translate model-based insights into implementable policies, assist in implementation, assess the results, and improve both the model and policies.

Experiments conducted in the virtual world of the model inform the design and implementation of experiments in the real world. That experience then leads to changes and improvements in the virtual world, in participants' mental models, and in actions taken in the real world, in an iterative process of continuous improvement.

Mathematically, the basic structure of a formal system dynamics computer simulation model is a system of coupled, nonlinear, first-order differential (or integral) equations:

$$\frac{d}{dt}\mathbf{X}(t) = \mathbf{f}(\mathbf{x},\mathbf{p}),$$

where $\mathbf{x}$ is a vector of levels (stocks or state variables), $\mathbf{p}$ is a set of parameters, and $\mathbf{f}$ is a nonlinear vector-valued function. Such a system has been variously called a *state-determined system* in the engineering literature, an *absolute system* (Ashby 1956), an *equifinal system* (von Bertalanffy 1968), and a *dynamical system* (Nicholis and Prigogine 1977).

Simulation of such systems is easily accomplished by partitioning simulated time into discrete intervals of length dt and stepping the system through time one dt at a time. Each state variable is computed from its previous value and its net rate of change $x'(t)$: $x(t) = x(t - dt) + dt \cdot x'(t - dt)$. In the earliest simulation language in the field (DYNAMO), this equation was written with time scripts K (the current moment), J (the previous moment), and JK (the interval between time J and K): $X_K = X_J + DT \cdot XRATE_{JK}$ (see, e.g., Richardson and Pugh 1981). The computation interval dt is selected small enough to have no discernible effect on the patterns of dynamic behavior exhibited by the model. In more recent simulation environments, more sophisticated integration schemes are available (although the equation written by the user may look like this simple Euler integration scheme), and time scripts may not be in evidence. Important current simulation environments include STELLA and iThink (isee Systems, http://www.iseesystems.com/), Vensim (Ventana Systems, http://www.vensim.com/), and Powersim (http://www.powersim.com/).

Forrester's original work stressed a continuous approach, but increasingly modern applications of system dynamics contain a mix of discrete difference equations and continuous differential or integral equations. Some practitioners associated with the field of system dynamics work on the mathematics of such structures, including the theory and mechanics of computer simulation, analysis and simplification of dynamic systems, policy optimization, dynamical systems theory, and complex nonlinear dynamics and deterministic chaos.

The main applied work in the field, however, focuses on understanding the dynamics of complex systems for the purpose of policy analysis and design. The conceptual tools and concepts of the field – including feedback thinking, stocks and flows, the concept of feedback loop dominance, and an endogenous point of view – are as important to the field as its simulation methods.

Feedback Thinking

Conceptually, the feedback concept is at the heart of the system dynamics approach. Diagrams of loops of information feedback and circular causality are tools for conceptualizing the structure of a complex system and for communicating model-based insights. Intuitively, a feedback loop exists when information resulting from some action travels through a system and eventually returns in some form to its point of origin, potentially influencing future action. If the tendency in the loop is to reinforce the initial action, the loop is called a *positive* or *reinforcing* feedback loop; if the tendency is to oppose the initial action, the loop is called a *negative*, *counteracting*, or *balancing* feedback loop. The sign of the loop is called its *polarity*. Balancing loops can be variously characterized as goal-seeking, equilibrating, or stabilizing processes. They can sometimes generate oscillations, as when a pendulum seeking its equilibrium goal gathers momentum and overshoots it. Reinforcing loops are sources of growth or accelerating collapse; they are disequilibrating and destabilizing. Combined, balancing and reinforcing circular-causal feedback loops can generate all manner of dynamic patterns.

Feedback loops are ubiquitous in human and natural systems and, under various names and representations, have been widely recognized in popular and scholarly literature. Feedback thought has been present implicitly or explicitly for hundreds of years in the social sciences and literally thousands of years in recorded history (Richardson 1991b). We have the vicious circle originating in classical logic and morphing into common usage, the bandwagon effect, the invisible hand of Adam Smith, Malthus's correct observation of population growth as a self-reinforcing process, Keynes's consumption multiplier, the investment accelerator of Hicks and Samuelson, compound interest or inflation, the biological concepts of proprioception and homeostasis, Festinger's cognitive dissonance, Myrdal's principle of cumulative causation, Venn's idea of a suicidal prophecy, Merton's related notion of a self-fulfilling prophecy, and so on. Each of these ideas can be concisely and insightfully represented as one or more loops of causal influences with positive or negative polarities. Great social scientists are feedback thinkers; great social theories are feedback thoughts. (For a full exposition of the intellectual history of the feed-back concept, see Richardson 1991b.)

Loop Dominance and Nonlinearity

The loop concept underlying feedback and circular causality by itself is not enough, however. The explanatory power and insightfulness of feedback understandings also rest on the notions of active structure and loop dominance. Complex systems change over time. A crucial requirement for a pow-erful view of a dynamic system is the ability of a mental or formal model to change the strengths of influences as conditions change, that is to say, the ability to shift *active* or *dominant structure*.

In a system of equations, this ability to shift loop dominance comes about endogenously from non-linearities in the system. For example, the S-shaped dynamic behavior of the classic logistic growth model ($dP/dt = aP - bP^2$) or similar structures like the Gompertz curve ($dP/dt = aP - bP\ln(P)$) can be seen as the consequence of a shift in loop dominance from a positive, self-reinforcing feed-back loop (aP) producing exponential-like growth to a negative feedback loop ($- bP^2$ or $- bP\ln(P)$) that brings the system to its eventual goal. The shift in loop dominance in these models comes about from the nonlinearity in the second term, which grows faster than the first term and eventually over-takes it. Only nonlinear models can endogenously alter their active or dominant structure and shift loop dominance.

Real systems are perceived to change their active or dominant structure over time, often because of the buildup of internal forces. Thus, from a feedback perspective, the ability of non-linearities to generate shifts in loop dominance is the fundamental reason for advocating nonlinear models of social system behavior.

Figures 1 and 2, abstracted from an early, clas-sic paper (Forrester 1968), illustrate these ideas. In Fig. 1, salesmen (in the blue reinforcing loop) book orders for the company; if enough revenue is generated, there is enough budget to hire more salesmen, and corporate growth ensues. Whether

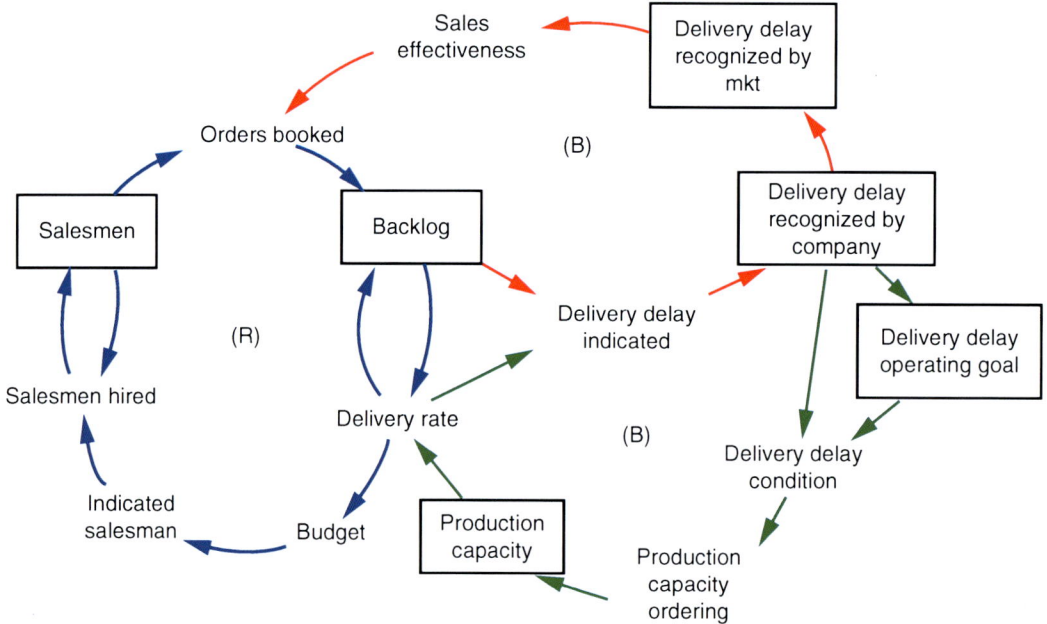

Core of System Dynamics, Fig. 1 Core structure of Forrester's market growth model (Forrester 1968), show-ing a *blue reinforcing loop* underlying the growth (or reinforcing decline) of Salesmen, Orders, and Revenue, a *red balancing loop* containing various delayed recogni-tions of the company's delivery delay, and a *green balancing loop* responsible for capacity ordering if the delivery delay drops too far below its operating goal

salesmen (in this simplified picture) book enough orders depends on the company's delivery delay for the product, as perceived by the market (red balancing loop). The company builds production capacity according to its perceived need, as indicated by its perceived delivery delay and its target for that (green balancing loop).

Figure 2 shows the dynamics this feedback structure endogenously generates. In the early phase, salesmen grow as orders and revenues grow; the system's exponential growth behavior in that phase is generated by the reinforcing salesmen loop. But then the feedback loop dominance soon shifts to the balancing delivery delay loop, which constrains sales effectiveness and brings a halt to growth. The system moves into an oscillatory phase generated by the various monitoring and perception delays around the now dominant red balancing loop. Salesmen eventual peak and decline, as the green production capacity ordering loop fails to keep production capacity sufficient to hold the delivery delays in check.

Thus, the dynamic behavior of this system is a consequence of its feedback structure and the nonlinearities that shift loop dominance endogenously over time. The particular decline scenario shown in Fig. 2 illustrates one of the deep insights of the model: the adaptive goal structure, in which the delivery delay operating goal moves slowly to accommodate changes in the company's delivery delay, weakens the green balancing loop trying to bring on capacity. The company never perceives its delivery delay is sufficiently higher than its (sliding) target, so it fails to order sufficient capacity to sustain growth. A fixed goal for the acceptable delivery delay sends a stronger signal, which can turn this corporate decline into oscillating growth (Forrester 1968).

Thus, nonlinearity is crucial to the system dynamics approach. However, it is crucial not merely because of its mathematical properties but because it enables the formalization of a profoundly powerful perspective on theory and policy – the *endogenous point of view*.

The Endogenous Point of View

The concept of endogenous change is fundamental to the system dynamics approach. It has both

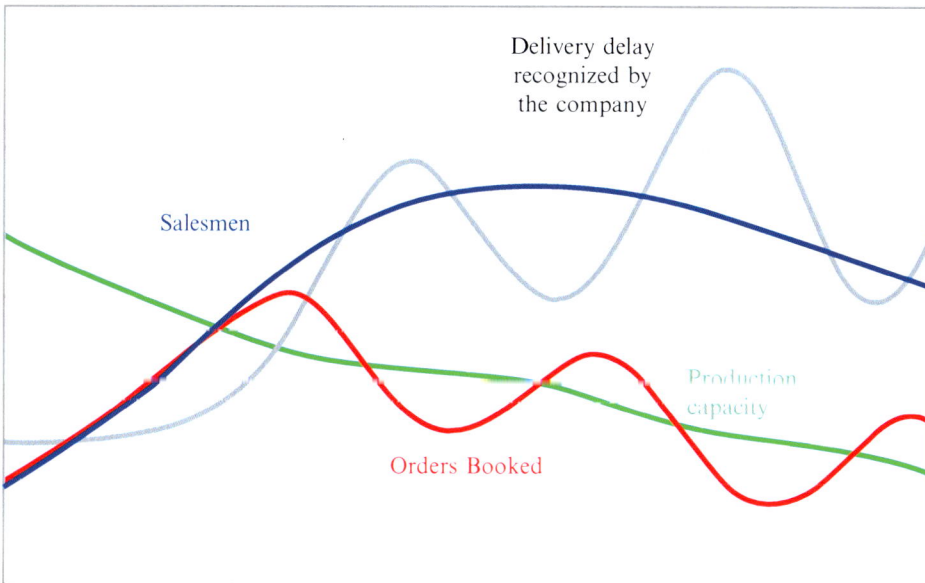

Core of System Dynamics, Fig. 2 The dynamic behavior of the model shown in Fig. 1, illustrating an early growth phase, which turns into an oscillatory phase as the feedback loop dominance shifts to the *red balancing* *delivery delay loop* and results in a long-term corporate decline as the *green capacity ordering loop* responds to a sliding operating goal for the acceptable delivery delay

philosophical and engineering origins. A deep and lasting insight of the earliest attempts at servo-mechanisms control is the realization that *the attempt to control a system generates dynamics of its own*, complicating the dynamics trying to be controlled. A governor imposed to control the speed of a steam engine can generate oscillatory "hunting behavior," as the control system over-shoots and undershoots the set point. As it becomes part of the system, the governing mechanism thus generates dynamics of its own.

The insight transfers readily, but with added significance, from engineering systems to people systems: attempts to control complex human systems – coercing, guiding, managing, and governing – generate dynamics of their own. Moreover, some of these endogenously gener-ated dynamics are created by the control mecha-nisms themselves (like the governor of a steam engine), and some are created by human creative responses to the management efforts (e.g., principal-agent interactions). These natural and human forces, creating counteracting and com-pensating pressures in response to system control efforts, emerge as complicated circular-causal feedback structures. The often complex, difficult-to-understand dynamics of such man-agement systems are to a great degree a conse-quence of their internal structures.

To capture and analyze such management complexities, one must look inward to see the ways a complex system naturally responds to system pressures. The endogenous point of view is thus central to the system dynamics approach. It dictates aspects of model formulation: exogenous disturbances are seen at most as *triggers* of system behavior (like displacing a pendulum); the *causes* are contained within the structure of the system itself (like the interaction of a pendulum's position and momentum that produces oscillations). Cor-rective responses are also not modeled as func-tions of time, but are dependent on conditions within the system. Time by itself is not seen as a cause in the endogenous point of view.

Theory building and policy analysis are signifi-cantly affected by this endogenous perspective. Tak-ing an endogenous view exposes the natural *compensating* tendencies in social systems that conspire to defeat many policy initiatives. Feedback and circular causality are delayed, devious, and deceptive. For understanding, system dynamics practitioners strive for an *endogenous point of view*. The effort is to uncover the sources of system behavior that exist within the structure of the system itself.

System Structure

These ideas are captured almost explicitly in Forrester's (1969) organizing framework for sys-tem structure:

Closed boundary
    Feedback loops
        Levels
        Rates
            Goal
            Observed condition
            Discrepancy
            Desired action

The *closed boundary* signals the endogenous point of view. The word *closed* here does not refer to open and closed systems in the general system sense, but rather refers to the effort to view a system as *causally* closed. The modeler's goal is to assemble a formal structure that can, *by itself*, without exogenous explanations, repro-duce the essential characteristics of a dynamic problem.

The causally closed system boundary at the head of this organizing framework identifies the endogenous point of view as the feedback view pressed to an extreme. Feedback thinking can be seen as a *consequence* of the effort to capture dynamics within a closed causal boundary. Without causal loops, all variables must trace the sources of their variation ultimately outside a system. Assuming instead that the causes of all significant behavior in the system are contained within some closed causal boundary forces causal influences to feed back upon themselves, forming causal loops. Feedback loops enable the endogenous point of view and give it structure.

Levels and Rates

Stocks (accumulations, or "levels" in early system dynamics literature) and the flows ("rates") that affect them are essential components of system structure. A map of causal influences and feedback loops is not enough to determine the dynamic behavior of a system. A constant inflow yields a linearly rising stock; a linearly rising inflow yields a stock rising along a parabolic path; a stock with inflow proportional to itself grows exponentially; two stocks in a balancing loop have a tendency to generate oscillations and so on. For example, the boxes in Fig. 1 represent accumulations in the company and its market; the three stocks in the red balancing loop (the order backlog and the two perceptions of the company's delivery delay) give that loop its tendency to generate oscillations which propagate throughout the system. Accumulations are the memory of a dynamic system and contribute to its disequilibrium and dynamic behavior.

Forrester (Forrester 1961) placed the operating policies of a system among its rates, the inflows and outflows governing change in the system. Many of these rates of change assume the classic structure of a negative feedback loop striving to take action to reduce the discrepancy between the observed condition of the system and a goal. The simplest such rate structure results in an equation of the form

$$rate = \frac{goal - level}{adjustment\ time},$$

where ADJUSTMENT TIME is the time over which the level adjusts to reach the goal. This simple formulation reflects Forrester's more general statement about rates in his hierarchy of system structure (above) which can be richly thought of as

$$
\begin{aligned}
rate \ &= \ f(desired\ action)\ desired\ action \\
&= \ g\big(desired\ condition, \\
& \qquad observed\ condition\big) observed\ condition \\
&= \ h(levels),
\end{aligned}
$$

for some functions f, g, and h representing particular system characteristics.

Operating policies in a management system can influence the *flows* of information, material, and resources, which are the only means of changing the accumulations in the system. While flows can be changed quickly, as a matter of relatively quick decision making, stocks change slowly – they rise when inflows are greater than outflows and decline when inflows are less than outflows.

The simple "tub dynamics" of stocks are clear even to children, yet can be befuddling in complex systems. The accumulation of greenhouse gases in the atmosphere, for example, affects the *flow* of heat energy radiated from the earth. To turn around global warming, the *accumulation* of greenhouse gases must drop far enough to raise radiant energy above the inflow of solar energy, a simple stock-and-flow insight. But to cause the accumulation of greenhouse gases to drop, their generation must fall below their natural absorption rate (another simple stock-and-flow observation). So turning around global warming is a process involving a chain of at least two significant accumulations, and people have trouble thinking it through reliably. The accumulations can only be changed by managing their associated flows. They will change only slowly even if we manage the technical and political pitfalls involved in lowering greenhouse gas production (see Sterman and Sweeney 2002).

The significance of stocks in complex systems is vivid in a resource-based view of strategy and policy. Resources that enable a corporation or government to function or flourish are stocks usually accumulated over long periods of time with significant investment of time, energy, and money. Reputations are also stocks, built over similarly long periods of time. While inadequate by themselves to give a full picture of the dynamics of a complex system, stocks and flows are vital components of system structure, without which fundamental understandings of dynamics are impossible (Warren 2002).

Behavior Is a Consequence of System Structure

The importance of stocks and flows appears most clearly when one takes a *continuous* view of structure and dynamics. Although a discrete view, focusing on separate events and decisions is entirely compatible with an endogenous feedback perspective; the system dynamics approach emphasizes a continuous view (Forrester 1961). The continuous view strives to look beyond events to see the dynamic patterns underlying them: model not the appearance of a discrete new housing unit in a city, but focus instead on the rise and fall of aggregate numbers of housing units. Moreover, the continuous view focuses not on discrete decisions but on the *policy structure* underlying decisions: not why this particular apartment building was constructed but what persistent pressures exist in the urban system that produce decisions that change housing availability in the city. Events and decisions are seen as surface phenomena that ride on an underlying tide of system structure and behavior. It is that underlying tide of policy structure and continuous behavior that is the system dynamicist's focus.

There is thus a *distancing* inherent in the system dynamics approach – not so close as to be confused by discrete decisions and myriad operational details, but not so far away as to miss the critical elements of policy structure and behavior. Events are deliberately blurred into dynamic behavior. Decisions are deliberately blurred into perceived policy structures. Insights into the connections between system structure and dynamic behavior, which are the goals of the system dynamics approach, come from this particular distance of perspective.

Suggestions for Further Reading on the Core of System Dynamics

The *System Dynamics Review*, the journal of the System Dynamics Society, published by Wiley, is the best source of current activity in the field, including methodological advances and applications.

The core of a vibrant field is difficult to discern in the flow of current work. However, the works that the field itself singles out as exemplary can give some reliable hints about what is considered vital to the core. In this sense, two edited volumes are noteworthy: an early, interesting collection of

applications is Roberts (1978); Richardson (1996b) is a two-volume edited collection in the same spirit, containing prize-winning work in philosophical background, dynamic decision making, applications in the private and public sectors, and techniques for modeling with management.

In addition, the following works, selected among the winners of the System Dynamics Society's *Jay Wright Forrester Award* (see www. systemdynamics.org/awards), can be considered insightful although implicit exemplars of the core of system dynamics. (Publications are listed beginning with the most recent; see the bibliography for full citations):

- J. Bradley Morrison, Jenny W. Rudolph, and John S. Carroll, "The Dynamics of Action-Oriented Problem Solving: Linking Interpretation and Choice"
- Mark Paich, Corey Peck, and Jason Valant, *Pharmaceutical Product Branding Strategies: Simulating Patient Flow and Portfolio Dynamics*
- Kimberly Thompson and Radboud J. Duintjer Tebbens, "Eradication *Versus* Control for Poliomyelitis: An Economic Analysis"
- David Lane and Elke Husemann, "Steering away from Scylla, Falling into Charybdis: The Importance of Recognising, Simulating and Challenging Reinforcing Loops in Social Systems"
- Thomas S. Fiddaman, "Exploring Policy Options with a Behavioral Climate-Economy Model"
- Kim D. Warren, *Competitive Strategy Dynamics*
- Eric F. Wolstenholme, "Towards the Definition and Use of a Core Set of Archetypal Structures in System Dynamics"
- Nelson P. Repenning, "Understanding Fire Fighting in New Product Development"
- John D. Sterman, *Business Dynamics: Systems Thinking and Modeling for a Complex World*
- Peter Milling, "Modeling Innovation Processes for Decision Support and Management Simulation"

- Erling Moxnes, "Not Only the Tragedy of the Commons: Misperceptions of Bioeconomics"
- Jac A. M. Vennix, *Group Model Building: Facilitating Team Learning Using System Dynamics*
- Jack B. Homer, "A System Dynamics Model of National Cocaine Prevalence"
- Andrew Ford, "Estimating the Impact of Efficiency Standards on Uncertainty of the Northwest Electric System"
- Khalid Saeed, *Towards Sustainable Development: Essays on System Analysis of National Policy*
- Tarek Abdel-Hamid and Stuart Madnick, *Software Project Dynamics: An Integrated Approach*
- George P. Richardson, *Feedback Thought in Social Science and Systems Theory*
- Peter M. Senge, *The Fifth Discipline*
- John D. W. Morecroft, "Rationality in the Analysis of Behavioral Simulation Models"
- John D. Sterman, "Modeling Managerial Behavior: Misperceptions of Feedback in a Dynamic Decision Making Experiment"

For texts on the system dynamics approach, see Alfeld and Graham (1976), Richardson and Pugh (1981), Wolstenholme (1990), Ford (1999), Maani and Cavana (2000), and the most comprehensive text to date, Sterman (2001).

Bibliography

Abdel-Hamid T, Madnick S (1991) Software project dynamics: an integrated approach. Prentice Hall, Englewood Cliffs

Alfeld LE, Graham AK (1976) Introduction to urban dynamics. Pegasus Communications, Waltham

Ashby H (1956) An introduction to cybernetics. Chapman & Hall, London

Fiddaman TS (2002) Exploring policy options with a behavioral climate-economy model. Syst Dyn Rev 18(2):243–267

Ford A (1990) Estimating the impact of efficiency standards on uncertainty of the northwest electric system. Oper Res 38(4):580–597

Ford A (1999) Modeling the environment: an introduction to system dynamics of environmental systems. Island Press, Washington, DC

Forrester JW (1961) Industrial dynamics. MIT Press, Cambridge (Reprinted by Pegasus Communications, Waltham)

Forrester JW (1968) Market growth as influenced by capital investment. Ind Manag Rev (MIT, now Sloan Manag Rev) 9(2):83–105 (Reprinted widely, e.g., Richardson 1991)

Forrester JW (1969) Urban dynamics. MIT Press, Cambridge (Reprinted by Pegasus Communications, Waltham)

Homer JB (1992) A system dynamics model of national cocaine prevalence. Syst Dyn Rev 9(1):49–78

John DW, Morecroft JDW (1985) Rationality in the analysis of behavioral simulation models. Manag Sci 31(7):900–916

Lane DC, Husemann E (2002) Steering away from Scylla, falling into Charybdis: the importance of recognising, simulating and challenging reinforcing loops in social systems. In: Milling PM (ed) Entscheiden in komplexen Systemen. Duncker & Humblot, Berlin, pp 27–68

Maani KE, Cavana RY (2000) Systems thinking and modelling: understanding change and complexity. Pearson Education, New Zealand

Milling P (1996) Modeling innovation processes for decision support and management simulation. Syst Dyn Rev 12(3):211–234

Morecroft JDW, Sterman JD (eds) (1994) Modeling for learning organizations, System dynamics series. Pegasus Communications, Waltham

Morrison JB, Rudolph JW, Carroll JS (2009) The dynamics of action-oriented problem solving: linking interpretation and choice. Acad Manag Rev 34(4):733–756

Moxnes E (1998) Not only the tragedy of the commons: misperceptions of bioeconomics. Manag Sci 44(9):1234–1248

Nicholis G, Prigogine I (1977) Self-organization in nonequilibrium systems: from dissipative structures to order through fluctuations. Wiley, New York

Repenning NR (2001) Understanding fire fighting in new product development. J Prod Innov Manag 18(5): 285–300

Richardson GP (1991a) System dynamics: simulation for policy analysis from a feedback perspective. In: Fishwick PA, Luker PA (eds) Qualitative simulation modeling and analysis. Springer, New York

Richardson GP (1991b) Feedback thought in social science and systems theory. University of Pennsylvania Press, Philadelphia (Reprinted by Pegasus Communications, 1999)

Richardson GP (1996a) System dynamics. In: Gass S, Harris C (eds) The encyclopedia of operations research and management science. Kluwer, New York

Richardson GP (ed) (1996b) Modelling for management: simulation in support of systems thinking. International library of management. Dartmouth, Aldershot

Richardson GP, Pugh AL III (1981) Introduction to system dynamics modeling with DYNAMO. MIT Press,

Cambridge (Reprinted by Pegasus Communications, Waltham)

Richmond B (1993) Systems thinking: critical thinking skills for the 1990s and beyond. Syst Dyn Rev 9(2):113–133

Roberts EB (ed) (1978) Managerial applications of system dynamics. MIT Press, Cambridge (Reprinted by Pegasus Communications, Waltham)

Saeed K (1991) Towards sustainable development: essays on system analysis of national policy. Progressive Publishers, Lahore

Senge PM (1990) The fifth discipline: the art and practice of the learning organization. Doubleday/Currency, New York

Sterman JD (1988) Modeling managerial behavior: misperceptions of feedback in a dynamic decision making experiment. Manag Sci 35(3):321–339

Sterman JD (2001) Business dynamics: systems thinking and modeling for a complex world. Irwin McGraw-Hill, Boston

Sterman JD, Sweeney LB (2002) Cloudy skies: assessing public understanding of global warming. Syst Dyn Rev 18(2):207–240

System Dynamics Review 1985-present. Wiley, Chichester

Thompson KM, Duintjer Tebbens RJ (2007) Eradication versus control for poliomyelitis: an economic analysis. The Lancet 369(9570):1363–1371

Vennix JAM (1996) Group model building: facilitating team learning using system dynamics. Wiley, Chichester

von Bertalanffy L (1968) General systems theory: foundations, development, applications. George Braziller, New York

Warren KD (2002) Competitive strategy dynamics. Wiley, Chichester, UK.

Wolstenholme EF (1990) System enquiry: a system dynamics approach. Wiley, Chichester

Wolstenholme EF (2003) Towards the definition and use of a core set of archetypal structures in system dynamics. Syst Dyn Rev 19(1):7–26

System Dynamics in the Evolution of the Systems Approach

Markus Schwaninger
Institute of Management and Strategy, University of St. Gallen, St. Gallen, Switzerland

Article Outline

Glossary
Definition of the Subject
Introduction
Emergence of the Systems Approach
Common Grounds and Differences
The Variety of Systems Methodologies
System Dynamics: Its Features, Strengths, and
  Limitations
Strengths of SD
Limitations of SD
Actual and Potential Relationships
Outlook
Appendix – Milestones in the Evolution of the
  Systems Approach in General and System
  Dynamics in Particular
Bibliography

Glossary

Cybernetics The science of communication and control in complex, dynamic systems. The core objects of study are information, communication, feedback, adaptation, and control or governance. In the newer versions of cybernetics, the emphasis is on observation, self organization, self reference, and learning

Dynamical system The dynamical system concept is a mathematical formalization of time-dependent processes. Examples include the mathematical models that describe the swinging of a clock pendulum, the flow of water in a river, or the evolution of a population of fish in a lake.

Law of requisite variety Ashby's law of requisite variety says "Only variety can destroy variety." It implies that the varieties of two interacting systems must be in balance, if stability is to be achieved.

Organizational cybernetics The science which applies cybernetic principles to organization. Synonyms are *management cybernetics* and *managerial cybernetics*.

System dynamics (SD) A methodology and discipline for the modeling, simulation, and control of dynamic systems. The main emphasis falls on the role of structure and its relationship with the dynamic behavior of systems, which are modeled as networks of informationally closed feedback loops comprising stock and flow variables.

System There are many definitions of *system*. Two examples: A portion of the world sufficiently well defined to be the subject of study; something characterized by a structure, for example, a social system (Anatol Rapoport). A system is a family of relationships between its members acting as a whole (International Society for the Systems Sciences).

Systems approach A perspective of inquiry, education, and management, which is based on systems theory and cybernetics.

Systems theory A formal science of the structure, behavior, and development of systems. In fact there are different systems theories. General systems theory is a transdisciplinary framework for the description and analysis of any kind of system. Systems theories have been developed in many domains, e.g., mathematics, computer science, engineering, sociology, psychotherapy, biology, and ecology.

Variety A technical term for *complexity* which denotes the amount of (potential or actual) states or modes of behavior of a system. *Repertoire of behavior* is a synonym.

© Springer Science+Business Media, LLC, part of Springer Nature 2020
B. Dangerfield (ed.), *System Dynamics*,
https://doi.org/10.1007/978-1-4939-8790-0_537

Originally published in
R. A. Meyers (ed.), *Encyclopedia of Complexity and Systems Science*, © Springer Science+Business
Media LLC 2019, https://doi.org/10.1007/978-0-387-30440-3_537-5

Definition of the Subject

The purpose of this chapter is to give an overview of the role of system dynamics (SD) in the context of the evolution of the systems movement. This is necessary because SD is often erroneously taken as the systems approach as such, not as part of it. It is also requisite to show that the processes of the evolution of both the systems movement as a whole and SD in particular are intimately linked and intertwined. Finally, in view of the purpose of the chapter, the actual and potential relationships between system dynamics and the other strands of the systems movement are evaluated. This way, complementarities and potential synergies are identified.

Introduction

The purpose of this contribution is to give an overview of the role of system dynamics in the context of the evolution of the systems movement. "Systems movement" – often referred to briefly as "systemics" also cyber systemics – is a broad term, which takes into account the fact that there is no single systems approach but a range of different ones. The diagram in Fig. 1 is not a complete representation but the result of an attempt to map the major threads of the systems movement and some of their interrelations. Hence, the schema does not cover all schools or protagonists of the movement. Why does the diagram show a dynamic and evolutionary systems thread and an additional cybernetics thread, if cybernetics is about dynamic systems? The latter embraces all the approaches that are explicitly grounded in cybernetics. The former relates to all other approaches concerned with dynamic or evolutionary systems. The streamlining made it necessary to somewhat curtail logical perfection for the sake of conveying a synoptic view of the different systems approaches, in a language that uses the categories common in current scientific and professional discourse. Overlaps exist, e.g., between dynamic systems and chaos theory, cellular automata, and agent-based modeling, etc.

The common denominator of the different systems approaches in our day is that they share a worldview focused on complex dynamic systems and an interest in describing, explaining, and designing or at least influencing them. Therefore, most of the systems approaches offer not only a theory but also a way of thinking ("systems thinking" or "systemic thinking") and a methodology for dealing with systemic issues or problems.

System dynamics (SD) is a discipline and a methodology for the modeling, simulation, and control of complex, dynamic systems. SD was developed by MIT professor Jay W. Forrester (e.g., Forrester 1961, 1968) and has been propagated by his students and associates. The terms "methodology" and "method" should not be used interchangeably (Jackson 2003). Our working definition for "methodology" embraces two kinds of use: (a) a set of principles that frame the way in which research (theoretical, empirical, practical) is to be undertaken and (b) a coherent group of complementary methods used in a given methodology. The term "method" derives from Ancient Greek: ὁδός (hodós) for path and μετά (metá) for in pursuit of. The working definition we are using for "method" refers to a technique or way toward a purpose or serving a specific paradigm. "Paradigm" is a mode of thinking or a set of assumptions shared by a (scientific) community. SD has grown to a school of numerous academics and practitioners all over the world. The particular approach of SD lies in representing the issues or systems in focus as meshes of closed feedback loops made up of stocks and flows, in continuous time, and subject to delays. Given these features, SD modeling is widely applicable to all kinds of systems – economic, social, ecological, etc. – and the construction of generic models is facilitated.

The SD view is focused on continuous as opposed to discrete processes. This makes it extraordinarily powerful in dealing with managerial issues. Other modeling and simulation approaches exist. One is the discrete event approach, which is focused on the modeling of systems conceived as queuing networks (Brailsford 2014). It is a widely used technique of operations research, particularly for fine-grained modeling, simulation, and control, in the

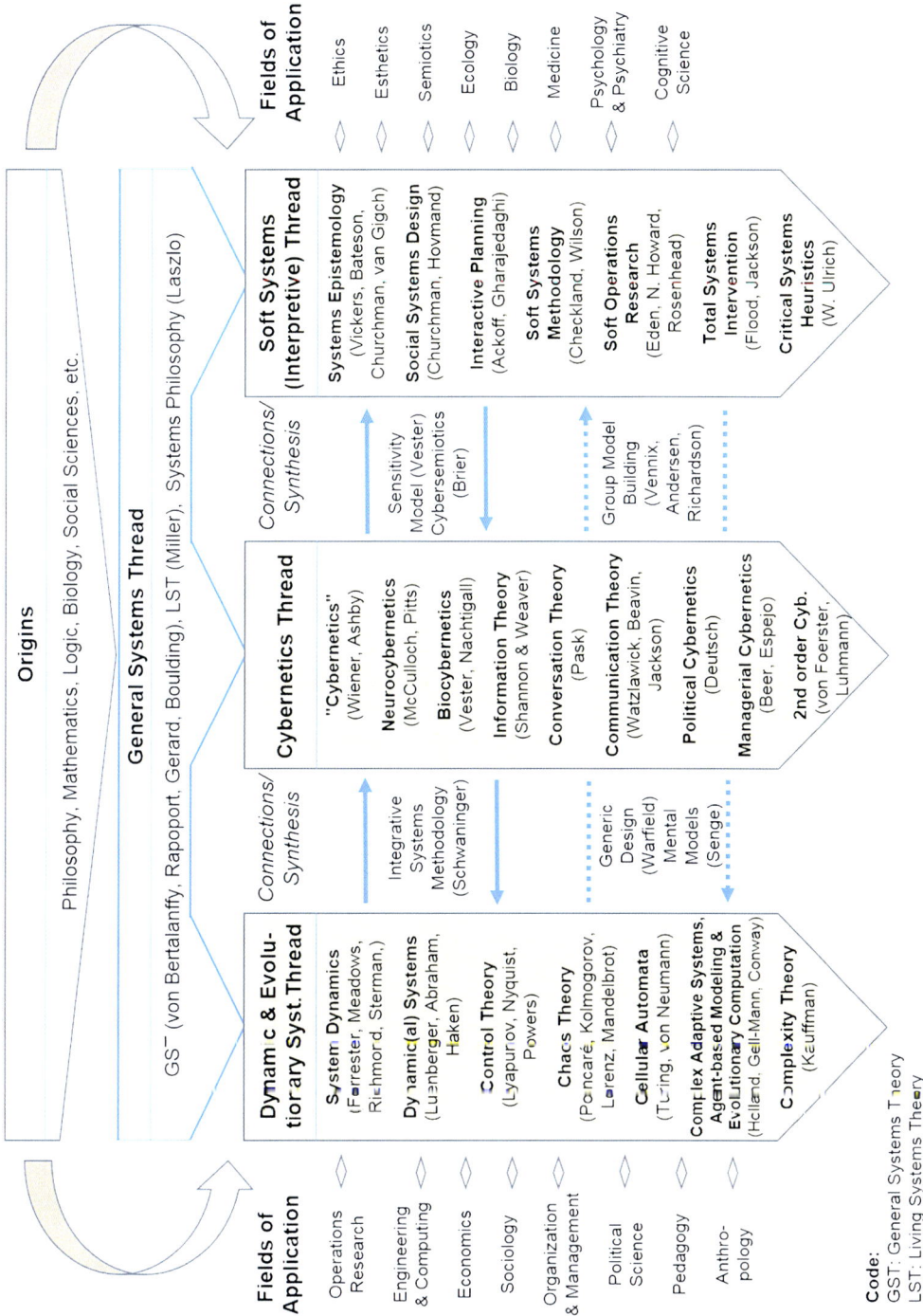

System Dynamics in the Evolution of the Systems Approach, Fig. 1   Systems approaches: An overview

domains of logistics, production, distribution systems, and transportation networks. SD is normally used for modeling and simulation at relatively higher levels of aggregation. Its perspective is top-down, which makes broad and long-term issues tangible and manageable. As a complement, a methodology has emerged which follows a bottom-up logic: the agent-based object approach (Holland 1995; Miller and Page 2007). Agent-based models are in discrete time. They represent sets of individual agents, whose interactions are simulated. Hence, the emergent behavior at macrolevel is generated from the bottom, by the individual behaviors at the microlevel. The three methodologies are complementary. Hybrid forms of modeling have been developed, which build bridges and enable synergies between the respective strengths of these different methodologies (Brailsford et al. 2014).

The development of the system dynamics methodology and the worldwide community that applies SD to modeling and simulation in highly varied contexts suggest that it is a "systems approach" on its own. Nevertheless, taking "system dynamics" as the (one and only) synonym for "systemic thinking" would be going too far, given the other approaches to systemic thinking as well as a variety of systems theories and methodologies, many of which are complementary to SD. In any case, however, the SD community has become the strongest "school" of the systems approach, if one takes the numbers of members in organizations representing the different schools as a measure (by early 2019, the System Dynamics Society has 1150 members).

The rationale and structure of this contribution are as follows. Starting with the emergence of the systems approach, the multiple roots and theoretical streams of systemics are outlined. Next, the common grounds and differences among different strands of the systems approach are highlighted, and the various systems methodologies are explored. Then the distinctive features of SD are analyzed. Finally comes a reflection on the relationships of SD with the rest of the systems movement and on potential complementarities and synergies.

In the Appendix, a timeline overview of some milestones in the evolution of the systems approach in general and system dynamics in particular is given. Elaborating on each of the sources quoted therein would reach beyond the purpose of this chapter. However, a synoptic view of the systems approaches has been provided in the diagram in Fig. 1.

Emergence of the Systems Approach

The systems movement has many roots and facets, with some of its concepts going back as far as ancient Greece. What we name as "the systems approach" today materialized in the first half of the twentieth century. At least two important components should be mentioned: those proposed by von Bertalanffy and by Wiener.

Ludwig von Bertalanffy, an American biologist of Austrian origin, developed the idea that organized wholes of any kind should be describable and, to a certain extent, explainable, by means of the same categories and ultimately by one and the same formal apparatus. His *General Systems Theory* (Von Bertalanffy 1950, 1968) triggered a whole movement which has tried to identify invariant structures and "mechanisms" across different kinds of organized wholes (e.g., hierarchy, teleology, purposefulness, differentiation, morphogenesis, stability, ultrastability, emergence, and evolution).

In 1948 Norbert Wiener, an American mathematician at the Massachusetts Institute of Technology, published his seminal book on *Cybernetics*, building upon interdisciplinary work carried out in cooperation with Bigelow, an IBM engineer, and Rosenblueth, a physiologist. Wiener's opus became the transdisciplinary foundation for a new science of capturing as well as designing control and communication mechanisms in all kinds of dynamic systems (Wiener 1948). Cyberneticists have been interested in concepts such as information, communication, complexity, autonomy, interdependence, cooperation and conflict, self-production ("autopoiesis"), self-organization, (self-) control, self-reference, and (self-) transformation of complex dynamic systems.

Along the genetic line of the tradition which led to the evolution of general systems theory (von Bertalanffy, Boulding, Gerard, Miller, Rapoport) and cybernetics (Wiener, McCulloch, Ashby, Powers, Pask, Beer), a number of roots can be identified, in particular:

- Mathematics (e.g., Newton, Poincaré, Lyapunov, Lotka, Volterra, Rashevsky).
- Logic (e.g., Epimenides, Leibniz, Boole, Russell and Whitehead, Goedel, Spencer-Brown).
- Biology, including general physiology and neurophysiology (e.g., Hippocrates, Cannon, Rosenblueth, McCulloch, Rosen).
- Engineering and computer science, including the respective physical and mathematical foundations (e.g., Heron, Kepler, Watt, Euler, Fourier, Maxwell, Hertz, Turing, Shannon and Weaver, von Neumann, Walsh).
- Social and human sciences, including economics (e.g., Hume, Adam Smith, Adam Ferguson, John Stuart Mill, Dewey, Bateson, Merton, Simon, Piaget).

In this last-mentioned strand of the systems movement, one focus of inquiry is on the role of feedback in communication and control in (and between) organizations and society, as well as in technical systems. The other focus of interest is on the multidimensional nature and the multilevel structures of complex systems. Specific theory building, methodological developments, and pertinent applications have occurred at the following levels:

- Individual and family levels (e.g., systemic psychotherapy, family therapy, holistic medicine, cognitive therapy, reality therapy)
- Organizational and societal levels (e.g., managerial cybernetics, organizational cybernetics, sociocybernetics, social systems design, social ecology, learning organizations)
- The level of complex (socio-)technical systems (systems engineering).

The notion of "socio-technical systems" has become widely used in the context of the design of organized wholes involving interactions of people and technology (for instance, Linstone's multiperspectives framework, known by way of the mnemonic TOP (*t*echnical, *o*rganizational, *p*ersonal/individual)).

As can be noted from these preliminaries, different kinds of systems theory and methodology have evolved over time. One of these is a theory of dynamic systems by Jay W. Forrester, which serves as a basis for the methodology of system dynamics. Two eminent titles are Forrester (1961) and (1968). In SD, the main emphasis falls on the role of structure and its relationship with the dynamic behavior of systems, modeled as networks of informationally closed feedback loops between stock and flow variables. Several other mathematical systems theories have been elaborated, for example, mathematical general systems theory (Klir, Pestel, Mesarovic, and Takahara), as well as a whole stream of theoretical developments which can be subsumed under the terms "dynamic systems theory" or "theories of nonlinear dynamics" (e.g., catastrophe theory, chaos theory, and complexity theory). Under the latter, branches such as the theory of fractals (Mandelbrot), geometry of behavior (Abraham), self-organized criticality (Bak), and network theory (Barabasi, Watts) are subsumed. In this context, the term "sciences of complexity" is used.

In addition, a number of mathematical theories, which can be called "systems theories," have emerged in different application contexts, examples of which are discernible in the following fields:

- Engineering, namely, information and communication theory (Shannon and Weaver), technology and computer-aided systems theory (e.g., control theory, automata, cellular automata, agent-based modeling, artificial intelligence, cybernetic machines, neural nets, machine learning)
- Operations research (e.g., modeling theory and simulation methodologies, Markov chains, genetic algorithms, fuzzy control, orthogonal sets, rough sets)
- Social sciences, economics in particular (e.g., game theory, decision theory)
- Biology (e.g., Sabelli's Bios theory of creation)
- Ecology (e.g., E. and H. Odum's systems ecology)

Most of these theories are transdisciplinary in nature, i.e., they can be applied across disciplines. The Bios theory, for example, is applicable to clinical, social, ecological, and personal settings (Sabelli 2005). Examples of essentially non-mathematical systems theories can be found in many different areas of study:

- Economics, namely, its institutional/evolution-ist strand (Veblen, Myrdal, Boulding, Dopfer)
- Sociology (e.g., Parsons' and Luhmann's social systems theories, Hall's cultural systems theory)
- Political sciences (e.g., Easton, Deutsch, Wallerstein)
- Anthropology (e.g., Levi Strauss's structuralist-functionalist anthropology, Margaret Mead)
- Semiotics (e.g., general semantics (Korzybski, Hayakawa, Rapoport), cybersemiotics (Brier))
- Psychology and psychotherapy (e.g., systemic intervention (Bateson, Watzlawick, F. Simon) and fractal affect logic (Ciompi))
- Ethics and epistemology (e.g., Vickers, Churchman, von Foerster, van Gigch).

Several system-theoretic contributions have merged the quantitative and the qualitative in new ways. This is the case, for example, in Rapoport's works in game theory as well as general systems theory, Pask's conversation theory, von Foerster's *Cybernetics of Cybernetics* (second-order cybernetics), and Stafford Beer's opus in managerial cybernetics. In all four cases, mathematical expression is virtuously connected to ethical, philosophical, and epistemological reflection. Further examples are Prigogine's theory of dissipative structures, Mandelbrot's theory of fractals, complex adaptive systems (Holland et al.), Kauffman's complexity theory, and Haken's synergetics, all of which combine mathematical analysis and a strong component of qualitative interpretation.

A large number of systems methodologies, with the pertinent threads of systems practice, have emanated from these theoretical developments. Many of them are expounded in detail in specialized publications (e.g., François 2004; Jackson 2019) and, under a specific theme, named *Systems Science and Cybernetics* of

the *Encyclopedia of Life Support Systems* (Encyclopedia of Life Support Systems 2002). In this chapter, only some of these will be addressed explicitly, in order to shed light on the role of SD as part of the systems movement.

Common Grounds and Differences

Even though the spectrum of systems theories and methodologies outlined in the preceding section may seem multifarious, all of them have a strong common denominator: They build on the idea of systems as organized wholes. An objectivist working definition of a system is that of a whole, the organization of which is made up by interrelationships. A subjectivist definition is that of a set of interdependent variables in the mind of an observer, or, a mental construct of a whole, an aspect that has been emphasized by the position of constructivism. *Constructivism* is a synonym for *second-order cybernetics*. While first-order cybernetics concentrates on regulation, information, and feedback, second-order cybernetics focuses on observation, self-organization, and self-reference. Heinz von Foerster established the distinction between "observed systems" for the former and "observing systems" for the latter (Von Foerster 1984).

From the standpoint of operational philosophy, a system is, as Rapoport says, "a part of the world, which is sufficiently well defined to be the object of an inquiry or also something, which is characterized by a structure, for example, a production system" (Rapoport 1953).

In recent systems theory, the aspect of relationships has been emphasized as the main building block of a system, as one can see from a definition published by the International Society for the Systems Sciences (ISSS), "A system is a family of relationships between its members acting as a whole" (Shapiro et al. 1996). Also, purpose and interaction have played an important part in reflections on systems. Systems are conceived, in the words of Forrester (1968), as "wholes of elements, which cooperate towards a common goal." Purposeful behavior is driven by internal goals, while purposive behavior rests on a function assigned from the outside. Finally, the aspects of open and closed functioning have been emphasized. Open systems are characterized by the import and export of matter, energy, and information. A variant of particular relevance in the case of social systems is the operationally closed system, that is, a system which is self-referential in the sense that its self-production (autopoiesis) is a function of production rules and processes by which order and identity are maintained, which cannot be modified directly from outside. As we shall see, this concept of operational closure is very much in line with the concept of circularity used in SD.

Another generic concept is that of complex adaptive system (CAS). The CAS is defined as a system that has a large number of components, often called agents, which interact and adapt or learn (John Holland 2006). It has features such as self-similarity (adaptation takes place at both levels – component and whole), complexity, emergence, and self-organization. The behavior of the whole system cannot be simply derived from the actions of its components; it emerges from their interaction.

At this point, it is worth elaborating on the specific differences between two major threads of the systems movement, which are of special interest because they are grounded in "feedback thought" (Richardson 1999): the cybernetic thread, from which organizational cybernetics has emanated, and the servomechanic thread in which SD is grounded. As Richardson's detailed study shows, the strongest influence on cybernetics came from biologists and physiologists, while the thinking of economists and engineers essentially shaped the servomechanic thread.

Consequently, the concepts of the former are more focused on the adaptation and control of complex systems for the purpose of maintaining stability under exogenous disturbances. Servomechanics, on the other hand, and SD, in particular, take an endogenous view, being mainly interested in understanding circular causality as the principal source of a system's behavior. Cybernetics is more connected with communication theory, the general concern of which can be summarized as how to deal with randomly varying input. SD, on the other hand, shows a stronger

link with engineering control theory, which is primarily concerned with behavior generated by the control system itself and by the role of nonlinearities. Managerial cybernetics and SD both share the concern of contributing to management science but with different emphases and with instruments that are distinct but in principle complementary. Finally, the mathematical foundations are generally more evident in the basic literature on SD than in the writings on organizational cybernetics, in which the formal apparatus underlying model formulation is confined to a small number of publications (e.g., Beer 1981, 1994), which are less known than the qualitative treatises. The terms *management cybernetics* and *managerial cybernetics* are used as synonyms for *organizational cybernetics*.

The Variety of Systems Methodologies

The methodologies that have evolved as part of the systems movement cannot be expounded in detail here. The two epistemological strands in which they are grounded, however, can be identified – the positivist tradition and the interpretivist tradition.

Positivist tradition denotes those methodological approaches that focus on the generation of "positive knowledge," that is, a knowledge based on "positively" ascertained facts. *Interpretivist tradition* denotes those methodological approaches that emphasize the importance of subjective interpretations of phenomena. This stream goes back to Greek art and science of the interpretation and understanding of texts.

Some systems methodologies have been rooted in the positivist tradition and others in the *interpretivist tradition*. As proposed in works on integrative systems methodology (see below), the differences between the two can be described along the following set of polarities:

- An objectivist versus a subjectivist position
- A conceptual-instrumental versus a communicational/cultural/political rationality
- An inclination to quantitative versus qualitative modeling

- A structuralist versus a discursive orientation

A positivistic methodological position tends toward the objectivistic, conceptual-instrumental, quantitative, and structuralist-functionalist in its approach. An interpretive position, on the other hand, tends to emphasize the subjectivist, communicational, cultural, political, ethical, and esthetic, that is, the qualitative and discursive aspects. It would be too simplistic to classify a specific methodology in itself as being "positivistic" or "interpretative." Despite the traditions they have grown out of, several methodologies have evolved and been reinterpreted or opened to new aspects (see below).

In the following, a sample of systems methodologies will be characterized and positioned in relation to these two traditions, beginning with those in the positivistic strand:

- *Hard OR methods.* Operations research (OR) uses a wide variety of mathematical and statistical methods and techniques, for example, of optimization, queuing, dynamic programming, graph theory, and time series analysis, to provide solutions for organizational and managerial problems, in the operational domains of production, logistics, and finance, among others.
- *Living systems theory.* In his LST, James Grier Miller (1978) identifies a set of 20 necessary components that can be discerned in living systems of any kind. These structural features are specified on the basis of a huge empirical study and proposed as the "critical subsystems" that "make up a living system." LST has been used as a device for diagnosis and design in the domains of engineering and the social sciences.
- *Viable system model.* To date, Stafford Beer's VSM is probably the best known product of organizational cybernetics. It specifies a set of management functions and their interrelationships as the sufficient preconditions for the viability of any human or social system (see Beer 1981). These are applicable in a recursive mode, for example, to the different levels of an

organization. The VSM has been widely applied in the diagnostic mode but also to support the design of all kinds of social systems (Espejo and Reyes 2011; Pérez Ríos 2012; Malik 2016). Specific methodologies for these purposes have been developed, for instance, for use in consultancy. The term viable system diagnosis (VSD) is also used.

The methodologies and models addressed up to this point have by and large been created in the positivistic tradition of science. Other strands in this tradition do exist, e.g., systems analysis and systems engineering, which together with OR have been called "hard systems thinking" (Jackson 2000, p. 127). Also, more recent developments such as mathematical complexity and network theories, agent-based modeling, game theory, and machine learning can be classified as hard systems approaches.

The respective approaches have not altogether been excluded from fertile contacts with the interpretivist strand of inquiry. In principle, all of them can be considered as instruments for supporting discourses about different interpretations of an organizational reality or alternative futures studied in concrete cases. In our time, most applications of the VSM, for example, are constructivist in nature. To put it in a nutshell, these applications are (usually collective) constructions of a (new) reality, in which observation and interpretation play a crucial part. In this process, the actors involved make sense of the system under study, i.e., the organization in focus, by mapping it on the VSM. At the same time, they bring forth "multiple realities rather than striving for a fit with one reality" (Harnden 1989, p. 299).

The second group of methodologies is part of the interpretive strand:

- *Interactive planning.* IP is a methodology, designed by Russell Ackoff (1981) and developed further by Jamshid Gharajedaghi (1999), for the purpose of dealing with "messes" and enabling actors to design their desired futures, as well as to bring them about. It is grounded in theoretical work on purposeful systems;

reverts to the principles of continuous, participative, and holistic planning; and centers on the idea of an "idealized design."

- *Soft systems methodology.* SSM is a heuristic designed by Peter Checkland (1981; Checkland and Poulter 2006) for dealing with complex situations. Checkland suggests a process of inquiry constituted by two aspects: a conceptual one, which is logic based, and a sociopolitical one, which is concerned with the cultural feasibility, desirability, and implementation of change.
- *Critical systems heuristics.* CSH is a methodology, which Werner Ulrich (1983, 1996) proposed for the purpose of scientifically informing planning and design in order to lead to an improvement in the human condition. The process aims at uncovering the interests that the system under study serves. The legitimacy and expertise of actors and particularly the impacts of decisions and behaviors of the system on others – the "affected" – are elicited by means of a set of boundary questions. CSH can be seen as part of a wider movement known as the "emancipatory systems approaches" which embraces, e.g., Freire's critical pedagogy, interpretive systemology, and community OR (see Jackson 2000, p. 291 ff).

All three of these methodologies (IP, SSM, and CSH) are positioned in the interpretive tradition. Other methodologies and concepts which can be subsumed under the interpretive systems approach are, e.g., Warfield's science of generic design, Churchman's social system design, Senge's soft systems thinking, Mason and Mitroff's strategic assumptions surfacing and testing (SAST), Eden and Ackermann's strategic options development and analysis (SODA), and other methodologies of soft operational research (for details, see Jackson 2000, p. 211 ff). The interpretive methodologies were designed to deal with qualitative aspects in the analysis and design of complex systems, emphasizing the communicational, social, political, and ethical dimensions of problem solving. Several authors mention explicitly that they do

not preclude the use of quantitative techniques, or they include such techniques in their repertoire (e.g., the biocyberneticist Frederic Vester).

In an advanced understanding of system dynamics, both of these traditions positivist and interpretivist are synthesized. The adherents of SD have conceived of model building and validation (the pursuit of model quality) as a semiformal, relativistic, and holistic social process. Validity is understood as usefulness or fitness in relation to the purpose of the model and validation as an elaborate set of procedures – including logico-structural, heuristic, algorithmic, statistical, and also discursive components – by which the quality of and the confidence in a model are gradually improved (see Barlas 1996; Barlas and Carpenter 1990; Schwaninger and Groesser 2009).

System Dynamics: Its Features, Strengths, and Limitations

The features, strengths, and limitations of the SD methodology are a consequence of its specific characteristics. In the context of the multiple theories and methodologies of the systems movement, some of the distinctive features of SD are (for an overview, see Richardson 1999; Jackson 2000, p. 142 ff):

- *Feedback as conceptual basis.* SD model systems are higher-order, multiple-loop networks of closed loops of information. Concomitantly, an interest in nonlinearities, long-term patterns, and internal structure rather than external disturbances is characteristic of SD (Meadows 1980, p. 31). However, SD models are not "closed systems," as sometimes is claimed, in the sense that (a) flows can originate from outside the system's boundaries, (b) representations of exogenous factors or systems can be incorporated into any model as parameters or special modules, and (c) new information can be accommodated via changes to a model. In other words, the SD view hinges on a view of systems which are closed in a causal sense but not materially (Richardson 1999, p. 297).

- *Focus on internally generated dynamics.* SD models are conceived as causally closed systems. The interest of users is in the dynamics generated inside those systems (Richardson 2011). Given the nature of closed feedback loops and the fact that delays occur within them, the dynamic behavior of these systems is essentially nonlinear.
- *Emphasis on understanding.* For system dynamicists, the understanding of the dynamics of a system is the first goal to be achieved by means of modeling and simulation. Conceptually, they try to understand events as embedded in patterns of behavior, which in turn are generated by underlying structures. Such understanding is enabled by SD as it "shows how present policies lead to future consequences" (Forrester 1971, Sect. VIII). Thereby, the feedback loops are "a major source of puzzling behavior and policy difficulties" (Richardson 1999, p. 300). SD models purport to test mental models, hone intuition, and improve learning (see Sterman 1994).
- *High degree of operationality.* SD relies on formal modeling. This fosters disciplined thinking; assumptions, underlying equations, and quantifications must be clarified. Feedback loops and delays are visualized and formalized; therewith the causal logic inherent in a model is made more transparent and discussable than in most other methodologies (Richmond 1997). Also, a high level of realism in the models can be achieved. SD is therefore apt to support decision-making processes effectively.
- *Far-reaching requirements (and possibilities) for the combination of qualitative and quantitative aspects of modeling and simulation.* This is a consequence of the emphasis on understanding. Qualitative modeling and visualization are carried out by means of causal loop diagrams and stock and flow diagrams, which are then underlaid quantitatively with equations. The focus is not on point-precise prediction but on the generation of insights into the patterns generated by the systems under study and the underlying structures which produce those patterns.

- *High level of generality and scale robustness.* The representation of dynamic issues in terms of stocks and flows is a generic form, which is adequate for a wide spectrum of potential applications. This spectrum is both broad as to the potential subjects under study and deep as to the possible degrees of resolution and detail (La Roche and Simon 2000). In addition, the SD methodology enables one to deal with large numbers of variables within multiple interacting feedback loops (Forrester 1969, p. 9). SD has been applied to the most diverse subject areas, e.g., global modeling, environmental issues, social and economic policy, corporate and public management, regional planning, medicine, psychology, and education in mathematics, physics, and biology.

The features of SD just sketched out result in both strengths and limitations. We start with the strengths.

Strengths of SD

1. Its *specific modeling approach* makes SD particularly helpful in gaining insights into the patterns exhibited by dynamic systems, as well as the structures underlying them. Closed-loop modeling has been found most useful in fostering understanding of the dynamic functioning of complex systems. Such understanding is especially facilitated by the principle of modeling the systems or issues under study in a continuous mode and at rather high aggregation levels (Forrester 1961; La Roche and Simon 2000). With the help of relatively small but insightful models, and by means of sensitivity analyses as well as optimization heuristics incorporated in the application software packages, decision-spaces can be thoroughly explored. Vulnerabilities and the consequences of different system designs can be examined with relative ease.

2. The *generality of the methodology* and its power to crystallize operational thinking in realistic models have triggered applications in

the most varied contexts. Easy-to-use software and the features of modeling via graphic user interfaces provide a strong lever for collaborative model building in teams (Andersen and Richardson 1997; Vennix 1996).

3. Its specific features make SD an exceptionally effective tool for *conveying systemic thinking* to anybody. Therefore, it also has an outstanding track record of classroom applications for which "learner-directed learning" (Forrester 1993a) or "learner-centered learning" is advocated (Forrester 1993b, 1997). Pertinent audiences range from schoolchildren at the levels of secondary and primary schools to managers and scientists.

4. The *momentum of the SD movement* is remarkable. Due to the strengths commented above this point, the community of users has grown steadily, being probably the largest community within the systems movement. Lane (2006, p. 484) has termed SD "one of the most widely used systems approaches in the world." Hence, a critical mass exists to spark future developments.

Given these strengths, the community of users has not only grown significantly but has also transcended disciplinary boundaries, ranging from the formal and natural sciences to the humanities and covering multiple uses from theory building and education to the tackling of real-world problems at almost any conceivable level. Applications to organizational, societal, and ecological issues have seen a particularly strong increase. This feeds back on the availability and growth of the knowledge upon which the individual modeler can draw.

The flip side of most of the strengths outlined here embodies the limitations of SD: we concentrate on those which can be relevant to a possible complementarity of SD with other systems methodologies.

Limitations of SD

1. The main point here is that SD does not provide a framework or methodology for the *diagnosis and design of organizational structures*

in the sense of interrelationships among organizational actors. This makes SD susceptible to completion from without – a completion which organizational cybernetics (OC), and the VSM in particular but also living systems theory (LST), especially can provide. The choice falls on these two approaches because of their strong heuristic power and their complementary strengths in relation to SD (Schwaninger 2006; Schwaninger and Pérez Ríos 2008).

2. Another limitation of SD is related to the *absorption of variety* (complexity) by an organization. *Variety* is a technical term for *complexity*, which denotes a (high) number of potential states or behaviors of a system (based on Ashby 1956; Beer 1966). SD offers an approach to the handling of variety which allows modeling at different scales of a problem or system (Odum and Odum 2000). It focuses on the identification, at a certain resolution level or possibly several resolution levels, of the main stock variables which will be affected by the respective flows. These, in turn, will be influenced by parameters and auxiliary variables. This approach, even though it enables thinking and modeling at different scales, does not provide a formal procedure for an organization to cope with the external complexity it faces, namely, for designing a structure which can absorb that complexity. In contrast, OC and LST offer elaborate models to enable the absorption of variety, in the case of the VSM based explicitly on Ashby's *law of requisite variety*. It says "Only variety can destroy variety," which implies that the varieties of two interacting systems must be in balance, if stability is to be achieved (Ashby 1956). The VSM has two salient features in this respect. Firstly, it helps design an organizational unit for viability, by enabling it to attenuate the complexity of its environment and also to enhance its eigen-variety, so that the two are in balance. The term *variety engineering* has been used in this context (Beer 1979). Secondly, the recursive structure of the VSM ensures that an organization with several levels will develop sufficient eigen-variety along the fronts on which the complexity it

faces unfolds. Similarly, LST offers the conditions for social systems to survive, by maintaining thermodynamically highly improbable energy states via continuous interaction with their environments. The difference between the two approaches is that the VSM functions more in the strategic and informational domains, while the LST model essentially focuses on the operational domain. In sum, both can make a strong contribution related to coping with the complexity faced by organizations and therefore can deliver a strong complement to SD.

3. The design of *modeling processes* confronts SD with specific challenges. The original SD methodology of modeling and simulation was to a large extent functionally and technically oriented. This made it strong in the domain of logical analysis, while the sociocultural and political dimensions of the modeling process were, if not completely out of consideration, at least not a significant concern in methodological developments. The SD community – also under the influence of the soft systems approaches – has become aware of this limitation and has worked on incorporating features of the social sciences into its repertoire. The following examples, which document this effort to close the gap, stand for many. Extensive work on group model building has been achieved, which explores the potential of collaborative model building (Vennix 1996). A new schema for the modeling process has been proposed, which complements logic-based analysis by cultural analysis (Lane and Oliva 1998). The social dimension of system dynamics-based modeling has become a subject to intensive discussion (Vriens and Achtenbergh 2006) and other contributions to the special issue of *Systems Research and Behavioral Science*, Vol. 51, No. 49. Finally, in relation to consultancy methodology, modeling has been framed as a learning process (Lane 1994a; Kineman 2017) and as second-order intervention (Schwaninger et al. 2006).

4. System dynamics modeling adopts a *top-down logic*. It operates with high-level aggregates

and is not convenient for models that represent the individual behaviors of high quantities of agents. In cases where microagents produce behavior at macrolevel, agent-based models (ABM) are the prevalent methodological solution. A combination of SD and ABM can leverage the complementarities of the two approaches (Duggan 2007; Martin and Schlüter 2015; Nava Guerrero et al. 2016). This way a great contribution should be possible, for example, to the modeling of processes of emergence and the understanding of emergent phenomena, an area where more work is needed (Kineman 2017).

As has been shown, there is a need to complement classical SD with other methodologies, when issues are at stake which it cannot handle by itself. VSM and LST are pertinent choices when issues of organizational diagnosis or design are to be tackled.

The limitations addressed here call attention to other methodologies which exhibit certain features that traditionally were not incorporated, or at least not explicit, in SD methodology. One aspect concerns the features that explicitly address the subjectivity of purposes and meanings ascribed to systems. In this context, support for problem formulation, model construction, and strategy design by individuals on the one hand and groups on the other are relevant issues. Also, techniques for an enhancement of creativity (e.g., the generation and the reframing of options) in both individuals and groups are a matter of concern. Two further aspects relate to methodological arrangements for coping with the specific issues of negotiation and alignment in pluralist and coercive settings.

As far as the modeling and simulation processes are concerned, group model building has become first a complement of SD and then a crucial feature of most modeling activities (Kineman 2017). In addition, there are other methodologies which should be considered as potentially apt to enrich SD analysis, namely, the soft systems approaches commented upon earlier, e.g., interactive planning, soft systems methodology, and critical systems heuristics.

On the other hand, SD can be a powerful complement to other methodologies which are more abstract or more static in nature. This potential refers essentially to all systems approaches which stand in the interpretive ("soft") tradition but also to approaches which stand in the positivist traditions, such as the VSM and LST. These should revert to the support of SD in the event that trade-offs between different goals must be handled, or if implications of long-term decisions on short-term outcomes (and vice versa) have to be ascertained, and whenever contingencies or vulnerabilities must be assessed.

Actual and Potential Relationships

It should be clear by now that the systems movement has bred a number of theories and methodologies, none of which can be considered all-embracing or complete. All of them have their strengths and weaknesses and their specific potentials and limitations.

Since Burrell and Morgan (1979) adverted to incommensurability between different paradigms of social theory, several authors have acknowledged or even advocated methodological complementarism. They argue that there is a potential complementarity between different methodologies and, one may add, models and methods, even if they come from distinct paradigms. Among these authors are, e.g., Brocklesby (1993), Jackson (1991), Midgley (2000), Mingers (1997), Schwaninger (1997), Yolles (1998), and García-Díaz and Olaya (2018). One of these protagonists proposes "that well-designed and well-executed multi-method research has inferential advantages over research relying on a single method" (Seawright 2016, p. 1). Seawright advocates a "well-constructed integrative multi-method design" (Ibidem: 9). In other words, a new perspective, in comparison with the non-complementaristic state of the art, has been opened.

In the past, the different methodologies have led to the formation of their own traditions and "schools," with boundaries across which not much dialogue has evolved. The methodologies have kept their protagonists busy testing them and developing them further. Also, the differences

between different language games and epistemological traditions have often suggested incommensurability and therewith have impaired communication. Prejudices and a lack of knowledge of the respective other side have accentuated this problem: Typically, "hard" systems scientists have been suspicious of "soft" systems scientists. For example, many members of the OR community, not unlike orthodox quantitatively oriented economists, adhere to the opinion that "SD is too soft." On the other hand, the protagonists of "soft" systems approaches, even though many of them have adopted feedback diagrams (causal loop diagrams) for the sake of visualization, are all too often convinced that "SD is too hard." Both of these judgments indicate a lack of knowledge, in particular of the SD validation and testing methods available, on the one hand, and the technical advancements achieved in modeling and simulation, on the other (see Barlas and Carpenter 1990; Schwaninger and Groesser 2009-2019; Sterman 2000).

In principle, both approaches are complementary. The qualitative view can enrich quantitative models, and it is connected to their philosophical, ethical, and esthetical foundations. However, qualitative reasoning tends to be misleading if applied to causal network structures without being complemented by formalization and quantification of relationships and variables. Furthermore, the quantitative simulation fosters insights into qualitative patterns and principles. It is thus a most valuable device for validating and honing the intuition of decision-makers, via corroboration and falsification.

Proposals that advocate mutual learning between the different "schools" have been formulated inside the SD community (e.g., Lane 1994b). The International System Dynamics Conference of 1994 in Stirling, held under the banner of "Transcending the Boundaries," was dedicated to the dialogue between different streams of the systems movement.

Also, from the 1990s onward, there were vigorous efforts to deal with methodological challenges, which traditionally had not been an important matter of scientific interest within the SD community. Some of the progress made in these areas is documented in a special edition of *Systems Research and Behavioral Science* (Vol. 21, No. 4, July–August 2004). The main point is that much of the available potential is based on the complementarity, not the mutual exclusiveness, of the different systems approaches.

In the future, much can be gained from leveraging these complementarities. Here are two examples of methodological developments in this direction, which appear to be achievable and potentially fertile: first, the enhancement of qualitative components in "soft" systems methodologies in the process of knowledge elicitation and model building (Vennix 1996) and second, the combination of cybernetics-based organizational design with SD-based modeling and simulation (Schwaninger and Pérez Ríos 2008). One must add that potential complementarities exist not only across the quality-quantity boundary but also within each one of the domains. For example, with the help of advanced software, SD modeling ("top-down") and agent-based modeling ("bottom-up") can be used in combination.

From a meta-methodological stance, generalist frameworks have been elaborated which contain blueprints for combining different methodologies where this is indicated. The following are two early and two more recent examples:

- *Total systems intervention* (TSI) is a framework proposed by Flood and Jackson (1991), which furnishes a number of heuristic schemes and principles for the purpose of selecting and combining systems methods/methodologies in a customized way, according to the issue to be tackled. SD is among the recommended "tools."
- *Integrative systems methodology* (ISM) is a heuristic for providing actors in organizations with requisite variety (Schwaninger 1997, 2004). It advocates (a) dealing with both content- and context-related issues during the process and (b) placing a stronger emphasis on the validation of qualitative and quantitative models as well as strategies, in both dimensions of the content of the issue under study and the organizational context into which that issue is embedded. For this purpose, the tools

of SD (to model content) and organizational cybernetics – the VSM (to model context) – are cogently integrated.

- *Evolutionary learning laboratory (ELLab)* denotes an approach to the management of complex issues developed by Bosch and coauthors (Bosch et al. 2013). The framework includes a process design for collective learning, involving participatory systems analysis, the interpretation of system structures to identify levers for systemic interventions, strategy design, and implementation. The methodology is based on several methods including qualitative system dynamics.
- *Social systems engineering* (SSE) is an application-oriented framework for coping with complexity, based on a nonmechanistic engineering viewpoint. It has a tradition going back to the 1970s and is now being reactivated (Garcia-Díaz and Olaya 2018). The approach considers technical, human, and social perspectives, which are crucial to solving complex problems. It focuses on systemic interventions aimed at the design, redesign, and transformation of social systems. Different approaches to multimethodology are proposed (Ibidem), e.g., actor-network theory as a heuristic for designing and building agent-based models (Méndez-Fajardo 2018) or a holistic approach to the design of healthcare systems, involving SD and the VSM (Schwaninger and Klocker 2018).

Recently, multimethodological frameworks have been developed, which are focused on integration (a systemic principle), but not explicitly tied to systemic methodologies (e.g., Morgan 2014; Seawright 2016). Given their meta-methodological position, these frameworks are in principle open to including systems methods or methodologies into their repertoire.

In principle, SD could make an important contribution in the context of any multi-methodological framework, far beyond the extent to which this has been the case. Systems methodologists and practitioners can potentially benefit enormously from including the SD methodology in their repertoires.

Outlook

There have been recurrent calls for an eclectic "mixing and matching" of methodologies. In light of the epistemological tendencies of our time toward radical relativism, it is necessary to warn against taking a course in which "anything goes." It is most important to emphasize that the desirable methodological progress can only be achieved on the grounds of scientific rigor. This postulate of "rigor" is not to be confused with an encouragement of "rigidity." The necessary methodological principles advocated here are disciplined thinking, a permanent quest for models of highest quality (i.e., thorough model validation), and full transparency in the formalizations as well as of the underlying assumptions and sources used. Scientific rigor, in this context, also implies that combinations of methodologies, methods, or models reach beyond merely eclectic add-ons so that genuine integration toward better adequacy to the issues at hand is achieved.

The contribution of system dynamics can come in the realms of the following:

- Fostering disciplined thinking
- Understanding dynamic behaviors of systems and the structures that generate them
- Exploring paths into the future and the concrete implications of decisions
- Assessing strategies as to their robustness, vulnerabilities, and impact, in ways precluded by other, more philosophical, and generally "soft" systems approaches

These latter streams can contribute to reflecting and tackling the meaning- and value-laden dimensions of complex human, social, and ecological systems. Some of their features should and can be combined synergistically with system dynamics, particularly by being incorporated into the repertoires of system dynamicists. From the reverse perspective, incorporating system dynamics as a standard tool will be of great benefit for the adopters of broad methodological frameworks. Model formalization and dynamic simulation may even be considered necessary components for the study of the concrete dynamics of complex systems.

There are also many developments in the "hard," i.e., mathematics-, statistics-, logic-, and informatics-based methods and technologies, which are apt to enrich the system dynamics methodology, namely, in terms of modeling and decision support. For example, the constantly evolving techniques of time series analysis, filtering, neural networks, control theory, machine learning, etc. can improve the design of system-dynamics-based systems of (self-)control. Also, a bridge across the divide between the top-down modeling approach of SD and the bottom-up approach of agent-based modeling has been built with the help of modern software.

Furthermore, a promising perspective for the design of genuinely "intelligent organizations" emerges if one combines dynamic modeling and cybernetic modeling, namely, system dynamics and the viable system model. Also, combining the "soft" systems methodologies and the "hard" methods from operations research could enable better, model-based management.

The approaches of integrating complementary methodologies, as outlined in this contribution, definitely mark a new phase in the history of the systems movement.

Appendix – Milestones in the Evolution of the Systems Approach in General and System Dynamics in Particular

The following table gives an overview of the systems movement's evolution, as shown in its main literature. This overview is not exhaustive.

Foundations of general systems theory		
Von Bertalanffy	*Zu einer allgemeinen Systemlehre*	1945
	An Outline of General System Theory	1950
	General System Theory	1968

(continued)

Bertalanffy, Boulding, Gerard, Rapoport	*Foundation of the Society for General Systems Research*	1953
Klir	*An Approach to General Systems Theory*	1968
Simon	*The Sciences of the Artificial*	1969
Pichler	*Mathematische Systemtheorie*	1975
Miller	*Living Systems*	1978
Mesarovic and Takahara	*Abstract Systems Theory*	1985
Rapoport	*General System Theory*	1986

Foundations of cybernetics

Macy conferences (Josiah Macy, Jr. foundation)	*Cybernetics, Circular Causal, and Feedback Mechanisms in Biological and Social Systems*	1946–1951
Wiener	*Cybernetics or Control and Communication in the Animal and in the Machine*	1948
Ashby	*An Introduction to Cybernetics*	1956
Pask	*An Approach to Cybernetics*	1961
Von Foerster, Zopf	*Principles of Self-Organization*	1962
McCulloch	*Embodiments of Mind*	1965

Foundations of organizational cybernetics

Beer	*Cybernetics and Management*	1959
	Towards the Cybernetic Factory	1962
	Decision and Control	1966
	Brain of the Firm	1972
Von Foerster	*Cybernetics of Cybernetics*	1974

Foundations of system dynamics

Forrester	*Industrial Dynamics*	1961
	Principles of Systems	1968
	Urban Dynamics	1969
	World Dynamics	1971
Meadows et al.	*Limits to Growth*	1972

(continued)

Richardson	*Feedback Thought in Social Science and Systems Theory*	1991

Systems methodology

Churchman	*Challenge to Reason*	1968
	The Systems Approach	1968
Vester and von Hesler	*Sensitivitätsmodell*	1980
Checkland	*Systems Thinking, Systems Practice*	1981
Ackoff	*Creating the Corporate Future*	1981
Ulrich	*Critical Heuristics of Social Planning*	1983
Warfield	*A Science of Generic Design*	1994
Schwaninger	*Integrative Systems Methodology*	1997
Gharajedaghi	*Systems Thinking*	1999
Sabelli	*Bios: A Study of Creation*	2005
Garcia and Olaya	*Social Systems Engineering*	2018

Selected recent works in system dynamics

Senge	*The Fifth Discipline*	1990
Barlas and Carpenter	*Model Validity*	1990
Vennix	*Group Model Building*	1996
Lane and Oliva	*Synthesis of System Dynamics and Soft Systems Methodology*	1998
Sterman	*Business Dynamics*	2000
Warren	*Strategy Dynamics*	2002, 2008
Wolstenholme	*Archetypal Structures*	2003
Morecroft	*Strategic Modeling*	2007
Schwaninger and Grösser	*Theory-Building with System Dynamics and Model Validation*	2008, 2009
Rahmandad, Oliva, Osgood	*Analytical Methods for Dynamic Modelers*	2015
Kunc	*Strategic Analytics*	2019

Bibliography

Primary Literature

Ackoff RL (1981) Creating the corporate future. Wiley, New York

Andersen DF, Richardson GP (1997) Scripts for group model building. Syst Dyn Rev 13(2):107–129

Ashby WR (1956) An introduction to cybernetics. Chapman & Hall, London

Barlas Y (1996) Formal aspects of model validity and validation in system dynamics. Syst Dyn Rev 12(3): 183–210

Barlas Y, Carpenter S (1990) Philosophical roots of model validity: two paradigms. Syst Dyn Rev 6(2):148–166

Beer S (1959) Cybernetics and management. English Universities Press, London

Beer S (1966) Decision and control. Wiley, Chichester

Beer S (1979) The heart of enterprise. Wiley, Chichester

Beer S (1981) Brain of the firm, 2nd edn. Wiley, Chichester

Beer S (1994) Towards the cybernetic factory. In: Harnden R, Leonard A (eds) How many grapes went into the wine. Stafford Beer on the art and science of holistic management. Wiley, Chichester, pp 163–225. (reprint, originally published in 1962)

Bosch OJH, Nguyen NC, Maeno T, Yasui T (2013) Managing complex issues through evolutionary learning laboratories. Syst Res Behav Sci 30(2):116–135

Brailsford S (2014) Theoretical comparison of discrete-event simulation and system dynamics. In: Brailsford S, Churilov L, Dangerfield B (eds) Discrete-event simulation and system dynamics for management decision making. Wiley, Chichester, pp 105–124

Brailsford S, Churilov L, Dangerfield B (eds) (2014) Discrete-event simulation and system dynamics for management decision making. Wiley, Chichester

Brocklesby J (1993) Methodological complementarism or separate paradigm development – examining the options for enhanced operational research. Aust J Manag 18(2):133–157

Burrell G, Morgan G (1979) Sociological paradigms and organisational analysis. Gower, Hants

Checkland PB (1981) Systems thinking, systems practice. Wiley, Chichester

Checkland PB, Poulter J (2006) Learning for action: a short definitive account of soft systems methodology, and its use practitioners, teachers and students. Wiley, Chichester

Churchman CW (1968a) Challenge to reason. McGraw-Hill, New York

Churchman CW (1968b) The systems approach. Delacorte Press, New York

Churchman CW (1979) The systems approach and its enemies. Basic Books, New York

Duggan J (2007) Modelling agent-based systems using system dynamics. In: Proceedings of the 2007 international conference of the system dynamics society. The System Dynamics Society, Boston

Encyclopedia of Life Support Systems (2002) published under: http://www.eolss.net/

Espejo R, Reyes A (2011) Organizational systems. Managing complexity with the viable system model. Springer, Berlin

Flood RL, Jackson MC (1991) Creative problem solving. Total systems intervention. Wiley, Chichester

Forrester JW (1961) Industrial dynamics. MIT Press, Cambridge

Forrester JW (1968) Principles of systems. MIT Press, Cambridge

Forrester JW (1969) Urban dynamics. MIT Press, Cambridge

Forrester JW (1971) World dynamics. Pegasus Communications, Waltham

Forrester JW (1993a) System dynamics and the lessons of 35 years. In: KB DG (ed) Systems-based approach to policy making. Kluwer, Boston

Forrester JW (1993b) System dynamics as an organizing framework for pre-college education. Syst Dyn Rev 9(2):183–194

Forrester JW (1997) System dynamics and K-12 teachers. A lecture at the University of Virginia School of Education, Massachusetts Institute of Technology. System Dynamics Group Paper D-4665-4

François C (2004) International encyclopedia of systems and cybernetics, 2nd edn. Saur, München

Garcia-Diaz C, Olaya C (eds) (2018) Social systems engineering: the design of complexity. Wiley, Chichester

Gharajedaghi J (1999) Systems thinking. Managing chaos and complexity. Butterworth-Heinemann, Boston

Harnden RJ (1989) Technology for enabling: the implications for management science of a hermeneutics of distinction. The University of Aston, Birmingham

Holland JH (1995) Hidden order: how adaption builds complexity. Addison-Wesley, Reading

Holland JH (2006) Studying complex adaptive systems. J Syst Sci Complex 19(1):1–8. https://doi.org/10.1007/s11424-006-0001-z. ISSN 1009-6124

Jackson MC (1991) Systems methodology for the management sciences. Plenum Press, New York

Jackson MC (2000) Systems approaches to management. Kluwer/Plenum, New York

Jackson MC (2003) The power of multi-methodology: some thoughts for John Mingers. J Oper Res Soc 54(12):1300–1301

Jackson MC (2019) Critical systems thinking and the management of complexity. Wiley Hoboken, NJ

Kauffman SA (1993) The origins of order. Self-organization and selection in evolution. Oxford University Press, New York

Kineman JJ (2017) Modeling and simulation. In: Edson MD, Buckle Henning P, Sankaran S (eds) A guide to systems research: philosophy, processes and practice. Springer, Singapore, pp 81–110

Klir GJ (1969) An approach to general systems theory. Nostrand, New York

Kunc M (2019) Strategic analytics: integrating management science and strategy. Wiley, Hoboken/Chichester

La Roche U, Simon M (2000) Geschäftsprozesse simulieren: flexibel und zielorientiert führen mit Fliessmodellen. Orell Füssli, Zürich

Lane DC (1994a) Modeling as learning: a consultancy methodology for enhancing learning in management in management teams. In: Morecroft J, Sterman JD (eds) Modeling for learning organizations. Productivity Press, Portland, pp 205–240

Lane DC (1994b) With a little help from our friends: how system dynamics and soft OR can learn from each other. Syst Dyn Rev 10(2–3):101–134

Lane DC (2006) IFORS' operational research hall of fame. Jay Wright Forrester. Int Trans Oper Res 13:483–492

Lane DC, Oliva R (1998) The greater whole: towards a synthesis of system dynamics and soft system methodology. Eur J Oper Res 107(1):214–235

Malik F (2016) Strategy for managing complex systems: a contribution to management cybernetics for evolutionary systems. Campus Verlag, Frankfurt/New York

Martin R, Schlüter M (2015) Combining system dynamics and agent-based modeling to analyze social-ecological interactions – an example from modeling restoration of a shallow lake. Front Environ Sci. https://doi.org/10.3389/fenvs.2015.00066. 29 Aug 2017

McCulloch WS (1965) Embodiments of mind. MIT Press, Cambridge

Meadows DH (1980) The unavoidable a priori. In: Randers J (ed) Elements of the system dynamics method. MIT Press, Cambridge, pp 23–57

Meadows DH, Meadows DL, Randers J, Behrens WW III (1972) Limits to growth. Universe Books, New York

Méndez-Fajardo S (2018) Using actor-network theory in agent-based modelling. In: Garcia-Díaz C, Olaya C (eds) Social systems engineering: the design of complexity. Wiley, Chichester, pp 157–196

Mesarovic MD, Takahara Y (1985) Abstract systems theory. Springer, Berlin

Midgley G (2000) Systemic intervention. Philosophy, methodology, and practice. Kluwer, New York

Miller JG (1978) Living systems. McGraw-Hill, New York

Miller JH, Page SE (2007) Complex adaptive systems: an introduction to computational models of social life. Princeton University Press, Princeton

Mingers J (1997) Multi-paradigm multimethodology. In: Mingers J, Gill A (eds) Multimethodology. Wiley, Chichester

Morecroft J (2007) Strategic modelling and business dynamics: a feedback systems approach. Wiley, Chichester

Morgan DL (2014) Integrating qualitative and quantitative methods: a pragmatic approach. Sage, Los Angeles

Nava Guerrero G, Schwarz P, Slinger JH (2016) A recent overview of the integration of system dynamics and agent-based modelling and simulation. In: Proceedings of the 34th international conference of the system dynamics society, July 17–21, 2016 Delft, Netherlands, p 1153

Odum HT, Odum EC (2000) Modeling for all scales: an introduction to system simulation. Academic, San Diego

Pask G (1961) An approach to cybernetics. Hutchinson, London

Pérez Ríos J (2012) Design and diagnosis for sustainable organizations: the viable system method. Springer, Berlin

Pichler F (1975) Mathematische Systemtheorie. de Gruyter, Berlin

Rapoport A (1953) Operational philosophy: integrating knowledge and action. Harper, New York

Rapoport A (1986) General system theory. Essential concepts and applications. Abacus Press, Turnbridge Wells

Richardson GP (1999) Feedback thought in social science and systems theory. Pegasus communications, Waltham. (Originally published in 1991)

Richmond B (1997) The "thinking" in systems thinking: how can we make it easier to master? Syst Think 8(2):1–5

Sabelli H (2005) Bios: a study of creation. World Scientific, Hackensack

Schwaninger M (1997) Integrative systems methodology: heuristic for requisite variety. Int Trans Oper Res 4(4):109–123

Schwaninger M (2004) Methodologies in conflict: achieving synergies between system dynamics and organizational cybernetics. Syst Res Behav Sci 21(4):1–21

Schwaninger M (2006) Theories of viability. A comparison. Syst Res Behav Sci 23:337–347

Schwaninger M, Groesser S (2008) System dynamics as model-based theory building. Syst Res Behav Sci 25:1–19

Schwaninger M, Groesser S (2009, new edition 2019) Model validation: the quest for quality in system dynamics modeling. Encyclopaedia of complexity and systems science. Springer, New York, pp 9000–9014. Reprinted in: Complex System in Finance and econometrics. Springer New York, 2011: 767–781

Schwaninger M, Klocker J (2018) Holistic system design: the oncology Carinthia study. In: Garcia-Díaz C, Olaya C (eds) Social systems engineering: the design of complexity. Wiley, Chichester, pp 235–266

Schwaninger M, Pérez Ríos J (2008) System dynamics and cybernetics: a synergetic pair. Syst Dyn Rev 24(2):145–174

Schwaninger M, Janovjak M, Ambroz K (2006) Second-order intervention: enhancing organizational competence and performance. Syst Res Behav Sci 23:529–545

Seawright J (2016) Multi-method social science: combining qualitative and quantitative tools. Cambridge University Press, Cambridge, UK

Senge PM (1990) The fifth discipline. The art and practice of the learning organization. Doubleday, New York

Shapiro M, Mandel T, Schwaninger M et al (1996) The primer toolbox. International Society for the Systems Sciences. http://www.isss.org/primer/toolbox.htm

Simon HA (1969) The sciences of the artificial. MIT Press, Cambridge

Sterman JD (1994) Learning in and about complex systems. Syst Dyn Rev 10(2–3):291–330

Sterman JD (2000) Business dynamics. Systems thinking and modeling for a complex world. Irwin/McGraw-Hill, Boston

Ulrich W (1983) Critical heuristics of social planning. Haupt, Bern

Ulrich W (1996) A primer to critical systems heuristics for action researchers. The Centre of systems studies. University of Hull, Hull

Vennix JAM (1996) Group model building. Facilitating team learning using system dynamics. Wiley, Chichester

Vester F, Von Hesler A (1980) Sensitivitätsmodell. Regionale Planungsgemeinschaft Untermain, Frankfurt am Main

Von Bertalanffy L (1949) Zu einer allgemeinen Systemlehre. Bl Dtsch Philos 18(3/4) (Excerpts: in Biol Gen 19(1):114–129 and in General System Theory, 1968, Chapter III)

Von Bertalanffy L (1950) An outline of general system theory. Br J Philos Sci 1:139–164

Von Bertalanffy L (1968) General system theory. Braziller, New York

Von Foerster H (1984) Observing systems, 2nd edn. Intersystems Publications, Seaside

Von Foerster H (ed) (1995) Cybernetics of cybernetics, 2nd edn. Future Systems, Minneapolis. Originally published in 1974

Von Foerster H, Zopf GW (eds) (1962) Principles of self-organization. Pergamon Press, Oxford

Vriens D, Achtenbergh J (2006) The social dimension of system dynamics-based modelling. Syst Res Behav Sci 23(4):553–563

Warfield JN (1994) A science of generic design: managing complexity through systems design, 2nd edn. Iowa State University Press, Ames

Warren K (2002) Competitive strategy dynamics. Wiley, Chichester

Warren K (2008) Strategic management dynamics. Wiley, Chichester

Wiener N (1948) Cybernetics: control and communication in the animal and in the machine. MIT Press, Cambridge

Wolstenholme E (2003) Towards the definition and use of a core set of archetypal structures in system dynamics. Syst Dyn Rev 19(1):7–26

Yolles MA (1998) Cybernetic exploration of methodological complement. Kybern 27(5):527–542

Books and Reviews

Dangerfield B (2014) Systems thinking and system dynamics: a primer. In: Brailsford S, Churilov L, Dangerfield B (eds) Discrete – event simulation and system dynamics for management decision-making. Wiley, Chichester, pp 26–51

Jackson MC (2003b) Systems thinking: creative holism for managers. Wiley, Chichester

Jackson MC (2019) Critical systems thinking and the management of complexity. Wiley, Chichester

Klir GJ (2001) Facets of systems science, 2nd edn. Kluwer/Plenum, New York

Kunc M (ed) (2018) System dynamics: soft and hard operational research. Palgrave Macmillan, London

Midgley G (ed) (2003) Systems thinking, vol vol 4. Sage, London

Ragsdell G, Wilby J (eds) (2001) Understanding complexity. Kluwer/Plenum, New York

Rahmandad H, Oliva R, Osgood ND (eds) (2015) Analytical methods for dynamic modelers. MIT Press, Cambridge, MA

Ramage M, Shipp K (2009) Systems thinkers. Springer, Dordrecht

Richardson GP (ed) (1996) Modelling for management. Simulation in support of systems thinking, vol 2 vols. Dartmouth, Aldershot

Richardson GP (2011) Reflections on the foundations of system dynamics. Syst Dyn Rev 27(3):219–243

Schwaninger M (2009) Intelligent organizations. Powerful model for systemic management, 2nd edn. Springer, Berlin

Seawright J (2016b) Multi-method social science. Cambridge University Press, Cambridge

Van Gigch JP (2003) Metadecisions. Rehabilitating epistemology. Kluwer/Plenum, New York

Scenario-Driven Planning with System Dynamics

Nicholas C. Georgantzas
Fordham University Business Schools,
New York, USA

Article Outline

Glossary

Mental model How one perceives cause and effect relations in a system, along with its boundary, i.e., exogenous variables, and the time horizon needed to articulate, formulate, or frame a decision situation; one's implicit causal map of a system, sometimes linked to the reference performance scenarios it might produce.

Product Either a physical good or an intangible service a firm delivers to its clients or customers.

Real option Right and obligation to make a business decision, typically a tangible investment. The option to invest, e.g., in a firm's store expansion. In contrast to financial "call" and "put" options, a *strategic real option* is not tradable. Any time it invests, a firm might be at once acquiring the *strategic real options* of expanding, downsizing, or abandoning projects in future. Examples include research and development (abbreviated R&D), merger and acquisition (abbreviated M&A), licensing abroad, and media options.

Scenario A postulated sequence or development of events through time; via Latin *scena* "scene," from Greek σκηνή, *skēnē* "tent, stage." In contrast to a forecast of what *will* happen in the future, a scenario shows what *might* happen. The term *scenario* must *not* be used loosely to mean situation. *Macro-environmental* as well as *industry-*, *task-*, or *transactional-environmental* scenarios are merely inputs to the *strategic objectives* and *real options* a firm must subsequently explore through *strategic scenarios, computed* or *simulated* with an explicit, formal *system dynamics* (abbreviated SD) model of its strategic situation. *Computed strategic scenarios* create the multiple perspectives that strategic thinkers need to defeat the tyranny of dogmatism that often assails firms, governments, and other social entities or organizations.

Scenario-driven planning (abbreviated SdP) To attain high performance through strategic flexibility, firms use the SdP *management technology* to create foresight and to anticipate the future with strategic real options, in situations where the business environment accelerates frequently and is highly complex or interdependent, thereby causing uncertainty.

Situation The set of circumstances in which a firm finds itself; its (strategic) state of affairs.

Strategic management process (abbreviated SMP) Geared at detecting environmental threats and turning them into opportunities, it *proceeds from* a firm's mission, vision, and environmental constraints *to* strategic goals and objectives *to* strategy design or formulation *to* strategy implementation or strategic action *to* evaluation and control

© Springer Science+Business Media, LLC, part of Springer Nature 2020
B. Dangerfield (ed.), *System Dynamics*,
https://doi.org/10.1007/978-1-4939-8790-0_465

Originally published in
R. A. Meyers (ed.), *Encyclopedia of Complexity and Systems Science*, © Springer Science+Business
Media LLC 2017, https://doi.org/10.1007/978-0-387-30440-3_465-4

to learning through feedback (background, Fig. 2).

SMP-1 environmental scanning Monitors, evaluates, and disseminates knowledge about a firm's internal and external environments to its people. The internal environment contains *strengths* and *weaknesses* within the firm; the external shows future *opportunities* and *threats* (abbreviated SWOT).

SMP-2 mission A firm's purpose, *raison d'être* or reason for being.

SMP-3 objectives Performance (*P*) goals that SMP often quantifies for some *P* metrics.

SMP-4 policy Decision-making guidelines that link strategy design or formulation to action or implementation tactics.

SMP-5 strategy A comprehensive plan that shows how a firm might achieve its mission and objectives. The three strategy levels are: corporate, business, and process or functional.

SMP-6 strategy design or formulation The interactive, as opposed to antagonistic, interplay of strategic content and process that creates flexible long-range plans to turn future environmental threats into opportunities; includes internal strengths and weaknesses as well as strategic mission and objectives and policy guidelines.

SMP-7 strategic action or implementation The process by which strategies and policies are put into action through the development of programs, processes, budgets, and procedures.

SMP-8 evaluation and control Sub-process that monitors activities and performance, comparing actual results with desired performance.

SMP-9 learning through feedback Occurs as knowledge about each SMP element enables improving previous SMP elements (background, Fig. 2).

System An organized group of interrelated components, elements, or parts working together for a purpose; parts might be either goal seeking or purposeful.

System dynamics (abbreviated SD) A lucid modeling method born from the need to manage business performance through time. Thanks to Forrester (Forrester 1958), who discovered that all change propagates itself through stock and

flow sequences, and user-friendly SD software (*iThink*®, *Vensim*®, etc.), SD models let managers see exactly how and why, like other biological and social organizations, business firms perform the way they do. Unlike other social sciences, SD shows exactly how *feedback loops*, i.e., circular cause and effect chains, each containing at least one time lag or delay, interact within a system to determine its performance through time.

Variable or metric Something that changes either though time or among different entities at the same time. An *internal change lever* is a decision or policy variable that a strategy-design modeling, or client, team controls. An *external change trigger* is an environmental or policy variable that a strategy-design modeling team does not control. Both *trigger* and *lever* variables can initiate change and be either endogenous or exogenous to a model of a system.

However certain our expectation, the moment foreseen may be unexpected when it arrives
—T.S. Eliot

Definition of the Subject

Many of us live and work in and about business ecosystems with complex structures and behaviors. Some realize that poor performance often results from our very own past actions or decisions, which come back to haunt us. So business leaders in diverse industries and firms, such as *Airbus, General Motors, Hewlett-Packard, Intel* and *Merck*, use scenario-driven planning (SdP) with system dynamics (SD) to help them identify, design, and apply high-leverage, sustainable solutions to dynamically complex strategic-decision situations. One must know, e.g., if the effect of an environmental change or strategic action gets magnified through time or is dampened and smoothed away. What may seem insignificant at first might cause major disruption in performance. SdP with SD show the causal processes behind such dynamics, so firms can respond to mitigate impacts on performance.

Accelerating change and complexity in the global business environment make firms and other social organizations abandon their *inactive, reactive* and *preactive* modes (Ackoff 1981). SdP with SD turns them *proactive*, so they can translate anticipation into action. To properly transform anticipation into action, computed with SD models, "strategic scenarios" must meet four conditions: *consistency, likelihood, relevance,* and *transparency* (Godet and Roubelat 1996). Combining SdP with SD for that purpose with other tools, like actor and stakeholder purposes, morphological methods, or probability, might help avoid entertainment and explore all possible scenarios. Indeed, SdP with SD

> does not stand alone... modeling projects are part of a larger effort... modeling works best as a complement to other tools, not as a substitute (see p. 80 in Sterman 2000).

SdP with SD is a systematic approach to a vital top-management job: leading today's firm in the rapidly changing and highly complex global environment. Anticipating a world where product life cycles, technology, and the mix of collective- and competitive-strategy patterns change at an unprecedented rate is hard enough. Moving ahead of it might prove larger than the talent and resources now available in leading firms. SdP with SD leads to a decisive integration of strategy design and operations, with the dividing line much lower than at present. As mid-level managers take on more responsibility, senior executives become free to give more time and attention to economic conditions, product innovation and the changes needed to enhance creativity toward strategic flexibility (Forrester 1958).

It is perhaps its capacity to reintegrate strategy content and process that turns SdP with SD into a new paradigm for competitive advantage (Istvan 1992), and simulation modeling in general (Georgantzas 2001a), into a critical fifth tool, in addition to the four tools used in science: observation, logical-mathematical analysis, hypothesis testing and experiment (Turner 1997). But full-fledged SD models also allow computing scenarios to assess possible implications of strategic situations. Strategic scenarios are not merely hypothesized plausible futures, but computed by simulating combined changes in strategy and in the business environment (Georgantzas and Acar 1995).

Computed scenarios help managers understand what they do not know, enabling strategy design and implementation through the coalignment of timely tactics to improve long-term performance. Through its judicious use of resources, scenario-driven planning with system dynamics makes the tactics required for implementation clear (Georgantzas 1995). And because computed scenarios reveal the required coalignment of tactics through time, SdP with SD helps firms become flexible, dependable, and efficient, and save time! Everyone's mind sees differently, but if there is truth in the adage "a picture is worth a thousand words," then the complex interrelations that SdP with SD unearth and show must be worth billions. In a world where strategic chitchat dominates, one can only hope that SdP with SD will play a central role in public and private dialogues about dynamically complex opportunities and threats.

> We shape our buildings; thereafter,
> our buildings shape us.
> —Winston Churchill

Introduction

Following on the heels of Ackoff and Emery (1972) and Christensen (1997), respectively, Gharajedaghi (1999) and Raynor (2007) show how strategies with the best chances for brilliant success expose firms to debilitating uncertainty. Firms fail as their recipes for success turn bad through time. Gharajedaghi (1999) shows, e.g., five strategy scenarios that convert success to failure. Each scenario plays a critically different role. Together, however, these scenarios form a dynamically complex system. Through time, as each scenario plays, it enables the context for the next:

1. *Noble ape* or *copycat* strategy imitates and replicates advantage. Also called "shadow marketing", it lets shadowy copycats instantly

shadow market product technology, often disruptively.

2. *Patchy* or *sluggish* strategy delays responses to new technology. When this *second* scenario plays, then patching up wastes time, enabling competitors to deliver new technology and to dominate markets. Worse, it causes costs to rise as it drives down product quality.

3. *Satisficing* or *suboptimal* strategy scenarios take many forms. One entails a false assumption: if a policy lever helps produce desired performance, then pulling or pushing on that lever will push performance further.

4. *Gambling* or *changing the game* strategy scenario transforms a strategic situation by playing the game successfully. While dealing with a challenge, firms gradually transform their strategic situation and change the basis for competition, so a whole new game and set of issues emerge. Success marked, e.g., the beginning of the *information era*. But competitive advantage has already moved away from having access to information. In our *systems era* (Ackoff 1981), creating new knowledge and generating insight is the new game (Zeleny 2005).

Lastly, the cumulative effects from these four strategy scenarios trickle down to the:

Archetypal swing or *paradigm shift* scenario. Both learning and unlearning can cause archetypal swings and paradigm shifts to unfold through time intentionally (Sterman and Wittenberg 1999). These also occur unintentionally when conventional wisdom fails to explain patterns of events that challenge prevailing mental models. The lack of a convincing explanation creates a twilight zone where acceptable ideas are not competent and competent ideas are not acceptable.

Beliefs about the future drive strategies. But the future is unpredictable. Worse, success demands commitments that make it impossible to adapt to a future that turns out surprising. So, strategies with great success potential also bear high failure probabilities. Raynor (2007) calls this

the *strategy paradox*. Dissolving it requires turning environmental uncertainty into strategic flexibility. To make it so, Raynor urges managers to: anticipate multiple futures with scenarios, formulate optimal strategies for each future, accumulate *strategic real options* (Anderson 2000) and manage the select options portfolio.

SdP with SD helps managers who operate in an uncertain world question their assumptions about how the world works, so they can see it more clearly. To survive, the human mind overestimates small risks and underestimates large risks. Likewise, it is much more sensitive to losses than to gains. So the capability to leverage opportunities and to mitigate risk might have become an economic value driver.

The purpose of computing scenarios is to help managers alter their view of reality, to match it up more closely with reality as is and as it might become. To become a leader, a manager must define reality. The SdP with SD purpose is *not*, however, to paint a more accurate picture of tomorrow, but to improve the quality of decisions about the future. Raynor says that the requisite strategic flexibility, which SdP with SD creates:

> is not a pastiche of existing approaches. Integrating these tools and grounding them in a validated theory of organizational hierarchy creates something that is quite different from any of these tools on its own, or in mere combination with the others (see p. 13 in Raynor 2007).

Indeed, knowledge of common purposes and the acceptable means of achieving them form and hold together a purposeful hierarchical system. Its members know and share values embedded in their culture, which knits parts into a cohesive whole. And because each part has a lot to say about the whole, consensus is essential to SdP with SD for the co-alignment of diverse interests and purposes.

Ackoff and Emery (1972), Gharajedaghi (1999), and Nicolis (1986) concur that purpose offers the lens one needs to see a firm as a multiminded social net. A purposeful firm produces either the same result differently in the same environment or different results in the same or different environments. Choosing among strategic real options is necessary but insufficient for purposefulness. Firms that behave differently but show only one result per environment are goal seeking, not purposeful. Servomechanisms are goal seeking but people are purposeful. As a purposeful system, the firm is part of purposeful subsystems, such as its *industry value chain* (Porter 1985) and the society. And firms have purposeful people as members. The result is a dynamically interdependent, i.e., complex, hierarchical purposeful system.

A firm's value chain is, along with its primary and support activities, at once a member of at least one industry value chain and of the society or *macroenvironment*. Industry analysis requires looking at value chains independently from the society (Porter 1985). But people, the society, and firm and industry value chains are so interdependent, so interconnected, that an optimal solution might not exist for any of them independently of the others. SdP with SD helps firms co-align the "plural rationality" of purposeful stakeholder groups with each other and that of the system as a whole.

Seeing strategic management as a *strategies and tactics net* (Georgantzas 1995) is in perfect syzygy with the *plural rationality* that SdP with SD accounts for among individuals, groups and organizations. Singer (1992, 1994) contrasts monothematic conventional universes of traditional rationality with the multiverse-directed view of plural rationality. In counterpoint, Morecroft's (1985) computed scenarios trace the dysfunctional interactions among sales objectives, overtime and sales force motivation to the intended, i.e., stated, singular rationality that drove action in a large sales organization.

Because their superordinate purpose is neither to compete nor to collaborate, but to develop new wealth-creating capabilities, in unique ways that serve both current and future stakeholder interests, customers and clients included (Moore 1991), firms can benefit from the multiverse-directed view of strategic management as a net of strategies and tactics. SdP with SD helps firms break free from the trade-offs tyranny of the mass-production era. Evidently, adherents to trade-offs-free strategy like *Bell Atlantic*, *Daimler-Benz*,

Hallmark, and *Motorola* "can have it all" (Pine et al. 1993).

A firm must serve the purposes of its people as well as those of its environment, not as a mindless mechanical system, but as a living, purposeful, *knowledge-bonded* hierarchical system (Ackoff and Emery 1972; Gharajedaghi 1999; Nicolis 1986; Zeleny 2005). To clarify, a bike always yields to its rider, e.g., regardless of the rider's desire; even if that entails running into a solid brick wall. Ouch! But riding a horse is an entirely different story. Horse and rider form a knowledge-bonded system: the horse must know the rider and the rider must know exactly how to lead the horse.

SdP with SD History: Always Back, Always in Style, Always Practical

Herman Kahn introduced scenarios to planning while at RAND Corporation in the 1950s (Kahn and Wiener 1967). Scenarios entered military strategy studies conducted for the US government. In the 1960s, Ozbekhan (1977) used urban planning scenarios in Paris, France. Organization theorists and even novelists were quick to catch on. The meaning of scenarios became literary. Imaginative improvisation produced flickering apocalyptic predictions of strikingly optimistic and pessimistic futures. Political and marketing experts use scenarios today to jazz up visions of favorable and unfavorable futures.

Wack (1985a, b) asserts it was Royal Dutch Shell that came up with the idea of scenarios in the early 1970s. Godet (1987) points to the French OTAM team as the first to use scenarios in a futures study by DATAR in 1971. Brauers and Weber (1988) claim that Battelle's scenarios method (Millet and Randles 1986) was originally a German approach. In connection with planning, however, most authors see scenario methods as typically American.

Indeed, during the 1970s, US researchers Olaf Helmer and Norman Dalkey developed scenario methods at RAND for eliciting and aggregating group judgments via Delphi and cross-impact matrices (Amara and Lipinski 1983). They extended cross impact analysis within statistical decision theory (Helmer 1983). A synthesis of scenario methods began in the 1970s that draws together multiple views, including those of professional planners, analysts and line managers.

Ansoff (1985) and other strategy theorists state that the 1970s witnessed the transformation of global markets. Today, changes in the external sociopolitical environment become pivotal in strategy making. Combined with the geographical expansion of markets, they increase the complexity of managerial work. As environmental challenges move progressively faster, they increase the likelihood of strategic surprises. So, strategic thinkers use scenarios to capture the nonlinearity of turbulent environments. Examples are Hax and Majluf (1996) and, more clearly so, Porter (1985) and Raynor (2007). They consider scenarios instrumental both in defining uncertainty and in anticipating environmental trends.

Huss and Honton (1987) see scenarios emerge as a distinct field of study, a hybrid of a few disciplines. They identify multiple scenarios methods that fall into three major categories:

1. Intuitive logics (Wack 1985a, b), now practiced by *SRI International*
2. Trend-impact analysis, practiced by the *Futures Group*
3. Cross-impact analysis, practiced by the *Center for Futures Research* using INTERAX (Interactive Cross-Impact Simulation) and by Battelle using BASICS (BAttelle Scenario Inputs to Corporate Strategies)

Similarly, after joining Ozbekhan to advocate reference scenarios, Ackoff (1981) distinguishes between:

1. Reference projections as piecemeal extrapolations of past trends and
2. The overall reference scenario that results from putting them together

Based on Acar's (1983) work under Ackoff, Georgantzas and Acar (1995) explore these distinctions with a practical managerial technology: *comprehensive situation mapping* (CSM). CSM is simple enough for MBA students to master in their capstone Business Policy course. With the help of *Vensim® PLE* (Eberlein 2002), CSM computes scenarios toward achieving a well-structured process of managing ill-structured strategic situations. In their introduction to SD, Georgantzas and Acar (see Chap. 10 in Georgantzas and Acar 1995) draw from the banquet talk that Jay Wright Forrester, Germeshausen Professor Emeritus, MIT, gave at the 1989 *International Conference of the System Dynamics Society* in Germany at the University of Stuttgart.

After attending the Engineering College, University of Nebraska, which included control dynamics at its core, Forrester went to MIT. There he worked for Gordon S. Brown, a pioneer in feedback control systems. During World War II, Brown and Forrester worked on servomechanisms for the control of radar antennas and gun mounts. This was research toward an extremely practical end, during which Forrester run literally from mathematical theory to the battlefield, aboard the US carrier *Lexington*.

After the war, Forrester worked on an analog aircraft flight simulator that could do little more than solve its own internal idiosyncrasies. So, Forrester invented *random-access magnetic storage* or *core memory*. His invention went into the heart of *Whirlwind*, a digital computer used for experimental development of military combat systems that eventually became the *semi*automatic *ground* environment (SAGE) air defense system for North America.

Alfred P. Sloan, the man who built *General Motors*, founded the *Sloan School of Management* in 1952. Forrester joined the school in 1956. Having spent fifteen years in the science and engineering side of MIT, he took the challenge of exploring what engineering could do for management.

One day, he found himself among students from *General Electric*. Their household appliance plants in Kentucky puzzled them: they would work with three or four shifts for some time and then, a few years later, with half the people laid off. Even if business cycles would explain fluctuating demand, that did not seem to be the entire reason. *GE*'s managers felt something was wrong.

After talking with them about hiring, firing, and inventory policies, Forrester did some simulation on a paper pad. He started with columns for inventories, employees, and customer orders. Given these metrics and *GE*'s policies, he could tell how many people would be hired or fired a week later. Each decision gave new conditions for employment, inventories, and production. It became clear that wholly determined internally, the system had potential for oscillatory dynamics. Even with constant incoming orders, the policies caused employment instability. That longform simulation of *GE*'s inventory and workforce system marked the beginning of system dynamics (Forrester 1958, 1961, 1987, 1992).

SdP with SD Use and Roadmap

Scenarios mostly help forecast alternative futures but, as firms abandon traditional forecasting methods for interactive planning systems, line managers in diverse business areas adopt scenario-driven planning with system dynamics. Realizing that a trade-offs-free strategy design requires insight about a firm's environment, both business and sociopolitical, to provide intelligence at *all* strategy levels, firms use SdP with SD to design *corporate, business*, and *process* or *functional* strategies. SdP with SD is not a panacea and requires discipline, but has been successful in many settings. Its transdisciplinary nature helps multiple applications, namely, capital budgeting, career planning, civil litigation (Georgantzas 2007), competitive analysis, crisis management, decision support systems (DSS), macroeconomic analysis, marketing, portfolio management, and product development (Repenning 2003). SdP with SD is a quest for managers who wish to be leaders, not just conciliators. They recognize that *logical incrementalism*, a piecemeal approach, is inadequate when the environment and their strategy change together.

Top management might see both divisional, i.e., business, and process or functional strategies as ways of implementing corporate strategy. But *active subsidiaries* (Jarillo 1988; Jarillo and Martínez 1990) provide both strategic ideas and results to their parent enterprise. Drawing too stiff a line between the corporate office and its divisions might be

an unhealthy side effect of our collective obsession with generating returns. The frameworks for developing competitive strategy that have emerged over the last thirty years have given us unparalleled insight into how companies can succeed. And competitive strategy remains enormously important, but it should be the preserve of divisional management... corporate strategy should be focused on the management of strategic uncertainty (see p. 11 in Raynor 2007).

Roadmap

It is material to disconnect scenarios from unproductive guesswork and to anchor them to sound practices for strategy design. This guided tour through the fascinating but possibly intimidating jungle of scenario definitions shows what the future might hold for SdP with SD. Extensive literature, examples, practical guidelines and two real-life cases show how computed scenarios help manage uncertainty, that necessary disciple of our open market system. Unlike extrapolation techniques, SdP with SD encourages managers to think broadly about the future.

The above sections clarify the required context and provide a glossary. Conceptual confusion leads to language games at best and to operational confusion at worst (Donaldson 1992). SdP with SD helps firms avert both types of confusion. Instead of shifting their focus away from actuality and rationality, managers improve their insight about fundamental assumptions underlying changes in strategy. The mind-set of SdP with SD makes it specific enough to give practical guidance to those managing in the real world, both now and in the future.

The sections below look at *three* SdP with SD facets linked to strategy design and implementation. The *first* facet involves the business environment, the forces behind its texture and future's requisite uncertainty (section "Environmental Turbulence and Future Uncertainty"). The *second* entails unearthing unstated assumptions about changes in the environment and in strategy, and about their potential combined effects on performance. The SdP with SD framework (section "SdP with SD: The Modeling Process ≡ Strategic Situation Formulation") builds on existing scenario methods. Its integrative view delineates processes that enhance institutional learning, bolster productivity, and improve performance through strategic flexibility. It shows why interest in computed scenarios is growing.

The *third* facet entails *computing* the combined or mixed effects on performance of changes both in the environment and in strategy. Even in mature economies, no matter how and how frequently said, decision-makers often forget how the same action yields different results as the environment changes. The result is often disastrous. Conversely, the tight coupling between computed scenarios and strategic results can create new knowledge. Linking a mixed environmental and decision scenario in a one-to-one correspondence to a strategic result suits the normative inclination of strategic management, placing rationalistic inquiry at par with purely descriptive approaches in strategy research.

The unified treatment of SdP with SD and the strategy-making process grants a practical bonus, accounting for the entry's peculiar nature. It is not only a conceptual or idea contribution, but also an application-oriented entry. Sections "Case 1: Cyprus' Environment and Hotel Profitability" and "Case 2: A Japanese Chemicals Keiretsu (JCK)" present two real-life cases of scenario-driven planning with system dynamics. Written with both the concrete and the abstract thinker in mind, the two cases show how firms and organizations build scenarios with a modest investment. SdP with SD provide an effective management technology that serves well those who adopt it. It saves them both time and energy.

Improvements in causal mapping (Eden 1994, 2004), and SD modeling and analysis (Mojtahedzadeh et al. 2004; Oliva and Mojtahedzadeh 2004) contribute to the SdP with SD trend (section "Future Directions"). Behavioral decision theory and cognitive science also help translate the knowledge of managers into SD models. The emphasis remains on small, transparent models of strategic situations and on dialogue between the managers' mental models and the computed scenarios (Morecroft 1988).

> All prognosticators are bloody fools.
> —Winston Churchill

Environmental Turbulence and Future Uncertainty

Environmental Turbulence

Abundant frameworks describe the business environment, but the one by Emery and Trist (1965), which Duncan (1972) abridged, has been guiding many a strategic thinker. It shows four business environments, each more complex and troublesome for the firm than the preceding one (Fig. 1a).

1. *Placid* or *independent-static environment*: infrequent changes are independent and randomly distributed, i.e., IID. Surprises are rare, but no new major opportunities to exploit either (*cell* 1, Fig. 1a).
2. *Placid-clustered* or *complex-static environment*: patterned changes make forecasting crucial. Comparable to the economist's idea of imperfect competition, this environment lets firms develop distinctive competencies to fit limited opportunities that lead to growth and bureaucracy (*cell* 2, Fig. 1a).
3. *Disturbed-reactive* or *independent-dynamic environment*: firms might influence patterned changes. Comparable to oligopoly in economics, this environment makes changes difficult to predict, so firms increase their operational flexibility through decentralization (*cell* 3, Fig. 1a).
4. *Turbulent field* or *complex-dynamic environment*: most frequent changes are also complex, i.e., interdependent, originating both from autonomous shifts in the environment and from interdependence among firms and conglomerates. Social values accepted by members guide strategic responses (*cell* 4, Fig. 1a).

Ansoff and McDonnell (1990) extend the dichotomous environmental uncertainty perceptions by breaking turbulent environments (*cell* 4, Fig. 1a) into *discontinuous* and *surprising*. This is a step in the right direction, but not as helpful as a causal model specific to the system structure of a firm's strategic situation. Assuredly, 2 × 2 typologies help clarify exposition and are most frequent

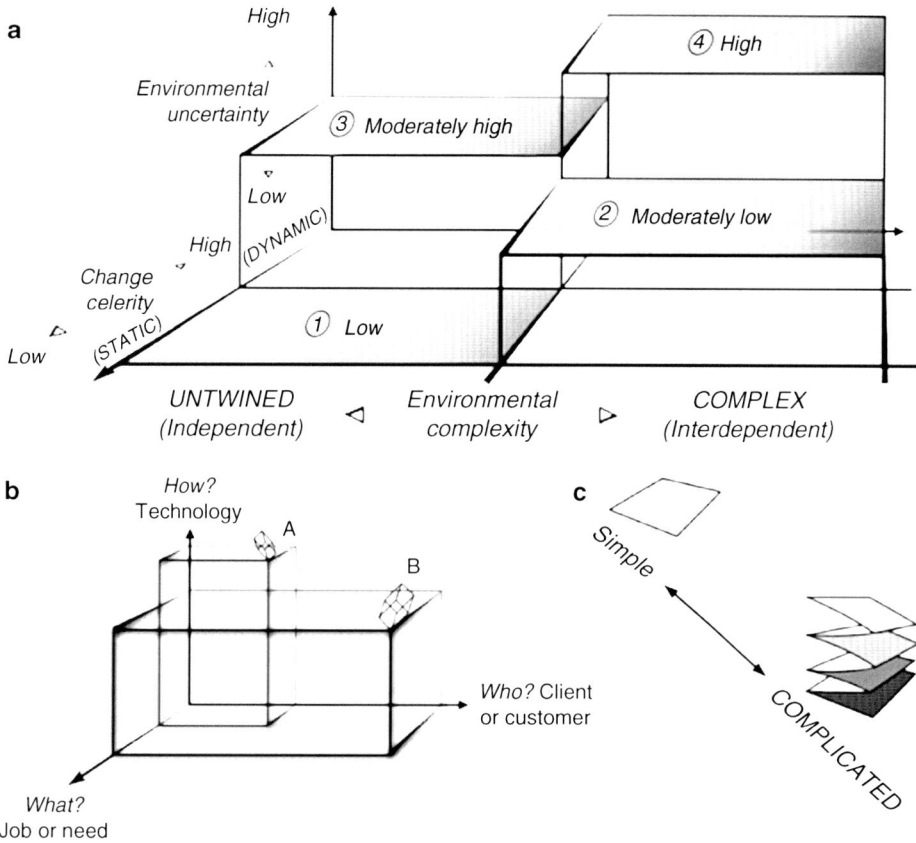

Scenario-Driven Planning with System Dynamics, Fig. 1 (**a**) Environmental complexity and change celerity dimensions that cause perceived environmental uncertainty (Adapted from Georgantzas and Acar (1995)). (**b**) Scenario-driven planning with system dynamics helps with strategy-design fundamentals, e.g., defining a business along the requisite client-job-technology three-dimensional grid. (**c**) The simple-complicated dimension must *not* be confused with the environmental complexity dimension (Adapted from Lissack and Roos (1999))

in the organization theory and strategy literatures. The mystical significance of duality affected even Leibniz, who associated one with God and zero with nothingness in the binary system. The generic solutions that dichotomies provide leave out the specifics that decision-makers need. No matter what business they are in (Fig. 1b), managers cannot wait until a better theory comes along; they must act now.

It is worth noting that people often confuse the term "complex" with "complicated." Etymology shows that *complicated* uses the Latin ending -*plic*: *to fold*, but *complex* contains the Greek root πλέξ - "*plēx-*": *to weave*. A complicated structure is thereby folded, with hidden facets stuffed into a small space (Fig. 1c). But a complex structure has interwoven parts with mutual interdependencies that cause dynamic complexity (Lissack and Roos 1999). Remember: complex is the opposite of independent or untwined (Fig. 1a) and complicated is the opposite of simple (Fig. 1c).

Daft and Weick's (1984) vista on firm *intrusiveness* and *environmental equivocality* is pertinent here. They see many events and trends in the environment as being inherently unclear. Managers discuss such events and trends, and form mental models and visions expressed in a fuzzy language and label system (Zeleny 1988). Within an *enactment* process, equivocality relates to managerial assumptions underlying the *analyzability* of the environment. A firm's *intrusiveness* determines how *active* or *passive* the firm is about environmental scanning. In this context, as the global environment gets turbulent, active firms and their subsidiaries construct SdP with SD models and compute scenarios to improve performance.

Managers of active firms combine knowledge acquisition with interpretations about the environment and their strategic situation. They reduce equivocality by assessing alternative futures through computed scenarios. In frequent meetings and debates, some by videoconferencing, managers use the dialectical inquiry process for *st*rategic *a*ssumption *s*urfacing and *t*esting (SAST), a vital strategic loop. Often ignored, the SAST loop gives active firms a strategic compass (Mason and Mitroff 1981).

Conversely, passive firms do not actively seek knowledge but reduce equivocality through rules, procedures and regular reports: reams of laser-printed paper with little or no pertinent information. Managers in passive firms use the media to interpret environmental events and trends. They obtain insight from personal contacts with significant others in their environment. Data are personal and informal, obtained as the opportunity arises.

Future Uncertainty

"If we were omnipotent," says Ackoff, then we could get "perfectly accurate forecasts" (see p. 60 in Ackoff 1981). Thank God the future is unpredictable and we must yet create it. If it were not, then life would have been so boring! Here are some facts about straight forecasting:

1. Forecasts are seldom perfect, in fact, they are always wrong, so a useful forecasting model is one that minimizes error.
2. Forecasts always assume underlying stability in systems.
3. Product family and aggregated forecasts are always more accurate than single product forecasts, so the large numbers law applies.
4. In the short term, managers can forecast but cannot act because time is too short; in the long term, they can act but cannot forecast.

To offset conundrum #4, SdP with SD juxtaposes the decomposition of performance dynamics into the growth and decline archetypes caused by *balancing* ($-$) and *reinforcing* ($+$) recursive causal-link chains or *feedback loops* (Georgantzas and Katsamakas 2007; Mojtahedzadeh et al. 2004). A thermostat is a typical example of a goal-seeking feedback loop that causes either balancing growth or decline. The gap between desired and room temperature causes action, which alters temperature with a time lag or delay. Temperature changes in turn close the gap between desired and room temperature.

Conversely, a typical loop that feeds on itself to cause either exponential growth or decline is that of an arms race. One side increases its arms. The other sides increase theirs. The first side then reacts by increasing its arms, and so on. Price wars between stores, promotional competition, shouting matches, one-upmanship, and the wild-card interest rates of the late 1970s are good examples too. Escalation might persist until the system explodes or outside intervention occurs or one side quits, surrenders, or goes out of business. In the case of wildcard interest rates, outside intervention by a regulatory agency can bring an end to irrationally escalating rates.

We've never been here before.

—Peter Senge

SdP with SD: The Modeling Process $\equiv$ Strategic Situation Formulation

The strategic management process (SMP, Fig. 2) starts with environmental scanning , in order to gauge environmental trends, opportunities, and threats. Examples include increasing rivalry among existing competitors and Porter's (1991) emphasis on the bargaining power of buyers and suppliers as well as on the threats of new entrants and substitutes. Even if some firms reduce environmental scanning to industry analysis in practice, changes in the environment beyond an industry's boundaries can determine what happens within the industry and its entry, exit and

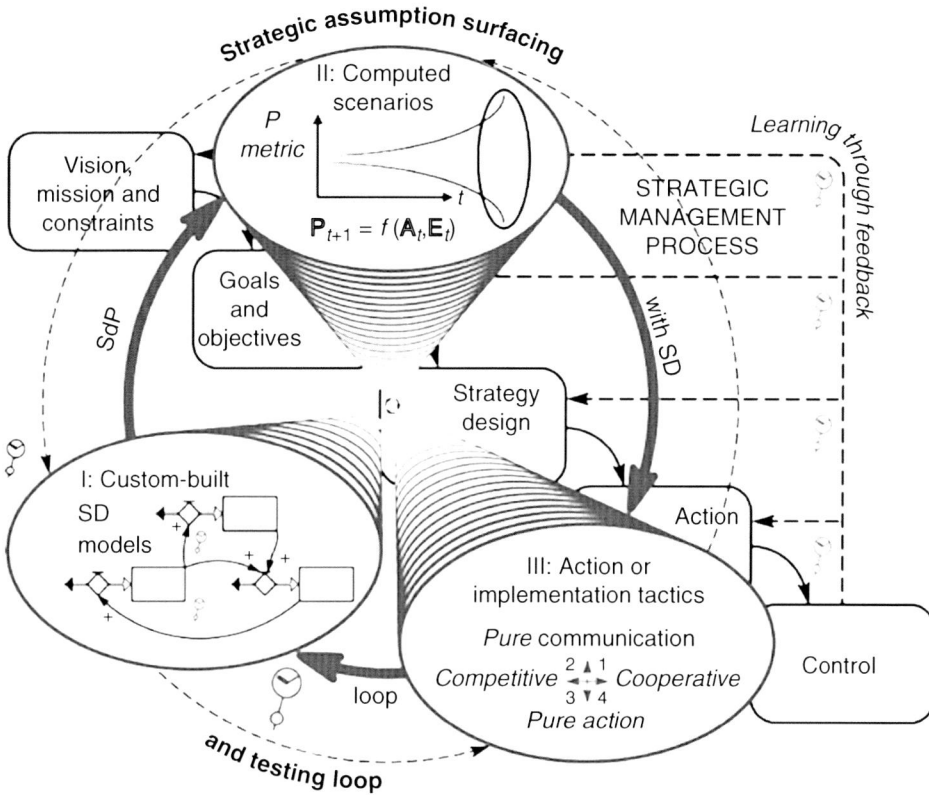

Strategic assumption surfacing

II: Computed scenarios

P metric

$P_{t+1} = f(A_t, E_t)$

Vision, mission and constraints

STRATEGIC MANAGEMENT PROCESS

Learning through feedback

SdP

Goals and objectives

with SD

Strategy design

I: Custom-built SD models

Action

III: Action or implementation tactics

Control

Pure communication

Competitive $\overset{2\ \blacktriangle\ 1}{\underset{3\ \blacktriangledown\ 4}{\rightleftarrows}}$ *Cooperative*

Pure action

loop

and testing loop

Scenario-Driven Planning with System Dynamics, Fig. 2 Cones of resolution show how scenario-driven planning with system dynamics enhances the strategy design component of the strategic management process (SMP; adapted from Georgantzas and Acar (1995))

inertia barriers. Internal capability analysis comes next. It examines a firm's past actions and internal policy levers that can both propel and limit future actions. The integrative perspective of the SdP with SD framework on Fig. 2 delineates processes that enhance institutional learning, bolster productivity, and improve performance through strategic flexibility.

Strategy design begins by identifying variables pertinent to a firm's strategic situation, along with their interrelated causal links. Changes in these variables can have profound effects on performance. Some of the variables belong to a firm's external environment. Examples are emerging new markets, processes and products, government regulations and international interest and currency rates. Changes either in these or their interrelated causal links determine a firm's performance through time.

It is a manager's job to understand the causal links underlying a strategic situation. SdP with SD helps anticipate the effects of future changes triggered in the external environment. Other variables are within a firm's control. Pulling or pushing on these internal levers also affects performance. To evaluate a change in strategy, one must look at potential results along with changes in the environment, matching resource capabilities, stakeholder purposes, and organizational goals and objectives (Fig. 2).

Most variables interact. Often, the entire set of possible outcomes is obscure, difficult to imagine. But if managers oversimplify, then they end up ignoring the combined effects of chain reactions. Even well-intended rationality often leads to oversimplification, which causes cognitive biases (CBs) that mislead decision-makers (Eilon 1984; Schwenk 1984; Simon 1979). Conversely, computing mixed environmental and decision scenarios that link internal and external metrics can reveal unwarranted simplification.

SdP with SD integrates business intelligence with strategy design, not as a narrow specialty, but as an admission of limitations and environmental complexity. It also uses multiperspective dialectics, crucial for strategic assumption surfacing and testing (SAST). Crucial because the language and labels managers use to coordinate strategic real options are imprecise and fuzzy. Fuzzy language is not only adequate *initially* for managing interdependence-induced uncertainty but required (Zeleny 1988). Decision-makers rely on it to overcome psychological barriers and Schwenk's (1984) groups of CBs.

The best-case scenario for a passive firm is to activate modeling on Fig. 2, sometimes unknowingly. When its managers boot up, e.g., electronic spreadsheets that contain inside-out causal models, with assumptions hidden deeply within many a formula. At bootup, only the numbers show. So passive-firm managers use electronic spreadsheets to laser-print matrices with comforting numbers. They

> twiddle a few numbers and diligently sucker themselves into thinking that they're forecasting the future (Schrage 1991).

And that is only when rapid changes in the environment force them to stop playing *blame the stakeholder*. They stop fighting the last war for a while, artfully name the situation a crisis, roll up their sleeves, and chat about and argue, but quickly agree on some arbitrary interpretation of the situation to generate strategic face-saving options. Miller and Friesen (see pp. 225–227 in Miller and Friesen 1983) show how for futile firms, rapid environmental changes lead to crisis-oriented decisions. Conversely, successful firms look far into the future as they counter environmental dynamism through strategy design with real options. Together, their options and interpretation of the environment, through the consensus that SdP with SD facilitates, enable a shared logic to emerge: a shared mental model that filters hidden spreadsheet patterns and heroic assumptions clean and clear.

Managers of active firms enter the SdP with SD loop of Fig. 2 both consciously and conscientiously. They activate strategic intelligence via computed scenarios and the SAST loop. Instead of twiddling spreadsheet numbers, *proactive* firm managers twiddle model assumptions. They stake, through SD model diagrams, their intuition about how they perceive the nature and structure of a strategic situation. Computed scenarios quantitatively assess their perceived implications. Having quantified the implications of shared visions and claims about the structure of the strategic situation, managers of active firms are likely to reduce uncertainty and equivocality. Now they can manage strategic interdependence. Because articulated perception is the starting point of all scenarios, computed scenarios give active firms a fair chance at becoming fast strategic learners.

The design of action or implementation tactics requires detailing how, when, and where a strategy goes into action. In addition to assuming the form of *pure communication* (III: 1 and 2, Fig. 2) or *pure action* (III: 3 and 4, Fig. 2), in a pragmatic sense, tactics can be either cooperative or competitive and defensive or offensive. Market location tactics, e.g., can be either offensive, trying to rob market share from established competitors, or defensive, preventing competitors from stealing one's market share. An offensive tactic takes the form of frontal assault, flanking maneuver, encirclement, bypass attack or guerilla warfare. A defensive tactic might entail raising structural barriers, increasing expected retaliation or lowering the inducement for future attack. Conversely, cooperative tactics try to gain mutual advantage by working with rather than against others. Cooperative tactics take the form of alliances, joint ventures, licensing agreements, mutual service consortia and value-chain partnerships, the co-location of which often creates industrial districts (Georgantzas 2001b).

The usual copycat strategy retort shows linear thinking at best and clumsy *benchmarking*, also known as shadow marketing, at worst. Its proponents assume performance can improve *incrementally*, with disconnected tactics alone, when strategy design is of primary concern. Piecemeal tactics can undermine strategy, but they are secondary. It might be possible to improve performance through efficient tactics, but is best to design strategies that expel counterproductive tactics. Counterproductive tactics examples are coercive moves that increase rivalry, without a real payoff, either direct or indirect, for the industry incumbent who initiates them. It is atypical of an industry or market leader to initiate such moves.

In strategy, superb action demands superior design. According to the design school, which Ansoff, Channon, McMillan, Porter, Thomas, and others lead, logical incrementalism may help implementation, but becomes just another prescription for failure when the environment shifts. Through its judicious use of corporate resources, SdP with SD makes the tactics required for action clear. Also, it reveals their proper coalignment through time, so a firm can build strategic flexibility and save time!

The Modeling Process ≡ Strategic Situation Formulation

SdP with SD (Fig. 2) begins by modeling a business or "social process" than a business or "social system." It is more productive to identify a *social process* first and then seek its causes than to slice a chunk of the real world and ask what dynamics it might generate. Distinguishing between a *social system* and a social process is roughly equivalent to distinguishing between a system's underlying causal structure and its dynamics. Randers (see p. 120 in Randers 1980) defines a social system as a set of cause and effect relations. Its structure is a causal diagram or map of a real-world chunk. A social process is a behavior pattern of events evolving through time. The simulation results of SdP with SD models show such chains of events as they might occur in the real world. An example of a social system (structure) is the set of rules and practices that a firm might enact when dealing with changes in demand, along with the communication channels used for transmitting information and decisions. A corresponding social process (dynamics) might be the stop-and-go pattern of capital investment caused by a conservative bias in a firm's culture.

In his model of a new, fast-growing product line, e.g., Forrester (1961) incorporates such a facet of corporate culture. Causing sales to stagnate, considerable back orders had to accumulate to justify expansion because the firm's president insisted on personally controlling all capital expenditures.

People often jump into describing system structure, perhaps because of its tangible nature as opposed to the elusive character of dynamics or social process fragments. Also, modelers present model structure first and then behavior. Ultimately, the goal in modeling a strategic situation is to link system structure and behavior. Yet, in the early stages of modeling it is best to start with system dynamics and then seek underlying causes. Indeed, SD is particularly keen on understanding system performance, "not structure per se" (see p. 331 in Oliva 2004), in lieu of SD's core tenet that structure causes performance.

The modeling process itself is recursive in nature. The path from real-world events, trends, and negligible externalities to an effective formal model usually resembles an expanding spiral (Fig. 3a). A useful model requires conceptualization; also focusing the modeling effort by establishing both the time horizon and the perspective from which to frame a decision situation. Typically, strategy-design models require a long-term horizon, over which computed scenarios assess the likely effects of changes both in strategy and in the environment.

Computer simulation is what makes SdP with SD models most useful. Qualitative cause-and-effect diagrams are too vague and tricky to simulate mentally. Produced through knowledge elicitation, their complexity vastly exceeds the human capacity to see their implications. Casting a chosen perspective into a formal SdP with SD model entails postulating a detailed structure; a diagramming description precise enough to propagate images of alternative futures, i.e., computed scenarios, "though not necessarily accurate" (see p. 118 in Randers 1980). But the modeling process must never downplay the managers' mental database and its knowledge content. Useful models always draw on that mental database (Forrester 1961).

Following Morecroft (1988), SdP with SD adopters might strive to replace the notion of

Scenario-Driven Planning with System Dynamics, Fig. 3 The recursive nature of the modeling process that scenario-driven planning with system dynamics entails (**a**) creates a sustainable, ever-expanding vortex of insight and wisdom, needed in strategic real-options valuation, and (**b**) saves both time and money as it renders negligible the cost of resistance (*R*) to change

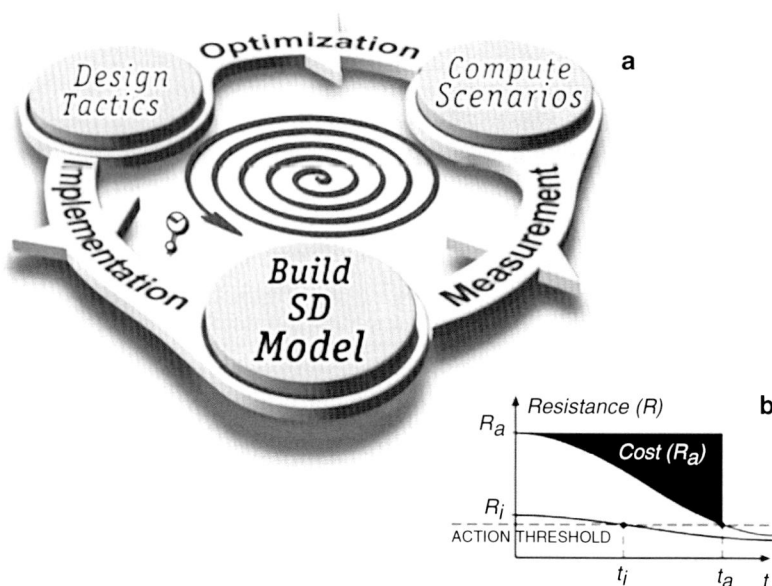

modeling an objectively singular world *out there*, with the much softer approach of building formal models to improve managers' mental models. The expanding spiral of Fig. 3a shows that the insight required for decisive action increases as the quantity of information decreases, by orders of magnitude. The required quantification of the relations among variables pertinent to a strategic situation changes the character of the information content as one moves from mental to written to numerical data. Perceptibly, a few data remain, but much more pertinent to the nature and structure of the situation. Thanks to computed scenarios, clarity rules in the end. And, if the modeling process stays *i*nteractive (i), as opposed to *a*ntagonistic (a), then clarity means low resistance to change ($R_i < R_a$, Fig. 3b), which helps reach a firm's action/implementation threshold quickly ($t_i < t_a$, Fig. 3b). This is how SdP with SD users build strategic flexibility while they save both time and money!

Case 1: Cyprus' Environment and Hotel Profitability

Cyprus' Hotel Association wished to test how Cyprus' year 2010 official tourism strategy might affect tourist arrivals, hotel bed capacity and profitability, and the island's environment (Georgantzas 2003). Computed with a system dynamics simulation model, four tourism growth scenarios show what might happen to Cyprus' tourism over the next 40 years, along with its potential effects on the sustainability of Cyprus' environment and hotel profitability. The following is a partial description of the system dynamics model that precedes its dynamics.

Model Description (Case 1)

The SD model highlights member interactions along Cyprus' hotel value chain. The model incorporates a generic value-chain management structure that allows modeling customer-supplier value chains in business as well as in physical, biological and other social systems. Although the structure is generic, its situation-specific parameters faithfully reproduce the dynamic behavior patterns seen in Cyprus' hotel value-chain processes, business rules, and resources.

Cyprus' Environment, Population, and Tourism Model Sectors

Within Cyprus' environment and population sector (Fig. 4a), the carbon dioxide (CO_2) pollution stock is the accumulation of Cyprus' anthropogenic emissions less the Mediterranean Sea region's self-clean-up rate. The clean-up rate that drains Cyprus' CO_2 pollution depends on the level of anthropogenic pollution itself as well as on the average clean-up time and its standard deviation (sd). Emissions that feed CO_2 pollution depend on Cyprus' population and tourism and on emissions per person (Brenkert 1998).

In SD models, rectangles represent stocks, i.e., level or state variables that accumulate through time, e.g., the Tourism stock on Fig. 4b. The double-line, pipe-and-valve-like icons that fill and drain the stocks, often emanating from cloud-like sources and ebbing into cloud-like sinks, represent material flows that cause the stocks to change. The arrive rate in Fig. 4b, e.g., shows tourists who flow into the tourism stock per month. Single-line arrows represent information flows, while plain text or circular icons depict auxiliary constant or converter variables, i.e., behavioral relations or decision points that convert information into decisions. Changes in the

tourism stock, e.g., depend on annual tourism, adjusted by tourism seasonality. Both the diagram on Fig. 4a and Table 1 are reproduced from the actual simulation model, first built on the glass of a computer screen using the diagramming interface of *iThink*® (Richmond et al. 2006), and then specifying simple algebraic equations and parameter values. Built-in functions help quantify policy parameters and variables pertinent to Cyprus' tourism situation.

There is a one-to-one correspondence between the model diagram in Fig. 4a and its equations (Table 1). Like the diagram, the equations are the actual output from *iThink*® too. The equations corresponding to Fig. 4b are archived in Georgantzas (2003). Together, Cyprus' population, local tourism, and monthly tourism determine the population and tourism sum (Eq. 1.11, Table 1). According to CYSTAT (2000), both *Cyprus' Tourism Organization* and its government attach great importance to local tourism. A study on domestic tourism conducted in 1995

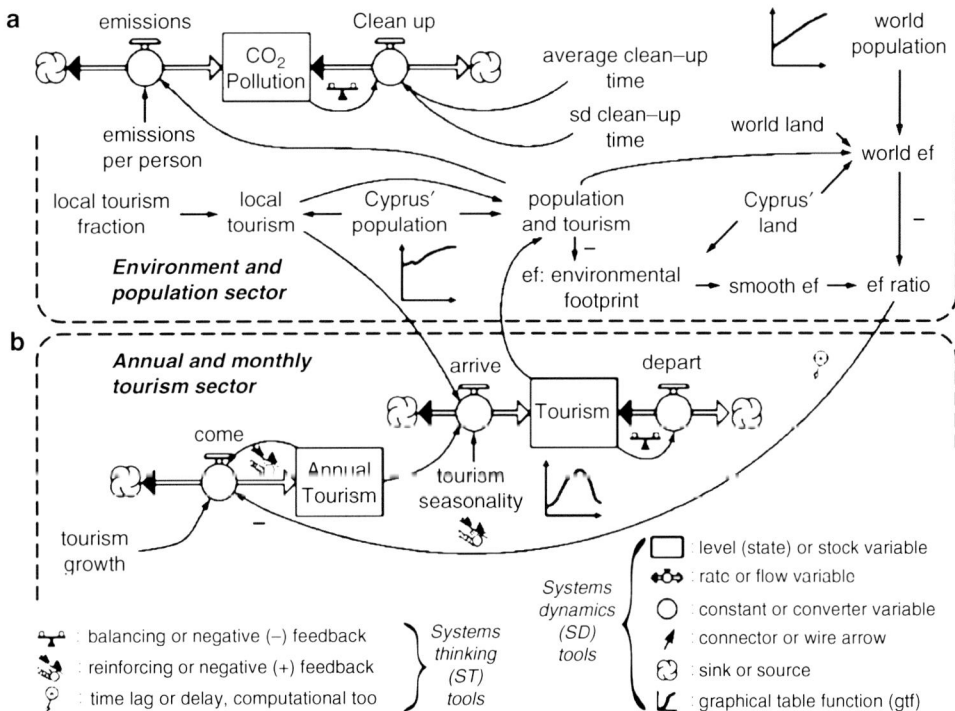

Scenario-Driven Planning with System Dynamics, Fig. 4 Cyprus' (**a**) environment and population and (**b**) annual and monthly tourism model sectors (Adapted from and extending (Georgantzas 2003))

Scenario-Driven Planning with System Dynamics, Table 1 Cyprus' environment and population (and local tourism) model sector (Fig. 4a) equations, with variable, constant parameter, and unit definitions

Level or state variables (stocks)	Eq. #
$CO_2Pollution(t) = CO_2Pollution(t - dt) + (\text{emissions-clean-up})^* dt$	(1.1)
INIT CO_2 Pollution = emissions (Based on 1995 gridded carbon dioxide anthropogenic emission data; unit: 1000 metric ton C per one degree latitude by one degree longitude grid cell)	(1.1.1)
Rate variables (**flows**)	
Emissions = emissions per person* population and tourism (unit: 1000 metric tons C/month)	(1.2)
Cleanup = max(0, $CO_2Pollution$/average clean-up time) (unit: 1000 metric tons C/month)	(1.3)
Auxiliary variables and constants (**converters**)	
Average clean-up time = 120 (Med Sea region average self-clean-up time = 100 years; unit: months)	(1.4)
Cyprus' land = If (time $\leq$ 168) then (9251 * 247.1052) else ((9251 3355) * 247.1052) (Cyprus' free land area; unit: acres; 1 km^2 = 247.1052 acres)	(1.5)
EF ratio = smooth EF/world EF (unit: unitless)	(1.6)
EF: environmental footprint = Cyprus' land/population and tourism (unit: acres/person)	(1.7)
Emissions per person = 1413.4/702000/12 (unit: anthropogenic emissions/person/month)	(1.8)
Local tourism = local tourism fraction * Cyprus' population (unit: persons/month)	(1.9)
Local tourism fraction = 0.46*(0.61 + 0.08) (Percentages based on a 1995 study on domestic tourism; unit: unitless)	(1.10)
Population and tourism = Cyprus' population + tourism − local tourism (Subtracts local tourists already included in Cyprus' population; unit: persons)	(1.11)
Sd clean-up time = 240 (clean-up time standard deviation = 20 years; unit: months)	(1.12)
Smooth EF = SMTH3 (EF: environmental footprint, 36) (Third-order exponential smooth of EF)	(1.13)
World EF = (world land − Cyprus' land)/(world population − population and tourism) (unit: acres/person)	(1.14)
World land = 36677577730.80 (unit: acres)	(1.15)
Cyprus' population = GRAPH (time/12) (Divided by 12 since these are annual data; unit: persons) (0.00, 493984), (1.00, 498898), (2.00, 496570), (3.00, 502001), (4.00, 505622), (5.00, 509329), (6.00, 512950), (7.00, 516743), (8.00, 520968), (9.00, 525364), (10.0, 529847), (11.0, 534330), (12.0, 539934), (13.0, 546486), (14.0, 552348), (15.0, 526313), (16.0, 516054), (17.0, 515881), (18.0, 518123), (19.0, 521657), (20.0, 526744), (21.0, 532692), (22.0, 538210), (23.0, 544675), (24.0, 551659), (25.0, 558038), (26.0, 560366), (27.0, 568469), (28.0, 572622), (29.0, 578394), (30.0, 587392), (31.0, 598217), (32.0, 609751), (33.0, 619658), (34.0, 626534), (35.0, 632082), (36.0, 636790), (37.0, 641169), (38.0, 645560), (39.0, 649759), (40.0, 653786), (41.0, 657686), (42.0, 661502), (43.0, 665246), (44.0, 668928), (45.0, 672554), (46.0, 676147), (47.0, 679730), (48.0, 683305), (49.0, 686870), (50.0, 690425), (51.0, 693975), (52.0, 697524), (53.0, 701056), (54.0, 704547), (55.0, 707970), (56.0, 711305), (57.0, 714535), (58.0, 717646), (59.0, 720613), (60.0, 723415), (61.0, 726032), (62.0, 728442), (63.0, 730629), (64.0, 732578), (65.0, 734280), (66.0, 735730), (67.0, 736928), (68.0, 737887), (69.0, 738627), (70.0, 739172), (71.0, 739540), (72.0, 739743), (73.0, 739792), (74.0, 739697), (75.0, 739472), (76.0, 739123), (77.0, 738658), (78.0, 738083), (79.0, 737406), (80.0, 737406)	(1.16)
World population = GRAPH (time/12) (Divided by 12 since these are annual data; unit: persons) (0.00, 3e+09), (1.00, 3.1e+09), (2.00, 3.1e+09), (3.00, 3.2e+09), (4.00, 3.3e+09), (5.00, 3.3e+09), (6.00, 3.4e+09), (7.00, 3.5e+09), (8.00, 3.6e+09), (9.00, 3.6e+09), (10.0, 3.7e+09), (11.0, 3.8e+09), (12.0, 3.9e+09), (13.0, 3.9e+09), (14.0, 4e+09), (15.0, 4.1e+09), (16.0, 4.2e+09), (17.0, 4.2e+09), (18.0, 4.3e+09), (19.0, 4.4e+09), (20.0, 4.5e+09), (21.0, 4.5e+09), (22.0, 4.6e+09), (23.0, 4.7e+09), (24.0, 4.8e+09), (25.0, 4.9e+09), (26.0, 4.9e+09), (27.0, 5e+09), (28.0, 5.1e+09), (29.0, 5.2e+09), (30.0, 5.3e+09), (31.0, 5.4e+09), (32.0, 5.4e+09), (33.0, 5.5e+09), (34.0, 5.6e+09), (35.0, 5.7e+09), (36.0, 5.8e+09), (37.0, 5.8e+09), (38.0, 5.9e+09), (39.0, 6e+09), (40.0, 6.1e+09), (41.0, 6.2e+09), (42.0, 6.2e+09), (43.0, 6.3e+09), (44.0, 6.4e+09), (45.0, 6.5e+09), (46.0, 6.5e+09), (47.0, 6.6e+09), (48.0, 6.7e+09), (49.0, 6.8e+09), (50.0, 6.8e+09), (51.0, 6.9e+09), (52.0, 7e+09), (53.0, 7e+09), (54.0, 7.1e+09), (55.0, 7.2e+09), (56.0, 7.2e+09), (57.0, 7.3e+09), (58.0, 7.4e+09), (59.0, 7.5e+09), (60.0, 7.5e+09), (61.0, 7.6e+09), (62.0, 7.6e+09), (63.0, 7.7e+09), (64.0, 7.8e+09), (65.0, 7.8e+09), (66.0, 7.9e+09), (67.0, 8e+09), (68.0, 8e+09), (69.0, 8.1e+09), (70.0, 8.1e+09), (71.0, 8.2e+09), (72.0, 8.3e+09), (73.0, 8.3e+09), (74.0, 8.4e+09), (75.0, 8.4e+09), (76.0, 8.5e+09), (77.0, 8.5e+09), (78.0, 8.6e+09), (79.0, 8.6e+09), (80.0, 8.7e+09)	(1.17)

revealed that about 46 percent of Cypriots take long holidays. Of these, 61 percent take long holidays exclusively in Cyprus and eight percent in Cyprus and abroad, while 31 percent chose to travel abroad only. These are precisely the percentages in the model (Eq. 1.10, Table 1).

In Fig. 4a, the world land and population data, minus Cyprus' land, population, and tourism co-determine the world EF (environmental footprint, Eq. 1.14, Table 1). Compared to Cyprus' smooth EF, i.e., the smooth ratio of the island's free land divided by its total population and tourism, the world EF gives a dynamic measure of Cyprus' relative attractiveness to the rest of the world. The EF ratio (Eq. 1.6, Table 1), i.e., the ratio of Cyprus' smooth EF (Eq. 1.13, Table 1) divided by the world EF (Eq. 1.14, Table 1), assumes that the higher this ratio is, the more attractive the island is to potential tourists, and vice versa. The EF ratio, which depends on Cyprus' total population and tourism, feeds back to the island's annual tourism via the inflow of foreign visitors who come to visit Cyprus every year (Fig. 4b).

The logistic or Verhulst growth model, after François Verhulst, who published it in 1838 (Richardson 1991), helps explain Cyprus' actual annual tourism, a quantity that cannot grow forever (Fig. 4b). Every system that initially grows exponentially eventually approaches the carrying capacity of its environment, whether it is food supply for moose, the number of people susceptible to infection, or the potential market for a good or a service. As an "autopoietic" system approaches its limits to growth, it goes through a nonlinear transition from a region where positive feedback dominates to a negative-feedback-dominated regime. S-shaped growth often results: a smooth transition from exponential growth to equilibrium.

The logistic model conforms to the requirements for S-shaped growth and the ecological idea of carrying capacity. The population it models typically grows in a fixed environment, such as Cyprus' foreign annual tourism has done since 1960 up to 2000. Initially dominated by positive feedback, Cyprus' annual tourism might soon reach the island's carrying capacity, with a nonlinear shift to dominance by negative feedback. While accounting for Cyprus' tourism lost to the summer of 1974 Turkish invasion, officially a very long "military intervention," further depleting annual tourism is the outflow of Cyprus' visitors (not shown here) who might go as the island's free area reaches its Carrying Capacity, estimated at seventy times the number of Cyprus' visitors in 1960 (Georgantzas 2003).

Cyprus' Hotel Association listed Cyprus tourism seasonality as one of its major concerns. At the time of this investigation, CYSTAT (2000) had compiled monthly tourism data for only 30 months. These were used for computing Cyprus' tourism seasonality (Fig. 4b). Incorporating both the foreign annual tourism and the monthly tourism stocks in the model allows both looking at the big picture of annual tourism growth and assessing the potential long-term effects of tourism seasonality on the sustainability of Cyprus' environment and hotel EBITDA, i.e., *e*arnings *b*efore *i*nterest, *t*axes, *d*epreciation, and *a*mortization. The publicly available actual annual tourism data allow testing the model's usefulness, i.e., how faithfully it reproduces actual data between 1960 and 2000 (Georgantzas 2003).

Cyprus' foreign visitors and local tourists arrive at the island's hotels and resorts according to Cyprus' tourism seasonality, thereby feeding Cyprus' monthly tourism stock. About 11.3 days later, according to CYSTAT's (2000) estimated average stay days, both foreign visitors and local tourists depart, thereby depleting the monthly tourism stock. By letting tourism growth = 0, and Cyprus' tourism seasonality continue repeating its established pattern, the model computes a zero-growth or *base-run* scenario. Subsequently, however, tourism growth values other than zero initiate different scenarios.

Cyprus' Tourism Growth Scenarios (Case 1)

What can Cyprus' hoteliers expect to see in terms of bottom-line dynamics? According to the four tourism-growth scenarios computed on Fig. 5, seasonal variations notwithstanding, the higher Cyprus' tourism growth is, the lower hotel EBITDA (smooth hE) is, in the short term. In

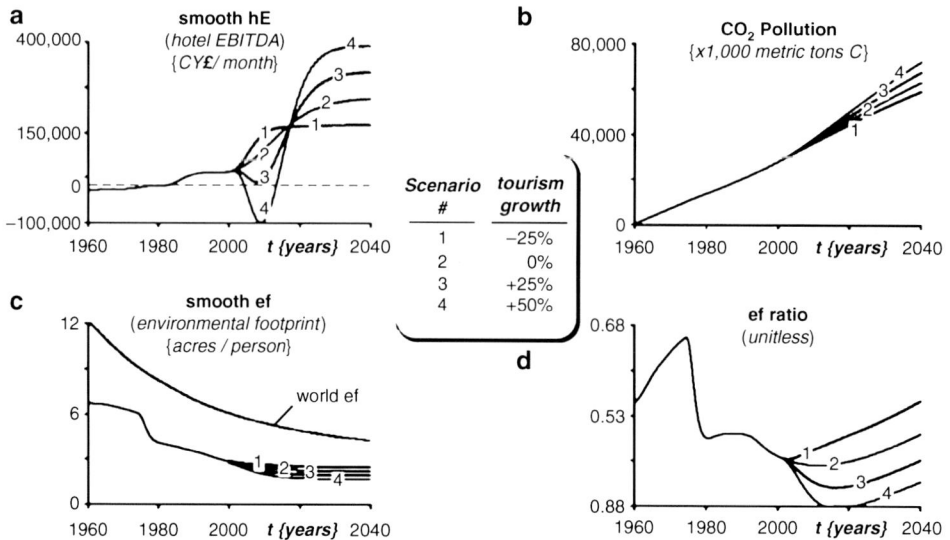

Scenario-Driven Planning with System Dynamics, Fig. 5 Four computed scenarios show how tourism growth might affect Cyprus' hotel EBITDA (smooth hE) and the island's environment, with carbon-dioxide (CO_2) pollution (Adapted from and extending (Georgantzas 2003))

the long term, however, higher tourism growth yields higher profitability in constant year 2000 prices.

High tourism growth implies accommodating overbooked hotel reservations for tourists who actually show up. Free cruises erode Cyprus' hotel EBITDA. The alternative is, however, angry tourists going off in hotel lobbies. Tourists have gotten angry at hotels before, but hotels have made the problem worse in recent years worldwide (Drucker and Higgins 2001). They have tightened check-in rules, doubled their renovations, and increased the rate of overbooking by about 30 percent. The results can be explosive if one adds the record flight delays that travelers endure. Anyhow, free cruises to nearby Egypt and Israel sound much better than simply training employees to handle unhappy guests that scream in hotel lobbies.

Eventually, as Cyprus' bed capacity increases and thereby catches up with tourism demand, there will be less overbooking and a few free cruises to erode Cyprus' hotel EBITDA. Given enough time for an initial bed capacity disequilibrium adjustment, in the long term, high tourism growth increases both hotel EBITDA (Fig. 5a) and cash (Georgantzas 2003).

In addition to the profound consequences for its hotel value-chain participants, Cyprus' tourism growth might also determine the fate of the island's environment. Depending on the island's population and emissions per person, high tourism growth implies high anthropogenic emissions feeding Cyprus' CO_2 pollution. Anthropogenic CO_2 emissions attributed to the upward and downward movements of recurring tourist arrivals create much more stress and strain for the island's natural environment than a consistent stream of tourism with low seasonality would. High tourism growth lowers Cyprus' environmental footprint (EF). The 1974 summer Turkish *military intervention* has had a drastic negative effect on Cyprus' relative attractiveness because it reduced the island's free land by 41 percent.

Although qualitatively similar to the world's average EF after the invasion, Cyprus' environmental footprint is lower than the world's average EF (Fig. 5c), rendering the island's free area relatively less attractive as more foreign tourists visit. Manifested in the EF ratio (Fig. 5d), Cyprus becomes relatively less attractive as more visitors choose to vacation on the island's free area.

Qualitatively, Cyprus' CO_2 pollution scenarios (Fig. 5b) look exactly like the A2 scenario family of harmonized anthropogenic CO_2 emissions, which the *Intergovernmental Panel on Climate Change* (IPCC) computed to access the risks of human-induced climate change (Nakicenovic et al. 2000). Like in the rest of the world, unless drastic changes in policy or technology alter the emissions per person ratio in the next 40 years, CO_2 pollution is expected to grow proportionally with Cyprus' tourism, degrading the island's environment.

Case 2: A Japanese Chemicals Keiretsu (JCK)

Home of NASA's *Johnson Space Center*, the Clear Lake region in Texas boasts strong high technology, biotechnology, and specialty chemicals firms. Among them is JCK, whose recent investment helps the Clear Lake region continue its stalwart role in Houston's regional economic expansion (Hodgin 2001).

An active member of a famous Japanese giant conglomerate, JCK's history began in the late 1800s. Despite its long history, it has not been easy for JCK to evade the feedback loop that drives Japanese firms to manufacture outside Japan. Since the 1950s, with Japan still recovering from WWII, the better Japanese companies performed, the better their national currency did. But the better Japan's currency did, the harder it became for its firms to export. The higher the yen, the more expensive and, therefore, less competitive Japan's exports become. This simple loop explains JCK's manufacturing lineage from Japan to USA (Georgantzas et al. 2002).

But the transition process behind this lineage is not that simple. JCK's use of SdP with SD reveals a lot about its strategy design and implementation tactics. The model below shows a tiny fragment of JCK's gigantic effort to re-perceive itself. The firm wants to see its keiretsu transform into an agile, virtual enterprise network (VEN) of active agents that collaborate to achieve its transnational business goals. Although still flying low under the media's collective radar screen, VENs receive increased attention by strategic managers (Georgantzas 2001b).

Sterman (see Chaps. 17 and 18 in Sterman 2000) presents a generic value-chain management structure that can unearth what VENs are about. By becoming a VEN, JCK is poised to bring the necessary people and production processes together to form *autopoietic*, i.e., self-organizing, customer-centric value chains in the specialty chemicals industry. JCK decided to build its own plant in the USA because the net present value (NPV) of the anticipated combined cash flow resulting from a merger with other specialty chemicals manufacturers in USA would have been less than the sum of the NPVs of the projected cash flows of the firms acting independently. Moreover, JCK's own technology transfer cost is so low that the internalization cost associated with a merger would far exceed supplier charges plus market transaction costs. To remain competitive (Porter 1991), JCK will not integrate the activity but offshoot it as a branch of its VEN-becoming keiretsu. The plant will be fully operational in January 2008. In order to maximize the combined *net present value of earnings before interest, taxes, depreciation, and amortization*, i.e., NPV(EBITDA), of its new USA plant and the existing one in Asia, JCK wishes to improve its US sales revenue before production starts in the USA.

JCK's pre-production marketing tactics aim at building a sales force to increase sales in the USA. Until the completion of the new plant (Dec. 2007), JCK will keep importing chemicals from its plant in Asia. Once production starts in the USA (Jan. 2008), then the flow of goods from Asia to the USA stops, the plant in the USA supplies the US market and the flow of goods from the USA to Asia begins.

Strategic scenarios are not new to the chemical industry (Zentner 1987). SdP with SD helps this specialty chemicals producer integrate its business intelligence efforts with strategy design in anticipation of environmental change. Modeling JCK's strategic situation requires a comprehensive inquiry into the environmental causalities and equivocalities that dictate its actions. Computed strategic and tactical scenarios probe the combined consequences of environmental trends, changes in JCK's own strategy, and the moves of its current and future competitors. The section below describes briefly how JCK plans to implement its transnational strategy of balanced marketing and production. This takes the form of a system dynamics simulation model, which precedes the interpretation of its computed scenarios.

Model Description (Case 2)

The entire model has multiple sectors, four of which compute financial accounting data. Figure 6a shows the production and sales, and Fig. 6b the total NPV(EBITDA) model sectors. The corresponding algebra is in Georgantzas et al. (2002). While JCK is building its US factory, its factory in Asia makes and sells all specialty chemicals the US market cannot yet absorb. This is what the *feed-forward* link from the production in Asia flow to the sales in Asia rate shows. The surplus demand JCK faces in Asia for its fine chemicals accounts for this rather unorthodox model structure. The surplus demand in Asia is the model's enabling *safety valve*, i.e., a major

strategic assumption that renders tactical implementation feasible.

With the plant in Asia producing at full capacity before the switch, sales in the USA both depletes the tank in Asian stock and reduces sales in Asia. The US sales depend on JCK's US sales force. But the size of this decision variable is just one determinant of sales in the USA.

Sales productivity depends on many parameters, such as the annual growth before the switch rate of specialty chemicals in the USA, the average expected volume a salesperson can sell per month, and on the diminishing returns that sales people experience after the successful calls they initially make on their industrial customers. B2B, or business to business, i.e., industrial marketing, can sometimes be as tough as B2C or business to customer, i.e., selling retail.

Time t = 30 months corresponds to January 2008, when the switch time converter cuts off the supply of JCK's chemicals from its plant in Asia.

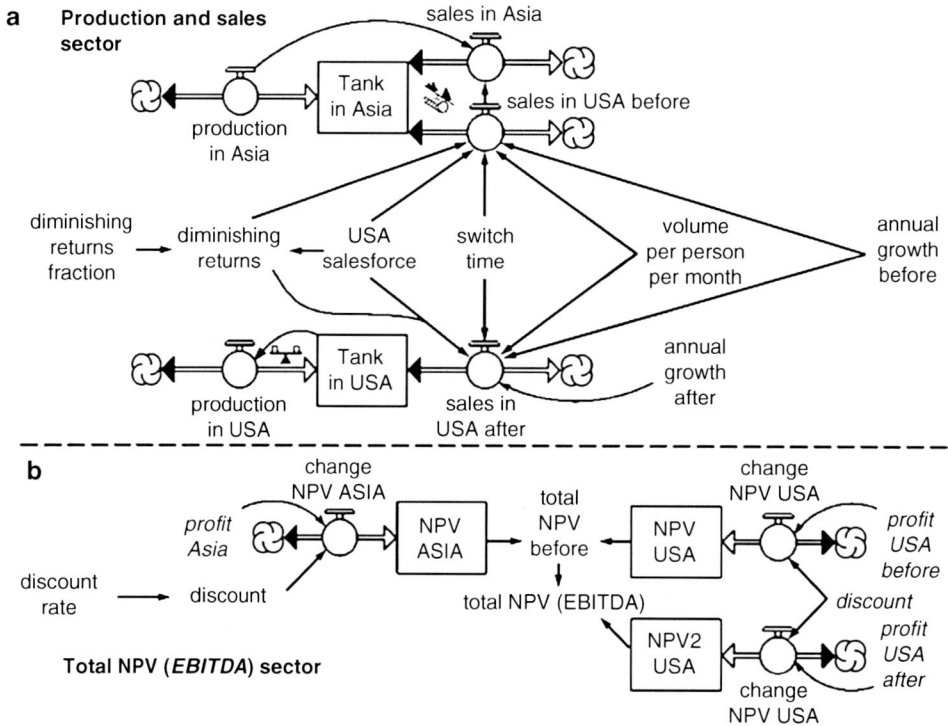

Scenario-Driven Planning with System Dynamics, Fig. 6 JCK's (**a**) production and sales and (**b**) total NPV (EBITDA) model sectors (Adapted from Georgantzas et al.

(2002); NPV = net present value, and EBITDA = earnings before interest, taxes, depreciation and amortization)

Ready by December 2007, the factory in the USA can supply the entire customer base its US sales force will have been building for 30 months. As production in the USA begins, the sales in the USA before flow stops draining the tank in Asia and sales in Asia resume to match JCK's surplus demand there. Acting both as a production flow and as a continuous-review inventory order point, after January 2008, production in the USA feeds the tank in US stock of the rudimentary value-chain management structure on Fig. 6a.

Value chains entail stock and flow structures for the acquisition, storage, and conversion of inputs into outputs, and the decision rules that govern the flows. The jet ski value chain includes, e.g., hulls and bows that travel down monorail assembly paths. At each stage in the process, a stock of parts buffers production. This includes the inventory of fiberglass laminate between hull and bow acquisition and usage, the inventory of hulls and bows for the jet ski lower and upper structures, and the inventory of jet skis between dealer acquisition and sales. The decision rules governing the flows entail policies for ordering fiber-glass laminate from suppliers, scheduling the spraying of preformed molds with layers of fiberglass laminate before assembly, shipping new jet skis to dealers and customer demand.

A typical firm's or VEN's value chain consists of cascades of supply chains, which often extend beyond a firm's boundaries. Effective value chain models must incorporate different agents and firms, including suppliers, the firm, distribution channels, and customers. Scenario-driven planning with system dynamics is well suited for value chain modeling and policy design because value chains involve multiple stock and flow chains, with time lags or delays, and the decision rules governing the flows create multiple feed-back loops among VEN members or value- and supply-chain partners (see Chaps. 17 and 18 in Sterman 2000).

Back to JCK, its tank in the USA feeds information about its level back to production in the USA. Acting first as a decision point, production in the USA compares the tank in the US level to the tank's capacity. If the tank is not full, then production in the USA places an order to itself and, once the USA factory has the requisite capacity, production in the USA refills the tank in the USA, but only until sales in the USA after the switch drains the tank. Then the cycle begins all over again.

Meanwhile, the profit in Asia, profit in the USA before, and profit in the USA after sectors (Georgantzas et al. 2002) perform all the financial accounting necessary to keep track of the transactions that take place in the value chain production and sales sector (Fig. 6a). As each scenario runs, the profit in Asia, the USA before and the USA after sectors feed the corresponding change in net present value (NPV) flows of the model's total NPV(EBITDA) sector (Fig. 6b). By adjusting each profit sector's EBITDA according to the discount rate, the change in NPV flows compute the total NPV(EBITDA) both in Asia and in the USA, both before and after JCK's January 2008 supply switch.

JCK's Computed Scenarios (Case 2)

Recall that the SdP with SD modeling-process spiral enabled our modeling team to crystallize JCK's strategic situation into the cyclical pattern that Fig. 3a shows. Although heavily disguised, the JCK measurement data and econometric sales functions let the system dynamics model compute scenarios to answer that razor-sharp optimization question the JCK executives asked:

> What size a USA sales force must we build in order to get a smooth switch in both sales and production in January 2008, and also to maximize the combined NPV(EBITDA) at our two plants in Asia and USA from now through 2012?

Treating the US sales force policy parameter in the "Sensi Specs..." menu item of *iThink*[®] allowed computing a set of 30 strategic scenarios. The 30 scenarios correspond to JCK's hiring from 1 to 30 sales people, respectively, to sell specialty chemicals to manufacturing firms in the USA, both before and after the January 2008 switch. Figures 7 and 8 show the 30 computed scenarios.

Figure 7c shows the response surfaces the production in US rate and tank in the US stock form after January 2008, in response to the 30 computed scenarios. The computed scenario

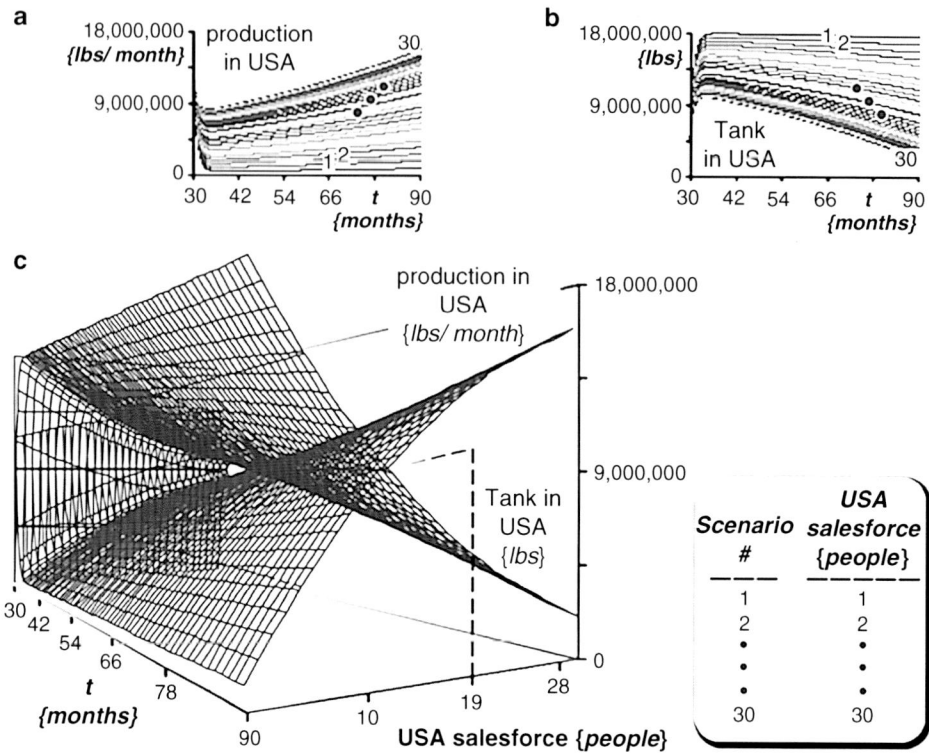

Scenario-Driven Planning with System Dynamics, Fig. 7 Thirty computed scenarios show JCK's dual, smooth-switch, and profitable purpose in production (Adapted from Georgantzas et al. (2002))

that corresponds to JCK's building a US sales force of 19 people achieves a smooth balance between sales in Asia and in the USA. Under this scenario, after January 2008, on the line where the two surfaces cross each other, not only the number of pounds of chemicals made and sold in Asia equals the number of pounds of chemicals made and sold in the USA, but as Fig. 7c shows, production in the USA also equals the tank in US stock. So hiring 19 sales people now meets JCK's smooth switch in sales and production objective. But how?

How does producing and selling in the USA at rates equal to the corresponding rates in Asia constitute a fair response to JCK's smooth switch objective? The JCK executives seemed to accept this at face value. But our team had to clearly explain the dynamics of JCK's rudimentary USA value chain (Fig. 6a), in order to unearth what the USA member of this transnational VEN-becoming keiretsu might be up to.

It looks simple, but the value chain of the production and sales sector in Fig. 6a can show the same amplification symptoms that much more elaborate value chains show when they fall prey to bullwhip effects. Locally rational policies that create smooth and stable adjustment of individual business units can, through their interaction with other functions and firms, cause oscillation and instability. Figure 8a shows the profound consequences of JCK's switch for its value chain in the USA. Because of the sudden switch in January 2008, the computed scenarios cause 30 sudden step changes. Both variables' adjustment rates increase, but the tank in the US stock's amplification remains almost constant below 50 percent. As customer demand steps up, so do both metrics' new equilibrium points, but in direct proportion to the step increase in customer demand in the USA.

The 30 computed scenarios confirm Sterman's argument that, while the magnitude of amplification depends on stock adjustment times and

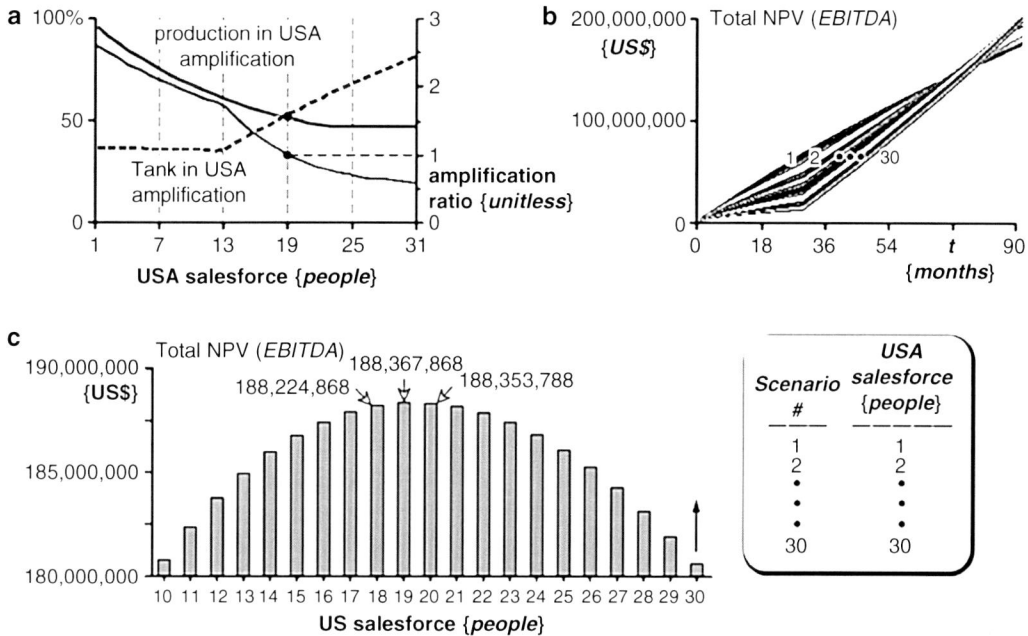

Scenario-Driven Planning with System Dynamics, Fig. 8 Thirty computed scenarios show how hiring a sales force of 19 people in the USA might maximize JCK's NPV(EBITDA), and thereby fulfill its dual, smooth-switch, and profitable purpose (Adapted from Georgantzas et al. (2002))

delivery lags, its existence does not. No matter how drastically customers and firms downstream in a value chain change an orders' magnitude, they cannot affect supply chain amplification. Value chain managers must never blame customers and downstream firms or their forecasts for bullwhip effects. The production in US amplification is almost double the tank in USA for a small US sales force, suggesting that JCK's US factory faces much larger changes in demand than its sales people do. Although temporary, during its disequilibrium adjustment, the tank in the USA consistently overshoots the new equilibrium points that it seeks after the switch (Fig. 7b), an inevitable consequence of stock and flow structure. Customers are innocent, but JCK's value chain structure is not.

First, the tank in the US stock adjustment process creates significant amplification of the production in the US rate. Though the tank in the US relative amplification is 36.18 percent under the US sales force = 1 scenario, e.g., the relative amplification of production in the USA (Fig. 8a) increases by a maximum of more than 90 percent

(the peak production in the US rate, after t = 30 months, divided by the minimum production in the US rate = 11.766.430.01/1 , 026 , 107.64 = 91.28 percent). The *amplification ratio*, i.e., the ratio of maximum change in output to maximum change in input, therefore, is 91.28% / 36.18% = 2.52. A one-percent increase in demand for JCK's chemicals causes a 2.52 percent surge in demand at JCK's USA plant. While the amplification ratio magnitude depends on the stock adjustment times and delivery lags, its existence does not (see p. 673 in Sterman 2000).

Second, amplification is temporary. In the long run, a one-percent increase in sales in the USA after leads to a one-percent increase in production in the USA. After two-adjustment times, i.e., two months, production in the USA gradually drops (Fig. 7a). During the disequilibrium adjustment, however, production in the USA overshoots its new equilibrium, an inevitable consequence of the stock and flow structure of customer-supplier value chains, no matter how tiny or simple they are. The only way the tank in the US stock can increase is for its inflow production in the US rate (order rate) to exceed its outflow rate sales in the USA after (Fig. 6a). Within a VEN's or keiretsu's customer-supplier value chain, supply agents face much larger changes in demand than finished-goods inventory, and the surge in demand is temporary.

The computed scenarios show that as the US sales force increases, the production in the USA's rate of amplification declines because its new long-term equilibrium point is closer to its initial jump in January 2008. Conversely, as the tank in the US stock's long-term equilibrium point remains consistently high because of the larger US sales force, its relative amplification begins to rise. Since the two variables' relative amplification moves in opposite directions, eventually, they meet. What a coincidence! They meet above the US sales force = 19 people. Now, is this not a much better interpretation of the word "smooth" in fair response to JCK's smooth-switch performance purpose? The answer to JCK is now pertinent to its balancing its value chain in the USA. With a US sales force = 19, JCK's value chain components show equal relative amplification to sudden changes in demand, attaining nothing less than a magnificent amplification ratio = 1. Now that is smooth!

But what of profitability? JCK's polite executives said: "maximize... combined... NPV." In the time domain (Fig. 8b), total NPV(EBITDA) creates intricate dynamics that obscures the US sales force effect. But the phase plot on Fig. 8c clearly shows a concave down behavior along the US sales force: US sales force = 19 maximizes the two plants' combined total NPV(EBITDA).

Future Directions

The above cases show how scenario-driven planning with system dynamics helps control performance by enabling organizational learning and the management of uncertainty. The strategic intelligence system that SdP with SD provides rests on the idea of a collective inquiry, which translates the environmental "macrocosm" and a firm's "microcosm" into a shared causal map with computed scenarios. Informed discussion then takes place. Seeing SdP with SD as an inquiry system might help the outcomes of the situation formulation-solution-implementation sequence, each stage built on successive learning.

Strategic situations are complex and uncertain Because planning is directed toward the future, predictions of changes in the environment are its indispensable components. Conventional forecasting by itself provides no cohesive way of understanding the effect of changes that might occur in the future. Conversely, SdP with SD and its computed scenarios provide strategic intelligence and a link from traditional forecasting to modern interactive planning systems. In today's quest for managers who are more leaders than conciliators, the strategists' or executives' interest in scenarios must be welcomed. A clearer delineation of SdP with SD might make it a very rich field of application and research.

The SdP with SD inquiry system on Fig. 2 includes several contributions. *First*, by translating the environmental *macrocosm* and the firm *microcosm* into a common context for conceptualization, the requisites of theory building can be addressed. Planning analysts no longer have to operate piecemeal. A theory and a dominant logic typically emerge from shared perceptions about a firm, its environment, and stakeholder purposes through model construction.

Second, the outputs of the strategic management process activities build on each other as successive layers. The SAST loop on Fig. 2 follows the counterclockwise direction of multiperspective dialectics (Mason and Mitroff 1981). This process allows adjustment of individual and organizational theories and logic, leading to an evolutionary interpretation of the real system that strategic decisions target.

Third, the inquiry system in Fig. 2 enables flexible support for all phases of strategy design. Problem finding or forming, or situation formulation receives equal attention as problem solving.

SdP with SD helps open up the black box of decision-makers' mental models, so they can specify the ideas and rules they apply. That in turn helps enrich their language and label system, organizational capability and knowledge, and the strategic decision processing system. Computed scenarios bring about transformation rules not previously thought of as well as new variables and interaction paths.

As an entity, each decision-maker has a local scope and deals only with specific variables and access paths to other entities. But success factors are not etched in stone. Often, we only observe a representative state of each entity, namely, locally meaningful variables and parts of a scenario. This representativeness changes dynamically in the process of computing scenarios. Beyond the purely technical advantages of computed scenarios, planning becomes interactive, and language and label systems render themselves more adequate, effective, and precise. Their associated organizational capability develops even more. In addition, the minor and major assumptions in decision-makers' mental models surface as computed scenarios specify the conditions under which performance changes.

A line of great immediate concern requires researchers and practitioners alike to explore the modeling process behind SdP with SD. For the sake of realism, to make negotiated perceptions of reality explicit, we need representations where strategic real options and self-interest projections mold the way in which managers incorporate their observations and interpretations into strategy models. This is an unavoidable, most challenging path to tread, if we want to build a dialectical debate into the strategy design process.

Do we really want to? Yes, because:

1. The traditional hierarchical organization dogma has been planning, managing, and controlling, whereas the new reality of the learning organization incorporates vision, values, and mental models. It entails training managers and teams in the IPRD learning cycle conceived by Dewey (1938) (cf. Senge and Sterman 1992):

2. In the strategic management process (SMP) evolution, planning is evolving too, from objective-driven to budget-driven to strategy-driven to scenario-driven planning with system dynamics (SdP with SD, see pp. 271–272 in Georgantzas and Acar 1995).

3. The inquiry system that mediates the restructuring of organizational *theory in use* (Schön 1983) determines the quality of organizational learning.

By looking into the dynamics of strategy design and the resulting performance of firms, the SdP with SD framework on Fig. 2 might let managers, planners, and business researchers see the tremendous potential of computed strategic scenarios. They might choose to build intelligence systems around SdP with SD to create insight for strategy design . They will be building real knowledge in the process, while developing capability for institutional learning. Both Pascale (1984) and de Geus (1992) see the capability to speed up institutional learning as a truly sustainable competitive advantage.

Bibliography

Primary Literature

Acar W (1983) Toward a theory of problem formulation and the planning of change: causal mapping and dialectical debate in situation formulation. UMI, Ann Arbor

Ackoff RL (1981) Creating the corporate future. Wiley, New York

Ackoff RL, Emery FE (1972) On purposeful systems. Aldine-Atherton, Chicago

Amara R, Lipinski AJ (1983) Business planning for an uncertain future: scenarios and strategies. Pergamon, New York

Anderson TJ (2000) Real options analysis in strategic decision making: an applied approach in a dual options framework. J Appl Manag Stud 9(2):235–255

Ansoff HI (1985) Conceptual underpinnings of systematic strategic management. Eur J Oper Res 19(1):2–19

Ansoff HI, McDonnell E (1990) Implanting strategic management, 2nd edn. Prentice-Hall, New York

Brauers J, Weber M (1988) A new method of scenario analysis for strategic planning. J Forecast 7(1):31–47

Brenkert AL (1998) Carbon dioxide emission estimates from fossil-fuel burning, hydraulic cement production, and gas flaring for 1995 on a one degree grid cell basis. Oak Ridge National Laboratory, Carbon Dioxide Information Analysis Center, Oak Ridge, TN (Database: NDP-058A 2-1998)

Christensen CM (1997) The innovator's dilemma: when new technologies cause great firms to fail. Harvard Business School Press, Cambridge

CYSTAT (2000) Cyprus key figures: tourism. The Statistical Service of Cyprus (CYSTAT), Nicosia

Daft RL, Weick KE (1984) Toward a model of organizations as interpretation systems. Acad Manag Rev 9:284–295

Dewey J (1938) Logic: the theory of inquiry. Holt, Rinehart and Winston, New York

Donaldson L (1992) The Weick stuff: managing beyond games. Organ Sci 3(4):461–466

Drucker J, Higgins M (2001) Hotel rage: losing it in the lobby. Wall Street J:W1–W7

Duncan RB (1972) Characteristics of organizational environments and perceived environmental uncertainty. Adm Sci Q 17:313–327

Eberlein RL (2002) Vensim® PLE software, V 5.2a. Ventana Systems Inc, Harvard

Eden C (1994) Cognitive mapping and problem structuring for system dynamics model building. Syst Dyn Rev 10(3):257–276

Eden C (2004) Analyzing cognitive maps to help structure issues or problems. Eur J Oper Res 159:673–686

Eilon S (1984) The art of reckoning: analysis of performance criteria. Academic, London

Emery FE, Trist EL (1965) The causal texture of organizational environments. Hum Relat 18:21–32

Forrester JW (1958) Industrial dynamics: a major breakthrough for decision makers. Harv Bus Rev 36(4):37–66

Forrester JW (1961) Industrial dynamics. MIT Press, Cambridge

Forrester JW (1987) Lessons from system dynamics modeling. Syst Dyn Rev 3(2):136–149

Forrester JW (1992) Policies, decisions and information sources for modeling. Eur J Oper Res 59(1):42–63

Georgantzas NC (1995) Strategy design tradeoffs-free. Hum Syst Manag 14(2):149–161

Georgantzas NC (2001a) Simulation modeling. In: Warner M (ed) International encyclopedia of business and management, 2nd edn. Thomson Learning, London, pp 5861–5872

Georgantzas NC (2001b) Virtual enterprise networks: the fifth element of corporate governance. Hum Syst Manag 20(3):171-188, with a 2003 reprint ICFAI J Corp Gov 2(4):67–91

Georgantzas NC (2003) Tourism dynamics: Cyprus' hotel value chain and profitability. Syst Dyn Rev 19(3):175–212

Georgantzas NC (2007) Digest® wisdom: collaborate for win-win human systems. In: Shi Y et al (eds) Advances in multiple criteria decision making and human systems management. IOS Press, Amsterdam, pp 341–371

Georgantzas NC, Acar W (1995) Scenario-driven planning: learning to manage strategic uncertainty. Greenwood-Quorum, Westport

Georgantzas NC, Katsamakas E (2007) Disruptive innovation strategy effects on hard-disk maker population: a system dynamics study. Inf Resour Manag J 20(2):90–107

Georgantzas NC, Sasai K, Schrömbgens P, Richtenburg K et al (2002) A chemical firm's penetration strategy, balance and profitability. In: Proceedings of the 20th international system dynamics society conference, 28 July–1 Aug, Villa Igiea

de Geus AP (1992) Modelling to predict or to learn? Eur J Oper Res 59(1):1–5

Gharajedaghi J (1999) Systems thinking: Managing chaos and complexity – a platform for designing business architecture. Butterworth-Heinemann, Boston

Godet M (1987) Scenarios and strategic management: prospective et planification stratégique. Butterworths, London

Godet M, Roubelat F (1996) Creating the future: the use and misuse of scenarios. Long Range Plan 29(2):164–171

Hax AC, Majluf NS (1996) The strategy concept and process: a pragmatic approach, 2nd edn. Prentice Hall, Upper Saddle River

Helmer O (1983) Looking forward. Sage, Beverly Hills

Hodgin RF (2001) Clear lake area industry and projections 2001. Center for Economic Development and Research, University of Houston-Clear Lake, Houston

Huss WR, Honton EJ (1987) Scenario planning: what style should you use? Long Range Plan 20(4):21–29

Istvan RL (1992) A new productivity paradigm for competitive advantage. Strateg Manag J 13(7):525–537

Jarillo JC (1988) On strategic networks. Strateg Manag J 9(1):31–41

Jarillo JC, Martinez JI (1990) Different roles for subsidiaries: the case of multinational corporations in Spain. Manag J 11(7):501–512

Kahn H, Wiener AJ (1967) The next thirty-three years: a framework for speculation. Daedalus 96(3):705–732

Lissack MR, Roos J (1999) The next common sense: the e-manager's guide to mastering complexity. Nicholas Brealey Publishing, London

Mason RO, Mitroff II (1981) Challenging strategic planning assumptions. Wiley, New York

Miller D, Friesen PH (1983) Strategy making and environment: the third link. Strateg Manag J 4:221–235

Millet SM, Randles F (1986) Scenarios for strategic business planning: a case history for aerospace and defence companies. Interfaces 16(6):64–72

Mojtahedzadeh MT, Andersen D, Richardson GP (2004) Using Digest® to implement the pathway participation method for detecting influential system structure. Syst Dyn Rev 20(1):1–20

Moore GA (1991) Crossing the chasm. Harper-Collins, New York

Morecroft JDW (1985) Rationality in the analysis of behavioral simulation models. Manag Sci 31:900–916

Morecroft JDW (1988) System dynamics and microworlds for policymakers. Eur J Oper Res 35:310–320

Nakicenovic N, Davidson O, Davis G et al (2000) Summary for policymakers-emissions scenarios: a special report of working group III of the intergovernmental panel on climate change (IPCC). World Meteorological Organization (WMO) and United Nations Environment Programme (UNEP), New York

Nicolis JS (1986) Dynamics of hierarchical systems: an evolutionary approach. Springer, Berlin

Oliva R (2004) Model structure analysis through graph theory: partition heuristics and feedback structure decomposition. Syst Dyn Rev 20(4):313–336

Oliva R, Mojtahedzadeh MT (2004) Keep it simple: a dominance assessment of short feedback loops. In: Proceedings of the 22nd international system dynamics society conference, 25–29 July 2004. Keble College, Oxford University, Oxford, UK

Ozbekhan H (1977) The future of Paris: a systems study in strategic urban planning. Philos Trans R Soc Lond 387:523–544

Pascale RT (1984) Perspectives on strategy: the real story behind Honda's success. Calif Manag Rev 26:47–72

Pine BJ-I, Victor B, Boynton AC (1993) Making mass customization work. Harv Bus Rev 71(5):108–115

Porter ME (1985) Competitive advantage: creating and sustaining superior performance. Free Press, New York

Porter ME (1991) Towards a dynamic theory of strategy. Strateg Manag J 12:95–117

Randers J (1980) Guidelines for model conceptualization. In: Randers J (ed) Elements of the system dynamics method. MIT Press, Cambridge, pp 117–139

Raynor ME (2007) The strategy paradox: why committing to success leads to failure [and what to do about it]. Currency-Doubleday, New York

Repenning NP (2003) Selling system dynamics to (other) social scientists. Syst Dyn Rev 19(4):303–327

Richardson GP (1991) Feedback thought in social science and systems theory. University of Pennsylvania Press, Philadelphia

Richmond B et al (2006) iThink® Software V 9.0.2. iSee Systems™, Lebanon

Schön D (1983) Organizational learning. In: Morgan G (ed) Beyond method. Sage, London

Schrage M (1991) Spreadsheets: bulking up on data. San Francisco Examiner

Schwenk CR (1984) Cognitive simplification processes in strategic decision making. Strateg Manag J 5(2):111–128

Senge PM, Sterman JD (1992) Systems thinking and organizational learning: acting locally and thinking globally in the organization of the future. Eur J Oper Res 59(1):137–150

Simon HA (1979) Rational decision making in business organizations. Am Econ Rev 69(4):497–509

Singer AE (1992) Strategy as rationality. Hum Syst Manag 11(1):7–21

Singer AE (1994) Strategy as moral philosophy. Strateg Manag J 15:191–213

Sterman JD (2000) Business dynamics: systems thinking and modeling for a complex world. Irwin McGraw-Hill, Boston

Sterman JD, Wittenberg J (1999) Path dependence, competition and succession in the dynamics of scientific revolution. Organ Sci 10(3, Special issue: Application of complexity theory to organization science, May-June):322–341

Turner F (1997) Foreword: chaos and social science. In: Eve RA, Horsfall S, Lee ME (eds) Chaos, complexity and sociology. Sage, Thousand Oaks, pp xi–xxvii

Wack P (1985a) Scenarios: uncharted waters ahead. Harv Bus Rev 63(5):73–89

Wack P (1985b) Scenarios: shooting the rapids. Harv Bus Rev 63(6):139–150

Zeleny M (1988) Parallelism, integration, auto-coordination and ambiguity in human support systems. In: Gupta MM, Yamakawa T (eds) Fuzzy logic in knowledge-based systems, decision and control. Elsevier Science, North Holland, pp 107–122

Zeleny M (2005) Human systems management: Integrating knowledge, management and systems. World Scientific, Hackensack

Zentner RD (1987) Scenarios and the chemical industry. Chem Marketing Manag:21–25

Books and Reviews

Bazerman MH, Watkins MD (2004) Predictable surprises: the disasters you should have seen coming and how to prevent them. Harvard Business School Press, Boston

Bower G, Morrow D (1990) Mental models in narrative comprehension. Science 247(4938):44–48

Mittelstaedt RE (2005) Will your next mistake be fatal? Avoiding the chain of mistakes that can destroy your organization. Wharton School Publishing, Upper Saddle River

Morecroft JDW (2007) Strategic modeling and business dynamics: a feedback systems approach. Wiley, West Sussex

Schnaars SP (1989) Megamistakes: forecasting and the myth of rapid technological change. The Free Press, New York

Schwartz P (1991) The art of the long view. Doubleday-Currency, New York

Tuchman B (1985) The March of folly: from troy to Vietnam. Ballantine Books, New York

Vennix JAM (1996) Group model building: facilitating team learning using system dynamics. Wiley, Chichester

Engineering of Strategy: A General, Unified Theory of Performance and Strategic Management

Kim Warren
Strategy Dynamics Limited, Princes Risborough, UK

Article Outline

Introduction
The Theory/Practice Disconnect in Strategic Management
Requirements of a General Unified Theory (GUT)
The Question to Be Answered
Explaining *Immediate* Performance
Explaining the Changing Quantities of Resources
Explaining Rates of Change to Resources
Causal Ambiguity, Intangible Factors, and Capabilities
Feedback and the Business System
Competition
Fulfilling the Requirements for a GUT
Future Directions
Bibliography

Introduction

Strategy scholars and others have long been concerned about the limited relevance and usage of strategy theories and methods by business executives and management consultants. Reflecting on many years spent designing strategy for organizations and then teaching the topic with the field's standard frameworks and theories, it seemed that this disconnect with real-world strategic management may arise from some serious, yet basically simple problems. First, most research into the causes of enterprise performance attempts to answer the wrong question – what explains profitability differences between firms? In reality, as has been axiomatic in the Finance field since the 1970s, investors invest in firms in order to access hoped-for growth in free cash-flow (the cash flow that remains after paying interest and taxes and financing *future* growth plans). Secondly, the field has looked in the wrong places for the answer – twice! During its early history, it was assumed that exogenous factors largely determined the profitability firms could achieve. When it was shown that firm-specific factors were more influential (Rumelt 1991), research turned to focus on firms' internal features but chose to neglect the tangible resources of which any enterprise consists – simple things like customers, staff, product-range, and physical capacity. Lastly, research has relied on the deployment of statistical methods (notably multi-variate regression analysis) that simply cannot handle the accumulation processes, threshold effects, interdependencies, and feedback that pervade any real-world case.

Fortunately, the science of System Dynamics, which applies the discipline of engineering control-theory to socioeconomic systems (and many others) *does* handle all of those mechanisms and thus overcomes the severe limitations of statistical methods when it comes to explaining the changing performance of interconnected systems (Forrester 1968). Its use results in working, quantified, time-based simulations that have an uncanny capacity to replicate real-world behavior, not only of performance outcomes but of all the elements of the enterprise that *give rise to* that performance. We will demonstrate the basic principles that form the foundation of such simulation models with an initial model of the low-fare airline Ryanair. Since system dynamics can provide the same rigorous explanation for the performance of any organization, or any part thereof, of any scale, in any sector, it offers the foundations for a general, unified theory of the link between strategic management (what organizations choose to do) and performance outcomes.

© Springer Science+Business Media, LLC, part of Springer Nature 2020
B. Dangerfield (ed.), *System Dynamics*,
https://doi.org/10.1007/978-1-4939-8790-0_660

Originally published in
R. A. Meyers (ed.), *Encyclopedia of Complexity and Systems Science*, © Springer Science+Business Media LLC 2019, https://doi.org/10.1007/978-3-642-27737-5_660-1

The Theory/Practice Disconnect in Strategic Management

Strategy writers have long expressed concern that the field is not sufficiently relevant to the real world (Hambrick 2004; Ghoshal 2005; Farjoun 2007; Whittington and Jarzabkowski 2008). There is also suggestive evidence that its tools, methods, and frameworks (hereafter, simply "methods") are not widely used (Rigby and Bilodeau 2013) and may be regarded by practitioners as being of limited value and relevance (Coyne and Subramaniam 2000).

Efforts to explain this patchy deployment of strategy methods have focused on sociological explanations (Jarzabkowski and Kaplan 2014) in which methods are treated as "technologies of rationality" that may prove inappropriate when confronted with the reality of experience. Consequently, rather than using such methods to inform rational decision-making (Cabantous and Gond 2011; Jarratt and Stiles 2010), those methods are either adapted by practitioners or used merely as conceptual frameworks to guide thinking in some undefined manner (Jarzabkowski and Kaplan 2014).

This strategy-as-practice perspective implicitly accepts that a mismatch exists between strategy methods and the practical experience of the real world that is impossible to eliminate. Yet a strong professional field would be expected to deploy methods grounded in theories that are sufficiently reliable to become embedded in real-world practice, and thus achieve high relevance; indeed, relevance so great that professional practice *requires* those methods to be employed. Among fields related to management, this is already observable across all engineering disciplines, and in Finance, where the principles of discounted cash-flow (Burr 1938) and the capital-asset pricing model (Treynor and Black 1976) have provided the bedrock for professional practice in corporate investment and valuation for many decades (Koller et al. 2010).

It follows, then, that the limited relevance and use of strategy methods may not be inevitable but reflect instead an absence of strong, core theories, which has left practitioners with little choice but to do the best they can with the methods available. It has also left the door open to the fads and fashions of journalistic writing on strategy that make no pretense to theoretical validity. It is therefore reasonable to conclude that the strategy field should seek stronger theoretical foundations – ideally, a general, unified theory – that can provide substantially improved technologies of rationality. The alternative is to accept that no such rigorous theory and no reliable methods are possible, and continue to hope that strategy-as-practice can be improved in some way that will result in better strategic management of organizations and better performance outcomes. Given the importance of sound strategic management of all kinds of enterprise, such surrender to hopelessness is unacceptable.

Requirements of a General Unified Theory (GUT)

As the most integrative of management disciplines, a strategy GUT would need to fulfil wide-ranging and demanding requirements. It should be consistent with, and preferably integrate, existing theories and principles recognized in the field and reflected in practice methods that have proved at all useful. It should be consistent with, and if possible integrate, accepted theories and principles in other disciplines. A GUT must handle organizations' diverse objectives, including trade-offs that may need to be made (such as that between current profitability and growth), and should explain fully the link from those outcomes back to their causes. Correlation alone between causes and outcomes is not adequate – even though that is the standard test required of published articles in the strategic management field – a GUT should also demonstrate complete and valid causal pathways between those causes and outcomes.

To achieve widespread, reliable use in practice, a GUT should employ factors, concepts and measures that are identifiable and measurable in the real world, avoiding abstract concepts that are not directly observable. It should apply not only to single-activity businesses in any industry, but also to sub-units of a business, and to distinct business functions. It should be extensible to corporate strategy challenges of multi-business organizations and to multiple participating entities, such as business networks.

A GUT should be applicable at all times and deal with significant changes occurring between different points in time. It should inform the *continuous* management of strategy from period to period, as well as occasional decisions, such as choice of strategic position, acquisitions and entry to new markets. Since public services, voluntary and other noncommercial organizations also pursue long-term aims, a GUT should be as applicable to those contexts as it is to commercial cases.

Given these already challenging requirements, it is too much to expect any GUT to deal with both strategy content (*what to do, when and how much, at all times and across all functions, with what impact on longer-term performance*) and strategy process (*how to influence whom to develop and deliver that strategy and performance*). This paper therefore focuses on strategy content alone. However, a substantially improved strategy method should also assist the strategy process, since increased confidence in the rational technology would degrade uncertainty and the scope for dissent between participants engaged in developing and managing strategy.

The Question to Be Answered

Although other objectives must be addressed, and will be considered below, financial aims for commercial firms are widespread, and so provide a useful start-point for developing a GUT. It is axiomatic in the Finance field that the value to investors of a business or of any significant investment reflects the expected stream of future earnings – specifically free cash-flows (*profit after interest and tax*, plus *depreciation*, minus *capital invested* and *increases in working capital*; Brealey et al. 2007). When the aim is to deliver value for investors, then, the answer required from evaluating a strategy or strategic initiative is a cash-flow forecast. The weakness of existing strategy methods is shown by the fact that no means has been

documented by which any such method or combination of methods might be used to achieve such a forecast.

A GUT must therefore explain *growth* of cash flows from period to period, rather than profitability ratios, and inform management how to sustain such growth over long periods (Penrose 1959; Rugman and Verbeke 2002). This is of practical importance because, while firms may differ by a few percentage points of profit-margin, strongly performing firms deliver orders-of-magnitude more growth in cash flow than weaker rivals. It also implies that research in strategic management has for many decades largely focused on the wrong question – sustained superior profitability ratios –which are of only indirect relevance to the sustained growth in free cash-flow that is the main concern of business owners and managers.

For non-financial objectives, such as market share, service quality or reputation, improving the indicator from period to period is also the focus, since step changes to such items are rarely possible. A GUT should explain, and provide future estimates for, two classes of objective – average *performance* over some *period of time*, such as profit per year or average service quality during a month, and targets for a *quantity of some asset* to be achieved by a *point in time*, such as customer numbers or trained staff by the end of a year.

Figure 1 shows an example of the question that an aspiring theory of strategic management and performance may be expected to answer – why the

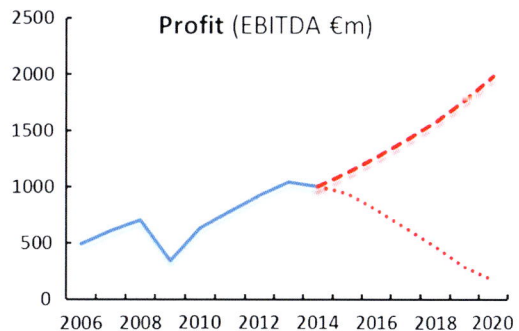

Engineering of Strategy: A General, Unified Theory of Performance and Strategic Management, Fig. 1 EBITDA profit for Ryanair plc, 2006 to 2020

cash operating profit of a company (Ryanair plc) has developed over time to its current value (€1014 m in the financial year ending in March of that year; hereafter "fiscal-2014"). The figure also shows an attractive future for the company's profit to 2020 that should result from pursuing a sound strategy in not-too-hostile market conditions, along with a less attractive trajectory that might arise from some combination of poor strategy or a challenging market environment. To be useful, a GUT should explain how the former outcome might be achieved, rather than the latter, and continue to do so from period to period.

The case of Ryanair also demonstrates the need for improved strategy methods. An assessment of industry forces (Porter 1980) might explain why profitability in the airline industry as a whole is low, and why profitability specifically among low-fare airlines has fallen since the sector's emergence in the 1980s and 1990s, as large numbers of new entrants competed away the sector's margins (60 new firms started up in Europe between 1995 and 2010, for example). However, since the strategic positioning and operating models of the many low-fare competitors have been virtually identical for over two decades, neither industry forces nor analysis of strategic resources offer much explanation for the diversity of performance between firms. That performance varies by *orders-of-magnitude* when viewed in terms of investors' interests – Ryanair generated over €1 billion in cash operating profits in fiscal-2014, about half of which was reinvested in new aircraft, while weaker competitors generated little or no free cash flow, or went out of business. Furthermore, not only have those strategic positions and operating models been largely indistinguishable, they have also not changed significantly during the same period. Neither industry-based nor resource-based methods, then, can assist with the critical decisions needed to *implement* strategy from period to period.

Explaining *Immediate* Performance

Developing a GUT to explain growth in profit and cash flow requires, first, an explanation for those values for a single period. And since investors value the receipt of absolute quantities of cash, that explanation should address absolute amounts, such as €millions/year, not any ratio. No research or statistical analysis is required to develop that explanation, since any period's operating profit and cash flows are completely explained by the accounting principles of the Income and Cash-Flow Statements. *Profit* is *revenue* minus *operating costs* (or, if value is added to bought-in items, by *gross profit* minus *operating costs*, and *gross profit* is *revenue* minus *cost of goods*). *Operating costs* are simply the total of the various cost categories required for the business to function and grow.

These relationships are set out for Ryanair's fiscal-2014 results in Fig. 2. The table at left presents the explanation for the year's profit in a conventional table, while the diagram at right displays exactly the same relationships in causal form. The arrows connecting items in this diagram have a specific meaning – that the dependent variable can be confidently estimated from the items linked to it by those arrows. In this case, each item can be exactly *calculated* from those on which it depends.

In virtually all commercial cases, revenue is caused by *customers* multiplied by their *average expenditure per period*, which in turn is caused by *transactions per period* multiplied by *average transaction value*. *Customers* are a tangible, somewhat reliable asset, in the sense that many customers are likely to continue purchasing from the business from period to period. (The gain and loss of customers is addressed below.) For Ryanair, the average customer buys flights about five times each year, information that allows the explanation for the company's revenue for fiscal-2014 to be completed, as shown in Fig. 3. The number of *Average customers* is depicted in a box to indicate that it is an accumulating asset-stock (Dierickx and Cool 1989).

With multiple customer segments, the relationship from customers to sales and revenue needs merely to be replicated and summed. An airline's customers, for example, will include some who take its flights frequently for commuting or similar purposes, others who are loyal but less-frequent

Revenue €m

Fare Revenues	3789.5
Ancillary Revenue	1247.2
Total Revenue	5036.7

Operating costs €m

Staff costs	463.6
Aircraft costs *	2230.7
Airport charges	617.2
Route costs	522.0
Marketing + other costs	192.8
Total (cash) operating costs	4026.3
**Profit EBITDA ** **	1010.4
Depreciation	351.8
Operating profit	658.6
Passenger Journeys Booked	81.7m
Average fare	€46.40

* excluding depreciation
** earnings before interest, tax, depreciation and amortization

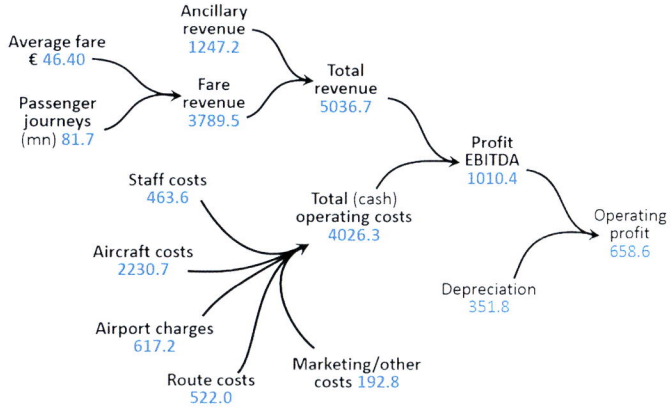

Engineering of Strategy: A General, Unified Theory of Performance and Strategic Management, Fig. 2 The causal explanation for Ryanair's profit in fiscal-2014 (includes rounding differences)

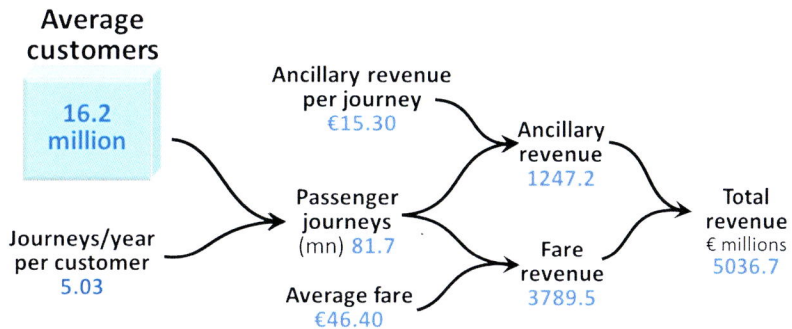

Engineering of Strategy: A General, Unified Theory of Performance and Strategic Management, Fig. 3 How customer numbers, purchase frequency, and price caused Ryanair revenue in fiscal-2014

travelers, and a few one-off customers who contribute a small fraction of sales and revenue.

Apart from the cost-of-goods embedded in the final product (insignificant in this case), costs are caused by costly assets, notably staff and capacity, and by the acquisition of those assets. *Staff costs*, for example, are caused by *staff numbers* multiplied by *average staff cost*, plus *staff hired per period* multiplied by *average hiring cost*. Some costs are simply decided upon by management but also relate to asset-building – marketing spend aims to acquire and retain customers, for example. For physical assets and some others, acquisition costs do not appear directly in the Income Statement, but are capitalized, and their depreciation (a non-cash item) is charged to the Income Statement instead.

Figure 4 shows how aircraft drive aircraft operating costs for Ryanair in fiscal-2014. The average

number of aircraft during the year is again depicted in a box because it, too, is an accumulating asset-stock. Other asset-stocks on the supply-side of the business are staff, the routes on which flights are offered (the company's product range) and the airports between which those routes are provided. The average operating costs of each aircraft is fully explained by additional items – the number of operating hours in each period and the costs of maintenance and fuel for each operating hour. The cost of fuel is an example of an exogenous variable over which management has little influence.

Supply-side assets may also comprise multiple segments, such as distinct staff groups, so these links between those assets and the resulting costs may need to be replicated and summed. Not all tangible assets need be owned, just available from

Average aircraft

301

Operating cost
per aircraft
€7.41m

Aircraft
operating costs
€ millions
2230.7

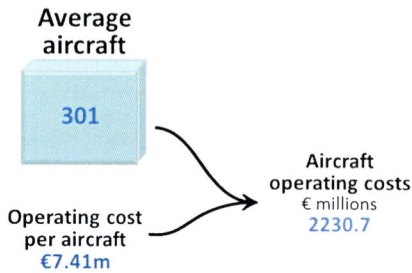

Engineering of Strategy: A General, Unified Theory of Performance and Strategic Management, Fig. 4 How capacity drives operating costs for Ryanair in fiscal-2014

period to period with some reliability – customers being the notable example, together with staff and leased equipment. In many cases, entire functions may be out-sourced to third-party providers, in which case costs are typically driven by the activity rates purchased by the firm.

Staff, capacity, and product range are simple, tangible resources, observable and measurable in virtually all cases. An airline's airports are an example of "points of presence" required to reach its customers – a function fulfilled in other contexts by dealers or distributors, or by physical assets such as vending machines.

Certain industries feature additional types of tangible resources, such as the mineral reserves that are extracted by natural resources firms, or the order-book for large-scale manufacturers. Together with customers, these supply-side categories of resources are virtually standard across any industry, and since they drive revenues and costs in a common manner captured by a company's Income Statement, they provide a completely reliable start-point for explaining the operating performance of any business in that sector, for any period.

Highly comparable assets arise in public-service and other noncommercial cases. "Demand" is driven by some population; numbers of children drive the need for schooling, and numbers of refugees drive the demand for food, water, and sanitation. "Supply" in such cases is again determined by the capacity of physical assets and by the staffing needed to fulfil that demand. Although profit may not be of primary concern,

such organizations must nevertheless be financially viable or operate within financial constraints. This means that very similar income statement and cash flow analyses will apply, and that the critical asset of "cash" will feature, just as in commercial cases.

None of the relationships set out so far are discovered by statistical analysis – they are simply arithmetical relationships that are universally recognizable. The causes of nonfinancial performance outcomes are not always so mechanical, but these too depend on quantities of assets. Service quality may be low if the current quantity of relevant staff is too small to support the demands from the current number of customers, for example.

For consistency with established terminology in the strategy field, the assets involved in these causal relationships are commonly referred to as "resources." However, since most of the resources directly involved in driving performance in any period are tangible, they may not fulfil the "strategic" criteria of the resource-based view of strategy (RBV – Wernerfelt 1984; Barney 1991; Amit and Schoemaker 1993; Montgomery 1995). Capacity, for example, may not be scarce or hard to imitate, whereas specialized staff are frequently valuable, rare, hard to imitate, and embedded in organizational processes – the so-called VRIO criteria for any resource to be regarded by scholars as "strategic." Intangible factors and capabilities more frequently fulfil these VRIO criteria. However, since current performance is directly explained by only a small number of readily identified, tangible resources, such VRIO factors can *only* affect performance by changing the quantities of those tangible factors over time. Consequently, no theory that relies solely on VRIO factors can adequately explain any measure of performance.

The reasoning to this point provides the first element of a GUT – that performance for any period of time depends on the average quantities of tangible resources that exist during that time. Two further types of factor directly affect current performance, the first being some management decisions. Raising marketing spend *immediately* cuts the profit for the current period, for example, regardless of its further impacts on changing customer numbers or purchase rates. Exogenous factors also affect current performance, such as changes in consumer disposable income, the costs of bought-in items, or the prices charged by competitors. The first principle of the GUT is thus as follows:

Performance, P, at time t is a function of the quantity of resources R_1 to R_n, discretionary management choices, M, and exogenous factors, E, at that time (Eq. 1).

$$P(t) = f\ [R_1(t), \ldots R_n(t), M(t), E(t)] \quad (1)$$

When an objective itself concerns the achievement of some quantity of resource by some point in time, the dependent variable is some $R_i(t)$, dealt with by the second element of the GUT, so Eq. 1 is not relevant.

In both business and other cases, performance depends not just upon the quantity of each resource but also on certain attributes or qualities they possess – the purchase-rate of customers, the salaries of staff, the output of production units, and so on. These attributes are already reflected in the causal relationships described thus far, but *changes* to those values over time also need to be explained.

Since performance is commonly monitored and reported for certain *periods* of time – a financial year or a month of operations, for example – Eq. 1 gives the *rate* of performance for such periods, so must be reported with correspondingly appropriate units, such as sales in *units per month* or cash flow in *€million per year*. Most nonfinancial performance outcomes feature similar "per time-period" measures or are reported on that basis, such as customer complaints per month, website visits per week, or production-plant yield per day.

In contrast, the resources that cause performance – being asset-stocks – are measured and reported at certain *points* in time, such as the start and end of a month or a financial quarter or year. Consequently, the quantities of resources R_1 to R_n in Eq. 1 must be the *average* quantities that exist during the reporting period to which the value of P relates. This is precisely analogous to the relationship between the quantity of cash in a bank account and the rate of

interest that it earns (*interest per period = average cash balance* multiplied by *interest rate per period*).

Since the quantities of resources and the performance they cause typically change significantly during a year, Eq. 1 becomes more meaningful and useful when calculated for shorter periods of time. This means that a GUT built upon Eq. 1 provides useful information for any required operating period, making it valuable for the implementation of strategy. In principle, Eq. 1 becomes most precise for the smallest possible periods of time. The rate of sales *right now*, for example, depends on the number of customers right now, and the fraction who are purchasing each minute. Although, at the limit, this relationship breaks down due to the stochastic nature of individual transactions.

Explaining the Changing Quantities of Resources

Since the objective is to explain *changes* in cash flow or other indicators from period to period, and these depend on quantities of resources, we must next explain how those quantities have arisen, and provide estimates for *future* changes to those resources. Since resources are accumulating asset-stocks (Dierickx and Cool 1989; Barney 1989), their behavior between any two points in time is readily specified.

The current quantity of resource R_i at time t is its quantity at time $t - 1$, plus or minus any gains or losses that have occurred between $t - 1$ and t (Eq. 2).

$$R_i(t) = R_i\ (t - 1) + / - \Delta R_i(t - 1 \ldots t) \quad (2)$$

This mathematical behavior, captured by integral calculus, is not a theory, opinion, or statistical observation but is axiomatic of asset-stock behavior. It is also a mathematical identity – cash today is *exactly* equal, to the cent, to the quantity yesterday, plus or minus any cash added or lost since that time. Numbers of customers or staff today are likewise exactly equal to the number yesterday, plus or minus any that have been won or lost. Since this is also always true for all time periods, back to the point in time when the asset-stock was first created, the current quantity of any asset-stock can only be *fully* explained by its entire history of gains and losses, but that explanation is absolute, with no error.

This mathematical property of asset-stocks has two critical implications. Since the quantity of any resource is precisely the sum of all quantities ever added, minus all quantities ever lost, there can be no *other* explanation for that quantity – the current quantity of cash cannot be worked out from current revenue and costs, today's number of customers is not "caused" by marketing or price, and current staff numbers are not explained by pay rates or working conditions.

Secondly, if the current quantity of a resource cannot be explained by anything except its own history of gains and losses, then neither can any value that *depends on* those resources – notably profit. This fatally damages efforts to confirm hypothetical explanations for financial or other outcomes through multivariate regression analysis. Current performance cannot be meaningfully correlated with either the current value of other variables, except those in Eq. 1. Nor can current performance be explained by time-lagged values of any variables, since there exist no causal mechanisms (in strategy or any other field) that can operate between remote points in time. For time-lagged influences to occur requires that some factor be stored through time, which is precisely what asset-stocks do.

Equation 2 is readily illustrated with resources from the Ryanair example:

Customers(end-March 2014)
    = *Customers* (end-March 2013)
    + *customers won* (in fiscal-2014)
    − *customers lost* (in fiscal-2014)

Aircraft(end-March 2014)
    = *Aircraft* (end-March 2013)
    + *aircraft bought* (in fiscal-2014)
    − *aircraft sold* (in fiscal-2014)

Similar relationships explain changes to numbers of staff, routes, and airports. The box in each diagram in Fig. 5 holds the stock-value for the relevant resource and displays the opening balance and closing balance at the start and end of fiscal-2014. The icons at the left and right of each stock indicate the rate at which resource flows into and out of that stock – hence the term "stock and flow" framework by which such diagrams are known. (*Since these icons can be thought of as pipes and pumps, the structure is also known as the "bath-tub metaphor."*)

The units for each flow-rate in any stock-flow relationship are *always* the same as those for the stock, with the addition of "... per period," so *customer* numbers are changed by flows of *customers per year* (or per week, month and so on). There is no exception to this rule, since it is axiomatic of asset-stocks' behavior.

It is rarely adequate to report the *net* change to the quantity of any resource during a period. The distinct values for the rate of gain and loss for any resource are nearly always important – winning 10 customers and losing none is not the same as winning 100 and losing 90. Moreover, the *causes* of gain and loss usually differ significantly – new customers are won for a different mix of reasons than those which cause current customers to be lost, and while the hiring of new staff is simply a management decision, losses of existing staff are caused by a variety of other factors. Since we are seeking a complete explanation for performance, each flow of resources into and out of a stock needs to be identified and explained. Note also that there can be more than one of each type of flow – staff may resign *and* be dismissed, for example.

Where chosen objectives concern achieving some resource-quantity by some point in time – $R_i(t)$ – rather than performance for a reporting period, the analysis starts with Eq. 2. For example, a target to grow a firm's customer-base from some current number today to a larger number at a future point in time can *only* be reached by achieving gains and losses over the intervening period that result in the required increase. The same principle applies for objectives to grow numbers of staff or other assets. Retailers, for example, frequently have objectives for the numbers of stores they operate, which can only be met by a new-store opening rate that exceeds any closures by a large enough difference to reach the goal.

Equation 2 exactly explains the change in the quantity of each resource over any desired period, and therefore explains, unambiguously, the change in performance over that same period. Explaining the changing trajectory of performance over multiple periods therefore requires only that the equation be evaluated repeatedly, for all resources across all time-periods. For example, if the changes to Ryanair's customer numbers in Fig. 5 applied to fiscal-2014, then they also applied for all previous years (and indeed for all periods within all of those years) and will continue to do so for all future years. The single-period causal relationships in Fig. 5 can therefore be replaced with time-charts for the same items, going back in history and forward into the future as far as required.

Figure 6 shows the relationships between the stock and flows of Ryanair customers from fiscal-2006 to 2014, and plausible projections to fiscal-2020 that would account for the preferred outcome shown in Fig. 1 (*No airline, Ryanair included, reports publicly the number of customers and their journey frequency, although they must know those values. The values shown in Figs. 5 and 6 are therefore plausible estimates. However, they must reconcile with the total quantity of passenger journeys sold in each period. If, therefore, actual journey frequency was lower than shown for any year, then customer numbers must have been correspondingly higher, and if the change in journey frequency between 2 years was different than shown, then gains and losses of customers must also have differed sufficiently to make up for that difference.*).

Figure 6 also shows the linkage to a commonly reported key performance indicator in the industry; sales of passenger journeys. Note that the stock displays the number of customers at the start of each fiscal year, but the average number for each year (and the year-end value) can be calculated from that value and the two rates-of-change.

Like earlier figures, Fig. 6 may again be an unfamiliar representation, although its relationships could readily be formulated in a spread-sheet and reported in a table. However, the diagram offers a more intuitive picture of how the dependencies play out over time, which proves to be of considerable value when teams seek to understand why performance is changing as it is, and plan activities and decisions to improve that performance. Also, since the relationships are totally reliable for any chosen period, the analysis may be made and presented in as much temporal detail as required. Figures 1 and 6 show changes over many years, expressed in annual rates, but could equally be examined over the same or shorter periods, for example, the last 3 years in quarterly or monthly rates. This enables use of the method for continuous strategy implementation, since changes can be highlighted over short periods of time and decisions adjusted accordingly.

Exactly equivalent relationships-over-time can be calculated and displayed for changes to the numbers of aircraft and all other resource-stocks, resulting in the complete explanation for Ryanair's profit trajectory shown in Fig. 7. This figure is more than a visual description of the causes for the company's profit – every value displayed is the actual value for fiscal-2014 (*start-of-year values for each resource-stock and average values for the year for performance outcomes and other items*), and every value can be

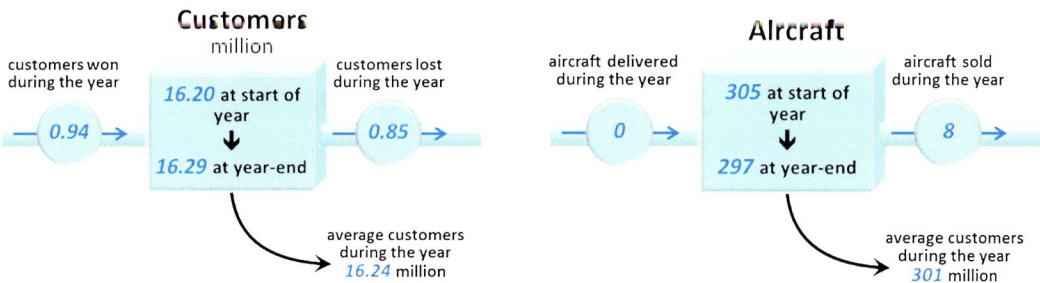

Engineering of Strategy: A General, Unified Theory of Performance and Strategic Management, Fig. 5 Changes to numbers of customers and aircraft at Ryanair during fiscal-2014

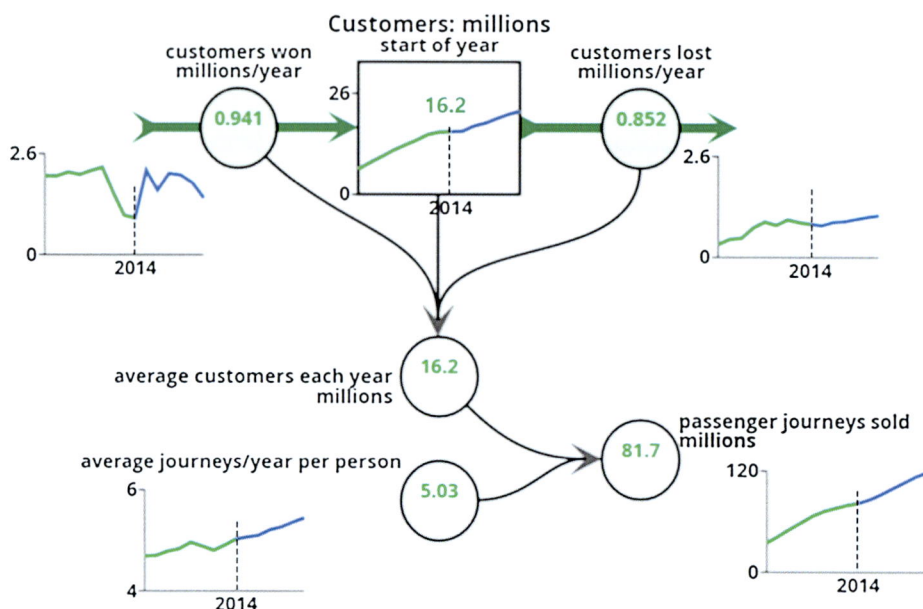

Engineering of Strategy: A General, Unified Theory of Performance and Strategic Management, Fig. 6 Changes to Ryanair customer numbers from fiscal-2006 to 2020 (rounded values)

accurately calculated using the causal links from items to their left. The same is true for every value indicated by the time-charts for every period prior to, and after, 2014.

Figure 7 therefore provides a visual, quantified explanation for the company's strategic management and performance over the analysis period. Ryanair could have suffered a severe loss of customers and fall in passenger journeys in fiscal-2009 and 2010 – the first financial years hit by the recession. However, a sharp cut in fares prevented both of these problems and enabled continued growth in sales and revenue. The loss of customers and sales might again have been expected to occur in fiscal-2011 and 2012, when Ryanair moved fares back to pre-recession levels. However, that problem too was prevented because the company opened services on many new routes in those years – additions that were possible because it had previously started operations at many new airports during fiscal-2009. Throughout the period, additions to aircraft and staff simply followed what was necessary to provide the physical and service capacity needed to support the route-services offered and sales of passenger journeys.

Explaining Rates of Change to Resources

The remaining unknowns from Eqs. 1 and 2 are the rates of change for each resource, $\Delta R_i(t - 1.. t)$. The summary explanation for all such flow-rates, described more fully below, is of exactly the same nature as the explanation for performance outcomes that we saw in Eq. 1:

The change in quantity of R_i from time $t - 1$ to time t depends on the quantity of resources R_1 to R_n at time $t - 1$ (*including that of resource R_i itself and its own potential*), on management choices, M, and on exogenous factors E at that time (Eq. 3).

$$\Delta R_i(t - 1 \ldots t) = f\,[R_1(t - 1), \ldots R_n(t - 1), \atop M(t - 1), E(t - 1)] \qquad (3)$$

Like Eqs. 1 and 2, Eq. 3 may be expanded to make explicit multiple segments of resource. Also, since the items on the right-hand side of Eq. 3 are likely to change significantly over a whole year, it becomes more accurate and useful if the time period t is short enough for the change $\Delta R_i(t - 1 \ldots t)$ to be small, relative to the scale of resource R_i. Again, monthly or weekly analysis makes it possible to use

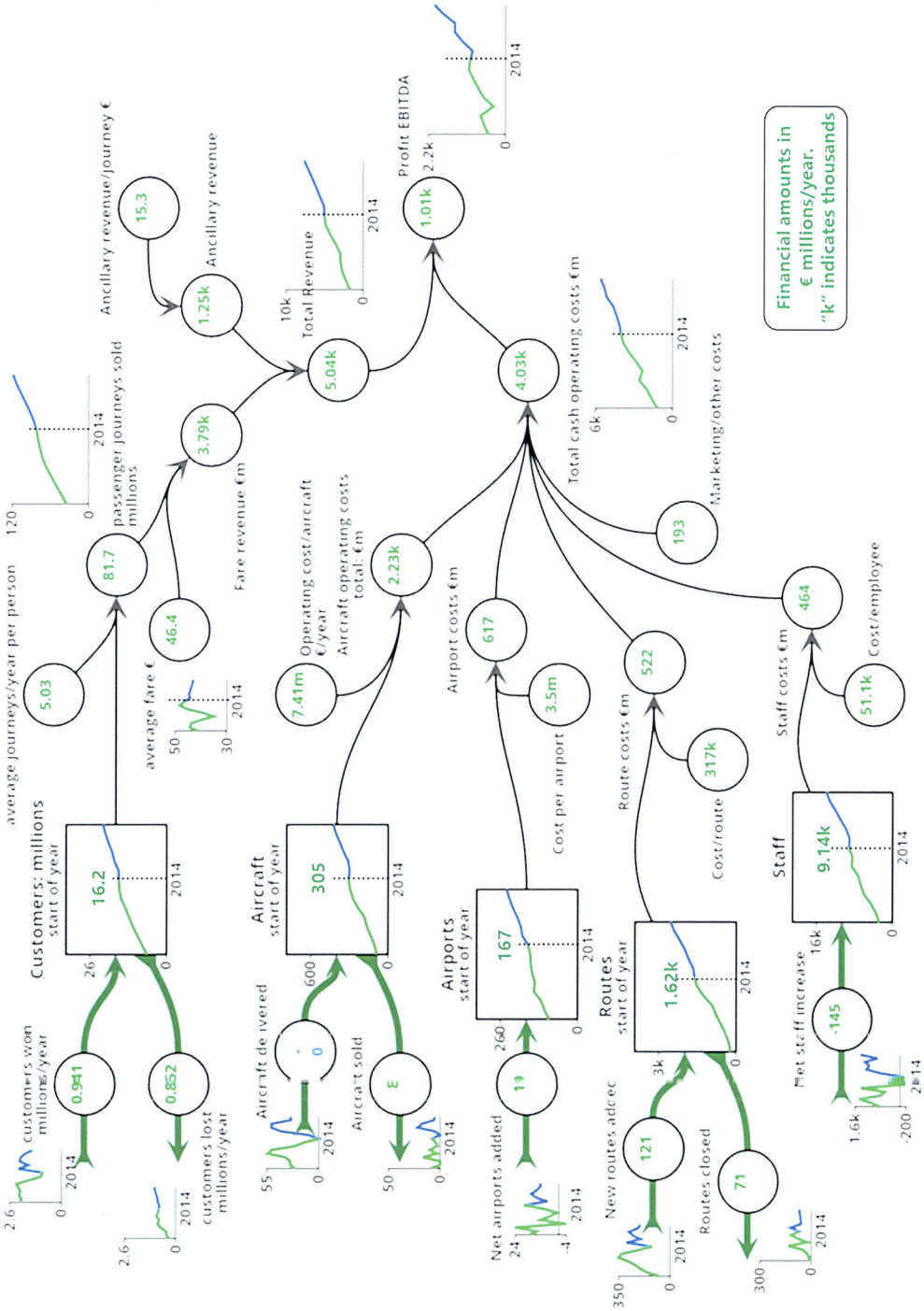

Engineering of Strategy: A General, Unified Theory of Performance and Strategic Management, Fig. 7 How changes to tangible resources drive changes to Ryanair EBITDA profit over time

the GUT for continuous strategic management – a capability not offered by any other method or theory.

The simplest instances of Eq. 3 concern rates of change that are simply chosen by management, such as capacity increases and closures, the launch and withdrawal of products, and the hiring and dismissal of staff. Not all such decisions necessarily result in exactly what management wishes, however. We may try to hire a certain number of staff, but cannot be certain to succeed in doing so because that hiring rate is also subject to other decisions, such as pay rates we offer, and exogenous factors such as competitors' pay rates.

The least certain flow rates concern the capture and loss of customers. Decisions on marketing expenditure, pricing and sales effort are set with the hope of winning customers at some rate and retaining existing customers (as well as changing customers' purchase rates), but the outcome is also affected by other factors, including the same efforts being made by competitors. The price decision plays a distinctive role. Management is free to make any price change that they wish, and if nothing else changes, this will immediately alter revenue and cash margin by an amount that is easily calculated. But other items *will* be changed by that same decision – customers' may purchase less often and will be won and lost at a different rate.

The second category of factor affecting rates of change to tangible resources concerns *existing* resource quantities. Both in-flows and out-flows are subject to such influences. A wider product range resource or a larger sales force may win customers more quickly. However, a too-small customer-support team may lead to poor service and result in higher customer losses that a larger team would have avoided.

The set of existing resource-quantities affecting the rate at which new resources can be acquired includes stocks of *potential* resources. Airlines can only capture potential customers who reside in the catchment area of the airports they serve, for example, and retailers can only capture potential consumers near their stores. Many firms struggle to grow numbers of customers because the potential populations have already been captured by the firm itself or by competitors. Firms may also be constrained by shortages of certain types of staff. The challenge is not limited to commercial organizations needing engineers, IT-specialists, experienced sales people, or accountants but applies equally to public service and other nonprofit organizations – shortages of teachers and medical staff constrain hiring rates for schools and health services, for example.

Potential resources do not feature in Eq. 1, since they do not directly drive current-period performance, but they do feature on the right-hand side of Eq. 3. But potential resources are themselves subject to Eq. 3 and exhibit the behavior defined by Eq. 2. This exposes them to influence by management decisions. Retailers create new potential consumers each time they open a store for example. Their product range, marketing, pricing and so on then determine whether those potential consumers flow into the stock of active consumers who generate sales and revenue. Likewise, for other business types, new products or services can be launched, new dealers appointed or new sales territories entered in order to reach new potential customers. The principle is not limited to customers – some organizations sponsor educational programs that train potential staff, precisely in order to create the stock of graduates from which they can then hire.

The third and last category of factors determining the gain and loss of resources concerns exogenous factors, including but not limited to the actions of competitors. Demographic, social, and economic changes may affect significantly both the rate at which new customers can be won and the rate at which existing customers are lost. For example, aging populations raise the numbers of customers who can be won by firms offering products and services for the elderly, and urbanization creates new customers for urban transport and other services. And many businesses acquired new customers because of the rising spending power of consumers that occurred during the economic boom of 2005–2008, only to lose them again as that spending power fell during the subsequent recession.

Competition, of course, is an exogenous factor with considerable influence. We already noted that competitors' activities and choices affect our own firm's *current* performance directly through Eq. 1 – their price being the most obvious example – but their activities and choices also affect the rate at which our firm can win and retain customers and staff. However, competitors cannot control all of our resource flow rates – they cannot usually control our ability to change our product range and capacity, for example.

Equation 3 exposes another challenge when seeking explanations for performance – the rate of accumulation or depletion of any particular resource is often highly nonlinear. Marketing expenditure, for example, may have little impact on a customer win-rate until it is sufficient to exceed some threshold of awareness or interest among potential customers at which they choose to respond. Staff losses, on the other hand, may continue at a modest rate up to a point of

unacceptable work-load, at which point the rate may escalate sharply. Such thresholds are a common source of tipping points: both helpful and harmful.

The implementation of Eq. 3 can again be demonstrated with the information on Ryanair plc in Fig. 8. The factors causing new customers to be won from the reachable potential are at upper-right – choice of routes, general rates of consumer spending (an exogenous factor), changes in the fares charged (a decision), and accessibility to new airports. The role of competition is implicit here, in that customers respond to the *relative* choice of routes and fares charged, as compared with those offered by rivals. The drivers of customer losses are at upper-right. In addition to normal losses arising from changing demographics and consumer habits, losses also arise due to falls in overall consumer spending and

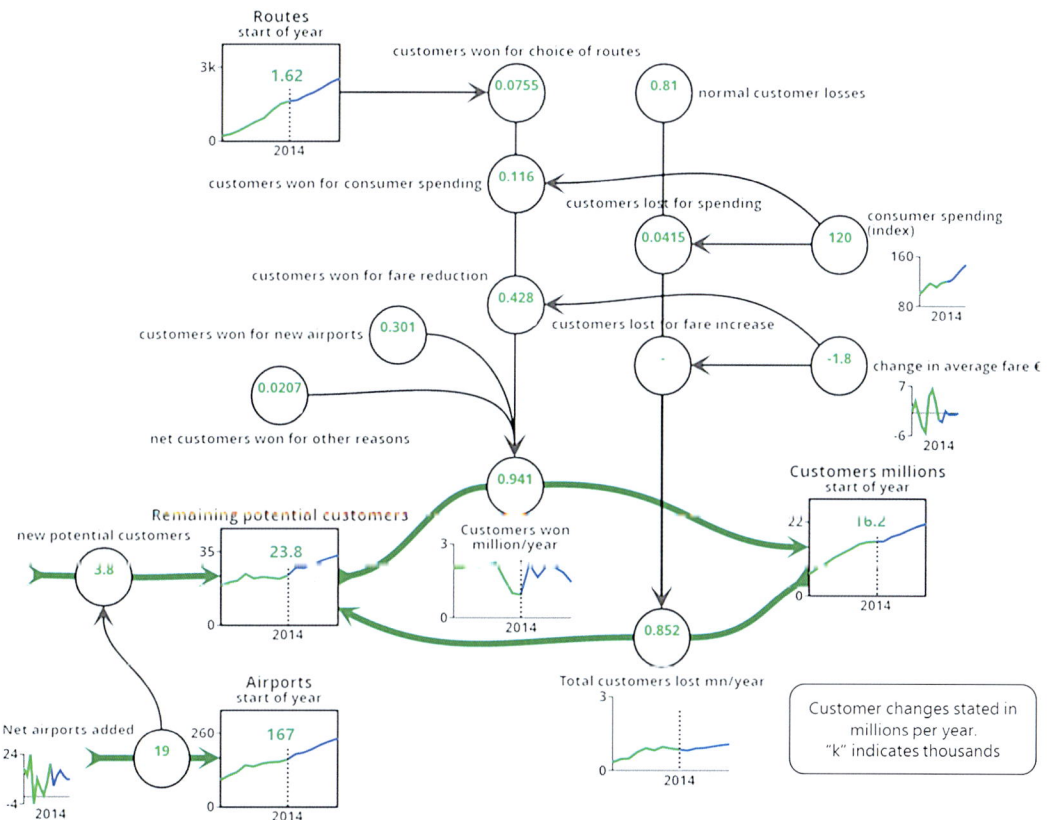

Engineering of Strategy: A General, Unified Theory of Performance and Strategic Management, Fig. 8 Causes of changes to Ryanair customer numbers from fiscal-2006 to 2020

fare increases. The population of potential customers from which the company draws active customers is shown at lower left, driven by the opening up of flight services from additional airports.

The exact causality for the win- and loss-rates of customers in this and other cases requires internal company information and extensive, repeated customer research, neither of which is possible with external analysis. However, the values shown in Fig. 8 reconcile with the customer numbers and sales rates shown in Figs. 6 and 7 (The item *net customers won for other reasons* is a small residual value in each year).

It was noted earlier that regression analysis cannot be meaningfully used to explain the value for any accumulating stock or, therefore, for any item that depends on such a stock. However, such methods can be used safely to explain rates of change. In Fig. 8, the customer win-rate is *some* function of items such as those shown, and statistical investigation of regular customer research such as conjoint analysis would provide the necessary estimates of those functions. This, incidentally, develops the popular strategy "value curve" framework (Kim and Mauborgne 1999, 2004) into a rigorous method that can be used continually for strategy implementation.

Causal Ambiguity, Intangible Factors, and Capabilities

It was noted in the introduction that any GUT should, so far as possible, build on existing knowledge. The strategy field has long been aware of resource accumulation and interdependence, so the contribution of Eqs. 1, 2, and 3 is, first, to operationalize those mechanisms in a form that makes them both researchable by scholars and usable by executives, and second, to embed them in a complete theory of organizational performance and a corresponding working model.

Dierickx and Cool (1989) note important consequences of resource accumulation. The function in Eq. 3 will frequently cause time compression diseconomies that limit a resource's growth rate,

and asset mass efficiencies are captured by Eqs. 2 and 3. Resource interconnectedness is explicit in Eq. 3. Erosion of a resource is a negative rate of change in Eq. 2 that arises if effort or expenditure is insufficient to sustain the level of that resource – physical assets degrade and staff skills become obsolete, for example.

The GUT also clarifies the location of the mechanisms that give rise to the causal ambiguity noted by Dierickx and Cool. Such ambiguity obfuscates the relationship between the strategic management of a business, the interaction with its market environment, and the resulting performance outcomes. There is no causal ambiguity in the relationship between tangible resources and performance – Eq. 1 is simply arithmetical and is readily confirmed in the case of other performance indicators. Eq. 2 expresses a mathematical identity, so provided that it is addressed with the correct tools (integral calculus), it too exhibits no causal ambiguity whatever.

Equation 3 is therefore the element of the GUT where causal ambiguity resides. The possibility that thresholds may arise in the function defining any resource's rate of accumulation or depletion has already been noted. The resulting nonlinear relationship between any change to elements on the right of Eq. 3 and the flow-rate they determine will certainly hamper any effort to discover that causal relationship. In addition, resource gains and losses other than direct management decisions often reflect choices made by others, notably customers and staff. The uncertainty in these behavioral responses clouds any explanation for rates of resource gains and losses, even for the firm itself, let alone for competitors or external analysts. This challenge is exacerbated by the fact that the resources on the right-hand side of Eq. 3 include intangible factors.

Extensive efforts have been made to clarify intangible resources and their impact in the RBV of strategy (Wernerfelt 1984; Barney 1991; Mahoney and Pandian 1992; Amit and Schoemaker 1993; Peteraf 1993; Collis and Montgomery 1995). However, it has proved hard to do this in a manner that makes these items tractable in practice. To understand the impact of intangible factors requires *both* that those factors be properly specified and measured *and* that their causal influence on the flow-rates of tangible resources in the business system be defined. Reputation, for example, undoubtedly influences the rate at which many firms win customers. But not only is that influence rarely precisely known, so too are the reasons why that reputation level itself rises or falls.

Intangibles, too, conform to Eqs. 2 and 3, and typically fall into one of three categories. Reputation is an example of a *state-of-mind factor*; others include customer annoyance, staff skills and motivation. Organizations increasingly measure and monitor such items, knowing them to be important, but need means by which to better manage both the intangibles themselves and their impact on the business system and performance. Besides state-of-mind factors, other intangible asset-stocks include certain *quality factors*, such as the reliability of equipment or systems, and *information-based factors*, ranging from the simple customer-information used to ensure good service to specialist technologies and the sophisticated knowledge-bases employed by professional-service firms.

Capabilities are a further source of causal ambiguity, but differ from resources, since they represent the ability of a firm to undertake strategically important activities, notably the building and retention of resources (Collis 1994; Dosi et al. 2000; Dutta et al. 2005; Chmielewski and Paladino 2007; Hoopes and Madsen 2008). Such strategic capabilities are themselves distinct from *dynamic* capabilities, which refer to the ability of an organization to change its strategy (Teece et al. 1997; Winter 2003).

Unfortunately, academic research in strategic management is ambiguous about such matters. Definitions of the terms "resource" and "capability" are inconsistent and abstract, and rarely is any attempt made to actually measure the items concerned, with research relying instead on proxies. R&D expenditure or numbers of patents, for example, may be assumed to be indicators of technological capability.

So, like the various intangible resources, it has proved hard to specify both capabilities and their influence mechanisms in a tractable form. This purpose is assisted by conforming to common language usage. Since a capability is about how well something is *done*, it is expressed in grammatical terms as a verb's present participle – market*ing*, hir*ing*, serv*ing* customers, develop*ing* products – or the noun describing these processes, such as recruit*ment* or product develop*ment*. Direct indicators of all such factors are possible. How well we serve customers, for example, may be the speed with which we resolve their enquiries, or the fraction of such enquiries we succeed in resolving. Exactly which indicator are relevant depends on exactly which factor affect observed customer behavior.

In order to incorporate capabilities in the GUT, the questions to be addressed are: (1) In what way do they affect the accumulation and retention of the tangible resources that drive performance? (2) What measurements indicate the scale of that influence? (3) What exactly constitutes each capability itself? Since previous reasoning has shown that the flow-rates building and sustaining levels of tangible resources are critical determinants of how performance changes over time, the clearest benefit that a capability offers is to enable either the faster accumulation or the slower depletion of some resource. In addition, a strong capability would enable those flows to occur at lower cost and ensure that resources built and retained were of high quality (high-value customers, skilled staff, appealing products and efficient and reliable capacity). Measurements of such a capability should therefore include indicators for each of these three features.

An organization's capability specified in this way clearly depends on the scale of tangible resources required; more sales people enable customers to be won faster and lost more slowly, and more product development staff and equipment enable more and better products to be developed more quickly. This element of capabilities is already captured by Eqs. 2 and 3. However each capability is enhanced by the skills of the relevant staff-groups, the data and other information available to them (an intangible resource), and the processes and procedures by which that information is deployed (a further intangible resource). A capability is therefore a composite asset-stock, reflecting the numbers of relevant people, together with their skills, available information and processes. In combination, these enable the resource-building activity to be done "well," as measured by the speed, cost and quality with which it is accomplished. Learning is also readily specified, as the rate at which any capability itself increases over time, driven by the experience that comes from having repeated the activity many times.

Specified in this way, organizational capabilities and their impact on firm performance are readily observable, measurable, and manageable. Ryanair, for example, opens new routes more quickly, at lower cost, and with greater impact on customer numbers than its rivals, and it can do so because it has more experience, built over many years, from undertaking that task many times. The same capability-enhancing learning applies to major strategic moves. General Electric, for example, acquires and integrates large numbers of companies quickly and effectively because it has done so many times. The development of both types of capability can be substantially enhanced by deliberate practice, codification and deployment of effective procedures, as has been the case for both of these examples. This deliberate formulation of effective procedures also lies at the heart of "The Toyota Way" which has made that company so strong in its industry.

Combining the arguments above with the specification of tangible resources earlier in this paper leads to the taxonomy of asset-stocks offered in Fig. 9, including for completeness operational asset-stocks that are not usually of strategic importance.

Feedback and the Business System

Note that Eq. 3 includes the current quantity of the same resource whose rate of change is to be explained – R_i is included in the set $R_1 \ldots R_n$. This means that any given resource may be involved in *its own* rate of growth or decline. This autodependence gives rise to feedback processes that add to the sources of causal ambiguity already identified (thresholds, the behaviors of system participants, intangible factors, and capabilities). Feedback may take two forms (Forrester 1961, 1968; Sterman 2000).

When *reinforcing* feedback drives growth, the current quantity of a resource stock causes an in-flow to that same stock and thus an increase in its quantity in the next period. This mechanism is colloquially known as a virtuous circle. An initial customer-base may win *new* customers through word-of-mouth, for example, so that the next period starts with more customers who drive a still greater win-rate in the next period. Such mechanisms need not rely on explicit recommendation but simply reflect individuals' tendency to follow role-models. Even then, the resulting explosion of growth can be spectacular, as in cases such as Facebook.

Reinforcing feedback can also drive decline, when an asset-stock's current quantity causes its own loss in what is known as a vicious cycle. For example, a staff group suffering over-load may lose individuals, leading to still greater pressure on those who remain and causing a further loss of that stock in the next period.

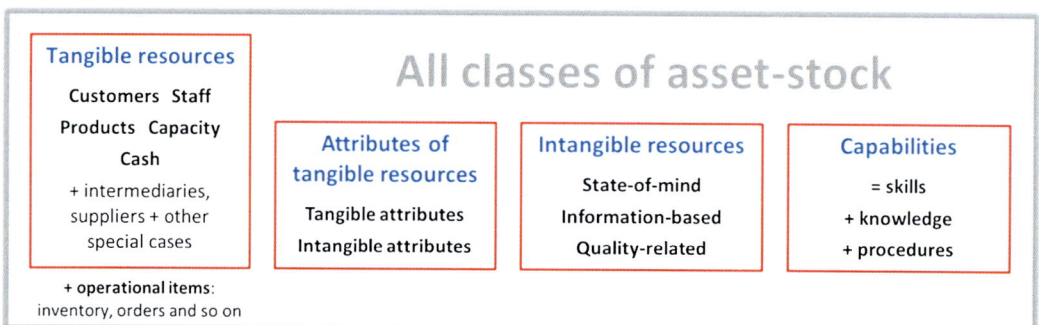

Engineering of Strategy: A General, Unified Theory of Performance and Strategic Management, Fig. 9 A taxonomy of resources (tangible and intangible) and capabilities

The second form of feedback is *balancing* feedback, in which a flow-rate gives rise to a change to its associated resource that then slows that flow-rate in the next period. Such feedback damps any tendency of the resource and the system of which it is a part to grow or decline, and commonly arises due to limited quantities of the resource itself or other stocks. For example, growth in a customer-base causes a decline in the number of remaining potential customers, and thus slows that same growth. Balancing feedback may also arise from limited availability of *other* resources, such as inadequate staffing or physical capacity. A resource stock may also limit its own decline. Customer or staff losses may slow, for example, simply because fewer customers or staff remain to be lost.

Each form of feedback can arise both from direct causal dependency of a flow rate on its own related stock, and indirectly through changes to other stocks. If more consumers (a first stock) purchase a product, for example, more stores may stock it (accumulation of a second stock), causing still more consumers to start purchasing it (the in-flow to the first stock). Whether feedback arises directly or indirectly, however, the functions required to describe it are already captured by Eqs. 1, 2, and 3, so the GUT requires no additional functionality in order to explain the performance consequences arising from such feedback.

The system-behaviors arising from feedback mechanisms, whether alone or in combination, are well known (Senge 1990). These include exponential growth, limits to growth, S-shaped growth when reinforcing and balancing feedback combine, and cyclicality as can be observed in the boom-bust cycles of many industries, from ship-building to construction to insurance.

Competition

Adding intangible factors and capabilities as defined above, along with their behavior and impact, to the system of tangible resources previously developed completes the GUT for a single enterprise. Competitors, too, operate business systems that are subject to exactly the same mechanisms specified by Eqs. 1, 2, and 3. Competitive interactions and relative performance can therefore be captured in a highly explicit manner, by confronting our own firm's resource-system with the corresponding systems of one or more competitors and the potential factors in the markets where they compete (notably potential customers and staff). Implementing these mechanisms gives rise to three characteristic rivalry mechanisms that may operate alone or in combination, depending on the nature and stage of development of the relevant market.

Type-1 competition occurs in its pure form with entirely new products or services, where no actual customers are yet purchasing, so only *potential* customers exist. The functionality, price and marketing of any firm therefore competes against others in seeking to win customers from that potential population at a faster rate than do those of competitors. This mechanism is clearly observable, for example, on every occasion when a new generation of cell-phone technology is introduced – existing users of the old technology (plus any remaining non-users) form the potential pool from which the new-generation rivals race to build a subscriber base. That potential pool may itself be added to by the collective actions of competitors, through the enhancement of the products and services offered, falling prices, and marketing spill-over that grows customer demand for the product/service *category*, not just for the specific item being marketed.

Type-2 competition occurs in its pure form only where all potential customers already exist – a fully saturated market – so any competitor can only grow its customer base if its functionality, price, and marketing can steal existing customers away from competitors. Competition amongst cell-phone operators with established technology comes close to this extreme case. However, in this and virtually all other cases, some residual element of Type-1 rivalry continues, if only to capture new potential customers who arise due to demographic changes or new-business formation in business-to-business cases.

Both Type-1 and Type-2 rivalry occur in their pure forms only in cases where customers must be exclusive to a single provider. This is generally the case for cell-phone services, but is also common in other contract-based sectors, such as mortgages, insurance, and utilities. It is also common in business-to-business markets, such as the supply of IT-services. Where disloyal behavior by customers is feasible – buying routinely from more than one provider, such as for consumer packaged goods –Type-3 rivalry is observed, in which any supplier's product/service functionality, price, and marketing seek to capture a larger share of those purchase decisions.

Competition in every circumstance can be captured by one or more of these three types of competition, enabling a full explanation for the relative performance of as many competitors as may be involved. New entry is simply the arrival of an additional firm's resource-system, initially populated with no active customers or revenue but seeking to grow that population by one or more of the mechanisms above. Firm failure and exit is simply the cessation of one rival's operations, at which point its products, staff, capacity, and customers may be absorbed by remaining rivals. These phenomena, together with the changing functionality and price of rival products and services explain the dynamics of entire industry sectors, a phenomenon that has proved difficult to explain in more than descriptive terms (McGahan 2004). None of these extended applications of the GUT, however, require any addition to the core principles captured in Eqs. 1, 2, and 3.

Fulfilling the Requirements for a GUT

Only a limited review of strategy principles and methods is possible here. Links with RBV have already been noted. The cost and margin make-up of strategy's value chain analysis (Porter 1985) is identifiable through Eq. 1, to any level of detail required. Value-curve factors (Kim and Mauborgne 1999, 2004) drive customer transaction rates in Eq. 1, as well as customer win and loss rates in Eq. 3 – variables for which other strategy tools and frameworks may be used. The experience curve, in which unit costs fall by some fraction as cumulative output rises, is precisely expressed with the three equations (Boston

Consulting Group 1972; Hax and Majluf 1982). Organizational procedures constitute a resource, so Eq. 3 applies.

The GUT assists choice of strategic position by identifying stocks of potential customers, feasible development of products that could be attractive to them, and the realism of building other supply-side resources of sufficient scale and quality to capture those customers and thus grow revenues and cash flows. It also informs continuous choices from period to period on all significant decisions and in response to changing competitive and other exogenous conditions. It offers more rigor for steering strategy than is typical in balanced scorecards (Kaplan and Norton 1996, 2004) and goes beyond such scorecards by including competitive and other exogenous factors.

Since all significant interdependencies are captured, the GUT handles multiple objectives and trade-offs between them. Properly applied, Eqs. 1 and 3 gain support from evidence rather than from abstract and ambiguous terminology, or from proxies. Adequate explanations for historic performance and estimates of future performance can be obtained from factors, concepts, and measures that are practical for management to identify and measure.

Industry structure is itself affected by its participants (Porter 1980; McGahan 2004). Each firm, new entrant, and substitute runs its own set of Eqs. 1, 2, and 3, with dynamic impacts on industry growth and levels of price and profitability. Pricing and capacity-building drive the growth in number of new potential customers that early firms then exploit, deterring new entry by others and slowing the loss of customers to substitutes. An industry's entire dynamics are captured by a stock of firms, the arrival of new entrants, and loss of exiting firms (Christensen 1997). In strategic groups (McGee and Thomas 1986), firms in each group operate near-identical architectures that differ from those operated by firms in other groups, whether in scale, segmentation, or featured resources.

Eqs. 1, 2, and 3, being highly generic, capture principles of other disciplines and business functions. Eq. 1 follows basic finance and accounting concepts, and Eq. 2 is the link between balance sheet and cash flow statements, where cash is the resource. Resources commonly move through stages – junior staff are promoted to senior levels, customers become aware, then informed, then start to buy, and so on. The equations describe such development chains, so can handle marketing's various customer development models (Palda 1966; Bass 1969; Kotler and Keller 2006), product development (Ulrich and Eppinger 2007), new product adoption (Utterback 1996; Christensen 1997; Rogers 2005), human resource strategy (Gratton et al. 1999), and the sustaining of physical assets and knowledge (Spender and Grant 1996; Alavi and Leidner 2001; Jardine and Tsang 2006).

These considerations make the GUT as applicable to departments and functions, as it is to whole organizations. For multi-business corporations, Eqs. 1, 2, and 3 can be replicated for each business unit and for each division serving distinct geographic or other market segments. Resources and capabilities developed collectively by, and for the benefit of several units, such as shared sales forces or IT services, simply appear repeatedly in the architectures of each unit. This makes explicit the concept of "relatedness" in corporate strategy (Porter 1985; Grant 1988; Bowman and Helfat 2001) and assists decisions on diversification, acquisition, alliances, and other corporate strategy choices. The same principles allow portrayal of vertical relationships between firms and other forms of business network.

The elements and concepts in Eqs. 1, 2, and 3 are applicable with little modification to public services, voluntary, and other nonprofit cases. Such organizations pursue objectives over time, serve identifiable constituencies with identifiable products or services, and develop capacity, people, and other supply-side resources for this purpose. The GUT is as applicable, then, to noncommercial organizations as it is to profit-oriented firms in competitive commercial markets – a purpose not currently well-served.

The GUT's equations have no meaning in the absence of time passing, so are inherently suited to the *continuous* management of strategy from period to period, responding to constantly changing circumstances – true "strategy dynamics."

They can incorporate a wide range of theories, concepts, and principles from strategy and other disciplines, and thus offer the basis for a rigorous, integrated and cumulative body of knowledge.

Taken together, implementing Eqs. 1, 2, and 3 creates what can be termed a "strategic architecture" of an organization's operating system, displaying all significant components, relationships and outcomes, to any level of detail required for confidence in the findings. That procedure brings the same rigor to the management of customers, staff and other resources that is taken as normal in other disciplines. The relationship between a stock and its flows is identical to that between a company's financial values reported on its Balance Sheet and the changes in those values laid out in its Cash Flow Statement. This is also how production and supply-chain operations are understood and managed. The method also extends that rigor to the interdependent relationships within the entire enterprise, and beyond, resulting in a true, working business model (Osterwalder and Pigneur 2010). Like any worthwhile theory, the GUT is amenable to falsification – the seeking of any case where it fails.

The equations' relevance is not limited by the location of firm boundaries – distribution facilities, IT services, or staff-recruitment, for example, are equally represented, whether owned and operated by the firm, or bought from third parties. Indeed, the GUT can help identify preferable choices on such issues.

The design and management of feedback mechanisms in physical systems is already well understood and addressed by engineering control theory. For a production system to produce goods efficiently and reliably, it must be engineered so as to make that performance possible and then controlled with information feedback systems to ensure its continual effective operation. A for-profit business is also a "designed system," whose desired output is rising cash-flow, so it too can be (and often is) engineered to be capable of strong performance and controlled with information feedback systems. The same core principles of engineering control theory can therefore be applied (Forrester 1961, 1968; Sterman 2000). Furthermore, since public-service and other nonprofit organizations are *also* designed systems intended to generate some output (albeit other than growing cash-flow), those too can be engineered and controlled using the same principles.

Future Directions

This chapter has summarized the theoretical foundations, informed by the science of system dynamics, for a wholly general explanation of the relationships between the strategic choices that management makes for an enterprise, through the working system of interdependent resources that constitute such an enterprise, to the performance it is able to achieve. There are no exceptions to these relationships, so there is little scope for the theory itself to be developed.

The method that embeds the theory has been deployed in most business sectors, for enterprises at every stage of life, as well as for specific plans and issues in all business functions. The method has been extended in scope to cover "integrated reporting" (adding environmental and social impacts to the purely financial performance outcomes). However, scope remains for certain aspects of the theory's application to be more precisely specified, notably in the measurement and capture of intangibles and capabilities. There is also scope to test and specify how the method is operationalized for more complex strategic issues, such as complex corporate structures of interdependent business units and business ecosystems.

Bibliography

Alavi M, Leidner DE (2001) Review: knowledge management and knowledge management systems: conceptual foundations and research issues. Manag Inf Syst Q 25:107–136

Amit R, Schoemaker P (1993) Strategic assets and organizational rent. Strateg Manag J 14:33–46

Barney JB (1989) Asset-stock accumulation and sustained competitive advantage: a comment. Manag Sci 35:1511–1513

Barney JB (1991) Firm resources and sustained competitive advantage. J Manag 17:99–120

Bass F (1969) A new product growth model for consumer durables. Manag Sci 15:215–227

Boston Consulting Group (1972) Perspectives on experience. Boston Consulting Group, Boston

Bowman EH, Helfat CE (2001) Does corporate strategy matter? Strateg Manag J 22:1–23

Brealey RA, Myers SC, Allen F (2007) Principles of corporate finance. McGraw-Hill, New York

Burr J (1938) The theory of investment value. Harvard University Press, Cambridge

Cabantous L, Gond JP (2011) Rational decision making as performative praxis: explaining rationality's eternal retour. Organ Sci 22:573–586

Chmielewski DA, Paladino A (2007) Driving a resource orientation: reviewing the role of resource and capability characteristics. Manag Decis 45:462–483

Christensen CM (1997) The Innovator's dilemma. Harvard Business School Press, Boston

Collis DJ (1994) How valuable are organizational capabilities? Strateg Manag J 15(Winter Special Issue):143–152

Collis D, Montgomery C (1995) Competing on resources. Harv Bus Rev July–August:119–128

Coyne KP, Subramaniam S (2000) Bringing discipline to strategy. McKinsey Q June:14–25

Dierickx I, Cool K (1989) Asset stock accumulation and sustainability of competitive advantage. Manag Sci 35:1504–1511

Dosi G, Nelson R, Winter S (2000) Introduction: the nature and dynamics of organizational capabilities. In: Dosi G, Nelson R, Winter SG (eds) Nature and dynamics of organizational capabilities. Oxford University Press, New York, pp 1–21

Dutta S, Narasimhan O, Rajiv S (2005) Conceptualizing and measuring capabilities: methodology and empirical application. Strateg Manag J 26:277–285

Farjoun M (2007) The end of strategy? Strateg Organ 5:197–210

Forrester J (1961) Industrial dynamics. Pegasus Communications, Waltham

Forrester J (1968) Industrial dynamics – after the first decade. Manag Sci 14:398–415

Ghoshal S (2005) Bad management theories are destroying good management practices. Acad Manag Learn Edu 4:75–91

Grant RM (1988) On 'dominant logic', relatedness and the link between diversity and performance. Strateg Manag J 9:639–642

Gratton L, Hope VH, Stiles P, Truss C (eds) (1999) Strategic human resource management. Oxford University Press, Oxford

Hambrick DC (2004) The disintegration of strategic management: It's time to consolidate our gains. Strateg Organ 2:91–98

Hax AC, Majluf NS (1982) Competitive cost dynamics: the experience curve. Interfaces 12:50–61

Hoopes D, Madsen T (2008) A capability-based view of competitive heterogeneity. Ind Corp Chang 17:393–427

Jardine AKS, Tsang AHC (2006) Maintenance, replacement & reliability: theory and applications. CRC/Taylor & Francis, London

Jarratt D, Stiles D (2010) How are methodologies and tools framing managers' strategizing practice in competitive strategy development? Br J Manag 21:28–43

Jarzabkowski P, Kaplan S (2014) Strategy tools-in-use: a framework for understanding 'Technologies of Rationality' in practice. Strateg Manag J. Early View: May 8

Kaplan R, Norton D (1996) The balanced scorecard. Harvard Business School Press, Boston

Kaplan R, Norton D (2004) Strategy maps. Harvard Business School Press, Boston

Kim C, Mauborgne R (1999) Creating new market space. Harv Bus Rev 77:83–93

Kim C, Mauborgne R (2004) Blue Ocean strategy: how to create uncontested market space and make the competition irrelevant. Harvard Business School Press, Boston

Koller T, Goedhart M, Wessels D (2010) Valuation – measuring and managing the value of companies, 5th edn. Wiley, Chichest

Kotler P, Keller K (2006) Marketing management, 12th edn. Prentice Hall, Upper Saddle River

Mahoney J, Pandian JR (1992) The resource-based view within the conversation of strategic management. Strateg Manag J 13:363–380

McGahan AM (2004) How industries evolve. Harvard Business School Press, Boston

McGee J, Thomas H (1986) Strategic groups: theory, research and taxonomy. Strateg Manag J 7:141–160

Montgomery CA (1995) Resource-based and evolutionary theories of the firm. Kluwer, Boston

Osterwalder A, Pigneur Y (2010) Business model generation. Wiley, Hoboken

Palda KS (1966) The hypothesis of a hierarchy of affects: a partial evaluation. J Mark Res 3:13–24

Penrose ET (1959) The theory of the growth of the firm. Oxford University Press, Oxford

Peteraf MA (1993) The cornerstones of competitive advantage: a resource-based view. Strateg Manag J 14:179–192

Porter ME (1980) Competitive strategy. Free Press, New York

Porter ME (1985) Competitive advantage. Free Press, New York

Rigby D, Bilodeau B (2013) Management tools and trends. Bain & Co, New York

Rogers E (2005) The diffusion of innovations, 5th edn. Free Press, New York

Rugman AM, Verbeke A (2002) Edith Penrose's contribution to the resource-based view of strategic management. Strateg Manag J 23:769–780

Rumelt R (1991) How much does industry matter? Strateg Manag J 12(3):167–185

Senge P (1990) The fifth discipline. Doubleday, New York

Spender JC, Grant RM (1996) Knowledge and the firm: overview. Strateg Manag J 17(Special Issue):5–9

Sterman J (2000) Business dynamics. Irwin/McGraw-Hill, New York

Teece DJ, Pisano G, Shuen A (1997) Dynamic capabilities and strategic management. Strateg Manag J 18:509–533

Treynor JL, Black F (1976) Corporate investment decisions. In: Myers SC (ed) Modern developments in financial management. Praeger, New York

Ulrich KT, Eppinger SD (2007) Product design and development. McGraw-Hill, New York

Utterback JM (1996) Mastering the dynamics of innovation. Harvard Business School Press, Boston

Wernerfelt B (1984) A resource-based view of the firm. Strateg Manag J 5:171–180

Whittington R, Jarzabkowski P (2008) Directions for a troubled discipline: strategy research, teaching, and practice – introduction to the dialogue. J Manag Inq 17:266–268

Winter SG (2003) Understanding dynamic capabilities. Strateg Manag J 24(Winter Special Issue):991–995

Group Model Building

Etiënne A. J. A. Rouwette and Jac A. M. Vennix
Institute for Management Research, Radboud
University, Nijmegen, The Netherlands

Article Outline

Glossary

Client Person (or agency) who buys a model

Facilitator Person who guides the group process in group model building

Gatekeeper Person who forms the linking pin between modelling team and management team

Knowledge elicitation Process of capturing the knowledge contained in the mental models of team members of the management team

Modeller Person who constructs the quantified model during group model building

Recorder Person who takes notes during group model building sessions and constructs workbooks

Workbook Booklet which contains summary of previous group model building sessions and prepares for subsequent sessions

Definition of the Subject and Its Importance

Computer (simulation) models have been used to support policy and decision-making since the decades after World War II. Over the years modellers learned that the application of these models to policy problems was not as straightforward as had been thought initially. As of the beginning of the 1970s, studies started to appear that questioned the use of large-scale computer models to support policy and decision-making (cf. Hoos 1972; Lee 1973). Lee's article bears the significant title: "Requiem for large scale models," a statement that leaves little room for ambiguity. Other authors who have studied the impact records of computer models also seem rather skeptical (e.g., Brewer 1973; Greenberger et al. 1976; Watt 1977; House 1982). It is interesting to note that Greenberger et al., after interviewing both modellers and policy makers (for whom the models were constructed), found that modellers generally pointed to the fact that they learned a lot from modelling a particular policy issue. Policy makers on the other hand indicated that they did not really understand the models nor had much confidence in them. The results of these studies pointed in the direction of learning from computer models, i.e., conceptual or enlightenment use rather than instrumental use, where policy recommendations could straightforwardly be deduced from the model analyses and outcomes. Thus it is in the process of modelling a policy problem where the learning takes place, which is required to (re)solve a problem. And it is also in this process that one needs to anticipate the implementation of policy changes. By the end of the 1970s, system dynamics modellers pointed out that implementation of model outcomes was a neglected area (e.g., Roberts 1978; Weil 1980) and that modellers sometimes naively assumed that implementation was straightforward, thereby neglecting organizational decision-making as a political arena.

© Springer Science+Business Media, LLC, part of Springer Nature 2020
B. Dangerfield (ed.), *System Dynamics*,
https://doi.org/10.1007/978-1-4939-8790-0_264

Originally published in
R. A. Meyers (ed.), *Encyclopedia of Complexity and Systems Science*, © Springer Science+Business
Media LLC 2019, https://doi.org/10.1007/978-3-642-27737-5_264-4

In other words client participation in the process of model construction and analysis is required for successful modelling and implementation of insights from the model into policy making or as Meadows and Robinson put it:

> Experienced consultants state that the most important guarantee of modelling success is the interested participation of the client in the modelling process. (Meadows and Robinson 1985, p. 408)

Over the years this has given rise to all kinds of experiments to involve clients in the process of model construction. In the 1990s the term group model building was introduced to refer to more or less structured approaches for client involvement in system dynamics model construction and analysis.

Introduction

From the early days of the field, the topic of client involvement in the process of model construction has raised attention in the system dynamics literature. Jay Forrester, the founder of the field of system dynamics, has repeatedly indicated that most of the knowledge needed to construct a system dynamics models can be found in the mental database of the participants of the system to be modelled (Forrester 1961, 1987). Over the years several system dynamics modellers have experimented with approaches to involve client (groups) in model construction and analysis. This development in the system dynamics community parallels a movement in the (behavioral) operational research and systems fields toward more attention for stakeholders' opinions (De Gooyert et al. 2017). A number of authors (e.g., Ackoff 1979) criticized traditional OR and systems approaches as unsuitable for ill-structured problems that arise from differences between stakeholders' views on the problem. For ill-structured problems, a range of new methods was developed (Mingers and Rosenhead 2004).

The developments in the system dynamics, operational research, and systems communities have given rise to a set of distinct methods and approaches. However, practitioners work on problems that have clear similarities to those encountered in other disciplines and frequently borrow techniques from one another. The boundaries between methods are therefore difficult to draw, and there is a degree of overlap between approaches in and between fields. Below we first describe the distinguishing characteristics of system dynamics, as this separates group model building most clearly from other approaches fostering client involvement. We then describe a number of distinct group model building approaches.

System Dynamics

System dynamics is most easily characterized by its emphasis on two ideas: (a) the importance of closed loops of information and action for social systems, i.e., social systems as information feedback systems, and (b) the need to use formal models to study these loops. System dynamicists assume that the dynamic behavior of a social system is the result of its underlying feedback structure. Actors use the information about the structure as input to their decisions and by implementing their decision influence system behavior. This creates an interlocked chain of action and information which is also known as a feedback loop. Richardson (1991: 1) describes a feedback loop as follows:

> The essence of the concept... is a circle of interactions, a closed loop of action and information. The patterns of behaviour of any two variables in such a closed loop are linked, each influencing, and in turn responding to the behaviour of the other.

As an illustration of the use of information on the system state in decisions, imagine a simple example on customer behavior. Let us assume that if customers perceive that a product's functionality increases, more products will be bought. This will increase profits which the manufacturing company may use to increase its design budget. An increased design budget can be used to improve the product's design, which will lead more customers to buy the product, and so on. Thus, decisions of actors within the system have an important influence on the system's behavior. If we continue to add other factors and relations to our example and capture these in a model, the following diagram may result.

As the figure shows, a causal loop diagram consists of variables, relationships, and feedback loops. Relations can be of two types: positive and negative. A positive relation indicates that both variables change in the same direction. In the model above, an increase in retail price will lead to an increase in profits, indicating a positive relationship. Variables in a negative relationship change in opposite directions. For instance, the lower part of Fig. 1 shows that an increase in costs will decrease profits. The snowball rolling down the slope in the right-hand side of the figure indicates a positive feedback loop. We assumed that an increase in profit results in a direct increase in the design budget. A higher budget allows for increased product functionality, which increases sales volume and finally profit. Starting from an increase in profit, the result is a further increase in profit. This is a so-called positive or self-reinforcing loop. However, if we assume that the design department uses its complete budget each year, an increased budget will contribute to design costs and lower profits. This is a negative or balancing loop, indicated by the balance symbol.

The second important idea in system dynamics is that formal models are necessary to understand the consequences of system structure. Since system dynamics models contain many (often nonlinear) relations and feedback loops, it becomes very difficult to predict their behavior without mathematical simulation. Systems are assumed to consist of interacting feedback loops, which may change in dominance over time. Diagrams such as the one depicted above are frequently used in the interaction with clients. Before the dynamic consequences of the structure captured in the figure can be studied, it is necessary to further specify both the variables and relations used in the model. Two categories of variables are distinguished: stocks and flows. Stocks are entities existing at a particular point in time, for example, supplies, personnel, or water in a reservoir. Flows are entities measured over a time period, such as deliveries, recruitments, or inflow of water. Relationships are separated into physical flows and information flows. If we capture differences between stocks and flows and information and physical flows in a diagram, a stock and flow diagram results.

As can be seen in Fig. 2, information links are depicted with a single arrow and physical flows with a double arrow. The physical human resources flow is separated in three stocks: number of rookies, number of junior researchers, and number of senior researchers. Recruitment will

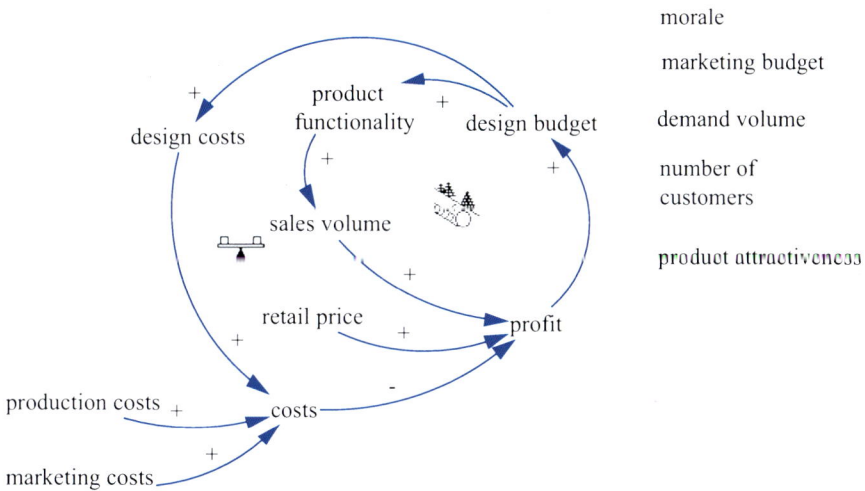

Group Model Building, Fig. 1 Example of a causal loop diagram (the unconnected variables on the right-hand side were elicited in the first phase of model construction and are still to be included in the model, see section "Group Model Building: Basic Ideas and Concepts")

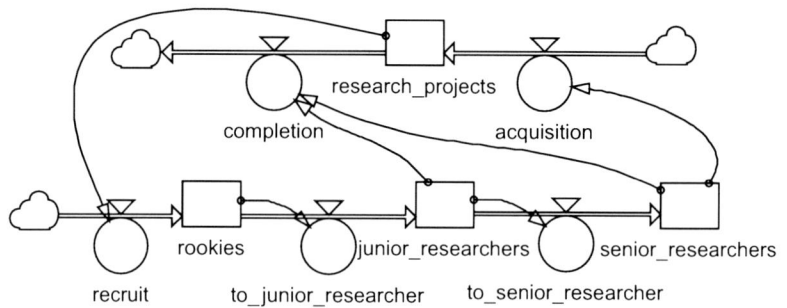

lead to an increase in the number of rookies. Two other flows influence the number of people in the stocks: rookies may promote to junior researcher, and junior researchers may promote to senior researchers. The human resources flow is related to the project flow with information links, for instance, indicating that acquisition of research projects is determined by the number of senior researchers.

Group Model Building Approaches

As pointed out before, client involvement has been important to system dynamics from the start of the field. System dynamics emphasizes feedback loops and the use of formal models. In this section we describe how models based on these ideas are built in interaction with actors and stakeholders in the problem at hand. A number of different participative model building formats can be identified, i.e., the reference group approach (Randers 1977; Stenberg 1980), the stepwise approach (Wolstenholme 1992), the Strategic Forum (Richmond 1997), modelling as learning (Lane 1992), strategy dynamics (Warren 2000, 2005), and Hines' "standard method" (Otto and Struben 2004). Below we describe each approach briefly.

In the reference group approach (Randers 1977; Stenberg 1980), participation takes the form of frequent interaction between the modelling team and a group of eight to ten clients. The approach starts with the identification of interest groups, of which representatives are invited to contribute to the modelling effort. The representatives are referred to as referents. In a series of interviews and meetings, the problem to be addressed is defined in more depth. On the basis

of this definition and the information gathered in the interviews and meetings, the modelling team develops a preliminary model. In the remainder of the project, the modellers are responsible for model improvements, while the referents function as critics. This model is elaborated in a series of meetings and is at the same time used as a tool for structuring the discussion. In later sessions, model output is used for developing scenarios. In a scenario discussion, the model is run, and results are described and analyzed by the modellers. The reference group is then asked to determine to what extent the model's behavior corresponds to their expectations about reality and, if it does not, to suggest changes. These suggestions can trigger changes in the model structure, initiating a new round in the discussions.

The stepwise approach (Wolstenholme 1992) is founded on the idea that full quantification of models is not always possible or desirable. The approach starts with a definition of the problematic behavior. If possible, this definition is given in the form of a reference mode of behavior. Modelling starts by roughly sketching the feedback loops responsible for this behavior. The key variables related to the cause for concern are identified, followed by the system resources connected to these key variables and their initial states. The resources are used to derive the central stocks in the system. From the resources, the resource flows can then be sketched with the associated rates of conversion. Delays are added to these flows if they are significant. Next, organizational boundaries, flows of information and strategies through which the stocks influence the flows, are added. Again, if there are significant delays, these are added to the information linkages. In the final step, information flows and strategies linking different resource flows are added. The steps are repeated until the relevant feedback loops have all been included. Wolstenholme indicates that these steps often provide the insights necessary to infer system behavior from the structure, which reduces the need for quantification. Models can also be analyzed in a qualitative manner.

The steps that make up the Strategic Forum (Richmond 1997) provide a detailed insight into how clients are encouraged to participate in modelling. The Strategic Forum consists of eight steps, of which the first two are conducted before the actual meeting (also called the forum) with the client group. The process begins with interviews based on a brief questionnaire, in which three issues are addressed: ideas on the current situation, a statement of the vision for the future, and agreement on a preliminary map of the problem. On the basis of the interview results, the modeller constructs an integrated map and accompanying computer model. In the second step, the project team designs a number of small group exercises that will be used during the forum. The exercises are aimed at discovering important structural and behavioral elements and are similar to the scenario discussions in the reference group approach. One important difference is that before simulation results are shown, participants have to "put a stake in the ground," i.e., they have to make a prediction of model behavior on the basis of a change in a policy variable and values for connected parameters. The model is then simulated, and results are compared with participants' expectations. Discrepancies between predictions and simulations are identified and might point to inconsistencies in participants' ideas or lead to model improvements. In the following steps, the participants meet in a series of workshops. Each workshop opens with an introduction and a big picture discussion. The heart of the session consists of exercises aimed at internal consistency checks, with an eye to increasing the correspondence between the group members' mental models and the computer model. In line with other approaches, model structure will be changed if inconsistencies with the participants' ideas on the problem are revealed. In the final phase of policy design, potential consequences of strategic policies are addressed resulting in an estimation of the existing capability for realizing the strategic objectives. A wrap-up discussion and identification of follow-up activities conclude the Strategic Forum. Richmond (1997: 146) emphasizes that the main purpose of the Strategic Forum is to check the consistency of strategy. The insights gained by the client therefore frequently lead to changes in strategy or operating policies, but less frequently to changes in objectives or the mission statement. One important element of ensuring an impact on participants' ideas is the (dis)confirmation of expectations on simulation outcomes.

Lane (1992) describes a modelling approach developed at Shell International Petroleum, known as "modelling as learning." Lane explicitly sets this approach apart from the widely used expert consultancy methodology (e.g., Schein 1987). His approach also puts strong emphasis on involving decision-makers in the modelling process. By showing decision-makers the benefits of participation early on in the process, an attempt is made to persuade them to spend time in direct interaction with the model. The approach centers on capturing and expressing the client's ideas, initiating a discussion on the issue with "no a-priori certainty regarding quantification, or even cause and effect" (1992: 70). The modellers also strive to include both hard and "soft" aspects of the problematic situation. Including both ensures that the clients' ideas are included in the model and contribute to creating ownership of the model. Ownership is also fostered by making models and model output transparent to participants, helping the client "to learn whichever techniques are used in a project" (1992: 71). Lane states that the focus throughout the approach is on a process of learning, using such elements as experimentation with the model, testing of assumptions, and representing and structuring ideas in a logical way.

Hines' approach (Otto and Struben 2004) starts off by diagnosing the problem. This step comes down to gathering and clustering problem variables. Problem behavior is visualized by drawing the reference mode of behavior. In the second step, the structure underlying the problematic behavior is captured in a causal diagram. This so-called dynamic hypothesis incorporates many of the problem variables identified earlier. The diagram helps to clarify the boundary of the problem that will be addressed and thus limits the project scope. The next step is to identify accumulations in the system, which will form the stocks in the system dynamics model. In the construction of the computer model, most work is done by the modellers, with client participation limited to providing data such as numerical values and details of the work processes relevant to the problem at hand. Model structure and behavior is then explained to the client. Discussions with the client lead to a series of model iterations, increasing confidence of the client in model calibration and validity. Similar to other participative approaches, policy runs are used to test proposed interventions in the problem.

Warren (2000, 2005) describes an approach to participative modelling that strongly focuses on identifying accumulations (stocks) in the system. In order to identify central accumulations, clients are asked to identify the strategic resources in the problem at hand. Increases and decreases in resources then lead to the identification of flows. Warren's approach differs from the ones described above in the sense that stocks and flows are differentiated from the outset. This means that causal loop diagrams are not used. In addition, graphs over time are recorded next to each variable in the model. By progressively adding elements to the model while visually relating structure and behavior, the clients' understanding of the problem is gradually increased.

As mentioned before, the boundaries around approaches are not easy to draw, and one method may "borrow" elements of another. Insights and practices from the operational research and system fields have been combined with those in system dynamics to develop combined methods. For example, modelling as learning is one of the approaches incorporating elements of soft operational research methodologies, more recently known as facilitated modelling (Franco and Montibeller 2010). Lane and Oliva (1998) describe the theoretical basis for integrating system dynamics and soft systems methodology. The cognitive mapping approach (e.g., Eden and

Ackermann 2001) also offers tools and techniques that are used in system dynamics studies (see Ackermann et al. 2011 for an integration of both methods).

In addition to combining different methods, approaches are sometimes also tailored to use in specific content areas. An example is Van den Belt's (2004) mediated modelling, which combines insights from participative system dynamics modelling and consensus building on environmental issues.

Group Model Building: Basic Ideas and Concepts

The separate approaches described in the last section continue to be developed and used in practical problems. Although we are not sure that all proponents of these approaches would characterize themselves as using "group model building," this term has been used more and more in the last decades refer to system dynamics approaches with client involvement in a general sense. The two approaches that coined the term group model building evolved more or less simultaneously, with considerable cross-fertilization of ideas, at SUNY at Albany and Radboud University of Nijmegen in the Netherlands (see Morecroft and Sterman 1992; Vennix et al. 1997). In an early application at Radboud University, participants were involved in a Delphi study consisting of mailed questionnaires and workbooks, followed by workshops (Vennix et al. 1990). In the dissertations by Verburgh (1994) and Akkermans (1995), a similar approach is used under the name of participative policy modelling and participative business modelling, respectively. In its latest version, group model building is a very open approach, which allows for the use of preliminary models or a start from scratch and uses individual interviews, documents, and group sessions, qualitative or quantitative modelling, and small as well as large models. Vennix (1996, 1999) provides a set of guidelines for choosing among these different approaches, building on and adding to the studies mentioned above. Andersen and Richardson (1997) and Hovmand and colleagues

(Hovmand et al. 2012) provide a large number of "scripts" that can help in setting up modelling projects. The procedures described are a long way from the earlier descriptions of a set of steps that seem to prescribe standard approaches applicable to most modelling projects. Instead, the guidelines offered have more the appearance of tool boxes, from which the appropriate technique can be selected on the basis of problem characteristics and the clients involved.

Group model building is generally conducted with a group of at least 6 and up to 15 people. The group is guided by at least two persons: a facilitator and a modeller/recorder. The group is seated in a semicircle in front of a whiteboard and/or projection screen, which serves as a so-called group memory. A projection screen is typically used when a model is constructed with the aid of system dynamics modelling software with a graphic interface (e.g., Vensim, Powersim, Ithink). This group memory documents the model under construction and is used as a parking lot for all kinds of unresolved issues which surface during the deliberations of the group (Fig. 3).

In the figure, the small circles indicate the persons present in the session. Apart from the participants, there is a facilitator and a recorder. The facilitator has the most important role in the session as he or she guides the group process. His/her task, as a neutral outsider, is to (a) elicit relevant knowledge form the group members, (b) to (help) translate elicited knowledge into

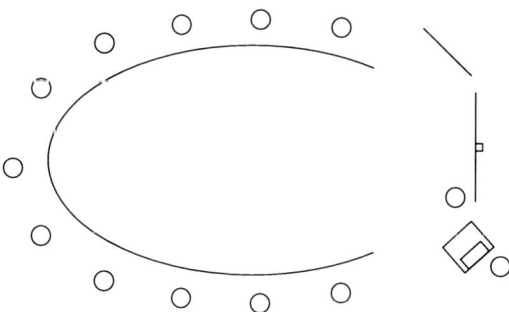

Group Model Building, Fig. 3 Typical room layout for group model building with participants seated in a semicircle, whiteboard and facilitator in front, and computer and overhead projector. (Adapted from Andersen and Richardson 1997)

system dynamics modelling terms, and (c) make sure that there is an open communication climate so that in the end consensus and commitment will result. The recorder keeps track of the elements of the model. In the figure (s)he is seated behind a computer, and the model is projected on the screen in front of the group. A separate whiteboard (upper right-hand corner) is used to depict the reference mode of behavior and record comments or preliminary model structure. As the model is visible to all participants, it serves as a group memory that at each moment reflects the content of the discussion up to that point. A group model building session is generally conducted in the so-called chauffeured style, where only the facilitator uses electronic support and projection equipment, while participants do not have access to electronic communication media (Nunamaker et al. 1991). The central screen or whiteboard will be used to depict the model, as shown in the figure.

The role of liaison between the organization and the modelling team is performed by the gatekeeper, who is generally also a member of the group. The gatekeeper is the contact between both parties and has an important role in the decision which participants to involve in the sessions. Apart from the gatekeeper, the facilitator and the recorder, two other roles may be important in a modelling session (Richardson and Andersen 1995): the process and the modelling coach. The process coach functions as an observer and primarily pays attention to the group process. The model coach needs to be experienced in system dynamics modelling but might also be an expert in the content area as well. As Richardson and Andersen (1995) point out, all roles are important in group model building, but not all of them have to be taken up by a single person. One person might, for instance, combine the roles of facilitator and process coach. Taken together, these different roles constitute the facilitation or modelling team.

In principle the group follows the normal steps in the construction of a system dynamics model. This means that the first step is the identification of the strategic issue to be discussed, preferably in the form of a so-called reference mode of behavior, i.e., a time series derived from the system to be modelled which indicates a (historical) undesirable development over time. As an example let us take copper production in Chile. An initial problem statement might be increasing and then levelling off of production. Typically the problematic behavior will be depicted in a graph over time as in figure below (Fig. 4).

The next step is to elicit relevant variables with which the model construction process can be started. Depending on the type of problem, this will take the form of either a causal loop diagram or a stock and flow diagram and is generally referred to as the conceptualization stage. The next step is to write mathematical equations (model formulation) and to quantify the model parameters. As described in the introductory section, most of the model formulation work is done backstage as it is quite time-consuming, and members of a management team generally are not very much interested in this stage of model construction. In this stage, the group is only consulted for crucial model formulations and parameter estimations. Experienced group model builders will start

to construct a simple running model as soon as possible and complicate it from there on if required. In the end the model should of course be able to replicate the reference mode of behavior (as one of the many validity tests) before it can be sensibly be used as a means to simulate the potential effects of strategies and scenarios.

Objectives of Group Model Building

As stated, the founder of system dynamics has repeatedly pointed out that much of the knowledge and information which is needed to construct a model can be found in the mental models of system participants. At first sight it may seem that the most important objective of building a system dynamics model is to find a robust strategy to solve the problem of the organization. In the end that is why one builds these models. From this perspective the most important issue in group model building is how to elicit the relevant knowledge from the group. However, as stated before, decision-making in organizations has its own logic, and in many cases, there is quite some disagreement about the problem and how it

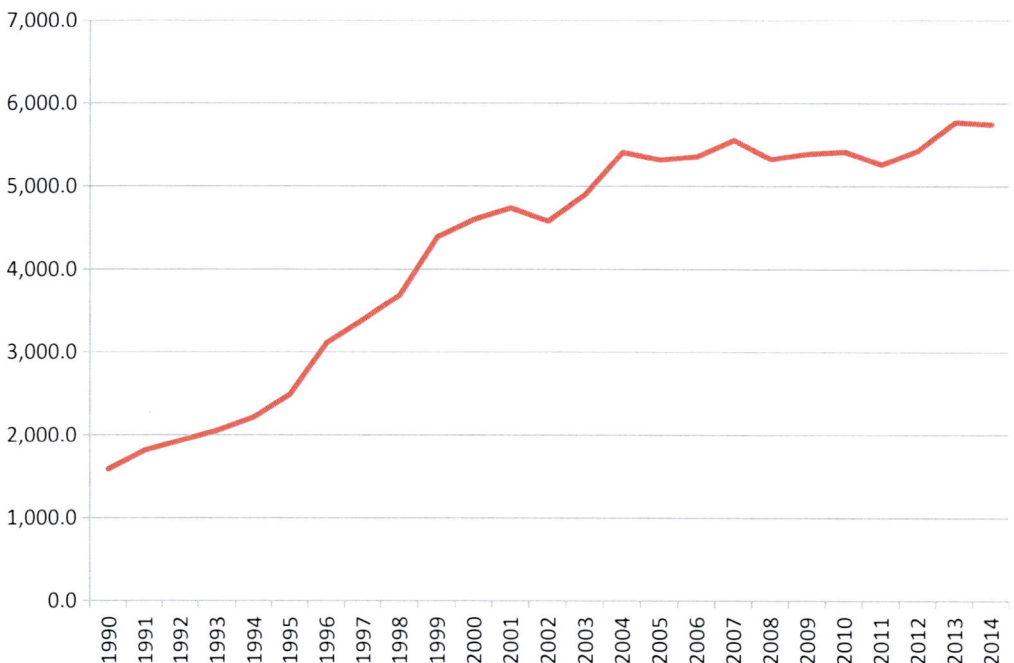

Group Model Building, Fig. 4 Problem description in graphical form: reference mode of behavior (example Copper production Chile, in 1000 metric tons, after Welting 2015)

should be tackled. No wonder that implementation of model outcomes is difficult if the model building process is not well integrated with decision-making processes in organizations, when it comes to creating agreement and commitment with a decision. From that perspective knowledge, elicitation is only one element in the process of model construction. It is not so much the model but to a greater extent the process of model construction which becomes important. Somewhat simplified one could say that in the traditional approach, when an organization is confronted with a strategic problem, it hires a modeller (or group of modellers) to construct a model and come up with recommendations to "solve" the problem. However, in most cases these recommendations become part of the discussion in the management team and get misunderstood or adapted or frequently just disappear as a potential solution from the discussion. Hence the title of Watt's (1977) paper: "Why won't anyone believe us?" becomes very much understandable from the point of view of the modeller. So rather than creating a situation where modellers "take away" the problem from the organization and (after considerable time) return with their recommendations, the model building process is now used to structure the problem and guide communication about it. Ultimately, it is used to test the robustness of strategies, taking into account other criteria and information which are not included in the model but do play a role for the organization when making the decision. Stated differently, the model building process now becomes intertwined with the process of decision-making in an organization. And this in turn means that other objectives than knowledge elicitation become important.

Simultaneously with the attempts to involve clients in the process of constructing system dynamics models, the objectives of group model building have been defined at several levels, i.e., the individual, the group, and the organizational level (cf. Andersen et al. 1997; Vennix 1999). The main goal at the individual level is change of mental models and learning. The idea is that participants should better understand the relationship between structure and dynamics and how their

interventions may create counterintuitive results. Unfortunately research has revealed that this is hardly the case. Even after extensive training, people have difficulty to understand the relationship between structure and dynamics (for a review, see Rouwette and Vennix 2006; Andersen et al. 2007). A second goal at the individual level is behavioral change. Frequently the conclusions of a modelling intervention point in the direction of behavioral change, for example, implementing a new job rotation scheme, or a change in purchasing policy. The question can then be asked how insights from the modelling intervention are translated to changes in behavior. Rouwette (e.g., Rouwette 2003; Rouwette and Vennix 2006; Rouwette et al. 2009) uses a framework from social psychology to understand the impact of modelling on behavior. The theory of Ajzen (1991, 2001) explains behavior on the basis of (a) attitude, (b) perceptions of norms, and (c) perceptions of control. It seems likely that each of these concepts is influenced in modelling sessions. When, for example, model simulations reveal unexpected levers for improving system behavior, we can expect that perceived control will increase. Another example: let's imagine that a manager is participating in a modelling session, where another participant reveals positive outcomes of a certain policy option. If these positive outcomes were previously not known to the manager, hearing them might make his/her attitude toward that option more positive (cf. Petty and Cacioppo 1986).

At the group level, objectives refer to mental model alignment (Huz et al. 1997) and fostering consensus (Rohrbaugh 1992; Vennix et al. 1993; Winch 1993). Creating consensus should not be confused with premature consensus, i.e., not discussing conflicting viewpoint. Here it concerns creation of consensus after critical debate, and discussion of opinions has taken place. This type of discussion which needs to take place in a cooperative communication climate is helpful to also create commitment with the resulting decision.

At the organizational level, goals have been discussed as system process change (are things done differently) and system outcome change (are customers impacted differently) (Cavaleri and Sterman 1997), although it has to be pointed out that in many cases system changes are the result of changes in attitude and behavior of participants in the system. An overview of group model building objectives is given in Table 1. In the table one finds a number of additional objectives such as positive reaction and creation of a shared language that are more fully reviewed by Huz et al. (1997), Rouwette et al. (2002), and Rouwette and Vennix (2006).

Designing Group Model Building Projects

When designing group model building projects, there are a number of questions that need to be addressed. The first concerns the suitability of system dynamics for the problem at hand. System dynamicists generally say that a problem needs to be dynamically complex in order to be suitable to model it through system dynamics. This means that one should at least hypothesize that there are positive and negative feedback process underlying the problem. From a more practical point of view, one could say that it should be possible to represent the problem in the form of a reference mode of behavior. If the latter is not possible, one should seriously question the use of system dynamics for the problem.

A second issue which needs to be given some thought is the question whether to use qualitative or quantitative modelling. Within the system dynamics community, there is still a debate

Group Model Building, Table 1 Objectives of group model building (cf. Rouwette and Vennix 2006)

Individual	Positive reaction
	Mental model refinement
	Commitment
	Behavioral change
Group	Increased quality of communication
	Creation of a shared language
	Consensus and alignment
Organization	System changes
	System improvement or results
Method	Further use
	Efficiency

about the question whether qualitative modelling (or mapping) can be considered system dynamics (see Coyle 2000, 2001; Homer and Oliva 2001). In short those who disagree point out that without quantification and simulation, one cannot reliably develop a robust policy simply because the human mind is not capable of predicting the dynamic effects of (interventions in) a dynamically complex structure. Those who do use mapping on the other hand point out that mapping in itself can have the beneficial effect of structuring the problem and at least will make managers aware of potential underlying feedback loops and their potential counterintuitive effects when intervening in a dynamically complex system. Basically the issue to quantify or not depends on the goals of the group model building intervention. If the goal is to find robust policies, then quantification is required. However, if the aim is to structure a problem and to create consensus on a strategic issue, then qualitative modelling may be all that is needed. This links up with Zagonel's (2004) distinction between the use of models as microworlds or as boundary objects. When used as a boundary object, the emphasis is on supporting negotiation and exchange of viewpoints in a group. This is clearly the case when problems are messy, i.e., connected to other problems and when there is much divergence of opinion on what the problem is and sometimes even whether there is a problem at all.

A third issue is the question who to involve in the sessions. There are a number of criteria which are generally employed. First it is important to involve people who have the power to make decisions and changes. A second criterion is to involve people who are knowledgeable about the problem at hand. A third criterion is to involve a wide variety of viewpoints, in order to make sure that all relevant knowledge about the problem is included. Of course these guidelines may create dilemmas. For example, involving more people in the process will make the group communication process more difficult and may thus endanger the creation of consensus and commitment.

Another issue is whether to use a preliminary model or to start from scratch (see Richardson 2013). Although using a preliminary model may speed up the process, the inherent danger is that it will be difficult to build group ownership over the model. Group ownership is clearly required to create consensus and commitment.

Finally, a range of methods and techniques is available to elicit relevant knowledge both from individuals and from groups. When it comes to individuals, well-known methods are interviews, questionnaires, and so-called workbooks. The latter are a kind of modified questionnaires, which are used in between sessions to report back to the group and ask new question in preparation of the next session. Interviews are being used routinely as a preparation for group model building sessions.

If a decision is made on the issues discussed above, the next important question is how to plan and execute the modelling sessions. This question is a central topic in the group model building literature, and its success heavily depends on the correct choice of available techniques and the quality of the facilitator.

Conducting Group Model Building Sessions

Although careful preparation of group model building sessions is a necessity, the most important part of the whole project is what happens in the group model building sessions themselves. During the sessions not only the analysis of the problem takes place (and the model is constructed) but also the interaction process between members of the management team unfolds. It is this interaction process which needs to be guided in such a way that consensus and commitment will emerge and implementation of results will follow. As pointed out the process is guided by the group facilitator, generally someone who is not only specialized in facilitation of group processes but also in system dynamics model construction. The facilitator is supported by a recorder or modeller who actually helps constructing the system dynamics model while the facilitator interacts with the management team.

The facilitator may choose from a wide variety of techniques in setting up and conducting a session. As a foundation for choosing techniques, Andersen and Richardson (1997) develop a set of guiding principles and so-called scripts for group model building sessions. Guiding principles capture basic ideas in the interaction with clients, such as break task/group structure several times each day, clarify group products, maintain visual consistency, and avoid talking heads. Scripts are more concrete instances of these principles and refer to small elements of the interaction process (Andersen and Richardson 1997; Luna-Reyes et al. 2006; Hovmand et al. 2012; https://en. wikibooks.org/wiki/Scriptapedia). The following table shows scripts described in Andersen and Richardson's (1997) original paper (Table 2).

In choosing a script, it is first important to be aware of the phase that is relevant in the project at that time. A common starting point, as we saw in

Group Model Building, Table 2 Group model building scripts (cf. Andersen and Richardson 1997)

Phase in modelling	Script
Defining a problem	Presenting reference modes
	Eliciting reference modes
	Audience, purpose, and policy options
Conceptualizing model structure	Sectors, a top-down approach
	Maintain sector overview while working within a sector
	Stocks and flows, by sector
	Name that variable or sector
Eliciting feedback structure	Direct feedback loop elicitation
	Capacity utilization script
	System archetype templates
	"Black box" means-ends script
Equation writing and parameterization	Data estimation script
	Model refinement script
	"Parking lot" for unclear terms
Policy development	Eliciting mental model-based policy stories
	Create a matrix that links policy levers to key system flows
	"Complete the graph" policy script
	Modeller/reflector feedback about policy implications
	Formal policy evaluation using multi-attribute utility models
	Scripts for "ending with a bang"

the description of group model building approaches, is to define the central problem of interest. The reference mode of behavior can function as a guideline for involving clients in this phase. Once the central problem is clear, a logical next step is to move toward model conceptualization. In this step again a number of options are available. Andersen and Richardson (1997) describe a script for identifying sectors that are important in the problem. An alternative is to start with more concrete variables in the problem, using a Nominal Group Technique (Delbecq et al. 1975).

Whatever scripts and techniques a facilitator employs, it is important that (s)he displays the right attitude and uses the correct skills. Several different aspects of attitude are important. First of all the facilitator is not the person who thinks (s)he knows the best solution, but needs to be helpful in guiding the group to find a solution to the strategic problem the organization is faced with. Second, a facilitator should be neutral with respect to the problem that is being discussed. Being too knowledgeable about a particular problem area may thus be dangerous, because it creates the tendency to participate in the discussions. Rather than being an expert, having an inquiry attitude (i.e., asking questions rather than providing answers) is more helpful to the group. Finally integrity and being authentic is important. Relying on tricks to guide the process will prove counterproductive, because people will look through them.

When it comes to skills, a thorough knowledge and experience in constructing system dynamics models is of course indispensable. Second, a facilitator needs to be knowledgeable about group process and has the skills to structure both the strategic problem and the group interaction process. For the latter, special group process techniques (e.g., brainstorming, Nominal Group Technique, Delphi) may be used, and knowledge about and skills in applying these techniques is a prerequisite for a successful group model building intervention. Finally, communication skills are important. Reflective listening is a skill which will help to prevent misunderstanding in communication, both between participants and the facilitator and between group members. For a more

thorough discussion of these attitudes and skills in the context of group model building, we refer to Vennix (1996, 1999).

Researching Group Model Building Effectiveness

In the above we described goals of group model building projects and principles and scripts for guiding the modelling process. In this section we consider the empirical evidence for a relation between modelling interventions and these intended outcomes. Empirical evidence can be gathered using a variety of research strategies, such as (field) experiments, surveys, or (in-depth) case studies. According to the review of modelling studies by Rouwette et al. (2002), the case study is the most frequently used design to study group model building interventions. We first report on the results found by these authors and then turn to other designs.

In the meta-analysis of Rouwette et al. (2002), the majority of group model building studies uses a case study design and assesses outcomes in a qualitative manner. Data are collected using observation, and a minority of studies employs individual group interviews. Case reports may be biased toward successful projects and are frequently not complete. The outcomes of the modelling projects were scored along the dimensions depicted in Table 1 in the section on modelling goals. The findings show positive outcomes in almost all dimensions of outcomes. Learning about the problem seems to be a robust outcome of group model building, for example:

- Of 101 studies that report on learning effects, 96 indicate a positive effect.
- Of 84 studies focusing on implementation of results, 42 report a positive effect.

Scott et al. (2015) look at studies published between 2001 and 2014. Results are in line with Rouwette et al. (2002), in that group model building achieves a range of outcomes such as communication, learning, consensus, behavioral change, and implementation.

Another set of studies, using quantitative assessment of results, is described by Rouwette and Vennix (2006). Although the research surveyed so far indicates positive effects of modelling on outcomes such as mental model refinement, consensus, and implementation of results, important challenges remain. Studies have paid little attention to the complexity of the intervention as described in the previous section. Pawson and Tilley (1997) urge us not to assume that intervention is similar and expect similar effects, since this would confuse meaningful differences between studies. Rouwette and Vennix (2006) describe two ways to learn more about the process through which outcomes of modelling are created: base research more on theory and/or to conduct research in more controlled settings. At present only few studies address elements of group model building in a controlled setting. Shields (2001, 2002) investigates the effect of type of modelling and facilitation on a group task. Most research on the use of system dynamics models concerns so-called management flight simulators. These studies aim to mimic the important characteristics of decision-making in complex, dynamic problems and test the effectiveness of various decision aids. Results are reviewed by Sterman (1994), Hsiao and Richardson (1999), and Rouwette et al. (2004).

Increased attention to theories may shed more light on the way in which modelling affects group decisions. Theories can help in specifying relations which can then be tested. Explanatory research is needed to connect the components and outcomes of group model building interventions (Andersen et al. 1997: 194). In the field of system dynamics modelling, three attempts at formulating theories on modelling components and outcomes are the work of Richardson et al. (1994), Rouwette (2003), and Black and Andersen (2012). The framework formulated by Rouwette (see Rouwette, Vennix, and Felling, 2009) builds on the theory of Ajzen (1991, 2001), which was explained before, and Petty and Cacioppo (1986). The latter add the idea that if people are motivated and able to process information, they will carefully scrutinize the arguments in any message they receive. As modelling usually addresses problems important to the participants invited to the sessions, we may assume they are highly motivated to process available information. Group model building's main role is then enabling participants to process information, for instance, by combining ideas into a coherent model of the situation, test the model, and determine the robustness of policies. Rouwette assumes that information exchanged in modelling sessions, if it provides arguments (so is new and relevant to the issue at hand), may change participants' evaluations and eventually actions. The theory of Richardson et al. (1994) separates mental models into means, ends, and means-ends models. The ends model contains goals, while the means model consists of strategies, tactics, and policy levers. The means-ends model contains the connection between the two former types of models and may contain either detailed "design" logic or more simple "operator" logic. On the basis of research on participants in a management flight simulator (Andersen et al. 1994), the authors conclude that operator logic, or high level heuristics, is a necessary condition for improving system performance. Therefore providing managers with operator knowledge is the key to implementation of system changes. Black and Andersen (2012) also look at the type of information that is exchanged in modelling. Whereas Rouwette and colleagues remain at a quite generic level by proposing that information contains "arguments," Black and Andersen are more specific. They propose that in a modelling session, information is added in "layers": first independent concepts, then relations between concepts, changes in the meaning of concepts, and finally clarification of actions that can be taken to resolve the issue. They illustrate these layers of information with clear examples from real-life cases, which drive the point home that each layer is the foundation for the next. It is clear that these theories do not contradict one another but instead have important similarities.

Future Directions

The success of group model building and problem structuring methodologies in general depends on a structured interaction between theory, methodology refinement, and application in practical project accompanied by systematic empirical evaluation.

Three areas for further development of theories stand out (cf. Rouwette and Vennix 2006):

– Review-related methodologies used in complex organizational problems, to determine which theories are used to explain effects. Examples that come to mind are insights from the operational research and systems fields (Morton et al. 2003; Franco and Montibeller 2010) and facilitated environmental modelling (Basco-Carrera et al. 2017).
– Forge a closer connection to research on electronic meeting systems. In this field, research is usually conducted in controlled settings (Pervan et al. 2004; Fjermestad and Hiltz 1999, 2001; Dennis et al. 2001), and theory development seems to be at a more advanced stage. Research on electronic meeting systems is interesting both because of the empirical results and explanatory theories used and because of insights on the intervention process. A recent development in the field is research on ThinkLets (Briggs et al. 2003). A ThinkLets is defined as a named, packaged facilitation intervention and thus seems very similar to the concept of a group model building script.
– A third source of theories is formed by research in psychology and group decision-making. Theories from these fields inform the definition of central concepts in group model building (see Table 1) and theories on modelling effectiveness. Rouwette and Vennix (2007) review literature on group information processing and relate this to elements of group model building interventions. Scott et al. (2015) and Rouwette (2016) provide a recent discussion of evidence for effects of group model building and avenues for further research.

From theories and evaluation research will come insights to further develop the methodology along the lines of (a) determining what kind of problem structuring methodology is best suited in what kind of situation, (b) refinement of procedures, (c) better understanding the nature of the intervention, and (d) better guidelines for facilitators how to work in different kinds of groups and situations.

Bibliography

Primary Literature

Ackermann F, Andersen DF, Eden C, Richardson GP (2011) ScriptsMap: a tool for designing multi-method policy-making workshops. Omega 39:427–434

Ackoff RA (1979) The future of operational research is past. J Oper Res Soc 30(2):93–104

Ajzen I (1991) The theory of planned behavior. Organ Behav Hum Decis Process 50:179–211

Ajzen I (2001) Nature and operation of attitudes. Annu Rev Psychol 52:27–58

Akkermans HA (1995) Modelling with managers: participative business modelling for effective strategic decision-making. PhD Dissertation: Technical University Eindhoven

Andersen DF, Richardson GP (1997) Scripts for group model-building. Syst Dyn Rev 13(2):107–129

Andersen DF, Maxwell TA, Richardson GP, Stewart TR (1994) Mental models and dynamic decision making in a simulation of welfare reform. In: Proceedings of the 1994 international system dynamics conference: social and public policy, Sterling, Scotland, pp 11–18

Andersen DF, Richardson GP, Vennix JAM (1997) Group model-building: adding more science to the craft. Syst Dyn Rev 13(2):187–201

Andersen DF, Vennix JAM, Richardson GP, Rouwette EAJA (2007) Group model building: problem structuring, policy simulation and decision support. J Oper Res Soc 58(5):691–694

Basco-Carrera L, Warren A, van Beek E, Jonoski A, Giardino A (2017) Collaborative modelling or participatory modelling? A framework for water resources management. Environ Model Softw 91:95–110

Black L, Andersen DA (2012) Using visual representations as boundary objects to resolve conflict in collaborative model-building approaches. Syst Res Behav Sci 29(2):194–208

Brewer GD (1973) Politicians, bureaucrats and the consultant: a critique of urban problem solving. Basic Books, New York

Briggs RO, de VGJ, Nunamaker JFJ (2003) Collaboration engineering with ThinkLets to pursue sustained success with group support systems. J Manag Inf Syst 19:31–63

Cavaleri S, Sterman JD (1997) Towards evaluation of systems thinking interventions: a case study. Syst Dyn Rev 13(2):171–186

Checkland P (1981) Systems thinking, systems practice. Wiley, Chichester/New York

Coyle G (2000) Qualitative and quantitative modelling in system dynamics: some research questions. Syst Dyn Rev 16(3):225–244

Coyle G (2001) Maps and models in system dynamics: rejoinder to Homer and Oliva. Syst Dyn Rev 17(4):357–363

De Gooyert V, Rouwette E, Kranenburg H v, Freeman E (2017) Reviewing the role of stakeholders in operational research: a stakeholder theory perspective. Eur J Oper Res 262(2):402–410

Delbecq AL, van de Ven AH, Gustafson DH (1975) Group techniques for program planning: a guide to nominal group and delphi processes. Glenview (Ill.). Foresman and Co, Scott

Dennis AR, Wixom BH, van den Berg RJ (2001) Understanding fit and appropriation effects in group support systems via meta-analysis. Manag Inf Syst Q 25:167–183

Eden C, Ackermann F (2001) SODA – the principles. In: Rosenhead J, Mingers J (eds) Rational analysis for a problematic world revisited. Problem structuring methods for complexity, uncertainty and conflict. Wiley, Chichester, pp 21–42

Fjermestad J, Hiltz SR (1999) An assessment of group support systems experimental research: methodology and results. J Manag Inf Syst 15:7–149

Fjermestad J, Hiltz SR (2001) A descriptive evaluation of group support systems case and field studies. J Manag Inf Syst 17:115–159

Forrester JW (1961) Industrial dynamics. MIT Press, Cambridge, MA

Forrester JW (1987) Lessons from system dynamics modelling. Syst Dyn Rev 3(2):136–149

Franco LA, Montibeller G (2010) Facilitated modelling in operational research. Eur J Oper Res 205(3):489–500

Greenberger M, Crenson MA, Crissey BL (1976) Models in the policy process: public decision making in the computer era. Russel Sage Foundation, New York

Homer J, Oliva R (2001) Maps and models in system dynamics: a response to Coyle. Syst Dyn Rev 17(4):347–355

Hoos IR (1972) Systems analysis in public policy: a critique. University of California Press, Berkeley/Los Angeles/London

House PW (1982) The art of public policy analysis: the arena of regulations and resources (2nd printing). Sage, Beverly Hills/London/New Delhi

Hovmand PS, Andersen DA, Rouwette EAJA, Richardson GP, Rux K, Calhoun A (2012) Group model building "scripts" as a collaborative planning tool. Syst Res Behav Sci 29:179–193

Hsiao N, Richardson GP (1999) In search of theories of dynamic decision making: a literature review. In: Cavana RY et al (eds) Systems thinking for the next millennium – Proceedings of the 17th international conference of the system dynamics society

Huz S (1999) Alignment from group model building for systems thinking: measurement and evaluation from a public policy setting. Unpublished doctoral dissertation. SUNY, Albany, New York

Huz S, Andersen DF, Richardson GP, Boothroyd R (1997) A framework for evaluating systems thinking interventions: an experimental approach to mental health system change. Syst Dyn Rev 13(2):149–169

Lane DC (1992) Modelling as learning: a consultancy methodology for enhancing learning in management teams. In: Morecroft JDW, Sterman JD (eds.) Modelling for learning, special issue of European journal of operational research. 59(1):64–84

Lane DC, Oliva R (1998) The greater whole: towards a synthesis of system dynamics and soft systems methodology. Eur J Oper Res 107(1):214–235

Lee DB (1973) Requiem for large-scale models. J Am Inst Plann 39(1):163–178

Luna-Reyes L, Martinez-Moyano I, Pardo T, Cresswell A, Andersen D, Richardson G (2006) Anatomy of a group model-building intervention: building dynamic theory from case study research. Syst Dyn Rev 22(4):291–320

Maxwell T, Andersen DF, Richardson GP, Stewart TR (1994) Mental models and dynamic decision making in a simulation of welfare reform. In: Proceedings of the 1994 international system dynamics conference. Stirling, Scotland, Social and Public Policy. System Dynamics Society, Lincoln, pp 11–28

Meadows DH, Robinson JM (1985) The electronic oracle: computer models and social decisions. Wiley, Chichester/New York

Mingers J, Rosenhead J (2004) Problem structuring methods in action. Eur J Oper Res 152:530–554

Morecroft JDW, Sterman JD (1992) Modelling for learning, special issue. Eur J Oper Res 59(1):1–230

Morton A, Ackermann F, Belton V (2003) Technology-driven and model-driven approaches to group decision support. Focus, research philosophy, and key concepts. Eur J Inform Sys 12(2):110–126

Nunamaker JF, Dennis AR, Valacich JS, Vogel DR, George JF (1991) Electronic meetings to support group work. Commun ACM 34(7):40–61

Otto PA, Struben J (2004) Gloucester fishery: insights from a group modeling intervention. Syst Dyn Rev 20(4):287–312

Pawson R, Tilley N (1997) Realistic evaluation. Sage, London

Pervan G, Lewis LF, Bajwa DS (2004) Adoption and use of electronic meeting systems in large Australian and New Zealand organizations. Group Decis Negot 13(5):403–414

Petty RE, Cacioppo JT (1986) The elaboration likelihood model of persuasion. Adv Exp Soc Psychol 19:123–205

Randers J (1977) The potential in simulation of macro-social processes, or how to be a useful builder of simulation models. Gruppen for Ressursstudier, Oslo

Richardson GP (1991) Feedback thought in social science and systems theory. University of Pennsylvania Press, Philadelphia

Richardson GP, Andersen DF (1995) Teamwork in group model-building. Syst Dyn Rev 11(2):113–137

Richardson GP (2013) Concept models in group model building. Syst Dyn Rev 29(1):42–55

Richardson GP, Andersen DF, Maxwell TA, Stewart TR (1994) Foundations of mental model research. In: Proceedings of the 1994 international system dynamics conference, Stirling, Scotland. System Dynamics Society, Lincoln, pp 181–192

Richmond B (1987) The Strategic Forum: from vision to strategy to operating policies and back again. High Performance Systems Inc., Lyme

Richmond B (1997) The strategic forum: aligning objectives, strategy and process. Syst Dyn Rev 13(2):131–148

Roberts EB (1978) Strategies for effective implementation of complex corporate models. In: Roberts EB (ed) Managerial applications of system dynamics. Productivity Press, Cambridge, MA, pp 77–85

Rohrbaugh JW (1992) Collective challenges and collective accomplishments. In: Bostron RP, Watson RT, Kinney ST (eds) Computer augmented team work: a guided tour. Van Nostrand Reinhold, New York, pp 299–324

Rouwette EAJA (2003) Group model building as mutual persuasion. Ph.D. dissertation Radboud University Nijmegen. Wolf Legal Publishers, Nijmegen

Rouwette EAJA (2016) The impact of group model building on behavior. In: Kunc M, Malpass J, White L (eds) Behavioral operational research: theory, methodology and practice. Palgrave Macmillan, London, pp 213–241

Rouwette EAJA, Vennix JAM (2006) System dynamics and organizational interventions. Syst Res Behav Sci 23(4):451–466

Rouwette EAJA, Vennix JAM (2007) Team learning on messy problems. In: London M, Sessa VI (eds) Work group learning: understanding, improving & assessing how groups learn in organizations. Lawrence Erlbaum Associates, Mahwah, pp 243–284

Rouwette EAJA, Vennix JAM, van Mullekom T (2002) Group model building effectiveness: a review of assessment studies. Syst Dyn Rev 18(1):5–45

Rouwette EAJA, Größler A, Vennix JAM (2004) Exploring influencing factors on rationality: a literature review of dynamic decision-making studies in system dynamics. Syst Res Behav Sci 21(4):351–370

Rouwette EAJA, Vennix JAM, Felling AJA (2009) On evaluating the performance of problem structuring methods: an attempt at formulating a conceptual model. Group Decis Negot 18(6):567–587

Schein EH (1987) Process consultation, vol II. Addison-Wesley, Reading

Scott RJ, Cavana RY, Cameron D (2015) Recent evidence on the effectiveness of group model building. Eur J Oper Res 249(3):908–918

Shields M (2001) An experimental investigation comparing the effects of case study, management flight simulator and facilitation of these methods on mental model development in a group setting. In: Proceedings of the 20th international conference of the system dynamics society 2001

Shields M (2002) The role of group dynamics in mental model development. In: Proceedings of the 20th international conference of the system dynamics society 2002

Stenberg L (1980) A modeling procedure for public policy. In: Randers J (ed) Elements of the system dynamics method. Productivity Press, Cambridge, MA, pp 292–312

Sterman JD (1994) Learning in and about complex systems. Syst Dyn Rev 10(2–3):291–330

van den Belt M (2004) Mediated modeling. A system dynamics approach to environmental consensus building. Island Press, Washington

Vennix JAM (1996) Group model-building: facilitating team learning using system dynamics. Wiley, Chichester/New York

Vennix JAM (1999) Group model building: tackling messy problems. Syst Dyn Rev 15(4):379–401

Vennix JAM, Gubbels JW, Post D, Poppen HJ (1990) A structured approach to knowledge elicitation in conceptual model-building. Syst Dyn Rev 6(2):31–45

Vennix JAM, Scheper W, Willems R (1993) Group model-building: what does the client think of it? In: Zepeda E, Machuca J (eds) The role of strategic modelling in international competitiveness. Proceedings of the 1993 International System Dynamics Conference Mexico, Cancun, pp 534–543

Vennix JAM, Akkermans HA, Rouwette EAJA (1996) Group model-building to facilitate organizational change: an exploratory study. Syst Dyn Rev 12(1):39–58

Vennix JAM, Andersen DF, Richardson GP (eds) (1997) Special issue on group model building. Syst Dyn Rev 13(2):187–201

Verburgh LD (1994) Participative policy modelling: applied to the health care industry. Ph.D. dissertation: Radboud University Nijmegen

Warren K (2000) Competitive strategy dynamics. Wiley, Chichester

Warren K (2005) Improving strategic management with the fundamental principles of system dynamics. Syst Dyn Rev 21(4):329–350

Watt CH (1977) Why won't anyone believe us? Simulation 28:1–3

Weil HB (1980) The evolution of an approach for achieving implemented results from system dynamic projects. In: Randers J (ed) Elements of the system dynamics method. Productivity Press, Cambridge, MA

Welting P (2015). Copper production and governmental funds in Chile: an initial effort towards the creation of a dynamic perspective on the developments in the Chilean copper mining industry and its impact on the Chilean government. Unpublished master thesis, Radboud University Nijmegen

Winch GW (1993) Consensus building in the planning process: benefits from a "hard" modelling approach. Syst Dyn Rev 9(3):287–300

Wolstenholme EF (1992) The definition and application of a stepwise approach to model conceptualisation and analysis. Eur J Oper Res 59:123–136

Zagonel AA (2004) Reflecting on group model building used to support welfare reform in New York state. Unpublished doctoral dissertation SUNY Albany, New York

Books and Reviews

Meadows DH, Richardson J, Bruckmann G (1982) Groping in the dark: the first decade of global modelling. Wiley, Chicester

Schwartz RM (1994) The skilled facilitator: practical wisdom for developing effective groups. Jossey-Bass, San Francisco

System Dynamics: Engineering Roots of Model Validation

Camilo Olaya
Universidad de Los Andes, Bogotá, Colombia

Article Outline

Glossary

Engineering The activity of designing and creating artifacts for improving a situation

Model An abstract or concrete artifact that is analog of another entity

Model Validity Judgment that determines whether a model fulfills the function for which it was created

Definition of the Subject

Engineering key elements that are present in System Dynamics (SD) give criteria for informing methodological questions. A good example is the issue of model validity. SD models are artifacts that can serve diverse purposes other than scientific goals. Hence, their validity rests on their effectiveness for accomplishing a specific purpose regardless of any type of justification, be it scientific or of other kind. However, the scope of that statement can be hard to appreciate. The belief that SD models are (or should be) scientific models is a popular one, and it is commonly used as the basis for developing and assessing models. However, this is not necessarily true. This entry underscores the meaning and scope of the engineering roots of SD for questions of model validation. In particular, the significance of the purpose of a model will be highlighted as the core issue to address.

Introduction

The origin of System Dynamics (SD) is placed in the works of the engineer Jay W. Forrester. System Dynamics modeling can be primarily described as a type of engineering activity, as opposed to a scientific activity, since it aims at designing artifacts, e.g. models, policies, plans, organizational schemes, etc., for concrete situations wanted to be improved under value-laden preferences and limited resources. However, the philosophy of science has been the usual place to identify the ground of System Dynamics (SD). In fact it is not uncommon to equate engineering with science or, similarly, to believe that engineering (or SD for that matter) is either applied science or some type of scientific activity. Those beliefs can complicate the discussion of methodological aspects of System Dynamics. The goal of designing improved systems means that a promising place to look for answers to methodological SD questions is in engineering. This entry shows how key engineering elements are present in the early writings of Forrester and how such elements inform methodological discussions regarding the validity of SD models. To consider defining elements of engineering that set it apart from science might be a required starting point, as follows.

Engineering

Despite of sharing many elements and mutual benefits, engineering and science, at their very

© Springer Science+Business Media, LLC, part of Springer Nature 2020

B. Dangerfield (ed.), *System Dynamics*,
https://doi.org/10.1007/978-1-4939-8790-0_544

Originally published in
R. A. Meyers (ed.), *Encyclopedia of Complexity and Systems Science*, © Springer Science+Business
Media LLC 2019, https://doi.org/10.1007/978-3-642-27737-5_544-2

core, are different activities that pursue different goals: scientists primarily *explain* phenomena; engineers *improve* a situation. The scientist deals mostly with the question "what is it?" The engineer deals with "how to change this situation?" Engineering is concerned "not with the necessary but with the contingent not with how things are but with how they might be" (Simon 1996, p. xii). Petroski puts it succinctly: scientists seek to explain the world; engineers try to change it (2010). Although engineers make the most of scientific knowledge, such different missions lead to different values, norms, rules, apparatus for reasoning, considerations, type of knowledge, methods, success criteria, and standards for evaluating results, in short, different epistemologies.

To illustrate those differences, the philosopher Sven Hansson (2007) establishes six defining characteristics of engineering:

1. Study objects are constructed by humans (rather than being objects from nature).
2. Design is an essential component. Objects are constructed by engineers.
3. The categories for classifying objects are usually specified according to functional rather than physical characteristics; e.g., to determine whether an object is a screwdriver requires determining whether it indeed drives screws.
4. Engineers operate in value-laden contexts that influence concepts and designs, e.g., "user friendly," "risk," "better," etc.
5. Engineering knowledge is difficult to generalize because of real-world restrictions and complexity attached to concrete contexts of application.
6. Exact mathematical precision and analytical solutions are not required if a sufficiently close approximation is available.

Instead of observing and explaining nature, engineers deal with the artificial world (Simon 1988, 1996) by adapting means to a preconceived end (Layton Jr 1974). In particular, engineers change the world through the design of artifacts that improve a situation and solve problems. *Design* is perhaps the key defining characteristic of engineering (Van De Poel 2010; Pitt 2011) – and hence the popular notion of engineering as "applied science" does not quite fit since design means a *creative* rather than merely an *applicative* (or *reproductive*) activity (Doridot 2008). Engineering design is in fact an activity intrinsically different from scientific research. The design of a bridge "does not derive from a set of equations expressing the laws of physics but rather from the creative mind of the engineer" (Petroski 2010, p. 47). The same applies for designing intangible artifacts such as plans, organizational or public policies, projects, laws, the way we organize ourselves into social units, etc. (Simon 1996; Romme 2003; Jelinek et al. 2008; Remington et al. 2012).

Moreover, engineering design is a contextual and intensely particular process (Goldman 2004) that requires concrete and practical considerations according to the specific situation to be improved. Engineering deals with particulars in its particularity and not as instantiations of a universal (Goldman 1990). That is why engineers have to resolve a variety of constraints that are the result of idiosyncratic values (economic, cultural, political, reliability, viability, ethical) that co-define and specify the design problem. In scientific research such constraints are absent in the definition of a scientific question (Kroes 2012). Scientists assume idealized situations. Engineers on the other hand deal with imperfect situations that do not allow for simplifications, e.g., the effects of friction or the way in which persons actually make decisions cannot be ignored (Hansson 2007). Furthermore, engineering exists because someone (a client, a corporation, a group of people, the State, etc.) is interested in improving a situation. These clients have both specific resources that also co-define the problem and subjective value-laden criteria for resolving trade-offs that will end up informing the artifact to be designed (Koen 2003), e.g., time vs. cost vs. environmental impact. Such orientation both to particulars (instead of universals) and to subjective criteria explains why there is no unique solution for an engineering problem; actually "an engineer who understands engineering will never claim to have found *the* solution... This is why there are so many different-looking airplanes and automobiles and why they operate differently... they are simply one engineers solution to a problem that has no unique solution" (Petroski 2010, p. 54). In contrast, scientific problems should be explained by the one best theory, at any given time – when a theory is shown to be erroneous, then it should be replaced for a better one (Goldman 1990, 1991).

Functional considerations distinguish engineering knowledge (Auyang 2009). That is, engineering does produce knowledge. But it is not scientific knowledge. Science is usually interested in the truthiness (or not) of statements according to evidence. A scientist wants to *know that* a description or an explanation is true or false. Engineers on the other hand show to *know how* things *should be done*. Engineering produces such a characteristic type of knowledge, *knowledge-how*, which is different from scientific *knowledge-that*. Knowledge-how cannot even be defined in terms of knowledge-that (Ryle 1945). "Engineering knowledge is... in the form of 'knowledge-how' to accomplish something, rather than 'knowledge-that' the universe operates in a particular way" (Schmidt 2012, p. 1162). Knowledge-how is not concerned with the truth or falsehood of statement, that is, "you cannot affirm or deny Mrs. Beetons recipes" (Ryle 1945, p. 12). A bridge is neither true nor false. Engineering knowledge is not descriptive but prescriptive.

The method is also different. The "scientific method" (in any of its versions) cannot produce knowledge-how. Instead, the method of engineers can be characterized by the term "heuristic." A heuristic is *"anything* that provides a plausible aid or direction in the solution of a problem but is in the final analysis *unjustified, incapable of justification, and potentially fallible*" (Koen 2010, p. 314, emphasis added). Koen highlights the distinctive nature of heuristics and the differences with scientific methods: a heuristic does not guarantee an answer or a solution, it may contradict other heuristics, it does not need justification, its relevance depends of the particular situation that the heuristic deals with, and its outcome is a matter of neither "truth" nor generalizability (Koen 2003, 2009). The engineering method –

as opposed to a scientific method – is unjustified, fallible, uncertain, context-defined, and problem-oriented.

Finally, to follow a heuristic method means that in order to get the job done, engineers use whatever works: "resolving engineering problems regularly requires the use of less than scientifically acceptable information" (Mitcham 1994); scientific "empirical evidence" might be useful, but it is not strictly necessary. Moreover, mathematical precision and analytical solutions are not needed; engineers require neither ideal conditions nor empirical justification. Because of that practical approach to create knowledge, unlike traditional science (perhaps with the exception of strict, and uncommon, Popperian science), engineering does not need epistemic justifications. The creation of artifacts is done by experimental methods that are more fundamental than (and not directly derived from) theories or scientific evidence (Doridot 2008). The justification for developing a design is ultimately irrelevant; an artifact does not have to be necessarily justified by evidence, theories, data, or similar. It can be freely generated with the help of any procedure, sourced from reason, or guided by previous expectations – "theoretic" or not (Stein and Lipton 1989), guided with the help of a model, or just based on imagination, intuition, or instincts, e.g. "the inventor or engineer... can proceed to design machines in ignorance of the laws of motion... These machines will either be successful or not" (Petroski 2010, p. 54). An artifact is not false or true (or closer to), instead, it works, or it does not. If it works, the engineer succeeds. The popular notion of knowledge that comes from the philosophy of science as "*justified* true belief" means little for the pragmatic engineer for whom knowledge is *unjustified*. In the words of Pitt: "If it solves our problem, then does it matter if we fail to have a philosophical justification for using it? To adopt this attitude is to reject the primary approach to philosophical analysis of science of the major part of the twentieth century, logical positivism, and to embrace pragmatism" (2011, p. 173). This is the yardstick for assessing an engineering artifact, not scientific justification but effectiveness according to purpose.

System Dynamics: Engineering "DNA"

The previous engineering trademarks can be identified in the origin of SD and its developments. Jay W. Forrester at M.I.T. wanted to bridge management education with his own engineering knowledge in feedback control systems and computers (Forrester 1975a). In 1956 he wrote a "note" to the Faculty Research Seminar. In this communication he sketched the worldview of what would be known as "System Dynamics" (Forrester 2003). He started with a criticism to economic models and instead favored techniques that were largely underused at that time: servomechanisms, differential equations, and what he called "the art of simulation." Anchored on the mentioned assessment and on these developments, Forrester conceived "a new avenue of attack for understanding the firm and the economy" (p. 336) envisaging *a new kind of models* that, unlike economic models that have science as a reference point, would have engineering elements such as:

Dynamic structure: detailed attention to the *operations* and *sequences of actions* which occur in the system being modeled and to the *forces* which trigger or temper such actions, with a particular concern on the controlling influences of lags and delays.

Information flows: explicit recognition of information flow channels and information transformation with time and transmission.

Decision criteria: reexamination of the proper decision criteria which must not be defined as depending only on current values of gross economic variables; instead, such criteria must be traced to the motivations, hopes, objectives, and optimism of the people involved, including as well what he calls business man's intuition which "represents a disordered accumulation of basic insights into how people and social systems react" (p. 342).

Nonlinear systems: economic systems present most – if not all – of the time highly nonlinear characteristics.

Differential equations: the behavior of economic systems should be better described by nonlinear differential equations since they have been developed to describe delays, momentum, elasticity, reservoirs, and accelerations, which are better suited quantities for describing the economic world. In practice these equations would be handled as incremental difference equations in order to obtain numerical solutions.

Model complexity: much complex and complete models can be developed with these techniques.

Solutions according to a particular situation: it is useless to look for explicit unique or "correct" solutions; instead, these models provide diverse solutions according to the different assumptions about the model structure and the initial values of the variables "out of a study of particular situations" (p. 344).

Symbolism and correspondence with real counterparts: the possibility of having a pictorial representation – a flow diagram – whose processes of information, money, goods, and people are moved, i.e., simulated, time-step-by-time-step from place to place.

Structure over coefficient accuracy: to prefer a structure in which we have confidence using intuitively estimated coefficients instead of using unlikely structures with accurately derived coefficients from statistical data.

A following step came in 1958 with the article "Industrial Dynamics: A Major Breakthrough for Decision Makers," published in the *Harvard Business Review* (Forrester 1975b), in which Forrester shaped the previous ideas with the concern that management should evolve from a highly fragmentized art to a profession capable of recognizing unified systems given that the task of management is to interrelate the flows of information, materials, labor, money, and capital equipment. He again emphasized features such as electronic data processing, decision-making, simulation, feedback control, and information flows. These elements were presented as the cornerstones of the innovative industrial dynamics program at MIT.

The definitive advance came in 1961 with his *magnum opus* "Industrial Dynamics" (Forrester 1961). Forrester presented the potential of this breakthrough in function of its possibilities to improve management effectiveness – for instance – in order handling and inventory adjustment. He made explicit the engineering spirit of the recently born "Industrial Dynamics":

> Physics has provided the foundation for a great upsurge in technology. But it is not adequate to specify the 'best' design of a space vehicle nor to guarantee our ability to make a roof that does not leak. Physics is a foundation of principles to explain underlying phenomena but not a substitute for invention, perception, and skill in applying the principles. (Forrester 1961, p. 5)

Forrester knew that the models in engineering differ "in principle" (p. 53) from the models of the physical sciences, "they arise in a different way, to be used for a different purpose" (p. 53). He thought that such engineering models provide a better precedent for management that deals with very complex systems. The purposes of this new type of model were for him different from the ones that are used in science:

> In engineering, models have been used for *designing* new systems; in economics, a common use is to *explain* existing systems... It appears that models that have been undertaken only for explanation have often had their goals set so low that they fail not only for design but also for explanatory purposes. (pp. 53–54, emphases original)

Hence, Forrester established the engineering task to address with Industrial Dynamics: "*designing* and *controlling* an industrial system" (p. 8). This new type of model would have as a main purpose to support the *design* of systems – as opposed to *explain*. Models should show how changes in policies or structure will produce better or worse behavior in real and concrete situations: the "search for useful solutions of *real* problems... The challenge to the future manager is to *design* improved enterprises" (pp. 3, 6, emphasis added).

Forrester also knew that each engineering model is unique; it corresponds to a particular situation, and its development depends on the question in hand: "there will be no such thing as *the* model of a social system, any more than there is *the* model of an aircraft...The factors that must be included arise directly from the questions that are to be answered... There may well be different models for different classes of questions about a particular system" (p. 60, emphasis original). In other words, such models do not pretend to provide generic, all-purpose explanations but to address specific problems. This stance defines, in turn, the question of validity:

> The validity (or significance) of a model should be judged by its suitability for a particular purpose. A model is sound and defendable if it accomplishes what is expected of it. This means that validity, as an abstract concept divorced from purpose, has no useful meaning. What may be an excellent model for one purpose may be misleading and therefore worse than useless for another purpose. (p. 116)

Forrester then made very clear the well-known validity principle of SD: suitability for a purpose. Since for him the purpose of these new models was "to aid in the *design* of improved industrial and economic systems" (p. 115), then "the ultimate test is whether or not better systems result from investigations based on model experimentation" (p. 115), that is, the effective improvement of a real and concrete situation. This is the pragmatic stamp of engineering.

Forrester knew his engineering, as this brief historical review has shown. Table 1 summarizes these trademarks in the engineering origins of SD.

Model Validity

System dynamicists produce knowledge through building models. But what is "a good model"? The answer will vary depending on the way we characterize SD models: are they scientific models, or are they instead closer to engineering models?

The belief that SD models are scientific models can be illustrated by the stance of Homer (1996)

System Dynamics: Engineering Roots of Model Validation, Table 1 Engineering underpinnings in Forrester's Industrial Dynamics (1961)

Goal	To improve a situation, to change it, to solve a problem (as opposed to explain)
Role of models	Aid for design, learning vehicles, multipurpose
Context of application	Decision processes, decisional nature
Domains of application	Specific (particular, concrete) problems
Validity of models	Effectiveness, usefulness, according to purpose

who defends what he calls "scientific modeling," a strict adherence to the "scientific method" which for him means that empirical evidence must support hypotheses and formulations. These preoccupations seem to be rooted in the justificationist paradigm of science that looks for epistemic support – in this case, observations and evidence (see, e.g., Homer 2014). Moreover, he takes the scientific method (if there is "one" scientific method) as "essential" for having reliable models. Such a position is a common one among SD modelers.

If we assume that an SD model is a scientific model, then the question of model validity will be driven by scientific criteria. An illustrative example illustrates this path. Barlas (1996) underlines the well-known criterion for judging SD models: the purpose of the model, i.e., "it is impossible to define an absolute notion of model validity divorced from its purpose" (p. 184). However, he seems to need a scientific reason for justifying that criterion. Barlas affirms that "judging the validity of the internal structure of a model is very problematic... because...the problem is directly related to the unresolved philosophical issue of verifying the truth of a (*scientific*) statement" (p. 186). Why? According to him:

> A valid system dynamics model embodies a theory about how a system actually works in some respect. Therefore, there has to be a strong connection between how theories are justified in the sciences (a major and unresolved philosophy of *science* question) and how such models are validated. This means that our conception of model validity depends on our philosophy (implicit or explicit) of how knowledge is obtained and confirmed. (p. 187)

Barlas explicitly suggests to use the theory of justification that comes from the philosophy of science, in order to deal with the question of how to secure SD knowledge, which leads to the question of model validity. Following the same path, with the aim of showing that SD models are "truly scientific" and that they produce "scientific knowledge," Barlas and Carpenter (1990) address the "philosophical roots of model validation" associating System Dynamics with what they call a "relativist philosophy of science" and by addressing the question on model validity as a question of

scientific knowledge: "The validity controversy is strongly tied to a fundamental problem in the philosophy of science: What is scientific knowledge, and what constitutes confirmation of a knowledge claim?" (p. 148). In this view, justification is pursued: such knowledge is seen as socially, culturally, and historically dependent, and it is defined as socially *justified* belief. Here "a valid model is assumed to be only one of many possible ways of describing a real situation... for every model carries in it the modeler's world view... validation is a matter of social conversation" (p. 157). Hence, the justification of what they call "a knowledge claim" is pursued through a social process relative to a frame of reference.

Actually, to equate engineering models with scientific models is not uncommon. Bissell and Dillon state why:

> Because engineering models use many of the same mathematical techniques as scientific models (differential equations, Fourier and Laplace transformations, vectors, tensors, for example) it is easy to assume that they are one and the same in essence. Yet in the case of engineering (and technology in general) such models are much more likely to be used to de- sign devices, equipment or industrial plant than for representing and analyzing natural objects or phenomena. (Bissell and Dillon 2012, p. v)

Models are artifacts. However, unlike scientific models that are built only (typically) for explaining or predicting phenomena, engineering models are multipurpose and are built for a wide variety of goals. For instance, Epstein (2008) highlights 16 reasons to build a model. Some of these (and further) purposes for building engineering models, including SD models, are:

- Design systems, policies, plans, courses of action.
- Drive conversations among designers, policymakers, communities, etc.
- Create shared understandings.
- Educate and learn, challenge mental models, change paradigms.
- Explore scenarios.
- Illuminate core dynamics.
- Suggest dynamical analogies.
- Discover new questions.
- Promote a scientific habit of mind.

- Bound (bracket) outcomes to plausible ranges.
- Illuminate core uncertainties.
- Offer crisis options in near real time.
- Demonstrate trade-offs/suggest efficiencies.
- Expose prevailing wisdom as incompatible with available data.
- Train practitioners.
- Discipline the policy dialogue.
- Educate the general public.
- Reveal the apparently simple (complex) to be complex (simple).

Many purposes, many uses, many criteria for assessing "usefulness." A System Dynamics model is hardly a scientific statement. If the purpose of an SD model is theory-building and/or produces scientific explanations, then scientific criteria surely will be needed for model validation; this promising path already has fruitful advances (Schwaninger and Grösser 2008; Lane 2008; Kopainsky and Luna-Reyes 2008; Sterman 1986; Rahmandad and Sterman 2012; Morrison et al. 2013). Nevertheless, as the list of diverse purposes above shows, the possibilities go far beyond. And as long as an SD model pursues goals different from scientific explanation, theorizing, or prediction, then its engineering roots become all the more important since they will widen the spectrum of possibilities for answering the question *what is a good model?* Norström (2013) perfectly illustrates this point by underlining that even engineering models can be unscientific and still be useful:

> Engineers commonly use rules, theories and models that lack scientific justification. Examples include rules of thumb based on experience, but also models based on obsolete science or folk theories. Centrifugal forces, heat and cold as substances, and sucking vacuum all belong to the latter group. These models contradict scientific knowledge, but are useful for prediction in limited contexts and they are used for this when convenient... The acceptance of the non-scientific models for action guidance could be experienced as contradictory... different epistemological frameworks must be used in science and technology. Technology is first and foremost what leads to useful results, not about finding the truth or generally applicable laws. (p. 377)

Technical assessments (extreme conditions, parameter assessment, sensitivity analysis, behavior reproduction, etc.), judgment, prototyping, inspection, internal consistency, social conversation, experiments, and observation, among other validation tests, are at disposal of the modeler for testing a model. However, the criteria for judging its usefulness come from its purpose. At the end, given that a model is built for a purpose (i.e., its reason for existing), then its capacity for accomplishing such a purpose should supersede other criteria. For instance, let us say that an SD model is built for training practitioners on a specific subject. Then the validity of the model will rest on its capacity and effectiveness for training those practitioners – certainly corresponding measurement tools will be needed – no matter, for example, how "realistic" or how "scientific" might the model be. Criteria for assessing the levels of expertise acquired by trainees are available in most of disciplinary domains, including SD training itself (Schaffernicht and Groesser 2016). For instance, a model can contradict the law of conservation of energy and still be useful for several purposes. Such a model will be problematic or unacceptable for a scientist. That same model could be used for a purpose other than describing or producing scientific knowledge, for example, as a trivial example, for testing the knowledge of physics students. In the latter case, the purpose is not scientific but educative, and if it proves to be useful for educating students, then the model succeeds. A different purpose will mean different criteria for "usefulness." For instance, for driving a policy dialogue, a model can be built and used for "creating shared understandings." Hence, those "shared understandings" will be either created or not (and hence the model will be useful or not), regardless of evidence, veracity, or the scientific accuracy of the model for representing the corresponding modeled situation. Does the model serve as a focal point for mediating and driving such a policy dialogue? Does it help to build consensus? Is it a reliable artifact for uncovering interests and positions? These would be primary questions to address. Or let us take, for instance, group model building in which participatory modeling might seek to empower communities through new systemic understandings. Did the modeling process and the model itself help to develop those new understandings? Several tools are at disposal of modelers and facilitators for assessing both the modeling process and the resulting models (Hovmand 2014; Andersen and Richardson 1997; Vennix 1996).

What is then "a good model"? The answer is straightforward (and not simplistic at all): a good model is a model that *works* for a given purpose. To establish if a model "works" is not an uncomplicated matter. There are several well-known tests that help to increase our confidence in a model (Sterman 2000; Barlas 1996; Schwaninger and Groesser 2018; Groesser and Schwaninger 2012). However, the ultimate razor for judging the "goodness" of SD models is, unmistakably, pragmatical and functional: effectiveness according to a purpose (and there are many possible purposes to choose from). This point cannot be emphasized enough. To enclose model validity under the principles of science could lead to a limiting way for developing and assessing SD models that might turn to be misleading. The question of truthiness might misinform the question of usefulness. To understand SD models from their engineering roots means that unrealistic SD models, that even may contradict scientific knowledge, should not be discarded based on such grounds, especially if those models prove to be useful. A model is a device. If the model works for the purpose in hand, then the model succeeds, regardless of justification, reasons, evidence, etc. The main challenge is then to develop adequate measures of effectiveness according to purpose. It might be inappropriate to frame the validation of SD models within the narrow question of justification that comes from the philosophy of science aiming at *explanation* (or prediction or representation) as a goal. Moreover, the epistemological tradition in what is known as "philosophy of science," being concerned with *knowing-that* and the truthiness of scientific statements, is unfit for addressing the type of knowledge that system dynamicists usually produce in large quantities: *knowledge-how*.

Future Directions

System Dynamics has been permeated by discussions that come from the philosophy of science. This has brought rigorous and systematic approaches that have boosted major developments. However, discussions on model validity might get complicated if SD models are understood as scientific models, not only because of the intricate issues regarding the validation of scientific models but also because of the implicit assumption that SD models, then, apparently would produce scientific knowledge, i.e., *knowledge-that*. This would be misleading. It is true that SD models can be used to *help* producing scientific knowledge. But that is not the same as saying that SD models *are* scientific models. The recognition of the engineering features of SD models helps to widen and clarify important SD questions. The issue of model validity is particularly central to SD modeling. Other questions related to the epistemology and methodology of System Dynamics, such as the type of knowledge that SD produces, model building methods, etc., are worth pursuing using engineering lenses. This would open possibilities for even broader, more creative, and innovative models.

Bibliography

Andersen DF, Richardson GP (1997) Scripts for group model building. Syst Dyn Rev 13:107–129

Auyang SY (2009) Knowledge in science and engineering. Synthese 168:319–331

Barlas Y (1996) Formal aspects of model validity and validation in system dynamics. Syst Dyn Rev 12:183–210

Barlas Y, Carpenter S (1990) Philosophical roots of model validation: two paradigms. Syst Dyn Rev 6:148–166

Bissell C, Dillon C (2012) Ways of thinking, ways of seeing. Mathematical and other modelling in engineering and technology. Springer, Berlin/Heidelberg

Doridot F (2008) Towards an engineered epistemology? Interdiscip Sci Rev 33:254–262

Epstein JM (2008) Why model? J Artif Soc Soc Simul 11:1–12

Forrester JW (1961) Industrial dynamics. Productivity Press, Cambridge, MA

Forrester JW (1975a) Industrial dynamics: after the first decade. Collected papers of Jay W. Forrester. Wright-Allen Press, Cambridge

Forrester JW (1975b) Industrial dynamics: a major breakthrough for decision makers. Collected papers of Jay W. Forrester. Wright-Allen Press, Cambridge

Forrester JW (2003) Dynamic models of economic systems and industrial organizations. Syst Dyn Rev 19:331–345

Goldman SL (1990) Philosophy, engineering, and western culture. In: Durbin PT (ed) Broad and narrow interpretations of philosophy of technology. Kluwer, Amsterdam

Goldman SL (1991) The social captivity of engineering. In: Durbin PT (ed) Critical perspectives on nonacademic science and engineering. Lehigh University Press, Bethlehem, PA

Goldman SL (2004) Why we need a philosophy of engineering: a work in progress. Interdiscip Sci Rev 29:163–176

Groesser SN, Schwaninger M (2012) Contributions to model validation: hierarchy, process, and cessation. Syst Dyn Rev 28:157–181

Hansson SO (2007) What is technological science? Stud Hist Phil Sci 38:523–527

Homer JB (1996) Why we iterate: scientific modeling in theory and practice. Syst Dyn Rev 12:1–19

Homer J (2014) Levels of evidence in system dynamics modeling. Syst Dyn Rev 30:75–80

Hovmand PS (2014) Community based system dynamics. Springer, New York

Jelinek M, Romme AGL, Boland RJ (2008) Introduction to the special issue. Organization studies as a science for design: creating collaborative artifacts and research. Organ Stud 29:317–329

Koen BV (2003) Discussion of the method. Oxford University Press, Oxford

Koen BV (2009) The engineering method and its implications for scientific, philosophical, and universal methods. Monist 92:357–386

Koen BV (2010) Quo vadis, humans? Engineering the survival of the human species. In: Van de Poel I, Goldberg DE (eds) Philosophy and engineering. An emerging agenda. Springer, Dordrecht

Kopainsky B, Luna-Reyes LF (2008) Closing the loop: promoting synergies with other theory building approaches to improve system dynamics practice. Syst Res Behav Sci 25:471–486

Kroes P (2012) Engineering design. In: Kroes P (ed) Technical artefacts: creations of mind and matter. Springer, Dordrecht

Lane D (2008) Formal theory building for the avalanche game: explaining counter-intuitive behaviour of a complex system using geometrical and human behavioural/physiological effects. Syst Res Behav Sci 25:521–542

Layton ET Jr (1974) Technology as knowledge. Technol Cult 15:31–41

Mitcham C (1994) Thinking through technology. the path between engineering and philosophy. The University of Chicago Press, Chicago

Morrison JB, Rudolph JW, Carroll JS (2013) Dynamic modeling as a multidiscipline collaborative journey. Syst Dyn Rev 29:4–25

Norström P (2013) Engineers' non-scientific models in technology education. Int J Technol Des Educ 23:377–390

Petroski H (2010) The essential engineer. Why science alone will not solve our global problems. Vintage Books, New York. (Ed. 2011)

Pitt JC (2011) Doing philosophy of technology. Springer, Dordrecht

Rahmandad H, Sterman J (2012) Reporting guidelines for simulation-based research in social sciences. Syst Dyn Rev 28:396–411

Remington R, Folk CL, Boehm-Davis DA (2012) Introduction to humans in engineered systems. Wiley, Hoboken

Romme AGL (2003) Making a difference: organization as design. Organ Sci 14:558–573

Ryle G (1945) Knowing how and knowing that. Proc Aristot Soc New Ser 46:1–16

Schaffernicht M, Groesser SN (2016) A competence development framework for learning and teaching system dynamics. Syst Dyn Rev 32:52–81

Schmidt JA (2012) What makes engineering, engineering? In: Carrato J, Burns J (eds) Structures congress proceedings. American Society of Civil Engineers, Reston

Schwaninger M, Groesser S (2018) System dynamics modeling: validation for quality assurance. In: Meyers RA (ed) Encyclopedia of complexity and systems science. Springer, Berlin/Heidelberg

Schwaninger M, Grösser S (2008) System dynamics as model-based theory building. Syst Res Behav Sci 25:447–465

Simon HA (1988) The science of design: creating the artificial. Des Issues,. IV 4:67–82

Simon HA (1996) The sciences of the artificial. MIT Press, Cambridge, MA

Stein E, Lipton P (1989) Where guesses come from: evolutionary epistemology and the anomaly of guided variation. Biol Philos 4:33–56

Sterman JD (1986) The economic long wave: Theory and evidence. Syst Dyn Rev 2:87–125

Sterman J (2000) Business dynamics. Systems thinking and modeling for a complex world. McGraw-Hill, Boston, MA

Van De Poel I (2010) Philosophy and Engineering: Setting the Stage. In: Van De Poel I, Goldberg DE (eds) Philosophy and engineering. An emerging agenda. Springer, Dordrecht

Vennix JAM (1996) Group model building. Wiley, Chichester

System Dynamics Modeling: Validation for Quality Assurance

Markus Schwaninger[1] and Stefan Groesser[2]
[1]Institute of Management and Strategy, University of St. Gallen, St. Gallen, Switzerland
[2]School of Engineering and Information Technology, Bern University of Applied Science, Bern, Switzerland

Article Outline

Glossary

Model/model system A model is an abstract representation of a concrete ("real") system. Models can be descriptive or prescriptive (normative). Their functions can be to enable explanation, anticipation, or design. A distinction used in this contribution is between causal and noncausal models, with System Dynamics models being of the former type. The term *model system* is used to stress the systemic character of a model; this serves to identify it as an organized whole of variables and relationships, on the one hand, and to distinguish it from the *real system* which is to be modeled, on the other.

Model validity A model's property of reflecting adequately the system modeled. Validity is the main feature of model quality. It is a matter of degree, not a dichotomized property.

Model purpose The goal for which a model is designed or the function it is supposed to fulfill. The model purpose adheres closely to the end-model user or model owner. Model purpose is the criterion for the choice of a model's boundary and design.

Modeling process The process involving phases such as problem articulation, boundary selection, development of a dynamic hypothesis, model formulation, model testing, policy formulation, and policy evaluation (Sterman 2000). The modeling process is followed by model use and implementation, i.e., the realization of actions designed or facilitated by the use of the model.

Validation process Validation is the process by which model validity is enhanced systematically. It consists in gradually building confidence in the usefulness of a model by applying validation tests as outlined in this entry. In principle, validation pervades all phases of the modeling process and, in addition, reaches into the phases of model implementation and use.

Introduction

The present entry addresses the question of building better models. This is crucial for coping with complexity in general and in particular for the management of dynamic systems (Schwaninger 2010). Both the epistemological and the methodological-technological aspects of model validation for the achievement of high-quality models are discussed. The focus is on formal models, i.e., those formulated in a stringent, logical, and mostly mathematical language.

The etymological root of "valid" is in the Latin word "validus," which denotes attributes such as strong, powerful, and firm. A valid model, then, is well founded and difficult to reject because it accurately represents the perceived real system which it is supposed to reflect. This system can be either one that already exists or one that is

Originally published in
R. A. Meyers (ed.), *Encyclopedia of Complexity and Systems Science*, © Springer Science+Business Media LLC 2018, https://doi.org/10.1007/978-3-642-27737-5_540-4

being constructed, or even anticipated, by a modeler or a group of modelers.

The validation standards in System Dynamics are more rigorous than those of many other methodologies. Let us distinguish between two types of mathematical models, which are fundamentally different: causal, theory-like models and noncausal, statistical (correlational) models (Barlas and Carpenter 1990). The former are explanatory, i.e., they embody theory about the functioning of a real system. The latter are descriptive and express observed associations among different elements of a real system. System Dynamics models are causal models.

Noncausal models are tested globally, in that the statistical fit between model and data series from the real system under study is assessed. If the fit is satisfactory, the model is considered to be accurate ("valid," "true"). In contrast, system dynamicists postulate that models be not only right but right for the right reasons. As the models are made up of causal interdependencies, accuracy is required for each and every variable and relationship. The following principle applies: In case only one component of the model is shown to be wrong, the whole model is rejected even if the overall model output fits the data (Barlas and Carpenter 1990). This strict standard is conducive to high-quality modeling practice.

A model is an abstract version of a perceived reality. Simulation is a way of experimenting with mathematical models to gain insights and to employ these to improve the real system under study. It is often said that System Dynamics models should portray problems or issues, not systems. This statement must be interpreted in the sense that one should not try to set the boundaries of the model too wide, but rather give the model a focus by concentrating on an object in accordance with the specific purpose of the model. In a narrower definition, even an issue or problem can be conceived of as a "system," i.e., "a portion of the world sufficiently well defined to be the subject of study" (Rapoport 1954). Validity then consists in a stringent correspondence between model system and "real system."

We will treat the issue of model validation as a means of assuring high-quality models. We

interject that validity is not the only criterion of model quality, the other criteria being parsimony, ease of use, practicality, importance, etc. (Schwaninger and Groesser 2008).

In the following, the epistemological foundations of model validity are reviewed (section "Epistemological Foundations"). Then, an overview of the methods for assuring model validity is given (section "Validation Methods"). Further, the survey includes an overview of the validation process (section "Validation Process") and our final conclusions (section "Synopsis and Outlook").

The substance of this entry will be made more tangible by means of the following frame of reference. We call it the validation cube. The diagram in Fig. 1 shows three dimensions of the validation topic:

Orders of reflection: The two layers addressed are *methodology* and *epistemology, at a meta-level.* These define the objects of the next two sections "Epistemological Foundations" and "Validation Methods."

Domains of validation: The three domains, *context, structure*, and *behavior*, refer to the groups of validation methods as described in the section on "Validation Methods."

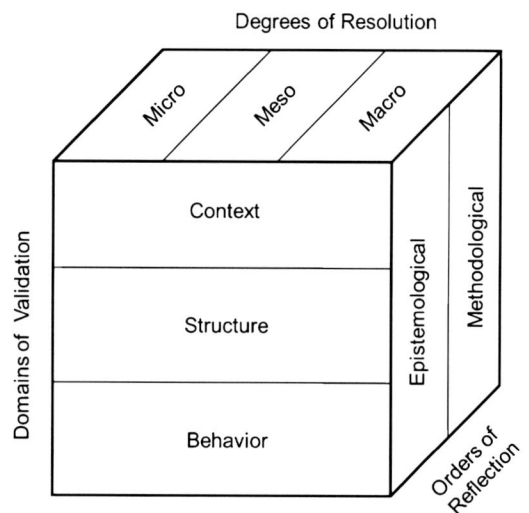

System Dynamics Modeling: Validation for Quality Assurance, Fig. 1 The validation cube – a frame of reference showing three dimensions of the validation topic

Degrees of resolution: We address the different granularities of models. *Micro* refers to the smallest building blocks of models (e.g., variables or small sets of variables), *meso* to modules which constitute a model, and *macro* to the model as a whole.

Epistemological Foundations

Epistemology is the theory that enquires into the nature and grounds of knowledge: "What can we know and how do we know it?" (Lacey 1996). These questions are of utmost importance when dealing with models and their validity, because a method of validation is only as good as its epistemological basis.

We can only briefly refer to the antecedents of the epistemological perspective inherent in the idea of model validation as commonly held today in the community of system dynamicists. One could go back to Socrates who, in Plato's Republic (fourth century BC), addressed the problematic relationship between reality, image, and knowledge. One could also refer to John Locke (seventeenth century), the first British empiricist who maintained that ideas could come only from experience, while admitting that our knowledge about external objects is uncertain. We will address the philosophical movements of the nineteenth and twentieth centuries, which are direct sources of the epistemology that is important for model validation. The reader may kindly excuse us for certain massive retrenchments that we are obliged to take.

What will be said here about theories applies equally to formal models. In System Dynamics, models either embody theories or they are considered essential components of theories. In addition, processes of modeling and theory building are of the same nature; a model, like any theory, is built and improved in a dialectic of propositions and refutations (Schwaninger and Groesser 2008).

Positivism and Critique

Positivism is a scientific doctrine founded by Auguste Comte (nineteenth century) which raises the *positive* to the principle of all scientific knowledge. "Positive," in this context, is not meant to be the opposite of negative, but the given, factual, or indubitably existent. The positive is associated with features such as being real, useful, certain, and precise. Positivism confines science to the observable and manipulable, drawing on the mathematical, empirical orientation of the natural sciences as its paragon. The objectivist claim of positivism is that things exist independently of the mind and that truths are detached from human values and beliefs. This stance calls for models that approximate an objective reality.

A younger development in this vein is the school of logical positivism, also logical empiricism (with Schlick, Neurath, Hempel, etc.), which concentrates on the problem of meaning and has developed the verifiability principle: Something is meaningful only if verifiable empirically, i.e., ultimately by observation through the senses. Verify here means to show to be true (Lacey 1996). For the logical positivists, the method of verification is the essence of theory building. Tests of theories hinge on the confirmation by facts. In SD, testing models on real-world data is a core component of validation.

Positivism has been criticized for being reductionist, i.e., for its tendency to reduce concepts to simpler or empirically more accessible ones and to conceive of learning as an accumulation of particular details. The critique has also interjected that there is no theory-independent identification of facts, and therefore different theories cannot be tested by means of the same data (Feyerabend 1993). Another objection maintains that social facts are not merely given, but produced by human action, and that they are subject to interpretation (Seiffert and Radnitzky 1994). These arguments introduce the principle of relativity, which is of crucial importance for the ambit of model validation: A model is a subjective construction by an observer.

Pragmatism: A Challenge to Positivism

Pragmatism, which arose in the second half of the nineteenth century, emphasizes action and the practical consequences of thinking. Its founder Charles Sanders Peirce was interested in the effects that the meaning of scientific concepts could have on human experience and action. He defined truth as "the opinion which is fated to be ultimately agreed to by all who investigate" (cit. Lacey 1996), whereby truth is linked to consensual validation. For pragmatists truth is in what works (Ferdinand Schiller) or satisfies us (John Dewey) and what we find believable and consistent: "'The true' ... is only the expedient in the way of our thinking," and "truth is *made* ... in the course of experience" (James 1987: 583, 581).

Pragmatism is often erroneously disdained for supposedly being a crass variety of utilitarianism and embodying a crude instrumentalist rationality. A more accurate view considers the fact that pragmatists are not satisfied with a mere ascertainment of truth; instead they ask: "If an idea or assumption is true, does this make a concrete difference to the life of people? How can this truth be actualized?" In other words, pragmatism does not crudely equalize truth and utility. It rather postulates that those truths which are useful to people ought to be put into practice (Seiffert and Radnitzky 1994).

Pragmatism introduces the criteria of confidence and usefulness, which are more operational as guides to the evaluation of experiments than is the notion of an absolute truth, which is unattainable in the realm of human affairs. At the same time, pragmatism triggers a crucial insight for the context of model building: The validity of a model depends not only on the absolute quality of that model but also hinges on its suitability with respect to a purpose (Forrester 1961). In the context of model validation, then, truth is a relative property; more exactly, a *truth* holds for a limited domain only.

More Challenges to Positivism

We discuss three more challenges to positivism in the twentieth century. The first is Thomas Kuhn's theory of scientific revolutions (Kuhn 1996): Kuhn shows, by means of historical cases, that in the sphere of science, generally accepted ways of looking at the world ("paradigms") change over time through fundamental shifts. Therefore, the activities of a scientist are largely shaped by the dominant scientific worldview. Second, Willard Van Orman Quine and Wilfrid Sellars argue that

knowledge creation and theory building is a holistic, conversational process, as opposed to the reductionist and confrontational views (Barlas and Carpenter 1990).

Both of these movements contribute to our understanding of how real systems are to be modeled and validated, as organized wholes, and consciously with respect to the values and beliefs underlying a given modeling process. This approach adheres to the spirit of models themselves, by means of which the behavior of whole systems can be simulated and tested on their inherent assumptions.

A third challenge are the interpretive streams of epistemology (for an overview, see Heracleous (2006)). Among them, a main force which expands the possibilities of scientific methodologies is the strand of hermeneutics. Derived from the Greek *hermeneuein* – to interpret or to explain – the term *hermeneutics* stands for a school, mainly associated with Hans-Georg Gadamer, which pursues the ideal of a human science of understanding. The emphasis is on interpretation in an interplay between a subject-matter and the interpreter's position. This emphasis introduces the subjective into scientific methodology. Hermeneutics denies both that a single "objective true interpretation" can transcend all individual viewpoints and that humans are forever confined within their own ken (Lacey 1996). This epistemology offers a necessary complement to a scientific stance, which exclusively hinges on "hard," quantitative methods in order supposedly to achieve absolute objectivity. The implication of hermeneutics for model validation is that it recognizes the pertinence of subjective judgment. In this connection, interpretive discourses play a crucial role in group model building and validation. Such discourses lead beyond the subjective, entailing the creation of intersubjective, shared realities. We will revert to this factor in section "Validation Process."

Critical Rationalism

Critical rationalism is a philosophical position founded by Karl R. Popper (1959, 1972). It grew out of positivism but rejected its verificationist stance. Critical rationalism posits that, in the social domain, theories can never be definitely proved, but can only reach greater or lesser levels of truth. Scientific proofs are confined to the realm of the formal sciences, namely logic and mathematics.

As Popper demonstrates, all theories are provisional. As a consequence, the main criterion for the assessment of a theory's truth status is *falsification* (Popper 1959). A theory holds as long as it is not refuted. Consequently, any theory can be upheld as long as it passes the test of falsification. In other words, the fertile approaches to science are not those of corroboration but the falsificationist efforts to test if the theory can be upheld. In the context of modeling this means that validation must undertake attempts to falsify a model, thereby testing its robustness.

Even Popper's theory of science is not unchallenged. For example, Kuhn has made the point that its principles are applicable only to normal science, which operates incrementally within a given paradigm, but not to anomalous science, which uncovers unsuspected phenomena in periods of scientific revolution (Kuhn 1996). This observation has an implication for model validation: Alternative and even multiple model designs should be assessed for their ability to account for fundamental change.

On the Meaning of Validity and Validation

One of the predominant convictions about science is the obsessive idea that proofs are the touchstone of the validity of both theories and models. We follow a different rationale, reverting to the philosophy of science as embodied in critical rationalism.

Popper's refutationist concept (as opposed to a verificationist concept) of theory testing implies both an evolutionist perspective and an empiricist stance. The evolutionist perspective is primary because it welcomes the challenges posed to a theory, since these attempts at falsification lead to an evolutionary process: Successful falsification efforts result in revisions and improvements of the theory. Correspondingly, empiricism is paramount in the social sciences, because the main source for the refutation of a theory is empirical evidence. However, falsification can also be grounded in logical arguments where empirical evidence cannot be obtained. In this sense, a structuralist approach as used in System Dynamics validation transcends the bounds of logical empiricism.

As a consequence of the evolutionist perspective, there is no such thing as absolute validity. Validity is always imperfect, but it can be improved over time. The empiricist aspect of theory building implies that theories must be validated by means of empirical data. However, logical assay, estimation, and judgment are complementary to this empiricist component (see below).

A validation process is about gradually building confidence in the model under study (Barlas 1996). This is both analytical and synthetic. It is directed at the model as a whole as much as it is at the components of the model. The touchstone of validity is less whether the model is right or wrong: As John Sterman states, ". . . all models are wrong" (Sterman 2000). Some models, however, fulfill the purpose ascribed to them, i.e., they are useful. Models are inherently incomplete; they cannot claim to be true in an absolute sense, but only to be relatively true (Barlas and Carpenter 1990). In this sense, *validation* is a *goal-oriented* activity and *validity* a *relative* concept.

Finally, the validation process often involves several people because the necessary knowledge is distributed. In these cases, the dialectics of propositions and refutations, as well as the interaction of different subjective viewpoints, and consensus-building, are integral. Validation processes, then, are semiformal, discursive social procedures with a holistic as opposed to a fragmentary orientation (ibidem).

On Objectivity

If subjective views and judgments are as prominent as alleged above, does objectivity play a role at all? Operational philosophy shows a way out of this dilemma: Anatol Rapoport defines objectivity as "invariance with respect to different observers" (Rapoport 1954). Popper has a similar stance in proposing that general statements must be formulated in a way that they can be criticized and, where applicable, falsified (Popper 1972). This concept of objectivity is a challenge to model validation: When defining concepts and functions, one must first of all strive for falsifiable statements. In principle, formal models meet this criterion: each variable and every function or relationship can be challenged. And they must be challenged, so that their robustness can be tested. The duty, then, is in finding the invariances that are intersubjectively accepted as the best approximations to truth. Frequently this is best achieved in group model-building processes (Vennix 1996). Finally, truth is something we search for but do not possess (Popper 1972), i.e., even an accepted model cannot guarantee truth with final certainty.

Validation Methods

For the enhancement of model validity, a considerable set of qualitative and quantitative tests has been developed. The state of the art has been documented in seminal publications (e.g., Forrester 1961, Forrester and Senge 1980, Barlas and Carpenter 1990, Petersen and Eberlein 1994, Lane 1995, Barlas 1996, Sterman 2000). Our purpose here is to present and exemplify the different tests to encourage and help those who strive to develop high-quality System Dynamics models.

In the following, an overview of the types of tests developed for System Dynamics models is given, without any claim to completeness. These tests have been documented extensively in Forrester (1961), Forrester and Senge (1980), Barlas (1996), and Sterman (2000). The descriptions of the tests adhere closely to the specifications of these authors (mainly Forrester and Senge). In addition, we have developed a new category for tests that concentrate on the context in which the model is to be developed. High-quality models can be created only if the relevant context is taken into consideration. To facilitate orientation, we have attached an overview of all described tests in the Appendix.

In this section, we are describing three groups of tests, those related to model-related context, tests of model structure, and tests of model behavior. Many of the tests described in the following

can be utilized for explanatory analysis which aims at an understanding of the problematic behavior of the issue under study. Others are suitable for normative ends, in analysis targeted on improvements of system performance with regard to a specified objective of the reference system. Also known as policy tests, or policy analysis, these "tests of policy implications differ from other tests in their explicit focus on comparing changes in a model and in the corresponding reality. Policy ... tests attempt to verify that response of a real system to a policy change would correspond to the response predicted by the model" (Forrester and Senge 1980). Policy testing can show the risk involved in adopting the model for policy making.

Tests About Model-Related Context
These tests deal with aspects related to the situation in which the model is to be developed and embedded. They imply metalevel decisions which have to be taken in the first place, before engaging in model building. Applied ex post facto, i.e., after modeling, they allow for assessing the utility of the modeling endeavor as such. These tests are extremely important, because they can help avoid ill-conceived models and the use of modeling methods which are inappropriate for dealing with the issues at hand. Model obsolescence is often due not only to flaws in the details of structure specifications but also and first of all because the wrong solution is applied to the problem under study.

Test of Model Framing
Framing means first of all to state an orientative purpose and clear goal(s) for the model. Related to this is the demarcation of the audience: mainly the user(s) of the model. Is it for managers, politicians, etc., and at what levels? What should be learned, and which kinds of insights should be gained from using the model? These decisions clarify the model's boundaries.

Issue Identification Test
The *raison d'être* of a System Dynamics model is its ability to adequately address an issue and to enhance stakeholders' understanding, an ability

which may lead to policy insights and system improvements. The issue identification test examines whether or not the identified issue or problem is indeed meaningful. Has the "right" problem been identified? Does the problem statement address the origins of an issue or only superficial *symptoms?* Whenever complex issues are addressed by a model, different perspectives (e.g., professional, economic, political) must be integrated for an accurate problem identification and modeling. This is not a "one-shot-only" test; it has to be applied recurrently during the modeling procedure. By reflecting regularly on the correctness of the identified issue, the modeler can increase the likelihood of capturing the origins of a suboptimal system behavior.

Adequacy of Methodology Test

Simulation models respond to the limitations of humans' mental ability to comprehend complex, dynamic feedback systems (Sterman 1989). The adequacy of methodology test scrutinizes whether the System Dynamics methodology is best-suited for dealing with the issue under study. One needs to clearly ascertain if that issue is characterized by dynamic complexity, feedback mechanisms, non-linear interdependency of structural elements, and delays between causes and effects. One needs to ask also if the issue under study could be better addressed by another methodology. For example, in a case where the question is to understand the difference in numerical outcomes between two configurations of a production system, it lets one determine whether discrete event simulation would fulfill this requirement more accurately than System Dynamics.

System Configuration Test

This test asks the fundamental question about whether the structural configuration chosen can be accepted. It challenges the assumption that the model represents the actual working of the system under study. The eventuality of a different design would be indicated to capture new conditions, such as different system configurations, phenomena, or rules of the game. Even revolutionary changes might be considered. Such an outlook may require a totally new model or an alternative model designed from a different vantage point. This would at least feasibly approximate the need to take paradigmatic change into account.

System Improvement Test

The purpose of modeling is to understand a part of reality and to resolve an issue. The system improvement test can be performed only after the modeling project (ex post facto test), once the insights derived from the model have already been implemented in the real system. This test reestablishes the connection between the abstract mathematical model and the real system. The system improvement test helps to evaluate whether or not the model development was successful. In operational terms, any improvements of the real system under study have to be compared with explicit objectives. In practice, the test might assess the impact of the modeling process or the model use either on the mental models of decision makers or on changes in organization structures. In principle, assessing the impact of a modeling endeavor is very difficult (one preliminary example is provided by Snabe and Grössler 2006).

Tests of Model Structure

Tests of model structure refer to the "nuts and bolts" of System Dynamics modeling, i.e., to the formal concepts and interrelationships which represent the real system. Model structure tests aim to increase confidence in the structure of the created theory about the behavior mode of interest. The model structure can be assessed by means of either direct or indirect inspection. Tests of model structure assess if the logic of the model is attuned to the corresponding structure in the real world. They do not yet compare the model behavior with time series data from the real system.

Direct Structure Tests

Direct structure tests assess whether or not the model structure conforms to relevant descriptive knowledge about the real system or class of systems under study. In principle they can be carried out without computers. By means of direct comparison, they qualitatively assess any disparities between the original system structure and the model structure.

Structure Examination Test Examination in this case means comparison in the sense just outlined. Qualitative or quantitative information about the real system structure can be obtained either empirically or theoretically. In the empirical version, structure examination tests compare the form of equations in the model with the relationships extant in the real system (Barlas 1996). They include reviews of model assumptions about system elements and their interdependencies, e.g., reviews made by highly knowledgeable experts of the real system. Theory-based tests compare the model structure with theoretical knowledge from literature about the type of system being studied.

To pass the structure examination test, the model must not contradict either the evidence or knowledge about the structure of the real system. This test ensures that the model contains only those structural elements and interconnections that are most likely extant in the real system. In this context, formal inspections of the model's equations, reviews of the syntax for the stock and flow diagram, and walkthroughs along the causal loop diagrams and their embodied causal explanations may be indicated. The experienced reader might recommend the use of statistical tests to identify and validate model structure. As Forrester and Senge (1980) indicate, a longstanding discussion exists about the application of inferential statistical tests for structure examination. After a series of experiments, Forrester and Senge conclude "that conventional statistical tests of model structure are not sufficient grounds for rejecting the causal hypotheses in a system dynamics model" (Forrester and Senge 1980). In the future, however, new statistical approaches might enrich the testing procedures.

Parameter Examination Test A parameter is a quantity that characterizes a system and is held constant in a case under study but may be varied in different cases (e.g., energy consumption per capita per day). The aim of parameter examination is to evaluate a model's parameters against evidence or knowledge about the real system. The test can utilize both empirical and theoretical information. Furthermore, the test can be conceptual or numerical. The conceptual parameter examination test is about construct validity; it identifies elements in the real system that correspond to the parameters of the model. Conceptual correspondence means that the parameters match elements of the real system's structure. Numerical parameter examination checks to see if the quantities of the conceptually confirmed parameters are estimated accurately. Techniques for the estimation of parameters are described in Graham (1980) and Struben et al. (2015).

Direct Extreme Condition Test Extreme conditions do not often occur in reality; they are exceptions. The validity of a model's equations under extreme conditions is evaluated by assessing the plausibility of the results generated by the model equations against the knowledge about what would happen under a similar condition in reality. Direct extreme condition testing is a mental process and does not involve computer simulation. Ideally, it is applied to each equation separately. It consists of assigning extreme values to the input variables of each equation. The values of the output variables are then interpreted in terms of what would happen in the real system under these extreme conditions. For example, if a population is zero, then neither births, deaths, nor consumption of resources can occur.

Boundary Adequacy Structure Test Boundary adequacy is given if the model contains the relevant structural relationships that are necessary and sufficient to satisfy a model's purpose. Consequently, the boundary adequacy test inquires whether the chosen level of aggregation is appropriate and if the model includes all relevant aspects of structure. It should ensure that the model contains the concepts that are important for addressing the problem endogenously. For instance, if parameters are likely to change over time, they should be endogenized (Forrester and Senge 1980). The pertinent validation question is: "Should this parameter be endogenized or not?" That question must be decided in view of the model's purpose.

The boundary adequacy test can be applied in three ways: as a structural test, as a behavioral test, and as a policy test. The names are correspondingly boundary adequacy structure test, boundary adequacy behavior test, and boundary adequacy policy test.

As a test of model structure, the boundary adequacy test involves developing a convincing hypothesis relating the proposed model structure to the particular issue addressed by the model. The boundary adequacy behavior/policy test (explained in the next section) continues this line of thinking.

Dimensional Consistency Test This test checks the dimensional consistency of measurement units of the expressions on both sides of an equation. The test is performed only at the equation level. In case all tests of the individual equations are passed, a large system of dimensionally consistent equations results. This test is only passed, if consistency is achieved without the use of parameters that have no meaning in respect to the real world. The dimensional consistency test is a powerful test to establish the internal validity of the model.

Indirect Structure Tests

Indirect structure tests assess the validity of the model structure indirectly by examining model-generated outcome behaviors. These tests require computer simulation. The comparative activities in these tests are based on logical plausibility considerations which in turn are based on the mental models of the analyst. Comparisons of model-generated data and time series about the real system are not yet involved. The tests can be applied to different degrees of model completeness, i.e., to the smallest "atomic" model components, to sub-models, as well as to the entire model.

Mass-Balance Check This powerful test was elaborated, as follows, by Brian Dangerfield (2014): In a model that contains flows of resources, inflows and outflows must balance out, i.e., the model must not gain or lose mass. The more flows a model includes, the greater is

the need for such a test. The procedure consists in accumulating all the inflows and outflows over time for each resource being modeled and then using the following balance or checksum equation:

$$\int_{t=0}^{t_{final}} [\text{Sum} of \text{ all} in flows - \text{Sum} of \text{ all} out flows$$
$$+ initial\ values\ of\ stocks$$
$$- current\ values\ of\ stocks] * dt$$
$$= 0$$

The correct result of the computation should be equivalent to zero throughout the run; anything else hints at a flawed model. (It is necessary to employ a double precision version of the software being utilized.) If more than one resource flow is involved (e.g., people and finance), then separate tests must be carried out for each flow. Figure 2 provides an example; in this model (Sterman 2000), a co-flow of labor requirements is tracked along an evolving structure of capital ("machines").

Indirect Extreme Condition Test For this test, the modeler assigns extreme values to selected model parameters and compares the generated model behavior to the observed or expected behavior of the real system under the same extreme condition. This test is the logical continuation of the direct extreme condition test, i.e., many of the extreme conditions mentally developed in the previous stage can now be deployed to evaluate the simulated behavioral consequences. This test can be used for the explanatory analysis phase of modeling but also for the normative phase of policy development. In the first instance, indirect extreme conditions are used to develop a structure that can reproduce the system behavior of interest and guard against developments impossible in reality. In the latter instance, the introduction of policies aims to improve the system's performance. The indirect extreme policy test introduces extreme policies to the model and compares the simulated consequences to what would be the most likely outcome of the real

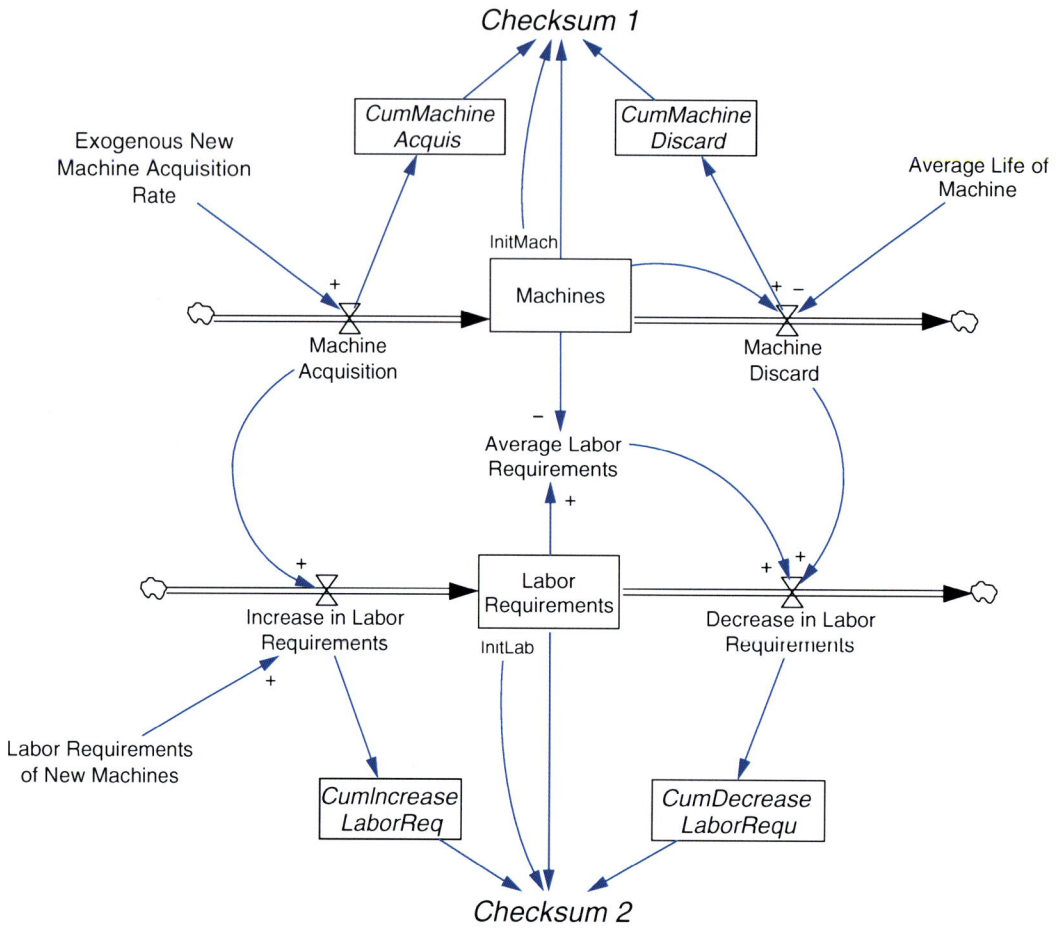

System Dynamics Modeling: Validation for Quality Assurance, Fig. 2 Mass-balance check – example (model in standard characters, testing gear in italics)

system if the same extreme policies would have been implemented.

Behavior Sensitivity Test Sensitivity analysis assesses changes of model outcome behavior given a systematic variation of input parameters. This test reveals those parameters to which the model behavior is highly sensitive and asks if the real system would exhibit a similar sensitivity to changes in the corresponding parameters. "The behavior sensitivity test examines whether or not plausible shifts in model parameters can cause a model to fail behavior tests previously passed. To the extent that such alternative parameter values are not found, confidence in the model is enhanced" (Forrester and Senge 1980). A model

can be numerically sensitive, i.e., the numerical values of variables change significantly, but the behavioral patterns are conserved. It can also exhibit behavioral sensitivity, i.e., the modes of model behavior change remarkably based on systematic parameter variations (Barlas (2006) defines several distinct patterns of model behavior).

As the test for indirect extreme conditions, the behavior sensitivity test can also be deployed to assess policy sensitivity. It can reveal the degree of robustness of model behavior and hence indicate to what degree model-based policy recommendations might be influenced by uncertainty in parameter values. If the same policies would be recommended regardless of parameter changes over a plausible range, risk in using the model would be less than if two plausible sets of parameters lead to distinct policy recommendations.

Integration Error Test Integration error is the deviation between the analytical solution of differential equations and the numerical solution of difference equations. This test ascertains whether the model behavior is sensitive to changes in either the applied integration method or the chosen integration interval (often referred to as simulation time step). Euler's method is the simplest numerical technique for solving ordinary differential and difference equations. For models that require more precise integration processes, the more elaborated Runge-Kutta integration methods can produce more accurate results, but they require more computational resources.

Boundary Adequacy Behavior Test/Boundary Adequacy Policy Test The logic for testing the boundary adequacy has already been developed under the aspect of direct structure testing in the preceding section. The indirect structure version of this test asks whether the model behavior would change significantly if the boundary were extended or reduced; i.e., the test involves conceptualizing additional structure or canceling unnecessary structure with regard to the purpose of the study. As one example of expanding the model boundary, this version of the test allows one to detail the treatment of model assumptions considered as unrealistically simple but still important for the model purpose. On the other hand, simplifying the model is also a way to reduce the model boundary. The loop-knockout analysis is a useful method to implement this two-sided test. Knockout analysis checks behavior changes induced by the connection and disconnection of a portion of the model structure and helps the modelers to evaluate the usefulness of those changes with respect to the model purpose.

The other version of this test is the boundary adequacy policy test. It examines whether policy recommendations would change significantly if the boundary were extended (or restricted): That is, what would happen if the boundary assumptions were relaxed (or confined)?

Loop Dominance Test Loop dominance analysis studies the internal mechanisms of a dynamic model and their temporal, relative contribution to the outcome behavior of the model. The relative contribution of a mechanism is a complex quantitative statement that explains the fraction of the analyzed behavior mode caused by the considered mechanism (see Kampmann and Oliva 2018). The analysis reveals the relative strengths of the feedback loops in the model. The loop dominance test compares these results with the modeler's or client's assumption about which are the dominant feedback loops in the real system. Since the results are analytical statements, interpretation and comparison with the real system require profound knowledge about the system under study.

Loop dominance analysis reveals insights about a model on a different level of analysis than the other validation tests discussed so far: It works not on the level of individual concepts or behaviors of variables but on the level of causal structure and compares the temporal significance of the different structures to each other. This test has developed from basic concepts (Richardson 1995) to sophistiated methods of analysis (Kampmann and Oliva op.cit.). If the relative loop dominances of the model map the relative loop dominances of the real system, confidence in the model is enhanced. In case the relative loop dominances of the real system are not known, it is still possible to evaluate whether or not the loop dominance logic in the model is reasonable.

Tests of Model Behavior

Tests of model behavior are empirical and compare simulation outcomes with data from the real system under study. On that basis, inferences about the adequacy of the model can be made. The empirical data can either be historical or refer to reasonable expectations about possible future developments.

Behavior Reproduction Tests

The family of behavior reproduction tests examines how well model-generated behavior matches the observed historical behavior of the real system. As a principle, models should be tested against data not only from periods of stability but also from unstable phases. Policies should not be designed or tested on the premise of normality, but rather should be validated with a view toward robustness and adaptiveness.

Symptom Generation Test This test indicates whether or not a model produces the symptom of difficulty that motivated the construction of the model. To pass the symptom generation test is a prerequisite for considering policy changes, because "unless one can show how internal policies and structures cause the symptoms, one is in a poor position to alter those causes" (Forrester and Senge 1980).

Summary statistics that measure and enable the interpretation of quantitative deviations provide the means to operationalize the symptom generation test. One known example is Theil inequality statistics, which measures the mean square error (MSE) between the model-generated behavior and historical time series data. It decomposes the deviation in three sources of error: bias (U_M), unequal variation (U_S), and unequal covariation (U_C) (Sterman 1984). An example taken from Schwaninger and Groesser (2008) illustrates the interpretation of the error sources.

The case in point, based on an example from an industrial firm, concerns the design of a model that replicates the historical product life-cycle

pattern, as observed in the enterprise, with high accuracy (Fig. 3). "Product revenue" is the main variable of interest and specifies the symptom (growth phase followed by rapid decay). The mean-square error for revenues is 0.35. The individual components of the inequality statistics are $U_M = 0.01$, $U_S = 0.01$, $U_C = 0.98$. The decomposition shows that the major part of the error is due to the unequal covariation component, while the other two sources of error are small. This signifies that the point-by-point values of the simulated and the historical data do not match, even though the model captures the dominant trend and the average values in the historical data. Such a situation indicates that the major part of the error is probably unsystematic and therefore that the model should not be rejected for failing to match the noise component of the data. The residuals of the historic and simulated time series show no significant trend. This strengthens the assessment that the model comprises of a structure that captures the fundamental dynamics of the issue under study.

Frequency Generation and Phase Relationship Tests These tests focus on the frequency and phase relationships between variables. An example is the pattern of investment cycles in an industry. These tests are superior to point-by-point

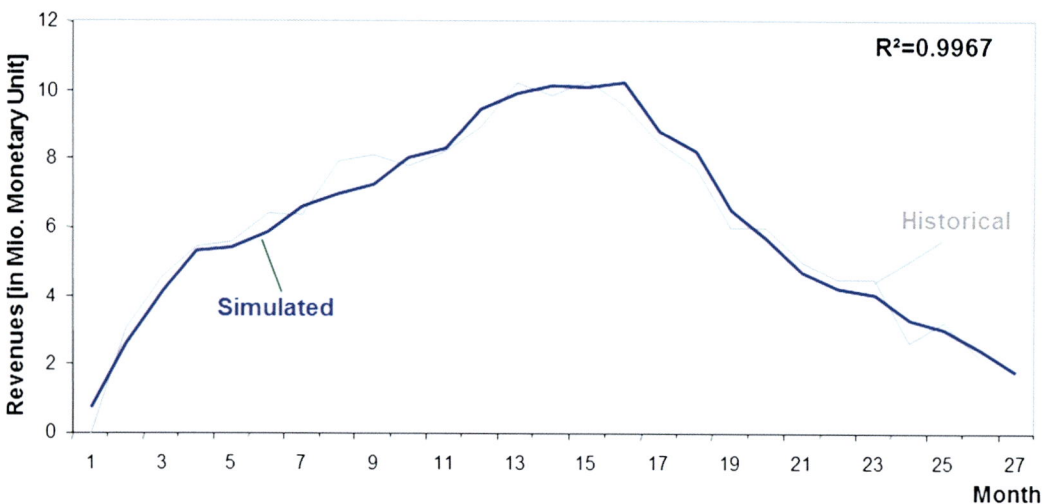

System Dynamics Modeling: Validation for Quality Assurance, Fig. 3 Comparison of historical and simulated time series for the product revenues – example. The explained variance is close to 100% ($R^2 = 0.9967$)

comparisons between model-generated and observed behavior (Forrester 1961).

Frequency refers to periodicities of fluctuation in a time series. Phase relationship is the relationship between the time series of at least two variables. In principle, three-phase relations are possible: preceding, simultaneous, and successive. The frequency generation test evaluates whether or not the periodicity of a variable is in accordance with the real system. The phase relationship test assesses the phase shifts of at least two variables by comparing their trajectories.

If the phase shift between the selected simulation variables contradicts the phase shift between the same variables as observed or expected in the real system, a structural flaw in the model might be diagnosed. The test can uncover failures in the model but offers only little guidance as to where the erroneous part of the model might be. The autocorrelation function test is one way to operationalize the frequency generation test (Barlas 1990). The function test consists in comparing the autocorrelation functions of the observed and the model-generated behavior outputs and can detect if significant errors between them exist.

Modified Behavior Test Modified behavior can arise from a modified model structure or changes in parameter values. This test concerns changes in the model structure. It can be performed if data about the behavior of a structurally modified version of the real system are available. "The model passes this test if it can generate similar modified behavior, when simulated with structural modifications that reflect the structure of the 'modified' real system" (Barlas 1996). The applicability of this test is rather limited since it requires specific data about the modified real system which must be similar in kind to the original real system. Only under this condition can additional insights into the suitability of the original model structure be obtained. If the modified real system deviates strongly from the original real system, the test does not result in any additional insights, because no stringent conclusions about the validity of the original system can be derived from a model that is dissimilar in its structure.

Multiple Modes Test A mode is a pattern of observed behavior. The multiple mode test considers whether a model is able to generate more than one mode of observed behavior, for instance, if a model about the production sector of an economy generates distinct patterns of fluctuations for the short term (production, employment, inventories, and prices) and for the long term (investment, capital stock) (Mass 1975). "A model able to generate two distinct periodicities of fluctuation observed in a real system provides the possibility for studying possible interaction of the mode and how policies differentially affect each mode" (Forrester and Senge 1980).

Behavior Characteristic Test Characteristics of a behavior are features of historical data that are clearly distinguishable, e.g., the peculiar shape of an oscillating time series, sharp peaks, long troughs, or such unusual events as an oil crisis. Since System Dynamics modeling is not about point prediction, the behavior characteristic test evaluates whether or not the model can generate the circumstances and behavior leading to the event. The creation of the exact time of the behavior is not part of the test.

Behavior Anticipation Tests

System Dynamics models do not strive to forecast future states of system variables. Nevertheless, given that the fundamental system structure is not subject to rapid and fundamental change, dynamic models might provide insights about the possible range of future behaviors. Hence, behavior anticipation tests are similar to behavior reproduction tests but possess a higher level of uncertainty.

Pattern Anticipation Test This test examines whether a model generates patterns of future behavior which are assumed to be qualitatively correct. The limits of anticipation are in that the structure of the system may change over time. The pattern anticipation test entails evaluation of periods, phase relationships, shape, or other characteristics of behavior anticipated by the model. One possibility for implementing this test is to split the historical time series into two data sets and introduce an artificial present time at the end of the first data series. The first set is then used for model development and calibration. The second data series is employed to perform the behavior anticipation test, i.e., to evaluate whether the model is able to anticipate the possible future behavior.

This test can also be used for policy considerations, in which case it is called "Changed Behavior Anticipation Test." It determines whether the model correctly anticipates how the behavior of the real system will change if a governing policy is changed.

Event Anticipation Test In respect to System Dynamics, the anticipation of events does not imply knowing the exact time at which the events occur; it rather means understanding the dynamic nature of events and being able to identify the antecedents leading to them. For instance, the event anticipation test is passed if a model has the ability to anticipate a steep peak in food prices based on the development of the conditioning factors.

Behavior Anomaly Test

In constructing and analyzing a System Dynamics model, one strives to have it behave like the real system under study. However, the analyst may detect anomalous features of the model's behavior which conflict with the behavior of the real system. Once the behavioral anomaly is traced to components of the model structure responsible for the anomaly, one often finds flaws in model assumptions. The test for recognizing behavioral anomalies is sporadically applied throughout the modeling process.

Family Member Test

A System Dynamics model often represents a family of social systems. Whenever possible, a model should be a general representation of the class of that systems to which the particular case belongs. One should ask if the model can generate the behavior in other instances of the same class. "The family-member test permits a repeat of the other tests of the model in the context of different special cases that fall within the general theory covered by the model. The general theory is embodied in the structure of the model. The special cases are embodied in the parameters. To make the test, one uses the particular member of the general family for picking parameter values. Then one examines the newly parameterized model in terms of the various model tests to see if the model has withstood transplantation to the special case" (Forrester and Senge 1980). The model should be calibrated so as to be applicable to the widest range of related systems. For the family member test, only the parameter values of the model are subject to alterations; changes in the model structure are part of the modified behavior test, as discussed in the preceding section.

Surprise Behavior Test

A surprising model behavior is a behavior that is not expected by the analysts. When such an unexpected behavior appears, the model analysts must first understand the causes of the unexpected behavior within the model. They then compare the behavior and its causes with those of the real system. In many cases, the surprising behavior turns out to be due to a formulation flaw in the model. However, if this procedure leads to the identification of behavior previously unrecognized in the real system, the confidence in the model's usefulness is strongly enhanced. Such a situation may signify a model-based identification of a counterintuitive behavior in a social system.

Turing Test

The Turing test is a qualitative test which uses the intuitive knowledge of system experts to evaluate model behavior. Experts are presented with a shuffled collection of real and simulated output behavior patterns. They are asked if they can distinguish between these two types of patterns. If they are unable to discern which pattern belongs to the real system and which to the simulation output, the Turing test is passed. Similar to the phase relationship test, the Turing test is powerful in its ability to indicate structural flaws but offers only little guidance for locating them in the model.

In another contribution, we have proposed heuristic principles for the choice of methods as a function of the complexity of the validation object, ranging from feedback loop, and combination of feedback loops, to whole model (Groesser and Schwaninger 2012).

Validation Process

The validation process pervades all phases of model building and reaches even beyond, into the phases of model implementation and use. The diagram in Fig. 4 visualizes the function of validation in the process of model building.

For the purpose of this contribution, validation is placed at the center of the scheme. From there it is dispersed through all steps of the modeling process, map (high-level modeling), model (building the formal model), simulate (scenarios, analysis), and design (of policies). We have limited the differentiation of these steps in order to highlight the structure of the process – a recursive structure drawn as a nested loop line. After the initial identification of issues and the articulation of model purpose, the simplified diagram denotes

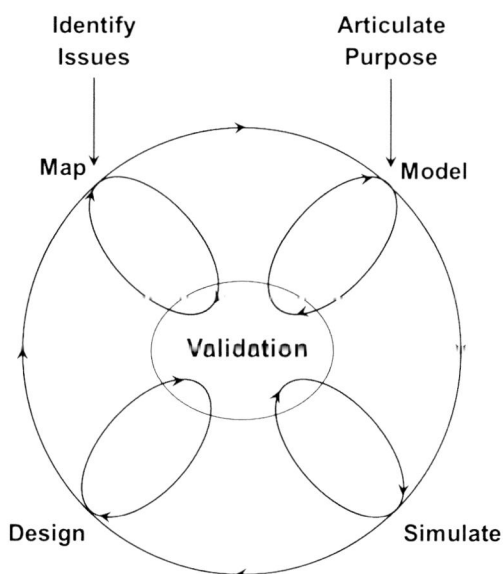

System Dynamics Modeling: Validation for Quality Assurance, Fig. 4 Validation in the context of the System Dynamics modeling procedure

the four phases, of mapping to modeling to simulation and design. The small loops symbolize the microprocesses in which, for example, a model is submitted to validation, e.g., a direct structure test, which may lead to its modification (two small arrows). The larger loop illustrates more comprehensive processes. For example, an indirect structure test of the model is carried out, in which the behavior is tested by means of simulation. Or a policy test by simulation leads to implications for design (large loop), and the design is validated in detail thereafter (small loop).

Now, we should note that the process scheme reminds us of a further aspect which is quite fundamental. If the results of the model's operation, e.g., a "prediction," diverge from the results of a test, then either the model is wrong or the test is inadequate (Smith 2008: 168). This meta-perspective lets us keep an eye on the adequacy of the tests: Is the logic of the test flawless? Are the data sources in order? (see "Adequacy of Methodology Test" in section "Validation Methods").

Model building is a process of knowledge creation, and model validation is an integral part of it. As the model is validated using the methods described in the former section, insights emerge, and a better understanding of the system under study keeps growing. But model building is also a construction of a reality in the minds of observers (von Glasersfeld 1991, von Foerster 1984) concerned with an issue. In this procedure, validation is supposed to be a "guarantor" for the realism of the model, a control function for preventing gross aberrations in individual and collective perceptions. Validation should encompass precautions against cognitive limitations and modeler blindness. The set of tests presented above is a system of partly heuristic and partly algorithmic devices for enhancing such provisions. A question not yet answered is how these tests should be ordered along the timeline. We have fleshed out three structural principles, which are illustrated in Fig. 5:

1. *Validation is a parallel process:* Validation at all three domains – context, structure, and behavior – is carried out in a synchronized

System Dynamics Modeling: Validation for Quality Assurance, Fig. 5 The interplay of validation activities

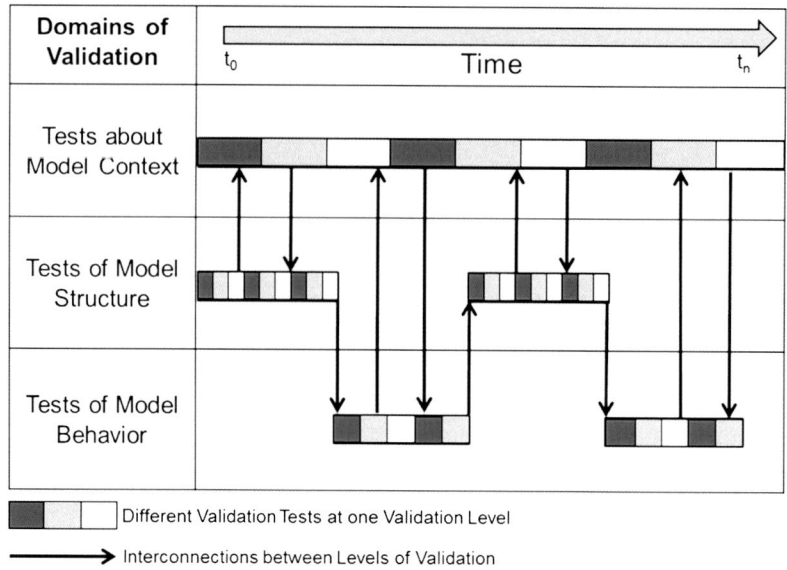

System Dynamics Modeling: Validation for Quality Assurance, Fig. 5 The interplay of validation activities

fashion. Contextual validation is perpetual, while the structural and behavioral validation are interrupted, from time to time by alternations.

2. *Parts of the validation process have a sequential structure:* This refers to the alternations between the components of structure and behavior validation. In principle, they occur alternately, with structural validation taking the lead and behavior validation following. After that, one might revert to structural validation again and so forth.

3. *Validation processes are polyrhythmic:* The length and accentuation of validation activities vary among the three levels. This fact is symbolized by the frequency of the vertical lines in the blocks of the chronogram.

A further important factor impinging on the validation process is the degree of resolution: micro, meso, or macro (as visualized in Fig. 1). The focus of validation is primarily on the micro-objects, the smallest building blocks of a model, for example, a stock or a subsystem containing a stock with its flows. One could call them meta-phorically *atoms* or *molecules*. Each building block should be validated individually, before it is integrated into the overall model structure. The reason is that at this atomic level,

dysfunctionalities or errors of thinking are discovered immediately, while at higher levels of resolution, the identification of structural flaws is more difficult and cumbersome. The same holds for the relation between modules (meso) and the whole model (macro). Before adding a module, it should have been validated in itself. This way, errors at the level of the whole system can be minimized, and also, it is very important to add, counterintuitive behavior of the model can be understood with more ease.

As far as details for the design of the validation process are concerned, along with the decision about when to stop the validation effort, we have dedicated a separate article to these issues (Groesser and Schwaninger 2012).

Until now we have treated what occurs in a validation process and how the process is structured. Finally, we raise the issue of who the actors are and why. In this context, we will concentrate on group processes in model validation.

Different observers associate diverse contents with a system, and they might even conceive the system distinctly, as far as its boundaries and structures are concerned. They might also succumb to erroneous inferences and therefore adhere to defective propositions. Consequently, error-correcting devices are needed. A powerful mechanism for this purpose is the practice of model building and validation in groups. We have already reverted to that concept in respect to several of the methods discussed in the section on "Validation Methods," and now we will briefly expand on it.

Group model building (GMB) is a methodology to facilitate team learning with the help of SD (Vennix 1996). The methodology consists in a set of methods and instruments as well as heuristic principles. These are meant to facilitate the elicitation of knowledge, the negotiation of meanings, the creation of a shared understanding of a problem in a team, as well as the joint construction and validation of models. The process of GMB is essentially a dialogue in which different interpretations of the real system under study are exposed, transformed, aligned, and translated into the concepts and relationships which make up the model system. In other words that process is about both modeling and validation.

Given its transdisciplinary approach, GMB enables an integration of different perspectives into one shared image of the system in focus. GMB is an important provision for attaining higher model quality: It can broaden the available knowledge base, inhibit errors, and show itself to be a cohesive force in the quest for consensual model validation. The opportunity for validation inheres in the broad knowledge base normally available in a modeling group. Much of this knowledge can be leveraged for validation purposes. Most validation tests are carried out in cadence with the model-building activities. Often the tests become a task to be accomplished between the workshops. However, the members of the model-building group can, in principle, be made available for knowledge input into and monitoring of the validation activities.

A functioning GMB process requires a number of necessary elements (Phillips 2007): commitment of key players (e.g., attendance of workshops); impartial facilitation; on-the-spot modeling at conversational pace, with continuous display of the developing model; as well as an interactive and iterative group process.

Let us not forget that there are many situations in which one single person is in charge of building and validating a model. In these cases the modeler must constantly challenge his or her own position. Normally it is indicated that one should also call for external judgment in reviews, walkthroughs, and the like. The same holds for knowledge supply. One-person modelers can find a lot of material in the media, libraries, the Internet, etc., but it is also usually beneficial to find experienced persons from whom to elicit relevant knowledge or even persons who join the modeling and validation venture.

Finally, validation processes cannot go on forever, because costs would become prohibitive. At what point should the process be terminated? This question is important but not easy to be answered. The validation cessation threshold is not a fixed value; it depends on a variety of contingency factors such as a target group's experience with modeling, the relative importance/risk of decision, model size, costs of validation, the target group's expectations, data availability, data intensity, potential degree of validity of the model, and a modeler's level of expertise. An extensive treatment of this issue was presented elsewhere (Groesser and Schwaninger 2012).

Synopsis and Outlook

Models should be relevant for coping with the complexity of the real world. At the same time, the methods by which they are constructed must be rigorous; otherwise the quality of the model suffers. Rigor and relevance are not entirely dichotomous, but given resource constraints, they are in competition to a certain extent. Lack of rigor in building a model is often worse than limitations to the model's relevance. One may say, cum *grano salis*: Incomplete validation entails complete irrelevance. Modelers must find a way to ensure both rigor and relevance, as both are necessary conditions for achieving the model purpose. Neither alone is sufficient, but one may assume that, taken together, rigor and relevance are sufficient conditions. The relative importance of these two dimensions of model building may vary over time as a function of the model quality achieved. At the beginning, relevance might be more important, while at high levels of model accomplishment rigor might become prevalent.

Investing in high model quality is indeed both worthwhile and imperative. It is impressive to register the fact that model validation has achieved higher levels of rigor not only in the academic ambit but also in the world of affairs: According to Coyle and Exelby, the need for orientating decisions about "real-world" affairs has also fueled strong efforts among commercial modelers and consultants for ensuring model validity (Coyle and Exelby 2000).

We have discussed two essential aspects of model validation, the epistemological foundations and methodological procedures for ensuring model validity. The main conclusion we have reached on epistemology is that crude positivism has been superseded by newer philosophical orientations that provide guidance for an adequate concept of validation in System Dynamics. Validation has been defined as a rich and well-defined process by which the confidence in a model is gradually enhanced. Validity, then, is always a matter of degree, never an absolute property.

Well defined here is not meant in the sense of a rigid algorithm, but as the rigorous application of a battery of validation methods which we have described in some detail.

Future research should account for the importance or priority of the different tests as perceived by practitioners and academics. This, of course, is relevant since resources for validation – i.e., time, monetary resources, expert knowledge, and appropriate data – are scarce (Groesser and Schwaninger 2012). We have included a number of new validation tests by which modelers' understanding of the relevant context can be scrutinized. These additional tests are rightly supposed to prevent wrong methodological choices. They should also trigger innovative approaches to the issues under study and foster the ability to think in terms of contingencies. Finally, they should liberate modelers from tunnel vision and open avenues to creativity. The imperative here is to cultivate a "sense of the possible" (Robert Musil's *Möglichkeitssinn*) and a skepticism against the supposedly impossible (see also Taleb 2007).

Simulation based on formal dynamic models is likely to become ever more important for both private and public organizations. It will continue to support managers at all levels in decision-making and policy design. The more that models are relied upon, the greater the importance of their high quality. Therefore, model validation is one of the big issues lying ahead in System Dynamics modeling.

Appendix: Overview of the Tests Described in This Entry

- Tests of Model-Related Context
  Test of Model Framing
  Issue Identification Test
  Adequacy of Methodology Test
  System Configuration Test
  System Improvement Test
- Tests of Model Structure
  Direct Structure Tests
  - Structure Examination Test
  - Parameter Examination Test
  - Direct Extreme Condition Test
  - Boundary Adequacy Structure Test
  - Dimensional Consistency Test
  Indirect Structure Tests
  - Mass-Balance Check
  - Indirect Extreme Condition Test
  - Behavior Sensitivity Test
  - Integration Error Test
  - Boundary Adequacy Behavior Test/ Boundary Adequacy Policy Test
  - Loop Dominance Test
- Tests of Model Behavior
  Behavior Reproduction Tests
  - Symptom Generation Test
  - Frequency Generation and Phase Relationship Test
  - Modified Behavior Test
  - Multiple Modes Test
  - Behavior Characteristic Test
  Behavior Anticipation Tests
  - Pattern Anticipation Test
  - Event Anticipation Test
  Behavior Anomaly Test
  Family Member Test
  Surprise Behavior Test
  Turing Test

Bibliography

Primary Literature

Barlas Y (1990) An autocorrelation function test for output validation. Simulation 55(1):7–16

Barlas Y (1996) Formal aspects of model validity and validation in system dynamics. Syst Dyn Rev 12(3):183–210

Barlas Y (2006) Model validity and testing in system dynamics: two specific tools. Paper presented at the 24th international conference of the system dynamics society, Nijmegen

Barlas Y, Carpenter S (1990) Philosophical roots of model validity – two paradigms. Syst Dyn Rev 6(2):148–166

Coyle G, Exelby D (2000) The validation of commercial system dynamics models. Syst Dyn Rev 16(1):27–41

Dangerfield B (2014) Systems thinking and system dynamics: a primer. In: Brailsford S, Churilov L et al (eds) Discrete-event simulation and system dynamics for management decision making. Wiley, Chichester, pp 26–51

Feyerabend P (1993) Against method, 3rd edn. Verso, London

Forrester JW (1961) Industrial dynamics. MIT Press, Cambridge, MA

Forrester JW, Senge PM (1980) Test for building confidence in system dynamics models. In: Legasto AA Jr, Forrester JW, Lyneis JM (eds) System dynamics. North-Holland Publishing Company, Amsterdam, pp 209–228

Graham AK (1980) Parameter estimation in system dynamics. In: Randers J (ed) Elements of the system dynamics method. MIT Press, Cambridge, MA, pp 143–161

Groesser SN, Schwaninger M (2012) Contributions to model validation: hierarchy, process, and cessation. Syst Dyn Rev 28(2):157–181

Heracleous L (2006) Discourse, interpretation, organization. Cambridge University Press, Cambridge, MA

James W (1987) Writings 1902–1910. Library of America, New York

Kampmann CE, Oliva R (2018) System dynamics: analytical methods for structural dominance analysis. In: Encyclopaedia of complexity and systems science. Springer, New York/London/Berlin

Kuhn T (1996) The structure of scientific revolutions, 3rd edn. University of Chicago Press, Chicago

Lacey AR (1996) A dictionary of philosophy, 3rd edn. Barnes and Noble, New York

Lane DC (1995) The folding star: a comparative reframing and extension of validity concepts in system dynamics. In: Simada T, Saeed K (eds) Proceedings of 1995 international system dynamics conference, 30 July–4 Aug, vol I. System Dynamics Society, Lincoln, pp 111–130

Mass NJ (1975) Economic cycles: an analysis of underlying causes. Productivity Press, Cambridge, MA

Mattheij RMM, Rienstra SW, ten Thije Boonkkamp JHM (2005) Partial differential equations: modeling, analysis, computation. Society for Industrial & Applied Mathematics (SIAM), Eindhoven

Petersen DW, Eberlein RL (1994) Understanding models with Vensim. In: JDW M, Sterman JD (eds) Modeling for learning organiziations. Productivity Press, Portland, pp 339–358

Phillips LD (2007) Decision conferencing. In: Edwards W, Miles RF, von Winterfeldt D (eds) Advances in decision analysis. From foundations to applications. Cambridge University Press, Cambridge, pp 375–399

Popper KR (1959) The logic of scientific discovery. Basic Books, New York (latest edition: 2002, Routledge, London)

Popper KR (1972) Objective knowledge: an evolutionary approach. Clarendon Press, Oxford

Rapoport A (1954) Operational philosophy. Integrating knowledge and action. Harper, New York

Richardson GP (1995, originally published in 1984) Loop polarity, loop dominance, and the concept of dominant polarity. Syst Dyn Rev 11(1): 67–88

Schwaninger M (2010) Model-based management (MBM): a vital prerequisite for organizational viability. Kybernetes 39(9/10):1419–1428

Schwaninger M, Groesser SN (2008) Model-based theory-building with system dynamics. Syst Res Behav Sci 25:1–19

Seiffert H, Radnitzky G (1994) Handlexikon der Wissenschaftstheorie, 2nd edn. DTV Wissenschaft, Munich

Smith VL (2008) Rationality in economics: constructivist and ecological forms. Cambridge University Press, Cambridge

Snabe B, Grössler A (2006) System dynamics modelling for strategy implementation – case study and issues. Syst Res Behav Sci 23(4):467–481

Sterman JD (1984) Appropriate summary statistics for evaluating the historical fit of system dynamics models. Dynamica 10(2):51–66

Sterman JD (1989) Misperceptions of feedback in dynamic decision making. Organ Behav Hum Decis Process 43(3):301–335

Sterman JD (2000) Business dynamics. Systems thinking and modeling for a complex world. Irwin/McGraw-Hill, Boston

Struben J, Sterman J, Keith D (2015) Parameter estimation through maximum likelihood and bootstrapping methods. In: Rahmandad H, Oliva R, Osgood ND (eds) Analytical methods for dynamic modelers. MIT Press, Cambridge, MA, pp 3–38

Taleb NN (2007) The black swan. The impact of the highly improbable. Random House, New York

Vennix JAM (1996) Group model building: facilitating team learning using system dynamics. Wiley, Chichester

von Foerster H (1984) Observing systems, 2nd edn. Intersystems Publications, Seaside

von Glasersfeld E (1991) Abschied von der Objektivität. In: Watzlawick P, Krieg P (eds) Das Auge des Betrachters. Piper, Munich, pp 17–30

Books and Reviews

Finlay PN (1997) Validity of decision support systems: towards a validation methodology. Syst Res Behav Sci 14(3):169–182

Forrester JW (1961) Industrial dynamics. MIT Press, Cambridge, MA

Law AM (2007) Simulation modeling and analysis, 4th edn. McGraw-Hill, New York

Legasto AA, Forrester JW, Lyneis JM (eds) (1980) System dynamics. North-Holland, Amsterdam

Morecroft J (2007) Strategic modelling and business dynamics: a feedback systems approach. Wiley, Chichester

Sargent RG (2004) Validation and verification of simulation models. In: Ingalls RG, Rossetti MD, Smith JS, Peters BA (eds) Proceedings of the 2004 winter simulation conference. ACM-Association for Computing Machinery, Washington, DC, pp 17–28

Schwaninger M (2011) System dynamics in the evolution of the systems approach. In: Meyers RA (ed) Complex systems in finance and econometrics, vol 2. Springer, New York, pp 753–766

Sterman JD (2000) Business dynamics. Systems thinking and modeling for a complex world. Irwing/Mc Graw-Hill, Boston

Warren K (2008) Strategic management dynamics. Wiley, Chichester

Optimization of System Dynamics Models

Brian Dangerfield[1] and Jim Duggan[2]
[1]School of Management, University of Bristol, Bristol, UK
[2]School of Computer Science, NUI Galway, Galway, Ireland

Article Outline

Glossary

Econometrics a statistical approach to economic modelling in which all the parameters in the structural equations are estimated according to a "best fit" to historical data.

Maximum likelihood a statistical concept which underpins calibration optimization and which generates the most likely parameter values; it is equivalent to the parameter set which minimizes the chi-square value.

Objective function see **Payoff** below.

Optimization the process of improving a model's results in terms of either an aspect of its performance or by calibrating it to fit reported time series data.

Payoff a formula which expresses the objective, say, maximization of profits, minimization of costs, or minimization of the differences between a model variable and historical data on that variable.

Zero-one parameter a parameter which is used as a multiplier in a policy equation and serves the effect of bringing in or removing a particular influence in determining the optimal policy.

Definition of the Subject and Its Importance

The term "optimization" when related to system dynamics (SD) models has a special significance. It relates to the mechanism used to improve the model vis-à-vis a criterion. This collapses into two fundamentally different intentions. Firstly one may wish to improve the model in terms of its performance. For instance, it may be desired to minimize overall costs of inventory while still offering a satisfactory level of service to the downstream customer. So the criterion here is cost, and this would be minimized after searching the parameter space related to service level. The direction of need may be reversed, and maximization may be desired as, for instance, if one had a model of a firm and wished to maximize profit subject to an acceptable level of payroll and advertising costs. Here the parameter space being explored would involve both payroll and advertising parameters. This type of optimization might be described generically as *policy optimization*.

Optimization of performance is also the *raison d'etre* of other management science tools, most notably mathematical programming. But such tools are usually static: they offer the "optimum" resource allocation given a set of constraints and a performance function to either maximize or minimize. These models normally relate to a single time point and may then need to be rerun on a weekly or monthly basis to determine a new optimal resource allocation. In addition, these models are often linear (certainly so in the case of linear

© Springer Science+Business Media, LLC, part of Springer Nature 2020
B. Dangerfield (ed.), *System Dynamics*,
https://doi.org/10.1007/978-1-4939-8790-0_542

Originally published in
R. A. Meyers (ed.), *Encyclopedia of Complexity and Systems Science*, © Springer Science+Business Media LLC 2019, https://doi.org/10.1007/978-3-642-27737-5_542-5

programming), whereas SD models are usually nonlinear. So the essential differences are that SD model optimization for performance involves both a dynamic and a nonlinear model.

A separate improvement to the model may be sought where it is required to fit the model to past time series data. Optimization here involves minimizing a statistical function which expresses how well the model fits a time series of data pertaining to an important model variable. In other words, a vector of parameters are explored with a view to determining the particular parameter combination which offers the best fit between the chosen important model variable and a past time series dataset of this variable. This type of optimization might be generically termed *model calibration*. If *all* the parameters in the SD model are determined in this fashion, then the process is equivalent to the technique of econometric modelling. A good comparison between system dynamics and econometric modelling can be found in Meadows and Robinson (1985).

Optimization as Calibration

In these circumstances, we wish to determine optimal parameters, those which, following a search of the parameter space, offer the best fit of a particular model variable to a time series

dataset on that variable taken from real-world reporting.

As an example consider a variation of one of the epidemic models which are made available with the Vensim™ software. The stock-flow diagram is presented as Fig. 1.

In this epidemiological system, members of a susceptible population become infected and join the infected population. Epidemiologists call this an S-I model. It is a simpler variation of the S-I-R model which includes recovered (R) individuals.

Suppose some data on new infections (at intervals of 5 days) is available covering 25 days of a real-world epidemic. The model is set with a time horizon of 50 days which is consistent with, say, a flu epidemic or an infectious outbreak of dysentery in a closed population such as a cruise ship. The "current" run of the model is shown in Fig. 2, with the real-world data included for comparison.

Clearly there is not a very good correspondence between the actual data and the model variable for the infection rate (infections). We wish to achieve a better calibration, and so there is a need to select relevant parameters through which the calibration optimization can be performed over. Referring back to Fig. 1, we can see that the *fraction infected from contact* and the *rate that people contact other people* are two possible parameters to consider. The initial infected and initial susceptible are also parameters

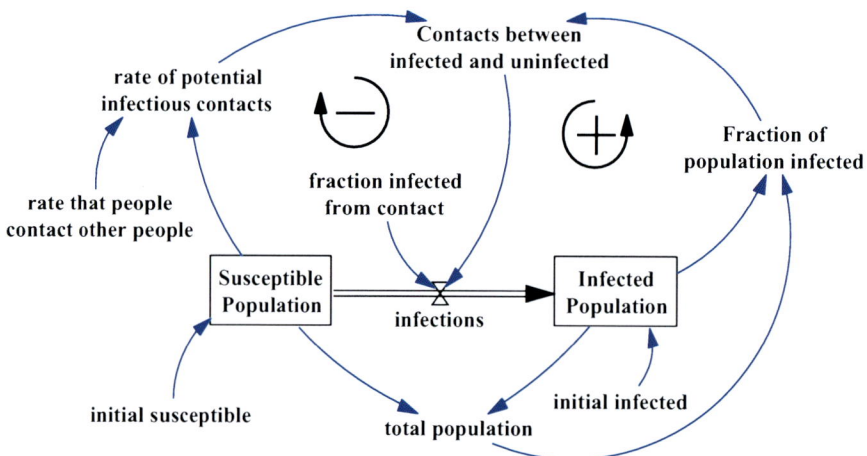

Optimization of System Dynamics Models, Fig. 1 Stock-flow diagram for a simple epidemic model

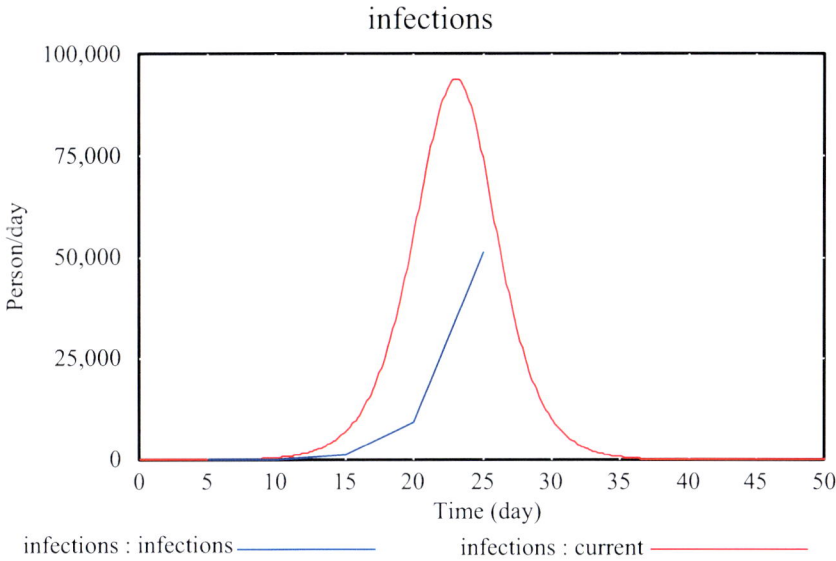

Optimization of System Dynamics Models, Fig. 2 Current (base) run of the model and reported data on infections

of the model in the strict sense of the term, but we will ignore them on this occasion. In this model, the *initial infected* is 10 persons and *initial susceptible* number 750,000 persons.

The chosen value for the *fraction infected from contact* is 0.1 while that for the *rate that people contact other people* is 5.0. The former is a dimensionless number, while the latter is measured as a fraction per day (1/day). This is obtained from consideration of the *rate of potential infectious contacts* (persons/day) as a proportion of the *susceptible population* (persons).

The optimization process for calibration involves reading into the model the time series data, in this case on new infections, and, secondly, determining the range for the search in parameter space. There is usually some basic background knowledge which allows a sensible range to be entered. For instance, a probability can only be specified between 0 and 1.0. In this case we have chosen to specify the ranges as follows:

0.03 < = fraction infected from contact <= 0.7
2 <= rate that people contact other people <= 10

A word of warning is necessary in respect of optimizing delay parameters. Because there is a risk of mathematical instability in the model if the

value of DT (the TIME STEP) is too large relative to the smallest first-order delay constant, it is important to ensure the TIME STEP employed in the model is sufficiently small to cope with delay constant values which may be reached during the search of the delay parameter space. In other words ensure the minimum number for the search range on the delay parameter is at least double the value of the TIME STEP.

Maximum Likelihood Estimation and the Payoff Function

The optimization process involves a determination of what are termed statistically as maximum likelihood estimates. In Vensim™, this is achieved by maximizing a payoff function. Initially this is negative, and the optimization process should ensure this becomes less negative. An ideal payoff value, after optimization, would be zero. A weighting is needed in the payoff function too, but for calibration optimization, this is normally 1.0. Driving the payoff value to be larger by making it less negative has parallels with the operation with the simplex algorithm common in linear programming. This algorithm was conceived initially for problems where the objective function was to be minimized. Its use on maximization problems is achieved by minimizing the negative of the objective function.

During the calibration search, Vensim™ takes the difference between the model variable and the data value, multiplies it by the weight, squares it, and adds it to the error sum. This error sum is minimized. Usually data points will not exist at every time point in the model. Here the model TIME STEP is 0.125 (1/8th), but let us assume that reported data on new infections have been made available only at times $t = 5$, 10, 15, 20, and 25 so the sum of squares operation is performed only at these five time points.

The data is shown as Table 1.

The Recording Point for Reported Data

System dynamics models differentiate between stock and flow variables and the software used for simulating such models advances by a small constant TIME STEP (also known as DT). This has implications for the task of fitting real-world reported data to each type of system dynamics model variable. The issue is: at what point in a continuum of time steps should the reported data be recorded at? This is important because the reported data has to be read into the model to be compared with the simulated data. The answer will be different for stock and flow variables.

Where the reported data relate to a stock variable, the appropriate time point for recording will be known. If it is recorded at the end of the day (say a closing bank balance), then the appropriate point for data entry in the model will be the beginning of the next day. Thus the first data point above is at time $t = 5$ (5.00) and would, if it were a stock, correspond to a record taken at the very end of time period 4.

However, if the data relate to a flow variable, as in the case of new infections here, the number is the total new infections which have occurred over the entire time unit (day, week, month, etc.), and so there is a decision to be reached as to which

time point the data are entered at. This is because the TIME STEP (DT) is hardly ever as large as the basic time unit which the model is calibrated in. The use of 5 (10, 15, etc.) above implies that the data on new infections over the period of time $t = 0$ to $t = 5$ is compared with the corresponding model variable at time 5 + 1*DT (and the new infections over the period $t = 5$ to $t = 10$ at time 10 + 1*DT, etc.). A more appropriate selection might be toward the end of the 5-day time period. Following the example above using a TIME STEP = 0.125, this might be at time 4 + 7*DT (i.e., at 4.875).

Calibration Optimization Results

Based upon the data on new infections shown above and the chosen ranges for the parameter search, the following output is obtained (Table 2). After 114 simulations, the optimized values for our two parameters are shown to be 0.08 and 5.12, and the payoff is over 2500 times larger (less negative). Replacing the original parameters with the optimized values reveals the result shown in Fig. 3. To take things further, we may wish to put confidence intervals on the estimated parameters. One way of accomplishing this is by profiling the likelihood and is described in Dangerfield and Roberts (1996a).

Avoid Cumulated Data

There might be a temptation to optimize parameters against cumulated data when the data is

Optimization of System Dynamics Models, Table 2 Results from the calibration optimization

Initial point of search
Fraction infected from contact = 0.1
Rate that people contact other people = 5
Simulations = 1
Pass = 0
Payoff = −2.67655e + 009
Maximum payoff found at
Fraction infected from contact = 0.0794332
*Rate that people contact other people = 5.11568
Simulations = 114
Pass = 6
Payoff = −1.06161e + 006

Optimization of System Dynamics Models, Table 1 Data used for calibration experiment

Time	5	10	15	20	25
Infections	30	230	1400	9500	51,400

Optimization of System Dynamics Models, Fig. 3 Reported data on infections and optimized (calibrated) model; the base case (current) is reproduced for reference

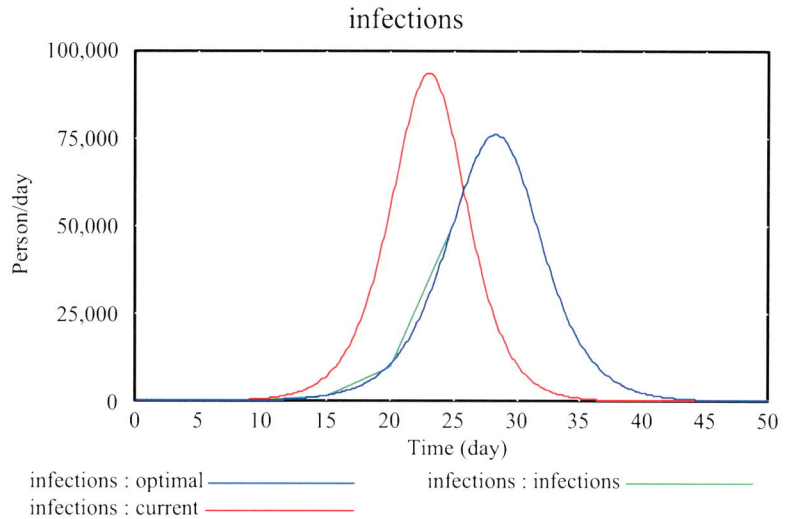

infections

infections : optimal ————
infections : current ————
infections : infections ————

reported essentially as a flow, as is the case here. The cumulated data obtained is shown in Table 3.

The results from this optimization are shown in Table 4. The ranges for the parameter space search are kept the same, but the payoff function now involves a comparison of the model variable *infected population* with the corresponding cumulated data. Figure 4 shows the resultant fit to infected population is good but that is manifestly not borne out when we consider the plot of infections obtained from the same optimization run (Fig. 5).

The reason for this is rooted in statistics. The maximum likelihood estimator is equivalent to the chi-squared statistic. This is turn assumes that each expected data value is independent. A cumulated data series would not exhibit this property of independence.

As an aside it is worth pointing out that this model, with suitable changes to the variable names and the time constants involved, could equally represent the diffusion of a new product into a virgin market. In systems terms, the structures are equivalent. The *fraction infected from contact* is the same as, say, the fraction reached by word of mouth or advertising, and the *rate that people contact other people* is a measure of the potential interactions at which new products might be mentioned among the members of the relevant market segment. *Infected population* is

Optimization of System Dynamics Models, Table 3 Cumulated reported data for the infected population

Time	5	10	15	20	25
Infected population	30	260	1660	11,160	62,560

Optimization of System Dynamics Models, Table 4 Results from the calibration using cumulated data

Initial point of search	
Fraction infected from contact = 0.1	
Rate that people contact other people = 5	
Simulations = 1	
Pass = 0	
Payoff = −2.48206e + 011	
Maximum payoff found at	
Fraction infected from contact = 0.0726811	
*Rate that people contact other people = 4.96546	
Simulations = 145	
Pass = 6	
Payoff = −212,645	
The final payoff is −212,645	

equivalent to customer base, the number of adopters of the relevant product. So it is possible to shed light on important real-world marketing parameters through a calibration optimization of models of this general structure.

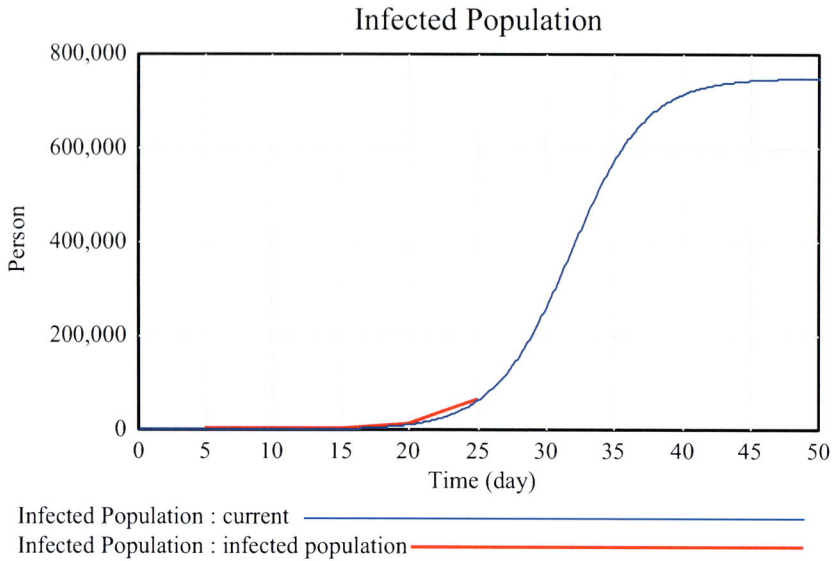

Optimization of System Dynamics Models, Fig. 4 The cumulative model variable (infected population) together with reported data

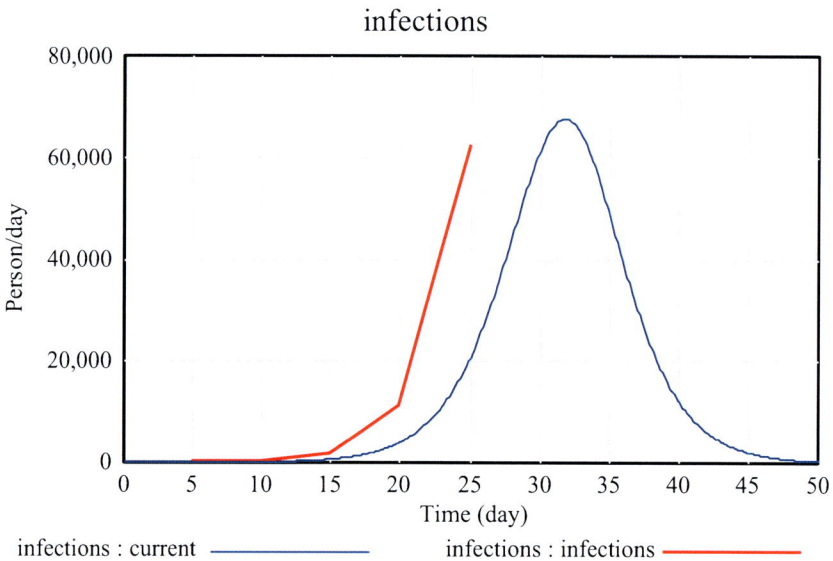

Optimization of System Dynamics Models, Fig. 5 The corresponding fit to infections is poor

Optimization of Performance (Policy Optimization)

An example model is to be used to illustrate the process of optimization to improve the performance of the system, and this is illustrated in

Fig. 6. It concerns the service requirements which can arise following the sale of a durable good. These items are typically sold with a 12-month warranty, and during this time the vendor is obliged to offer service if a customer calls for it. In this particular case, the vendor is not being responsive

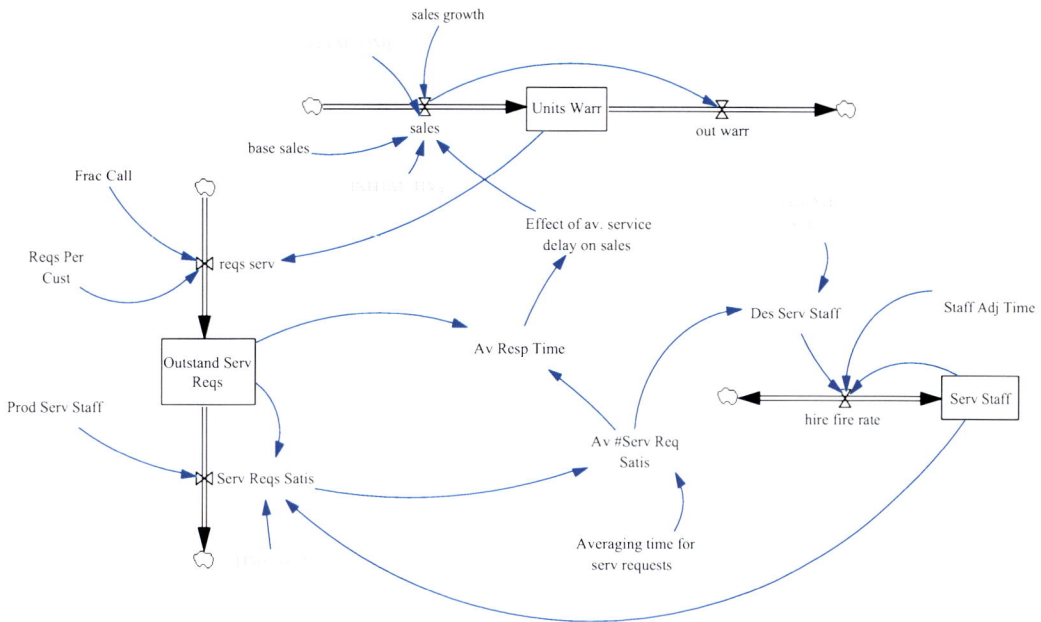

Optimization of System Dynamics Models, Fig. 6 Model of service delays for durable goods under warranty

in terms of staffing the service section. The result is that as sales grow the increasing number of service requests is putting pressure on the service personnel. The delay in responding to service calls also increases, and the effect of this is that future sales are depressed because of the vendor's acquired reputation for poor service response. The basic behavior mode is overshoot and collapse.

In the model depicted in Fig. 6, the growth process is achieved by a RAMP function which causes sales of the good to increase linearly by 20 units per month from a base of 500 units per month.

The payoff function is restricted to the variable *Sales*. However, this need not be the case. Where a number of variables might be options in a payoff function, it is possible to assign weights to each such that the sum of the weights is 1.0 (or 100). The optimization process will then proceed with the software accumulating a weighted payoff which it will attempt to maximize. Weights are positive when more is better and negative when less is better.

Policy Experiment No. 1

Here it is decided to try to improve the productivity of the service staff. Currently they manage, on

average, to respond to 120 calls per operative per month. It may be an option to improve their productivity by, say, providing them with handheld devices which direct each operative from one call to the next – calls which may have arisen since setting out from their base. In this way, their call routing is improved.

The optimization parameter is *Prod Serv Staff*, and we select an upper limit for the search range of 240 calls per person per month. The chosen performance variable is *Sales* since we wish to maximize this – or at least not have it overly depressed by poor response times. The results are shown in Table 5. We see that the payoff is increased and that the optimum productivity is a modest increase of 2.6 requests per month, on average. This should be easily achievable and perhaps without expenditure on high-tech devices. The graphical output for sales is shown in Fig. 7.

For comparison, the effect of increasing the productivity to as high as 150 calls per month, on average, is also shown. This would represent an increase of 25% and would be much more difficult to accomplish. Here the benefit of optimization is highlighted. A modest increase in productivity returns a visibly improved sales performance (although the basic behavior mode is unchanged), while a much greater productivity increase offers little extra benefit for the effort and cost involved in improving productivity.

Policy Experiment No. 2

Another approach to policy optimization involves the use of a zero-one parameter which has the effect of either including or excluding an influence on policy. Suppose it was thought that the quantity of product units in warranty should exert an influence on the numbers of service personnel hired (or fired). The equation for the desired

number of service staff (*Des Serv Staff*) can be expressed as:

Des Serv Staff = "Av #Serv Req Satis"/Prod Serv Staff* trigger + ("Av #Serv Req Satis"/Prod Serv Staff)*(Units Warr/initial units in warranty)* (1-trigger) (Units: Persons)

The *trigger* variable is initially set to 1.0, and so the more sophisticated policy is not active. The optimization run results are shown in Table 6. Clearly there is benefit from including the more sophisticated policy which takes into account the current numbers of product units in warranty.

The graphical output is unequivocal (Fig. 8). Sales are continuously increasing when the recruitment policy for service personnel takes into account the number of product units in warranty. The depressive effect on sales of poor service performance is nonexistent.

While this might seem an obvious policy, it is surprising how easily the naïve alternative might be accepted without question. The number of calls a typical operative can manage each month is well known along with the (historical) number of service requests satisfied. Hence, the desired number of staff is more or less fixed. This comes undone when there is a growth in the number of products sold. In this different

Optimization of System Dynamics Models, Table 5 Optimization results for the productivity of the service staff

Initial point of search
Prod Serv Staff = 120
Simulations = 1
Pass = 0
Payoff = 27743.5
Maximum payoff found at
*Prod Serv Staff = 122.647
Simulations = 27
Pass = 3
Payoff = 29,915

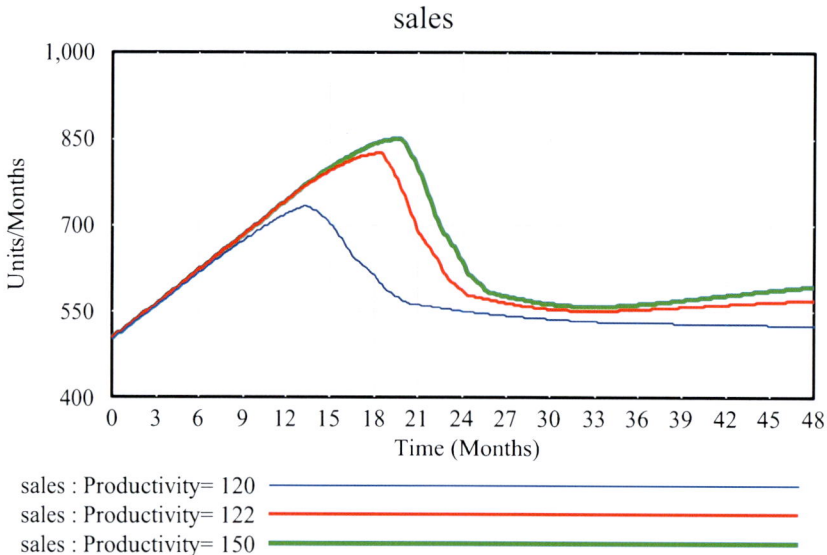

Optimization of System Dynamics Models, Fig. 7 Plots of sales achieved for differing productivities

environment such a simplistic policy can, as shown, lead to overshoot and collapse. Notice needs to be taken of the changing number of product units in warranty in order that a more effective system performance is achieved.

The above experiments are illustrative only, and there is no intention of overworking a simple teaching model in order to uncover an ideal policy. In the case of policy optimization, a wide range of possible alternatives exists. Indeed, a process of learning naturally arises through carrying out repeated optimization experiments with the model (Coyle 1996).

Optimization of System Dynamics Models, Table 6 Optimization results from selection of policy drivers

Initial point of search
Trigger = 1
Simulations = 1
Pass = 0
Payoff = 27743.5
Maximum payoff found at
*Trigger = 0
Simulations = 13
Pass = 3
Payoff = 47090.9

Examples of SD Optimization Reported in the Literature

Among the earliest work in this area, the writings of Keloharju are worthy of mention. He contributed several papers on the topic in the pages of *Dynamica*. See, for example, Keloharju (1977). His work brought the concept to prominence, but he did not employ the method on anything other than problems described in text books or postulated by himself. For instance, an application of optimization to the project model contained in Richardson and Pugh's (1981) text is contained in Keloharju and Wolstenholme (1989). A statement of the method together with some textbook examples is also available (Keloharju and Wolstenholme 1988). Additionally, an overview of the methods and their deployment on textbook examples has been contributed by one of the current authors (Dangerfield and Roberts 1996a). Finally, there is an example of optimization applied to defense analysis. Again, though, it is a standard defense model – the armored advance model – rather than any real-world study (Wolstenholme and Al-Alusi 1987).

Retaining the emphasis on textbook problems for the moment, Duggan (2005) employs Coyle's (1996) model of the *Domestic Manufacturing*

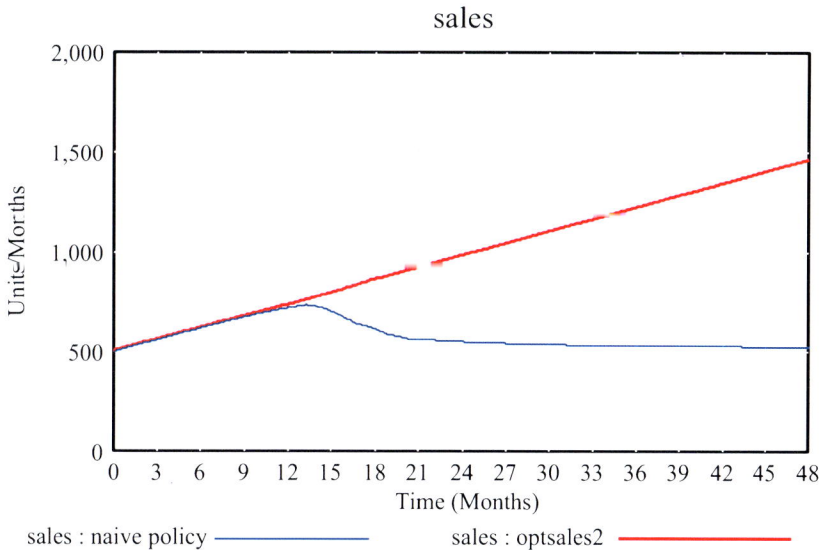

sales

Optimization of System Dynamics Models, Fig. 8 Comparison of sales from two different policy drivers

Company to illustrate the methods of multi-objective optimization – an advance over standard SD optimization with its single objective function. The concept of multiple objectives arises from multi-criteria decision-making where a situation can be judged on more than one performance metric. While a multi-objective payoff function can be formulated using a set of weights, it is argued that the selection of the weights is very individual-specific. The multi-objective approach – underpinned by the methods of genetic algorithms – rests upon determining a Pareto-optimal situation, defined as one where no improvement is possible without making some other aspect worse. In other words the method strives for an optimal solution which is not dominated by any other solution. The author demonstrates the approach by combining two objectives in the model: one for the differences between desired stock and actual and another between desired backlog and actual.

In terms of applications to real-world problems, one of the current authors has also used the methods of optimization in research conducted in connection with modelling the epidemiology of HIV/AIDS. Fitting a model of AIDS spread to data was carried out for a number of European countries (Dangerfield and Roberts 1994; Dangerfield and Roberts 1996b). The optimized parameters furnished support for some of the features of AIDS epidemiology which, at the time, were being uncovered by other branches of science. For example, the optimized output revealed that a U-shaped profile of infectiousness in a host was necessary in order to achieve a best fit to data on new AIDS cases. This infectiousness profile was also evidenced by virologists who had analyzed patients' blood and other secretions on a longitudinal basis.

Within this strand of research, a much more complex optimization was performed using American data on transfusion-associated AIDS cases (Dangerfield and Roberts 1999). The purpose here was to estimate the parameters for a number of plausible statistical HIV incubation distributions. Given the nature of the data, the point of infection could be quite accurately determined, but two difficulties were evident: the data

were right-censored, and the number receiving infected transfusions in each quarter was unknown. However, the SD optimization could estimate this number as part of the process, in addition to estimating parameters of the incubation distribution. The best fit was found to be a three-stage distribution similar to the gamma and one which accorded with the high-low-high U-shaped infectiousness profile which was receiving support from a number of sources.

In the marketing domain, Graham and Ariza (2003) carried out an optimization on a system dynamics model which was designed to shed light on the allocations to make from a marketing budget in a high-tech client firm. Assuming the budget was fixed, the task was to optimize the allocations across more than 90 "buckets" – combinations of product lines, marketing channels, and types of marketing. However, these were not discrete: advertising on one product line might have crossover effects on another, and the impacts could propagate over a period of time. One major conclusion for this firm was that the advertising allocation should be increased markedly. In general, intuitive allocations were shown to fall short of the ideal: they were directionally correct, but magnitudes fell short often by factors of three or four.

As a further counterbalance to the preponderance of the use of optimization solely on textbook or artificial problems, one of the current authors has contributed to research on target setting for weight reduction in the context of childhood overweight and obesity in the UK (Abidin et al. 2014). A UK government-set target in 2008 aimed to reverse the prevalence of obesity back to 2000 levels by 2020. The model-based analysis showed that the earliest this target might be achieved was 2026. It was founded upon an optimization of the required parameter changes associated with children's eating behavior to reverse the (then) present increase in average weight and move the relevant metrics toward the government target. As this behavioral change needed to happen progressively over time, the relevant parameters were reduced by optimization of the slope of an assumed linear decay function (Fig. 9). A separate optimization on the parameters

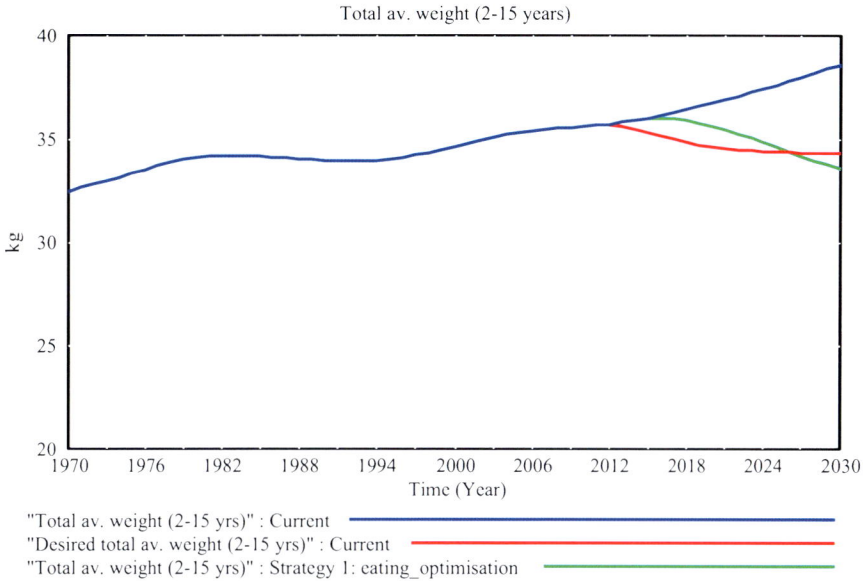

Optimization of System Dynamics Models, Fig. 9 Simulated projection for the average weight across three age groups spanning 2–15 years (blue line); a plausible decline which would achieve the government's 2020 target (red line); and the resultant trajectory following an optimization (green line) to try to achieve the red line trajectory through changes in parameters associated with eating behavior (target not achieved until 2026)

associated with physical activity showed policy directed here was even less effective in achieving the target; the policy precept is that interventions need to be made in child energy intake rather than energy expenditure.

Hesamamiri and Bourouni (2016) combine particle swarm optimization (PSO) with system dynamics and demonstrate this approach through a customer satisfaction model. The PSO method is an evolutionary computation approach inspired by bird flocking or fish schooling where potential solutions "fly" through a hyperspace and accelerate toward better or more optimum solutions. Their method starts with developing a detailed SD model of a system and proceeds to deploy the PSO algorithm in order to optimize the parameter values, subject to an objective function. Liu et al. (2012) also apply evolutionary computation methods to system dynamics and provide a sector-based approach for optimization, where fitness functions can be developed for a local sector (e.g., a retailer in a supply line). Evolutionary optimization approaches, based on game theory, are then used to explore the impact of different ordering heuristics and provide a testing ground

for exploring impact that locally rational behavior has on overall system performance. Ng et al. (2012) introduce a robust goal-seeking approach for the parameter design of SD models, which synthesizes developments in robust optimization technology, eigenvalue analysis, and the goal-seeking behavior of decision agents. An advantage of their approach is the ease of computation, although the initial SD model must be linearized to obtain a state-space model. The authors also highlight the importance of open-source software and how available optimization modules have the potential to encourage adoption in practice.

In the area of model calibration, Struben et al. (2015) discuss the importance of grounding models in data if modellers are to provide reliable advice to policymakers, and emphasize that the ability of a model to replicate the relevant system history is an important test. They present the technique of parameter estimation through maximum likelihood and bootstrapping methods and provide a process for confidence interval estimation that aligns both approaches. Osgood and Liu (2015) present a worked example of the Markov chain Monte Carlo (MCMC) method for model calibration, which – using Bayesian statistics – provides a rigorous way to derive posterior distributions of model parameters. A benefit of this method is that it can be used to *dynamically recalibrate* models when new data becomes available. Although computationally expensive, the growing availability of hardware resources (e.g., parallel computation) should see an increased use of this method within the system dynamics methodology.

Hosseinichimeh et al. (2016) address the challenge that is common within the social sciences, whereby there are few data points available for analysis. They introduce indirect inference, which is a simulation-based estimation method that finds model parameters by ensuring that simulation data and available empirical observations produce similar auxiliary statistics. Their results, based on a model of depression and rumination, also indicate that this approach can be applied to SD models in other areas.

Future Directions in SD Optimization

A primary aim must be to see more published work which describes optimization studies carried out on real-world SD applications. There may be frequent use of optimization in consulting assignments, but such activities are rarely published. The references herein suggest that, thus far, outside of unpublished work, the number may be around half a dozen and is certainly less than ten. While software requirements may have inhibited use of SD optimization in the past, there are now fewer computational barriers to its use, and it is to be hoped that in the future this quite powerful analytical tool in the SD methodological repertoire will feature in more application studies.

An advance in the methodology itself has been developed by Duggan (2008), and this is a promising pointer for the future. Based on genetic algorithms, it is best suited to the class of SD problems that are agent-based, and this highlights a slight limitation. Traditional optimization takes the policy equations as given and explores the parameter space to determine an optimal policy.

Instead he has offered an approach which searches over both parameter space and policy (strategy). Theoretically there is no limit to the number of strategies which can be evaluated in this approach, although the user has to define a set of such strategies in advance of the runs. Under a conventional optimization approach, a limited tilt at this is possible using the zero-one parameter method suggested above, although this would restrict the enumerated strategies to two only. Duggan demonstrates the new approach using a classic SD problem: the four-agent beer game. Game theory approaches can also be successfully integrated with system dynamics, for example, the work on *differential games*, which has the potential to empower system dynamics modellers to address new challenges where multi-stakeholder competition is a defining characteristic of the system under study (Rahmandad and Spiteri 2015).

The emergence of big data (Struben et al. 2015) provides a further opportunity to align system dynamics with modern data mining and machine learning methods and tools. In the system dynamics literature, data-driven "black box" methods (Barlas 1996) and technologies such as R (Duggan 2016) are becoming more influential, as modellers explore ways to combine the increasing availability of evidential data with the power of stock and flow modelling, in order to support policy analysis. Hamarat et al. (2013) combine a computational model-based approach to support decision-making under deep uncertainty. Thousands of scenarios are analyzed using system dynamics models, data mining techniques (Pruyt and Islam 2015), and optimization methods, and they demonstrate the effectiveness of this hybrid approach through the exploration of energy transitions. Yücel and Barlas (2011) also deploy machine learning methods to provide a tool that supports model calibration. Their pattern-identification method enables a parameter search process based on qualitative features of a desired behavior pattern. The suggested search process obtains parameter values that match qualitative patterns of the desired behavior.

This promising synthesis of mathematical modelling and data science methods is also impacting important policy domains, particularly where the construction and implementation of complex nonlinear models are essential for informed decision support. For example, the use of direct optimization of the values of policy functions through white-box methods can measurably improve optimization performance (Vierhaus et al. 2017). They demonstrate this with benchmarks provided on the classic World2 model. In public health, a key challenge is how to robustly weigh evidence from diverse datasets and handle the dependence between them, in order to make more informed responses to epidemic outbreaks (Funk et al. 2015). These new approaches have high potential, particularly if models can be constructed that can be rapidly calibrated by new data sources. For example, these could include real-time data feeds based on the latest incidence outbreaks, weather data, and evidence from internet search queries and social media feeds to support situational awareness (Signorini et al. 2011). We look forward to the further integration of these optimization methods with system dynamics, in order to enhance decision support for policymakers.

Bibliography

Abidin NZ, Mustafa M, Dangerfield BC et al (2014) Combating obesity through healthy eating behaviour: a call for system dynamics optimisation. PLoS One 9(12):e114135, 17pp. https://doi.org/10.1371/journal.pone.0114135

Barlas Y (1996) Formal aspects of model validity and validation in system dynamics. Syst Dyn Rev 12(3):183–210

Coyle RG (1996) System dynamics modelling: a practical approach. Chapman & Hall, London

Dangerfield BC, Roberts CA (1994) Fitting a model of the Spread of AIDS to data from five European countries. In: O.R. Work in HIV/AIDS, 2nd edn. Operational Research Society, Birmingham, pp 7–13

Dangerfield BC, Roberts CA (1996a) An overview of strategy and tactics in system dynamics optimisation. J Oper Res Soc 47(3):405–423

Dangerfield BC, Roberts CA (1996b) Relating a transmission model of AIDS spread to data: some international comparisons. In: Isham V, Medley G (eds) Models for infectious human diseases: their structure and relation to data. Cambridge University Press, Cambridge, pp 473–476

Dangerfield BC, Roberts CA (1999) Optimisation as a statistical estimation tool: an example in estimating the AIDS treatment-free incubation period distribution. Syst Dyn Rev 15(3):273–291

Duggan J (2005) Using multiple objective optimisation to generate policy insights for system dynamics models. In: Proceedings of the international system dynamics conference. System Dynamics Society, Boston

Duggan J (2008) Equation-based policy optimisation for agent-oriented system dynamics models. Syst Dyn Rev 24(1):97–118

Duggan J (2016) An introduction to R. In: System dynamics modeling with R. Springer International Publishing, Cham, pp 25–47

Funk S, Bansal S, Bauch CT, Eames KTD, Edmunds WJ, Galvani AP, Klepac P (2015) Nine challenges in incorporating the dynamics of behaviour in infectious disease models. Epidemics 10:21–25. https://doi.org/10.1016/j.epidem.2014.09.005

Graham AK, Ariza CA (2003) Dynamic, hard and strategic questions: using optimisation to answer a marketing resource allocation question. Syst Dyn Rev 19(1):27–46

Hamarat C, Kwakkel JH, Pruyt E (2013) Adaptive robust design under deep uncertainty. Technol Forecast Soc Chang 80(3):408–418. https://doi.org/10.1016/j.techfore.2012.10.004

Hesamamiri R, Bourouni A (2016) Customer support optimization using system dynamics: a multi-parameter approach. Kybernetes 45(6):900–914. https://doi.org/10.1108/K-10-2015-0257

Hosseinichimeh N, Rahmandad H, Jalali MS, Wittenborn AK (2016) Estimating the parameters of system dynamics models using indirect inference. Syst Dyn Rev 32(2):156–180. https://doi.org/10.1002/sdr.1558

Keloharju R (1977) Multi-objective decision models in system dynamics. Dynamica 3(1):3–13; 3(2):45–55

Keloharju R, Wolstenholme EF (1988) The basic concepts of system dynamics optimisation. Syst Pract 1:65–86

Keloharju R, Wolstenholme EF (1989) A case study in system dynamics optimisation. J Oper Res Soc 40(3):221–230

Liu H, Howley E, Duggan J (2012) Co-evolutionary analysis: a policy exploration method for system dynamics models. Syst Dyn Rev 28(4):361–369. https://doi.org/10.1002/sdr.1482

Meadows DM, Robinson JM (1985) The electronic oracle. Wiley, Chichester. (Now available from the System Dynamics Society, Albany NY)

Ng TS, Sy CL, Lee LH (2012) Robust parameter design for system dynamics models: a formal approach based on goal-seeking behavior. Syst Dyn Rev 28(3):230–254. https://doi.org/10.1002/sdr.1475

Osgood ND, Liu J (2015) Combining Markov Chain Monte Carlo approaches and dynamic modeling. In: Analytical methods for dynamic modelers. MIT Press, Cambridge, MA, p 125

Pruyt E, Islam T (2015) On generating and exploring the behavior space of complex models. Syst Dyn Rev 31(4):220–249. https://doi.org/10.1002/sdr.1544

Rahmandad H, Spiteri RJ (2015) Modeling competing actors using differential games. In: Analytical methods for dynamic modelers. MIT Press, Cambridge, MA, p 373

Richardson GP, Pugh AL (1981) An introduction to system dynamics modelling with DYNAMO. MIT Press, Cambridge, MA

Signorini A, Segre AM, Polgreen PM (2011) The use of Twitter to track levels of disease activity and public concern in the U.S. during the Influenza A H1N1 Pandemic. PLoS One 6(5):e19467. https://doi.org/10.1371/journal.pone.0019467

Struben J, Sterman J, Keith D (2015) Parameter estimation through maximum likelihood and bootstrapping methods. In: Analytical methods for dynamic modelers. MIT Press, Cambridge, MA, pp 3–38

Vierhaus I, Fügenschuh A, Gottwald R, Grösser S (2017) Using white-box nonlinear optimization methods in system dynamics policy improvement. Syst Dyn Rev 33(2):138–168. https://doi.org/10.1002/sdr.1583

Wolstenholme EF, Al-Alusi AS (1987) System dynamics and heuristic optimisation in defence analysis. Syst Dyn Rev 3(2):102–115

Yücel G, Barlas Y (2011) Automated parameter specification in dynamic feedback models based on behavior pattern features. Syst Dyn Rev 27(2):195–215. https://doi.org/10.1002/sdr.457

Analytical Methods for Structural Dominance Analysis in System Dynamics

Christian Erik Kampmann[1] and Rogelio Oliva[2]
[1]Department of Innovation and Organizational Economics, Copenhagen Business School, Copenhagen, Denmark
[2]Mays Business School, Texas A&M University, College Station, TX, USA

Article Outline

Glossary
Definition of the Subject
Introduction
Characterizing Linear and Nonlinear Systems
Traditional Control Theory Approaches
Pathway Participation Metrics
Eigenvalue Elasticity Analysis
Eigenvectors and Dynamic Decomposition
  Weights (DDW)
Future Directions
Bibliography

Glossary

Behavior mode The traditional meaning of the term is the qualitative nature of the observed system behavior, such as damped or expanding oscillations, overshoot and collapse, exponential growth or adjustment to equilibrium, or limit cycles. In linear systems theory, the term has a more specific meaning, cf. the explanation for eigenvalues.

Bode plot (phase and gain plot) A tool used in classical control theory to characterize the frequency response, i.e., the amplification A and phase shift ϕ in the system output variable of interest $x(t) = A \sin(\omega t + \phi)$ compared to the input variable $u(t) = \sin(\omega t)$, as a function of the frequency ω of the input.

Chaos A type of behavior exhibited by nonlinear systems that appears to be approximately periodic but with a seemingly random element. A hallmark of chaotic behavior is that it is sensitive to initial conditions.

Dominant structure A general term for the feedback loops (or possibly external driving forces) that are "most important" in generating a behavior pattern of interest. In nonlinear models, particularly single-transient models, there is frequently a shift in structural dominance, i.e., in the strength and significance of certain feedback loops.

Dynamic decomposition weights (DDW) An application of eigenvector elasticity analysis (EVA) that focuses on how parameter changes influence the relative weights (DDW) of the system behavior modes in a particular variable.

Eigenvalue elasticity analysis (EEA) A method of analyzing the significance of a structural element, say a loop or a link in the model with a gain g, in terms of its marginal effect upon the eigenvalues λ of the system. There are several such measures, such as the *influence measure* $\partial\lambda/\partial g \cdot g$, the *elasticity* $\partial\lambda/\partial g \cdot (g/\lambda)$, or, in the case of complex-valued eigenvalues, the effect upon the damping ratio, natural frequency, damping time, etc., as illustrated in Fig. 8. See also *Loop Eigenvalue Elasticity Analysis (LEEA)*.

Eigenvalue An eigenvalue for a square matrix A is a value λ for which the equation $Ar = \lambda r$ has a nonzero solution $r \neq 0$. The column vector r is called the (right) eigenvector corresponding to the eigenvalue λ. The eigenvalues and eigenvectors determine the behavior modes (components) in the solution to the linear dynamical system $\dot{x} = Ax$. A real eigenvalue λ leads to an exponential behavior mode $\exp(\lambda t)$, while a complex eigenvalue $\lambda = \tau \pm i\omega$ leads to oscillatory behavior modes $\exp(\tau t) \sin(\omega t + \phi)$. The eigenvectors determine the weight, or the degree to which a particular

© Springer Science+Business Media, LLC, part of Springer Nature 2020
B. Dangerfield (ed.), *System Dynamics*,
https://doi.org/10.1007/978-1-4939-8790-0_535

Originally published in
R. A. Meyers (ed.), *Encyclopedia of Complexity and Systems Science*, © Springer Science+Business Media LLC 2017, https://doi.org/10.1007/978-3-642-27737-5_535-2

behavior mode is expressed in a particular system variable.

Eigenvector elasticity analysis (EVA) A complement to eigenvalue elasticity analysis (EEA) that looks explicitly at the expression or relative weight of each behavior mode in each system variable. These weights are related to the eigenvectors of the system matrix.

Eigenvector See explanation for *Eigenvalue*.

Frequency domain A term used to describe the analysis of signals with respect to frequency. While a time domain graph shows the behavior of the signal over time, the frequency domain graphs show how much of the signal variance lies within each given frequency band.

Independent loop set (ILS) Although the number of feedback loops in a model can be very large (theoretically astronomically large), there is a much smaller *independent loop set* that can be considered independent structural elements. For a strongly connected system (where any pair of variables are connected via causal chain in both directions) with N links and n variables, there are exactly $N - n + 1$ independent loops. Simple algorithms exist for constructing independent loop sets, in particular *shortest independent loop sets (SILS)*. See also explanation for *Loop Eigenvalue Elasticity Analysis (LEEA)*.

Linear dynamical system A system where the rates $\dot{x} = (dx_1/dt, \ldots, dx_n/dt)$ are a linear function of the state variables $x = (x_1, \ldots, x_n)$ and exogenous or control variables $u = (u_1, \ldots, u_p)$, expressed by the equation $\dot{x} = Ax + Bu$ where A is an $n \times n$ matrix and B is an $n \times p$ matrix. Unlike nonlinear systems of the general form $\dot{x} = f(x, u)$, linear systems have analytical solutions based on the eigenvalues and eigenvectors of the matrix A (cf. explanation for *Eigenvalues*).

Linear systems theory The mathematical theory of *linear dynamical systems*.

Loop eigenvalue elasticity analysis (LEEA) A form of eigenvalue elasticity analysis (EEA) that uses graph theory to express structural changes in terms of change in the strength of individual feedback loops. *Independent* loops can be assigned individual (loop) eigenvalue

elasticities or influence measures just like other structural elements (see explanation for *Eigenvalue Elasticity Analysis (EEA)* and *Independent Loop Set (ILS)*).

Model simplification approach A way of attributing dynamic behavior to particular pieces of structure by replacing the full model with a simplified structure. See also *Structure contribution approach*.

Nonlinear systems Systems of the form $\dot{x} = f(x, u)$ where f is a nonlinear function. See explanation for *Linear Dynamical Systems*.

Pathway participation metric A measure that decomposes the curvature $(\ddot{x} = d^2x/dt^2)$ of a variable x_i into the individual driving components, $\ddot{x}_i = \sum \frac{\partial f_i}{\partial x_j} \dot{x}_j$. By considering the sign of the curvature relative to the slope, i.e., $\ddot{x}/\dot{x}$, one may define behavior as (apparently) dominated by positive $\ddot{x}/\dot{x} > 0$ or negative $\ddot{x}/\dot{x} < 0$ feedback loops. The component (pathway) with the largest absolute value and the same sign as $\ddot{x}/\dot{x}$ is then defined as the dominant structure.

Quasilinear models Models that are almost linear in structure around the operating point of interest so that they may be well approximated by a linear model.

Quasiperiodic behavior A behavior that is a sum of oscillations of incommensurate frequencies so that the system never returns to exactly the same point (which would be the case for periodic behavior).

Shortest independent loop set (SILS) An *independent loop set (ILS)* that consists of the shortest possible loops (in terms of the number of nodes and links in each loop). Since the choice of ILS is far from unique, a SILS provides a more focused choice of loops, which are typically also easier to interpret due to their short length.

Single-transient models Models where the behavior of interest is the transition toward an equilibrium or constant growth rate. Models are typically nonlinear, exhibit patterns such as smooth transition, overshoot and collapse, growth, or stagnation.

Structure contribution approach A way of linking model structure to dynamic behavior by considering how individual pieces of structure (feedback loops or subsystems) contribute to the behavior pattern of interest by turning the structure on or off (in traditional simulation experiments) or by considering the marginal effect of small changes in structure (the eigenvalue approach). See also *Model Simplification Approach*.

Definition of the Subject

The link between system structure and dynamic behavior is one of the defining elements in the system dynamics paradigm, yet it is only recently that systematic, mathematically rigorous methods for exploring this link have started to become available. In a sense, a simulation model can be viewed as an explicit and consistent theory of the behavior it exhibits. Although this point of view has certain merits, not least the fact that it lifts the discussion from outcomes to causes of these outcomes and from events to underlying structure (Forrester 1961; Sterman 2000), we are concerned here with a more compact explanation of the system's behavior. In fact, most system dynamics modeling projects report their results in terms of simpler explanations of the observed results, typically in terms of dominant feedback loops that produce the salient features of the behavior.

Most often, dominant structure is thought of in terms of feedback loops and, occasionally, external driving forces to the system. For simple systems with relatively few variables, it is usually easy to use intuition and trial-and-error simulation experiments to explain the dynamic behavior as resulting from particular feedback loops. In larger systems, this method becomes increasingly difficult and the risk of incorrect explanations rises accordingly. There is a need, therefore, for analytical methods that provide some consistency and rigor to this process.

These analytical tools are important to the practitioner because the structure-behavior link is the key to finding leverage points for policy initiatives. And they are important to the theorist because a system dynamics theory of a particular phenomenon is an account of how certain feedback loops cause certain dynamic patterns of behavior to appear. The qualitative understanding of the model behavior is often at least as important as the particular numerical predictions obtained, even in applied studies. Yet the rigor of such an account depends directly on the rigor with which structure-behavior link can be made in a given model.

The classical disciplines of linear systems theory and control engineering have provided a set of concepts and tools, particularly system eigenvalues and eigenvectors, that can also be applied under many circumstances to the nonlinear models found in system dynamics, not as a complete theory but as a pragmatic aid. This entry reviews the recent advances in analytical tools based on linear systems theory and discusses its future potential for both the system dynamics practitioner and the theorist.

Though we strongly believe in the utility of these methods, it is important to realize that advances in nonlinear dynamics and complexity theory in recent decades have shown that it is not possible to construct a complete theory of dominant structure because nonlinear systems are capable of exceedingly complex and intricate behavior that is impossible to predict without actually simulating the system. Furthermore, applications of graph theory to system dynamics models have revealed that the concept of feedback loops has some inherent problems and limitations because there are potentially many different loop descriptions of the same system (see Kampmann 2012; Oliva 2004). Thus, the analytical tools should be viewed as pragmatic aids to model analysis that can guide the modeler's intuition, rather than universal methods that provide automatic answers.

We first provide a brief historical introduction to the different ways scholars have thought about the notion of dominant structure, including an example of the traditional approach to structural analysis. In the next section we present the formal mathematical representation of linear and nonlinear systems and how one may describe the dynamic behavior in terms of behavior modes and system eigenvalues. In the four following sections we present alternative approaches to performing this analysis. We conclude with a summary of the current state of research and a discussion of future directions.

Introduction

Understanding model behavior is closely related to the process of model testing and validation, for which there is a well-established tradition and an extensive literature in the field (e.g., Barlas 1989; Ford 1999; Forrester and Senge 1980; Morecroft 1985; Richardson1986, 1995). Indeed there is no sharp line between model building, testing, validation, and analysis – in practice, the analyst undertakes all these processes simultaneously (Forrester and Senge 1980).

Of particular concern is whether one can identify pieces of structure that are in some sense "important" in generating the observed behavior of interest. Traditionally, system dynamics analysts have relied on trial-and-error simulation to discover these structures, by changing parameter values or switching individual links and feedback loops on and off. The tradition is well developed and includes a set of principles for partial model formulation and testing based on the organizational theory of bounded rationality (Homer 2012; Morecroft 1985).

The intuition guiding this effort often relies on simple feedback systems with one or a few state variables, where the behavior is fully documented and understood. In particular, the modeler uses well-understood "generic structures" that seem to appear again and again in system dynamics models, such as "overshoot and carrying capacity collapse," and "drifting goal structure" (see Lane and Smart 1996; Senge 1990; Wolstenholme 2004 for an account of these structures). Clearly such structures can be a useful aid to understanding if the model is sufficiently simple to allow such simple structures to be identified.

A simple example of a generic structure is the classical model of diffusion, sometimes known as the Bass model (Bass 1969, see also Chap. 9 in Sterman 2000). The model structure is illustrated in Fig. 1, and the resulting behavior, an s-shaped growth curve, is illustrated in Fig. 2. The idea behind the model is that the adoption of a new

technology is driven by the number of users that have already adopted it, through a word-of-mouth effect. One may interpret the s-shaped behavior as the interaction of two feedback loops, namely loop no. 2, the positive "word-of-mouth," and loop no. 1, the negative "exhaustion" loop (see Fig. 1). In the beginning, the positive loop dominates, leading to exponential growth in the number of adopters. Later, however, the negative loop gains strength, and the behavior shifts to an exponential adjustment toward the eventual market saturation. Thus, the traditional feedback loop analysis helps give an intuitive understanding of the dynamics of the model.

In large-scale models with perhaps hundreds of state variables, however, the traditional approach shows significant limitations. In practice, model building and analysis is often done using a "nested" partial model testing approach where one goes from the level of small pieces of structure to entire subsystems of the model, with frequent reuse of known formulations and partial models. Although this approach does carry a long way, it can be very difficult to discover feedback mechanisms that transcend model substructures in ways not anticipated by the modeler in the original dynamic hypothesis. Thus, there is a danger that observed behavior is falsely attributed to certain feedback mechanisms when in fact another set of feedbacks is driving the outcome. Likewise, one may make false inferences about how a particular feedback mechanism modifies the behavior, e.g., whether it attenuates or amplifies a particular oscillation.

Modern software packages can run extensive tests for sensitivity and "reality checks" where a large number of parameters are varied simultaneously (Peterson and Eberlein 1994). This is clearly a significant improvement over "manual" trial-and-error methods, particularly when these methods are combined with statistical inference methods such as Kalman filtering or Monte Carlo maximum likelihood estimation (Eberlein 1986; Eberlein and Wang 1985; Oliva 2003; Peterson 1980; Radzicki 2004; Schweppe 1973). A variant of this approach involves using statistical experimental design and correlation methods to screen for significant model structure (parameters), as

**Analytical Methods for
Structural Dominance
Analysis in System
Dynamics, Fig. 1** The
Bass model of diffusion

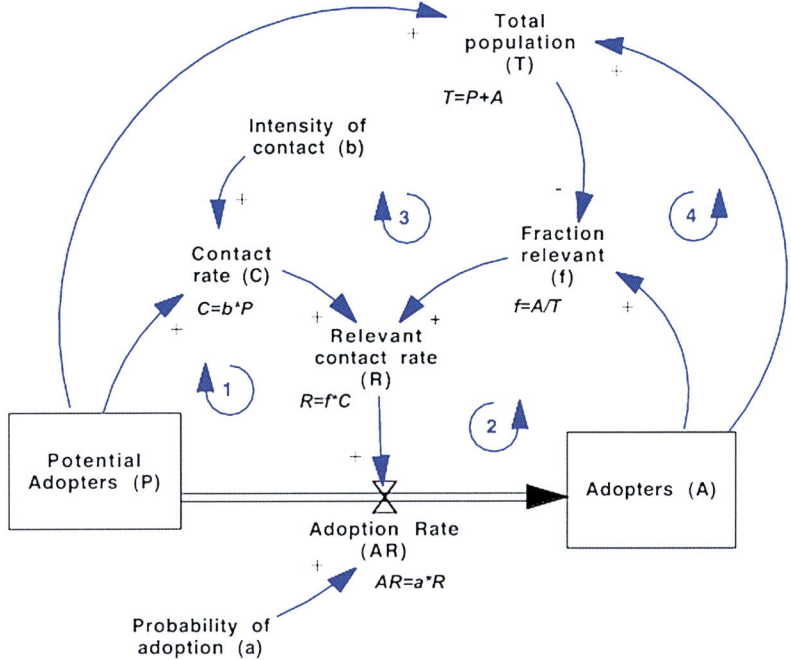

suggested by Ford and Flynn (2005). Indeed, the prospects of marrying such methods with modern search and optimization methods like classifier systems (Holland 1992) or genetic algorithms (Goldberg 1989) seem very promising. However, these methods are more addressing issues in estimation, validation, and testing than inferences about or understanding how (dominant) structure is causing behavior.

Richardson (1986) suggested a taxonomy of approaches to the notion of dominant structure, where he distinguishes along three dimensions, namely linear vs. nonlinear systems, model reduction vs. loop contribution, and the characterization of behavior in terms of time graphs vs. eigenvalues or frequency response. Of these, the distinction between model reduction and loop contribution is the most important.

In the model reduction approach, the idea is to replace a large complicated model with a simplified smaller model that captures the "essence" of the dynamics. A good example of this is Sterman's simple model of the economic long wave (Sterman 1985), which was distilled from the much larger System Dynamics National Model (Forrester et al. 1976). Eberlein (1984,

1989) attempted to tackle model simplification in a systematic way in linear systems by focusing on retaining specific behavior modes. In large part his results were negative: it is generally not possible to build simpler models that reproduce the salient behavior without sacrificing either the accuracy of the behavior or the ability to relate the simplified model variables to those in the full model. It is fair to say that this line of inquiry has largely been abandoned as a result. Extracting the "essence" of a model remains an art more than a science.

The focus here will be on Richardson's second category, the loop contribution or, more generally, the *structure contribution* approach. It reflects the intuitive idea that if one removes the element under consideration, e.g., by weakening a link or switching off a feedback loop, and the behavior then "disappears," one would say that the element in some sense "causes" the observed behavior.

This notion underlies the traditional trial-and-error simulation approach, sometimes supplemented with methods from the classical control engineering, which focuses on how structural elements modify the behavior of the system, viewed in terms of the frequency response. Typically, the

**Analytical Methods for
Structural Dominance
Analysis in System
Dynamics,
Fig. 2** Behavior of the
Bass model

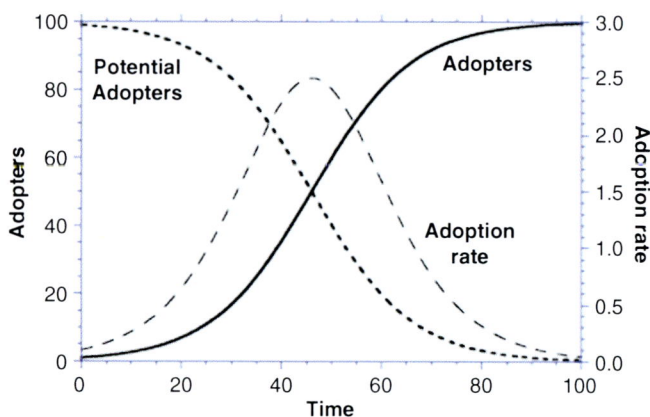

method works "backwards" by starting with simple feedback systems of single loops and then considering the marginal effect of adding links and loops. We discuss this approach in section "Traditional Control Theory Approaches."

If, instead, one considers marginal (infinitesimal) changes in structure, e.g., in the strength of a particular link, it is possible to derive rigorous analytical results for the resulting change in behavior expressed as the eigenvalues of the linearized model. One would then say that if a change in a system element has a relatively large effect upon the behavior pattern of interest, this element is "significant" in "causing" the behavior. This is what underlies the *eigenvalue elasticity* and *eigenvector* approaches discussed in sections "Eigenvalue Elasticity Analysis" and "Eigenvectors and Dynamic Decomposition Weights (DDW)." The marginal and experimental approaches may supplement each other well, where a marginal analysis may identify elements that can then be tested experimentally for their significance.

Unlike the traditional control method and the eigenvalue method that work in the structural and *frequency domain*, the *pathway participation* method (PPM) relates directly to the time path of particular system variables and is more concerned with the qualitative nature of the time path, expressed in terms of signs of the slope (whether growing or declining) and curvature (whether convex or concave) than with numerical measures of degree of influence. Briefly stated, the PPM traces the causal links in the variables influencing

the system variable in question and then identifies the most important chain of links. We discuss this method in section "Pathway Participation Metrics."

Common to the approaches discussed here is that they all build upon a precise mathematical characterization of the system behavior (see Duggan and Oliva 2013 for an overview of other approaches to perform structural dominance analysis). In the next section, we demonstrate how the concepts from linear systems theory may be used to give a precise characterization of behavior in terms of component *behavior modes*.

Characterizing Linear and Nonlinear Systems

A system dynamics model can be represented mathematically as a set of ordinary differential equations

$$\frac{d\boldsymbol{x}(t)}{dt} \equiv \dot{\boldsymbol{x}}(t) = \boldsymbol{f}(\boldsymbol{x}(t), \boldsymbol{u}(t)), \qquad (1)$$

where $\boldsymbol{x}(t)$ is a (column) vector of n state variables (levels) $x_1(t), \ldots, x_n(t)$, $\boldsymbol{u}(t)$ is a column vector of p exogenous variables or control variables $u_1(t), \ldots, u_p(t)$, $\boldsymbol{f}$ is a corresponding vector function, and t is simulated time. In this entry, we restrict our attention to the state variables (levels) of the model for notational convenience, ignoring the auxiliary variables. Mathematically, a model can always be brought to the *reduced* form (1), but in practice, the auxiliary variables give a more intuitive account of the analysis. Likewise, we do not consider time-varying systems (where time t enters as an explicit argument in the function f), since these can usually be accommodated by an appropriate definition of the exogenous variables u. In general, f is a nonlinear function of its arguments, and we speak of a *nonlinear* system. Conversely, if f is a linear function, we speak of a *linear* system.

Figure 3 and Table 1 show a well-known example, the inventory-workforce model. It has three state variables, inventory (INV), workforce (WF), and expected demand (ED), and one exogenous variable, demand (DEM), i.e.,

$$x(t) = \begin{pmatrix} \text{INV} \\ \text{WF} \\ \text{ED} \end{pmatrix}, u(t) = (\text{DEM}), \qquad (2)$$

and the function f is determined by the equations in Table 1.

Given the model structure (1), knowledge of the initial conditions $x(0)$, and the path of the input variables $u(t)$, the behavior of the model is completely determined. It is in this sense that the model structure (1) constitutes a "theory" of the time behavior $x(t)$, as mentioned in the introduction. Yet, we are interested in methods that yield a more compact explanation, short of having to simulate the entire model structure.

It turns out that in its ultimate form, this dream is beyond reach: Since the days of Henri Poincaré, mathematicians have known that it is impossible to find general analytical solutions to nonlinear systems. Furthermore, the development of nonlinear dynamics and chaos theory has proven that such systems, even when they have very few state variables, can produce highly complex and intricate behavior that goes beyond general analytic methods (e.g., Ott 1993; Richardson 1988). Thus, we will never find a final general theory where we can infer the behavior of the system directly from its structure; instead, we will always have to rely on simulation to discover the dynamics implied by the structure. (This is not to say that no general analytical results exist in nonlinear systems. The field of chaos theory has uncovered a number of universal features, e.g., relating to the transition from periodic or quasiperiodic behavior to chaos, where the transitions show both qualitative and quantitative similarities that are independent of the specific forms of the model equations (see, e.g., Ott 1993). However, these universal features relate to specific situations such as period-doubling or intermittency routes to chaos).

Analytical Methods for Structural Dominance Analysis in System Dynamics, Fig. 3 Flow diagram of the inventory workforce model

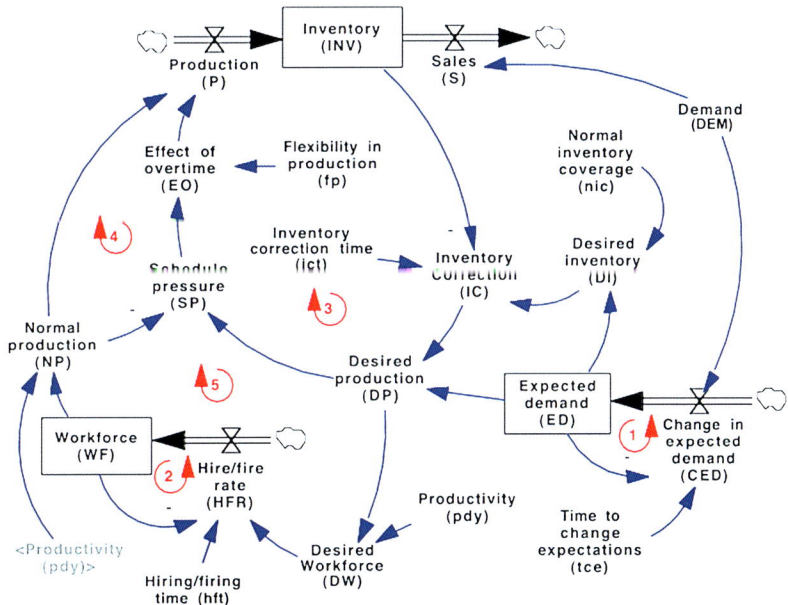

Analytical Methods for Structural Dominance Analysis in System Dynamics, Table 1 Equations of the inventory workforce model

Equation	Name	Units
$d/dt(\text{ED}) = (\text{DEM} - \text{ED})/\text{tce}, \text{ED}_0 = \text{DEM}^*\text{df}$	Expected demand	[Units/Month]
$d/dt(\text{INV}) = \text{P} - \text{S}, \text{INV}_0 = \text{DI}$	Inventory	[Units]
$d/dt(\text{FW}) = \text{HFR}, \text{FW}_0 = \text{DWF}$	Workforce	[Workers]
S = DEM	Shipments	[Units/Month]
P = NP*EO	Production	[Units/Month]
EO = fp*(1 − (1 − 1/fp)^SP)	Effect of overtime	[Dimensionless]
NP = WF*pdy	Normal production	[Units/Month]
DI = ED*nic	Desired inventory	[Units]
SP = DP/NP	Schedule pressure	[Dimensionless]
DEM = 1(Exogenous)	Demand	[Units/Month]
DP = ED + IC	Desired production	[Units/Month]
IC = (DI − INV)/ict	Inventory correction	[Units/Month]
HFR = (DWF − WF)/hft	Hire/fire rate	[Workers/Month]
DWF = DP/pdy	Desired workforce	[Workers]
hft = 5	Hire/fire time	[Month]
pdy = 1	Productivity	[Units/Month/Worker]
tce = 4	Time to change expectations	[Month]
fp = 1.05	Flexibility in production	[Dimensionless]
nic = 3	Normal inventory coverage	[Months]
ict = 2	Inventory correction time	[Months]
df = 0.5	Disequilibrium fraction	[Dimensionless]

The best we can hope for, therefore, is a set of tools that will guide intuition and help identify *dominant structure* in the model. By dominant structure we mean particular feedback loops that are in some sense "important" in shaping the behavior of interest. To the extent that we can identify such dominant structures, we may say that we have found a "theory" of the observed behavior.

Although the term "behavior" may appear rather loose, experience and reflection tells us that there is a limited number, perhaps a dozen or so, of relevant behavior patterns that dynamical systems can exhibit. Some of these behaviors, like exponential growth, exponential adjustment, and damped or expanding oscillations, are typical of linear systems. Others, like limit cycles, quasiperiodic motion, mode-locking, and chaos, can only be exhibited by nonlinear systems.

Common to the approaches considered in this entry is that they are based on tools from linear systems theory, i.e., they approximate the

nonlinear model (1) with a linearized version, using first-order Taylor expansion around some operating point x_0, u_0, i.e.,

$$\dot{x}(t) \approx f(x_0, u_0) + \frac{\partial f}{\partial x}(x - x_0) + \frac{\partial f}{\partial u}(u - u_0),$$

$$(3)$$

or, by redefinition of the variables $x \rightarrow x - x_0$, $t \rightarrow t - t_0$ and $u \rightarrow u - u_0$,

$$\dot{x}(t) \approx Ax(t) + Bu(t), \qquad (4)$$

where A is a constant $n \times n$ matrix of partial derivatives $\partial f_i / \partial x_j$ and B is a constant $n \times p$ matrix of partial derivatives $\partial f_i / \partial u_j$, where all partial derivatives are evaluated at the operating point.

For the linear system (4), there is a well-developed and extensive theory of the system behavior as a function of its structure, expressed in the matrices A and B. One may broadly distinguish two parts of the theory, named classical control theory (e.g., Ogata 1990) and modern linear systems theory (e.g., Chen 1970; Luenberger 1979). We return to the classical control theory in the next section.

Modern control theory or linear systems theory (LST) is concerned with the dynamical properties of the system as a direct function of the system matrices A and B. A key element in this theory is the notion of the system *eigenvalues*, i.e., the eigenvalues of the matrix A. If, for simplicity, we restrict ourselves to the endogenous dynamics of the system (set $u = 0$), we can write the solution to (4) as

$$x_i(t) = c_{i1}\exp(\lambda_1 t) + c_{i2}\exp(\lambda_2 t) \\ + \cdots + c_{in}\exp(\lambda_n t), \quad i = 1, \ldots, n \quad (5)$$

where $\lambda_1, \ldots, \lambda_n$ are the n eigenvalues of the matrix A and c_{ij} are constants that depend upon the eigenvectors and the initial condition of the system. In other words, the resulting behavior is a weighted sum of distinct *behavior modes*, $\exp(\lambda t)$. If an eigenvalue is real, the corresponding behavior mode is exponential growth (if $\lambda > 0$) or exponential decay (if $\lambda < 0$). Complex-valued eigenvalues come in complex conjugate pairs $\lambda = \tau \pm i\omega$ which give rise to oscillations $\exp(\tau t) \sin(\omega t + \phi)$ of frequency ω that are either expanding (if $\tau > 0$) or damped (if $\tau < 0$). In this manner, the eigenvalues serve as a compact and rigorous characterization of the behavior (of linear systems).

At any point in time, any system, linear or nonlinear, may be approximated by the expression (5). Whether it remains a good approximation depends upon how much and how quickly the eigenvalues change due to the nonlinearities in the function f. If they are more or less constant for significant periods of time, we may speak of *quasilinear systems* that are well approximated by the linear system. In some cases, however, the eigenvalues change so rapidly that it makes little sense to characterize the behavior by Eq. 5 (See Kampmann and Oliva 2006 for further discussion).

Traditional Control Theory Approaches

The first set of methods, which we call the traditional approach, has been used for decades and is part of the standard curriculum in system dynamics teaching at the graduate level. It involves using the concepts from classical control theory (Ogata 1990) to very simple systems with only a few state variables.

The starting point is the simple first- and second-order positive and negative feedback loops found in any introductory treatment of system dynamics. The advantage of the approach is its simplicity. Although it serves as a guide to intuition, however, the obvious shortage is that it applies rigorously only to simple systems. There have been some attempts to treat higher-order systems by adding a few feedback loops (Graham 1977), but the step to large-scale models is beyond this method given its inherent limitations.

Graham (1977) distills a number of principles that are based on the metaphor of a "disturbance" traveling along the chain of causal links in a feedback loop and getting amplified, damped, and possibly delayed in the process. For major negative feedback loops, which are known to tend to produce oscillation, adding minor negative loops and cross-links, or shortening the delay times increases the damping. Conversely, adding positive loops into the oscillatory system tends to lengthen the period of oscillation whereas the effect on the damping depends upon the delays in the positive loop. Using the metaphor of pushing a child on a swing, it becomes clear that the timing of the propagation of a disturbance has as much importance for its effect on the damping as its strength.

For analyzing the behavior of positive feedback loops, Graham suggested calculating the open-loop steady-state gain (OLSSG), a measure of the amplification around the loop. A gain greater than unity will result in exponential growth, while gains less than 1 will give exponential adjustment (leveling off or decay). The intuition is perhaps best illustrated by an example: sales-driven growth. Suppose a salesperson can eventually pull in $100,000 per month in orders (probably with a several-month long delay) and assume that the company allocates 10% of revenue to marketing. This eventually leads to $10,000 per month for sales efforts. If the cost of a salesperson (salary, overhead, expenses, etc.) is, say, $8,000 dollars per month, then the efforts of the current sales force will provide enough revenue to support $10,000/8,000 = 1.25$ persons per current person. Thus, the OLSSG of the positive loop from salespersons $\rightarrow$ orders $\rightarrow$ revenues $\rightarrow$ marketing budget $\rightarrow$ salespersons is 1.25, and the system will grow exponentially (until other factors limit the growth). Conversely, if the gain is less than 1, one salesperson will not sell enough to support their own cost, and the loop will lead to exponential decay. Graham showed how the actual rate of growth is partly determined by the OLSSG, and partly by the time constants (delays, etc.) involved (See also Sect. 15.3 in Sterman 2000).

In the context of oscillating systems, system dynamics has also employed a concept from classical control theory, frequency response. The frequency response is determined from the transfer function of the system, $G(i\omega)$, which is a complex-valued function that specifies how an input signal $u(t)$ with frequency ω results in an output signal $x(t)$ that may be phase shifted (delayed), and either amplified or attenuated. For linear systems, G can be calculated directly from the system matrices in (4) – the transfer function (matrix) is $G(i\omega) = B(i\omega I - A)^{-1}$, where I is the identity matrix (see e.g., Chen 1970). For nonlinear systems, G may be found through simulation experiments.

Usually, G is represented in a *Bode* or *phase-and-gain* diagram. For instance, Fig. 4 shows a Bode diagram of the inventory variable INV(t) relative to the exogenous demand input variable DEM(t) in the inventory workforce model in Fig. 3. The diagram shows how the relative amplitude of the oscillation and the relative phase shift (in radians) between input and output varies as a function of the frequency of the input.

It is clear from the diagram that there is a certain frequency range, around the system's own natural frequency, where fluctuations in demand are greatly amplified compared to other frequencies. Indeed, it is a general phenomenon in systems that they will tend to amplify certain frequencies while attenuating other frequencies. This may be used to explain or understand the role of particular structures in the model in generating oscillation at certain frequencies, even when there are no oscillations coming in from the outside world. (External random noise is enough to produce oscillations in the system because random noise contains fluctuations at all frequencies.) In this manner, the approach nicely demonstrates the "endogenous viewpoint" that behavior (oscillations) is generated internally by the system. As an analytic tool for large-scale systems, however, the method does not seem to produce any additional insights. Thus, we may conclude that the classical approaches serve mostly as intuitive metaphors to guide the analyst rather than as full analytical tools.

Pathway Participation Metrics

The *pathway participation* method (Mojtahedzadeh 1996; Mojtahedzadeh et al. 2004) represents a further development of an original suggestion by Richardson (1995) to provide a rigorous definition of loop polarity and loop dominance. Richardson motivated this with the common confusion associated with positive feedback loops, which may exhibit a wide range of behaviors (Graham 1977), as Barry Richmond noted with wonderful humor:

> Positive loops are ... er, well, they give rise to exponential growth ... or collapse ... but only under certain conditions ... Under other conditions they behave like negative feedback loops .... (Richmond 1980)

Richardson proposed that the polarity of a loop be defined as the sign of the expression

$$\frac{\partial \dot{x}_i}{\partial x_i} = \frac{\partial f_i(x, u)}{\partial x_i}, \qquad (6)$$

in the model (1), with a positive sign indicating a positive loop and vice versa. When several loops operate simultaneously, the sign of the expression indicates whether the positive or negative loops dominate. Note, however, that the definition only applies to minor loops (i.e., loops involving a single level). Put differently, it only considers the diagonal elements of the matrix A in the linearized system (4). Richardson (1995) demonstrates how

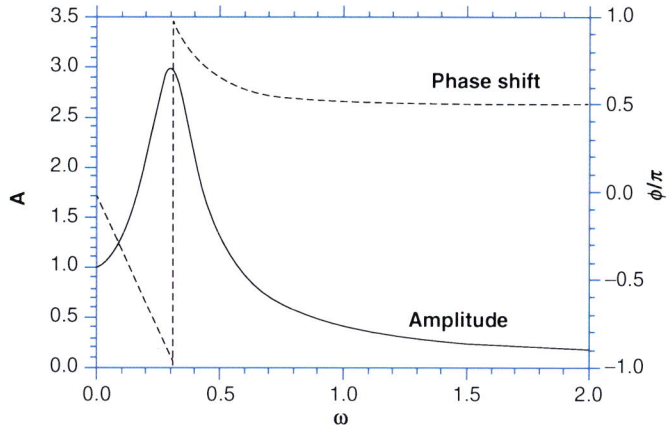

Analytical Methods for Structural Dominance Analysis in System Dynamics, Fig. 4 Phase-and-gain diagram (Bode diagram) showing the inventory $A \sin(\omega t + \phi)$ with amplitude A and phase shift ϕ, relative to a sinusoidal demand $\sin(\omega t)$, for varying values of the frequency ω of the demand fluctuation

even with this limitation, analyzing the system with this metric can (sometimes) yield insights into behavior of higher-order systems.

The expression (6) hints that it is relevant to consider the curvature, i.e., the second time derivative, $\ddot{x}$, of a variable when looking for dominant structure. Although he does not say so explicitly, this is effectively the focus of Mojtahedzadeh's pathway method. Figure 5 shows how one may classify behavior by comparing the first and second time derivatives of a variable. As seen in the figure, the sign of the expression $\ddot{x}/\dot{x}$, which Mojtahedzadeh denotes the total *pathway participation metric* or *PPM*, indicates whether the behavior appears dominated by positive or negative loops, much in line with Richardson's definition of dominant polarity. A zero curvature indicates a shift in loop dominance (cf. the middle column in the figure). Note, however, that the interpretation of the middle row in the figure where the slope $\dot{x}$ is zero has no clear interpretation in terms of loop dominance. Indeed this hints at one of the weaknesses of the approach that we will return to below.

Mojtahedzadeh's method proceeds by decomposing the PPM into its constituent terms as follows,

$$PPM_i = \frac{\ddot{x}_i}{\dot{x}_i} = \sum_j \frac{\partial f_i}{\partial x_j} \frac{\dot{x}_j}{\dot{x}_i}, \qquad (7)$$

where, for brevity, we have chosen to ignore the exogenous variables u. One might say that each of the terms in the sum in (7) represents the separate influence of each of the systems' state variables on the behavior of x_i. Mojtahedzadeh in fact uses a normalized measure for the terms,

$$\frac{(\partial f_i/\partial x_j)\dot{x}_j}{\sum\limits_{k=1}^{n} |(\partial f_i/\partial x_k)\dot{x}_k|}, \qquad (8)$$

which can vary between -1 and $+1$, to measure the relative importance of the pathway from variable j. By explicitly considering auxiliary variables y in the model, one may further decompose each term $\partial f_i/\partial x_j$ into a sum of terms

$$\frac{\partial f_i^k}{\partial x_j} = \frac{\partial f_i}{\partial y_1} \cdot \frac{\partial y_1}{\partial y_2} \cdots \cdot \frac{\partial y_{m-1}}{\partial y_m} \cdot \frac{\partial y_m}{\partial x_j}, \qquad (9)$$

corresponding to a causal chain or pathway $\pi_k = \{x_j \rightarrow y_m \rightarrow \cdots y_2 \rightarrow y_1 \rightarrow \dot{x}_i\}$. Mojtahedzadeh now considers each possible pathway (9) and defines the "dominant" pathway as the one with the largest numerical value and the same sign as PPM_i. Having selected this dominant pathway, $\pi_{ij}^* = \{x_j \rightarrow y_m \rightarrow \cdots y_2 \rightarrow y_1 \rightarrow \dot{x}_i\}$, which originates in the state variable x_j, the procedure is repeated for that state variable x_j, and so forth, until one either reaches one of the already "visited" state variables (in which case a loop has been found) or an exogenous variable (in which case an external driving force has been found). Thus, the procedure may result in three alternative forms of dominant

Analytical Methods for Structural Dominance Analysis in System Dynamics, Fig. 5 Characteristic behavior patterns based on the first and second time derivatives

structure illustrated in Fig. 6, namely a "pure" minor or major feedback loop, a pathway from a feedback loop elsewhere in the system, or a pathway from an exogenous variable.

By dividing the observed model behavior into different phases according to the taxonomy in Fig. 5 and then applying the method just described at different points in during these phases, one can reveal how the dominant structure changes over time. For illustration, the PPM method is applied to the Bass model and the results are presented in Fig. 7. The figure shows the metrics of four alternative pathways (four feedback loops) and the results accord nicely with the informal analysis done earlier (cf. Fig. 1): The method identifies two phases, exponential growth, exponential adjustment, and identifies the "word-of-mouth" positive loop (loop 2) as dominant in the first phase and the "exhaustion" loop (loop 1) as dominant in the second phase.

The PPM method is still mostly used at an early explorative stage on rather simple models, where

it does appear to aid insight into the dynamics (e.g., Oliva and Mojtahedzadeh 2004), and has been implemented in a software package, *Digest* (Mojtahedzadeh et al. 2004).

From the studies performed so far, it is clear that the main strength of this method is its relative computational simplicity (it does not require computing eigenvalues, which is a numerically demanding task), and the intuitive and direct connection it makes between the observed behavior and the influencing structural elements. Unlike the other approaches which operate in the "frequency domain," the method considers the time path of a specific variable directly.

There are, however, some important outstanding issues that remain to be clarified. First, the method is not suitable for oscillatory systems. The problem is easy to recognize when one considers how the PPM measure will vary over the course of a sinusoidal outcome: The sign of the PPM will shift twice during each cycle, indicating

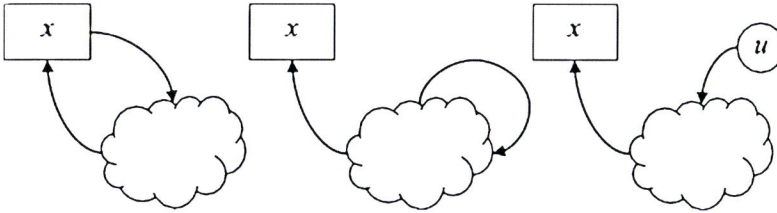

Analytical Methods for Structural Dominance Analysis in System Dynamics, Fig. 6 Three alternative forms of dominant structure in the PPM method

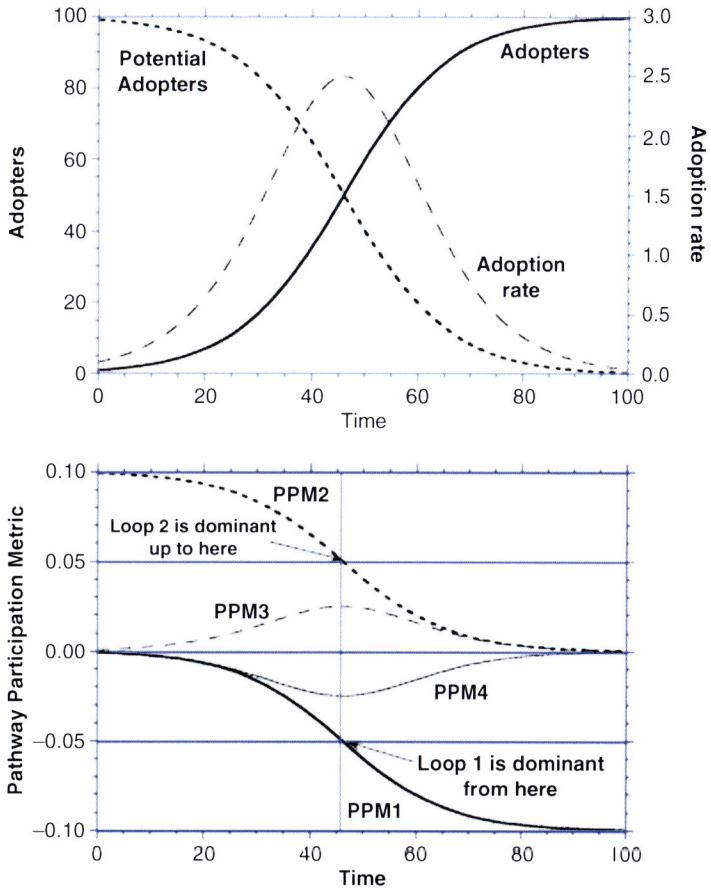

Analytical Methods for Structural Dominance Analysis in System Dynamics,
Fig. 7 Pathway participation measures in the Bass model

that the behavior is alternately dominated by positive and negative loops, even though the system structure, and hence the loop dominance, may remain unchanged all the time. Richardson (1995) already alluded to this problem by noting that the measure only considers the diagonal elements in the system matrix in (4), yet we know that the structure causing oscillation is the major

negative loop that involve the off-diagonal elements. This is a significant limitation, given the prevalence and importance of oscillation in system dynamics analysis.

A second limitation of the current implementation of PPM is that it uses a depth-first search for the single most influential pathway for a variable. This strategy does not capture the situation where more than one structure may contribute significantly to the model behavior and, through the depth-first algorithm, may miss alternative paths that could prove to yield a larger total value of the metric. This problem could be addressed by modifying the search algorithm and is most likely of minor importance.

Another issue is how to treat the case when $\dot{x} = 0$ since it appears in the denominator of the terms in (6). However, it is not clear that it is necessary to do this division, given that it is easy to identify the nine cases in the figure by simply examining its sign. Thus, the issue is probably not of much significance.

The fourth issue, on the other hand, is more significant, namely the emphasis on identifying a single "dominant" structure. In reality, of course, the behavior of a variable is influenced by many loops and pathways at once. Reducing the consideration to a single one of these may miss important features of the structure-behavior relationships. For instance, a variable may be influenced by two negative loops and one positive, with the sum of the two negative loops dominating the influence of the positive loop, even though that loop by itself has the strongest influence on the behavior. It is more appropriate to consider the relative importance of alternative pathways, yet the method does not address how one would partition the behavior among pathways (the three structures in Fig. 6) – only among individual links. Recently, Hayward and Boswell (2014) have worked around this limitation by simplifying the PPM method and developing an algorithm that identifies the minimum combination of loops of like polarity that is larger than the combination of all loops in the opposite polarity. Thus, while the notion of pathways seems an interesting and useful idea, it may be that it will ultimately be more effective to use a list, ranked in order of magnitude, of the pathways that influence a variable.

Finally, the method shares a weakness with the traditional method in that it considers primarily partial system structures rather than global system properties. In contrast, the two eigenvalue methods to which we now turn are based on a rigorous characterization of the entire system (at a given point in time).

Eigenvalue Elasticity Analysis

The third method may be termed *eigenvalue elasticity analysis* (or EEA for short) and builds upon the tools from modern linear systems theory (LST), applied to the linearized model (4). The method is concerned with the structural elements that significantly affect the system eigenvalues or behavior modes – the values λ in (5). Specifically, it measures the influence of a structural model parameter g on an eigenvalue λ by the elasticity, defined as $\varepsilon = (\partial\lambda/\partial g)(g/\lambda)$, i.e., the fractional change in the eigenvalue relative to the fractional change in the parameter. The advantage of this fractional measure is that it is dimensionless, i.e., independent upon the choice of units, including the time scale unit. Sometimes, the influence measure is used instead, defined as $\mu = (\partial\lambda/\partial g)g$. This measure has dimension [1/*time*] and so depends upon the choice of its time unit, but it is generally easier to interpret for complex-valued eigenvalues and avoids numerical problems with very small or zero eigenvalues (see Kampmann and Oliva 2006; Saleh et al. 2010).

The idea behind EEA was first introduced in system dynamics by Forrester (1982) in the context of economic stabilization policy. For purposes of policy analysis in oscillating systems, one may define a number of criteria from engineering control theory, all of which relate to the eigenvalues of the system, as summarized in Table 2. Figure 8 provides a graphical characterization of the eigenvalues and policy criteria in the complex plane. Though these measures are not new, the EEA method is unique in its attempt to use them to gain qualitative intuitive understanding of the system. A significant step in this direction was first suggested by Forrester (1983) with the notion that the elasticities of any links in the model (corresponding to elements of the matrix A in the linearized system (4)) can be interpreted as the sum of elasticities of all feedback loops containing that link. We have chosen to name this approach *loop eigenvalue elasticity analysis (LEEA)*.

Kampmann (2012) provided a rigorous definition of LEEA and also pointed to the fact that

Analytical Methods for Structural Dominance Analysis in System Dynamics, Table 2 Stabilization policy criteria and corresponding effects on eigenvalues and DDW of a policy change in a system element g

Policy criterion	Description	Change in eigenvalue $\lambda = \delta \pm i\omega$, $\omega > 0$	Change in DDW w	Appropriate measure in time path
Damping	Increases the rate of decay of oscillation (or decreases the rate of expansion)	$\frac{\partial \delta}{\partial g} \frac{g}{\delta} < 0$	N/A	$\frac{x(t+T)}{x(T)}$
Frequency	Decreases the frequency of oscillation (or lengthens the period T)	$\frac{\partial \omega}{\partial g} \frac{g}{\omega} < 0$	N/A	T
Variance	Reduces the variance of a target variable (or the weighted average variances of several variables)	No simple relation	$\frac{\partial w}{\partial g} \frac{g}{w} < 0$	$\int x(t)^2 dt$
Auto-spectrum	Reduces variance of target variable(s) within a target frequency range	No simple relation	$\frac{\partial w}{\partial g} \frac{g}{w} < 0$	Filter in frequency domain
Frequency response gain	Reduces the gain (amplification) in the target frequency range for a particular combination of disturbance exogenous and output variables	Based upon transfer function $G(i\omega)$		

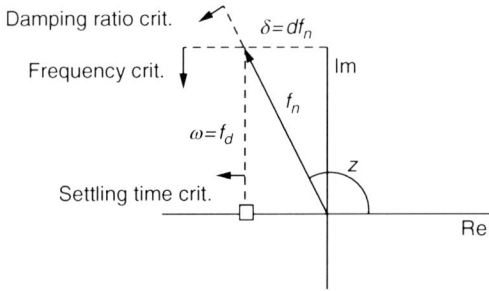

Analytical Methods for Structural Dominance Analysis in System Dynamics, Fig. 8 Characterization of eigenvalues plotted in the complex plane

feedback loops are not independent. In other words, given the possibly very large number of loops in a given model (Kampmann demonstrated how the theoretical maximum number of loops grows combinatorially with the number of variables), it only makes sense to speak of individual contributions of a limited set of *independent* loops. He proved that a fully connected system (where there is a feedback loop between any pair of variables – the typical case in system dynamics models) with N links and n variables has a total of $N - n + 1$ independent loops and provided a procedure for constructing this set and calculating the loop elasticities.

Kampmann's analysis points to a fundamental issue relating to the notion of feedback loops as a way to explain behavior: the significance assigned

to a particular loop depends upon the context (the chosen independent loop set). In other words, feedback loops are derived and relative concepts rather than fundamental independent building blocks of systems. Oliva (2004) further refined the definition of independent loop sets by introducing the *shortest independent loop set (SILS)* along with a procedure for constructing the set. Although a SILS is not generally unique, experience seems to suggest that it is easier to interpret (Oliva and Mojtahedzadeh 2004). Yet the issue remains that independent feedback loop sets are relative concepts.

In Table 3, we show how the LEEA analysis applies to the simple inventory-workforce model in Fig. 3. The model contains a total of five feedback loops, all of which are independent. The loops are listed in Table 3, including their constituent variables (nodes), and the gain of the loop (defined in a similar manner to the pathway participation metrics above). We see that there are three minor negative loops, related to the exponential smoothing of expected demand (loop 1) and the adjustment of workforce to desired workforce (loop 2). The minor loop 3 is the "overtime shortcut" that allows production to adjust part way to desired production immediately so one does not have to wait for the workforce to adjust. Loop 4 is the main major negative loop that adjusts inventory to desired levels via workforce

Analytical Methods for Structural Dominance Analysis in System Dynamics, Table 3 Loops and their influences in the inventory workforce model. Values are measured at time $t = 0$. The model contains three

eigenvalues, $\lambda_1 = -0.250$ and λ_2, $\lambda_3 = -0.138 \pm i\,0.285$. The influence measure is defined as $g \cdot \partial \lambda / \partial g$. For the imaginary part, a positive influence measure means that the frequency is increased

Loop	Nodes	Gain	Influence on Re[λ_1]	Influence on Re[λ_2]	Influence on Im[λ_2]
1	ED > CED	−0.250	−0.250	0.000	0.000
2	W > HFR	−0.200	0.000	−0.100	−0.022
3	INV > IC > DP > SP > EO > P	−0.076	0.000	−0.038	0.008
4	INV > IC > DP > DW > HFR > W > NP > P	−0.100	0.000	0.000	0.176
5	INV > IC > DP > DW > HFR > W > NP > SP > EO > P	0.015	0.000	0.000	−0.027

adjustment. Finally, loop 5 (the only positive loop) is a fairly weak loop that moderates the effect of loop 4 by adjusting the overtime effect "back to normal" when the workforce is brought in line with desired production.

Although the model is nonlinear (due to the overtime function), the eigenvalues do not change very much over the course of its behavior. The model contains one real eigenvalue ($\lambda_1 = -0.250$) and one pair of complex conjugate eigenvalues (λ_2, $\lambda_3 = -0.138 \pm i\,0.285$). The first eigenvalue corresponds to the adjustment of expected demand (ED). The other pair produces a damped oscillation in inventory and workforce.

Table 3 also shows the loop influences upon the three eigenvalues. Note how there is a one-to-one correspondence between loop 1 (the adjustment of expected demand) and the first eigenvalue. This is due to fact that the ED level constitutes a single strongly connected component of the model (see Fig. 3), i.e., there is no feedback between this level and the rest of the model. We also note that the workforce adjustment and the overtime loops have a stabilizing influence upon the behavior (they make the real part of the oscillatory eigenvalues more negative and have relative little effect upon the frequency of oscillation). Conversely, the major negative loop 4 has a destabilizing influence, since strengthening it will increase the frequency of oscillation and not increase the damping. The effects of loop 5 are fairly weak.

From this analysis, one would therefore expect parameters that strengthen loop 2 (shortening hire/fire time) or loop 3 (increase overtime effect) would stabilize the system while strengthening loop 4 (shorter inventory adjustment time) will

destabilize the system. Indeed this is what happens, as illustrated in the simulations in Fig. 9.

The EEA/LEEA method has been applied in a number of contexts (e.g., Abdel-Gawad et al. 2005; Gonçalves 2003; Gonçalves et al. 2000; Güneralp 2006; Kampmann and Oliva 2006; Saleh and Davidsen 2001; Saleh et al. 2010) but remains a tool employed only by specialists in fundamental research, not least because it has not been incorporated into standard software packages. Thus, the potential of the method for widespread practice remains unexplored.

One might be skeptical that a method derived from linear systems theory may have any use for the nonlinear models found in system dynamics. Kampmann and Oliva (2006) considered what types of models the method would be particularly suited for. They defined three categories of models, based upon the behavior they are designed to exhibit: (1) linear and quasilinear models, (2) nonlinear single-transient models, and (3) nonlinear periodic models. The first category encompasses models of oscillations, possibly combined with growth trends, with relatively stable equilibrium points (e.g., the classical industrial dynamics models Forrester 1961). Nonlinearities may modify behavior (particularly responses to extreme shocks), but the instabilities and growth trends can be analyzed in terms of linear relationships. Kampmann and Oliva concluded that LEEA showed the most promise and potential for this class of models because the analytical foundations are solid and valid, and because the method has the ability to find high-elasticity loops even in large models very quickly without much intervention on the part of the analyst.

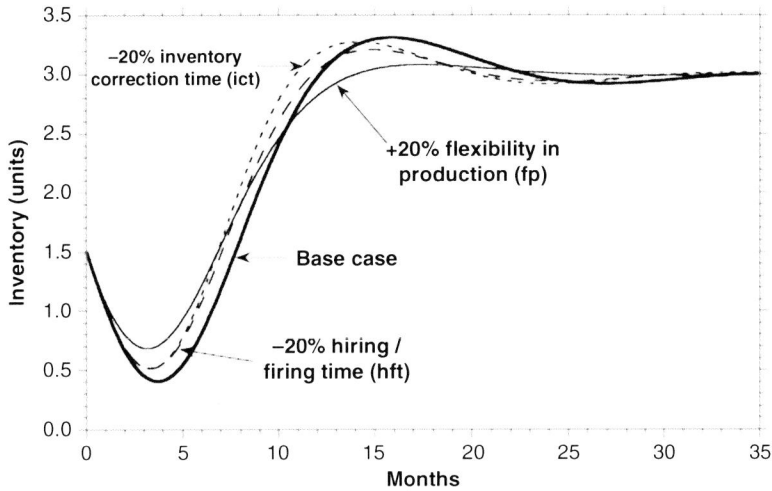

Analytical Methods for Structural Dominance Analysis in System Dynamics, Fig. 9 Simulated behavior of inventory-workforce model, showing the effect on inventory of parameter changes for overtime (flexibility of production), inventory adjustment time (ict), and labor hiring/firing time (hft), respectively

The second class is typical of scenario models like the World Model (Forrester 1971; Meadows et al. 1972), the Urban Dynamics Model (Forrester 1969), or the energy transition model in (Sterman 1981), to name a few, that show a single transient behavior pattern, like overshoot and collapse or a turbulent transition to a new equilibrium. In these models, nonlinearities usually play an essential role in the dynamics. Yet it is possible to divide the behavior into distinct phases where certain loops tend to dominate the behavior. In this class of models LEEA also shows promise by measuring shifts in structural dominance by the change in elasticities. But it requires more input from the analyst (e.g., in defining the different phases of the transition) and it has no obvious advantage over other methods, like PPM.

The third class, nonlinear periodic models, are those that exhibit fluctuating behavior in which nonlinearities play an essential role, such as like limit cycles, quasiperiodic behavior, or chaos (see, e.g., Richardson 1988). Here the utility of the method is much less clear and depends upon the specifics of the model in question. For example, the classic Lorenz model that exhibits limit cycles, period doubling, and deterministic chaos does not lend itself to any insight using LEEA (Kampmann and Oliva 2006). This is particularly

the case in systems with strong nonlinearities such as min and max functions. In these systems, the behavior may change abruptly (eigenvalues suddenly shift) in what is called border-collision bifurcations (Mosekilde and Laugesen 2006; Zhusubaliyev and Mosekilde 2003). In other cases, the method of breaking the behavior into phases with different dominant structures may yield significant insight from LEEA. For instance, Sterman's simple long wave model (Sterman 1985) lends itself well to this approach (e.g., Güneralp 2006; Kampmann 2012).

In the present entry, we add a fourth category of models or behavior for which the method has not been explored yet. We name this category nonlinear multimodal models. These encompass the cases where one behavior mode interacts with and therefore modifies another behavior mode – something that can only happen in nonlinear systems. The most common example is mode-locking or entrainment, in which oscillations become synchronized (e.g., Haxholdt et al. 1995). Another example is mode modification, where one behavior mode (growth or oscillation) affects the character of another (typically oscillation). An example of this is the interaction of the business cycle with the economic long wave, where the former tends to get more severe during long wave downturns (Forrester 1993). Whether LEEA can contribute to this class of models remains to be seen.

Compared to the former two methods, the EEA/LEEA is mathematically more general and rigorous, though many of the mathematical issues in the method remain to be addressed, as we summarize below. This rigor is also the main strength of the method, since it provides an unambiguous and complete measure of the influence of the entire feedback structure on all behavior modes.

A weakness or challenge that is starting to show up is the computational intensity in calculating eigenvalues and elasticities. This is not so much an issue of computer time and memory space as of the stability of numerical methods. Kampmann and Oliva (2006) found that the numerical method used sometimes proved unstable, yielding meaningless results. Clearly, there is a need to explore this issue further, possibly building upon the developments in control engineering.

A more serious weakness is the difficulty in interpreting the results: Eigenvalues do not directly relate to the observed behavior of a particular variable. The concepts of eigenvalues and elasticities are rather abstract and unintuitive (Ford 1999). There is a need for tools and methods that can translate them into visible, visceral, and salient measures. Here, the measures in Table 2 may provide a guide. In particular, it is possible to use (linear) filtering in the frequency domain to define a behavior of interest. For example, an analyst may be concerned with structures causing a typical business cycle (3–4-year oscillation) and, by specifying a filter that "picks out" that range of fluctuation, could obtain measures for structures that have elasticities in that range. Because filters are typically linear operators, all the analytical machinery of the LEEA method will also apply in this case – a significant advantage.

Using filters will also solve an issue that appears in large-scale models, namely the presence of several identical or nearly identical behavior modes. Saleh et al. (2010) do consider the analytical problems associated with repeated eigenvalues, where it becomes necessary to use generalized eigenvectors, and where other behavior modes appear involving power functions of time. A filter essentially constitutes a weighted average of behavior modes and in this fashion avoids the "identity problem" of nondistinct eigenvalues.

The most serious theoretical issue, in our view, is how the results are interpreted using the feedback loop concept. As mentioned, the concept is relative (to a choice of an independent loop set). Moreover, practice reveals that the number of loops to consider is rather large and that the loops elasticities often do not have an easy or intuitive explanation. A lot of care must be taken when interpreting the results. For instance, Kampmann and Oliva (2006) found that "phantom loops" – loops that cancel each other by logical necessity and are essentially artifacts of the equation formulations used in the model – could nonetheless have large elasticities and thus seriously distort the interpretation of the results. An example of "phantom loops" is found in the Bass model in Fig. 1, where loops 3 and 4 are artifacts of the way the model is formulated. If the variable *total population* (T) was eliminated from the equations, the loops would disappear and in fact they exactly cancel each other out (since T is constant). Nonetheless, they appear on the list of loops and appear to have a separate influence on behavior. These kinds of problems may not be intractable, but their resolution will require careful mathematical analysis.

Finally, a problem with EEA and LEEA is that it only considers changes to behavior modes, not the degree to which these modes are expressed in a system variable of interest. This issue is addressed by also considering the eigen*vectors* of the system, which is the foundation for the analysis in the next section.

Eigenvectors and Dynamic Decomposition Weights (DDW)

The last set of methods, which are still in early development, we have termed the *eigenvector-based* approach (EVA). EVA attempts to improve the EEA/LEEA method by considering how much an eigenvalue or behavior mode is expressed in a particular system variable. The logic of the method and how EEA and EVA complement each other is shown in Fig. 10. As shown by Kampmann (2012), in a sense there is a one-to-one correspondence between eigenvalues and loop gains whereas the eigenvectors arise from the remaining "degrees of freedom" in the system. The observed behavior of the state variables in the model is then the combined outcome of the behavior modes (from the loop gains) and the weights for each mode (from the eigenvectors) in the respective state variable.

A number of researchers have attempted to develop EVA methods. Some emphasize the curvature (second time derivative) of the behavior, similar to the starting point of the PPM method (Güneralp 2006; Saleh 2002; Saleh and Davidsen 2001). The slope or rate of change $\dot{x}(t)$ of a given variable x in the linearized system may be written by

$$\dot{x}(t - t_0) = w_1 \exp(\lambda_1 (t - t_0)) + \cdots$$
$$+ w_n \exp(\lambda_n (t - t_0)), \qquad (10)$$

where the weights w_i are related to the eigenvectors. Then the curvature at time t_0 is

$$\ddot{x}(t_0) = w_1 \lambda_1 + \cdots + w_n \lambda_n. \qquad (11)$$

One may therefore interpret (11) has the sum of contribution from individual behavior modes. Güneralp (2006) suggested using the terms on the right-hand side of (11) as weights to combine elasticities of individual behavior modes ε_i with respect to some system element (like a link gain or a loop gain) into a weighted sum

$$\bar{\varepsilon} = \frac{\sum\limits_{i=1}^{n} w_1 \lambda_1 \varepsilon_i}{\sum\limits_{i=1}^{n} |w_1 \lambda_1|}, \qquad (12)$$

as a measure of the overall significance of that system element. He further normalized the elasticity measure by the elasticity measure for other system elements, i.e., assuming there are K such elements (loops or links), the relative importance ρ_k of the kth element is defined as

$$\rho_k = \frac{\bar{\varepsilon}_k}{\sum\limits_{j=1}^{K} |\bar{\varepsilon}_j|}, \qquad (13)$$

with the motivation that elasticities may vary greatly in numerical values, making comparisons at different points in time difficult, whereas ρ_k is a relative measure varying between $+1$ and -1. His results shed an alternative light on the behavior of these models, but the mathematical meaning, consistency, and significance of the doubly normalized measure (13) remains to be clarified. It is still too early to tell what the most useful approach will be, but one may note that the emphasis on the curvature shares the basic weakness in the PPM approach in dealing with oscillations.

Other researchers have looked directly at the *dynamic decomposition weights (DDW)* w_i in (10), i.e., the relative weight of the modes for a particular variable, from a policy criterion perspective, similar to Forrester's original focus and the starting point for the EEA analysis (Gonçalves 2009; Saleh et al. 2010).

For instance, Saleh et al. (2010) look at how alternative stabilization policies affect the behavior of business cycle models, using a simple inventory-workforce model (2000). Using the procedure in Fig. 10, they decompose the net stabilizing effect of a policy into its effect on the behavior mode itself (LEEA) and its effect on the expression of that mode in the variable of interest, measured by the dynamic decomposition weights (DDW). Oliva (2015) furthered develop the integration of the LEEA and DDW approaches and demonstrated that the methods can identify high-elasticity loops and leverage points in large models that incorporate autocorrelated noise (Oliva 2016).

To illustrate the approach we perform the computations for the inventory-workforce model (Fig. 3 and Table 1). We find that the following equations describe the behavior of the state variables

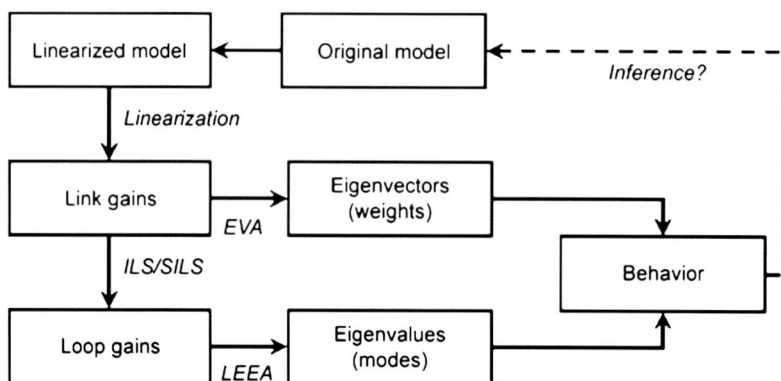

Analytical Methods for Structural Dominance Analysis in System Dynamics, Fig. 10 Schematic view of eigenvalue and eigenvector analysis approach

$$\text{ED} = 1 - 0.500e^{-0.250t}$$
$$\text{INV} = 3 - 2.167e^{-0.250t}$$
$$\quad + 3.403e^{-0.138t} \sin{(2.945 + 0.285t)}$$
$$\text{WF} = 1 + 0.669e^{-0.250t}$$
$$\quad - 1.169e^{-0.138t}(1.553 - 0.285t).$$

$$(14)$$

As expected from the structure of the model, the behavior of *expected demand* does not have an oscillatory component and only shows a short transient exponential adjustment for the stock to match *demand*. On the other hand, *inventory* and *workforce*, in addition to having the transient behavior to reach equilibrium captured by the first eigenvalue, have an oscillatory component represented by the second eigenvalue. Note that each state variable has a different *dynamic decomposition weight* (w) for each reference mode, i.e., each eigenvalue contributes differently to the overall behavior of each state variable.

An exploration of the policy design space can be achieved by assessing the influence of model parameters on the dynamic decomposition weight. By focusing on the weights of the behavior modes for the variable of interest, we can identify leverage points to increase or decrease the presence of a behavior mode in that variable. The weight elasticity column in Table 4 reports the parameter elasticity of w_2 (the weight of eigenvalue 2, the oscillatory behavior mode) on *inventory* ($\varepsilon_w = (dw/dp)(p/w)$). The magnitude of the elasticity quantifies the impact that changes in the

parameter value have on the weight of the oscillatory behavior model on *inventory*. The table is sorted in descending order of absolute value of elasticity.

Changes in parameters, however, not only impact the dynamic decomposition weights but also change the eigenvalues themselves. This dual impact of parameter changes introduces a challenge in developing policy recommendations. The last two columns of Table 4 report the influence on the eigenvalue (real and imaginary part) for each parameter. These measures of influence should be interpreted in a similar way as the weight elasticities. The influence measure is defined as $\mu_\lambda = (\partial\lambda/\partial p)p$. A positive real-part measure indicates that increasing the parameter will destabilize the system by lengthening the settling time and vice versa. A positive imaginary-part measure indicates that increasing the parameter will increase the frequency of oscillation – normally considered a destabilizing influence – and vice versa.

Three parameters, *demand*, *disequilibrium fraction*, and *time to change expectations*, have no influence on the oscillatory behavior mode. *Demand* and *disequilibrium fraction* are initialization constants that do not participate in any of the feedback loops in the model. While *time to change expectations* is involved in loop 1, it does not participate in the oscillatory behavior observed in the model since, as discuss above,

Analytical Methods for Structural Dominance Analysis in System Dynamics, Table 4 Elasticity to parameters of weight of eigenvalue 2 $(-0.138 + i\ 0.285)$ on inventory and influence of parameters on eigenvalue 2 – inventory-workforce model

Parameter	w_2 on INV Elasticity	Influence on $\mathrm{Re}[\lambda_2]$	Influence on $\mathrm{Im}[\lambda_2]$
Demand	2.000	0.000	0.000
Inventory correction time	0.656	0.038	−0.157
Flexibility in production	0.364	−0.549	−0.266
Productivity	−0.353	−0.116	−0.056
Time to change expectations	−0.240	0.000	0.000
Normal inventory coverage	0.239	0.173	0.084
Hiring/firing time	0.238	0.100	−0.127
Disequilibrium fraction	0.000	0.000	0.000

expected demand is in a separate strongly connected component of the model.

In accordance with LEEA, the *flexibility in production* parameter, which strengthens overtime loop 3 (cf. Fig. 3), has a strong stabilizing influence, by both increasing the damping and lowering the frequency of the oscillatory mode. Likewise, as predicted by LEEA, a shorter *hiring/firing time* will increase damping by strengthening the labor adjustment loop 5 but, again in accordance with LEEA, also increases the frequency of adjustment because it also strengthens the major loop 4. Finally, lowering the *inventory correction time* will strengthen the link from *inventory* to *desired production*, and consequently the three loops 3, 4, and 5, with the net effect that the adjustment is a little faster (a more negative real part). Reducing the *normal inventory coverage* also reduces the oscillation settling time. as *desired inventory* is much more stable. Note, however, that reducing the *normal inventory coverage* also reduces the frequency of the oscillation, while the reduction in *inventory correction time* increases the frequency significantly, it is a less effective way of stabilizing the system (cf. Fig. 9).

Note that *productivity* is essentially a scaling parameter defining the relationship between units of labor and goods. Redefining units should not affect the model dynamics. However, because we compute the DDWs at the current state of the system (in this case at $t=0$), the change in *productivity* does affect the gain of loop 3. Higher *productivity* reduces the settling time and reduces the frequency of the oscillations.

As an alternative approach, Fig. 11 shows what happens to the frequency response of the state variables (*inventory* INV, *workforce* WF, and *expected demand* ED) when the parameter hiring/firing time (hft) is reduced by 2% from 5 to 4.9. There are a number of things to notice in the figure. First, there is no effect whatsoever on the ED variable, which should not be surprising, given that there is no feedback to this variable from the rest of the system. Second, the effect on the amplitude, like the amplitude itself, is strongly dependent upon the frequency of variation. We see that there is a significant amount of dampening on the *inventory* fluctuation around the resonant frequencies in the range 0.1–0.3. On the other hand, there is a small amount of amplification of inventory in the higher frequency ranges. The effect on *workforce* is very different: though there is a small attenuation in the resonant frequencies, there is a significant increase in variance in the higher frequency range. In other words, although the LEEA analysis showed a faster hiring policy to be stabilizing (by strengthening loop 2, cf. Table 3), the DDW analysis shows that it depends – both upon the variable in question and the context (frequency of variation).

Future Directions

As mentioned above, it is not possible to construct a complete theory that will automatically provide modelers with "the" dominant structure. Given the analytical intractability of nonlinear high-order systems found in our field, the most we can hope for is a set of tools that will guide the analysis and aid the development of the modeler's intuition.

Analytical Methods for Structural Dominance Analysis in System Dynamics, Fig. 11 Effect on frequency response of the inventory workforce model of reducing the parameter hiring/firing time (hft) from 5.0 to 4.9. The diagram shows the gain of the base case (*upper graph*) for the three state variables, and the resulting change in the gain, measured as the ratio (A'/A) from the parameter change (*lower graph*)

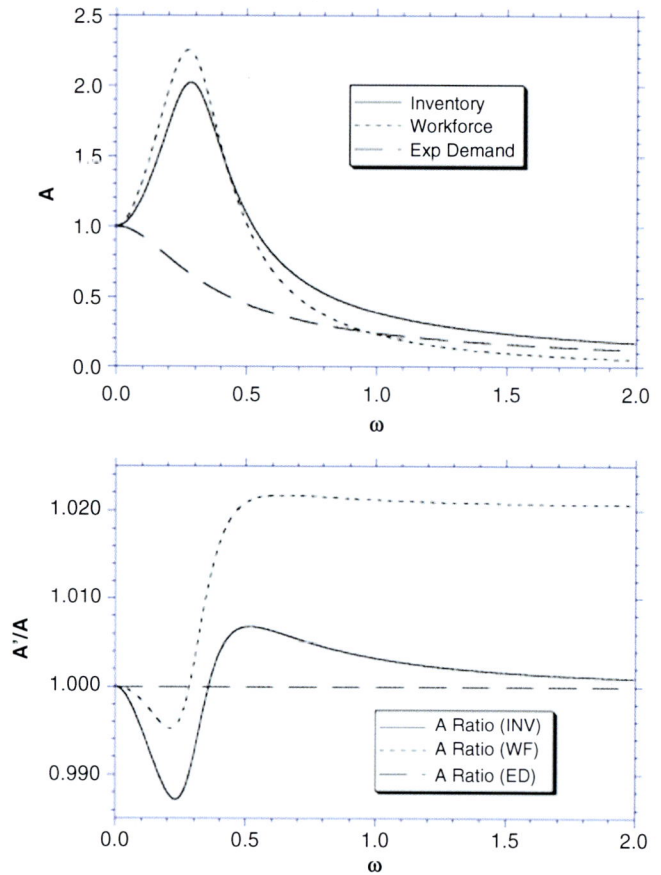

That said, however, we are left with an impression that the analytical foundation for these tools is in need of further development before one rushes into implementing them into software packages. We are quite satisfied with the current state of affairs in this regard, where code, models, and documentation are made freely to download (most of the cited papers provide a URL to their code and models). Understanding *how* and *why* the tools work the way they do is crucial, and this will require that a number of puzzles, uncertainties, and technical problems be addressed. Only then will the time come to submit the methods for wider application to test their real-world utility.

While the classical method remains a useful intuitive guide and teaching tool for graduate students, there are no signs that it may be developed further. (That said, it is possible that the classical control transfer function method may be employed in the eigensystem approaches to explore nested canonical systems, though this is purely speculative.) The pathway method would benefit from a firmer mathematical foundation. In particular, it would be important to compare how its results and conclusions compare to those found in the LST. It is possible that the pathway method may eventually be merged with the LST approaches as a subset of a general analytical toolbox. We believe that there is a great deal of promise in combining the eigenvalue and eigenvector analysis in the LST approaches. This combination will yield a complete system characterization and an understanding of both how particular feedback loops are involved in generating a behavior mode and how system elements determine the expression of that behavior mode in a particular variable. Significant progress has been made over the last 5 years (e.g., Duggan and Oliva

2013; Hayward and Boswell 2014; Oliva 2015, 2016) to develop a unified LST approach along the lines suggested in Fig. 10.

It will probably be a while, however, before these methods will find their way into widely available and use-friendly software packages. Apart from the theoretical issues alluded to above, a number of technical issues related to numerical calculations, various "pathological cases" (such as nondistinct eigenvalues), and special cases of feedback loops ("figure-eight" loops, for instance) will need to be addressed.

On the more creative side, it would be interesting to explore alternative forms of visualizing the various influence measures developed. For instance, one could imagine that links between variables in a model diagram "glow" in different colors and intensities depending upon their effect on a behavior pattern in question. This is not just a question of fancy user interfaces: as mentioned in the introduction, the function of these tools will be as intuitive consistent aids to understanding, not analytical "answering machines." In this light, the visualization is as important as the analytical principles behind it. Given the power of the human eye in finding patterns in visual data and the recent developments in visualization by data scientists, this could be a significant next step.

Bibliography

Abdel-Gawad A, Abdel-Aleem B, Saleh M, Davidsen P (2005) Identifying dominant behavior patterns, links and loops: automated eigenvalue analysis of system dynamics models. In: Proceedings of the Int system dynamics conference., July 2005. System Dynamics Society, Boston/Albany

Barlas Y (1989) Multiple tests for validation of system dynamics type of simulation models. Eur J Oper Res 42(1):59 87

Bass FM (1969) A new product growth for model consumer durables. Manag Sci 15(5):215–227

Chen CT (1970) Introduction to linear system theory. Holt, Rinnehart and Winston, New York

Duggan J, Oliva R (2013) Methods for identifying structural dominance. Syst Dyn Rev 29(virtual special issue). http://onlinelibrary.wiley.com/journal/10.1002/(ISSN)1099-1727/homepage/VirtualIssuesPage.html.

Eberlein R (1984) Simplifying models by retaining selected behavior modes. PhD Thesis, Sloan School of Management, MIT, Cambridge, MA

Eberlein RL (1986) Full feedback parameter estimation. In: Proceedings of the international systems dynamics conference. System Dynamics Society, Sevilla/Albany, pp 69–83

Eberlein RL (1989) Simplification and understanding of models. Syst Dyn Rev 5(1):51–68

Eberlein RL, Wang Q (1985) Statistical estimation and system dynamics models. In: Proceedings of the Int systems dynamics conference. System Dynamics Society, Keystone/Albany, pp 206–222

Ford DN (1999) A behavioral approach to feedback loop dominance analysis. Syst Dyn Rev 15(1):3–36

Ford A, Flynn H (2005) Statistical screening of system dynamics models. Syst Dyn Rev 21(4):273–303

Forrester JW (1961) Industrial dynamics. Productivity Press, Cambridge

Forrester JW (1969) Urban dynamics. Productivity Press, Cambridge

Forrester JW (1971) World dynamics. Productivity Press, Cambridge

Forrester N (1982) A dynamic synthesis of basic macroeconomic policy: implications for stabilization policy analysis. PhD Thesis, Sloan School of Management, MIT, Cambridge, MA.

Forrester N (1983) Eigenvalue analysis of dominant feedback loops. In: Proceedings of the Int system dynamics conference. System Dynamics Society, Chestnut Hill/Albany

Forrester JW (1993) System dynamics and the lessons of 35 years. In: DeGreene KB (ed) Systems-based approach to policymaking. Kluwer, Norwell, pp 199–240

Forrester JW, Senge PM (1980) Tests for building confidence in system dynamics models. TIMS Stud Manag Sci 14:209–228

Forrester JW, Mass NJ, Ryan CJ (1976) The system dynamics National Model: understanding socioeconomic behavior and policy alternatives. Technol Forecast Soc Chang 9(1–2):51–68

Goldberg DE (1989) Genetic algorithms in search, optimization and machine learning. Addison-Wesley, Reading

Gonçalves P (2003) Demand bubbles and phantom orders in supply chains. PhD Thesis, Sloan School of Management, MIT, Cambridge, MA

Gonçalves P (2009) Behavior modes, pathways and overall trajectories: eigenvalue and eigenvector analysis in dynamic systems. Syst Dyn Rev 25(1):35–62

Gonçalves P, Lerpattarapong C, Hines JH (2000) Implementing formal model analysis. In: Proceedings of the international system dynamics conference. August 2000. System Dynamics Society, Bergen/Albany

Graham AK. Principles on the relationship between structure and behavior of dynamic systems. PhD Thesis, Sloan School of Management, MIT, Cambridge; 1977.

Güneralp B (2006) Towards coherent loop dominance analysis: progress in eigenvalue elasticity analysis. Syst Dyn Rev 22(3):263–289

Haxholdt C, Kampmann CE, Mosekilde E, Sterman JD (1995) Mode locking and entrainment of endogenous economic cycles. Syst Dyn Rev 11(3):177–198

Hayward J, Boswell GP (2014) Model behaviour and the concept of loop impact: a practical method. Syst Dyn Rev 30(1–2):29–57

Holland JH (1992) Adaptation in natural and artificial systems. MIT Press, Cambridge

Homer JB (2012) Partial-model testing as a validation tool for system dynamics (1983). Syst Dyn Rev 28(3):281–294

Kampmann CE (2012) Feedback loop gains and system behavior (1996). Syst Dyn Rev 28(4):370–395

Kampmann CE, Oliva R (2006) Loop eigenvalue elasticity analysis: three case studies. Syst Dyn Rev 22(2): 146–162

Lane DC, Smart C (1996) Reinterpreting 'generic structure': evolution, application and limitations of a concept. Syst Dyn Rev 12(2):87–120

Luenberger DG (1979) Introduction to dynamic systems: theory, models and applications. Wiley, New York

Meadows DH, Meadows DL, Randers J, Behrens WW III (1972) The limits to growth: a report for the Club of Rome's project on the predicament of mankind. Universe Books, New York

Mojtahedzadeh MT. A path taken: computer-assisted heuristics for understanding dynamic systems. PhD Thesis, Rockefeller College of Pubic Affairs and Policy, State University of New York at Albany, Albany; 1996.

Mojtahedzadeh MT, Andersen D, Richardson GP (2004) Using digest to implement the pathway participation method for detecting influential system structure. Syst Dyn Rev 20(1):1–20

Morecroft JDW (1985) Rationality in the analysis of behavioral simulation models. Manag Sci 31(7): 900–916

Mosekilde E, Laugesen JL (2006) Nonlinear dynamic phenomena in the BEER model. Department of Physics, The Technical University of Denmark, Kongens Lyngby

Ogata K (1990) Modern control engineering, 2nd edn. Prentice Hall, Englewood Cliffs

Oliva R (2003) Model calibration as a testing strategy for system dynamics models. Eur J Oper Res 151(3): 552–568

Oliva R (2004) Model structure analysis through graph theory: partition heuristics and feedback structure decomposition. Syst Dyn Rev 20(4):313–336

Oliva R (2015) Linking structure to behavior using eigenvalue elasticity analysis. In: Rahmandad H, Oliva R, Osgood ND (eds) Analytical methods for dynamics modelers. MIT Press, Cambridge, MA, pp 207–239

Oliva R (2016) Structural dominance analysis of large and stochastic models. Syst Dyn Rev 31(1):56–51

Oliva R, Mojtahedzadeh M (2004) Keep it simple: dominance assessment of short feedback loops. In: Proceedings of the international system dynamics conference. July 2004. System Dynamics Society, Oxford, UK/Albany

Ott E (1993) Chaos in dynamical systems. Cambridge University Press, New York

Peterson DW (1980) Statistical tools for system dynamics. In: Randers J (ed) Elements of the system dynamics method. Productivity Press, Cambridge, pp 224–241

Peterson DW, Eberlein RL (1994) Reality checks: a bridge between systems thinking and system dynamics. Syst Dyn Rev 10(2/3):159–174

Radzicki MJ (2004) Expectation formation and parameter estimation in uncertain dynamical systems: the system dynamics approach to post Keynesian-institutional economics. In: Proceedings of the Int system dynamics conference. July 2004. System Dynamics Society, Oxford, UK/Albany

Richardson GP (1995) Loop polarity, loop dominance, and the concept of dominant polarity (1984). Syst Dyn Rev 11(1):67–88

Richardson GP (1986) Dominant structure. Syst Dyn Rev 2(1):68–75

Richardson GP (1988). Chaos Special Issue (ed). Syst Syn Rev 4:1–2

Richmond B (1980) A new look at an old friend. Plexus, Resource Policy Center, Thayer School of Engineering, Dartmouth College, Hanover

Saleh M. The characterization of model behavior and its causal foundation. PhD Thesis, Department of Information Science, University of Bergen, Bergen; 2002.

Saleh M, Davidsen P (2001) The origins of behavior patterns. In: Proceedings of the Int system dynamics conference., July 2001. System Dynamics Society, Atlanta/Albany

Saleh M, Oliva R, Kampmann CE, Davidsen P (2010) A comprehensive analytical approach for policy analysis of system dynamics models. Eur J Operat Res 203(3):673–683

Schweppe F (1973) Uncertain dynamical systems. Prentice-Hall, Englewood Cliffs

Senge PM (1990) The fifth discipline: the art & practice of the learning organization. Doubleday Currency, New York

Sterman JD (1981) The energy transition and the economy: a system dynamics approach. PhD Thesis, Sloan School of Management, MIT, Cambridge

Sterman JD (1985) A behavioral model of the economic long wave. J Econ Behav Organ 6(1):17–53

Sterman JD (2000) Business dynamics: systems thinking and modeling for a complex world. Irwin/McGraw-Hill, Boston

Wolstenholme E (2004) Using generic system archetypes to support thinking and modelling. Syst Dyn Rev 20(4):341–356

Zhusubaliyev ZT, Mosekilde E (2003) Bifurcation and chaos in piecewise-smooth dynamical systems. World Scientific, Singapore

Business Policy and Strategy, System Dynamics Applications to

James M. Lyneis
System Design and Management Program,
Massachusetts Institute of Technology,
Cambridge, MA, USA

Article Outline

Glossary

Business Policy and Strategy A firm's business strategy defines how and where it competes and its approach to doing so. A business strategy typically specifies a firm's goals, the products and services offered and the markets served, and the basis for competing (price, service, quality, etc.). A strategy may also define the organization structure, systems, and policies which implement the strategy. In addition, firms will have systems and policies which focus on operations and functions and are not truly "strategic" in nature. Nevertheless, these operational policies can be important in determining business performance.

Business Dynamics Business dynamics is the study of how the structure of a business (or a part of the business), the policies it follows, and its interactions with the outside world (customers, competitors, suppliers) determine its performance over time. Business structure consists of feedback loops surrounding the stocks and flows of resources, customers, and competitive factors that cause change over time; business policies are important components of these feedback loops. Business dynamics is a means of determining the likely performance that will result from alternative business policies and strategies.

Definition of the Subject

System dynamics has long been applied to problems of business performance. These applications range from operational/functional performance to overall strategic performance. Beginning with its founding at MIT's Sloan School of Management in 1957, an important focus of research, teaching, and application has been on understanding why companies and markets exhibit cycles or underperform competitors in terms of growth or profitability. The original publication in the field was Forrester's Industrial Dynamics (Forrester 1961), which not only laid the theoretical foundations for the field but also provided an understanding of the causes of instability in supply chains. Since that initial work, research and application has been widespread. It has addressed the dynamics underlying instability in manufacturing and service organizations, the processes which encourage or inhibit growth, the dynamics of research organizations, and the causes of cost and schedule overruns on individual projects. It has been applied in many industries, from manufacturing to high-tech to financial services and utilities, both by academics and consultants. Business theory and applications are taught at many universities, including but not limited to MIT, London Business School, and others in England, Bergen (Norway), Manheim, and Stuttgart (Germany) (see (Milling 2007; Morecroft and Wolstenholme 2007) for more details). Business policy and strategy has and will continue to be one

© Springer Science+Business Media, LLC, part of Springer Nature 2020
B. Dangerfield (ed.), *System Dynamics*,
https://doi.org/10.1007/978-1-4939-8790-0_45

Originally published in
R. A. Meyers (ed.), *Encyclopedia of Complexity and Systems Science*, © Springer Science+Business
Media New York 2013, https://doi.org/10.1007/978-3-642-27737-5_45-3

of the major application areas for system dynamics.

Introduction

Business strategy, sometimes called simply "policy" or "strategy," is primarily concerned with how and where firms choose to compete. It includes such decisions as setting goals; selecting which products and services to offer in which markets; establishing the basis for competing (price, service, quality, etc.); determining the organization structure, systems, and policies to accomplish the strategy; and designing policies for steering that strategy continually into the future. Academic and applied research on business strategy developed separately from system dynamics. That research, while widely disparate, has largely focused on static assessments and tools. For example, cross-sectional studies of many companies attempt to identify key differences that determine success or failure as a guide to management; "strategic frameworks" (e.g., learning curves, growth share matrices, Porter's five forces (Porter 1980, 1985)) assist managers in framing strategy and intuitively assessing performance over time; scenario planning helps managers visualize alternative futures; and the resource-based view of the firm and core competencies (Prahalad and Hamel 1990) help managers identify how resources and capabilities determine the best way to compete. While these tools provide valuable insights and frameworks, they leave the connection between a firm's business strategy and the evolution of its performance over time to the intuition of managers – while traditional business strategy addresses the starting point and the desired end point and the mechanisms that might allow the firm to transition between the two, the ability of those mechanisms to achieve that transition, and the path for getting between the two, is left unanswered.

Academic and applied research on operational and functional performance has similarly developed separately from system dynamics. Although it is difficult to generalize, this research is again typically static in nature and/or focused on the detailed management of a part of the organization over a relatively short period of time (e.g., optimization of production scheduling during a month, quarter, or year; optimal inventory management during a quarter or year). While this detailed management is necessary for running a business, it often overlooks the longer run implications of the policies established to manage the business in the short run and of the impacts of one part of the business on other parts.

In contrast, system dynamics addresses how structure (feedback loops, stocks, and flows) and policies determine performance over time – how does the firm, or a part of the firm, get from its current state to some future state. Evolving from this structural theory, system dynamicists have studied why firms and industries exhibit instability and cycles and why firms grow or decline. Two real examples of problematic behavior over time are shown in Fig. 1. The example on the left shows the pattern of orders for commercial jet aircraft – a system dynamicist would try to understand why the orders are cyclical and what can be done to make them less so (or to take advantage of the cycles by forecasting that cyclicality); the example on the right shows market shares of major players in a recently deregulated telecom market – a system dynamicist, working for the incumbent telecom, would try to understand the causes of market share loss and what can be done to reverse that loss.

From its beginnings in the late 1950s, system dynamics has been used to progressively develop structural theories to explain instability in supply chains (Forrester 1961), cycles of growth (Forrester 1968), boom and bust in product sales (Sterman 2000), and cost and schedule overrun on projects (Lyneis et al. 2001), to mention just a few. While system dynamics has at times borrowed, or in some cases reinvented, concepts from business policy and strategy, this structural theory development has until recently evolved largely independently of traditional business research and practice. There is, however, a great deal of potential synergy between traditional business research, particularly strategy research, and system dynamics that is increasingly being exploited.

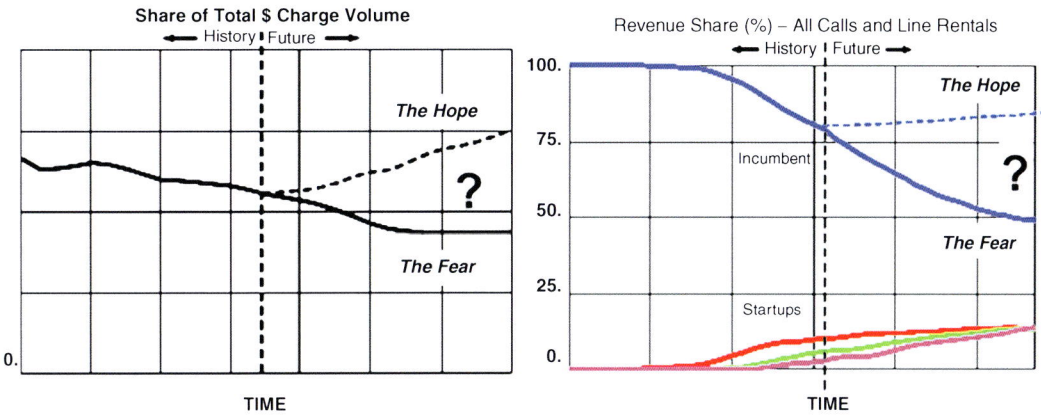

Business Policy and Strategy, System Dynamics Applications to, Fig. 1 Examples of problematic behavior over time

This entry surveys the application of system dynamics to business policy and strategy. The next section discusses the role of system dynamics models in policy and strategy formulation and implementation. Topics include how system dynamics fits into the typical policy and strategy formulation process, how system dynamics offers synergies with more traditional approaches, and how to conduct a modeling effort in order to enhance the implementation of any changes in policy or strategy. In section "Theory Development: Understanding the Drivers of Business Dynamics," system dynamics contribution to theory development is discussed – what are the structures underlying common business problems, and how can performance be improved? For example, what creates instability in supply chains, or "boom and bust" in product sales, and how can these behaviors be changed? Finally, in section "Applications and Case Examples," applications to real-world situations are presented – case studies that illustrate the value and impact of using system dynamics for policy and strategy development and implementation in specific firms and industries. As there has been a substantial amount of work done in this area, I must be selective, trying to touch on major themes and a representative sampling of work. Inevitably this will reflect my personal experiences, and my apologies to others that I have omitted either intentionally or unintentionally.

Using System Dynamics Models in Policy and Strategy Formulation and Implementation

Role of System Dynamics Models in Policy and Strategy Formulation

There is general agreement among system dynamics modelers on the role that models play in the policy and strategy formulation process. This role has been depicted diagrammatically and described by Morecroft (1984), Sterman (2000), and Dyson et al. (2007). The role of the model in policy and strategy formulation is to act as a "virtual world" – a simpler and transparent version of the "real world." The model serves as a vehicle for testing our understanding of the causes of behavior in the real world and as a laboratory for experimentation with alternative policies and/or strategies.

One version of that role is shown in Fig. 2. In many cases, the process starts with the definition of a problem – an aspect of behavior that is problematic or threatening. This might be a decline in market share or profitability or the threat posed by a new competitive product or service, as illustrated in the example of Fig. 1. Sometimes the problem can be expressed in terms of achieving business goals or objectives in the future. As illustrated in Fig. 2, the overall policy/strategy management process can be divided into three components: analysis, planning, and control. "Analysis" is usually triggered by a significant and/or persistent deviation between actual and expected performance. It

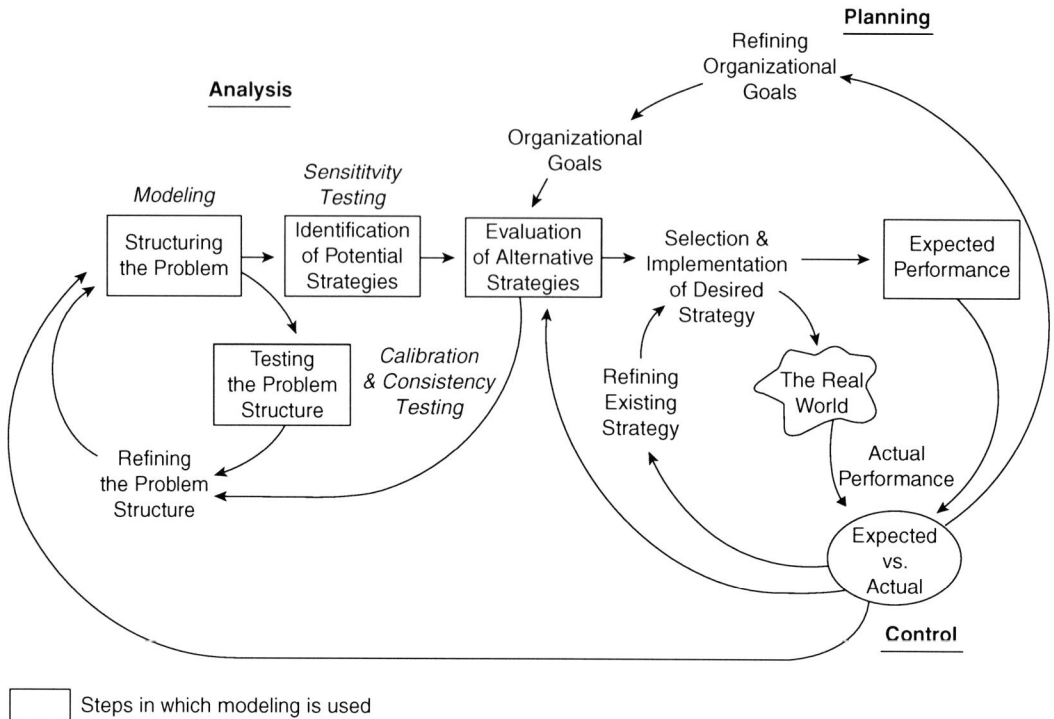

Business Policy and Strategy, System Dynamics Applications to, Fig. 2 Policy/strategy management process and the role of system dynamics models (Lyneis 1999)

involves the iterative structuring, testing, and refinement of an organization's understanding of its operational or strategic problems and of the options open to it to deal with the performance gap. The model is the vehicle for this analysis – does our understanding of the system as reflected in model equations in fact produce behavior consistent with the observed problem, and if not, how can our understanding of structure be made more consistent with reality? The process of evaluating alternative policies/strategies often sheds new light on the problems faced by an organization or reveals the need for further analyses. Note that the "modeling" cycle is iterative and compares simulated behavior to actual performance – the scientific method applied to strategy. Simulation of the model shows how business structure, policies, and external events together caused the past performance of the firm and how future performance will evolve if structure, policies, and external events differ. The next phase, "planning," is also an iterative process; it involves the

evaluation, selection, and implementation of policies/strategies – some authors refer to this as "rehearsing" strategy. Evaluation of alternative policies/strategies depends not only on projected accomplishment of organizational goals but also on the realities of current performance. The existing operational policies or existing strategy (vision, mission, strategic objectives) and goals are subject to refinement, as required, based on the successes and problems encountered and in response to changing conditions.

A third phase of the policy/strategy formulation process is here called "control." Ongoing policy/strategy *management* involves the continual, systematic monitoring of performance and the effective feeding back of successes, problems, threats, opportunities, experiences, and lessons learned to the other components of the policy/strategy management process. The control phase is where organizations continue to learn. *The model provides an essential element to the control process – a forecast of expected performance against which actual performance can be monitored on a regular basis.* Deviations provide a signal for additional analysis: Has the policy/strategy been implemented effectively? Have conditions about the external environment changed? Are competitors acting differently than expected? Has the structure of the system changed? The model provides a means of assessing the likely causes of the deviation and thereby provides an early warning of the need to act.

Synergy Between Traditional Strategy and System Dynamics

While Fig. 2 illustrates how system dynamics models fit into the policy/strategy formulation *process*, there is also a synergy between system

dynamics models and traditional *strategy frameworks and concepts*. Figure 3 illustrates the factors that drive business performance from a system dynamics perspective. Starting with resources, a firm's resources determine its product attractiveness; a firm's market share is based on that attractiveness compared to the attractiveness of competitor products; market share drives customer orders, which in turn generates profits and cash flow to finance the acquisition of additional resources for further growth – thereby completing a growth-producing feedback around the outside of the figure (or as in the example of the telecom in Fig. 1, "growth" in the downward direction for the incumbent). However, the acquisition of

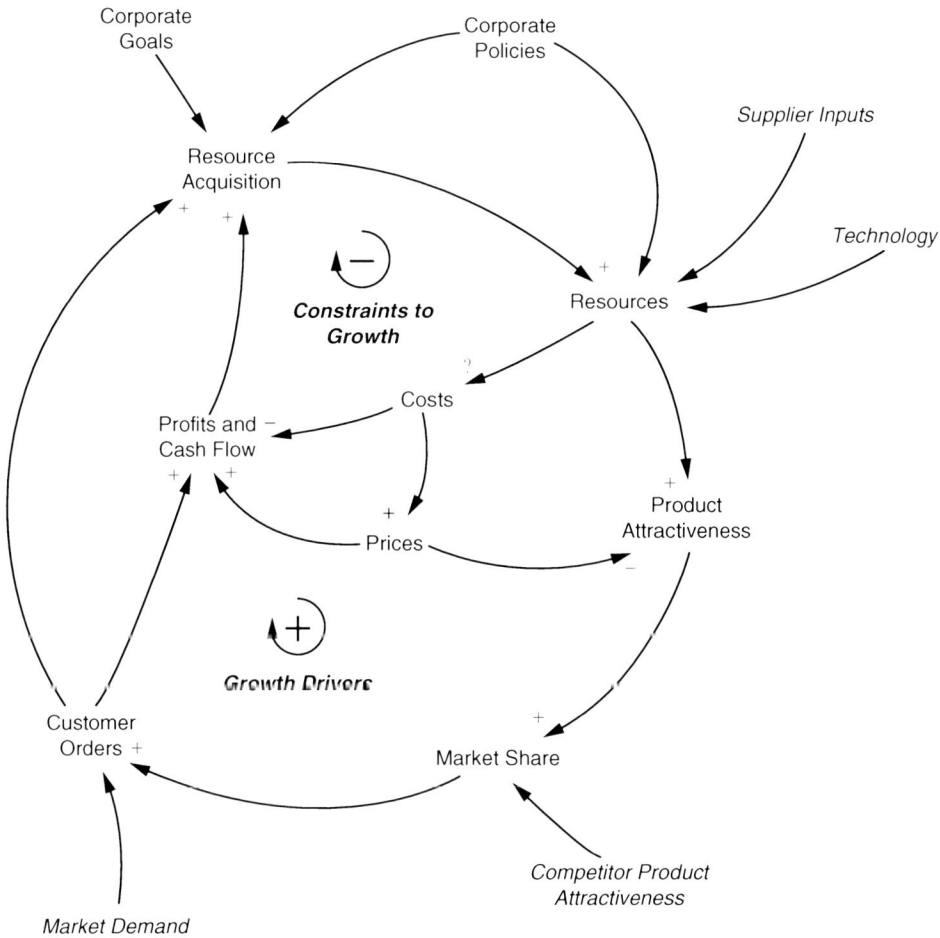

Business Policy and Strategy, System Dynamics Applications to, Fig. 3 Drivers of business performance (Adapted from Lyneis 1980)

additional resources can constrain future growth. To the extent increased resources increase costs, then profits and cash flow are reduced, and/or prices may need to increase. Both constrain growth (as might happen for the startups in the telecom example of Fig. 1).

There are a number of places in Fig. 3 where the system dynamics approach can be, and has been, connected to traditional strategy research and practice. For example, concepts such as learning curves, economics of scale, and economies of scope define possible connections between resources and costs – system dynamics models typically represent these connections. Figure 3 shows a number of factors external to the firm: market demand, competitor product attractiveness, technology, and supplier inputs.

Strategy frameworks such as "five forces" and visioning approaches such as "scenario-based planning" (Winch 1999) provide methods for thinking through these inputs – system dynamics models determine the consequences of alternative assumptions for the performance of the firm (note that system dynamics models also often internally represent the structural dynamics of competitors, suppliers, and the market as appropriate to explain the behaviors and issues of interest, rather than specifying them as exogenous inputs). For example, Fig. 4 shows a "sector" diagram of the major components of a strategy model developed by a consulting company for a telecom company dealing with loss of market share as in Fig. 1 above (from Graham and Walker 1998, originally developed by

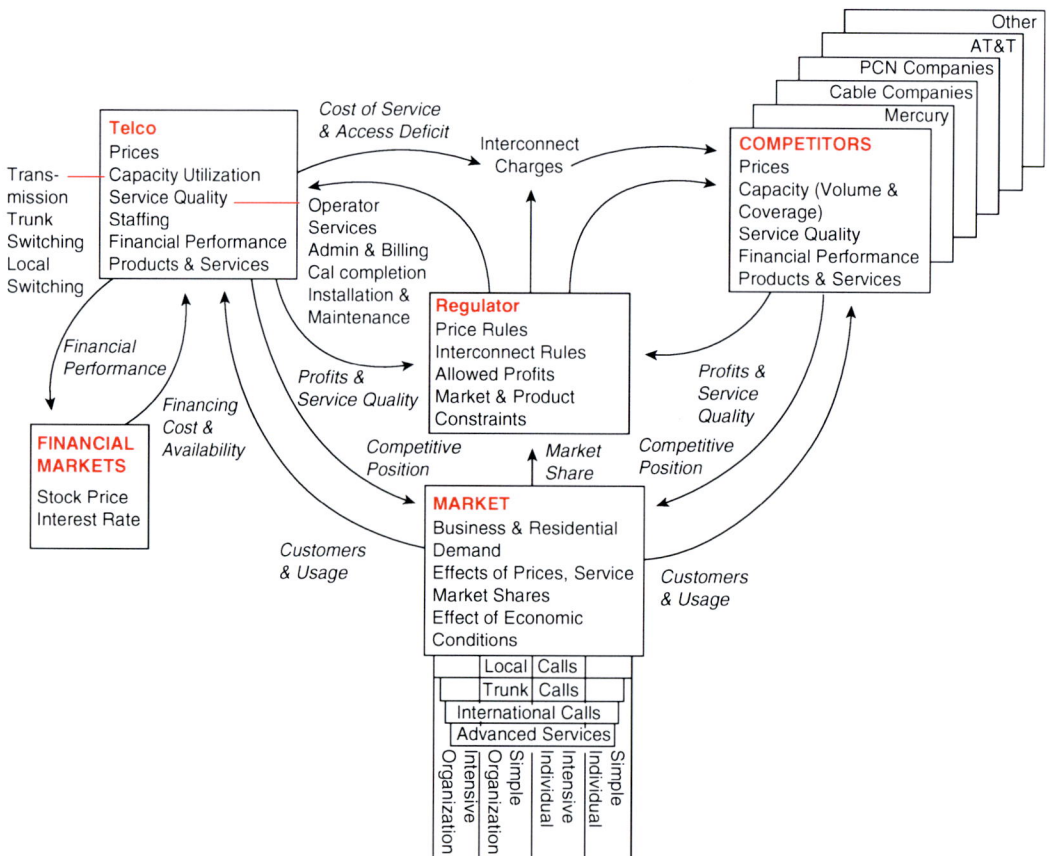

Business Policy and Strategy, System Dynamics Applications to, Fig. 4 "Sector" diagram from a typical strategy model

Lyneis). The model not only represents factors internal to the dynamics of the telecom firm but also factors related to the internal dynamics of competitors, regulatory responses to telecom and competitor performance, financial market responses to telecom performance, and market reactions to telecom and competitor competitive position. This sector diagram connects to a number of the traditional strategy frameworks. In addition to these general connections, a number of system dynamics papers have detailed specific connections to strategy concepts: learning curves and product portfolios (Merten et al. 1987), duopoly competition (Sice et al. 2000), diversification (Gary 2005; Morecroft 1999), and industry structure and evolution (Rockart et al. 2007).

Many of the connections between system dynamics and traditional strategy practice and research are discussed in Warren's *Competitive Strategy Dynamics* (Warren 2002) and *Strategic Management Dynamics* (Warren 2007). More importantly, Warren's book details and expands on the connections between the system dynamics concepts of structure, particularly the concepts of stocks and flows, and the well-established resource-based view (RBV) of strategy and performance. (See Montgomery 1995 and Foss 1997 for explanations of the development of RBV; a managerial explanation of how this theoretical perspective can be applied can be found in Chap. 5 in Grant 2005). Again, system dynamics provides a means of simulating the consequences of alternative resource acquisition and allocation strategies on firm performance. As such, there would seem to be a strong synergy between system dynamics and this strategy approach.

In summary, while system dynamics and traditional strategy approaches developed largely independently, the potential synergies between the two are significant. Until recently, few researchers and practitioners have made the effort to cross disciplines and exploit this synergy. More effort and publication are needed to demonstrate areas of synergy and get system dynamics into the mainstream of business strategy research and ultimately practice.

Working with Management Teams to Achieve Implementation

While Fig. 2 depicts the overall role of the model in strategy management, the approach to developing and using the model itself involves an iterative, multi-phased process. That process has evolved over time as practitioners and researchers have learned from experience. Since its inception, system dynamics has been concerned with having an impact on business decisions. Jay Forrester, the founder of the field, stressed the importance of working on important problems – those that affect the success or failure of firms – and generating solutions that are relevant to those problems. Ed Roberts, one of the first researchers and practitioners in the field, was also involved in early research and experimentation on organizational change and how models can fit into that process (Roberts 1977, 2007). The emphasis on having an impact has remained a central tenet of system dynamics, and over the years system dynamics practitioners have developed and refined methods of working with managers to not only solve problems but also enhance the likelihood of those solutions being implemented.

Starting with Roberts and his consulting firm Pugh-Roberts Associates (now a part of the PA Consulting Group), a number of researchers and practitioners have contributed to the evolving approach of working with managers to affect organizational change. These include, but are not limited to, Coyle and his group originally at Bradford University in the UK (Coyle 1996, 1997), Morecroft and his group at the London Business School (Morecroft 1984; Morecroft and van der Heijden 1994), Richmond (1997), Sterman (2000) and Hines (Hines and Johnson 1994) at MIT, the "group model building" approach (Richardson and Andersen 1995; Vennix 1996; Group Model Building) and Peter Senge's organizational learning (Senge 1990). While there are some differences in emphasis and details, there is in general agreement on the high-level process of using system dynamics to affect corporate strategy development and change. In this section, I describe that process, discuss some specific areas where there is some divergence in practice, and end with some examples to the approach in practice.

In the early years, the approach of most system dynamicists to consulting was heavy on "product" and light on "process." Like management science in general, many in system dynamics took the view that as experts we would solve the client's problem for him and present him with the solution. Practitioners gradually recognized that elegant solutions did not necessarily lead to implementation, and consulting styles changed to include increased client involvement (Roberts 1977; Weil 1980). At the same time, the "product" was evolving to meet the needs of clients (and to take advantage of the increased power of computers). Practitioners evolved from the smaller, policy-based models which characterized the original academic approach to more detailed models, along with the use of numerical time series data to calibrate these models and determine the expected numerical payoff to alternative strategies (Lyneis 1999). In addition, during the 1980s academic research began to focus more on "process": the use of models in support of business strategy and on more effective ways to involve the client in the actual building of the model (Morecroft 1984, 1985; Richardson and Andersen 1995; Richmond 1997; Vennix 1996).

System dynamics practitioners now generally agree on a four-phased approach to accomplish these objectives:

1. Structuring the problem
2. Developing an initial model and generating insight
3. Refining and expanding the model and developing a strategy
4. Ongoing strategy management and organizational learning

In practice, there is sometimes a fifth phase of work. Often modelers simplify a final project model in order to capture the core feedback loops that lie behind observed dynamics. Many of the generic structures discussed later in this paper arose in this way. In other cases, modelers will create management "games" and/or learning labs from the final project model.

As discussed below, the relative mix between "product" (detail complexity and calibration of model) and "process" (degree of involvement of client in model development and use) is perhaps the main difference in the style and approach of different practitioners of system dynamics. For those new to system dynamics, or for those seeking good examples, there is an excellent example of this process, including model development, by Kunc and Morecroft (2007).

Phase 1: Structuring the Problem

The purpose of the first phase of analysis is to clearly define the problem of interest (either a past problem behavior or a desired future trajectory), the likely causes of that problem (or desired trajectory), and any constraints that may arise in implementing a solution. It identifies the performance objectives of the organization and possible solutions – all to be rigorously tested in later phases. Similar to more traditional policy and strategy approaches, during this phase the consultant team reviews company documents, the business press, and available company data and interviews company managers and, possibly, customers and competitors. During this phase, the hypothesized drivers of business performance are identified: what compels customers to buy this product, what compels them to buy from one supplier versus another, what drives the internal acquisition and allocation of resources, and what major externalities affect the business (e.g., the economy, regulations, etc.), perhaps drawing on frameworks such as "five forces" and SWOT analysis. More importantly, these drivers are linked in a cause-effect model to form a working hypothesis of the reasons for company behavior. This hypothesis formation builds heavily on the tools and techniques of what is now commonly called "systems thinking":

1. Behavior-over-time graphs (reference modes) – Graphs of problematic behavior over time, often with objectives for future performance highlighted (using actual data where readily available).

2. Causal loop and mixed causal, stock-flow diagramming as a diagrammatic hypothesis of the causes of problematic behavior.
3. System archetypes, or common generic problem behaviors and structures observed over and over again in different businesses, as a means of identifying structure (see, e.g., Senge 1990; Kim and Lannon 1997).
4. Mental simulation – Does the hypothesis embodied in the conceptual model seem capable of explaining the observed problem(s)? Mental simulation is also used to identify the possible impact of alternative courses of action.

Note that the exercise to this point, as commonly practiced, is almost entirely qualitative. Warren (2002, 2007) introduces quantitative dimensions even in this phase.

Phase 2: Developing an Initial Model and Generating Insight

The power of system dynamics comes from building and analyzing formal computer models. This is best done in two steps. In the first, a small, insight-based model is developed to understand the dynamics of the business so as to generate insights into the direction of actions needed to improve behavior. The small, insight-based model is also the next logical progression beyond "systems thinking" in the education of the client in the methods and techniques of system dynamics modeling. In the second quantitative modeling step (Phase 3 below), a more detailed version of the first model is developed and is often calibrated to historical data. Its purpose is to quantify the actions needed, to assure that the model accurately reflects all relevant knowledge, and to sell others.

Small models (anywhere from 20 to 100 equations) make it much easier to understand the relationship between structure and behavior: How is it that a particular set of positive feedback loops, negative feedback loops, and stocks and delays interact to create the behavior shown in the simulation output? This can only be determined by experimentation and analysis, which is very difficult with large models. The focus of the first model is on insight generation, communication, and learning, rather than determining a specific

shift in strategic direction or investment. These models can help managers improve their intuition (mental models) about the nature of their business and thereby to better understand the rationale behind more detailed strategies that evolve in later phases of model development.

In summary, Phase 2 delivers:

- A small model which recreates the observed pattern of behavior or hypothesized future behavior (and is roughly right quantitatively)
- Analysis and understanding of the principal causes of that pattern of behavior
- Ideas of high leverage areas that could improve behavior into the future
- Recommendations as to where additional detail will improve the strategy advice or will make the results of the model more usable and/or easier to accept by others

Phase 3: Refining and Expanding the Model and Developing a Strategy

The final phase of model development entails the iterative expansion of the model to include more detail and often calibration to historical data, as deemed appropriate for the situation. One progressively adds detail and structure, initially to make the process manageable and then as necessary to correct discrepancies between simulated output and data or to add policy handles and implementation constraints. Further, model development is likely to continue in the "ongoing learning" phase as additional structure and/or detail is required to address new issues that arise. The purpose of this more elaborate modeling phase is to:

1. *Assure that the model contains all of the structure necessary to create the problem behavior.* Conceptual models, and even small, insight-based models, can miss dynamically important elements of structure, often because without data, the reference mode is incomplete or inaccurate (see Lyneis 1999 for examples of this).
2. *Accurately price out the cost-benefit of alternative choices.* Strategic moves often

require big investments and "worse-before-better" solutions. Knowing what is involved, and the magnitude of the risks and payoff, will make sticking with the strategy easier during implementation. Understanding the payoff and risks requires quantifying as accurately as possible the strengths of relationships.

3. *Facilitate strategy development and implementation*. Business operates at a detail level – information is often assembled at this level, and actions must be executed at that level. Therefore, the closer model information inputs and results can be made to the normal business lines and planning systems of the company, the easier strategy development and implementation will be.

4. *Sell the results to those not on the client's project team*. Few, if any, managers can dictate change – most often, change requires consensus, cooperation, and action by others. The "selling" of results may be required for a number of reasons. If, as in the optimal client situation, consensus among key decision-makers is achieved because they are all a part of the project team, then the only "selling" may be to bring on board those whose cooperation is needed to implement the change. Under less optimal client circumstances where the project is executed by advisors to key decision-makers, or by a support function such as strategy or planning, then selling to decision-maker(s) and to other functions will be required.

There are two important elements in this phase of work: adding detail to the model and, possibly, calibrating it to historical data. Adding detail to the small, insight-based model usually involves some combination of (1) disaggregation of products, staff, customers, etc.; (2) adding cause and effect relationships and feedback loops, often where the more detailed disaggregation requires representing allocations to products and markets, but also representing additional feedback effects that may seem secondary to understanding key dynamics, but may come into play under alternative scenarios, or may later help to "prove" the

feedbacks were not important; (3) including important external inputs, typically representing the economy, regulatory changes, etc.; and (4) adding detailed financial sectors, which entail numerous equations, with important feedback from profitability and cash flow to ability to invest, employment levels, pricing, and so on. Calibration is the iterative process of adjusting model parameters and revising structure, to achieve a better correspondence between simulated output and historical data. Whereas the Phase 2 model primarily relies on our store of knowledge and information about cause-effect structure, the Phase 3 model relies on our store of information about what actually happened over time.

In summary, Phase 3 delivers:

- An internally consistent data base on strategic information
- A detailed, calibrated model of the business issue
- A rigorous explanation and assessment of the causes of performance problems
- Analyses in support of strategic and/or tactical issues
- Specific recommendations for actions
- Expectations regarding the performance of the business under the new strategy and the most likely scenario

Phase 4: Ongoing Strategy Management System and Organizational Learning

True strategy management ("control") involves the ongoing, systematic monitoring of performance and the effective feeding back of successes, problems, threats, opportunities, experiences, and lessons learned to the other components of the strategy management process. The control phase is where organizations continue to learn. *The model provides an essential element to the control process – a forecast of expected performance against which actual performance can be monitored on a regular basis.* Deviations provide a signal for additional analysis: Has the strategy been implemented effectively? Have conditions about the external environment changed? Are competitors acting differently than expected?

Has the structure of the system changed? The model provides a means of assessing the likely causes of the deviation and thereby provides an early warning of the need to act. This feedback is only possible with a detailed, calibrated model.

Differences in Emphasis and Style

While there is general agreement among system dynamics practitioners regarding the role of models in the strategy development process and of the basic steps in that process as described above, there are some differences in emphasis and style regarding (1) the use of causal loop diagrams (CLDs) versus stock-flow (SF) diagrams, (2) whether you can stop after Phase 1 (i.e., after the "qualitative" phase of work), (3) whether calibration is necessary and/or cost-effective, and (4) how much model detail is desirable.

CLDs Versus SF

There are disagreements within the field about the value of causal loop diagramming (versus stock-flow diagrams). Causal loop diagrams focus on the feedback loop structure that is believed to generate behavior; stock-flow diagrams also include key stocks and flows and in the extreme correspond one to one with complete model equations. In my view, there is no "right" answer to this debate. The most important point is that in Phase 1 diagramming one is trying to develop a dynamic hypothesis that can explain the problem behavior and that can form the basis of more detailed diagramming and modeling – whether that dynamic hypothesis is a CLD, a stock-flow diagram with links and loops labeled, or some combination depends in part on:

- Personal style and experience – some people, Jay Forrester perhaps being the best example, seem to always start with the key stocks and flows and work from there; Kim Warren (2004) also argues for this approach as an effective means of connecting to the way managers view the problem and to the data.
- The structure of the system – some systems have "obvious" key chains of stocks and flows, and so starting there makes the most

sense (e.g., the aging chains in the urban dynamics model (Forrester 1969), the rework cycle on projects, and the inventory control systems); other systems, without critical chains of stocks and flows, may be easier to address starting with CLDs.

- Whether or not you are doing the model for yourself or with a group, especially if the group is not conversant with the basics of system dynamics, starting with CLDs is easier, and it is also easier to brainstorm with CLDs (which is different than developing a dynamic hypothesis); but again, it is personal preference and nature of system as well. In practice, I have found that CLDs alone, or a mixed stock-flow/causal diagram, are extremely valuable for eliciting ideas in a group setting about the cause-effect structure of the business and later for explaining the dynamics observed in simulation output. However, one cannot build a model literally from a causal diagram, and either explicit or implicit translation is required.

Qualitative Versus Quantitative Modeling

Some practitioners of system dynamics believe that strategic insights can sometimes be obtained after the first phase of work, after the dynamic hypothesis and mental simulation (and note that much of traditional strategy practice relies on such qualitative insights). Coyle (2000, 2001) argues for this; the popularity of "systems thinking" engendered by Senge's work (Senge 1990) has spawned a number of practitioners that use only qualitative modeling (Lyneis 1999). Coyle's views generated a strong counterresponse from (Homer and Oliva 2001). Wolstenholme (1999) provides a history and discusses his view on the advantages and disadvantages of each approach. My own view is that while Phase 1 and the systems thinking that is a key part of it are a necessary start, it should not be the end point. Two problems limit its effectiveness in supporting business strategy. First, simple causal diagrams represented by system archetypes, while useful pedagogically, take a very narrow view of the situation (typically, one or two feedback loops). In reality, more factors are likely to affect performance, and it is therefore dangerous to draw policy conclusions from such a limited view of the system. A more complete representation of the problem considers more feedback effects and distinguishes stocks from flows but introduces the second problem: research has shown that the human mind is incapable of drawing the correct dynamic insights from mental simulations on a system with more than two or three feedback loops (Paich and Sterman 1993; Sterman 1989a). In fact, without the rigor and check of a formal simulation model, a complex causal diagram might be used to argue any number of different conclusions. In addition to overcoming these limitations, as discussed below, formal modeling adds significant value to the development and implementation of effective business strategies. Warren (p. 347 in Warren 2005) also stresses need to focus on quantitative behavior to achieve management consensus.

Need for Data/Validation

The necessity of obtaining numerical data and calibrating model output to that data is also questioned by some practitioners. While I agree that curve fitting via exogenous variables is not a useful endeavor, proper calibration is an important part of the scientific method that involves systematically comparing simulation output to data, identifying causes of error, and correcting discrepancies by improving first the structure of the model and then its parametrization. In some cases, discrepancies are ignored because they are deemed to be caused by factors irrelevant to the problem of interest or may be "fixed" by exogenous factors if these are deemed significant by the client and are consistent with the remaining model structure and calibration. As Homer (1996, 1997) argues, the use of historical data and calibration is essential to scientific modeling.

In some cases, organizations lack the data on key factors felt to be essential to the dynamic performance of the business and by implication essential to sound strategic management of the business. The modeling process can highlight these shortcomings and, in the short term, substitute educated assumptions for this data. In the longer term, companies can be encouraged to acquire this important data (and substitute it for much of the unimportant information companies generally pore over).

Accurate calibration can greatly enhance confidence in a model. This can be especially important when trying to convince others of the appropriateness of actions a management team is going to take or to demonstrate to others the need to take action themselves based on the results of the model. Calibration can also be important for other reasons: (1) numerical accuracy is often necessary to evaluate the relative cost and benefits of changes in strategy or to assess short-term costs before improvements occur; (2) calibration often uncovers errors in the data or other models, especially incomplete or incorrect mental models that form the basis for the dynamic hypothesis (see (Lyneis 1999) for examples); and (3) the "control" feedback in the fourth phase of the strategy management process is only possible with a detailed, calibrated model.

Level of Detail and Model Complexity

Some practitioners argue that large, complex models should be avoided, for a number of reasons: they can be even more like black boxes; they can be difficult to understand (not only for the non-modelers, but even the modelers); and they are costly to develop. Morecroft argues that a detailed model "loses its agility and becomes less effective as a basis for argument" (p. 227 in Morecroft 1984). In practice, the first two issues can be avoided and/or minimized by executing the model development in three phases as discussed above. This allows the client to grow slowly with the concepts, and it allows the modeling team to develop a solid understanding of the model. The third problem is generally not an issue if you are working on significant problems – in my view the cost of the consulting engagement is trivial relative to the expected payoff. While I believe that the client obtains value, regardless of when you stop, strategy consulting is one case where the "80/20 rule" does not apply – the client does not get 80% of the value for 20% of the cost (which would be essentially at the end of Phase 1). In part this is a function of what I view as the objective of the project – providing tools, strategic analyses, and advice in support of an important strategic and/or investment decision. In this situation, the "value" is back-end loaded. Finally, effective strategy management is only possible with a detailed, calibrated model.

In addition, detail and calibration are often necessary to sell the model to others. In many situations, everyone who may have an input to a strategic decision or be necessary for successful implementation cannot be a part of the client team. As surprising as it may seem, the selling of results (as opposed to understanding) is easier to accomplish with a detailed, calibrated model than with a small model. First, the numerical accuracy gives the model face validity. Second, a detailed model more often allows the modeler to counter the "have you considered (insert pet theory)?" criticism. I have often found that when you start explaining the model to others, they respond by asking "Have you considered this feedback? Or this effect?" And if you have not, that ends the discussion. Even though you may think that feedback or that effect may not have any impact, if it is not included in the model, you cannot say "Yes, we looked at that and it did not have any impact" and explain why. If it is not in the model, the critic can argue that your results would be changed by the inclusion of their pet theory. One has a hard time countering that assertion without a convincing argument based on simulation results. Finally, a detailed, calibrated model helps tell a convincing story. The simulation output, which corresponds closely to the data, can be used to explain (again with output) why, for example, a loss of market share occurred. Why price relative to the competitions' price was the key factor, and/or why the factors affecting share changed over time. The simulation output can and should be tied to specific events. We have found that an explanation like this is compelling and is important in enhancing the credibility of the model and the modeler.

The benefits of large, complex models in a consulting setting are also noted by Winch (1993). He specifically finds that "For the executive team to have confidence in the impartiality of the model, each person must feel it captures the detailed pressures and processes of his or her own sphere of responsibility yet produces a holistic view of the organization" (pp. 295–296 in Winch 1993) and that the model was essential to getting everyone to agree: "The process of building system dynamics models, in each case ostensibly as a forecasting and evaluation tool, enabled the managers eventually to develop a shared view, which formed the basis for formulating and agreeing upon a final strategy" (p. 298).

Process Examples

There are a number of published examples that support the four-phase process of applying system dynamics to business strategy:

- Lyneis (1999) not only provides a more fully developed description of the detailed, calibrated model Pugh-Roberts approach but also illustrates its application to the credit card and airline manufacturing industries.
- Morecroft et al. (1991) describe how a model was created and used to stimulate debate and discussion about growth management in a biotechnology startup firm. The paper highlights several novel features about the *process* used for capturing management team knowledge. A heavy emphasis was placed on mapping the operating structure of the factory and distribution channels. Qualitative modeling methods (structural diagrams, descriptive variable names, "friendly" algebra) were used to capture the management team's descriptions of the business. Simulation scenarios were crafted to stimulated debate about strategic issues such as capacity allocation, capacity expansion, customer recruitment, customer retention, and market growth and to engage the management team in using the computer to design strategic scenarios. The article concludes with comments on the impact of the project.
- Winch (1993) examines the role that building and using a system dynamics model plays in developing consensus within management teams facing key strategic decisions: a shared view emerges within the team as individual views of the company, its industry, and the socioeconomic climate are articulated and compared. Examples are given based on two

actual consulting assignments in which differing views concerning the competitive environment and the general business outlook initially pointed to quite different strategies. The emergence of consensus was considered a major benefit in addition to the forecasts and quantitative evaluations the model provided. In its analyses and examples, this article emphasizes both the "hard" benefits of forecasts and an objective framework for quantitative evaluations and the "soft" benefits of building consensus within management teams.

- Coyle (1997, 1998) also has an approach that he discusses, with emphasis on CLDs (he terms these "influence diagrams," and his group was instrumental in the initial use of this technique).
- Snabe and Grossler (2006) show how modeling can be supportive for strategy implementation in organizations and illustrate with a detailed case study from a high-tech company.
- A special issue of the *Journal of the Operational Research Society* on System Dynamics for Policy, Strategy, and Management, edited by Coyle and Morecroft (1999), contains a number of papers which in part discuss consulting process issues (Delauzun and Mollona 1999; Wolstenholme 1999; Winch 1999) among others.
- The special issue Fall 2001 of *System Dynamics Review* on consulting practice contains papers by Thompson (1999), Campbell (2001), and Backus et al. (2001) that focus on the consulting process.

Theory Development: Understanding the Drivers of Business Dynamics

Another important contribution of system dynamics to business policy and strategy formulation is the development of structural theories to explain commonly observed patterns of behavior. Theory development provides us with an understanding of the basic drivers of business dynamics; insights, enhanced mental models, and policy guidelines for improved performance; and building blocks of tested model equations for real

applications (equations for the model must be provided for models to add to our base of theory). System dynamicists have developed structural theories to explain the basic patterns of business dynamics: (1) cycles and instability, (2) productivity and eroding performance, (3) life cycles, and (4) growth. Each is discussed in turn below.

Cycles and Instability: Stock Management, Supply Chains, and Manufacturing Systems

The very first applications of system dynamics were to understanding the tendencies of production-distribution systems, or "supply chains," toward cycles and instability; these applications remain important to this day (Forrester 1958, 1961). For example, the "Beer Game," now distributed by the System Dynamics Society, was developed and refined at MIT beginning in the early 1960s and remains one of the most popular introductions to both system dynamics principles and to supply chain issues.

Supply chains are an important component of all industrialized societies. They exist in any industry where goods are produced and distributed to consumers, for example, food and beverage production and distribution, or manufactured goods such as automobiles and appliances. Supply chains exhibit a classic behavior pattern which has impacts not only on the individual company but also on the economy as a whole: as one moves up the supply chain from the end user, any variation in orders from the end user is progressively *amplified* and delayed at each additional stage in the chain – factory variations are greater than customer variations; raw material production variations are greater than factory variations (see Chaps. 17 and 20 in Sterman 2000 for real-world examples of this behavior). This behavior is also sometimes referred to as the "bull-whip" effect.

Figure 5 illustrates the structure of one stage of a typical supply chain and the causes of amplification: a stock of inventory is depleted by shipments (here assumed equal to demand) and replenished by production completions (or more generally, shipments from a supplier); the stocks of goods in production (or goods being assembled and shipped by a supplier) are increased by

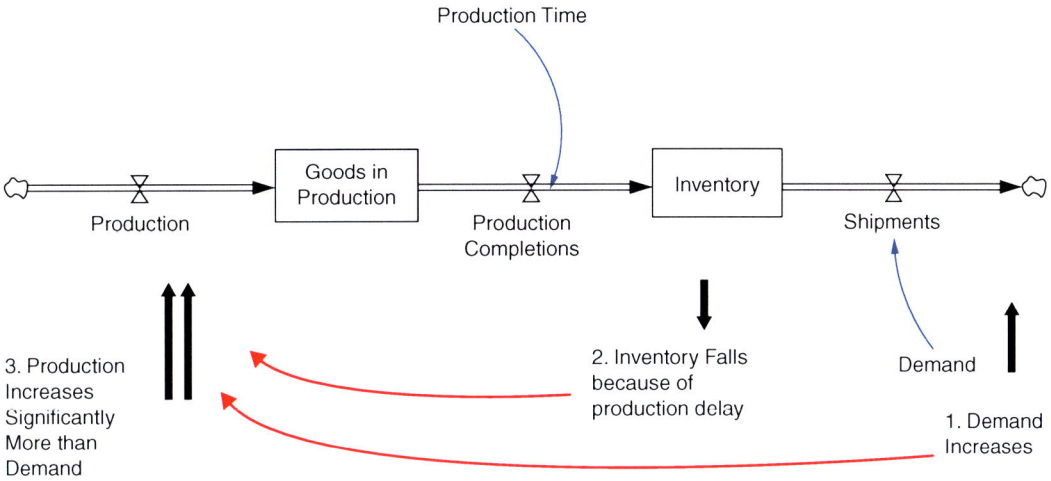

Business Policy and Strategy, System Dynamics Applications to, Fig. 5 Structure and causes of amplification in one stage of a supply chain

production and reduced, after the production (and/or shipping) delay, by production completions. This structure has a tendency to "amplify" any changes in demand – that is, "production" (or orders and reorders, depending on the system) increases or decreases more than any increase or decrease in demand and tends to lag changes in demand. For example, in Fig. 5, when demand increases, even if production increases immediately, inventory falls because production completions are delayed by the production (and/or shipping) delay. Therefore, production must increase higher than demand (amplification) in order to rebuild inventories. In addition to production and shipping delays, inventory might also fall because of delays caused by smoothing information about demand (such that production changes lag changes in demand). Production further increases above demand because of the need to increase inventories and production or supply lines to higher target levels. Intuitively, and verified by simulations, amplification is greater if desired inventories are larger, production/transit delays are longer, and/or responses to inventory gaps are more aggressive (smaller adjustment time constant, as discussed below).

This basic structure in Fig. 5 also illustrates the "stock management" problem. In Fig. 5, the stock of finished goods inventory must be managed in order to serve customer demand in a timely fashion. Figure 6 details the structure typically used to control stocks, one of the most used and important structures in system dynamics:

Production = Expected Demand
+Inventory Correction + Goods In Process
Correction Inventory Correction
= (Desired Inventory – Inventory)/Time to
Correct Inventory Desired Inventory
= Expected Demand × Months Coverage
Goal Goods In Process Correction =
(Desired Goods In Production –
Desired In Production)/Time to Correct Inventory
= Expected Demand × Production Time

Stock management is complicated by delays in replenishing the stock, here a production delay. Depending on the pattern of demand, there is often a tradeoff between amplification and variations in inventory – less aggressive responses (longer time to correct inventory) generally reduce amplification but cause greater variations in inventory (and therefore may necessitate higher target levels to reduce the likelihood of stockouts); more aggressive responses (shorter time to correct inventory) increase amplification and demands on manufacturing and suppliers, and potentially costs, but can result in more stable inventory

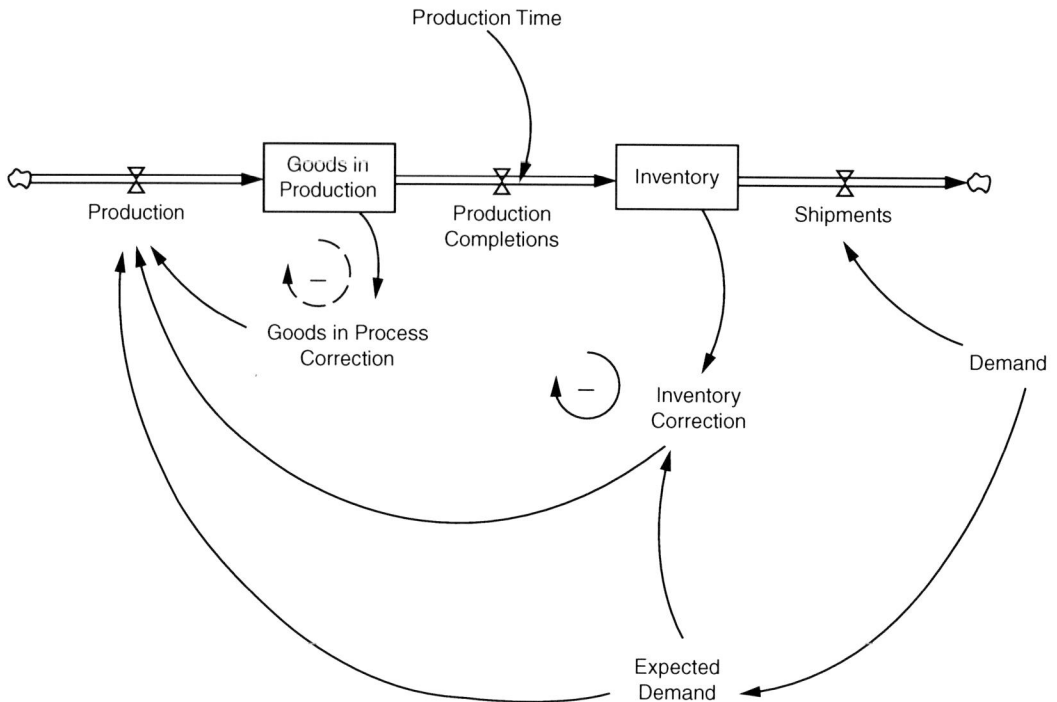

Business Policy and Strategy, System Dynamics Applications to, Fig. 6 Stock management structure

levels. However, under some demand patterns and production conditions, aggressive responses can increase both amplification and inventory instability. While as noted below structural changes can significantly improve the overall performance of stock management and supply chain systems, nevertheless this fundamental tradeoff between amplification and inventory levels will remain. The "optimal" solution will vary by firm and over time as the inherent pattern of demand changes. These dynamics and tradeoffs are discussed in depth in (Lyneis 1980; Sterman 2000).

In a typical supply chain, there are multiple stages connected in series, for example, in the automotive industry: dealers, car manufacturers/ assemblers, machine tool producers, parts manufacturers, and raw material suppliers (with potentially several stock management stages in some of these main categories). The upstream stages suffer greater amplification than the downstream stages. In the main, this occurs because each stage uses as its demand signal the orders from the prior stage, which are amplified by that stage's stock management policies as discussed above. Other reasons for increased upstream amplification include (Akkermans and Vos 2003) the following: (1) in determining "expected demand," each stage tends to extrapolate trends in orders from the prior stage; (2) order batching; (3) price fluctuations (in response to inventory levels); and (4) shortage gaming (ordering more than you really need to get a higher share of the rationed goods from the supplier; see (Lyneis 1999) for an example in the aircraft industry). System dynamics analyses have identified a number of structural changes which can improve supply chain and stock management performance: (1) reduce delays, (2) reduce inventory, and (3) share information (e.g., if upstream stages are aware of the downstream end user customer demand pattern, they can use that information rather than the amplified orders from the prior stage as the basis for their decisions and at least partially avoid amplification (Croson and Donohue 2005)).

Applications of system dynamics to supply chain management, and production management, remain an important area of research and applications. Akkermans and Daellaert (2005), in an article entitled "The rediscovery of industrial dynamics: the contribution of system dynamics to supply chain management in a dynamic and fragmented world," provide an excellent survey of supply chain management and system dynamics potential role in moving that field forward. Additional work in this area includes Morecroft's original analysis of the dynamics created by MRP systems (Morecroft 1983); Gonçalves doctoral dissertation (Gonçalves 2003), some of which is summarized in (Gonçalves 2006; Gonçalves et al. 2005); Anderson and Fine (1999) on capital equipment supply cycles; and Zahn et al. (1998) on flexible assembly systems. Each of these discusses variations on the basic stock/production/supply chain management systems and provides references for further research.

In addition to inventories, firms need to manage other stocks and resources, including raw materials, employees, capital equipment, and so on; the stock management structure described above for inventory applies to these other stocks as well. The management of stocks and resources is central to dynamic and strategy problems in many industries. First, the management of one stock often influences the ability to manage other stocks (e.g., capital equipment and employees determine production). Not only does this interdependency create constraints, the additional negative feedback control in managing resources is another source of cyclical behavior (see Chap. 19 in Sterman 2000 and Chap. 5 of Morecroft 2007). Second, in addition to the stocks of resources, production is affected by the productivity of those resources. Dynamic drivers of productivity, such as experience and fatigue, are discussed in the next section.

While the negative control feedbacks described above are central to the observed cyclical behavior of supply chains and resource-based firms, an additional negative feedback through the market adds a further source of instability. This dynamic is perhaps clearest in commodity-based industries, which have also been extensively modeled by system dynamicists as first summarized by Meadows (1970). As illustrated in Fig. 7, these models integrate the supply chain with the dynamics created by supply and demand – in a typical commodity system, there are three major negative feedback loops: two supply feedbacks (one through production, often representing the resource labor, and one through the resource capacity) and one demand feedback (e.g., an increase in inventory causes prices to fall, which increases demand and leads to a decrease in inventory from what it otherwise would be). Commodity industries typically exhibit behaviors that include cycles of two periodicities, one determined primarily by the shorter production feedback loop and another longer cycle driven by the capacity loop (see Chap. 20 in Sterman 2000 for both detailed equations and for examples of the structure applied to the livestock and paper industries). The demand feedback loop, however, can play a role in the dynamics as well – if the demand feedback is strong and with a short delay, then demand corrections occur before the supply feedbacks operate and system stability is improved; however, if the magnitude of the delay in the demand loop is similar to the magnitude of the delays in either of the supply loops, the intensity of the corresponding cycle is increased as two negative feedback loops are both independently acting to "solve" the inventory problem. In addition, commodity industries, where they involve a depletable resource such as oil, can experience long-term resource depletion dynamics (Sterman et al. 1988).

In conclusion, manufacturing and supply chain dynamics are central to many of the behaviors observed in businesses (see Chap. 20 in Sterman 2000 for real-world examples of these cycles). The supply chain, stock management, resource management, and commodity structures discussed above are therefore important components of many system dynamics models developed to support business policy and strategy. In some cases, the firm can change policies to reduce the severity of these cycles; in other cases, especially where the cycles are driven primarily by industry dynamics, the individual firm can use

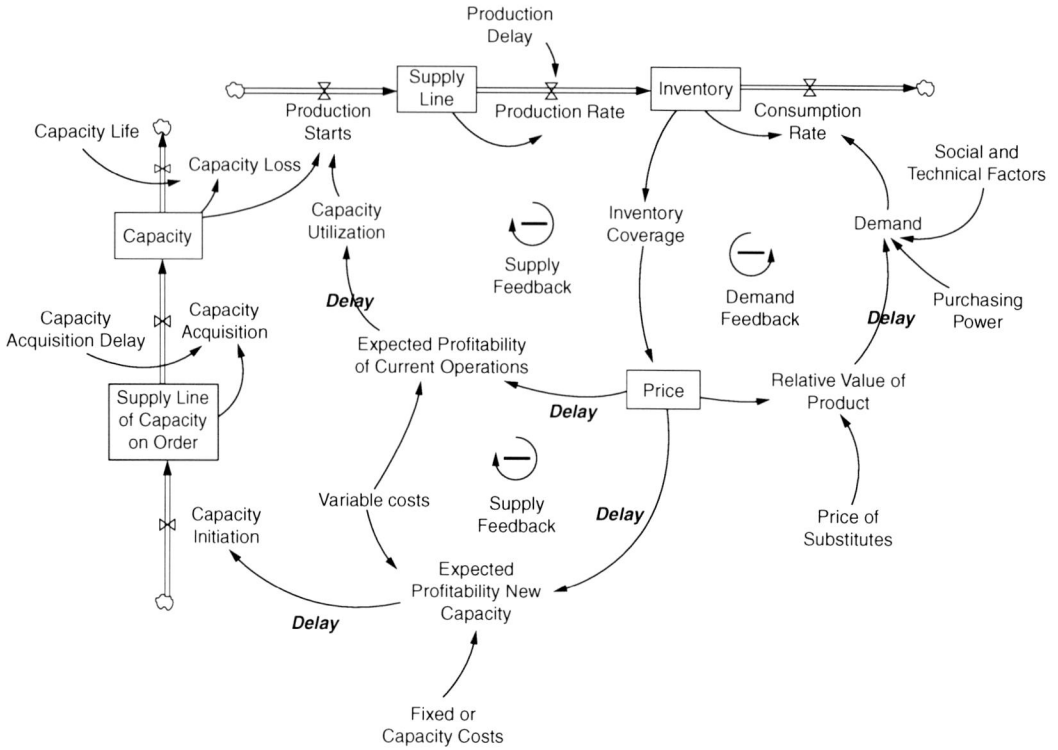

Business Policy and Strategy, System Dynamics Applications to, Fig. 7 Commodity industry dynamics showing three controlling feedbacks (Adapted from Sterman 2000)

the enhanced understanding and forecasting of these cycles for more strategic decisions such as new product introduction and capacity planning (as in the commercial aircraft market case illustrated in Fig. 1 and discussed in section "Applications and Case Examples").

Productivity and Eroding Performance: Service Industry Dynamics

Service-based firms (e.g., professional services, transportation, catalog and online shopping, etc.) and the service arms of manufacturing-based organizations have a somewhat different set of structural dynamics. Firms in these industries have a number of characteristics that make them more difficult to manage than more manufacturing intensive industries: (1) their product is difficult if not impossible to inventory, and so something else must buffer changes in demand; (2) they are particularly dependent on the performance of people (although the productivity of

resources is also important to manufacturing-based businesses as well); and (3) the performance of the system can be harder to detect, and so they are much more subject to a gradual erosion in performance and goals. "The major recurring problems observed in service industry – erosion of service quality, high turnover, and low profitability – can be explained by the organization's response to changes in work pressure" (see p. 28 of Oliva 2001).

One of the primary distinguishing features of service-based firms is that their end product is people dependent and cannot be inventoried. While there may be inventories of products that support delivery of the service, that delivery must be performed based on current resources. As a result, work or order backlogs are the stock that buffers demand from "production." A simplified example is shown in Fig. 8. Customer orders (demand) fill an order backlog, which is depleted by order fulfillment. Order fulfillment is based on

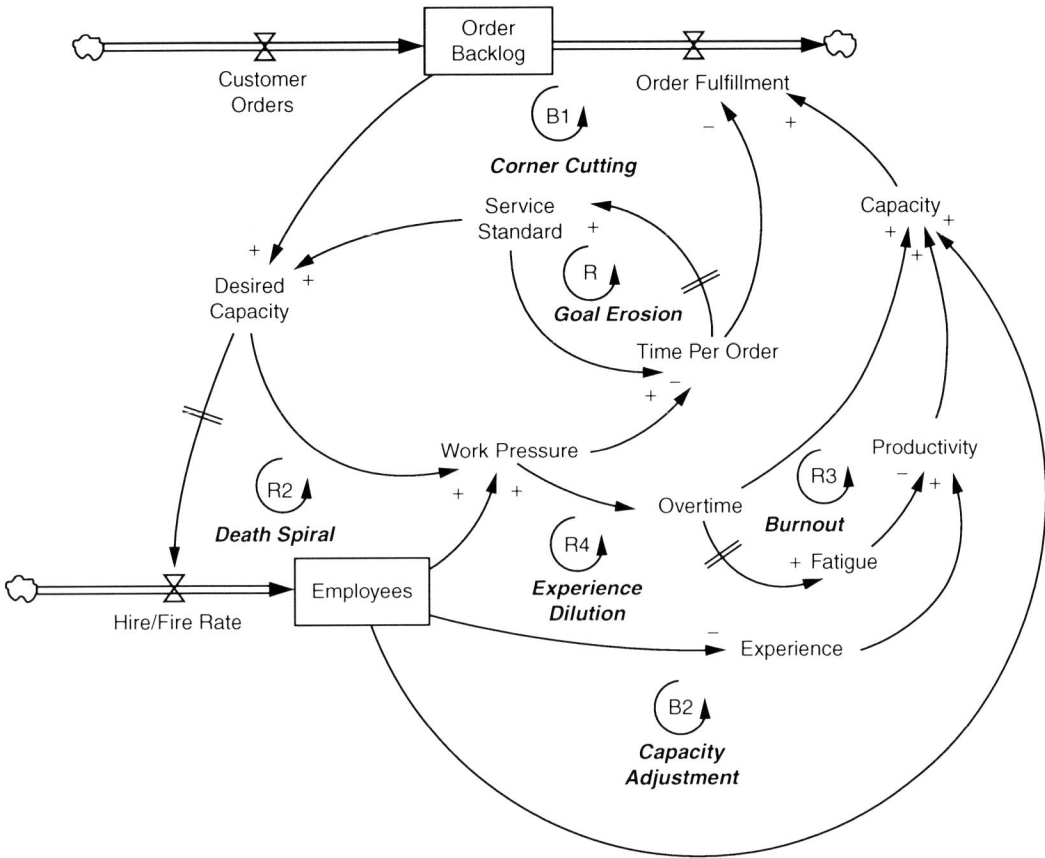

Business Policy and Strategy, System Dynamics Applications to, Fig. 8   Drivers of service business dynamics

the firm's service "capacity" and the amount of time spent per order – for the same capacity, order fulfillment will be greater if less time is spent per order (although service quality may suffer). Capacity is dependent upon people, overtime, and productivity, as discussed further below. Employees are increased or decreased based on desired capacity, which in turn depends on order backlog relative to the firm's service standard (time per order). Work pressure depends on desired capacity relative to the current stock of employees – to the extent the number of employees does not increase as needed, work pressure builds which can result in increases in effective capacity via overtime or reductions in time spent per order. Time spent per order depends on the firm's service standard for time spent per order, modified by work pressure – if work

pressure is high, time spent can be reduced. The service standard often responds to actual performance.

Another important characteristic of service supply firms shown in Fig. 8 is that their capacity is particularly dependent on the performance of people, both in numbers and in productivity. While productivity is also a factor in manufacturing systems, the sensitivity of performance to people factors is generally less than in service-based firms. Therefore, models of such firms generally represent in some detail the factors that drive productivity, including (1) skill and experience, often using an aging chain or "rookie-pro" structure (Sterman 2000; Winch 2001); (2) fatigue from sustained amounts of overtime; and (3) work intensity increasing productivity, but with "haste-makes-waste" impacts on errors and rework (not shown in

Fig. 8). In these situations, when demand is growing there are considerable short-term forces which reduce productivity and cause a deterioration in service quality: adding people reduces productivity because of "experience dilution"; working overtime increases fatigue and reduces productivity; pressures to work more intensely increase errors and cause additional work. As shown in Fig. 8, these productivity effects form reinforcing feedback loops which can drive down a service system's performance: an increase in work backlog and desired capacity causes the firm to hire more people; experience levels and productivity decline as a result, thereby reducing order fulfillment below what it otherwise would have been; order backlog does not fall as much as expected, necessitating additional capacity, further hires, and decreased experience; this completes the "experience dilution" R4 loop. The "burnout" loop through overtime and fatigue is similarly a reinforcing loop (R3).

Oliva (2001) shows that how management responds to these work pressure problems can determine the long-term success or failure of a service-based organization, largely as a result of the third characteristic of such systems: performance of the system can be harder to detect, and so they are much more subject to a gradual erosion in performance and goals. Oliva demonstrates that if the firm reduces its service standards (goals) in response to deteriorating performance (loop R1 goal erosion), a death spiral can ensue in which the declining goals cause the firm to add fewer people, which locks in a situation of excessive work pressure and further declining performance (thereby completing the "death spiral" loop R2). Unless there is a cyclical downturn in demand which alleviates the pressure, a firm's service performance will gradually erode until competition captures the market. He further discusses solutions to these problems, including buffers and faster response. Oliva's work, together with applications noted below, suggests that a service company should hire steadily rather than in spurts to avoid problems of inexperience, should hire enough workers to avoid overwork and a drift to low standards, and (in the case of equipment service) should give preventive maintenance high priority to avoid a spiral of equipment failures.

The resultant financial pressures engendered by the dynamics described above often drive service organizations to investments in process improvement and other cost containment initiatives to seek efficiency gains. Such investments, while offering perhaps the only long-term solution to remaining competitive, cause short-term workloads that further increase the demands on service personnel. This is demonstrated in the work of Repenning and Kofmann (1997) and Repenning and Sterman (2001, 2002).

Akkermans and Vos (2003) and Anderson et al. (2005) have studied the extent to which service industries have multistage supply chains similar to manufacturing industries, albeit with backlogs rather than inventories. Akkermans and Vos demonstrate that "inventory" cycles in service chains manifest themselves in terms of order backlog and workload cycles and that while some of the causes of amplification existent in product supply chains apply to service supply chains (demand signaling and pricing), others, particularly those related to inventory management, do not (order batching, shortage gaming). They find that the real drivers of amplification in service supply chains come from the interactions of workloads, process quality, and rework. Because of delays in hiring and firing, capacity is slow to respond to changes and is likely to exacerbate cycles. Anderson et al. (2005) find that the bullwhip effect may or may not occur in service supply chains, depending on the policies used to manage each stage. However, when it does occur, they find that the systemic improvements that can often be achieved in physical supply chains by locally applied policies (e.g., reducing delay times and sharing information) do not have as many parallels in service chains. Instead service supply chains are characterized by numerous tradeoffs between improving local performance and improving system performance.

The modeling of service delivery has also had a long history in system dynamics, though the number of published works is more modest than in other areas. Much of the early work was in the area of health care and education (Levin et al. 1976). Later works of note include models of People Express Airlines (Sterman 1989b), Hanover Insurance claims processing (Senge and

Sterman 1992), NatWest Bank lending (Oliva 1996), and DuPont chemical plant equipment maintenance (Carroll et al. 1998). Homer (1999) presents a case application for a major producer of equipment for semiconductor manufacturing that demonstrates many of the structures and policy issues enumerated above. These works incorporate the basic dynamic theory discussed above and illustrated in Fig. 8 and add another set of important structural theories to the building blocks for business strategy applications (note that the modeling of various effects on productivity is much more extensive in the area of project modeling, as discussed in (Lyneis and Ford 2007)).

Life Cycles of Products and Diffusion

Another important pattern of behavior characteristic of many firms (or subsets of firms) is that of a life cycle (for the flow) and S-shaped pattern for the stock, as illustrated in Fig. 9: a gradual increase from a low level up to a peak, followed by a gradual decline either to zero or to some replacement level (sometimes referred to as "boom and bust" behavior). Sterman (2000) and Oliva et al. (2003) provide some real-world examples of this behavior. The example shown is common for the sales of new products: the flow represents people becoming customers, and the stock, customers.

The structure which creates this "boom and bust" dynamics is shown in Fig. 10. In the marketing literature this structure is referred to as the "Bass Diffusion Model" after its original proponent (Bass 1969). The structure consists of three feedback loops: a reinforcing "word-of-mouth" loop that dominates behavior in the first half of customer sales growth; a balancing "market saturation" loop that constrains and eventually shuts down growth as the number of potential customers falls to zero; and another balancing loop "advertising saturation," which represents other means of stimulating awareness, such as advertising, direct sales efforts, and media reports. These other channels are usually assumed to be proportional to the size of the pool of potential customers and therefore initially stimulate the flow of "becoming customers" but then decline over time as the pool is depleted.

The dynamics of this structure, extensions to it (e.g., loss of customers, competition, repeat sales), and policy implications are discussed in depth in Chap. 9 in (Sterman 2000) and Chap. 6 in (Morecroft 2007). This structure forms the basis of many system dynamics models that represent product sales, customer development, and the diffusion of innovations. Published examples include the work of Milling (1996, 2002) and Maier (1998) on the management of innovation diffusions and Oliva et al. (2003) on boom and bust in e-commerce. Milling discusses the ▸ "System Dynamics Analysis of the Diffusion of Innovations" in more depth.

Growth Dynamics

Growth is fundamentally a dynamic process, and therefore it is no surprise that since its

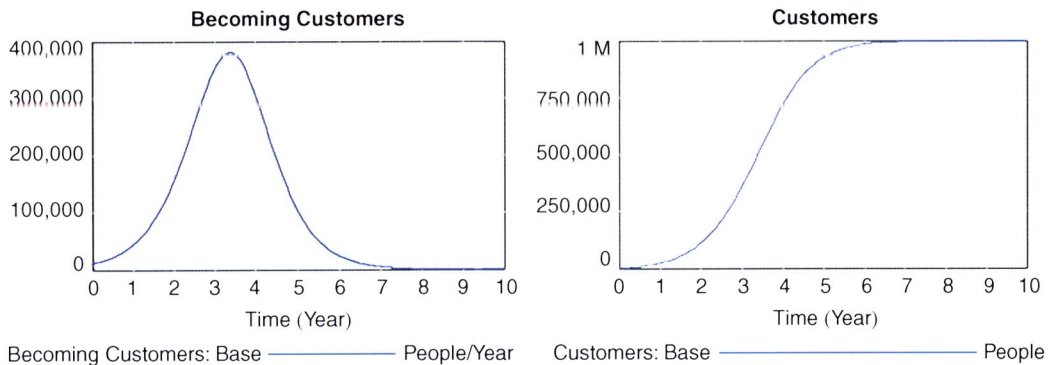

Business Policy and Strategy, System Dynamics Applications to, Fig. 9 Life cycle behavior mode ("boom and bust")

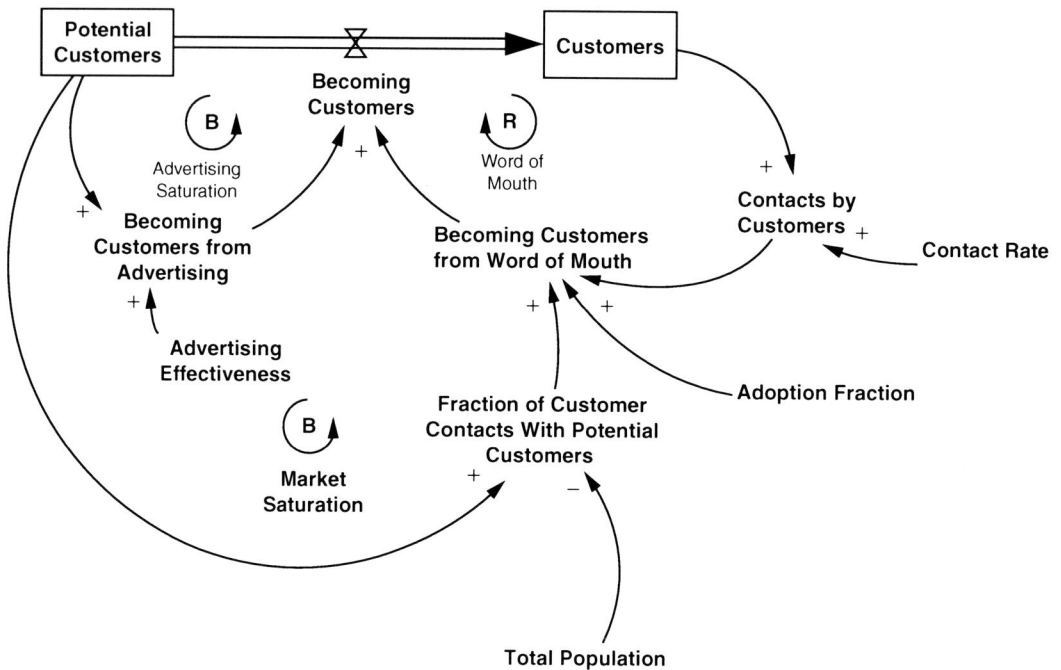

Business Policy and Strategy, System Dynamics Applications to, Fig. 10 Basic structure generating boom and bust dynamics

early days system dynamicists have shown an interest in the dynamics of corporate growth. Forrester (1964, 1968), Packer (1964), Lyneis (1975, 1980), Morecroft (1986), Morecroft and Lane (1991), and others studied corporate growth in the field's early years. More recently, the People Express (Sterman 1989b) and B&B ("boom and bust") flight simulators (Paich and Sterman 1993) illustrate the field's interest in growth dynamics.

In his 1964 article, Forrester identified the range of possible growth patterns (see Fig. 11): smooth, steady growth; growth with repeated setbacks; stagnation; and decline. Examples of these patterns can be found in many real-world industries, as illustrated for the computer industry in Fig. 11 (see also Lyneis 1998 for examples). In his classic article "Market growth as influenced by capital investment," Forrester detailed the three types of feedback loops which can create the range of possible growth patterns. These are illustrated in Fig. 11 (the equations for this model are provided in the original Forrester article; the model is also

presented and discussed and analyzed in detail in Chap. 15 in Sterman 2000 and Chap. 7 in Morecroft 2007).

On the left in Fig. 12 is the reinforcing "salesforce expansion" loop: the salesforce generates sales; a portion of those sales are allocated to future marketing budgets, which allows an increase in the size of the salesforce and a further increase in sales. The salesforce expansion loop in isolation can create smooth growth forever (until the market is saturated). However, assuming a fixed capacity, the balancing "capacity constraints" loop activates: if sales exceed capacity, delivery delay increases such that, after a delay, sales effectiveness falls and sales decline. The goal of the loop is to equate sales and capacity, and the two loops together can produce growth followed by stagnation (with fluctuations caused by delays in the balancing loop). In response to increasing delivery delay, however, firms often increase capacity ("capacity expansion" loop): when delivery delay exceeds the firm's delivery delay goal, capacity orders increase, which after a delay increases capacity and thereby reduces

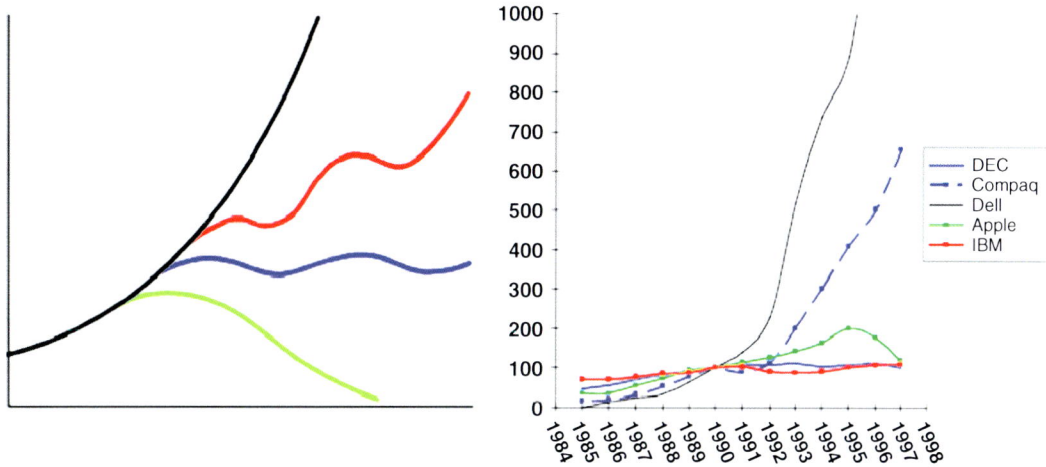

Business Policy and Strategy, System Dynamics Applications to, Fig. 11 Stylized patterns of growth and examples from the computer hardware industry

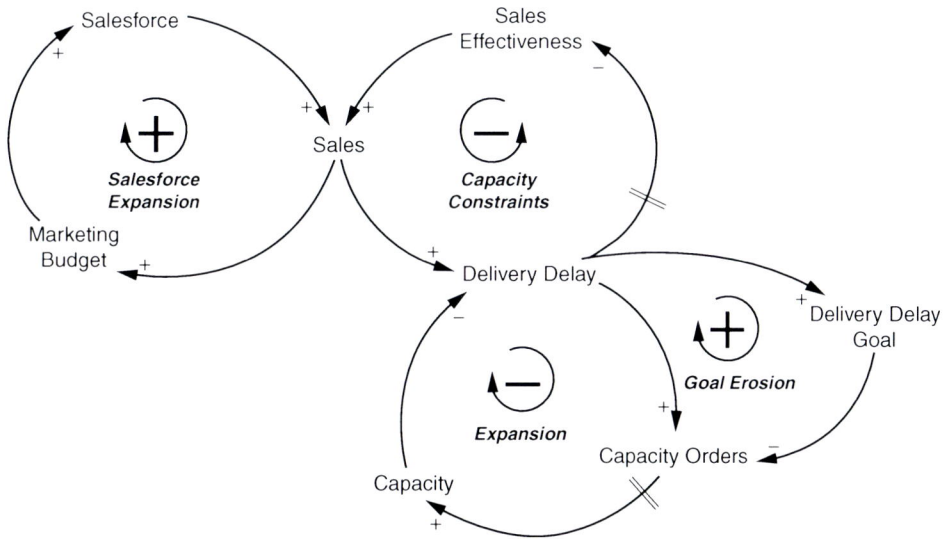

Business Policy and Strategy, System Dynamics Applications to, Fig. 12 Feedback loops creating observed patterns of growth (Adapted from Forrester 1968)

delivery delay; the goal of this loop is to equate delivery delay to the firm's delivery delay goal. However, once delivery delay is reduced, sales effectiveness and sales increase, thereby stimulating additional salesforce expansion, such that the growth with setbacks pattern of behavior can result. The final loop shown in Fig. 12 is the reinforcing "goal erosion" loop: if the firm's delivery delay goal responds to the actual delivery delay performance, a downward spiral can ensue – the goal increases, less expansion occurs than had before, capacity is less than needed, delivery delay increases, the goal is further increased, and so on (this loop is similar to the service standard goal erosion loop discussed above). The goal erosion loop can create the decline dynamics illustrated in Fig. 11 (although in actual practice the decline would likely occur over a much more extended period than shown).

In actual practice, there are numerous positive feedback loops through resources that might stimulate growth. These loops are listed below, and many are discussed and diagrammed in Chap. 10 in (Sterman 2000). In each of these loops, an increase in sales causes management actions and/or investments that further increase the resource and sales:

- Sales channels – sales capability (which might include salesforce as discussed above or retail stores), advertising, word-of-mouth contagion (as in the diffusion model), and media hype (sales create media exposure which attracts potential customers and more sales)
- Price – product attractiveness channels (operationalizing the link between resources and costs in Fig. 2 – spreading of fixed costs over more units, thereby lowering unit costs; learning curves; economies of scale; economies of scope; investments in process improvements)
- Market channels which increase the pool of potential customers – network effects (the more people using cell phones, the greater their value) and development of complementary goods (software applications for computers)
- Product investment channels – product improvement, new products
- Market power channels – over suppliers, over labor, over customers, and cost of capital

With all these positive feedback loops, how can anyone fail? In fact, there are also numerous constraints to growth, including depletion of the pool of potential customers as discussed in the last section, growth of competition, delays in acquiring production capacity and/or service capacity, limits to financial capital (which can increase delays or limit acquiring productive assets), and increases in organizational size, complexity, and administrative overheads (which might make the resources-costs loop in Fig. 2 revert to a positive connection, thereby constraining growth). Structures for representing

these constraints are provided in the earlier references to this section, especially (Lyneis 1980; Morecroft 2007; Sterman 2000), and form the building blocks for many of the practical applications of system dynamics to business growth strategy.

As system dynamicists have long recognized, managing growth is one of the more challenging management tasks. It entails fostering the positive, reinforcing feedback loops while simultaneously relaxing the constraining, negative feedback loops. While it is difficult to generalize without sounding platitudinous, a number of important lessons have emerged from studies of growth. Lyneis (1998) discusses these in more detail and references other work:

Lesson 1	You won't achieve what you don't try for (if a firm is timid in its growth objectives, it will be timid in the acquisition of resources – balancing loops through, e.g., delivery delay will then drive sales growth to the firm's resource growth). Corollary 1: don't mistake forecasts for reality (a firm may be continually surprised by how accurate their sales forecasts are, because if resources are based on these forecasts, the balancing loops will drive sales to those resources). Corollary 2: provide sufficient buffers and contingencies (these help minimize the risks that the balancing loops will become dominant)
Lesson 2	Avoid the temptation to reduce objectives in the face of performance problems (the "goal erosion" loop in Figs. 8 and 12)
Lesson 3	In a world of limited resources, something must limit growth. Proactively managing these limits is a key factor affecting performance. For example, if financial constraints are limiting expansion, with resultant delivery constraints on sales, why not increase prices to limit sales, and use the extra cash to finance expansion?
Lesson 4	Make an effort to account for delays, especially in market response (e.g., improvements in service will take a while to manifest themselves in improved sales, so avoid the temptation to cut back productive resources that will later be needed)
Lesson 5	Account for short-term productivity losses such as fatigue and experience dilution in resource expansion decisions (in the short term, you may be getting less capacity than you think)

(continued)

Lesson 6	Account for likely competitor responses in taking actions (it's easy for competitors to follow price changes and trigger a price war; improvements in other components of product attractiveness are harder to detect and replicate)

Applications and Case Examples

The process and structural developments discussed in the last sections have formed the basis of numerous applications of system dynamics in support of business strategy. In turn, these applications have provided the practice field through which the process has been refined and the structural models, insights, and tools validated. While most of the published "real-world" applications of system dynamics do not provide details of the models, they do nevertheless provide diagrams which show the nature of the model, representative results and policy conclusions, and the role the models played in business strategy formulation.

Space limitations preclude covering specific applications in depth. Therefore, I have chosen to reference applications in a number of different areas so readers can find references to the literature in their particular area of interest. Before getting to that, however, there are a couple of general references worthy of note: Roberts (1978) covers many of the early published applications of system dynamics, with sections on manufacturing, marketing, research and development, and management and financial control. Coyle (1996, 1997, 1998) touches on many of the applications initiated by his team at the University of Bradford, including work in the defense, natural resources, and utility industries. Richardson (1996) provides an edited collection of academic journal articles containing some of the best work in system dynamics for business (and public) policy from its early years to the 1990s. Beyond these general compendiums, business strategy applications can perhaps best be classified by the industry of interest.

First, there have been a number of industry-wide models. The purpose of these models is typically to understand the drivers of change in the industry and to forecast demand for use in other planning models. These include:

- Aircraft market as illustrated in Fig. 1 above (Liehr et al. 2001; Lyneis 2000)
- Health-care market (Homer et al. 2004; Thompson 2006)
- Oil market (Morecroft 2007; Morecroft and van der Heijden 1994; Sterman et al. 1988)
- Shipping market (Randers and Göluke 2007)

Second, in addition to these industry-wide models, multiple applications to specific firms (which might also include some industry and/or competitive modeling) have been done in the following industry sectors:

- Utilities/regulation – In the electric industry, work by Ford (1997), Lyneis (1997), and Bunn and Larsen (1997; Larsen and Bunn 1999) is covered elsewhere in System Dynamics Models of Environment, Energy and Climate Change.
- Telecoms (Graham and Walker 1998)
- Financial services – Work for MasterCard (Lyneis 1999) and in the insurance industry (Barlas et al. 2000; Doman et al. 1995; Senge and Sterman 1992; Thompson 1999)

Finally, applications to specific types of firms, particularly small- and medium-sized enterprises (Bianchi 2002).

For particular examples of where system dynamics has had an impact in changing or forming business strategy, the MasterCard application described by Lyneis (1999) and the General Motors OnStar application described by Barabba et al. (2002) are noteworthy. These applications provide some detail about the model structure and describe how the modeling process changed management intuition and thereby led to significant shifts in business strategy. The MasterCard model represents growth and competitive dynamics in some detail and is used to illustrate the multistage development process detailed in section "Using System Dynamics Models in Policy and Strategy

Formulation and Implementation" above. The OnStar example describes the process of modeling an industry that does not yet exist. The model itself builds from the diffusion model discussed above, with significant elaboration of potential customers, provision of service, alliances, dealers, and financial performance. The entry details the significant role that the system dynamics model played in reshaping GM's strategy.

Future Directions

System dynamics has made significant theoretical and practical contributions to business strategy. These contributions fall into two general categories: first, the process through which models are developed, working with management teams to enhance model validity and implementation, and, second, understanding of the business structures and policies which cause observed problem behavior within firms and industries. Nevertheless, the impact of system dynamics on business strategy has been modest – relatively few firms use system dynamics. In my view, several factors contribute to this slow uptake. These are listed below, with possible actions that could be taken to speed up the process:

1. Knowledge of system dynamics and its potential in strategy formulation is limited. Senior executives of firms are unaware of system dynamics, or its potential (Warren 2004). In part this is because system dynamics is taught at relatively few business schools. Over time, this problem will be solved, but it will take a long time. But more importantly, researchers and practitioners of system dynamics rarely publish their work in publications which are widely read by senior management, such as the Harvard Business Review. This is a problem which can be addressed in the near future if researchers and practitioners made the effort to communicate with this market.
2. System dynamics is not well connected with traditional strategy research and practice. As noted earlier, system dynamics and traditional strategy developed largely independently. While the potential synergies between the two

are significant, until recently few researchers and practitioners have made the effort to cross disciplines. More effort and publication are needed to demonstrate areas of synergy and get system dynamics into the mainstream of business strategy research and ultimately practice. Warren (2007) makes a start on this.
3. System dynamics is hard. Building system dynamics models, calibrating them to data, and analyzing their behavior to improve business strategy requires significant skill and experience. This has traditionally been developed via an apprenticeship program, either in university or consulting firms. Much more can be done to hasten this process. First, tried and true model structures that can be used as building blocks for models must be better documented and gaps closed. While the theoretical basis for business dynamics described above is a good start, and reflects structures which all practicing system dynamicists should know, the underlying models and building blocks are widely dispersed and difficult to access. For example, the models in (Sterman 2000) are contained within a 900-page introductory textbook; the models in (Lyneis 1980) are in old software, and the book is out of print. In both cases, the models are only starting points, and much unpublished work has occurred that expands these introductory models to make them more relevant and directly applicable to real business strategy problems. Efforts could and should be made to expand the library of business strategy building blocks. In addition, the process of building models and working with managers needs to be better documented and disseminated, both in formal courses and in published works, so as to facilitate the apprenticeship learning process.

Bibliography

Akkermans HA, Daellaert N (2005) The rediscovery of industrial dynamics: the contribution of system dynamics to supply chain management in a dynamic and fragmented world. Syst Dyn Rev 21(3):173–186

Akkermans HA, Vos B (2003) Amplification in service supply chains: an exploratory case study from the telecom industry. Prod Oper Manag 12(2):204–223

Anderson E, Fine C (1999) Business cycles and productivity in capital equipment supply chains. In: Tayur S et al (eds) Quantitative models for supply chain management. Kluwer, Norwell

Anderson E, Morrice D, Lundeen G (2005) The 'physics' of capacity and backlog management in service and custom manufacturing supply chains. Syst Dyn Rev 21(3):187–216

Backus G, Schwein MT, Johnson ST, Walker RJ (2001) Comparing expectations to actual events: the post mortem of a Y2K analysis. Syst Dyn Rev 17(3): 217–235

Barabba V, Huber C, Cooke F, Pudar N, Smith J, Paich M (2002) A multimethod approach for creating new business models: the general motors on star project. Interfaces 32(1):20–34

Barlas Y, Cırak K, Duman E (2000) Dynamic simulation for strategic insurance management. Syst Dyn Rev 16(1):43–58

Bass FM (1969) New product growth model for consumer durables. Manag Sci 15:215–227

Bianchi C (2002) Editorial to special issue on systems thinking and system dynamics in small-medium enterprises. Syst Dyn Rev 18(3):311–314

Bunn DW, Larsen ER (eds) (1997) Systems modeling for energy policy. Wiley, Chichester

Campbell D (2001) The long and winding (and frequently bumpy) road to successful client engagement: one team's journey. Syst Dyn Rev 17(3):195–215

Carroll JS, Sterman JD, Marcus AA (1998) Playing the maintenance game: how mental models drive organizational decisions. In: Halpern JJ, Stern RN (eds) Nonrational elements of organizational decision making. Cornell University Press, Ithaca

Coyle RG (1996) System dynamics modelling: a practical approach. Chapman and Hall, London

Coyle RG (1997) System dynamics at Bradford University: a silver jubilee review. Syst Dyn Rev 13(4): 311–321

Coyle RG (1998) The practice of system dynamics: milestones, lessons and ideas from 30 years experience. Syst Dyn Rev 14(4):343–365

Coyle RG (2000) Qualitative and quantitative modeling in system dynamics: some research questions. Syst Dyn Rev 16(3):225–244

Coyle RG (2001) Rejoinder to Homer and Oliva. Syst Dyn Rev 17(4):357–363

Coyle RG, Morecroft JDW (1999) System dynamics for policy, strategy and management education. J Oper Res Soc 50(4):291–449

Croson R, Donohue K (2005) Upstream versus downstream information and its impact on the bullwhip effect. Syst Dyn Rev 21(3):187–216

Delauzun F, Mollona E (1999) Introducing system dynamics to the BBC world service: an insider perspective. J Oper Res Soc 50(4):364–371

Doman A, Glucksman M, Mass N, Sasportes M (1995) The dynamics of managing a life insurance company. Syst Dyn Rev 11(3):219–232

Dyson RG, Bryant J, Morecroft J, O'Brien F (2007) The strategic development process. In: O'Brien FA, Dyson RG (eds) Supporting strategy. Wiley, Chichester

Ford AJ (1997) System dynamics and the electric power industry. Syst Dyn Rev 13(1):57–85

Forrester JW (1958) Industrial dynamics: a major breakthrough for decision makers. Harv Bus Rev 36 (4):37–66

Forrester JW (1961) Industrial dynamics. MIT Press, Cambridge (now available from Pegasus Communications, Waltham)

Forrester JW (1964) Common foundations underlying engineering and management. IEEE Spectr 1(9):6–77

Forrester JW (1968) Market growth as influenced by capital investment. Ind Manag Rev 9(2):83–105 (Reprinted in Forrester JW (1975) Collected papers of Jay W Forrester. Pegasus Communications, Waltham)

Forrester JW (1969) Urban dynamics. Pegasus Communications, Waltham

Foss NJ (ed) (1997) Resources, firms and strategies. Oxford University Press, Oxford

Gary MS (2005) Implementation strategy and performance outcomes in related diversification. Strateg Manag J 26:643–664

Gonçalves PM (2003) Demand bubbles and phantom orders in supply chains. Unpublished dissertation Sloan School of Management, MIT, Cambridge

Gonçalves PM (2006) The impact of customer response on inventory and utilization policies. J Bus Logist 27 (2):103–128

Gonçalves P, Hines J, Sterman J (2005) The impact of endogenous demand on push-pull production systems. Syst Dyn Rev 21(3):187–216

Graham AK, Walker RJ (1998) Strategy modeling for top management: going beyond modeling orthodoxy at Bell Canada. In: Proceedings of the 1998 international system dynamics conference, Quebec

Grant RM (2005) Contemporary strategy analysis, 5th edn. Blackwell, Cambridge

Hines JH, Johnson DW (1994) Launching system dynamics. In: Proceedings of the 1994 international system dynamics conference, business decision-making, Stirling

Homer JB (1996) Why we iterate: scientific modeling in theory and practice. Syst Dyn Rev 12(1):1–19

Homer JB (1997) Structure, data, and compelling conclusions: notes from the field. Syst Dyn Rev 13 (4):293–309

Homer JB (1999) Macro- and micro-modeling of field service dynamics. Syst Dyn Rev 15(2):139–162

Homer J, Oliva R (2001) Maps and models in system dynamics: a response to Coyle. Syst Dyn Rev 17 (4):347–355

Homer J, Hirsch G, Minniti M, Pierson M (2004) Models for collaboration: how system dynamics helped a community organize cost-effective care for chronic illness. Syst Dyn Rev 20(3):199–222

Kim DH, Lannon C (1997) Applying systems archetypes. Pegasus Communications Inc, Waltham

Kunc M, Morecroft J (2007) System dynamics modelling for strategic development. In: O'Brien FA, Dyson RG (eds) Supporting strategy. Wiley, Chichester

Larsen ER, Bunn DW (1999) Deregulation in electricity: understanding strategic and regulatory risks. J Oper Res Soc 50(4):337–344

Levin G, Roberts EB, Hirsch GB (1976) The dynamics of human service delivery. Ballinger, Cambridge

Liehr M, Großler A, Klein M, Milling PM (2001) Cycles in the sky: understanding and managing business cycles in the airline market. Syst Dyn Rev 17(4):311–332

Lyneis JM (1975) Designing financial policies to deal with limited financial resources. Financ Manag 4(1)

Lyneis JM (1980) Corporate planning and policy design: a system dynamics approach. M.I.T Press, Cambridge

Lyneis JM (1997) Preparing for a competitive environment: developing strategies for America's electric utilities. In: Bunn DW, Larsen ER (eds) Systems modeling for energy policy. Wiley, Chichester

Lyneis JM (1998) Learning to manage growth: lessons from a management flight simulator. In: Proceedings of the 1998 international system dynamics conference (plenary session), Quebec City, 1998

Lyneis JM (1999) System dynamics for business strategy: a phased approach. Syst Dyn Rev 15(1):37–70

Lyneis JM (2000) System dynamics for market forecasting and structural analysis. Syst Dyn Rev 16(1):3–25

Lyneis JM, Ford DN (2007) System dynamics applied to project management: a survey, assessment, and directions for future research. Syst Dyn Rev 23(2/3):157–189

Lyneis JM, Cooper KG, Els SA (2001) Strategic management of complex projects: a case study using system dynamics. Syst Dyn Rev 17(3):237–260

Magee J (1958) Production planning and inventory control. McGraw-Hill, London

Maier FH (1998) New product diffusion models in innovation management – a system dynamics perspective. Syst Dyn Rev 14(4):285–308

Meadows DL (1970) Dynamics of commodity production cycles. Pegasus Communications, Waltham

Merten PP, Loffler R, Wiedmann KP (1987) Portfolio simulation: a tool to support strategic management. Syst Dyn Rev 3(2):81–101

Milling PM (1996) Modeling innovation processes for decision support and management simulation. Syst Dyn Rev 12(3):211–223

Milling PM (2002) Understanding and managing innovation processes. Syst Dyn Rev 18(1):73–86

Milling PM (2007) A brief history of system dynamics in continental Europe. Syst Dyn Rev 23(2/3):205–214

Montgomery CA (ed) (1995) Resource-based and evolutionary theories of the firm. Kluwer, Boston

Morecroft JDW (1983) A systems perspective on material requirements planning. Decis Sci 14(1):1–18

Morecroft JDW (1984) Strategy support models. Strateg Manag J 5:215–229

Morecroft JDW (1985) The feedback view of business policy and strategy. Syst Dyn Rev 1(1):4–19

Morecroft JDW (1986) The dynamics of a fledgling high-technology growth market. Syst Dyn Rev 2(1):36–61

Morecroft JDW (1999) Management attitudes, learning and scale in successful diversification: a dynamic and behavioural resource system view. J Oper Res Soc 50(4):315–336

Morecroft JDW (2007) Strategic modelling and business dynamics: a feedback systems view. Wiley, Chichester

Morecroft JDW, van der Heijden KAJM (1994) Modeling the oil producers: capturing oil industry knowledge in a behavioral simulation model. In: Morecroft JDW, Sterman JD (eds) Modeling for learning organizations. Productivity Press, Portland

Morecroft JDW, Wolstenholme E (2007) System dynamics in the UK: a journey from Stirling to Oxford and beyond. Syst Dyn Rev 23(2/3):205–214

Morecroft JDW, Lane DC, Viita PS (1991) Modeling growth strategy in a biotechnology startup firm. Syst Dyn Rev 7(2):93–116

O'Brien FA, Dyson RG (eds) (2007) Supporting strategy. Wiley, Chichester

Oliva R (1996) A dynamic theory of service delivery: implications for managing service quality. PhD thesis, Sloan School of Management, Massachusetts Institute of Technology

Oliva R (2001) Tradeoffs in responses to work pressure in the service industry. Calif Manag Rev 43(4):26–43

Oliva R, Sterman JD, Giese M (2003) Limits to growth in the new economy: exploring the 'get big fast' strategy in e-commerce. Syst Dyn Rev 19(2):83–117

Packer DW (1964) Resource acquisition in corporate growth. M.I.T Press, Cambridge

Paich M, Sterman JD (1993) Boom, bust, and failures to learn in experimental markets. Manag Sci 39(12):1439

Porter ME (1980) Competitive strategy. The Free Press, New York

Porter ME (1985) Competitive advantage. The Free Press, New York

Prahalad CK, Hamel G (1990) The core competence of the corporation. Harv Bus Rev (May–June):79–91

Randers J, Göluke U (2007) Forecasting turning points in shipping freight rates: lessons from 30 years of practical effort. Syst Dyn Rev 23(2/3):253–284

Repenning N, Kofmann F (1997) Unanticipated side effects of successful quality programs: exploring a paradox of organizational improvement. Manag Sci 43(4):503–521

Repenning NP, Sterman JD (2001) Nobody ever gets credit for fixing problems that never happened: creating and sustaining process improvement. Calif Manag Rev 43(4):64–88

Repenning NP, Sterman JD (2002) Capability traps and self-confirming attribution errors in the dynamics of process improvement. Adm Sci Q 47:265–295

Richardson GP (1996) Modelling for management: simulation in support of systems thinking. Dartmouth, Aldershot

Richardson GP, Andersen DF (1995) Teamwork in group model building. Syst Dyn Rev 11(2):113–138

Richmond B (1997) The strategic forum: aligning objectives, strategy and process. Syst Dyn Rev 13(2): 131–148

Risch J, Troyano-Bermúdez L, Sterman J (1995) Designing corporate strategy with system dynamics: a case study in the pulp and paper industry. Syst Dyn Rev 11(4):249–274

Roberts EB (1977) Strategies for the effective implementation of complex corporate models. Interfaces 8(1):26–33

Roberts EB (ed) (1978) Managerial applications of system dynamics. The MIT Press, Cambridge

Roberts EB (2007) Making system dynamics useful: a personal memoir. Syst Dyn Rev 23(2/3):119–136

Rockart SF, Lenox MJ, Lewin AY (2007) Interdependency, competition, and industry dynamics. Manag Sci 53(4): 599–615

Senge PM (1990) The fifth discipline: the art and practice of the learning organization. Doubleday, New York

Senge PM, Sterman JD (1992) Systems thinking and organizational learning: acting locally and thinking globally in the organization of the future. Eur J Oper Res 59(1):137–150

Sice P, Mosekilde E, Moscardini A, Lawler K, French I (2000) Using system dynamics to analyse interactions in duopoly competition. Syst Dyn Rev 16(2):113–133

Snabe B, Grossler A (2006) System dynamics modelling for strategy implementation-case study and issues. Syst Res Behav Sci 23:467–481

Sterman JD (1989a) Modeling management behavior: misperceptions of feedback in a dynamic decision making experiment. Manag Sci 35:321–339

Sterman JD (1989b) Strategy dynamics: the rise and fall of people express. Memorandum D-3939-1 Sloan School of Management, M.I.T, Cambridge, MA.

Sterman JD (2000) Business dynamics: systems thinking and modeling for a complex world. McGraw-Hill, New York

Sterman JD, Richardson G, Davidsen P (1988) Modeling the estimation of petroleum resources in the United States. Technol Forecast Soc Chang 33(3): 219–249

Thompson JP (1999) Consulting approaches with system dynamics: three case studies. Syst Dyn Rev 15(1): 71–95

Thompson JP (2006) Making sense of US health care system dynamics. In: Proceedings of the 2006 international system dynamics conference, Nijmegen

Vennix JAM (1996) Group model building: facilitating team learning using system dynamics field. Wiley, Chichester

Warren KD (2002) Competitive strategy dynamics. Wiley, Chichester

Warren KD (2004) Why has feedback systems thinking struggled to influence strategy and policy formulation? Suggestive evidence, explanations and solutions. Syst Res Behav Sci 21:331–347

Warren KD (2005) Improving strategic management with the fundamental principles of system dynamics. Syst Dyn Rev 21(4):329–350

Warren KD (2007) Strategic management dynamics. Wiley, Chichester

Weil HB (1980) The evolution of an approach for achieving implemented results from system dynamics projects. In: Randers J (ed) Elements of the system dynamics method. MIT Press, Cambridge

Winch GW (1993) Consensus building in the planning process: benefits from a "hard" modeling approach. Syst Dyn Rev 9(3):287–300

Winch GW (1999) Dynamic visioning for dynamic environments. J Oper Res Soc 50(4):354–361

Winch GW (2001) Management of the "skills inventory" in times of major change. Syst Dyn Rev 17(2):151–159

Wolstenholme EF (1999) Qualitative vs. quantitative modelling: the evolving balance. J Oper Res Soc 50 (4):422–428

Zahn E, Dillerup R, Schmid U (1998) Strategic evaluation of flexible assembly systems on the basis of hard and soft decision criteria. Syst Dyn Rev 14(4): 263–284

System Dynamics Applications to Health Care in the United States

Gary B. Hirsch[1] and Jack Homer[2]
[1]Creator of Learning Environments, Wayland, MA, USA
[2]Homer Consulting, New York, USA

Article Outline

Glossary
Definition of the Subject
Introduction
Models of Specific Chronic Illnesses
Models for Exploring Health System Reform
Future Directions
Bibliography

Glossary

Chronic illness A disease or adverse health state that persists over time and cannot in general be cured, although its symptoms may be treatable.

Feedback loop A closed loop of causality that acts to counterbalance or reinforce prior change in a system state.

Flow A rate-of-change variable affecting a stock, such as births flowing into a population or deaths flowing out.

Stock An accumulation or state variable, such as the size of a population.

Definition of the Subject

Healthcare involves a complex system of interactions among patients, providers, payers, and other stakeholders. This system is difficult to manage in the United States because of its free market approach and relative lack of regulation. System dynamics (SD) simulation modeling is an effective method for understanding and explaining causes of dysfunction in US healthcare and for suggesting approaches to improving health outcomes and slowing rising costs. Applications since the 1970s have covered diverse areas in healthcare including the epidemiology of diseases and substance abuse as well as the dynamics of health-care capacity and delivery and their impacts on health. Many of these applications have dealt with the mounting burden of chronic illnesses, such as diabetes. In this entry several such applications are described. Several applications to health reform are also described.

Introduction

Despite remarkable successes in some areas, the health enterprise in the United States faces difficult challenges in meeting its primary goal of reducing the burden of disease and injury. These challenges include the growth of the underinsured population, epidemics of obesity and asthma, drug-resistant infectious diseases, ineffective management of chronic illness, long-standing racial and ethnic health disparities, and an overall decline in the health-related quality of life (Zack et al. 2004). Many of these complex problems have persisted for decades, often proving resistant to attempts to solve them (Lee and Paxman 1997).

It has been argued that these interventions fail because they are made in piecemeal fashion, rather than comprehensively and from a whole-system perspective (Heirich 1999). This compartmentalized approach is engrained in the financial structures, intervention designs, and evaluation methods of most health agencies. Conventional analytic methods are generally unable to satisfactorily address situations in which population needs change over time (often in response to the interventions themselves) and in which risk factors, diseases, and health resources are in a continuous state of interaction and flux (Schorr 1997).

© Springer Science+Business Media, LLC, part of Springer Nature 2020
B. Dangerfield (ed.), *System Dynamics*,
https://doi.org/10.1007/978-1-4939-8790-0_270

Originally published in
R. A. Meyers (ed.), *Encyclopedia of Complexity and Systems Science*, © Springer Science+Business Media LLC 2017, https://doi.org/10.1007/978-3-642-27737-5_270-3

The term *dynamic complexity* has been used to describe such evolving situations (Sterman 2000). Dynamically complex problems are often characterized by long delays between causes and effects and by multiple goals and interests that may in some ways conflict with one another. In such situations, it is difficult to know how, where, and when to intervene, because most interventions will have unintended consequences and will tend to be resisted or undermined by opposing interests or as a result of limited resources or capacities.

The SD systems modeling methodology is well suited to addressing the challenges of dynamic complexity in public health. The methodology involves the development of causal diagrams and policy-oriented computer simulation models that are unique to each problem setting. The approach was developed by computer pioneer Jay W. Forrester in the mid-1950s and first described at length in his book *Industrial Dynamics* (Forrester 1961) with some additional principles presented in later works (Forrester 1980; Forrester and Senge 1980). The International System Dynamics Society was established in 1983, and within the society a special interest group on health issues was organized in 2003.

SD modeling has been applied to health and health-care issues in the United States since the 1970s (Hirsch et al. 2015). Topic areas in recent years have included:

Disease epidemiology including work in cardiovascular disease (Hirsch et al. 2010; Homer et al. 2004, 2010, 2014; Lich et al. 2014), diabetes (Homer et al. 2004; Jones et al. 2006; Milstein et al. 2007), obesity (Homer et al. 2006), polio (Thompson and Tebbens 2007), drug-resistant infections (Homer et al. 2000), pandemic influenza (Hirsch 2006), domestic violence. (Hovmand et al. 2009), and social determinants of health (Mahamoud et al. 2012).

Substance abuse epidemiology including smoking (Tengs et al. 2001) and abuse of prescription opioids (Wakeland et al. 2011).

Health-care capacity and delivery in such areas as dental care (Edelstein et al. 2015; Hirsch et al. 2012a), organ transplantation (Hirsch et al. 2012b), and as affected by natural disasters or terrorist acts (Hirsch 2004; Hoard et al. 2005; LeClaire et al. 2009; Manley et al. 2005).

Interactions between health-care or public health capacity and disease epidemiology (Hirsch and Homer 2004; Hirsch and Immediato 1999; Homer et al. 2007, 2016; Homer and Milstein 2004).

Most of these modeling efforts have been done with the close involvement of clinicians and policymakers who have a direct stake in the problem being modeled. Established SD techniques for group model building (Vennix 1996) can help to harness the insights and involvement of those who deal with public health problems on a day-to-day basis.

It is useful to consider how SD models compare with those of other simulation methods that have been applied to public health issues, particularly in epidemiological modeling. One may characterize any population health model in terms of its degree of aggregation, that is, the extent to which individuals in the population are combined together in categories of disease, risk, or age and other demographic attributes. At the most aggregate end of the scale are lumped contagion models (Anderson 1994; Kaplan et al. 2002); the more disaggregated are Markov models (Gunning-Schepers 1989; Honeycutt et al. 2003; Naidoo et al. 1997); the most disaggregated are microsimulations at the level of individuals (Halloran et al. 2002; Schlessinger and Eddy 2002; Wolfson 1994).

Most SD population health models are relatively high in aggregation, with a broad model boundary sufficient to include a variety of realistic causal factors, policy levers, and feedback loops. Although it is possible to build models that are both broad in scope and highly disaggregated, experience suggests that such very large models nearly always suffer in terms of their ability to be easily and fully tested, understood, and maintained. In choosing between broader scope and finer disaggregation, SD modelers tend to opt for the former, because a broad scope is generally needed for diagnosing and finding effective solutions to dynamically complex problems (Sterman 1988, 2000).

The remainder of this entry describes in detail several specific examples of real-world SD application, some addressing specific chronic illnesses and others addressing health system reform more broadly. The entry concludes with a discussion of promising areas for future work.

Models of Specific Chronic Illnesses

Chronic illness is responsible for at least 70% of all deaths and 75% of all health-care costs in the United States (Centers for Disease Control and Prevention, National Center for Chronic Disease Prevention and Health Promotion (CDC/NCCDPHP) 2007). Four applications are presented here dealing with diabetes and cardiovascular disease at the local and national levels.

Diabetes and Heart Failure Management at the Community Level

Two hours north of Seattle in the state of Washington lies Whatcom County, with a population of about 170,000. The county embarked on a major effort to address chronic illness care and was selected by the Robert Wood Johnson Foundation as one of seven sites in a larger chronic care program called Pursuing Perfection (Homer et al. 2004). The program initially concentrated on two chronic illnesses as prototypes for improved care: diabetes and congestive heart failure. Both of these illnesses affect millions of people in the United States and other countries and exact a heavy toll in terms of direct medical expenditures as well as indirect costs due to disability and premature mortality (American Heart Association (AHA) 2009; National Institute of Diabetes and Digestive and Kidney Diseases (NIDDK) 2004; O'Connell and Bristow 1993). The prevalence of both diseases is growing rapidly as the numbers of people above age 65 increase and also due to the epidemic rise in obesity, which is a risk factor for both diabetes and heart disease (Flegal et al. 2002).

Leaders of the Whatcom County program had two critical needs for making decisions about potential interventions for improving the care of chronic illnesses such as diabetes and heart failure. First, they wanted to get a sense of the overall impact of these interventions on incidence and prevalence of diabetes and heart failure, health-care utilization and cost, and mortality and disability rates in the community. Second, they wanted to understand the impact of the various interventions on health-care providers, insurers, employers, and patients in the community and on those who pay for care insurers, employers, and patients themselves. There was a concern that the costs and benefits of the program be shared equitably and that providers who helped produce savings should not suffer a resulting loss of revenue to their businesses.

These analytic needs could not be met with spreadsheet and other models that project impacts in a simple, linear fashion. Interventions in chronic illness do not have simple direct impacts. The aging of the population, incidence of new cases, disease progression, deaths, and interventions themselves all create a constantly changing situation. Interventions ideally reduce mortality rates, but this leaves more people with the disease alive and requiring care.

Figure 1 presents a simplified view of the stock-and-flow structure used in modeling non-insulin-dependent (type 2) diabetes. The actual model has two separate structures like those shown in Fig. 1, one for the 18–64 age group and one for the 65 and older age group, which are linked by flows of patients turning 65. The model also calculates an inflow of population turning 18, death outflows from each stock based on patient age and stage of illness, and flows of migration into and out of the county. The rectangular boxes in Fig. 1 represent subpopulations with particular characteristics. The arrows signify flows of people from one population group to another (e.g., from uncontrolled to controlled diabetes at a particular stage). Lines from ovals (programmatic interventions such as disease management) to population flows indicate control of or influence on those flows.

The three stages of diabetes portrayed in this figure were identified through discussions with clinicians in Whatcom County. The population at risk includes those with family history, the obese, and, most directly, those with a condition of moderately elevated blood sugar known as predi-abetes. Further increases in blood sugar lead to stage 1 diabetes, in which blood vessels suffer degradation, but there is not yet any damage to organs of the body nor typically any symptoms of the encroaching disease. More than half of stage 1 diabetics are undiagnosed. If stage 1 diabetics go untreated, most will eventually progress to stage 2, marked by organ disease. In stage 2 diabetes, blood flow disturbances impair the functioning of organ systems and potentially lead to irreversible damage. A patient who has suffered irreversible organ damage, or organ failure, is said to be in stage 3; this would include diabetics who suffer heart attacks, strokes, blindness, amputations, or end-stage renal disease. These patients are at the greatest risk of further complications leading to death.

Studies show that the incidence, progression, complications, and costs of diabetes can be reduced significantly through concerted intervention (American Diabetes Association (ADA) and National Institute of Diabetes and Digestive and Kidney Diseases (NIDDK) 2002; Diabetes Prevention Program Research Group (DPPRG) 2002; U.K. Prospective Diabetes Study Group (UKPDSG) 1998; Wagner et al. 2001). Such intervention may include primary prevention or disease management. As indicated in Fig. 1, primary prevention would consist of efforts to screen the at-risk population and educate them about the diet and activity changes they need to prevent progression to diabetes. Disease management, on the other hand, addresses existing diabetics. A comprehensive disease management approach, such as that employed by the Whatcom County program, can increase the fraction of patients who are able to keep their blood sugar under effective control from the 40% or less typically seen without a program up to perhaps 80% or more.

The SD model of diabetes in Whatcom County was first used to produce a 20 year status quo or baseline projection, which assumes that no intervention program is implemented. In this projection, the prevalence of diabetes among all adults gradually increases from 6.5% to 7.5%, because

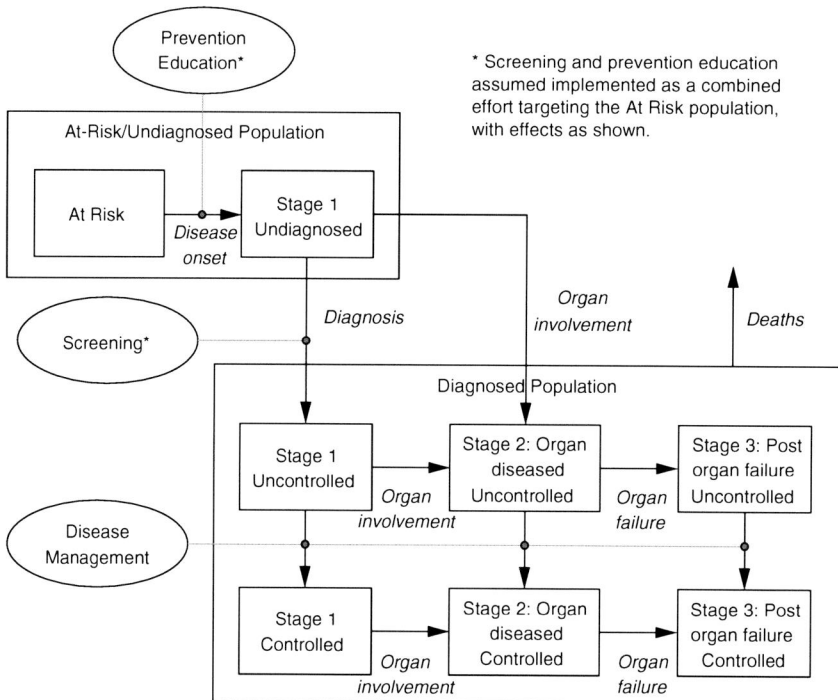

System Dynamics Applications to Health Care in the United States, Fig. 1 Disease stages and intervention points in the Whatcom County Diabetes Model

of a growing elderly population; the prevalence of diabetes among the elderly is 17%, compared with 5% among the non-elderly. Total costs of diabetes, including direct costs for healthcare and pharmaceuticals and indirect economic losses due to disability, grow substantially in this baseline projection.

The next step was to use the model to examine the impact of various program options. These included (1) a partial approach enhancing disease management but not primary prevention, (2) a full implementation approach combining enhancement of both disease management and primary prevention, and (3) an approach that goes beyond full implementation by also providing greater financial assistance to the elderly for purchasing drugs needed for the control of diabetes.

Simulations of these options projected results in terms of various outcome variables, including deaths from complications of diabetes and total costs of diabetes. Figure 2 shows typical simulation results obtained by projecting these options, in this case, the number of deaths over time that might be expected due to complications of diabetes. Under the status quo projection, the number of diabetes-related deaths grows continuously along with the size of the diabetic population. The partial (disease management only) approach is effective at reducing deaths early on but becomes increasingly less effective over time. The full program approach (including primary prevention)

overcomes this shortcoming and, by the end of the 20-year simulation, reduces diabetes-related deaths by 40% relative to the status quo. Addition of a drug purchase plan for the elderly does even better, facilitating greater disease control and thereby reducing diabetes-related deaths by 54% relative to the status quo.

With regard to total costs of diabetes, the simulations indicate that the full program approach can achieve net savings only 2 years after the program is launched. By the end of 20 years, the full program approach results in a net savings amounting to 7% of the status quo costs, two-thirds of that savings coming from reduction in disability-related costs. The model suggests that these anticipated net savings are the result of keeping people in the less severe stages of the diseases for a longer period of time and reducing the number of diabetes-related hospitalizations.

The simulations provided important information and ideas to the Whatcom County program planners as well as supporting detailed discussions of how various costs and benefits could be equitably distributed among the participants. This helped to reassure participants that none of them would be unfairly affected by the proposed chronic illness program. Perhaps the most important contribution of modeling to the program planning process was its ability to demonstrate that the program, if implemented in its full form, would likely reduce total costs, even though it would

System Dynamics Applications to Health Care in the United States, Fig. 2 Typical results from policy simulations with the Whatcom County Diabetes Model

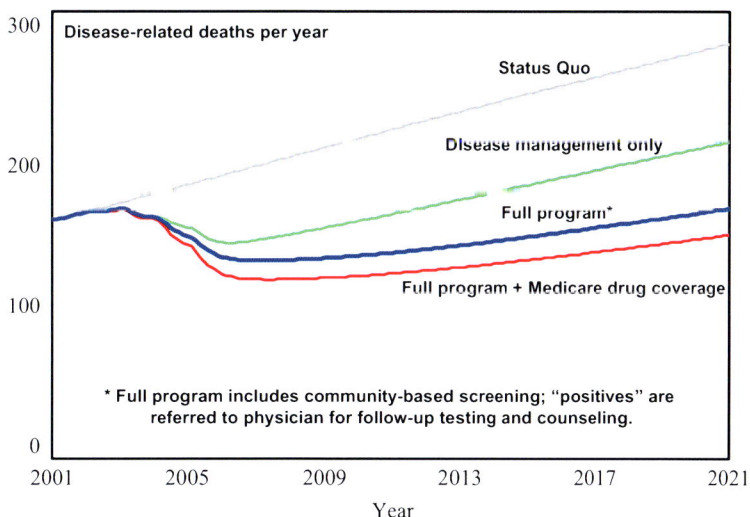

extend the longevity of many diabetics requiring costly care. Given the sensitivity of payers who were already bearing high costs, this finding helped to motivate their continued participation in the program.

Diabetes Prevention and Management at the National Level

Another SD model of diabetes in the population was developed for the Division of Diabetes Translation of the US Centers for Disease Control and Prevention (CDC) (Jones et al. 2006). This model, a structural overview of which is presented in Fig. 3, builds upon the Whatcom County work but looks more closely at the drivers of diabetes onset, including the roles of prediabetes and obesity. The core of the CDC model is a chain of population stocks and flows portraying the movement of people among the stages of normal blood glucose, prediabetes, uncomplicated diabetes, and complicated diabetes. The prediabetes and diabetes stages are further divided among stocks of people whose conditions are diagnosed or

undiagnosed. Also shown in Fig. 3 are the potentially modifiable influences in the model that affect the rates of population flow. These flow-rate drivers include obesity and the detection and management of prediabetes and of diabetes.

The model's parameters were calibrated based on historical data available for the US adult population as well as estimates from the scientific literature. The model is able to reproduce historical time series, some going as far back as 1980, on diagnosed diabetes prevalence, the diagnosed fraction of diabetes, prediabetes prevalence, the obese fractions of people with prediabetes and diabetes, and the health burden (specifically, the mortality, morbidity, and costs) attributable to diabetes. The model suggests that two forces worked in opposition to affect the diabetes health burden from 1980 to 2004. The first force is a rise in the prevalence of obesity, which led to a greater incidence and prevalence of prediabetes and diabetes through the chain of causation seen in Fig. 3. The second and opposing force is a significant improvement in the control of diabetes, achieved

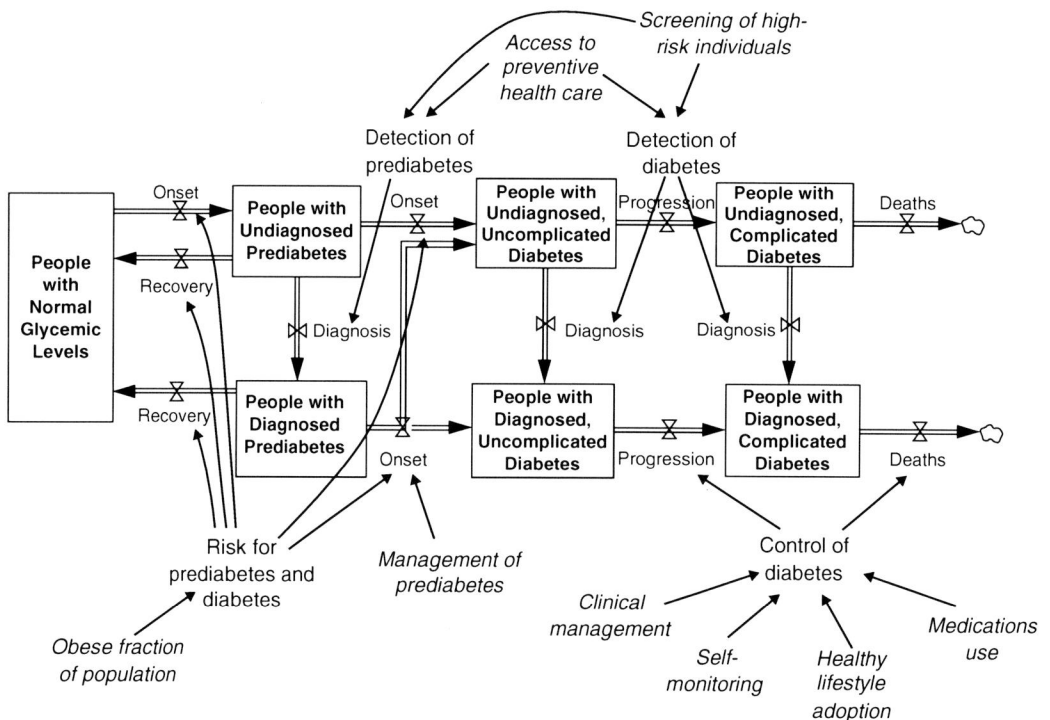

System Dynamics Applications to Health Care in the United States, Fig. 3 Structure of the CDC diabetes model

through greater efforts to detect and manage the disease. The second force managed to hold the health burden of diabetes more or less flat during 1980–2004.

Looking toward the future, a baseline scenario assumes that no further changes occur in obesity prevalence after 2006 and that inputs affecting the detection and management of prediabetes and diabetes remain fixed at their 2004 values through 2050. This fixed-inputs assumption for the baseline scenario is not meant to represent a forecast of what is most likely to in the future but does provide a useful and easily understood starting point for policy analysis.

The CDC diabetes model has been used to examine a variety of future scenarios involving policy interventions intended to limit growth in the burden of diabetes. These include scenarios improving the management of diabetes, increasing the management of prediabetes, or reducing the prevalence of general population obesity over time. Enhanced diabetes management can significantly reduce the burden of diabetes in the short term, but does not prevent the growth of greater burden in the longer term due to the growth of diabetes prevalence. Indeed, the effect of enhanced diabetes management on diabetes prevalence is not to decrease it at all but rather to increase it somewhat by increasing the longevity of people with diabetes.

Increased management of prediabetes does, in contrast, reduce diabetes onset and the growth of diabetes prevalence. However, it does not have as much impact as one might expect. This is because many people with prediabetes are not diagnosed and also because the policy does nothing to reduce the growth of prediabetes prevalence in the general population.

A reduction in prediabetes can be achieved only by reducing population obesity. Significant obesity reduction may take 20 years or more to accomplish fully, but the model suggests that such a policy can be quite a powerful one in halting the growth of diabetes prevalence and burden even before those 20 years are through.

Overall, the CDC model suggests that no single type of intervention is sufficient to limit the growth of the diabetes burden in both the short

term and the long term. Rather, what is needed is a combination of disease management for the short term and primary prevention for the longer term. The model also suggests that effective primary prevention may require obesity reduction in the general population and a focus on managing diagnosed prediabetes.

At the state and regional level, the CDC diabetes model has become the basis for a model-based workshop called the "Diabetes Action Lab." Participants have included state and local public health officials along with nongovernmental stakeholders including health-care professionals, leaders of not-for-profit agencies, and advocates for people living with diabetes. The workshops have helped the participants improve their intervention strategies and goals and become more hopeful and determined about seeing their actions yield positive results in the future.

The CDC diabetes model has led to other SD modeling efforts at the CDC emphasizing disease prevention, including studies of obesity (Homer et al. 2006) and cardiovascular risk (see PRISM model below). The obesity study involved the careful analysis of population survey data to identify patterns of weight gain over the entire course of life from childhood to old age. It explored likely impacts decades into the future of interventions to reduce or prevent obesity that may be targeted at specific age categories. Key findings included (1) that the average amount of caloric reduction necessary to reverse the growth of obesity in the population is less than 100 calories per day and (2) that a focus of intervention efforts on school-age children would likely have only a small impact on future obesity in the adult population.

Cardiovascular Disease (CVD) Program Planning at the Local Level

More than 27 million Americans have experienced an ischemic heart event, stroke, or hospitalization for heart failure or peripheral arterial disease. These cardiovascular events are the leading cause of death in the United States, responsible for over 700,000 deaths each year, $155 billion of direct health-care costs, and another $92 billion of indirect costs reflecting lost productivity (American Heart Association (AHA) 2009). Much of the responsibility for organizing local programs and systems of care falls on county health departments and other local agencies. An SD model was built to support planning programs for CVD in El Paso County (EPC), Colorado. The model was used for projecting future CVD incidence and prevalence and evaluating alternative strategies for reducing the county's CVD burden (Hirsch et al. 2010).

The model includes several proximal risk factors that affect the incidence of first-time events directly, namely, high blood pressure, high cholesterol, diabetes, smoking, secondhand smoke, and small particulate air pollution. Obesity, physical activity, and quality of diet act indirectly through high blood pressure, high cholesterol, and diabetes. Psychosocial stress acts indirectly through high blood pressure and obesity. Receiving quality primary care affects the likelihood of diagnosing and controlling high blood pressure, high cholesterol, and diabetes as well as guiding patients to quit smoking, lose weight, and manage their stress. Getting people to quit smoking also reduces the likelihood that others will be exposed to secondhand smoke, another risk factor.

The model also contains various programmatic interventions for reducing the prevalence of risk factors and, as a result, the incidence of cardiovascular events. Interventions may increase the accessibility of relevant services, products, especially for those people who cannot afford them, or promote their utilization through social marketing. These services include primary care, smoking cessation, weight loss, and stress management as well as opportunities for physical activity and healthy eating. Another intervention would increase the quality of primary care by educating providers as to current guidelines for risk factor management. Other interventions are regulatory in nature rather than service oriented. These include tobacco taxes and sales regulations, workplace smoking bans, air pollution regulations, and junk food taxes and sales regulations.

The model was used to assess the impact of a number of different interventions including three major strategies:

1. *Community-wide lifestyle change*: All of the model's lifestyle and environment interventions are implemented, but none of the medical and mental health-care ones.
2. *Community-wide lifestyle plus post-CVD care*: In addition to the lifestyle and environment interventions, the model's medical and mental health-care interventions are implemented but only for the post-CVD population.
3. *All programs*: In addition to the lifestyle and environment interventions, the medical and mental health-care interventions are implemented for the entire adult population, not only for the post-CVD population.

The results of the simulations helped the health department get a sense of which strategy would best fit the county's needs and resources. Of the three strategies, *all programs* has the greatest ability to reduce CVD events and deaths as well as total consequence costs. It is at least twice as effective at reducing CVD events and deaths as *lifestyle and environment* and also somewhat more effective than the *community-wide lifestyle plus post-CVD care* strategy. However, the *all programs* strategy also implies a large increase in the need for primary care capacity, 30% relative to the status quo. The lack of availability of primary care in El Paso County was already a serious problem and expected to get worse as the population grows and ages. In this context, a strategy that depends so heavily on the expansion of primary care may not be realistic. The *community-wide lifestyle plus post-CVD care* strategy does quite well in terms of outcomes (about 80% as well as *all programs*) while requiring a primary care expansion of only 13%. Health planners in EPC felt that this more modest primary care expansion was achievable, though still a challenge.

CVD Strategy at the National Level

Another cardiovascular disease model is the Prevention Impacts Simulation Model (PRISM) (Hirsch et al. 2014; Homer et al. 2010, 2014). PRISM is an evidence-based tool that simulates the multiyear impacts of a wide variety of interventions aimed at reducing risks for CVD (i.e., coronary heart disease, stroke, heart failure, and peripheral arterial disease) and other chronic conditions and diseases (e.g., hypertension, diabetes, renal disease, obstructive pulmonary disease, certain cancers).

PRISM was sponsored by the CDC and the US National Heart, Lung, and Blood Institute (NHLBI) and developed over several years starting in 2007. The model simulates from 1990 to 2040 and has been calibrated to represent the United States overall as well as some particular local areas.

In PRISM, the population is segmented by six childhood and adult age categories, by gender, and by CVD event status (not-yet or "non-CVD" and already-had or "post-CVD"). The model depicts changes in the population through birth, death, migration, aging, and movement from non-CVD to post-CVD status. It also depicts flows into and out of three blood pressure categories, three blood cholesterol categories, three blood sugar categories, four smoking categories, and two body mass index categories. The end points of the model include risk-related mortality and cost. The model simulates by quarter-year increments

through more than 4,000 intermediate and final output variables.

PRISM includes more than 50 intervention levers that may prevent or mitigate certain well-established cardiovascular risks – high blood pressure, high cholesterol, diabetes, obesity, poor nutrition, inactivity, psychological distress, smoking, secondhand smoke, small particulate air pollution, sleep apnea, periodontal disease, and inadequate use of aspirin. The interventions fall into four groups: (1) clinical (e.g., treating high blood pressure), (2) behavioral support (e.g., weight loss), (3) health promotion and access (against smoking and for better diet and exercise), and (4) taxes and regulation (cigarette taxes, workplace smoking rules). All interventions in the model are based on peer-reviewed literature and discussions with experts working in the field who helped to specify and quantify their causal effects. Figure 4 shows how these elements relate to each other.

Simulations with PRISM indicated that clinical interventions could have the greatest impact on CVD deaths, but clinical interventions are also the most costly to implement and not necessarily the

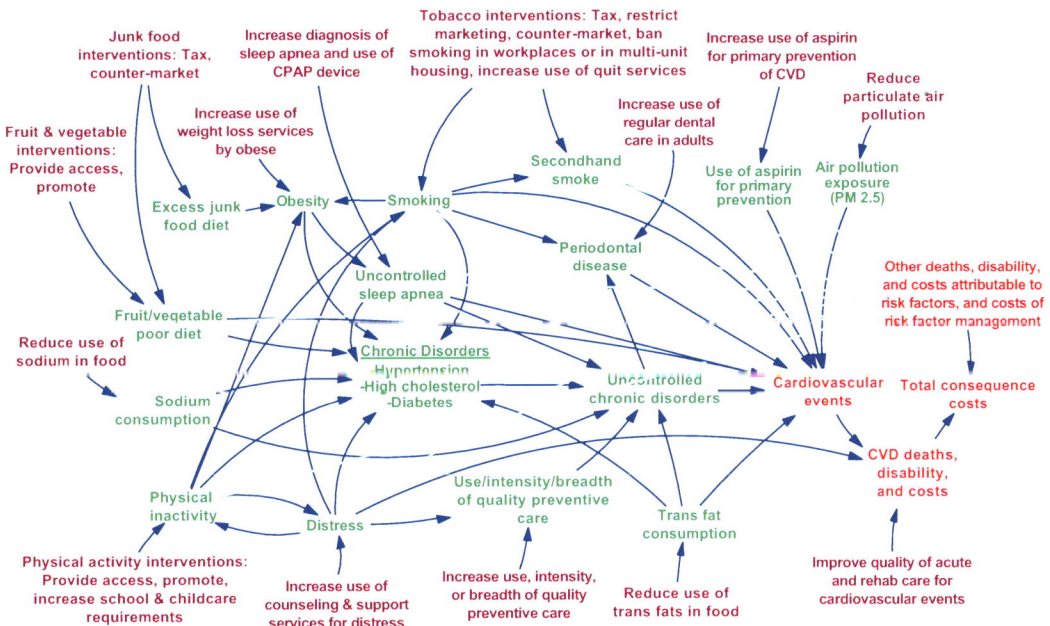

System Dynamics Applications to Health Care in the United States, Fig. 4 Overview of the Prevention Impacts Simulation Model (PRISM) for studying cardiovascular disease

most cost-effective. Taxes and regulations that act to suppress risk factors for CVD cost much less than clinical interventions and have significant benefit in reducing morbidity and mortality, especially in the longer term (i.e., after a decade or more.) Combining all of the interventions would, of course, have the greatest impact in both the short and long term.

PRISM has also been used extensively at the local level for planning CVD prevention programs supported by CDC grants.

Models for Exploring Health System Reform

Health leaders in the United States face many challenges beyond the prevention and control of specific chronic illnesses and would like to better understand the overall health system and how it may be best improved for the sake of patients, providers, payers, and the general economy. Four applications are presented here dealing with entire health systems at the local and national levels.

Three of these applications (all but the second one below) were developed not only as analytic tools but also as simulation-based games available for direct use by health leaders and other interested parties. A simulation-based game combines an underlying model, a user-friendly interface, and a well-thought out learning experience to engage decision makers so that they may try their own ideas and reach their own conclusions. In a situation such as health reform where people have entrenched attitudes and tend to only see parts of the problem, the learning experience provided by a game can be especially useful for unfreezing attitudes and expanding frameworks.

Health-Care and Illness Prevention at a Community Level

Hirsch and Immediato (1999) describe a comprehensive view of health at the level of a community. Their "Healthcare Microworld," depicted in simplified form in Fig. 5, simulates the health status and health-care delivery for people in the community. The microworld was created for a

consortium of health-care providers who were facing a wide range of changes in the mid-1990s and needed a means for their staffs to understand the managerial implications of these changes for how they managed. The underlying SD model consists of hundreds of equations and was designed to reflect with realistic detail a typical American community and its providers, with data taken from public sources as well as proprietary surveys.

Users of the microworld have a wide array of options for expanding the capacity and performance of the community's health-care delivery system such as adding personnel and facilities, investing in clinical information systems, and process redesign. They have a similar range of alternatives for improving health status and changing the demand for care including screening for and enhanced maintenance care of people with chronic illnesses, programs to reduce behavioral risks such as smoking and alcohol abuse, environmental protection, and longer-term risk reduction strategies such as providing social services, remedial education, and job training.

The microworld's comprehensive view of health status and health-care delivery can provide insights not available from approaches that focus on one component of the system at a time. For example, users can play roles of different providers in the community and get a better understanding of why attempts at creating integrated delivery systems can fail when providers care more about their own bottom lines and prerogatives than about creating a viable system. When examining strategies for improving health status, users can get a better sense of how a focus on enhanced care of people with chronic illnesses provides short-term benefits in terms of reduced deaths, hospital admissions, and costs, but how better long-term results can be obtained by also investing in programs that reduce social and behavioral health risks.

Chronic Illness and Growth of the National Health-Care Economy

Despite rapid growth in health-care spending in the United States in recent decades, the health of Americans has not noticeably improved. An SD

Health Care Delivery System

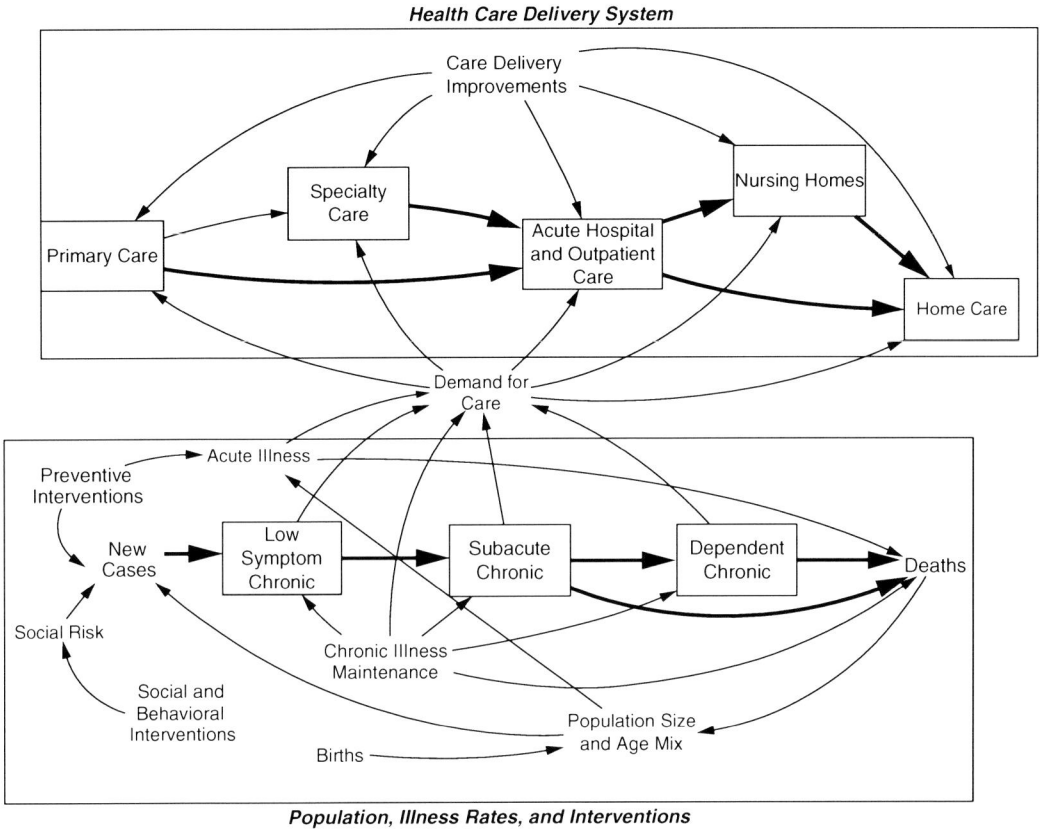

System Dynamics Applications to Health Care in the United States, Fig. 5 Overview of the healthcare microworld

model addresses the question of why the United States has not been more successful in preventing and controlling chronic illness and related costs (Homer et al. 2007). This model reproduces patterns of change in disease prevalence and mortality in the United States, but its structure is a generic one and should be applicable to other countries. The model examines the growing prevalence of disease and responses to it, responses which include the treatment of complications as well as disease management activities designed to slow the progression of illness and reduce the occurrence of future complications. The model shows how progress in complication treatment and disease management has slowed since 1980 in the United States, largely due to a behavioral tug-of-war between health-care payers and providers that has resulted in price inflation and an unstable climate for health-care investments. The

model is also used to demonstrate the impact of moving "upstream" by managing known risk factors to prevent illness onset and moving even further upstream by addressing adverse behaviors and living conditions linked to the development of these risk factors in the first place.

An overview of the model's causal structure is presented in Fig. 6. The population stock of disease prevalence is increased by disease incidence and decreased by deaths. The death rate can be reduced by a greater extent of disease care, including urgent care and disease management. Disease incidence draws from a stock of risk prevalence, where risk refers to physical or psychological conditions or individual behaviors that may lead to disease. The risk prevalence stock is increased by adverse behaviors and living conditions. Adverse behaviors may include poor diet, lack of physical activity, or substance abuse. Adverse

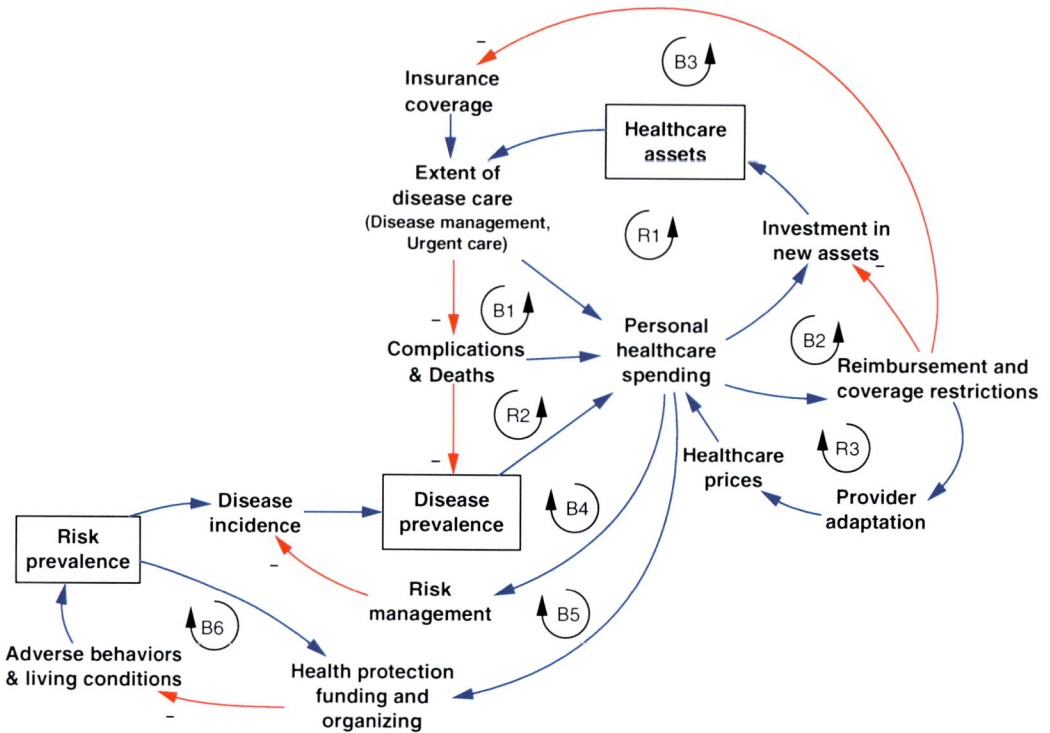

System Dynamics Applications to Health Care in the United States, Fig. 6 Overview of a national-level model of healthcare and illness prevention. Key to feedback loops ("R" denotes self-reinforcing, "B" denotes counterbalancing). *R1*: Health-care revenues are reinvested for further growth. *B1*: Disease management reduces need for urgent care. *R2*: Disease care prolongs life and further increases need for care. *B2*: Reimbursement restriction limits spending growth. *B3*: Insurance denial limits spending growth. *R3*: Providers circumvent reimbursement restrictions, leading to a tug-of-war with payers. *B4*: Risk management proportional to downstream spending can help limit it. *B5*: Health protection proportional to downstream spending can help limit it. *B6*: Health protection (via sin taxes) proportional to risk prevalence can help limit it

living conditions can encompass many factors, including crime, lack of access to healthy foods, inadequate regulation of smoking, weak social networks, substandard housing, poverty, or poor educational opportunities.

The extent of care is explained in the model by two key factors: the abundance of health-care assets (structures, fixed equipment, personnel) and insurance coverage. The uninsured are less likely than the insured to receive health-care services and especially disease management.

The stock of assets is increased by investments, which may be viewed as the reinvestment of some fraction of health-care revenues. Such reinvestment drives further growth of care and revenue, and the resulting exponential growth process is identified as loop R1 in Fig. 6. The data indicate,

however, that the reinvestment process has slowed significantly since 1980. This decline in the reinvestment rate has likely been the response by potential investors to various forms of cost control, including the restriction of insurance reimbursements, which affect the providers of health-care goods services, Health-care costs and cost controls have also led to elimination of private health insurance coverage by some employers, although some of the lost coverage has been replaced by publicly funded insurance.

One additional part of the downstream health-care story portrayed in Fig. 6 is the growth of health-care prices. Health-care prices are measured in terms of a medical care consumer price index (CPI), which since 1980 has grown much more rapidly than the general CPI for the overall economy. One reason for this price inflation is shown in Fig. 6 as "provider adaptation." This is the idea that, in response to cost containment efforts, providers may increase fees, prescribe more services, prescribe more complex services (or simply bill for them), order more follow-up visits, or do a combination of these. By one estimate, unnecessary and inflationary care represented 29% of all personal health-care spending in 1989 (Homer et al. 2007).

Simulations of the model suggest that there are no easy downstream fixes to the problem of an underperforming and expensive health-care system in the US mold. More effective fixes appear to lie upstream. There are two broad categories of such efforts represented in the model: risk management for people already at risk and health protection for the population at large to change adverse behaviors and mitigate unhealthy living conditions. While spending on population-based health protection and risk management programs has grown somewhat, it still represents a small fraction of total US health-care spending, on the order of 5% in 2004 (Homer et al. 2007).

Figure 6 includes three balancing loops to indicate how, in general terms, efforts in risk management and health protection might be funded or resourced more systematically and in proportion to indicators of capability or relative need. Loop B4 suggests that funding for programs promoting risk management could be made proportional to spending on downstream care, so that when downstream care grows funding for risk management would grow as well. Loop B5 suggests something similar for health protection, supposing that government budgets and philanthropic investments for health protection could be set in proportion to recent health care spending. Loop B6 takes a different approach to the funding of health protection, linking it not to health-care spending but to risk prevalence, the stock which health protection most directly seeks to reduce. The linkage to risk prevalence can be made fiscally through "sin taxes" on unhealthy items, such as cigarettes (already taxed throughout the United States to varying extents) and sugary drinks. In theory, the optimal magnitude of such taxes may be rather large in some cases, as the taxes can be

used both to discourage unhealthy activities and promote healthier ones (O'Donoghue and Rabin 2006).

Simulations of the model suggest that whether the approach to upstream action is risk management or health protection, such actions can reduce illness prevalence and ultimately save money. Focusing only on downstream health-care costs, the payback time of these efforts may be on the order of 20 years. However, if one also included productivity losses, the payback period on upstream action could shrink to only a few years, as suggested by the Whatcom County and PRISM models described above.

National Health System Reform

The HealthBound policy simulation game was developed with support from the CDC for those wanting to experience the possibility of transforming the troubled US health system (Milstein et al. 2010). The game's simulator tracks movement of the US population among states of health, risky behavior, environmental exposures, and socioeconomic status. The model is quantified based on publicly available data from the early 2000s as well as studies from the literature on health-care utilization and programmatic impact. Players try to steer the country's health system toward greater levels of health, equity, and cost-effectiveness. The goals are difficult to achieve, in part, because the game includes resource constraints, time delays, and side effects of intervention similar to those of the actual health system.

Figure 7 provides an overview of the model underlying HealthBound. The basic causal logic is as follows: many health problems have their genesis in unhealthy behaviors and hazardous and stressful environments. Socioeconomic disadvantage worsens all of these social determinants of health. The disadvantaged also have worse access to regular office-based healthcare than do the advantaged, due to less insurance coverage and less sufficiency of primary care providers (PCPs) to meet patient demand. This insufficiency of office-based care leads to the increased use of hospital emergency departments for nonurgent care. The insufficiency is largely related to relatively low PCP incomes, especially for PCPs who

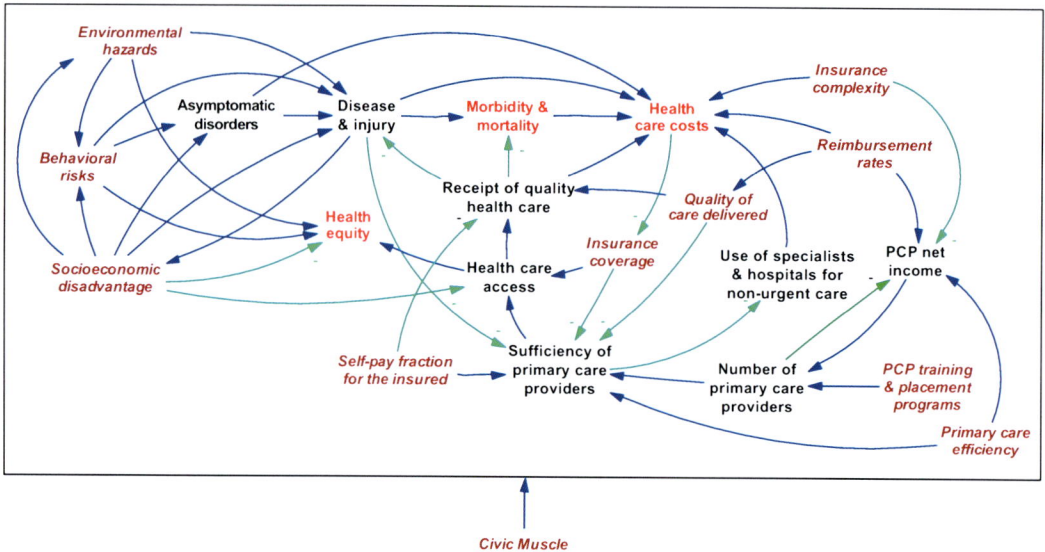

System Dynamics Applications to Health Care in the United States, Fig. 7 Major elements in the healthBound model

serve the disadvantaged. Another factor affecting health outcomes, for both the disadvantaged and the advantaged, is the quality of care delivered, describing the extent to which providers follow guidelines for best practice with regard to screening, monitoring, and treatment. Quality of care may suffer when insurance reimbursement rates are not adequate.

Users of HealthBound attempt to achieve the best results across four criteria simultaneously. They must (1) save lives, (2) improve well-being, (3) achieve health equity, and (4) lower health-care costs per capita, all the while being conscious of total intervention spending and the number of simultaneous interventions. The game tracks these scorecard variables and many others over 30 years: beginning with a 5-year comparison period, followed by 25 years during which players may intervene.

Users may employ several types of intervention, alone or in combination, to achieve their goals. These include (1) expanding insurance coverage; (2) improving quality of care; (3) reducing insurance complexity; (4) expanding the supply of primary care providers, particularly for disadvantaged populations; (5) improving primary care efficiency; (6) changing reimbursement rates to

physicians or hospitals; (7) requiring gatekeeper approval for specialist services; (8) changing the self-pay fraction for those who have insurance; (9) enabling healthier behaviors (e.g., reducing tobacco use); (10) building safer environments (e.g., reducing air pollution); (11) creating pathways to advantage (e.g., through education, job training, living wage policies); and (12) strengthening civic muscle to enable more effective implementation of the other interventions. Many of these general interventions can be further tailored by focusing on particular demographic subgroups.

The model underlying HealthBound has been used to do some extensive policy analyses (Milstein et al. 2010, 2011). Findings include:

Universal insurance coverage reduces morbidity and mortality but does not reduce overall health-care costs. Also, it can lead to a shortage of primary care capacity for the disadvantaged, in which case the disadvantaged do not benefit as much as they might otherwise.

Increasing the quality of chronic and preventive care by office-based primary care physicians and specialists can produce an even greater reduction in morbidity and mortality than that achieved by universal coverage, because the quality improvement applies to a much larger segment of the population. But like universal coverage, quality improvement does not reduce health-care costs, and it can worsen health inequity.

Reducing reimbursements to office-based providers initially lowers health-care costs but also diminishes the adequacy and quality of care. These lead to greater morbidity and more office and hospital visits, which cost money. The initial reduction in costs is ultimately negated.

The model points to two strategies that could work to improve health equity and reduce health-care costs. One of them is to increase PCP capacity for the disadvantaged, which can be done by making PCP office operations more efficient, by improving PCP insurance reimbursements, or by offering more scholarships and other incentives for medical students who commit to work with the disadvantaged after they graduate. With greater PCP capacity, the health-care demands of the disadvantaged may be better met in provider offices rather than in hospital emergency departments, thereby reducing health-care costs.

A second promising approach is an upstream one, either creating safer environments and enabling healthier behaviors or helping more people move out of their disadvantaged position, for example, through a mix of training/educational reforms and family income supports. Such a strategy takes several years to generate significant benefits, but it ultimately can reduce morbidity and mortality quite significantly and thereby reduces health-care costs.

Local Health System Reform

Health system reform is a national priority in the United States but is also being pursued through local initiatives. The Rippel Foundation's ReThink Health Dynamics Model is a realistic, yet simplified, representation of a local health system over time (Homer et al. 2016; Milstein et al. 2013). Its main purpose is to estimate the likely health and economic consequences over

time for a variety of interventions that may be enacted at a local or regional level. Simulated scenarios may be studied as a prelude to action in the real world.

Figure 8 shows an overview of the model's main elements. The model simulates changes in health, care, and cost over time and also tracks worker productivity as well as health equity between advantaged and disadvantaged groups. The model also accounts for changing levels of risk or vulnerability stemming from aging, unhealthy behaviors, crime, and environmental hazards as well as poverty and lack of insurance.

Together, those drivers affect health status over time, including the prevalence of physical illness (mild and severe), mental illness (treated and untreated) as well as acute episodes (urgent and nonurgent), and deaths. Health status, in turn, determines the demand and cost for healthcare in different locations, including physician offices, hospitals, nursing homes, and at home.

A sophisticated interface allows users to interact directly with the model and test a wide variety of initiatives. These include downstream initiatives for higher-valued care through cost reduction, expanded access to care, improved quality of care, and payment reform as well as upstream interventions to address adverse living conditions and promote healthier behavior. Other upstream interventions simulate the effect of providing opportunities that reduce socioeconomic disadvantage.

The model estimates the program cost for each simulated initiative over time. If the initiatives succeed in saving health-care costs, then users may choose to capture and reinvest those savings in an effort to sustain the selected initiatives over time and avoid running out of funding.

The ReThink Health Dynamics Model has been calibrated and customized for a number of communities and has helped in local health reform planning. A version of the model called *Anytown* that reflects a cross-section of the US population has been used extensively by communities and in academic settings to help people understand the complexities of local health system reform. One of the model's important benefits has been to help users identify, in advance, major pitfalls that

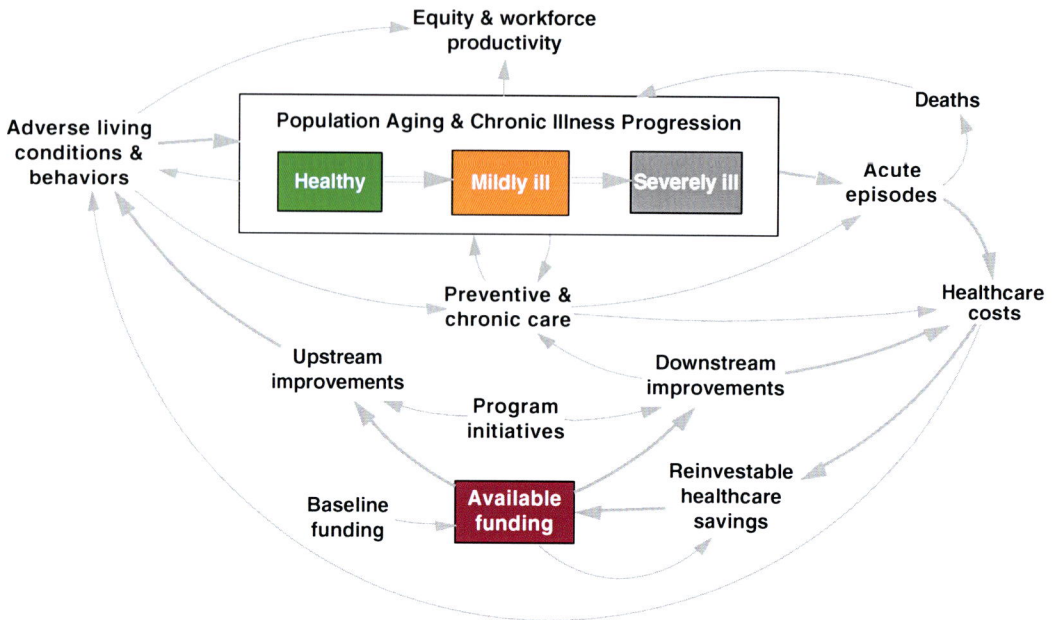

System Dynamics Applications to Health Care in the United States, Fig. 8 Overview of the rethink health dynamics model

undermine or derail regional change ventures. These pitfalls include:

Attempting too much without adequate funding

Exacerbating bottlenecks impeding access to primary care

Failing to anticipate "supply-push" responses by providers that can undercut health-care cost reduction efforts

Attempting near-term improvements but losing sight of the long term (and comparing alternative strategies using only a short time horizon)

Attempting long-term improvements but losing sight of the near term

Improving health, care, or cost overall but failing to address inequities – and possibly making them worse

Simulations with the model indicate that the greatest reduction in per capita health-care costs is likely to result from a combination of higher-value care, a shift to global payment, reinvestment of savings that are produced, and investment in upstream initiatives to promote healthier behavior. Antipoverty initiatives can further reduce the

prevalence of severe chronic illness and improve health equity and workforce productivity but do not help to reduce health-care costs (Homer et al. 2016).

Future Directions

As long as there are dynamically complex health issues in search of answers, the SD approach will have a place in the analytic armamentarium. There is still much to be learned about the population dynamics of individual chronic conditions like hypertension and risk factors like obesity. SD models could also address multiple interacting diseases and risks, giving a more realistic picture of their overall epidemiology and policy implications, particularly where the diseases and risks are mutually reinforcing. For example, it has been found that substance abuse, violence, and AIDS often cluster in the same urban subpopulations and that such "syndemics" are resistant to narrow policy interventions (Singer and Clair 2003; Wallace 1988). This idea could also be extended to the case of mental depression, which is often exacerbated by other chronic illnesses and may, in turn, interfere with the proper management of those illnesses. An exploratory simulation model has indicated that SD can usefully address the concept of syndemics (Homer and Milstein 2002).

There is also more to be learned about health-care delivery systems and capacities, with the inclusion of characteristics specific to selected real-world cases. Models combining delivery systems and risk and disease epidemiology could help policymakers and health-care providers understand the nature of coordination required to put ambitious public health and risk reduction programs in place without overwhelming delivery capacities. Such models could reach beyond the health-care delivery system per se to examine the potential roles of other delivery systems, such as schools and social service agencies, in health risk reduction.

The more complete view of population health dynamics advocated here may also be extended to address persistent challenges in the United States that will likely require policy changes at a national and state level and not only at the level of local communities. Examples include racial and ethnic health disparities and the shortage of nurses. SD modeling can help to identify the feedback structures responsible for these problems and point the way to policies that can make a lasting difference.

Bibliography

American Diabetes Association (ADA) and National Institute of Diabetes and Digestive and Kidney Diseases (NIDDK) (2002) The prevention or delay of type 2 diabetes. Diabetes Care 25:742–749

American Heart Association (AHA) (2009) Heart disease and stroke statistics 2010 update: a report from the American Heart Association. Circulation 121: e46–e215

Anderson R (1994) Populations, infectious disease and immunity: a very nonlinear world. Philos Trans R Soc Lond Ser B Biol Sci B346:457–505

Centers for Disease Control and Prevention, National Center for Chronic Disease Prevention and Health Promotion (CDC/NCCDPHP) (2007) Chronic disease overview. Available at http://www.cdc.gov/nccdphp/overview.htm

Diabetes Prevention Program Research Group (DPPRG) (2002) Reduction in the incidence of type 2 diabetes with lifestyle intervention or metformin. N Engl J Med 346:393–403

Edelstein B, Hirsch G, Frosh M, Kumar J (2015) Reducing early childhood caries in a Medicaid population: a systems model analysis. J Am Dent Assoc 146(4):224–232

Flegal KM, Carroll MD, Ogden CL, Johnson CL (2002) Prevalence and trends in obesity among US adults, 1999–2000. JAMA 288:1723–1727

Forrester JW (1961) Industrial dynamics. MIT Press, Cambridge

Forrester JW (1980) Information sources for modeling the national economy. J Am Stat Assoc 75:555–574

Forrester JW, Senge PM (1980) Tests for building confidence in system dynamics models. In: System dynamics, TIMS studies in the management sciences. North-Holland, Amsterdam, pp 209–228

Gunning-Schepers LJ (1989) The health benefits of prevention: a simulation approach. Health Policy Spec Issue 12:1–255

Halloran EM, Longini IM, Nizam A, Yang Y (2002) Containing bioterrorist smallpox. Science 298:1428–1432

Heirich M (1999) Rethinking health care: innovation and change in America. Westview Press, Boulder

Hirsch GB (2004) Modeling the consequences of major incidents for health care systems. In: 22nd international system dynamics conference. System Dynamics Society, Oxford. http://www.systemdynamics.org/publications.htm

Hirsch G (2006) A generic model of contagious disease and its application to human-to-human transmission of avian influenza. In: Proceedings of the international conference of the System Dynamics Society, Nijmegen. http://www.systemdynamics.org/publications.htm

Hirsch G, Homer J (2004) Modeling the dynamics of health care services for improved chronic illness management. In: 22nd international system dynamics conference. System Dynamics Society, Oxford. http://www.systemdynamics.org/publications.htm

Hirsch GB, Immediato CS (1999) Microworlds and generic structures as resources for integrating care and improving health. Syst Dyn Rev 15:315–330

Hirsch GB, Homer JB, Evans E, Zielinski A (2010) A system dynamics model for planning cardiovascular disease interventions. Am J Public Health 100(4):616–622

Hirsch GB, Edelstein BL, Frosh M, Anselmo T (2012a) A simulation model for designing effective interventions in early childhood caries. Prev Chronic Dis 9:110219. https://www.cdc.gov/pcd/issues/2012/11_0219.htm

Hirsch G, McCleary K, Meyer K, Saeed K (2012b) Deceased donor potential for organ transplantation: a system dynamics framework. International Conference of the System Dynamics Society, St. Gallen. http://www.systemdynamics.org/publications.htm

Hirsch G, Homer J, Wile K, Trogdon JG, Orenstein D (2014) Using simulation to compare four categories of intervention for reducing cardiovascular disease risks. Am J Public Health 104(7):1187–1195

Hirsch G, Homer J, Tomoaia-Cotisel A (eds) (2015) System dynamics applications to health and health care. Virtual issue of *System Dynamics Review* (15 previously published articles with new Introduction and Extended Bibliography). http://onlinelibrarywiley.com/journal/10.1002/(ISSN)1099-1727/homepage/VirtualIssuesPage.html

Hoard M, Homer J, Manley W, Furbee P, Haque A, Helmkamp J (2005) Systems modeling in support of evidence-based disaster planning for rural areas. Int J Hyg Environ Health 208:117–125

Homer J, Milstein B (2002) Communities with multiple afflictions: a system dynamics approach to the study and prevention of syndemics. In: 20th international system dynamics conference. System Dynamics Society, Palermo. http://www.systemdynamics.org/publications.htm

Homer J, Milstein B (2004) Optimal decision making in a dynamic model of community health. In: 37th Hawaii international conference on system sciences. IEEE, Waikoloa. http://csdl.computer.org/comp/proceedings/hicss/2004/2056/03/2056toc.htm

Homer J, Ritchie-Dunham J, Rabbino H, Puente LM, Jorgensen J, Hendricks K (2000) Toward a dynamic theory of antibiotic resistance. Syst Dyn Rev 16:287–319

Homer J, Hirsch G, Minniti M, Pierson M (2004) Models for collaboration: how system dynamics helped a community organize cost-effective care for chronic illness. Syst Dyn Rev 20:199–222

Homer J, Milstein B, Dietz W, Buchner D, Majestic E (2006) Obesity population dynamics: exploring historical growth and plausible futures in the U.S. In: 24th international system dynamics conference. System Dynamics Society, Nijmegen. http://www.systemdynamics.org/publications.htm

Homer J, Hirsch G, Milstein B (2007) Chronic illness in a complex health economy: the perils and promises of downstream and upstream reforms. Syst Dyn Rev 23(2–3):313–334

Homer J, Milstein B, Wile K, Trogdon J et al (2010) Simulating and evaluating local interventions to improve cardiovascular health. Prev Chronic Dis 7(1):080231. http://www.cdc.gov/pcd/issues/2010/jan/08_0231.htm

Homer J, Wile K, Yarnoff B, Trogdon JG et al (2014) Using simulation to compare established and emerging interventions to reduce cardiovascular disease risks in the United States. Prev Chronic Dis 11(E195):1–14. http://www.cdc.gov/pcd/issues/2014/14_0130.htm

Homer J, Hirsch G, Milstein B, Fisher E (2016) Combined regional investments could substantially enhance health system performance and be financially affordable. Health Aff 35(8):1435–1443

Honeycutt AA, Boyle JP, Broglio KR et al (2003) A dynamic Markov model for forecasting

diabetes prevalence in the United States through 2050. Health Care Manag Sci 6:155–164

Hovmand P, Ford D, Flom I, Kyriakakis S (2009) Victims arrested for domestic violence: unintended consequences of arrest policies. Syst Dyn Rev 25(3):161–181

Jones AP, Homer JB, Murphy DL, Essien JDK, Milstein B, Seville DA (2006) Understanding diabetes population dynamics through simulation modeling and experimentation. Am J Public Health 96(3):488–494

Kaplan EH, Craft DL, Wein LM (2002) Emergency response to a smallpox attack: the case for mass vaccination. Proc Natl Acad Sci 99:10935–10940

LeClaire R, Hirsch G, Bandlow A (2009) Learning environment simulator: a tool for local decision makers and first responders. In: Proceedings of the international conference of the System Dynamics Society, Albuquerque. http://www.systemdynamics.org/publications.htm

Lee P, Paxman D (1997) Reinventing public health. Ann Rev Public Health 18:1–35

Lich KH, Tian Y, Beadles CA, Williams L et al (2014) Strategic planning to reduce the burden of stroke among veterans: using simulation modeling to inform decision making. Stroke 45:2078–2084

Mahamoud A, Roche B, Homer J (2012) Modelling the social determinants of health and simulating short-term and long-term intervention impacts for the city of Toronto, Canada. Soc Sci Med 93:247–255

Manley W, Homer J et al (2005) A dynamic model to support surge capacity planning in a rural hospital. In: 23rd international system dynamics conference, Boston. http://www.systemdynamics.org/publications.htm

Milstein B, Hirsch G, Minyard K (2013) County officials embark on new collective endeavors to rethink their local health systems. J County Adm 1–10

Milstein B, Jones A, Homer J, Murphy D, Essien J, Seville D (2007) Charting plausible futures for diabetes prevalence in the United States: a role for system dynamics simulation modeling. Prev Chronic Dis 4(3). http://www.cdc.gov/pcd/issues/2007/jul/06_0070.htm

Milstein B, Homer J, Hirsch G (2010) Analyzing national health reform strategies with a dynamic simulation model. Am J Public Health 100:811–819

Milstein B, Homer J, Briss P, Burton D, Pechacek T (2011) Why behavioral and environmental interventions are needed to improve health at lower cost. Health Aff 30(5):1–11

Naidoo B, Thorogood M, McPherson K, Gunning-Schepers LJ (1997) Modeling the effects of increased physical activity on coronary heart disease in England and Wales. J Epidemiol Community Health 51:144–150

National Institute of Diabetes and Digestive and Kidney Diseases (NIDDK) (2004) National Diabetes Statistics. http://diabetes.niddk.nih.gov/dm/pubs/statistics/index.htm

O'Connell JB, Bristow MR (1993) Economic impact of heart failure in the United States: time for a different approach. J Heart Lung Transplant 13(Suppl): S107–S112

O'Donoghue T, Rabin M (2006) Optimal sin taxes. J Public Econ 90:1825–1849

Schlessinger L, Eddy DM (2002) Archimedes: a new model for simulating health care systems – the mathematical formulation. J Biomed Inform 35:37–50

Schorr LB (1997) Common purpose: strengthening families and neighborhoods to rebuild America. Doubleday/Anchor Books, New York

Singer M, Clair S (2003) Syndemics and public health: reconceptualizing disease in bio-social context. Med Anthropol Q 17:423–441

Sterman JD (1988) A skeptic's guide to computer models. In: Grant L (ed) Foresight and national decisions. University Press of America, Lanham, pp 133–169

Sterman JD (2000) Business dynamics: systems thinking and modeling for a complex world. Irwin/McGraw-Hill, Boston

Tengs TO, Osgood ND, Chen LL (2001) The cost-effectiveness of intensive national school-based anti-tobacco education: results from the tobacco policy model. Prev Med 33:558–570

Thompson KM, Tebbens RJD (2007) Eradication versus control for poliomyelitis: an economic analysis. Lancet 369:1363–1371

U.K. Prospective Diabetes Study Group (UKPDSG) (1998) Tight blood pressure control and risk of macrovascular and microvascular complications in type 2 diabetes. Lancet 352:703–713

Vennix JAM (1996) Group model-building: facilitating team learning using system dynamics. Wiley, Chichester

Wagner EH, Sandhu N, Newton KM et al (2001) Effect of improved glycemic control on health care costs and utilization. JAMA 285(2):182–189

Wakeland W, Schmidt T, Gilson A, Haddox J, Webster L (2011) System dynamics modeling as a potentially useful tool in analyzing mitigation strategies to reduce overdose deaths associated with pharmaceutical opioid treatment of chronic pain. Pain Med 12(Suppl 2): S49–S58

Wallace R (1988) A synergism of plagues. Environ Res 47:1–33

Wolfson MC (1994) POHEM: a framework for understanding and modeling the health of human populations. World Health Stat Q 47:157–176

Zack MM, Moriarty DG, Stroup DF, Ford ES, Mokdad AH (2004) Worsening trends in adult health-related quality of life and self-rated health-United States, 1993–2001. Public Health Rep 119:493–505

System Dynamics Applications to Health and Social Care in the United Kingdom and Europe

Eric Wolstenholme
Symmetric Scenarios, Edinburgh, UK

Article Outline

Glossary

Terms Related to System Dynamics

(A combination of adaptations from the System Dynamics Society website, system dynamics software, and personal views)

A System A collection of elements working together as parts of a complex whole whose behavior is greater than the sum of the parts. The word systems is often used to refer to complex phenomena existing in the world (such as financial systems, health systems, and computer systems), but the true meaning of systems refers more to an abstract or model of such phenomena existing only in a conceptual world we construct to think about systems.

Agent-based modeling A style of computer simulation based on the actions and interactions of autonomous agents. This type of modeling can be thought of as a combination of continuous and discrete simulation methods.

Causal loop maps A means of exposing and articulating the causal feedback processes at work in a complex system.

Complexity Having a large number of interconnected elements which allow systems to be self-organizing and adaptive, especially where the connections themselves form feedback loops (a influences b, b influences c, and c influences a).

Detailed complexity The number of elements contained in a system

Discrete event simulation modeling A method of computer simulation based on the movement of individual entities through systems over time represented as activities, sampled from statistical distributions, and queues.

© Springer Science+Business Media, LLC, part of Springer Nature 2020
B. Dangerfield (ed.), *System Dynamics*,
https://doi.org/10.1007/978-1-4939-8790-0_269

Originally published in
R. A. Meyers (ed.), *Encyclopedia of Complexity and Systems Science*, © Springer Science+Business
Media LLC 2019, https://doi.org/10.1007/978-3-642-27737-5_269-4

Dynamic complexity The number of interconnections contained in a system.

Dynamical systems Systems where movement and flow is important and where evolution is as a result of internal (endogenous) rules based on feedback within and between sub-organizations.

Expert groups and group model building A working group constituted to oversee the construction and validation of a system dynamics model and consisting of all representatives of all organizations and departments impacting on and responsible for a particular issue. Any model needs primarily to be a combination of the mental models of the group and to be owned by the group.

Feedback loops Feedback loops exists when information resulting from some action travels through a system and eventually returns in some form to its point of origin, potentially influencing future action. If the tendency in the loop is to reinforce the initial action, the loop is called a positive or reinforcing feedback loop; if the tendency is to oppose the initial action, the loop is called a negative or balancing feedback loop. The sign of the loop is called its polarity which is derived from the polarities of its individual causal links. Balancing loops are goal-seeking loops which tend to stabilize systems, whereas reinforcing loops are growth or collapse loops which tend to destabilize systems.

Modeling as learning The process of experimenting with virtual models in a risk-free environment to learn about the behavior of the real system.

Reference modes of behavior An observed past trend and future projected trends over time used to assist in defining the scope, time frame, and validation of a system dynamics model.

STELLA architect A particular version of system dynamics software used in this entry.

Stocks and flows A language by which to visualize and map the structure of a system in terms of accumulations and flows.

System archetypes System archetypes are generic combinations of reinforcing and balancing feedback loops which give rise to specific dynamic patterns of behavior over time. Since there are only two types of feedback loops, there are only four basic types of system archetypes. System archetypes are a concise means of articulating the problems caused by *unintended consequences* arising from *well-intended* actions and designing solutions necessary to mitigate these.

System behavior The term used in system dynamics to refer to the behavior over time of a particular system structure.

System dynamics System dynamics is a systems-based computer simulation approach to policy analysis and design based on a language of stocks and flows. The approach can be considered as a continuous, aggregate simulation method analogous to water flows through bathtubs and pipes and combines elements of fluid dynamics and control engineering. It applies to dynamic problems arising in complex social, managerial, economic, or ecological systems – any dynamic systems characterized by interdependence, mutual interaction, information feedback, and circular causality. It facilitates the emulation of such systems for the purpose of communication, sharing, and redesign.

System dynamics software Purpose-built simulation software used to model in system dynamics.

Systems thinking Generally, the interpretation of the world as a complex, self-regulating, and adaptive system based on taking a holistic, as distinct from a reductionist, perspective. Specifically, in system dynamics, a way of using the feedback structures at the heart of system dynamics to hypothesize system behavior over time without using formal simulation (or to help the conceptualization of a system dynamics model or to explain its behavior).

Unintended consequences The side effects of well-intended policies.

Terms Related to Health and Social Care

Acute hospitals	Hospital dealing with short-term conditions requiring mainly one-off treatments.

Adult social care	The provision of care services in residential care homes and at home (domiciliary care)		be centered on the person, not the provider.
Better care fund	A pooled budget between local authorities and the NHS in England to better integrate health and social care services.	Intermediate care	Short-term care to expedite the treatment of complex conditions.
Clinical Commissioning Groups (CCGs) [Formerly Primary Care Trusts (PCTs)]	The local operating agencies of the NHS in England, which both commission (buy) and deliver health services.	LGA	The body in England responsible for representing local government and hence social care.
		National Health Service (NHS)	The organization in the UK responsible for the delivery of health care. It takes different forms in each of the four countries of the UK.
Coping strategies	The strategies employed by service providers to attempt to maintain patient flow when faced with excessive demand beyond their delivery capacity.	NHS Confederation	The body in England responsible for representing individual NHS organizations.
		NHS Continuing Care	The provision of health services inside and outside hospitals and in nursing homes
General practitioners (GPs)	Locally based general clinicians who deliver primary care services and control access to specialist secondary health and social care services.	Nursing homes	Private and public residential establishments providing health care for older people.
		Outliers/ boarders	Patients located in hospital wards not related to their condition, due to bed shortages.
Health conditions	States or stages of an illness such as mild, moderate, and severe.	Scheduled or elective care	Medical or surgical procedures where treatment is chosen by the patient.
Health treatment	The application of drugs, therapies, and medical/ surgical interventions to treat illness.	Social services	In the UK, services which provide nonhealth-related care, mainly for children and older people, located within local government.
Hospital-delayed transfers of care – Delayed hospital discharges	Patients in hospital who have finished treatment and are fit for discharge but cannot be discharged, usually as a result of needing continuing heath or social care packages.	State-provided health services	Health services provided by the state through taxation as distinct from private health care funded through insurance.
Integrated care	A means of coordinating the delivery of diverse health and social care services to the same person, based on the belief that services should	Unscheduled or emergency care	Medical or surgical procedures arising from accidents or emergencies.

Introduction

All too often, complexity issues are ignored in decision making simply because they are just too difficult to represent. Managers feel that to expand the boundaries of the decision domain to include intricate, cross-boundary interconnections and feedback will detract from the clarity of the issue at stake. This is particularly true when the interconnections are behavioral and hard to quantify. Hence, the focus of decision making is either very subjective or based on simple, linear, easy-to-quantify components. However, such a reductionist stance, which ignores information feedback and multiple-ownership issues, can result in unsustainable, short-term benefits with major unintended consequences.

System dynamics is a particular way of thinking and analyzing situations, which makes visible the dynamic complexity of human activity systems for decision support. It is particularly important in the health and social care field where there are major issues of complexity associated with the incidence and prevalence of disease, an aging population, and a profusion of new technologies and multiple agencies responsible for the prevention and treatment of illness along very long patient pathways. Health is also linked at every stage to all facets of life and health policy has a strong social and political dimension in most countries, which can constrain implementation of solutions (El-Jardali et al. 2014).

This entry describes and reviews work in applying system dynamics to issues of health and social care in the UK and Europe which, despite cultural differences, have many similarities. One of the significant commonalities is the focus on treatment rather than prevention. There are many reasons for this, not the least being that illness prevention is in fact the province of a field much bigger than health, which includes economics, social deprivation, drugs, poverty, power, and politics.

The field of system dynamics in health in the UK and Europe reflects this dilemma and contrasts somewhat with the use of system dynamics in the USA (Hirsh and Homer 2019; Bishai et al. 2014; Tretter 2002). However, work often includes both disease and treatment states. For example, work on AIDS covers both prevention and drug treatment and work on long-term conditions, particularly mental health conditions, covers condition progression as well as alternative therapies. Also, there is a growing trend in public health strategies to keep people out of acute hospitals, and recent work on alcohol abuse, dementia, and obesity relates to this issue.

It is important to emphasize what this entry does not cover. By definition, system dynamics is a strategic approach aimed at assisting with the understanding of high-level feedback effects at work in organizations. It is therefore separate from the many applications of spreadsheets and discrete entity simulation methods applied to answer short-term operational-level issues in health. More will be said on the way these fields are converging in the later section on method consolidation.

It is also important to note where the knowledge base of this entry is located. System dynamics applications in health in Europe began in the 1980s and are expanding rapidly. However, as will be seen from the bibliography to this entry, much of the work is applied by internal and external consultants and academic staff in universities for the benefit of health-care managers and is reported in management, operational research, and system dynamics journals. Little of the work so far has been published in mainstream health literature. Indeed, an important trend in communication is not to write about models, but to allow interested people to actually use them. Many models are now available over the internet as learning environments (http://www.symmetriclab.com; http://www.thewholesystem.co.uk/simulation/examples.aspx). This allows individuals or groups to experiment with them and to use their own data.

It should also be pointed out that this entry should be seen as one of a number of recent publications explaining system dynamics and its role in health (Wolstenholme and McKelvie 2019; McKelvie 2013; Miller and McKelvie, 2016; Atkinson et al. 2015; Dangerfield 2016; Brailsford and Vissers 2010). There is also a section on the web site of the UK Chapter of the

System Dynamics Society entitled "Making an impact in the health and social care system" that describes a number of papers in health which won the Society's Steer Davies Gleave prize (http://systemdynamics.org.uk).

Initially the fundamentals of system dynamics will be described, followed by an overview of current health issues and responses in the UK and Europe. This is followed by a case study to demonstrate how effective and apposite system dynamics studies can be. There then follows a review of applications. Mention is also made of health workforce planning studies. Lastly, a review of future directions is described.

The History of System Dynamics

System dynamics was conceived at the Massachusetts Institute of Technology (MIT), Boston, in the late 1950s and has now grown into a major discipline (Forrester 1961; Sterman 2000, 2008), which was formally celebrated and reviewed at its 60th anniversary in 2017. It is widely used in the private business sector in production, marketing, oil, asset management, financial services, pharmaceuticals, and consultancy. It is also used in the public sector in defense, health, and criminal justice.

System dynamics has a long history in the UK and Europe. The first formal university group was established at the University of Bradford in England in 1970. Today many university departments and business schools offer courses in system dynamics and numerous consultancies of all types using the method in one form or another. Thousands of people have attended private and university and there is a well-established UK System Dynamics Society (http://systemdynamics.org.uk), which is the largest national grouping outside the USA.

The Need for System Dynamics

Most private and public organizations are large and complex. They exhibit both "detailed" complexity (the number of elements they contain) and plexity (the number of elements they contain) and more importantly "dynamic" complexity (the number of interconnections and interactions they embrace). They have long processes, which transcend many sectors, each with their own accounting and performance measures. In the case of health and social care organizations, this translates into long patient pathways across many agencies. Complexity and decision making in the public sector are also compounded by a multitude of planning time horizons and the political dimension.

Long processes mean that there are many opportunities for intervention, but that the best levers for overall improvement are often well away from symptoms of problems. Such interventions may benefit sectors other than those making the investments and require an open approach to improving patient outcomes rather than single agency advantage.

The management of complex organizations is complicated by the fact that human beings have limited cognitive ability to understand interconnections and consequently have limited mental models about the structure and dynamics of organizations.

A characteristic of complex organizations is a tendency for management to be risk averse, policy resistant, and quick to blame. This usually means they prefer to stick to traditional solutions and reactive, short-term gains. In doing this, managers ignore the response of other sectors and levels of the organization. In particular, they underestimate the role and effect of behavioral feedback.

Such oversight can result in unintended consequences in the medium term that undermine well-intended actions. Self-organizing and adaptive responses in organizations can lead to many types of informal coping actions, which in turn inhibit the realization of improvement attempts and distort data. A good example of these phenomena, arising from studies described here, is the use of "length of stay" in health and social care services as a policy lever to compensate for capacity shortages.

Planning within complex organization reflects the above characteristics. The core of current planning tends to be static in nature, sector based, and reliant on data and financial spreadsheets with limited transparency of assumptions. For example, the planning of new hospital services can quickly progress to detailed levels without assessment of trends in primary and post-acute care, that is, where hospital patients come from and go to.

In contrast, sustainable solutions to problems in complex organizations often require novel and balanced interventions over whole processes, which seem to defy logic and may even be counterintuitive. However, in order to realize such solutions, it requires a leap beyond both the thinking and planning tools commonly used today. In order to make significant changes in complex organizations, it is necessary to think differently and test ideas before use. System dynamics provides such a method.

The Components of System Dynamics

System dynamics is based on the idea of resisting the temptation to be over-reactive to events, learning instead to view patterns of behavior in organizations and ground these in the structure (operational processes and policies) of organizations. It uses purpose-built software to map processes and policies at a strategic level, to populate these maps with data, and to simulate the evolution of the processes under transparent assumptions, polices, and scenarios.

System dynamics is founded upon:

- Nonlinear dynamics and feedback control developed in mathematics, physics, and engineering
- Human, group, and organizational behavior developed in cognitive and social psychology and economics
- Problem solving and facilitation developed in operational research and statistics

System dynamics provides a set of *thinking* skills and a set of *modeling* tools which underpin the current trend of "whole systems thinking" in health and social care.

System Dynamics Thinking Skills for the Management of Complex Organizations

In order to understand and operate in complex organizations, it is necessary to develop a wide range of thinking skills (Senge 1990). The following are summarized after Richmond (Richmond 1994):

Dynamic thinking – the ability to conceptualize how organizations behave over time and how we would like them to behave

System-as-cause thinking – the ability to determine plausible explanations for the behavior of the organization over time in terms of past actions

Forest thinking – the ability to see the "big picture" (transcending organizational boundaries)

Operational thinking – the ability to analyze the contribution made to the overall behavior by the interaction of processes, information feedback, delays, and organizational boundaries

Closed-loop thinking – the ability to analyze feedback loops, including the way that results can feedback to influence causes

Quantitative thinking – the ability to determine the mathematical relationships needed to model cause and effect

Scientific thinking – the ability to construct and test hypotheses through modeling

System Dynamics Modeling Tools for Planning in Complex Organizations

A useful way to appreciate the tool set of system dynamics is by a brief comparison with other computer-based management tools for decision support. System dynamics is, by definition, a strategic rather than operational tool. It can be used in a detailed operational role, but is first and foremost a *strategic* tool aimed at integrating policies across organizations, where behavioral feedback is important. It is unique in its ability to address the strategic domain, and this places it apart from more operational tool sets such as process mapping, spreadsheets, data analysis, discrete entity simulation, and agent-based simulation.

System dynamics is based on representing process flows by "stock" and "rate" variables. Stocks are important measurable accumulations of physical (and nonphysical) resources in the world. They are built and depleted over time as input and output rates to them change under the influence of feedback from the stocks and outside factors. Recognizing the difference between stocks and rates is fundamental to understanding the world as a system and to understanding the concept of sustainability. Only when depletion rates of resources are less than their discovery rate can stocks of those resources be sustained. The superimposition of organizational sectors and boundaries on the processes is also fundamental to understanding the impact of culture and power on the flows. System dynamics also makes extensive use of causal maps both to highlight feedback processes within models and, of course, to make extensive use of purpose-built computer simulation tools, many of which are described in other entries of this encyclopedia.

Applying System Dynamics with Management Teams

However, the success of system dynamics lies as much in its process of application as in the tool set and hence demands greater skill in conceptualization and use than spreadsheets. Figure 1 shows the overall process of applying system dynamics. A key starting point is the definition of an initial significant issue of managerial concern and the establishment of a set of committed and consistent management teams from all agencies involved in the issue. Another requirement is a set of facilitators experienced in both conceptualizing and formulating system dynamics models. The models created must be shared extensions of the mental models of the management teams, not the facilitators, and importantly owned by the team.

The next step is the analysis of existing trends in major performance measures and the range of possible future trajectories, desired and undesired. This is referred to as the reference model of behavior of the issue and helps with the establishment of the time scale of the analysis. The key

contribution of system dynamics is then to formulate causal maps and high-level stock-flow maps, linking operations across organizations. These maps are then populated with the best data available and subjected to computer simulation. Initially this is to validate them against the mental models of the management team and against past data and shown that they are capable of reproducing the reference mode of behavior of the issue ("what is"). The model can then be used to design policies to realize desired futures ("what might be"). Maps and models are constructed in inexpensive purpose-built software (e.g., *Stella Architect*, *Vensim*, and *Powersim*) with very transparent graphical interfaces. The key is to produce the simplest model possible consistent with maintaining its transparency and having confidence in its ability to cast new light on the issue of concern.

An Overview of Health and Social Care in the UK and Europe

Ensuring that all residents have access to health and social care services is an important goal in all European countries. Most of the services are publicly financed, with taxation and social insurance providing the main sources of funding. Taxation is generally levied on both employees and employers at either the national level or local level or both. The role of private insurance varies between countries and generally private insurance is used as a supplement to, rather than as a substitute for, the main care system. The exceptions to this are Germany and the Netherlands. Further, people are increasingly required to pay part of the cost of medical care.

This entry is primarily concerned with health and social care service supply issues. Although the structure and terminology associated with supply varies across Europe, the underlying issues tend to be similar between countries. Hence the major issues will be described for England.

Health in England is primarily managed and delivered by the National Health Service (NHS) and is seemingly perpetually at the center of an improvement agenda, whereby the government

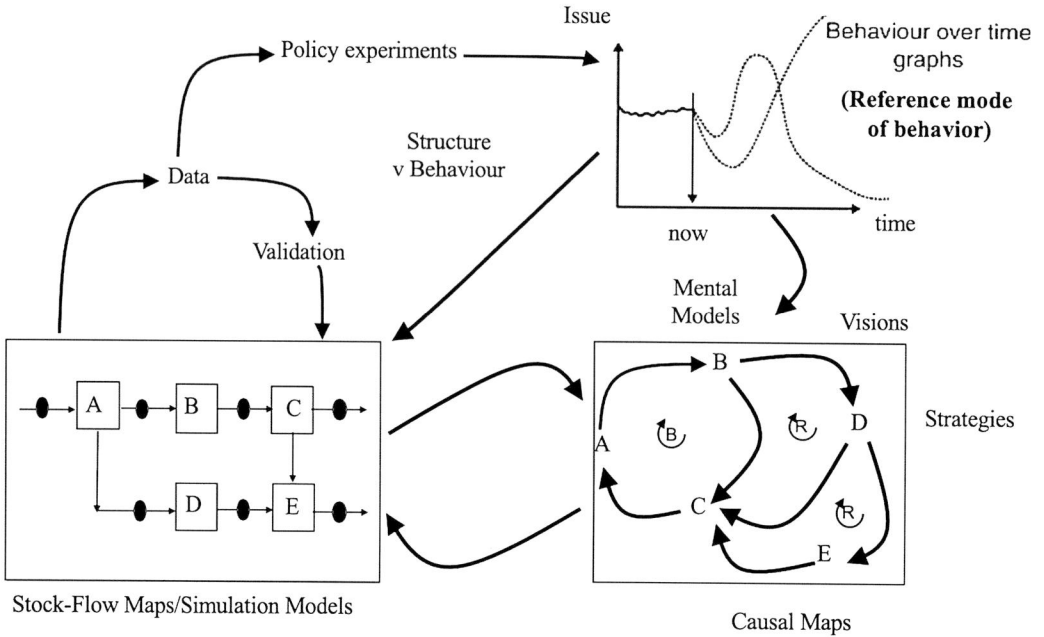

System Dynamics Applications to Health and Social Care in the United Kingdom and Europe, Fig. 1 The systems thinking/system dynamics method

sets out a program of change and targets against which the public may judge improved services.

A major mechanism for reform tends to be via frequent changes to the organizational structure. The current structure consists of large clinical commissioning groups (CCGs), formerly primary care trusts (PCTs), which purchase (commission) primary and secondary services from other agencies, both public and private. A key driver of structural change is to enhance primary care and to take the pressure off hospitals. Initiatives center on providing new services, such as diagnostic and treatment centers and shorter-term "intermediate" care. Emphasis is on bringing the services near to the users, patient choice, payment by results (rather than through block contracts), and service efficiency, the latter being driven by target setting and achievement. The government has made reform of public services a key plank in its legislative program, and pressure to achieve a broad range of often-conflicting targets is therefore immense. However, despite continual increases in funding, new initiatives are slow to take effect, and the performance and viability of the service are problematic with money often being used to clear deficits, rather than generate new solutions.

Social care in England is delivered both by a public sector located with local government social services directorates and a private sector. It consists of numerous services to support children and older people. The latter consists of care homes, nursing homes, and domiciliary (at home) care.

Many patient processes, particularly for older people, transcend health and social care boundaries and hence create a serious conflict of process structure and organizational structure, where the relative power of the different agencies is a major determinant of resource allocation (Wolstenholme and Monk 2004). Additionally, local government and hence social care is at a crisis with severe financial restrictions due to successive austerity measures.

A Case Study: Using System Dynamics to Influence Health and Social Care Policy Nationally in the UK, Delayed Transfers of Care

In order to give a flavor of the relevance and impact of applying system dynamics to health and social care issues, a concise case study will be presented (Wolstenholme et al. 2004a, c, 2007a; Wolstenholme and McKelvie 2019).

Issue

Delayed transfer of care (delayed hospital discharge) to social care is a long-standing issue wherever hospital discharge is government funded and financially constrained. In systems terms, it is a problem of a single patient pathway passing through different organizations and having inevitable boundary issues.

The issue centers on patients occupying hospital beds although they have been declared as "medically fit" for discharge and is most prevalent in the case of older people in hospital waiting for places in adult social care and continuing health care. It is basically an artifact of a caring society where care places are subsidized to try to ensure everyone is looked after post-discharge. Although small compared with the total number of hospital patients, delayed discharges constitute significant loss of hospital capacity for crucial elective procedures, significant risk of infection associated with longer hospital stays, increased cost, and low staff morale.

In the UK, the main organizations involved in delayed discharges are health and adult social care – health being funded by central government and social care funded (and delivered in part) by local government authorities (except in Northern Ireland). Social care also consists of many different services broadly classified as nursing homes (private and local authority controlled), which provide nursing care; care homes, which provide social care; domiciliary care (which means care at home); and NHS continuing care, which embraces services such as district nurses.

The scale of the problem is significant. In 2016/2017, there were 2.3 million delayed transfer days in England, an average of around 6200 per day and 25% higher than the previous year. Approximately, half of these transfers was attributable to waiting for further health services and half to waiting for social care. Estimates of the cost of delayed transfers to NHS providers vary but all are significant sums of money. The National Audit Office in its 2016 report found the total annual cost to the NHS was £820 m (National Audit Office 2016). Delays attributable

to social care have been estimated as £173 million in 2017 (Bate 2017) and £289 m in 2018 (Age UK).

Additionally, local government social care is reported to be on the brink of crisis with budgets having shrunk by £7bn since 2010. It is estimated there will be a £2.5 billion funding gap by 2019/ 2020, which is considerably more than can be met by the current government offer of £2bn over the next 3 years from 2017. Further, a government target for delayed discharges of 3.5% of hospital bed days by September 2017 has not been met. It is suggested that, unless additional funding is found, more people will be denied access to care and pressures will increase for service users, their families, and carers. The situation is not helped by the low esteem in which social care staff have been traditionally held. Staff turnover was recently reported as being 28% (Kings Fund/ Nuffield Trust 2017).

Proposed Solutions

Delayed hospital discharges were made part of the legislative agenda with the Delayed Discharge Act of 2003 and this was replaced by the Care Act 2014. The underlying principles embedded in these acts ranged from fining local authorities for delayed hospital discharges, to improving joint working between health and social care. The latter has taken the form of combining health and social care budgets (NHS Wales and Scotland and small parts of England now operate in this way), to the recently proposed establishment of a "Better Care Fund" (NHS England 2016). This is a pooled budget between Local Authorities and the NHS to better integrate health and social care services. A further proposed solution has been to move medically fit patients to separate pathways for discharge assessment. In order to emphasize their commitment to better integration in 2017, the UK government recently renamed the post of the England Secretary of State for Health to the Secretary of State for Health and Social care.

However, Health and Local Government Social Care are vastly different organizations with their own budgets, power, culture, and information systems and there remains distrust between the two. In addition to social care, local government is responsible for local services of all kinds, including budgets for education and waste disposal, and in times of austerity has to prioritize its own spending. These other sectors of local government are also in crisis, with education, in particular, suffering significant cuts to teacher numbers.

The government's approach to this issue at the time of the study was to essentially blame social care and fine them for delayed transfers of care. The idea of fines was challenged by the Local Government Association (LGA), which represents the interests of all local government agencies at the national level who suggested that a "system" approach should be undertaken to look at the complex interaction of factors affecting delayed transfer of care. This organization, together with the NHS Confederation (the partner organization representing the interests of the National Health Service organizations at a national level), then commissioned a system dynamics study to support their stance.

The remit was for consultants working with the representatives of the two organizations to create a system dynamics model of the "whole patient pathway" extending upstream and downstream from the stock of people delayed in hospital, to identify and test other interventions affecting the issue.

Model Overview

A system dynamics model was developed interactively with managers from the LGA and NHS, using national data to simulate pressures in a sample health economy covering primary, acute, and post-acute care over a 3-year period. The model was driven by variable demand including three winter pressure "peaks" when capacity in each sector was stretched to the limit. Figure 2 shows an overview of the sectors and patient pathways of the model.

The patient flows through the model were represented in hospital by two "stocks" – a "treatment stock" and an "'await discharge" stock – and broken down into medical and surgical flows with access to medical and surgical treatment being constrained by bed capacity. The medical flows were mainly emergency patients and the surgical

flows mainly nonemergency "elective" patients, who came via referral processes and wait lists.

Further, medical patients were broken down into "fast" and "slow" streams. The former were the normal patients who had a short stay in hospital and needed few post-acute services, and the latter the more complex cases (mainly older people), who require a longer stay in hospital and complex onward care packages from social care and community health services. This split was because although the slow patients were few in number, they constituted most of the people who caused delayed transfers of care. The post-hospital health and social care services of intermediate care, nursing/residential home care, and domiciliary care were included in the model and were also constrained in terms of assessment capacity and the number of care packages they could provide. The model incorporated a number of mechanisms by which hospitals coped during periods of high demand, for example, moving medical patients to surgical beds (outliers) and early discharges with allowance for readmissions.

Configuration of the Model

The model was set up to simulate a typical sample health economy over a 3-year period when driven by a variable demand (including three winter "peaks"). The capacity-constrained sectors of the model were given barely sufficient capacity to cope. This situation was designed to create shocks against which to test alternative policies for performance improvement. Major performance measures in use in the various agencies were incorporated. These included:

1. Cumulative episodes of elective surgery
2. Elective wait list size and wait time
3. Numbers of patients in hospital having completed treatment and assessment but not yet discharged (delayed transfers of care/delayed hospital discharges)
4. Number of "outliers"

The model was initially set up with a number of fixed experiments to introduce them to the concept of policy testing. From there, they were encouraged to devise their own experiments and

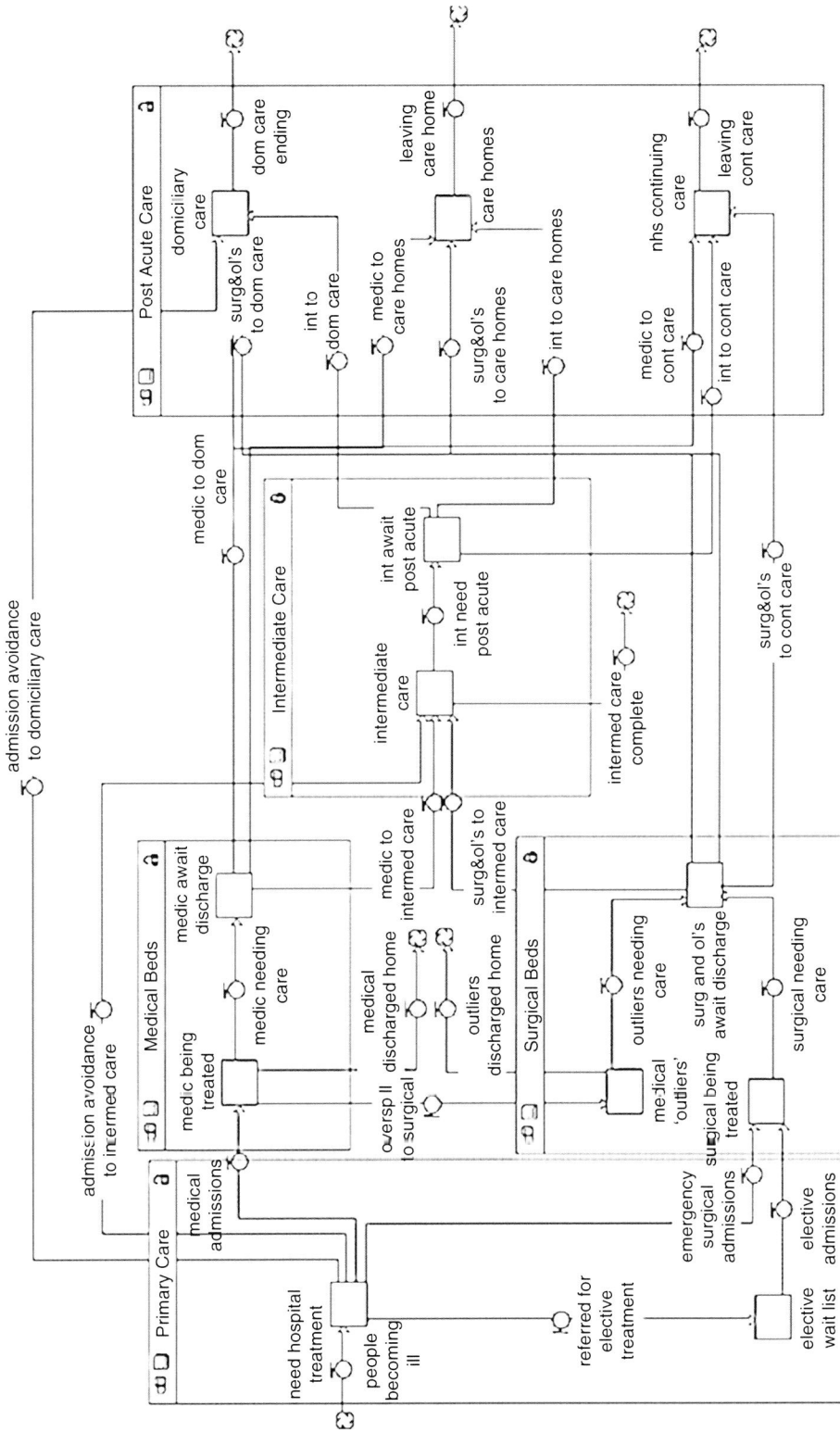

System Dynamics Applications to Health and Social Care in the United Kingdom and Europe, Fig. 2 A simplified overview of the stock/flow structure of the delayed discharge model ("normal" hospital discharges, capacity feedback constraints, and cost sectors are removed for clarity)

develop their own theories of useful interventions and commissioning strategies.

The three main polices tested in the fixed runs were:

1. Adding additional acute hospital bed capacity. This is the classic response used over many years by governments throughout the world to solve any patient pathway capacity problems and was a favorite "solution" here.
2. Adding additional post-acute capacity, both nursing and residential home beds, but also more domiciliary capacity.
3. Diverting more people away from hospital admission by the use of pre-hospital intermediate capacity and also expansion of treatment in primary care GP surgeries.

Example Results from the Delayed Transfers of Care Model

Figures 3, 4, and 5 show some typical outputs for the delayed hospital discharge model. Figure 3 captures the way capacity utilization was displayed (actual beds occupied vs. total available for both medical and surgical sectors of the hospital) and shows the occurrence of "outliers"

(transfers of patients from medical to surgical beds) whenever medical capacity was reached.

Figures 4 and 5 show comparative graphs of three policy runs for two major performance measures for two sectors of the patient pathway – delayed discharges for post-acute social services and cumulative elective procedures for acute hospitals. In each case, the base run is line 1. Line 2 shows the effect of increasing hospital beds by 10%, and line 3 shows the effect of increasing post-acute capacity by 10%.

The interesting feature of this example output is that the cheaper option of increasing post-acute capacity gives lower delayed discharges and higher elective operations, whereas the more expensive option of increasing acute hospital beds benefits the hospital, but makes delayed transfers of care worse. The key to this counterintuitive effect is that increasing post-acute capacity results in higher hospital discharges which in turn reduces the need for the "outlier" coping policy in the hospital, hence freeing up surgical capacity for elective operations.

Outcomes

The obvious common-sense solution of adding more hospital capacity was shown to exacerbate

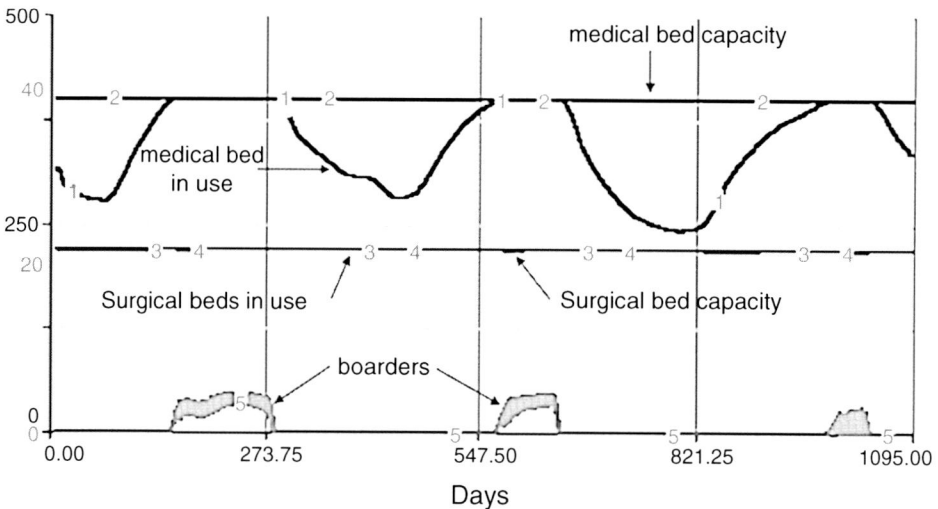

System Dynamics Applications to Health and Social Care in the United Kingdom and Europe, Fig. 3 Medical and surgical bed utilization in hospital and "outliers"

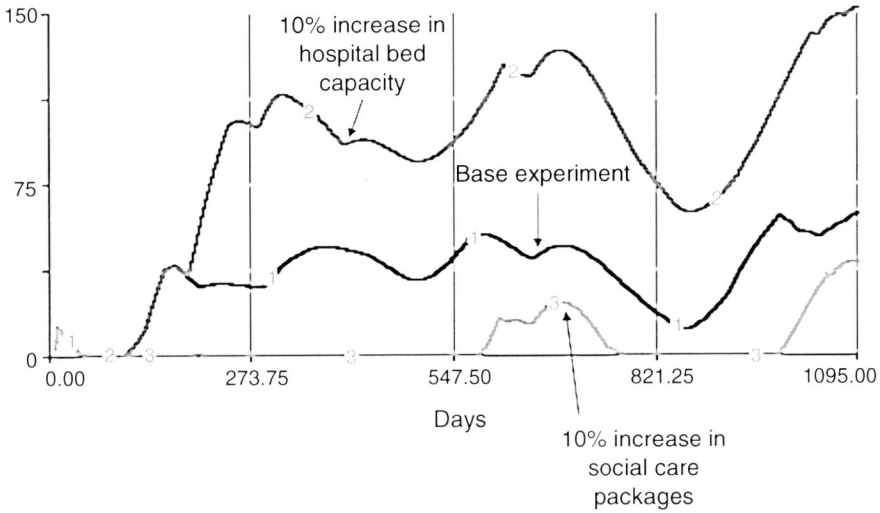

System Dynamics Applications to Health and Social Care in the United Kingdom and Europe, Fig. 4 Delayed hospital discharges for three policy runs of the model

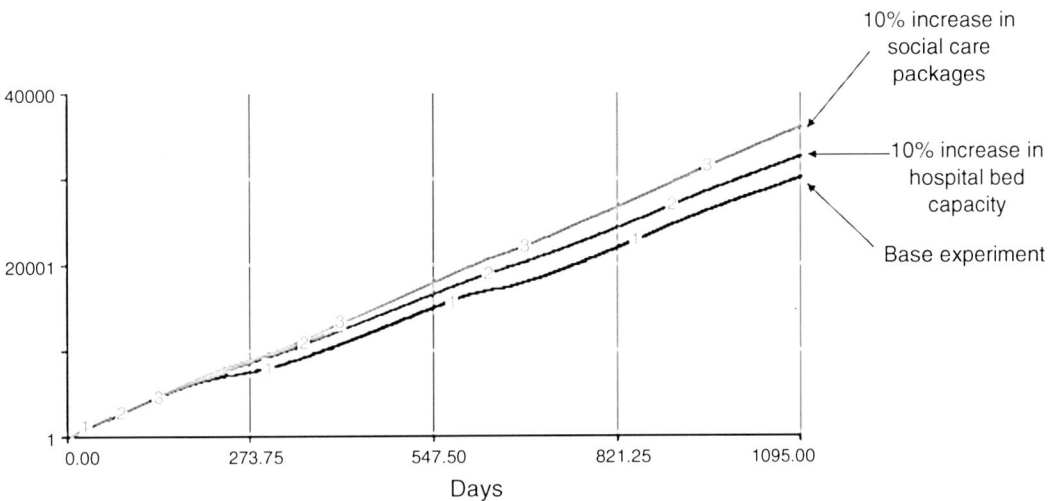

System Dynamics Applications to Health and Social Care in the United Kingdom and Europe, Fig. 5 Cumulative elective operations for three policy runs of the model

the delayed discharge situation. Increasing hospital capacity means facilitating more hospital admissions, but with no corresponding increase in hospital discharges. Hence, the new capacity will simply fill up and then more early discharges and outliers will be needed.

The use of fines on social care was shown to have numerous unintended consequences. If the money levied from social services was given to the hospitals to finance additional capacity, it would make delayed discharges worse. It would be worse still if it caused social care to cut services. The effects of service cuts may also then spill over into the other areas of local government including housing and education.

It was demonstrated that there were some interventions that could help:

1. Increasing social care capacity gives a win-win solution to both health and social care because

it increases all acute and post-acute sector performance measures. Such action allows hospital discharges to increase and eliminates the need for the hospitals to apply coping policies such as boarding emergency patients in elective beds. This, in turn, increases elective operations and reduces elective wait times. Further, counterintuitively, increasing medical capacity in hospital is more effective than increasing surgical capacity for reducing elective wait times as this directly reduces boarding.

2. Reducing assessment times and lengths of stay in all sectors is beneficial to all performance measures as it acts as a proxy for capacity increases.

3. Increasing diversion from hospitals into pre-admission intermediate care is almost as beneficial as increasing social care capacity.

4. If fines are levied, they need to be reinvested from a whole system perspective. This means rebalancing resources across all the sectors (not just adding to hospital capacity).

In general, the model showed that keeping people out of the hospital is more effective than trying to get them out faster. This is compounded by the fact that in-patients are more prone to infections, so the longer the patients are in hospital, the longer they will be in hospital. Also improving the quality of data was shown to be paramount to realizing the benefits of all policies. This is an interesting conclusion associated with many system dynamics studies, where explicit representation of the structure of the organization can lead to a review and redesign of the information needed by systems to really manage the organization.

An interesting generalization of the findings was that increasing stock variables where demand is rising (such as adding capacity) is an expensive and unsustainable solution, whereas increasing rate variables (throughput), by reducing delays and lengths of stay, is cheaper and sustainable.

The original model was shown at the political conferences of 2002 and generated considerable interest. It was instrumental in causing re-thinking of the intended legislation, so that social services were provided with investment funding to address capacity issues, and the implementation of fines was put on hold. Reference to the model was made in the House of Lords.

Moving the main amendment, Liberal Democrat health spokesperson Lord Clement-Jones asked the House to agree that the Bill failed to tackle the causes of delayed discharges and would create perverse incentives which would undermine joint working between local authorities and the NHS and distort priorities for care of elderly people by placing the requirement to meet discharge targets ahead of measures to avoid hospital admission. He referred to the whole system approach being put forward by the Local Government Association, health service managers, and social services directors involving joint local protocols and local action plans prepared in cooperation.

This case study demonstrates the ability of system dynamics to be applied quickly and purposefully to shed rigor and insight on an important issue. The study enabled the development of a very articulate and compelling case for the government to move from a reactive position of blaming social services to one of understanding and acting on a systemic basis and variants of the model have been used extensively over many years. However, delayed hospital discharges have remained a persistent problem, and at the time of writing, the use of a systemic approach is still at the embryonic stage. However, there are signs of a step forward with the establishment of integrated care partnerships between health and social care (Ham 2018) to smooth the movement of older people from hospitals. There are also demonstrations within recent system dynamics studies that small changes to the role, funding, and flexibility of the social care workforce can produce large reductions in delayed hospital discharges (Wolstenholme and McKelvie 2019 – Chaps. 7, 8, and 17).

System Dynamics Studies in Epidemiology

The potential for system dynamics in population health and disease control began in the UK in the late 1980s/early 1990s with the extensive studies carried out on AIDS modeling (Dangerfield and Roberts 1989a, b, 1990, 1992, 1994a, 1996; Dangerfield et al. 2001).

The earlier studies (Dangerfield and Roberts 1990) used a transition model to portray the nature of the disease and to better specify the types of data collection required for further developments of the model. The model was then developed further (Dangerfield and Roberts 1994a, 1996) and was fed with time-series data of actual cases. This enabled projections of future occurrence to be forecast. The latter models were more concerned with examining the resource and cost implications of treatments given to HIV-positive individuals and at their varying stages up until the ensuing onset of AIDS. Later models assessed the effects of triple antiretroviral therapy given to HIV patients.

Dangerfield explains some of the reasons (Dangerfield 1999a) why system dynamics acts as an excellent tool for epidemiological modeling. The positive and negative feedback loops help imitate the natural disposition of the spread and containment of diseases among the general population. Further, system dynamics allows delays associated with the incubation period of infectious diseases to be accurately and easily modeled (Dangerfield 1999c) without the need for complicated mathematical representations.

The work in the UK was complemented by the work in Holland on simulation as a tool in the decision-making process to prevent HIV incidence among homosexual men (Dijkgraaf et al. 1998) and on models for the analysis and evaluation of strategies for preventing AIDS (Lagergren 1992).

Other epidemiological studies in system dynamics in the UK have been related to BSE and Ebola. The outbreak out of BSE and the subsequent infection of humans with its human form nvCJD was modeled in 2003 (Curram and Coyle 2003) and the outbreaks of Ebola were modeled in 2015/2016 (Giuseppe 2015, 2016). More specific work on the spread of diseases has been undertaken around infections in hospitals (Auditor General 2009).

These epidemic models are all characterized by representing the growth of an infection as a reinforcing feedback loop and then intervening in this growth by introducing a balancing feedback loop based on treatment of the condition. This combination of feedback loops creates a system archetype (limits to growth) (Senge 1990; Wolstenholme 2003, 2004). These feedback effects result in the flow of people through different stocks over time, representing the different stages of the disease progression. The purpose of system dynamics modeling is then to test the effects of interventions, such as inoculation, aimed at slowing down the rate of progression of the condition or indeed moving people "upstream" to less severe states of the condition.

System Dynamics Studies in Health and Social Care Management

By far, the greatest number of studies and publications in the use of system dynamics in health and social care is associated with patient flow modeling for health-care planning, that is, the flow of patients through multiple service delivery channels. Patient pathway definition has been an area of health modernization, and these pathways lend themselves to represent as stock/flow resource flows in system dynamics. The purpose of this type of modeling is to identify bottlenecks, plan capacity, reduce wait lists, and improve the efficiency of patient assessments and times and the design of alternative pathways with shorter treatment times (e.g., intermediate care facilities both pre- and post-hospital treatment).

A characteristic of patient pathways is that they are long and pass through multiple agencies and hence confront the major health issues of working across boundaries and designing integrated policies. Studies in this area have examined the flow of many different populations of patients and often resulted in arrayed models to represent the flow of different "populations" or "needs groups" through several parallel service channels.

The most common set of models are associated with the flow of patients from primary care through acute hospitals and onward into post-acute care such as social services provisions for home care, nursing care, and residential care. The populations have often been split between the simple everyday cases and the complex cases associated with older people needing greater degrees of care. They have also involved medical and surgical splits. There are a number of books and review papers which supplement the work described below (Wolstenholme and McKelvie 2019; Abdul-Salam 2006; Dangerfield and Roberts 1994a. 1996; Maliapen and Dangerfield 2009; Bechberger et al. 2011; Lane et al. 2010).

In addition to the work in the 1990s on the interface between health and social care (Wolstenholme 1993, 1996, 1999) and the national-level UK work on older people flows through hospitals (Wolstenholme et al. 2004a, c, 2007b, c), Wolstenholme has reported that system dynamics applications are currently under way in numerous local health communities around the UK with the objectives of modeling patient flows across agency boundaries to provide a visual and quantitative stimulus to strategic multi-agency planning (Wolstenholme et al. 2004a).

Lane has reported a work in accident and emergency departments (Lane 2000) and in mapping acute patient flows (Lane and Husemann 2008), while Royston worked with the NHS to help develop and implement policies and programs in health care in England (Royston and Dost 1999). Taylor has undertaken an award winning modeling of the feedback effects of reconfiguring health services (Taylor 2002; Taylor and Dangerfield 2004; Taylor et al. 2005), while Lacey has reported numerous UK studies to support the strategic and performance management roles of health service management, including provision of intermediate care and reduction of delayed hospital discharges (Lacey 2005; Lacey 2011). Studies in providing intermediate care and social care are described by Bayer (Bayer 2002; Bayer and Barlow 2003) and further hospital capacity studies by Coyle (Coyle 1996). Elsewhere, there have been specific studies on bed blocking (El-Darzi and Vasilakis 1998) and screening (Osipenko 2006).

In Norway, system dynamics-based studies have focused on mapping the flows of patients in elderly nonacute care settings (Chen 2003). The purpose of this study according to Chen is to differentiate between acute and nonacute settings and thereby increase the understanding of the complexity and dynamics caused by influencing elements in the system. Also it is to provide a tool for local communities in Norway for their long-term budget planning in the nonacute health sector for the older people. Work on reducing waiting lists has been reported in Holland (Kim and Gogi 2003; Van Ackere and Smith 1999; Van Dijkum and Kuijk 1998; Verburgh and Gubbels 1990). Also in Holland, Vennix has reported a comprehensive work on modeling a regional Dutch health-care system (Vennix and Gubbels 1992).

Work has been undertaken to balance capacities in individual hospitals in Italy (Sedehi 2001) and in Norway (Petersen 2000; Ravn and Petersen 2007). Furthermore, normally the realm of more operational types of simulation, system dynamics has proved very effective here. There has also been work to assess the impact on health and social care of technological innovation, particularly telecare (Bayer 2001; Bayer and Barlow 2004). Additionally, systems thinking has been undertaken by doctors to examine the European time directive (Ratnarajah 2005). A highly significant body of emergent work is associated with the use of system dynamics modeling to try to reduce the impact of disease on communities. These tend to relate to prevention, but are currently driven by their impact on services. Examples are the conditions of alcohol use and abuse, dementia, obesity (Dangerfield 2011; Zainal et al. 2014), and norovirus (Lane et al. 2019).

With costs to the UK NHS escalating, the reduction of alcohol-related harm is seen as a priority, and models have been rolled out for regional use across the UK to assess the net impact on hospital admissions and cost from implementing interventions such as identification and brief advice, the employment of alcohol health workers, and specialized treatment (Wolstenholme and McKelvie 2019, McKelvie 2011, http://www.symmetriclab.com). The number of people with dementia is forecast to double in the UK and the cost of their services to treble. Modeling is being used by NHS commissioners to review the impact of a range of dementia service interventions, decide an appropriate service strategy, avoid a significant element of the expected growth in cost, and, in some cases, achieve actual savings, and improve service quality by avoiding unnecessary hospital and/or care home admission (http://www.symmetriclab.com). In particular, a variety of studies across the UK have been reported by Lacy in 2017. These cover such issues as determining staff mix and costs for alternative policies in the management of neurological patients, designing capacities for re-enabling dementia patients towards independent living, and developing understanding of why best practice is commonly too difficult to achieve for many services. The latter is an excellent case of the insights available from system dynamics and focuses on understanding the delays and psychology between establishing targets for best practice and its adoption and communication (http://www.thewholesystem.co.uk/simulation/examples.aspx).

Wolstenholme reports many general insights from the foregoing work. He suggests a hypothesis that the "normal" mode of operation for many health and social care organizations today is often well beyond their safe design capacity. This situation arises from having to cope with whatever demand arrives at their door irrespective of their supply capability. Risk levels can be high in these organizations and the consequences could be catastrophic for patients (Wolstenholme et al. 2005; Wolstenholme and McKelvie 2019).

Evidence for the hypothesis has emerged at many points along patient pathways in health and social care from a number of studies carried out using system dynamics simulation to identify and promote systemic practice in local health communities. The rigor involved in knowledge-capture and quantitative simulation model construction and running has identified mismatches between how managers claim their organizations work and the observed data and behavior. The discrepancies can only be explained by surfacing informal coping policies. Examples of such policies and their costly unintended consequences are:

1. Transferring medical emergency patients to surgical wards when medical wards are full. This results in cancelled elective procedures as in the case study in this entry (also reported by Lane et al. (2003)) and consultant time lost from sometimes not knowing where patients are.

2. Stemming admissions in times of high demand. This results in placing more responsibility on society and families.

3. Buying care on the private spot market to enable discharges – a very costly solution.

4. Reducing the length of stay for patients easy to discharge to compensate for lack of beds. This results in increased re-admissions.

All of these policies are reactive rather than thought out. Indeed, they make sound policy making more difficult since they produce questionable data, which reflects more the actions of managers than the true characteristics of patients.

The result of capacity pressure can mean that managers are unable, physically and financially, to break out from a fire-fighting mode to implement better resource investment and development policies for systemic and sustainable improvement. The insights reported are important for health and social care management, for the meaning of data, and for modeling. The key message here is that much needed systemic solutions and whole systems thinking can never be successfully implemented until organizations are allowed to articulate and dismantle their worst coping strategies and funded to return to working within best practice capacities.

Modeling to assist mental health reform has developed as a separate strand of health work in the UK (Smith and Wolstenholme 2004; Wolstenholme et al. 2006, 2007c). Mental health services in the UK over the past 50 years have undergone numerous major reforms. The National Institute for Health and Care Excellence has published a series of extensive research-based guidelines on the way stepped care might be best achieved (for example, NICE 2004). These involved moves towards a balanced, mixed community/institutional provision of services set within a range of significant reforms to the National Health Service. The latest and perhaps most significant reform is that associated with the introduction of "stepped care." Stepped care is aimed at bringing help to more patients more cheaply by developing intermediate staff, services, and treatments between GPs and the specialist health hospitals.

Having decided on the new treatments at each step and having designed the basic patient pathways, modeling has been used in the northwest of England to help with the communication of the benefits and to overcome the anticipated problems with resource reallocation issues (Wolstenholme et al. 2006). The software tools created to support this work were used by eight Care trusts across England.

Further work in Lincolnshire, UK (Wolstenholme et al. 2008), reports the increasing use of "matrix" modeling in mental health to capture the dynamics of both patient needs and treatments. This work also demonstrates the counterintuitive dangers of overinvestment in situations where demand is perhaps more latent in accrued backlogs due to past supply problems than real. In fact, demand could be decreasing due to better and more successful interventions.

A novel use of system dynamics modeling has also been to support static cost-benefit analyses in mental health (Wolstenholme 2007c, 2010). A planned expansion of cognitive behavioral therapy by recruiting large numbers of additional therapists in the UK was underpinned by dynamic cost-benefit analysis of the numbers of people who might benefit over time. By focusing the analysis on the dynamics of people flows over time, simulation was shown to assist the understanding of the investment proposal and its potential benefits. The number of patients expected to benefit was shown to depend on the treatment capacity, where people come from, the number and type of treatment channels and their parameters, the success of treatment, the provision for patients moving between treatment channels, the dynamics of the labor market, and employment opportunities. The work questions the magnitude of the potential benefits, introduces phasing issues, surfaces structural insights, takes account of the dynamics of the labor market, and forces linkages between the plan and other initiatives to get people back to work. The paper suggests that

cost-benefit analysis and system dynamics are very complementary and should be used together in strategic planning. This work is an example of the value that system dynamics can add to the conventional cost-benefit analysis.

Other mental health capacity planning studies have been carried out for individual mental health hospitals and trusts. One such study (Wolstenholme et al. 2007b) describes the application of system dynamics to assist decision making in the reallocation of resources within a specialist mental health trust in South London. Mental health service providers in the UK are under increasing pressure both to reduce their own costs and to move resources upstream in mental health patient pathways to facilitate treating more people while not compromising service quality.

The investigation here focused on the consequences of converting an existing specialist service ward in a mental health hospital into a "triage" ward, where patients are assessed and prioritized during a short stay for either discharge or onward admission to a normal ward. Various policies for the transition were studied together with the implications for those patients needing post-hospital services and relocation within the community. The model suggested that the introduction of a triage ward could meet the strategic requirement of a 10% shift away from institutional care and into community services. The paper includes a number of statements from the management team involved on the benefits of system dynamics and the impact of its application on their thinking.

Against a background of increasing exposure of child abuse in the UK, there has been recent work within the field of child social care to use the ideas of systems thinking to help shed light on reviewing child protection. The "Munro Review of Child Protection" was a high-profile examination of child protection activities in England, conducted for the Department for Education (Munro 2011) and this study employed a blend of systems thinking approaches to examine the activities, culture, effectiveness, and social relations of the child protection sector.

Within this work, systems thinking mapping was used to both visualize current operations and the effects of past policies on them and to give shape to the range of issues the review addressed. The maps acted as an organizing framework for the systemically determining a coherent set of recommendations. The analysis helped explain how a "compliance culture" had evolved within social care staff leading to less time with clients, less staff commitment, and increased absenteeism and staff turnover (Lane et al. 2016).

It is also important to mention that work has been carried out in a number of countries in the field of workforce planning related to health. In the UK, the NHS has deployed sophisticated workforce planning models to determine the training and staffing needs associated with numerous alternative service configurations. In the Spanish health system, modeling has been used to determine the number of doctors required for a number of specialist services and to attempt to explore solutions for the current imbalance among the supply and demand of physicians (Alonso Magdaleno 2002a, c; Bayer 2001). Elsewhere the factors affecting staff retention have been studied (Holmstroem and Elf 2004), and in the Netherlands, an advisory body of the Dutch government was given the responsibility of implying a new standard for the number of rheumatologists (Posmta and Smits 1992). One of the main factors that were studied in the scenario analysis stage was the influence of changing demographics on the demand for workforce in the health system. Other studies have covered time reduction legislation on doctor training (Derrick et al. 2005).

The problem of how to ensure current doctor recruitment rates match targets for future numbers in each grade is an on-going issue in the UK. System dynamics is particularly suited to this analysis, taking into account training delays, times in post, retirement, intake of pre-trained foreign labor and dropout rates (http://www.symmetriclab.com). Similar complexities exist in developing strategies for delivering safe, effective, and sustainable acute care in pediatrics and in nursing (http://www.thewholesystem.co.uk/simulation/examples.aspx).

System dynamics is one of a number of systems methods being applied to health issues in Europe and there are numerous initiatives to try to contrast and consolidate these methods (Brailsford and Lattimer 2004; Brailsford and Howick, Davies 1985; Jun et al. 1999; Luke and Stamatakis 2012; Sadsad et al. 2014).

In particular, The Cumberland Initiative (http://cumberland-initiative.org) has been established and has a movement to encourage all types of simulation and modeling of healthcare scenarios to improve NHS quality of care delivery. The idea is to transform the quality of care through radically better processes and systems and produce far superior outcomes and huge cost savings compared to traditional exercises centered on better selection of drugs or technology. Here the emphasis is directly on experimenting with models rather than on the real operations.

Participative Model Building

An important contribution to the process of model building can be seen in the work of Vennix. A characteristic of this work has been group model building (Vennix 1996). The main objectives (Heyne and Geurts 1994) are to improve communication, learning, and integration of multiple perspectives (Vennix and Gubbels 1992). Vennix brought together strategic managers and important stakeholders to participate in the process of building a system dynamics model of the Dutch health-care system. The policy problem which is modeled in Vennix's 1992 study is related to the gradual, but persistent, rise in health-care costs in the Netherlands. Vennix (Vennix and Gubbels 1992) attempts to find the underlying causes of those increases that emanate from within the health-care system itself rather than focusing on exogenous factors. By doing so, Vennix stands to identify potential levers within the health-care system that can be practically and appropriately adjusted to reduce cost increases. Vennix attempts to extract important assumptions from the key players by posing three straightforward questions:

1. What factors have been responsible for the increase in health-care costs?
2. How will health-care costs develop in the future?
3. What are the potential effects of several policy options to reduce these costs?

Participants are asked if they agreed or disagreed with the statements and why they thought the statements were true or not. The most frequently given reasons for the verbal statements were then incorporated into the statements to create causal arguments from the participant's mental models.

Similar methods were adopted to identify policies which represent the aggregate of many individual actions: for example, why a GP may decide on such matters as frequency of patients' appointments, drug choice, referral to other medical specialists, or a combination of all these. Vennix's model was subsequently formalized and quantified and converted into a computer-based learning environment for use by a wider range of health personnel. The idea of using system dynamics as a means of participative modeling for learning is also inherent in other work (Lane et al. 2003).

Future Directions

System dynamics has already made a significant impact on health and social care thinking across Europe. Many policy insights have been generated and organizations are increasingly being recognized as complex adaptive systems. However, true understanding and implementation of the messages require much more work, and too many organizations are still locked into a pattern of short termism which leads them to focus on the things they feel able to control – usually variables within their own individual spheres of control. There are also a few aspects of system reform in some countries that are producing perverse incentives, which encourage organizations to apply short-term policies.

The use of internet-based models is providing wider communication of studies and making models available for use by local health and social care teams and communities. This is an important development in demonstrating to a wider audience of managers and clinicians that health and social care organizations can add value to the whole while remaining autonomous. An important element is to train more people capable of modeling and facilitating studies and to simplify further the process and software of system dynamics. It would also be beneficial to broaden the use of system dynamics in European health and social care to more preventative studies.

Bibliography

Books
Senge P (1990) The fifth discipline. Doubleday, New York
Sterman J (2000) Business dynamics: system thinking and modelling for a complex world. McGrawHill, Boston
Vennix AM (1996) Group model building. Wiley, Chichester
Wolstenholme EF, McKelvie D (2019) The dynamics of care. Springer, (in press)

Primary Literature
Abdul-Salam O (2006) An overview of system dynamics applications in health care. In: A dissertation submitted to the University of Salford Centre for Operational Research and Applied Statistics for the degree of MSc Centre for Operational Research and Applied Statistics, University of Salford
Age UK Press release 9/7/2018
Alonso Magdaleno MI (2002a) Administrative policies and MIR vacancies: impact on the Spanish Health System. In: Proceedings of the 20th international conference of the System Dynamics Society, Palermo
Alonso Magdaleno MI (2002b) Dynamic analysis of some proposals for the management of the number of physicians in Spain. In: Proceedings of the 20th international conference of the System Dynamics Society, Palermo
Alonso Magdaleno MI (2002c) Elaboration of a model for the management of the number of specialized doctors in the spanish health system. In: Proceedings of the 20th international conference of the System Dynamics Society, Palermo
Atkinson JM, Wells R, Page A, Dominello A, Haines M, Wilson A (2015) Applications of system dynamics modelling to support health policy. Public Health Res Pract 25(3):c2531531
Bate, A (2017) Delayed transfers of care in the NHS, briefing paper no 7415, June 2017. House of Commons Library
Bayer S (2001) Planning the implementation of telecare services. In: The 19th international conference of the System Dynamics Society, Atlanta

Bayer S (2002) Post-hospital intermediate care: examining assumptions and systemic consequences of a health policy prescription. In: Proceedings of the 20th international conference of the System Dynamics Society, Palermo

Bayer S, Barlow J (2003) Simulating health and social care delivery. In: Proceedings of the 21st international conference of the System Dynamics Society, New York

Bayer S, Barlow J (2004) Assessing the impact of a care innovation: telecare. In: 22nd international conference of the System Dynamics Society, Oxford

Bishai D, Paina L, Li Q, Peters DH, Hyder AA (2014) Advancing the application of systems thinking in health: why cure crowds out prevention, vol 12. Health Research Policy and Systems 2014, p 2

Brailsford SC, Lattimer VA (2004) Emergency and on-demand health care: modelling a large complex system. J Oper Res Soc 55:34–42

Brailsford S, Vissers J (2010) OR in healthcare: a European perspective. Eur J Oper Res 212(3):223–234. https://doi.org/10.1016/j.ejor.2010.10.026

Chen Y (2003) A system dynamics-based study on elderly non-acute service in Norway. In: Proceedings of the 21st international conference of the System Dynamics Society, New York

Coyle RG (1996) A systems approach to the management of a hospital for short-term patients. Socio Econ Plan Sci 18(4):219–226

Curram S, Coyle JM (2003) Are you mad to go for surgery? Risk assessment for transmission of vCJD via surgical instruments: the contribution of system dynamics. In: Proceedings of the 21st international conference of the System Dynamics Society, New York

Dangerfield BC (1999a) System dynamics applications to European health care issues. System dynamics for policy, strategy and management education. J Oper Res Soc 50(4):345–353

Dangerfield BC (1999b) System dynamics applications to European health care issues. J Oper Res Soc 50(4):345–353

Dangerfield B (1999c) Optimisation as a statistical estimation tool: an example in estimating the AIDS treatment-free incubation period distribution. System Dynamics Review 15(3):273–291. (BC Dangerfield & CA Roberts)

Dangerfield B (2016) System dynamics applications to European healthcare issues. In: Mustafee N (ed) Operational research for emergency planning in healthcare: vol 2. Palgrave Macmillan, London, pp 296–315

Dangerfield BC, Roberts CA (1989a) A role for system dynamics in modelling the spread of AIDS. Trans Inst Meas Control 11(4):187–195

Dangerfield BC, Roberts CA (1989b) Understanding the epidemiology of HIV infection and AIDS: experiences with a system dynamics model. In: Murray-Smith D, Stephenson J, Zobel RN (eds) Proceedings of the 3rd European simulation congress. Simulation Councils Inc, San Diego, pp 241–247

Dangerfield BC, Roberts CA (1990) Modelling the epidemiological consequences of HIV infection and AIDS: a contribution from operational research. J Oper Res Soc 41(4):273–289

Dangerfield BC, Roberts CA (1992) Estimating the parameters of an AIDS spread model using optimisation software: results for two countries compared. In: Vennix JAM, Faber J, Scheper WJ, Takkenberg CA (eds) System dynamics. System Dynamics Society, Cambridge, pp 605–617

Dangerfield BC, Roberts CA (1994a) Fitting a model of the spread of AIDS to data from five european countries. In: Dangerfield BC, Roberts CA (eds) O.R. work in HIV/AIDS, vol 2. Operational Research Society, Birmingham, pp 7–13

Dangerfield BC, Roberts CA (1996) Relating a transmission model of AIDS spread to data: some international comparisons. In: Isham V, Medley G (eds) Models for infectious human diseases: their structure and relation to data. Cambridge University Press, Cambridge, pp 473–476

Dangerfield BC, Roberts CA, Fang Y (2001) Model-based scenarios for the epidemiology of HIV/AIDS: the consequences of highly active antiretroviral therapy. Syst Dyn Rev 17(2):119–150

Davies R (1985) An assessment of models in a health system. J Oper Res Soc 36:679–687

Derrick S, Winch GW, Badger B, Chandler J, Lovett J, Nokes T (2005) Evaluating the impacts of time-reduction legislation on junior doctor training and service. In: Proceedings of the 23rd international conference of the System Dynamics Society, Boston

Dijkgraaf MGW, van Greenstein GJP, Gourds JLA (1998) Interactive simulation as a tool in the decision-making process to prevent HIV incidence among homosexual men in the Netherlands: a proposal. In: Jager JC, Rotenberg EJ (eds) Statistical analysis and mathematical modelling of AIDS. Oxford University Press, Oxford, pp 112–122

El-Darzi E, Vasilakis C (1998) A simulation modelling approach to evaluating length of stay, occupancy, emptiness and bed blocking in hospitals. Health Care Manag Sci 1(2):143–149

El-Jardali F, Adam T, Ataya N, Jamal D, Jaafar M (2014 Dec) Constraints to applying systems thinking concepts in health systems: a regional perspective from surveying stakeholders in eastern Mediterranean countries. Int J Health Policy Manag 3(7):399–407

Giuseppe N, 2015 Modeling the effect of information feedback for the management of the Ebola Crisis "advances in business management. Towards systemic approach". In: 3rd business systems laboratory international symposium Perugia – January 21–23, 2015. http://bslab-symposium.net/Perugia.2015/Online-Proceedings-Book-Abstrac ts-BSLAB-2015.pdf

Giuseppe N, 2016 Modelling the Ebola Crisis with system dynamics, Boletín de Dinámica de Sistemas, June 2016. http://dinamica-de-sistemas.com/revista/0616b.htm

Gonzalez B, Garcia R (1999) Waiting lists in spanish public hospitals: a system dynamics approach. Syst Dyn Rev 15(3):201–224

Ham C (2018) Making sense of integrated care systems, integrated care partnerships and accountable care organisations in the NHS in England. https://www.kingsfund.org.uk/publications/making-sense-integrated-care-systems

Heyne G, Geurts JL (1994) Diagnost: a microworld in the healthcare for elderly people. In: 1994 international system dynamics conference. System Dynamics Society, Stirling

Hirsch G, Homer J (2019) System dynamics applications to health care in the United States. In: Meyers R (ed) Encyclopedia of complexity and system science. Springer-Verlag, Berlin. ISBN 978-0-387-75888-6.

Holmstroem P, Elf M (2004) Staff retention and job satisfaction at a hospital clinic: a case study. In: 22nd international conference of the System Dynamics Society, Oxford

Institute for Health Research (NIHR). http://www.nihr.ac.uk/

Jun JB, Jacobson SH, Swisher JR (1999) Application of discrete-event simulation in health care clinics: a survey. J Oper Res Soc 50(2):109–123

Kim DH, Gogi J (2003) System dynamics modeling for long term care policy. In: Proceedings of the 21st international conference of the System Dynamics Society, New York

Kings Fund/Nuffield Trust (2017) The Autumn Budget: joint statement on health and social care. Nuffield Trust, Health Foundation, Kings Fund

Lacey P (2005) Futures through the eyes of a health system simulator. Paper presented to the international system dynamics conference, Boston

Lacey P (2011) Integrated care development using systems modelling – a case study of intermediate care. In: Jain S, Creasey RR, Himmelspach J, White KP, Fu M (eds) Proceedings of the 2011 winter simulation conference, Boston

Lagergren M (1992) A family of models for analysis and evaluation of strategies for preventing AIDS. In: Jager JC (ed) Scenario analysis. Elsvier, Amsterdam, pp 117–145

Lane DC (2000) Looking in the wrong place for healthcare improvements: a system dynamics study of an accident and emergency department. J Oper Res Soc 51(5):518

Lane DC, Husemann E (2008) System dynamics mapping of acute patient flows. J Oper Res Soc 59:213–224

Lane DC, Monefeldt C et al (2003) Client involvement in simulation model building: hints and insights from a case study in a London hospital. Health Care Manag Sci 6(2):105–116. https://doi.org/10.1057/palgrave.jors.2600892

Lane DC, Munro E, Husemann E (2016) Blending systems thinking approaches for organisational analysis: reviewing child protection in England. Eur J Oper Res 251:613–623

Luke DA, Stamatakis KA (2012) Systems science methods in public health: dynamics, networks, and agents. Annu

Rev Public Health 33:357–376. https://doi.org/10.1146/annurev-publhealth-031210-101222. Epub 2012 Jan 3

Maliapen M, Dangerfield B (2009) A system dynamics based simulation study for managing clinical governance and pathways in a hospital. J Oper Res Soc 61:255–264

McKelvie D (2013) Modelling social care complexity: the potential of system dynamics methods review no 14, NIHR School for social care research. sscr.nihr.ac.uk/PDF/MR/MR14.pdf

Miller R, McKelvie D (2016) Commissioning for complexity: exploring the role of system dynamics in social care, 2016, NIHR School for Social Care Research Scoping Review. sscr.nihr.ac.uk/PDF/Findings/RF58.pdf (The School for Social Care Research The School for Social Care Research is a partnership between the London School of Economics and Political Science, King's College London and the Universities of Kent, Manchester and York, and is part of the National

Munro E (2011) The Munro review of child protection final report: a child-centred system. TSO, London

National Audit Office Annual Report (2016)

National Institute for Health and Care Excellence (2004) Depression: management of depression in primary and secondary care – NICE guidance, national clinical practice guideline 23

NHS England (2016) Better care operating guide. NHS England, Leeds

Osipenko L (2006) System dynamics model of a new prenatal screening technology (screening pregnant women). In: 24th international conference of the System Dynamics Society, Nijmegen, 23–27 July 2006

Petersen LO (2000) How should the capacity for treating heart decease be expanded? In: 18th international conference of the System Dynamics Society, Bergen

Posmta TJBM, Smits MT (1992) Personnel planning in health care: an example in the field of rheumatology. In: Proceedings of the 1992 international system dynamics conference of the System Dynamics Society. Utrecht

Ratnarajah M (2005) European union working time directive. In: Proceedings of 7th annual gathering, UK System Dynamics Chapter, Harrogate, Feb 2005

Ravn H, Petersen LO (2007) Balancing the surgical capacity in a hospital. Int J Healthcare Technol Manag 14:4023–4089

Royston G, Dost A (1999) Using system dynamics to help develop and implement policies and programmes in health care in England. Syst Dyn Rev 15(3):293–315

Sadsad R, McDonnell G, Viana J, Desai SM, Harper P, Brailsford S (2014) Hybrid modelling case studies. In: Brailsford S, Churilov L, Dangerfield B (eds) Discrete-event simulation and system dynamics for management decision making. Wiley, Chichester. https://doi.org/10.1002/9781118762745.ch14

Sedehi H (2001) HDS: health department simulator. In: The 19th international conference of the System Dynamics Society, Atlanta

Smith G, Wolstenholme EF (2004) Using system dynamics in modeling health issues in the UK. In: 22nd international conference of the System Dynamics Society, Oxford

Sterman T (ed) (2008) Exploring the next frontier: system dynamics at 50. Syst Dyn Rev 23:89–93

Taylor KS (2002) A system dynamics model for planning and evaluating shifts in health services: the case of cardiac catheterisation procedures. Unpublished PhD thesis, London School of Economics and Political Science

Taylor KS, Dangerfield BC (2004) Modelling the feedback effects of reconfiguring health services. J Oper Res Soc 56:659–675. (Published on-line Sept 2004)

Taylor KS, Dangerfield BC, LeGrand J (2005) Simulation analysis of the consequences of shifting the balance of health care: a system dynamics approach. J Health Serv Res Policy 10(4):196–202

Tretter F (2002) Modeling the system of health care and drug addicts. In: Proceedings of the 20th international conference of the System Dynamics Society, Palermo

Van Ackere A, Smith PC (1999) Towards a macro model of national health service waiting lists. Syst Dyn Rev 15(3):225

Van Dijkum C, Kuijk E (1998) Experiments with a non-linear model of health-related actions. In: 16th international conference of the System Dynamics Society, Quebec

Vennix JAM, Gubbels JW (1992) Knowledge elicitation in conceptual model building: a case study in modeling a regional Dutch health care system. Eur J Oper Res 59(1):85–101

Verburgh LD, Gubbels JW (1990) Model-based analyses of the Dutch health care system. In: System dynamics '90: proceedings of the 1990 international systems dynamics conference, Chestnut Hill

Wolstenholme EF (1993) A case study in community care using systems thinking. J Oper Res Soc 44(9):925–934

Wolstenholme EF (1996) A management flight simulator for community care. In: Cropper S (ed) Enhancing decision making in the NHS. Open University Press, Milton Keynes

Wolstenholme EF (1999) A patient flow perspective of UK health services: exploring the case for new intermediate care initiatives. Syst Dyn Rev 15(3):253–273

Wolstenholme EF (2003) Towards the definition and use of a core set of archetypal structures in system dynamics. Syst Dyn Rev 19(1):7–26

Wolstenholme EF (2004) Using generic system archetypes to support thinking and learning. Syst Dyn Rev 20(2):341–356

Wolstenholme EF (2010) Dynamic cost benefit analysis for mental health reform (with David Monk and David Todd). Kybernetes 39(9–10):1645–1658

Wolstenholme EF, McKelvie D (2004) Using system dynamics in modeling health and social care commissioning in the UK. In: 22nd international conference of the System Dynamics Society, Oxford

Wolstenholme EF, Monk D (2004) Using system dynamics to influence and interpret health and social care policy

in the UK. In: 22nd international conference of the System Dynamics Society, Oxford

Wolstenholme EF, Monk D, Smith G, McKelvie D (2004a) Using system dynamics in modelling health and social care commissioning in the UK. In: Proceedings of the 2004 system dynamics conference, Oxford

Wolstenholme EF, Monk D, Smith G, McKelvie D (2004b) Using system dynamics in modelling mental health issues in the UK. In: Proceedings of the 2004 system dynamics conference, Oxford

Wolstenholme EF, Monk D, Smith G, McKelvie D (2004c) Using system dynamics to influence and interpret health and social care policy in the UK. In: Proceedings of the 2004 system dynamics conference, Oxford

Wolstenholme EF, Arnold S, Monk D, Todd D, McKelvie D (2005) Coping but not coping in health and social care – masking the reality of running organisations well beyond safe design capacity. Syst Dyn Rev 23(4):371–389

Wolstenholme EF, Arnold S, Monk D, Todd D, McKelvie D (2006) Reforming mental health services in the UK – using system dynamics to support the design and implementation of a stepped care approach to depression in North-West England. In: Proceedings of the 2006 international system dynamics conference, Nijmegen

Wolstenholme EF, Monk D, McKelvie D, Gillespie P, O'Rourke D, Todd D (2007a) Reallocating mental health resources in the borough of Lambeth, London, UK. In: Proceedings of the 2007 international system dynamics conference, Boston

Wolstenholme EF, Monk D, McKelvie D, Todd D (2007b) The contribution of system dynamics to cost benefit analysis – a case study in planning new mental health services in the UK. In: Proceedings of the 2007 international system dynamics conference, Boston

Wolstenholme EF, Monk D, McKelvie D (2007c) Influencing and interpreting health and social care policy in the UK. In: Qudrat-Ullah H, Spector MJ, Davidsen PI (eds) Complex decision making: theory and practice. Springer, New York

Wolstenholme E, McKelvie D, Monk D, Todd D, Brad C (2008) Emerging opportunities for system dynamics in UK health and social care – the market-pull for systemic thinking. Paper submitted to the 2008 system dynamics conference, Athens

Zainal Abidin N, Mamat M, Dangerfield B (2014) Combating obesity through healthy eating behaviour: a call for system dynamics optimization. PLoS One 9(12): e114135. https://doi.org/10.1371/journal.pone.0114135

Reviews and Reports

Bechberger E, Lane DC, McBride T, Morton A, Quintas D, Wong CH (2011) The National Audit Office uses OR to assess the value for money of public services. Interfaces 41(4):365–374

Comptroller and Auditor General (2009) Reducing healthcare associated infections in hospitals in

England. Report HC 560 2008–2009. National Audit Office, London

Dangerfield B (2011) The role of behaviour change in eating and physical activity in the battle against childhood obesity. In: Proceedings of the international system dynamics conference, Washington, DC. Online at www.systemdynamics.org with N Zainal Abidin

Dangerfield BC, Roberts CA (eds) (1994b) Health and health care dynamics. Special issue of Syst Dyn Rev 15(3):526–567

Forrester JW (1961) Industrial dynamics. MIT Press/Cambridge Press, Cambridge, MA/Cambridge

Lane DC, Munro E (2010) Appendix 2: applying systems thinking ideas to child protection. In: Munro E (ed) The Munro review of child protection part one: a systems analysis. UK Government Publication, London, pp 47–52

Lane DC, Bianchi C, Bivona E (eds) (2010) Public sector applications of the system dynamics approach: selected papers from the fourth European system dynamics workshop, University of Palermo, Italy. Special Edition of Syst Res Behav Sci 27(4)

Lane DC, Husemann E, Holland D, Khaled, A (2019) Two approaches to understanding norovirus foodborne transmissions: a study for the food standards association, Presentation to the System Dynamics Chapter Annual Conference

McKelvie D (2011) Reducing alcohol related admissions to hospital. International system dynamics conference, Washington, DC. Online at www.systemdynamics.org

Richmond B (1994) iThink technical manual, ISEE Systems, Hanover

System Dynamics Applications to Public Policy

David F. Andersen[1], Eliot Rich[2] and
Roderick Macdonald[3]
[1]Rockefeller College of Public Affairs and Policy,
University at Albany, Albany, USA
[2]School of Business, University at Albany,
Albany, USA
[3]Initiative for System Dynamics in the Public
Sector, University at Albany, Albany, USA

Article Outline

Glossary

Causal loop diagram A diagrammatic artifact that captures the causal model and feedback structure underlying a problem situation. Commonly used as a first-cut tool to identify major stakeholder concerns and interactions. These diagrams are often precursors to formal models.

Dynamic modeling Formal examination of the behavior of a system over time. Contrast with point-estimation, which attempts to predict an average outcome.

Feedback A relationship where two or more variables are linked over time so that the influence of one variable on a second will later affect the state of the first. If the influence is such as to increase the state of the first over time, the feedback is termed reinforcing. If the influence is such as to decrease the state of the first, it is termed balancing.

Formal model The representation of a system structure in mathematical form. Contrast with causal model, which represents structure without the underlying mathematics.

Mental model The representation of a problem's structure as possessed by an expert in a particular domain. Mental models are often intangible until explicated by the expert.

Public policy Any and all actions or non-actions, decisions or non-decisions taken by government, at all levels, to address problems. These actions, non-actions, decisions or non-decisions are implemented through laws, regulations and the allocation of resources.

Group model building (GMB) An approach to problem definition that asks multiple experts and major stakeholders to provide collective insights into the structure and behavior of a system through facilitated exercises and artifacts. GMB is often used to explicate the contrasting mental models of stakeholders.

Stakeholder An individual or group that has significant interest or influence over a policy problem.

System dynamics An analytic approach to problem definition and solution that focuses on endogenous variables linked through feedback, information and material delays, and non-linear relationships. The structure of

Originally published in
R. A. Meyers (ed.), *Encyclopedia of Complexity and Systems Science*, © Springer-Verlag 2009,
https://doi.org/10.1007/978-0-387-30440-3_421

these linkages determines the behavior of the modeled system.

Definition of the Subject

System dynamics is an approach to problem understanding and solution. It captures the complexity of real-world problems through the explication of feedback among endogenous variables. This feedback, and the delays that accompany it, often drive public sector programs towards unanticipated or unsatisfactory results. Through formal and informal modeling, System Dynamics-based analysis explicates and opens these feedback structures to discussion, debate and consensus building necessary for successful public sector policymaking.

Introduction

In the 50 years since its founding, System Dynamics has contributed to public policy thought in a number of areas. Major works, such as *Urban Dynamics* (Forrester 1971) and *Limits to Growth* (Meadows et al. 1972) have sparked controversy and debate. Other works in the domains of military policy, illegal drugs, welfare reform, health care, international development, and education have provided deep insight into complex social problems. The perspective of System Dynamics, with its emphasis on feedback, changes over time, and the role of information delays, helps inform policy makers about the intended and unintended consequences of their choices. The System Dynamics method includes a problem-oriented focus and the accommodation of multiple stakeholders, both crucial to the development of sound policy. Through the use of formal simulation, decision makers may also use System Dynamics models to consider the effects of their choices on short- and long-term outcomes. We illustrate this process with real life examples, followed by a review of the features of System Dynamics as they relate to public policy issues. We then describe the conjunction of System Dynamics and Group Model Building as a mechanism for policy ideation and review. We identify some of the historical and current uses of System Dynamics in the public sector, and discuss techniques for evaluating its effects on policy and organizations.

Medical Malpractice: A System Dynamics and Public Policy Vignette

The year was 1987 and New York's medical malpractice insurance system was in a state of crisis. Fueled by unprecedented levels of litigation, total settlements were soaring as were the malpractice insurance rates charged to hospitals and physicians. Obstetricians stopped taking on new patients. Doctors threatened to or actually did leave the state. Commercial insurance carriers had stopped underwriting malpractice insurance policies, leaving state-sponsored risk pools as the only option. The Governor and the Legislature were under pressure to find a solution and to find it soon. At the center of this quandary was the state's Insurance Department, the agency responsible for regulating and setting rates for the state's insurance pools. The agency's head found himself in just the kind of media hot seat one seeks to avoid in the public service.

An in-house SWAT team of actuaries, lawyers, and analysts had been working to present a fiscally sound and politically viable set of options for the Agency to consider and recommend to the legislature. They had been working with a team of System Dynamics modelers to gain better understanding of the root causes of the crisis. Working as a group, they had laid out a whole-system view of the key forces driving malpractice premiums in New York State. Their simulation model, forged in the crucible of group consensus, portrayed the various options on a "level playing field," each option being analyzed using a consistent set of operating assumptions. One option stood out for its ability to offer immediate malpractice insurance premium relief, virtually insuring a rapid resolution to the current crisis. An actuarial restructuring of future liabilities arising from future possible lawsuits relieved immediate pressure on available reserve funds. Upward pressure on premiums would vanish; a showdown in the legislature would be averted. Obviously, the Commissioner was interested in this option–who would not be?

"But what happens in the later years, after our crisis is solved?" he asked. As the team pored over the simulation model, they found that today's solution sowed the seeds for tomorrow's problems. Ten, fifteen, or maybe more years into the future, the deferred liabilities piled up in the system creating a secondary crisis, quite literally a second crisis caused by the resolution of the first crisis.

"Take that option off the table – it creates an unacceptable future," was the Commissioner's snap judgment. At that moment a politically appointed official had summarily dismissed a viable and politically astute "silver bullet" cure to a current quandary because he was thinking dynamically, considering both short-term and long-term effects of policy.

The fascinating point of the medical malpractice vignette is that the option taken off the table was indeed, in the short run, a "silver bullet" to the immediate crisis. The System Dynamics model projected that the solution's unraveling would occur long after the present Commissioner's career was over, as well as after the elected life span of the Governor who had appointed him and the legislators whose votes would be needed to implement the solution. His decision did not define the current problem solely in terms of the current constellation of stakeholders at the negotiations, each with their particular interests and points of view. His dynamic thinking posed the current problem as the result of a system of forces that had accumulated in the past. Symmetrically, his dynamic thinking looked ahead in an attempt to forecast what would be the future dynamic consequences of each option. Might today's solution become tomorrow's problem?

This way of thinking supported by System Dynamics modeling invites speculation about long-run versus short-run effects. It sensitizes policy makers to the pressure of future possible stakeholders, especially future generations who may come to bear the burden of our current decisions. It draws attention into the past seeking causes that may be buried at far spatial and temporal distances

from current symptoms within the system. It seeks to understand the natural reaction time of the system, the period during which problems emerge and hence over which they need to be solved. System Dynamics-based analysis in the public sector draws analytic attention away from the riveting logic of the annual or biannual budget cycle, often focusing on options that will play themselves out years after current elected officials have left office. Such work is hard to do, but critical if one wants to think in systems terms.

What Is System Dynamics Modeling?

While other papers in this series may provide a more expanded answer to this basic question, it may be useful to begin this discussion of System Dynamics and public policy with a brief description of what System Dynamics is.

System Dynamics is an approach to policy analysis and design that applies to problems arising in complex social, managerial, economic, or ecological systems (Ford 1999; Forrester 1961a; Richardson and Pugh 1981; Sterman 2000). System Dynamics models are built around a particular problem. The problem defines the relevant factors and key variables to be included in the analysis. This represents the model's boundary, which may cross departmental or organizational boundaries. One of the unique advantages of using System Dynamics models to study public policy problems is that assumptions from a variety of stakeholders are explicitly stated, can be tested through simulation, and can be examined in context.

System Dynamics models rely on three sources of information: numerical data, the written database (reports, operations manuals, published works, etc.), and the expert knowledge of key participants in the system (Forrester 1980). The numerical database of most organizations is very small, the written database is larger, and the expert knowledge of key participants is vast. System Dynamicists rely on all three sources, with particular attention paid to the expert knowledge of key participants because it is only through such expert knowledge that we have any knowledge of the structure of the system. The explicit capturing of accumulated experience from multiple stakeholders in the model is one of the major differences between System Dynamics models and other simulation paradigms. An understanding of the long term effects of increased vigilance on the crime rate in a community needs to account for the reaction of courts, prisons, and rehabilitation agencies pressed to manage a larger population. This knowledge is spread across experts in several fields, and is not likely to be found in any single computer database. Rather, insight requires a process that makes these factors visible and explicit. For public sector problems, in particular, this approach helps move conflict out of the realm of inter-organizational conflict and towards a problem-solving focus.

Through the use of available data and by using the verbal descriptions of experts to develop mathematical relationships between variables, we expose new concepts and/or previously unknown but significant variables. System Dynamics models are appropriate to problems that arise in closed-loop systems, in which conditions are converted into information that is observed and acted upon, changing conditions that influence future decisions (Richardson 1991).

This idea of a "closed loop" or "endogenous" point of view on a system is really important to all good System Dynamics models. A simple example drawn from everyday life may help better to understand what an endogenous (versus exogenous) point of view means. If a father believes that his teenage daughter is always doing things to annoy him and put him in a bad mood, then he has an exogenous or "open loop" view of his own mood because he is seeing his mood as being controlled by forces outside of or exogenous to his own actions. However, if the father sees that his daughter and her moods are reacting to his own actions and moods while in turn his daughter's actions shape and define his moods, then this father has an endogenous point of view on his own mood. He understands how his mood is linked in a closed loop with another member of his family. Of course, the father with an endogenous view will be in a better position to more fully understand family dynamics and take actions that

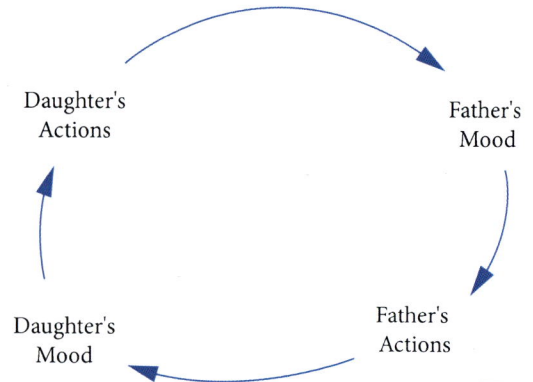

System Dynamics Applications to Public Policy, Fig. 1 Closed loop diagram of fathers and daughters

can prevent bad moods from spreading within the family (Fig. 1).

Using an SD Model to Develop a Theory

A System Dynamics model represents a theory about a particular problem. Since models in the social sciences represent a theory, the most we can hope for from all these models, mental or formal, is that they be useful (Sterman 1996). System Dynamics models are useful because the mathematical underpinning needed for computer simulation requires that the theory be precise. The process of combining numerical data, written data, and the knowledge of experts in mathematical form can identify inconsistencies about how we think the system is structured and how it behaves over time (Forrester 1994).

In policymaking it is often easy and convenient to blame other stakeholders for the problem state. Often, though, the structure of the system creates the problem by, for example, shifting resources to the wrong recipient or by inclusion of policies that intervene in politically visible but ineffective ways. The use of inclusive SD models educates us by identifying these inconsistencies through an iterative process involving hypotheses about system structure and tests of system behavior. Simulation allows us to see how the complex interactions we have identified work when they are all active at the same time. Furthermore, we can test a variety of policies quickly to see how they play out in the long run. The final result is a model that represents our most insightful and tested theory about the endogenous sources of problem behavior.

Behavior over Time Versus Forecasts

People who take a systems view of policy problems know that behavior generated by complex organizations cannot be well understood by examining the parts. By taking this holistic view, System Dynamicists capture time delays, amplification, and information distortion as they exist in organizations. By developing computer simulation models that incorporate information feedback, systems modelers seek to understand the internal policies and decisions, and the external dynamic phenomena that combine to generate the problems observed. They seek to predict dynamic implications of policy, not forecast the values of quantities at a given time in the future.

System Dynamics models are tools that examine the behavior of key variables over time. Historical data and performance goals provide baselines for determining whether a particular policy generates behavior of key variables that is better or worse, when compared to the baseline or other policies. Furthermore, models incorporating rich feedback structure often highlight circumstances where the forces governing a system may change in a radical fashion. For example, in early phases of its growth a town in an arid region may be driven by a need to attract new jobs to support its population. At some future point in time, the very fact of successful growth may lead to a water shortage. Now the search for more water, not more jobs, may be what controls growth in the system. Richardson (1991) has identified such phenomena as shifts in loop dominance that provide endogenous explanations for specific outcomes. Simulation allows us to compress time (Sterman 2000) so that many different policies can be tested, the outcomes explained, and the causes that generate a specific outcome can be examined by knowledgeable people working in the system, before policies are actually implemented.

Excellent short descriptions of System Dynamics methodology are found in Richardson (1991, 1996) and Barlas (2002). Furthermore, Forrester's (1961a) detailed explanation of the field in

Industrial Dynamics is still relevant, and Richardson and Pugh (1981), Roberts et al. (1983), Coyle (96), Ford (1999), Maani and Cavana (2007a), croft (2007) and Sterman (2000) are books describe the field and provide tools, techniques and modeling examples suitable for the novice as well as for experienced System Dynamics modelers.

An Application of System Dynamics – The Governor's Office of Regulatory Assistance (GORA) Example

When applied to public policy problems, the "nuts and bolts" of this System Dynamics process consist of identifying the problem, examining the behavior of key variables over time, creating a visualization of the feedback structure of the causes of the problem, and developing a formal simulation model. A second case illustration may assist in understanding the process. The New York State Governor's Office of Regulatory Assistance (GORA) is a governmental agency whose mission it is to provide information about government rules and regulations to entrepreneurs who seek to start up new businesses in the state. The case was described by Andersen et al. (2006) and is often used as a teaching case introducing System Dynamics to public managers.

Figure 2 below illustrates three key feedback loops that contribute both to the growth and eventual collapse of citizen service requests at GORA. The reinforcing feedback loop labeled "R1" illustrates how successful completion of citizen orders creates new contacts from word-of-mouth by satisfied citizens which in turn leads to more requests for service coming into the agency. If only this loop were working, a self-reinforcing process would lead to continuing expansion of citizen requests for services at GORA. The balancing loop labeled "B2" provides a balancing effect. As workers within the agency get more and more work to complete, the workload within the agency goes up with one effect being a possible drop in the quality in the work completed. Over time, loop B2 tells a story of how an increased workload can lead to a lower quality of work, with the effect of that lower quality being fewer incoming requests in the future. So over time, too many

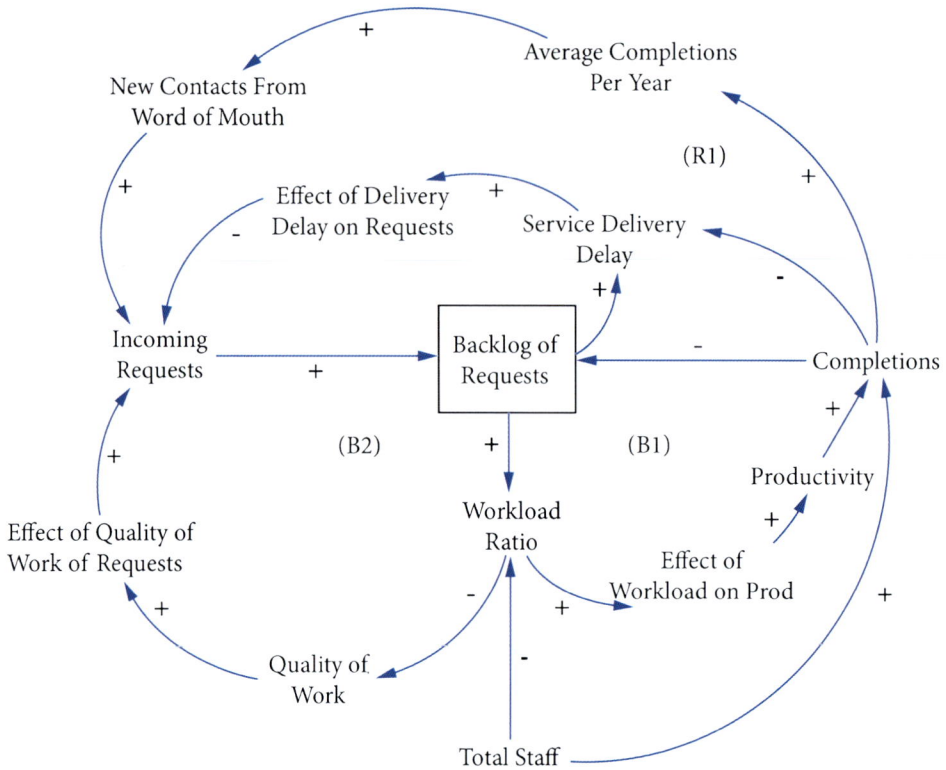

System Dynamics Applications to Public Policy, Fig. 2 Key feedback loops in a simulation of workflow in the Governor's Office of Regulatory Administration (GORA)

incoming requests set off a process that limits future requests by driving down quality. Many public managers who have worked with the GORA model find these two simple feedback loops to be realistic and powerful explanations of many of the problems that their agencies face on a day-to-day basis. The full GORA model has many other feedback loops and active variables not shown in the aggregated Fig. 2.

Once all of the variables have been represented by mathematical equations, a computer simulation is able to recreate an over time trajectory possible future values for all of the variables in the model. Figure 3 shows a graph over time of simulated data for key indicators in the GORA case study. The simulation begins when GORA comes into existence to provide services to the public and runs for 48 months. Initially, there is adequate staff and the amount of work to do is low, so the Workload Ratio, shown as part of loops B1 and B2 in the previous figure, is very

low. With a low Workload Ratio GORA employees are able to devote additional time to each task they perform and the Quality of Work[1] is thus relatively high. The Backlog of Requests and the Average Completions Per Year begin at 0 and then increase and level off over time to approximately 4,500 and 41,000 respectively. The Fraction Experienced Staff measures what portion of the overall workers are experienced and hence more efficient at doing their jobs. As shown in Fig. 3, the Fraction

---

[1]The Workload Ratio and Quality of Work are normalized variables. This means that they are measured against some predetermined standard. Therefore, when these two variables are equal to 1 they are operating in the desired state. Depending on the definition of the variable, values below or above 1 indicate when they are operating in a desired or undesired state. For example, Quality of Work above 1 indicates that quality is high, relative to the predetermined normal. However, Quality of Work below 1 indicates an undesirable state.

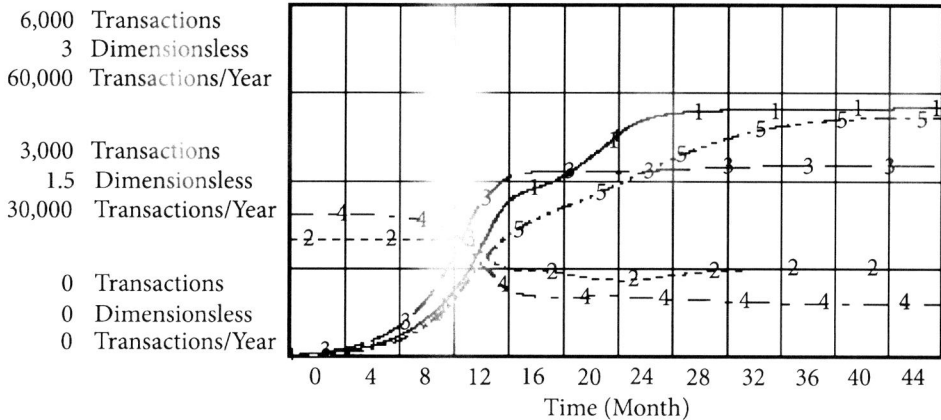

6,000	Transactions	
3	Dimensionless	
60,000	Transactions/Year	
3,000	Transactions	
1.5	Dimensionless	
30,000	Transactions/Year	
0	Transactions	
0	Dimensionless	
0	Transactions/Year	

Time (Month)

Backlog of Requests : Base Run — 1 — 1 — 1 — 1 — 1 — Transactions
Fraction Experienced : Base Run - - - 2 - - - - - 2 - - - - -2 - - - - 2 - - - - 2 - - Dimensionless
Workload Ration : Base Run — 3— 3— 3— 3— 3— 3 Dimensionless
Quality of Work : Base Run — 4 — - 4— - -4- — 4 — - 4— - — Dimensionless
Average Completions Per Year : Base Run — - - —5- - —5— - - -5- - Transactions/Year

System Dynamics Applications to Public Policy, Fig. 3 Simulated performance of key variables in the GORA case study

Experienced begins at 1 and then falls and increases slightly to .75 indicating that GORA is having a harder time retaining experienced staff and is experiencing higher employee turnover. (The full GORA model has a theory of employee burnout and turnover not shown in Fig. 2.)

The combination of the visualization in Fig. 2 with a formal model capable of generating the dynamic output shown in Fig. 3 illustrates the power of System Dynamics modeling for public policy issues. Linking behavior and structure helps stakeholders understand why the behavior of key variables unfolds over time as it does. In the GORA case, the program is initially successful as staff are experienced, are not overworked, and the quality of the services they provide is high. As clients receive services the R1 feedback loop is dominant and this attracts new clients to GORA. However, at the end of the first year the number of clients requesting services begins to exceed the ability of GORA staff to provide the requested services in a timely manner. The Workload Ratio increases, employees are very busy, the Quality of Work falls, and the B2 feedback loop works to limit the number of people seeking services.

Furthermore, people are waiting longer to receive services and some are discouraged from seeking services due to the delay. The initial success of the program cannot be sustained and the program settles down into an unsatisfactory situation where the Workload Ratio is high, Quality of Work is low, clients are waiting longer for services and staff turnover is high as indicated by the Fraction Experienced.

The model tells a story of high performance expectations, initial success and later reversal, all explained endogenously. Creating and examining the simulation helps managers consider possible problems before they occur – before staff are overtaxed, before turnover climbs, and before the agency has fallen behind. Having a model to consider compresses time and provides the opportunity for a priori analysis. Finally, having a good model can provide managers with a test bed for asking "what if" questions, allowing public managers to spend simulated dollars and make simulated errors all the while learning how to design better public policies at relatively low cost and without real (only simulated) risk.

How Is System Dynamics Used to Support Public Policy and Management?

The Medical Malpractice vignette that opened this chapter involving the New York State Commissioner of Insurance is more fully documented by Reagan-Cirincione et al. (1991) and is one of the first published examples of the results of a team of government executives working in a face-to face group model building session to create a System Dynamics model to support critical policy decisions facing the group. The combined group modeling and simulation approach had a number of positive effects on the policy process. Those positive effects are:

Make Mental Models of Key Players Explicit

When the Commissioner drew together his team, the members of this group held different pieces of information and expertise. Much of the most important information was held in the minds, in the mental models, of the Commissioner's staff, and not in data tabulations. The System Dynamics modeling process made it possible for managers to explicitly represent and manipulate their shared mental models in the form of a System Dynamics simulation model. This process of sharing and aligning mental models, as done during a System Dynamics modeling intervention, is an important aspect of a "learning organization" as emphasized by Senge (1990).

Create a Formal and Explicit Theory of the Public Policy Situation Under Discussion

The formal model of malpractice insurance contained an explicit and unambiguous theory of how the medical malpractice system in New York State functions. The shared mental models of the client team implied such a formal and model-based theory, but the requirements of creating a running simulation forced the group to be much more explicit and clear about their joint thinking. As the modeling team worked with the group, a shared consensus about how the whole medical malpractice system worked was cast, first into a causal-loop diagram, and later into the equations of the formal simulation model (Richardson and Pugh 1981; Sterman 2000).

Document all Key Parameters and Numbers Supporting the Policy Debate

In addition to creating a formal and explicit theory, the System Dynamics model was able to integrate explicit data and professional experience available to the Department of Insurance. Recording the assumptions of the model in a clear and concise way makes possible review and examination by those not part of the model's development. Capturing these insights and their derivation provides face validity to the model's constructs.

Building confidence in the utility of a System Dynamics model for use in solving a public policy problem involves a series of rigorous tests that probe how the model behaves over time as well as how available data, both numerical and tacit structural knowledge, have been integrated and used in the model. Forrester and Senge (1980) detail 17 tests for building confidence in a System Dynamics model. Sterman (2000) identifies 12 model tests, the purpose or goal of each test, and the steps that modelers should follow in undertaking those tests. Furthermore, Sterman (2000) also lists questions that model consumers should ask in order to generate confidence in a model. This is particularly important for public policy issues where the ultimate goal or outcome for different stakeholders may be shared, but underlying assumptions of the stakeholders may be different.

Create a Formal Model that Stimulates and Answers Key "what if" Questions

Once the formal model was constructed, the Commissioner and his policy team were able to explore "what if" scenarios in a cost-free and risk-free manner. Significant cost overruns in a simulated environment do not drive up real tax rates, nor do they lead to an elected official being voted out of office, nor to an appointed official losing her job. Quite the contrary, a simulated cost overrun or a simulated failed program provides an opportunity to learn how better to implement or manage the program or policy (or to avoid trying to implement the policy). Public managers get to experiment quickly with new policies or programs in a risk-free simulated environment until they "get it right" in the simulated world. Only then should

System Dynamics Applications to Public Policy, Fig. 4 A team of public managers working together to build a System Dynamics model of welfare reform policies

they take the risk of implementation in a high stakes policy environment.

Bringing a complex model to large groups sometimes requires the development of a more elaborate simulation, so that those who were not part of the initial analysis can also derive insight from its results. Iterative development and discussion provides an additional validation of the constructs and conclusions of the model, Zagonel et al. (2004) have described a case where local managers responsible for implementing the 1996 federal welfare reform legislation used a simulation model to explore such "what if" futures before taking risks of actual implementation.

Public policy problems are complex, cross organizational boundaries, involve stakeholders with widely different perspectives, and evolve over time. Changes in police procedures and/or resources may have an effect on prison and parole populations many years into the future. Health care policies will determine how resources are allocated at local hospitals and the types of treatments that can be obtained. Immigration policies in one country may influence the incomes and jobs of people in a second country. Miyakawa (1999) has pointed out that public policies are systemically interdependent. Solutions to one problem often create other problems. Increased enforcement of immigration along the U.S. borders has increased the workload of courts (Finely 2006). Besides being complex these examples also contain stakeholders with different sets of goals. In solving public policy problems, how diverse stakeholders work out their differences is a key component of successful policy solutions. System Dynamics modeling interventions, and in particular the techniques of group model building (Akkermans and Vennix 1997; Andersen and Richardson 1997; Richardson and Andersen 1995; Vennix 1996), provide a unique combination of tools and methods to promote shared understanding by key stakeholders within the system.

System Dynamics and Models: A Range of Analytic Scope and Products

In the malpractice insurance example, the Commissioner called his advisors into a room to explicitly engage in a group model building session. These formal group model-building sessions involve a specialized blend of projected computer support plus professional facilitation in a face-to-face meeting of public managers and policy analysts. Figure 4 is an illustration of a team of public managers working together in a group model building project. In this photograph, a facilitator is working on a hand drawn view of a simulation model's structure while projected views of computer output can be used to look at first cut simulation runs or refined images of the model being built by the group. Of course, the key feature of this whole process is facilitated face-to-face conversations between the key stakeholders responsible for the policy decisions being made.

Richardson and Andersen (1995), Andersen and Richardson (1997), Vennix (1996), and Luna-

Reyes et al. (2007) have provided detailed descriptions of how this kind of group model building process actually takes place. In addition to these group model building approaches, the System Dynamics literature describes five other ways that teams of modelers work with client groups. They are (1) the Reference Group approach (Stenberg 1980), (2) the Strategic Forum (Richmond 1997), (3) The stepwise approach (Wolstenholme 1992), (4) strategy dynamics (Warren 1999, 2002, 2005), and (5) the "standard method" of Hines (Otto and Struben 2004).

Some System Dynamics-oriented analyzes of public policies completed by groups of public managers and policy analysts stop short of building a formal simulation model. The models produced by Wolstenholme and Coyle (1983), Cavana et al. (2007) and the system archetypes promoted by Senge (1990) have described how these qualitative system mapping exercises, absent a formal running simulation model, can add significant value to a client group struggling with an important public policy problem. The absence of a formal simulation limits the results to a conceptual model, rather than a tool for systematic experimentation.

Finally, a number of public agencies and Non Governmental Organizations are joining their counterparts in the private sector by providing broad-based systems thinking training to their top leaders and administrative staff. A number of simulation-based management exercises such as the production-distribution game (also known as the "beer game") (Sterman 1992) and the People's Express Flight Simulator (Sterman 1988) have been developed and refined over time to support such training and professional development efforts. In addition, Cavana and Clifford (2006) have used GMB to develop a formal model and flight simulator to examine the policy implications of an excise tax policy on tobacco smoking.

What Are the Arenas in Which System Dynamics Models Are Used?

The malpractice insurance vignette and the GORA example represented cases where a model was developed for a single problem within one agency. Naill (1977) provides an example of how a sustained modeling capability can be installed within an agency to support a range of ongoing policy decisions (in this case the model was looking at transitional energy policies at the federal level). Barney (1982) developed a class of System Dynamics simulation models to support economic development and planning in developing nations. Wolstenholme (1993) reported on efforts to support health planning within the British Health Service.

Addressing a tactical problem within a single public sector agency, while quite common, is only one of the many types of decision arenas in which System Dynamics models can be and are used to support public policy. Indeed, how a model is used in a public policy debate is largely determined by the unique characteristics of the specific decision-making arena in which the model is to be used. Some of the more common examples follow.

Models Used to Support Inter-Agency and Inter-Governmental Collaborative Efforts

A quite different arena for the application of System Dynamics models to support the policy process occurs when an interagency or intergovernmental network of program managers must cooperate to meet a common mission. For example, Rohrbaugh (2000) and Zagonel et al. (2004) report a case where state and local officials from social services, labor, and health agencies combined their efforts with private and non-profit managers of day care services, health care services, and worker training and education services to plan for comprehensive reform of welfare policies in the late 1990s. These teams were seeking strategies to blend financial and program resources across a myriad of stovepipe regulations and reimbursement schemes to provide a seamless system of service to clients at the local level. To complete this task, they created a simulation model containing a wide range of system-level interactions and tested policies in that model to find out what blend of policies might work. Policy implementation followed this model-based and simulation-supported policy design.

Models Used to Support Expert Testimony in Courtroom Litigation

Cooper (1980) presented one of the first published accounts of a System Dynamic model being used as a sort of expert witness in courtroom litigation. In the case he reported, Litton Industries was involved in a protracted lawsuit with the U. S. Navy concerning cost and time overruns in the construction of several naval warships. In a nutshell, the Navy contended that the cost overruns were due to actions taken (or not taken) by Litton Industries as primary contractor on the project and as such the Navy should not be responsible for covering cost overruns. Litton maintained that a significant number of change orders made by the Navy were the primary drivers of cost overruns and time delays. A simulation model was constructed of the shipbuilding process and the simulation model then built two simulated ships without any change orders. A second set of "what if" runs subsequently built the same ships except that the change orders from the Navy were included in the construction process. By running and re-running the model, the analysts were able to tease out what fractions of the cost overrun could reasonably be attributed to Litton and what fraction should be attributed to naval change orders. Managers at Litton Industries attribute their receipt of hundreds of millions dollars of court-sanctioned payments to the analysis supported by this System Dynamics simulation model. Ackermann et al. (1997) have used a similar approach involving soft systems approaches combined with a System Dynamics model in litigation over cost overruns in the channel tunnel project.

Models Used as Part of the Legislative Process

While System Dynamics models have been actively used to support agency-level decision making, inter-agency and inter-governmental task forces and planning, and even courtroom litigation, their use in direct support of legislative processes has a more uneven track record. For example, Ford (1997) reports successes in using System Dynamics modeling to support regulatory rule making in the electric power industry, and Richardson and Lamitie (1989) report on how System Dynamics modeling helped redefine a legislative agenda relating to the school aid formula in the U.S. state of Connecticut. However, Andersen (1990) remains more pessimistic about the ability of System Dynamics models to directly support legislative decision making, especially when the decisions involve zero-sum tradeoffs in the allocation of resources (such as formula-driven aid involving local municipal or education formulas). This class of decisions appears to be dominated by short-term special interests. A longer-term dynamic view of such immediate resource allocation problems is less welcome. The pathway to affecting legislative decision making appears to be by working through and with public agencies, networks of providers, the courts, or even in some opinions, by directly influencing public opinion.

Models Used to Inform the Public and Support Public Debate

In addition to using System Dynamics modeling to support decision making in the executive, judicial, and legislative branches of government (often involving Non-Governmental Organizations and private sector support), a number of System Dynamics studies appeal directly to the public. These studies intend to affect public policy by shaping public opinion in the popular press and the policy debate. In the 1960s, Jay Forrester's *Urban Dynamics* (Forrester 1969) presented a System Dynamics model that looked at many of the problems facing urban America in the latter half of the twentieth century. Several years later in response to an invitation from the Club of Rome, Forrester put together a study that led to the publication of *World Dynamics* (Forrester 1971), a highly aggregate System Dynamics model that laid out a feedback-oriented view of a hypothesized set of relationships between human activity on the planet, industrialization, and environmental degradation. Meadows et al. (1972) followed on this study with a widely hailed (and critiqued) System Dynamics simulation study embodied in the best-selling book, *Limits to Growth*. Translated into over 26 languages, this volume coalesced a wide range of public opinion leading to a number of pieces of environmental reform in the decade of the 1970s. The debate engendered by that volume continues even 30 years later (Meadows et al. 2004). Donella Meadows continued in this tradition of appealing directly to public opinion through her syndicated column, The Global Citizen, which was nominated for the Pulitzer Prize in 1991. The column presented a System Dynamics-based view of environment matters for many years (http://www.pcdf.org/meadows/).

What Are some of the Substantive Areas Where System Dynamics Has Been Applied?

The International System Dynamics Society (http://www.systemdynamics.org) maintains a comprehensive bibliography of over 8,000 scholarly books and articles documenting a wide variety of applications of System Dynamics modeling to applied problems in all sectors. MacDonald et al. (2007) have created a bibliography extracted from this larger database that summarizes some of the major areas where System Dynamics modeling has been applied to public policy. Below, we summarize some of the substantive areas where System Dynamics has been applied, giving one or two sample illustrations for each area.

Health Care

System Dynamicists have been applying their tools to analyze health care issues at both the academic and practitioner level for many years. *The System Dynamics Review*, the official journal of the System Dynamics Society, devoted a special issue to health care in 1999 due to the importance of health care as a critical public policy issue high on the political agenda of many countries and as an area where much System Dynamics work has been performed. The extensive System Dynamics work performed in the health care area fell into three general categories: patient flow management, general health policy, and specific health problems.

The patient flow management category is exemplified by the work of Wolstenholme (1999), Lane et al. (1998), and van Ackere and Smith (1999). The articles written by these authors focused on issues and policies relating to patient flows in countries where health care service is universal.

The general health policy category is rather broad in that these articles covered policy and decision making from the micro level (Taylor and Lane 1998) to the macro level (Senge and Asay 1988). There were also many articles that showed how the process of modeling resulted in better understanding of the problem and issues facing health care providers and policy makers (Cavana et al. 1999).

The last category dealt with specific health-related problems such as the spread of AIDS (Homer and St Clair 1991; Roberts and Dangerfield 1990), smoking (Homer et al. 1982), and malaria control (Flessa 1999), as well as many other health-related conditions.

Education

The education articles touched on various topics relating to education ranging from using System Dynamics in the classroom as a student-centered teaching method to models that dealt with resource allocations in higher education. Nevertheless, many of the articles fell into five categories that could be labeled management case studies or flight simulators, teaching technology, research, teaching, and education policy.

The management case study and flight simulator articles are best exemplified by Sterman's (1992) article describing the Beer Game and Graham et al. (1992) article on "Model Supported Case Studies for Management Education." The emphasis of these works is on the use of case studies in higher education, with the addition of games or computer simulations. This is related to the teaching technology category in that both emphasize using System Dynamics models/tools to promote learning. However, the teaching technology category of articles stresses the introduction of computer technology, specifically System Dynamics computer technology, into the classroom. Steed (1992) has written an article that discusses the cognitive processes involved while using Stella to build models, while Waggoner (1984) examined new technologies versus traditional teaching approaches.

In addition to teaching technology are articles that focus on teaching. The teaching category is very broad in that it encompasses teaching System Dynamics in K-12 and higher education as subject matter (Forrester 1993; Roberts 1983) as well as ways to integrate research into the higher-education classroom (Richardson and Andersen 1979). System Dynamics models are also used to introduce advanced mathematical concepts through simulation and visualization, rather than through equations (Fisher 2001, 2004). In addition, lesson plans for the classroom are also part of this thread (Hopkins 1992). The Creative Learning Exchange (http://www.clexchange.com) provides a central repository of lessons and models useful for pre-college study of System Dynamics, including a selfstudy roadmap to System Dynamics principles (Creative Learning Exchange 2000).

There are also a number of articles that pertain to resource allocation (Chen et al. 1981) at the state level for K-12 schools along with articles that deal with resource-allocation decisions in higher education (Forsyth et al. 1976; Galbraith 1989). Saeed (1996) and Mashayekhi (1977) cover issues relating to higher education policy in developing countries.

The last education category involved research issues around education. These articles examined whether the System Dynamics methodology and simulation-based education approaches improved learning (Davidsen 1996; Keys and Wolfe 1996; Mandinach and Cline 1993).

Defense

System Dynamics modeling work around the military has focused on manpower issues, resource allocation decisions, decision making and conflict. Coyle (1992) developed a System Dynamics model to examine policies and scenarios involved in sending aircraft carriers against land-based targets. Wils et al. (1998) have modeled internal conflicts as a result of outbreaks of conflict over allocation and competition of scarce resources.

The manpower articles focused on recruitment and retention policies in the armed forces and are represented in articles by Lopez and Watson (1979), Andersen and Emmerichs (1982), Clark (1993), Clark et al. (1980) and Cavana et al. (2007). The resource allocation category deals with issues of money and materials, as opposed to manpower, and is represented by Clark (1981, 1987). Decision making in military affairs from a System Dynamics perspective is represented in the article by Bakken and Gilljam (2003).

Environment

The System Dynamics applications dealing with environmental resource issues can be traced back to when the techniques developed in *Industrial Dynamics* were beginning to be applied to other fields. The publication of Forrester's *World Dynamics* in 1971 and the follow-up study *Limits to Growth* (Meadows et al. 1972, 1992, 2004) used System Dynamics methodology to address the problem of continued population increases on industrial capital, food production, resource consumption and pollution. Furthermore, specific studies dealing with DDT, mercury and eutrophication of lakes were part of the Meadows et al. (1974) project and appeared as stand-alone journal articles prior to being published as a collection in Meadows and Meadows (1977).

The environmental applications of System Dynamics have moved on since that time. Recent work has combined environmental and climate issues with economic concerns thorough simulation experiments (Fiddaman 2002) as well as stakeholder participation in environmental issues (Stave 2002). In 2004, the *System Dynamics Review* ran a special issue dedicated to environmental issues. Cavana and Ford (2004) were the editors and did a review of the System Dynamics bibliography in 2004, identifying 635 citations with the key words "environmental" or "resource." Cavana and Ford broke the 635 citations into 11 categories they identified as resources, energy, environmental, population, water, sustainable, natural resources, forest, ecology, agriculture, pollution, fish, waste, earth, climate and wildlife.

General Public Policy

The System Dynamics field first addressed the issue of public policy with Forrester's *Urban Dynamics* (Forrester 1969) and the follow-up work contained in *Readings in Urban Dynamics* (Mass 1974) and Alfeld and Graham's *Introduction to Urban Dynamics* (Alfeld and Graham 1976). The field then branched out into the previously mentioned *World Dynamics* and the followup studies related to that work. Moreover, the application of System Dynamics to general public policy issues began to spread into areas as diverse as drug policy (Levin et al. 1975), and the causes of patient dropout from mental health programs (Levin and Roberts 1976), to ongoing work by Saeed (Saeed 1994, 1998, 2003) on development issues in emerging economies. More recently, Saysel et al. (2002) have examined water scarcity issues in agricultural areas, Mashayckhi (1998) reports on the impact on public finance of oil exports in countries that export oil and Jones et al. (2002) cover the issues of sustainability of forests when no single entity has direct control.

This brief review of the literature where System Dynamics modeling has been used to address public policy issues indicates that the field is making inroads at the micro level (within government agencies) and at the macro level (between government agencies). Furthermore, work has been performed at the international level and at what could truly be termed the global level with models addressing public policy issues aimed at climate change.

Evaluating the Effectiveness of System Dynamics Models in Supporting the Public Policy Process

System Dynamics modeling is a promising technology for policy development. But does it really work? Over the past several decades, a minor cottage industry has emerged that purports to document the successes (and a few failures) of System Dynamics models by reporting on case studies. These case studies report on successful applications and sometimes analyze weaknesses, making suggestions for improvement in future practice. Rouwette et al. (2002) have compiled a meta-analysis of 107 such case-based stories.

However, as compelling as such case stories may be, case studies are a famously biased and unsystematic way to evaluate effectiveness. Presumably, failed cases will not be commonly reported in the literature. In addition, such a research approach illustrates in almost textbook fashion the full litany of both internal and external threats to validity, making such cases an interesting but unscientific compilation of war stories. Attempts to study live management teams in naturally occurring decision situations can have high external validity but almost always lack internal controls necessary to create scientifically sound insights.

Huz et al. (1997) created an experimental design to test for the effectiveness of a controlled series of group-based System Dynamics cases in the public sector. They used a wide battery of pre- and post survey, interview, archival, administrative data, and qualitative observation techniques to evaluate eight carefully matched interventions. All eight interventions dealt with the integration of mental health and vocational rehabilitation services at the county level. Four of the eight interventions contained System Dynamics modeling sessions and four did not. These controlled interventions were designed to get at the impact of System Dynamics modeling on the public policy process.

Overall, Huz et al. (1997) envisioned that change could take place in nine domains measured across three separate levels of analysis as illustrated in Table 1 below.

Using the battery of pre- and post test instruments, Huz found important and statistically significant results in eight of the nine domains measured. The exceptions were in domain 9 where they did not measure client outcomes, in domain 5 where "participants were not significantly more aligned in their perceptions on strategies for changes" (but were more aligned in goals), and in domain 7 where "no significant change was found with respect to structural conditions within the network" (but two other dimensions of organizational relationships did change).

**System Dynamics Applications to Public Policy,
Table 1** Domains of measurement and evaluation used
to assess impact of systems-dynamics interventions (see
p. 151 in Huz et al. 1997)

Level I	Reflections of the modeling team
Domain 1	Modeling team's assessment of the intervention
Level II	Participant self-reports of the intervention
Domain 2	Participants' perceptions of the intervention
Domain 3	Shifts in participants' goal structures
Domain 4	Shifts in participants' change strategies
Domain 5	Alignment of participant mental models
Domain 6	Shifts in understanding how the system functions
Level III	Measurable system change and "bottom line" results
Domain 7	Shifts in network of agencies that support services integration
Domain 8	Changes in system-wide policies and procedures
Domain 9	Changes in outcomes for clients

In their meta-analysis of 107 case studies of
System Dynamics applications, Rouwette et al.
(2002) coded case studies with respect to eleven
classes of outcomes, sorted into individual level,
group level, and organizational level. The
107 cases were dominated by for-profit examples
with 65 such cases appearing in the literature
followed by 21 cases in the non-profit sector,
18 cases in governmental settings, and 3 cases in
mixed settings. While recognizing possible high
levels of bias in reported cases as well as difficulties in coding across cases and a high number of
missing categories, they found high percentages
of positive outcomes along all 11 dimensions of
analysis. For each separate dimension, they analyzed between 13 and 101 cases with the fraction
of positive outcomes for each dimension ranging
from a low of 83% to several dimensions where
100% of the cases reporting on a dimension found
positive results. At the individual level, they
coded for overall positive reactions to the work,
insight gained from the work, and some level of
individual commitment to the results emerging
from the study. At the group level, they coded
for increased levels of communication, the

emergence of shared language, and increases in
consensus or mental model alignment. Organizational level outcomes included implementation of
system level change. With respect to this important overall indicator they "found 84 projects
focused on implementation, which suggests that
in half (42) of the relevant cases changes are
implemented. More than half (24) of these
changes led to positive results" (see p. 20 in
Rouwette et al. 2002).

Rouwette (2003) followed this meta-analysis
with a detailed statistical analysis of a series of
System Dynamics-based interventions held
mostly in governmental settings in the Netherlands. He was able to estimate a statistical model
that demonstrated how System Dynamics group
model building sessions moved both individuals
and groups from beliefs to intentions to act, and
ultimately on to behavioral change.

In sum, attempts to evaluate System Dynamics
interventions in live settings continue to be
plagued by methodological problems that
researchers have struggled to overcome with a
number of innovative designs. What is emerging
from this body of study is a mixed, "good news
and bad news" picture. All studies that take into
account a reasonable sample of field studies show
some successes and some failures. About one-
quarter to one-half of the System Dynamics studies investigated showed low impact on decision
making. On the other hand, roughly half of the
studies have led to system-level implemented
change with approximately half of the
implemented studies being associated with positive measures of success.

Summary: System Dynamics – A Powerful Tool to Support Public Policy

While recognizing and respecting the difficulties
of scientific evaluation of System Dynamics studies in the public sector, we remain relentlessly
optimistic about the method's utility as a policy
design and problem-solving tool. Our glass is half
(or even three-quarters) full. A method that can
deliver high decision impact up to three-quarters
of the time and implement results in up to half of

the cases examined (and in a compressed time frame) is a dramatic improvement over alternative approaches that can struggle for months or even years without coming to closure on important policy directions.

System Dynamics-based modeling efforts are effective because they join the minds of public managers and policy makers in an emergent dialog that relies on formal modeling to integrate data, other empirical insights, and mental models into the policy process. Policy making begins with the pre-existing mental models and policy stories that managers bring with them into the room. Policy consensus and direction emerge from a process that combines social facilitation with technical modeling and analysis. The method blends dialog with data. It begins with an emergent discussion and ends with an analytic framework that moves from "what is" baseline knowledge to informed "what if" insights about future policy directions.

In sum, we believe that a number of the process features related to building System Dynamics models to solve public policy problems contribute to their appeal for frontline managers:

- **Engagement** Key managers can be in the room as the model is evolving, and their own expertise and insights drive all aspect of the analysis.
- **Mental models** The model-building process uses the language and concepts that managers bring to the room with them, making explicit the assumptions and causal mental models managers use to make their decisions.
- **Complexity** The resulting nonlinear simulation models lead to insights about how system structure influences system behavior, revealing understandable but initially counterintuitive tendencies like policy resistance or "worse before better" behavior.
- **Alignment** The modeling process benefits from diverse, sometimes competing points of view as stakeholders can have a chance to wrestle with causal assumptions in a group context. Often these discussions realign thinking and are among the most valuable portions of the overall modeling effort.

- **Refutability** The resulting formal model yields testable propositions, enabling managers to see how well their implicit theories match available data about overall system performance.
- **Empowerment** Using the model managers can see how actions under their control can change the future of the system.

System Dynamics modeling projects merge managers' causal and structural thinking with the available data, drawing upon expert judgment to fill in the gaps concerning possible futures. The resulting simulation models provide powerful tools to develop a shared understanding and to ground what-if thinking.

Future Directions

While the field of System Dynamics has reached its half-centenary in 2007, its influence on public policy continues to grow. Many of the problems defined by the earliest writers in the field continue to challenge us today. The growing literature base of environmental, social, and education policy is evidence of continued interest in the systems perspective. In addition, System Dynamics modeling is growing in popularity for defense analysis, computer security and infrastructure planning, and emergency management. These areas have the characteristic problems of complexity and uncertainty that require the integration of multiple perspectives and tacit knowledge that this method supports. Researchers and practitioners will continue to be attracted to the open nature of System Dynamics models as a vehicle for consensus and experimentation.

We anticipate that the tool base for developing and distributing System Dynamics models and insights will also grow. Graphical and multimedia-based simulations are growing in popularity, making it possible to build clearer models and disseminate insights easily. In addition, the development of materials for school-age learners to consider a systems perspective to social problems gives us optimism for the future of the field, as well as future policy.

Bibliography

Primary Literature

Ackermann F, Eden C, Williams T (1997) Modeling for litigation: mixing qualitative and quantitative approaches. Interfaces 27(2):48–65

Akkermans H, Vennix J (1997) Clients' opinions on group model-building: an exploratory study. Syst Dyn Rev 13(1):3–31

Alfeld L, Graham A (1976) Introduction to urban dynamics. Wright-Allen Press, Cambridge

Andersen D (1990) Analyzing who gains and who loses: the case of school finance reform in New York State. Syst Dyn Rev 6(1):21–43

Andersen D, Emmerichs R (1982) Analyzing US military retirement policies. Simulation 39(5):151–158

Andersen D, Richardson GP (1997) Scripts for group model building. Syst Dyn Rev 13(2):107–129

Andersen D, Bryson J, Richardson GP, Ackermann F, Eden C, Finn C (2006) Integrating modes of systems thinking into strategic planning education and practice: the thinking persons' institute approach. J Public Aff Educ 12(3):265–293

Bakken B, Gilljam M (2003) Dynamic intuition in military command and control: why it is important, and how it should be developed. Cogn Technol Work (5):197–205

Barlas Y (2002) System dynamics: systemic feedback modeling for policy analysis. In: Knowledge for sustainable development, an insight into the encyclopedia of life support systems, vol 1. UNESCO-EOLSS, Oxford, pp 1131–1175

Barney G (1982) The global 2000 report to the President: entering the twenty-first century. Penguin, New York

Cavana RY, Clifford L (2006) Demonstrating the utility for system dynamics for public policy analysis in New Zealand: the case for excise tax policy on tobacco. Syst Dyn Rev 22(4):321–348

Cavana RY, Ford A (2004) Environmental and resource systems: editor's introduction. Syst Dyn Rev 20(2):89–98

Cavana RY, Davies P et al (1999) Drivers of quality in health services: different worldviews of clinicians and policy managers revealed. Syst Dyn Rev 15(3):331–340

Cavana RY, Boyd D, Taylor R (2007) A systems thinking study of retention and recruitment issues for the New Zealand Army electronic technician trade group. Syst Res Behav Sci 24(2):201–216

Chen F, Andersen D et al (1981) A preliminary system dynamics model of the allocation of state aid to education. Dynamica 7(1):2–13

Clark R (1981) Readiness as a residual of resource allocation decisions. Def Manag J 1:20–24

Clark R (1987) Defense budget instability and weapon system acquisition. Public Budg Financ 7(2):24–36

Clark R (1993) The dynamics of US force reduction and reconstitution. Def Anal 9(1):51–68

Clark T, McCullough B et al (1980) A conceptual model of the effects of department of defense realignments. Behav Sci 25(2):149–160

Cooper K (1980) Naval ship production: a claim settled and framework built. Interfaces 10(6):20

Coyle RG (1992) A system dynamics model of aircraft carrier survivability. Syst Dyn Rev 8(3):193–213

Coyle RG (1996) System dynamics modelling: a practical approach. Chapman and Hall, London

Creative Learning Exchange (2000) Road maps: a guide to learning system dynamics. http://sysdyn.clexchange.org/road-maps/home.html

Davidsen P (1996) Educational features of the system dynamics approach to modelling and learning. J Struct Learn 12(4):269–290

Fiddaman TS (2002) Exploring policy options with a behavioral climate-economy model. Syst Dyn Rev 18(2):243–267

Finely B (2006) Migrant cases burden system: rise in deportations floods detention centers, courts. Denver Post, 2 Oct 2006. http://www.denverpost.com/immigration/ci_4428563

Fisher D (2001) Lessons in mathematics: a dynamic approach. iSee Systems, Lebanon

Fisher D (2004) Modeling dynamic systems: lessons for a first course. iSee Systems, Lebanon

Flessa S (1999) Decision support for malaria-control programmes – a system dynamics model. Health Care Manag Sci 2(3):181–191

Ford A (1997) System dynamics and the electric power industry. Syst Dyn Rev 13(1):57–85

Ford A (1999) Modeling the environment: an introduction to system dynamics modeling of environmental systems. Island Press, Washington, DC

Forrester J (1961a) Industrial dynamics. Pegasus Communications, Cambridge

Forrester J (1969) Urban dynamics. Pegasus Communications, Waltham

Forrester J (1971) World dynamics. Pegasus Communications, Waltham

Forrester J (1980) Information sources for modeling the national economy. J Am Stat Assoc 75(371):555–566

Forrester J (1993) System dynamics as an organizing framework for pre-college education. Syst Dyn Rev 9(2):183–194

Forrester J (1994) Policies, decisions, and information sources for modeling. In: Morecroft J, Sterman J (eds) Modeling for learning organizations. Productivity Press, Portland, pp 51–84

Forrester J, Senge P (1980) Tests for building confidence in system dynamics models. In: Legasto AA Jr et al (eds) System dynamics. North-Holland, New York, 14, pp 209–228

Forsyth B, Hirsch G et al (1976) Projecting a teaching hospital's future utilization: a dynamic simulation approach. J Med Educ 51(11):937–939

Galbraith P (1989) Mathematics education and the future: a long wave view of change. Learn Math 8(3):27–33

Graham A, Morecroft J et al (1992) Model supported case studies for management education. Eur J Oper Res 59(1):151–166

Homer J, St Clair C (1991) A model of HIV transmission through needle sharing. A model useful in analyzing public policies, such as a needle cleaning campaign. Interfaces 21(3):26–29

Homer J, Roberts E et al (1982) A systems view of the smoking problem. Int J Biomed Comput 13 69–86

Hopkins P (1992) Simulating hamlet in the classroom. Syst Dyn Rev 8(1):91–100

Huz S, Andersen D, Richardson GP, Boothroyd R (1997) A framework for evaluating systems thinking interventions: an experimental approach to mental health system change. Syst Dyn Rev 13(2)149–169

Jones A, Seville D et al (2002) Resource sustainability in commodity systems: the sawmill industry in the Northern Forest. Syst Dyn Rev 18(2):171–204

Keys B, Wolfe J (1996) The role of management games and simulations in education research. J Manag 16(2):307–336

Lane DC, Monefeldt C, Rosenhead JV (1998) Emergency – but no accident – a system dynamics study of an accident and emergency department. OR Insight 11(4):2–10

Levin G, Roberts E (eds) (1976) The dynamics of human service delivery. Ballinger, Cambridge

Levin G, Hirsch G, Roberts E (1975) The persistent poppy: a computer aided search for heroin policy. Ballinger, Cambridge

Lopez T, Watson J Jr (1979) A system dynamics simulation model of the U.S. Marine Corps manpower system. Dynamica 5(2):57–78

Luna-Reyes L, Martinez-Moyano I, Pardo T, Creswell A, Richardson GP, Andersen D(2007) Anatomy of a group model building intervention: building dynamic theory from case study research. Syst Dyn Rev 22(4):291–320

Maani KE, Cavana RY (2007a) Systems thinking, system dynamics: managing change and complexity. Pearson Education (NZ), Auckland

MacDonald R et al (2007) System dynamics public policy literature. Syst Dyn Soc. http://www.systemdynamics.org/short_bibliography.htm

Mandinach E, Cline H (1993) Systems, science, and schools. Syst Dyn Rev 9(2):195–206

Mashayekhi A (1977) Economic planning and growth of education in developing countries. Simulation 29(6):189–197

Mashayekhi A (1998) Public finance, oil revenue expenditure and economic performance: a comparative study of four countries. Syst Dyn Rev 14(2–3):189–219

Mass N (ed) (1974) Readings in urban dynamics. Wright-Allen Press, Cambridge

Meadows D, Meadows D (1977) Towards global equilibrium: collected papers. MIT Press, Cambridge

Meadows D, Meadows D, Randers J (1972) The limits to growth: a report for the club of Rome's project on the predicament of mankind. Universe Books, New York

Meadows D, Beherens W III, Meadows D, Nail R, Randers J, Zahn E (ed) (1974) Dynamics of growth in a finite world. Pegasus Communications, Waltham

Meadows D, Meadows D, Randers J (1992) Beyond the limits. Chelsea Green Publishing Company, Post Mills

Meadows D, Randers J, Meadows D (2004) Limits to growth: the 30-year update. Chelsea Green Publishing Company, White River Junction

Miyakawa T (ed) (1999) The science of public policy: essential readings in policy sciences 1. Routledge, London

Morecroft J (2007) Strategic modelling and business dynamics: a feedback systems approach. Wiley, West Sussex

Naill R (1977) Managing the energy transition. Ballinger Publishing Company, Cambridge

Otto P, Struben J (2004) Gloucester fishery: insights from a group modeling intervention. Syst Dyn Rev 20(4):287–312

Reagan-Cirincione P, Shuman S, Richardson GP, Dorf S (1991) Decision modeling: tools for strategic thinking. Interfaces 21(6):52–65

Richardson GP (1991) System dynamics: simulation for policy analysis from a feedback perspective. In: Fishwick P, Luker P (eds) Qualitative simulation modeling and analysis. Springer, New York

Richardson GP (1996) System dynamics. In: Gass S, Harris C (eds) Encyclopedia of operations research and management science. Kluwer Academic, Norwell

Richardson G, Andersen D (1979) Teaching for research in system dynamics. Dynamica 5(3)

Richardson G, Andersen D (1995) Teamwork in group model building. Syst Dyn Rev 11(2):113–137

Richardson G, Lamitie R (1989) Improving Connecticut school aid: a case study with model-based policy analysis. J Educ Financ 15(2):169–188

Richardson G, Pugh J (1981) Introduction to system dynamics modeling. Pegasus Communications, Waltham

Richmond B (1997) The strategic forum: aligning objectives strategy and process. Syst Dyn Rev 13(2):131–148

Roberts N (1983) An introductory curriculum in system dynamics. Dynamica 9(1):40–42

Roberts C, Dangerfield B (1990) Modelling the epidemiological consequences of HIV infection and AIDS: a contribution from operational research. J Oper Res Soc 41(4):273–289

Roberts N, Andersen DF, Deal RM, Grant MS, Schaffer WA (1983) Introduction to computer simulation: a system dynamics modeling approach. Addison Wesley, Reading

Rohrbaugh J (2000) The use of system dynamics in decision conferencing: implementing welfare reform in

New York State. In: Garson G (ed) Handbook of public information systems. Marc Dekker, New York, pp 521–533

Rouwette E (2003) Group model building as mutual persuasion. Wolf Legal Publisher Nijmegen

Rouwette E, Vennix J, Van Mullekom T (2002) Group model building effectiveness: a review of assessment studies. Syst Dyn Rev 18(1):5–45

Saeed K (1994) Development planning and policy design: a system dynamics approach. Ashgate, Aldershot

Saeed K (1996) The dynamics of collegial systems in the developing countries. High Educ Policy 9(1):75–86

Saeed K (1998) Towards sustainable development, 2nd edn. Essays on system analysis of national policy. Ashgate, Aldershot

Saeed K (2003) Articulating developmental problems for policy intervention: a system dynamics modeling approach. Simul Gaming 34(3):409–436

Saysel A, Barlas Y et al (2002) Environmental sustainability in an agricultural development project: a system dynamics approach. J Environ Manag 64(3):247–260

Senge P (1990) The fifth discipline: the art and practice of the learning organization. Doubleday/Currency, New York

Senge P, Asay D (1988) Rethinking the healthcare system. Healthc Forum J 31(3):32–34, 44–45

Stave K (2002) Using SD to improve public participation in environment decisions. Syst Dyn Rev 18(2):139–167

Steed M (1992) Stella, a simulation construction kit: cognitive process and educational implications. J Comput Math Sci Teach 11(1):39–52

Stenberg L (1980) A modeling procedure for the public policy scene. In: Randers J (ed) Elements of the system dynamics method. Pegasus Communications, Waltham, pp 257–288

Sterman J (1988) People express management flight simulator: simulation game, briefing book, and simulator guide. http://web.mit.edu/jsterman/www/SDG/MFS/PE.html

Sterman J (1992) Teaching takes off: flight simulators for management education. OR/MS Today (October):40–44

Sterman J (1996) A Skeptic's guide to computer models. In: Richardson GP (ed) Modelling for management. Dartmouth Publishing Company, Aldershot

Sterman J (2000) Business dynamics: systems thinking and modeling for a complex world. Irwin/McGraw-Hill, Boston

Taylor K, Lane D (1998) Simulation applied to health services: opportunities for applying the system dynamics approach. J Health Serv Res Policy 3(4):226–232

van Ackere A, Smith P (1999) Towards a macro model of national health service waiting lists. Syst Dyn Rev :225–253

Ve (1996) Group model building: facilitating team using system dynamics. Wiley, Chichester

W er M (1984) The new technologies versus the tradition in higher education: is change possible Educ Technol 24(3):7–13

W K (1999) Dynamics of strategy. Bus Strateg Rev (3):1–16

Warren K (2002) Competitive strategy dynamics. Wiley, Chichester

Warren K (2005) Improving strategic management with the fundamental principles of system dynamics. Syst Dyn Rev 21(4):329–350

Wils A, Kamiya M et al (1998) Threats to sustainability: simulating conflict within and between nations. Syst Dyn Rev 14(2–3):129–162

Wolstenholme E (1992) The definition and application of a stepwise approach to model conceptualization and analysis. Eur J Oper Res 59(1):123–136

Wolstenholme E (1993) A case study in community care using systems thinking. J Oper Res Soc 44(9):925–934

Wolstenholme E (1999) A patient flow perspective of UK health services: exploring the case for new intermediate care initiatives. Syst Dyn Rev 15(3):253–273

Wolstenholme E, Coyle RG (1983) The development of system dynamics as a methodology for system description and qualitative analysis. J Oper Res Soc 34(7):569–581

Zagonel A, Andersen D, Richardson GP, Rohrbaugh J (2004) Using simulation models to address "what if" questions about welfare reform. J Policy Anal Manag 23(4):890–901

Books and Reviews

Coyle RG (1998) The practice of system dynamics: milestones, lessons and ideas from 30 years of experience. Syst Dyn Rev 14(4):343–365

Forrester J (1961b) Industrial dynamics. Pegasus Communications, Waltham

Maani KE, Cavana RY (2007b) Systems thinking, system dynamics: managing change and complexity. Pearson Education (NZ), Auckland

MacDonald R (1998) Reducing traffic safety deaths: a system dynamics perspective. In: 16th International conference of the System Dynamics Society, Quebec '98. System Dynamics Society, Quebec City

Morecroft JDW, Sterman JD (eds) (1994) Modeling for learning organizations, system dynamics series. Productivity Press, Portland

Wolstenholme EF (1990) System enquiry: a system dynamics approach. Wiley, Chichester

System Dynamics and Operations Management

Andreas Größler
Department of Operations Management,
University of Stuttgart, Stuttgart, Germany

Article Outline

Glossary

Model A simplified and idealized (formal or informal) representation of a real or hypothetical system

Operations management The subfield of business administration that deals with the efficient management of operations, i.e., an organization's value creation processes

Operations strategy The subfield of business administration that deals with the effective reconciliation of operations' capabilities and external (e.g., of customers and competitors) requirements

Scenario Potential future development of a system, often derived from running computer simulations

Simulation Producing the behavior-over-time of a system, often based on a formal model and with the help of a computer

System dynamics Method for developing formal models and running computer simulations for changing systems, mostly in the socio-economic domain

Introduction

Operations management (OM) is concerned with planning, organizing, implementing, and controlling value generation through either the manufacturing of products or the provision of services. In a broad view, this creation of value can happen within the organization or in its supply chain (thus, supply chain management is considered here a subset of operations management and included in the following discussion). Operations has been named the "nexus of systems, people, processes and procedures" (Hill et al. 1999, 148) and "the activity of managing the resources which produce and deliver products and services" (Slack et al. 2010, 4). From a strategic point of view, operations management tries to reconcile external demands with internal capabilities (Slack and Lewis 2015). Within this context, OM must deal with changing requirements from customers and markets, the creation and depletion of organizational resources over time, and the continuous alignment with other functions in an organization. Operations is a primary function of for-profit organizations and firms but also exists and is relevant in not-for-profit organizations where similar decisions must be made (Slack et al. 2010).

System dynamics (SD) is a method to visualize, to model formally, and to simulate dynamic systems, for instance, the operations of manufacturing or service firms. In 1958, MIT professor Jay W. Forrester, who started the System Dynamics Group at the Sloan School 2 years before, published his seminal article "Industrial Dynamics: A Major Breakthrough for Decision Makers." SD is designed to study complex feedback systems and to provide decision support by model-based scenario analysis and policy testing. It aims to support (1) the development of "a better

© Springer Science+Business Media, LLC, part of Springer Nature 2020
B. Dangerfield (ed.), *System Dynamics*,
https://doi.org/10.1007/978-1-4939-8790-0_661

Originally published in
R. A. Meyers (ed.), *Encyclopedia of Complexity and Systems Science*, © Springer Science+Business
Media LLC 2018, https://doi.org/10.1007/978-3-642-27737-5_661-1

intuitive feel for the time-varying behaviour of industrial and economic systems," (2) the provision of "a background showing how the major aspects of a company are related to one another," (3) predicting "the future course of an existing organization," and (4) possibilities for improvement of "the future prospects of a company" (Forrester 1958).

As the term "industrial dynamics" implies, it was initially applied to industrial companies only. However, it became soon obvious that the goals and the methodology of industrial dynamics could be applied to all kinds of dynamic systems. For instance, applications to systems such as a city (Forrester 1969) and the whole world (Forrester 1971) followed, and the field was renamed to "system dynamics." Today, system dynamics is used to study a wide variety of systems that change over time, particularly in the socioeconomic domain (Morecroft 2007; Sterman 2000), including operations management (see Größler et al. 2008 for an article arguing that system dynamics constitutes a structural theory of operations; this chapter modifies and extends some parts of that study without replicating the discussion about the theory-ness of SD). For the history of system dynamics, see, e.g., Forrester (1989) and Lane (1999, 2007).

System dynamics is an extension of servomechanism theory (Richardson 1991). It proposes that the behavior of a dynamic system is the result of the system's structure. SD models are supposed to produce the "right behavior for the right reasons" (Barlas 1996), employing "operational thinking" when modelling real-world systems (Olaya 2015; Richmond 1993). The main structural concepts used in system dynamics are (i) feedback loops, (ii) accumulation processes, and (iii) delays between cause and effect, which are all characteristics of most operations management situations. For instance, the successful implementation of a program to increase efficiency in production might be characterized by a feedback loop: its successful implementation can influence worker motivation which, eventually, might itself affect the implementation level of the program, when worker motivation decreases because of fear of being redundant with the

program being in place. An example for accumulations in operations settings are the dynamics of increasing and decreasing inventories within a supply chain or the ramp-up and phaseout of production capacity. A change of operations strategy from being a niche producer to becoming a volume manufacturer is characterized by a substantial delay caused, for instance, by the time it takes to build up capacity, employee skills, and company reputation with (potential) customers.

With the help of system dynamics, new insights in OM issues are possible. For instance, Gonçalves et al. (2003) show how considering the feedback relation between product availability and customer demand ("endogenizing demand") renders conventional prescriptions of lean inventory and fast-responsive capacity utilization as not appropriate. In another example, a study by Anderson et al. (2003) demonstrates that lead-time reduction does not necessarily result in a dampening of the bullwhip effect (Lee et al. 1997), but can also amplify it, if it is not coordinated with capacity policies.

In the following section, the prominent role of modelling and simulation in operations management research is described. Furthermore, it discusses in which way system dynamics adds to this body of literature. The third section explains the system dynamics methodology in more detail. In the section thereafter, exemplary applications of system dynamics to operations' issues are presented. This chapter closes with an outlook to future research directions of system dynamics within the field of operations management.

Modelling and Simulation in Operations Management

System dynamics has been used in a variety of operations management studies. A recent collection of papers applying system dynamics in an operations management context can be found in a special issue of the Journal of Operations Management (Sterman et al. 2015). Bertrand and Fransoo (2002) distinguish model-based articles in operations management along two dimensions: "empirical" versus "axiomatic" and "descriptive" versus "normative." Concerning this

classification of modelling approaches pera-
tions management, system dynamics can
be classified as "empirical." In contra axiomatic" uses o modelling, system dyn tud-
ies are driven by empirical evidenc by
abstract concepts. Regarding the descr versus normative dimension, the focus stem
dynamics studies is not on deriving an a cally
solvable model, but on investigating ystem
with its relevant complexity (Akkerma 993).
Thus, system dynamics models are more descriptive" than "normative." Nevertheless, the
improvement of real systems is a goal in system
dynamics but without the aim of achieving an
optimal solution. In typical OM settings, the aim
usually is "to compromise rather than optimise"
(Hill et al. 1999, 142).

In the operations management field, modelling
and simulation has often been viewed as a "rationalist" and purely "deductive" way of generating
new knowledge (Meredith 1998; Swamidass
1991). The basis for this perspective is grounded
in the operations research type of models that are
frequently employed and theories that result from
them (e.g., parts of inventory theory). As
Swamidass (1991) states, fuzzy and messy concepts (like manufacturing strategy) are unsuitable
for such deductive research (cf. also Hill et al.
1999). However, this does not imply that modelling and simulation cannot be applied in general
for strategic issues. In system dynamics, models
are used to analyze dynamically complex and
messy concepts (mostly in the strategic area).
When addressing such strategic issues, it is regularly "possible to construct aggregate relationships that describe, in a simplified but adequate
manner, the overall behavior of an in itself highly
complex operations system without modelling
this complex system in detail" (Akkermans and
Bertrand 1997, 953). However, this has the draw-
back that some mathematical elegance is
sacrificed (the analytical derivation of optimal
solutions) for the attainability and relevance of
"just" robust solutions.

The validity of system dynamics models is
shown by the transferability of insights generated
by the model to reality, thus fulfilling a requirement of Hill et al. (1999) for the usefulness of

models in OM. Starting with Forrester's *Industrial Dynamics* model, results from SD studies
have been implemented and have led to improvements in real operations systems. Concerning the
information bases of system dynamics modelling,
it relies on empirical studies like surveys (Gupta
et al. 2006; Rungtusanatham et al. 2003; Scudder
and Hill 1998). In addition, more qualitative
methods of knowledge generation are considered
appropriate and useful for building simulation
models – like case studies and interviews (Hill
et al. 1999; Meredith 1998) – following a
grounded theory approach (Glaser and Strauss
1967; Yin 1994). See Oliva and Sterman (2001)
for a prototypical example of this kind of work.
System dynamics claims to depict policies and
structures as realistic as possible, to test potential
improvements with models and simulations, and
to implement improvements in the real system
(Snabe and Größler 2006). The actual implementation of a potential improvement is considered
more important than the search for an abstract
optimal system state (Sterman 1988).

System Dynamics as a Method to Understand Time–Dependent Aspects of Operations Management

Feedback

The central notion of system dynamics is that
feedback loops are the structural building blocks
of all social systems (Forrester 1961). This means
that any activity of an element in such a system
ultimately has an effect back on that element.
Corollaries of this definition are that action cannot
occur without reaction, that side effects are inevitable in social systems, and that decisions trigger
effects that are remote regarding time and space.
The concept of feedback is so prominent in system
dynamics that it is also used to define the borders
of any issues analyzed: all feedback loops that
have a substantial impact on the behavior of a
system must be included in a system dynamics
study (Forrester 1968a, ch. 4).

Two types of feedback loops exist: negative
(goal seeking) and positive (reinforcing) feedback
loops. Negative feedback loops are self-

correcting, in the sense that they counteract the change of the system state, while positive feedback loops amplify the change and are therefore considered self-reinforcing (Sterman 2000). Examples from operations management of negative feedback loops are inventory or capacity adjustment systems: whenever actual inventory or capacity deviates from a predefined target, adjustments are made, for instance, by putting additional items into stock, by reducing production (to lower the inventory level), or by investing or disinvesting in machines and personnel to adjust capacity. An example of a reinforcing feedback loop are operational improvement programs which result in increased profits that subsequently allow to conduct further improvement programs that lead to even more profits. Of course, these examples are simplified in a way that in real-world systems – as well as in system dynamics models – usually more than just one feedback loop exists. For instance, for the positive feedback loop described above, we must assume that limits to

this otherwise indefinite growth process exist (Meadows 1982), which in this case, for example, could be total market demand or capacity constraints of underlying resources. Feedback structures can be depicted with causal loop diagrams (for an example, see Fig. 1). In these diagrams, causal linkages between variables are shown together with the polarities of the linkages (a positive polarity indicating a change in the same direction, a negative polarity indicating a change in the opposite direction). Cycles of causal linkages form feedback loops that can be positive or negative as just discussed.

Accumulation

In system dynamics tradition, feedback loops are the major structural element of social systems. Nevertheless, some authors have argued that the concept of accumulation is even more important for system dynamics than feedback loops (Warren 2002, 2008). Every feedback loop contains at

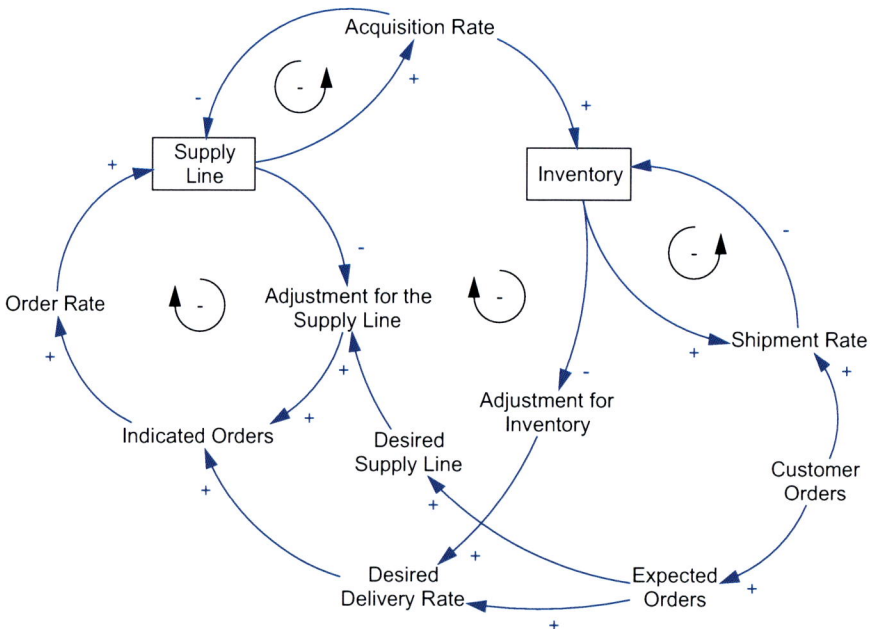

System Dynamics and Operations Management, Fig. 1 Example of causal loop diagram; here, feedback loops controlling the (delayed) adjustment of an inventory level based on the shipment rate affected by customer orders; the various feedback loops form a net of complex

causal relations. +/– indicate positive/negative link polarities; – in circles indicate goal-seeking feedback loops (by chance, this causal loop diagram does not contain reinforcing feedback loops)

least one level variable that conserves the state of the system and that can be used as an initial value for calculating the simulated behavior. Accumulations, or stocks as they are often called in system dynamics literature, preserve the path of a system through time: they incorporate the history of a system and determine the future of a system. Flow variables, which contain the mechanisms for state changes, and stock variables represent the two main types of variables in system dynamics (Forrester 1968a, ch. 4). In organizations, a stock is an accumulation of a resource, whereas a flow represents a change of a resource. Examples from OM for the widespread occurrence of stock-flow processes are the accumulation of raw materials, modules, and finished goods in sequences of production stages and inventories (Cachon and Terwiesch 2013); accumulative strategic capabilities in production (Ferdows and De Meyer 1990); and the accumulation of production experience (Henderson 1984).

Delays

Closely connected to the concepts of feedback loops and accumulating stocks is the idea that delays are ubiquitous in social systems. No business process can be conducted instantaneously. Decisions lead to effects that are distant in time and space. With the help of simulation, system dynamics compresses such delays, to analyze their effects within an acceptable time frame (Kim and Senge 1994). Examples from operations management are the delays connected with the effectiveness of improvement programs, the time it takes until new employees are fully productive, or the delays that are inherent in organizational reporting and controlling processes. The resulting behavior modes of delayed processes are often oscillations, which frequently have been analyzed in different forms with the help of system dynamics, for example, business cycles (Forrester 1976; Lorenz 1992; Liehr et al. 2001; Meadows 1970; Randers and Göluke 2007) and Kondratieff cycles (Ryzhenkov 2000; Sterman 1985).

Nonlinearity and "Soft" Variables

The combination of feedback loops, accumulation processes, and delays frequently results in nonlinear behavior of systems. The behavior modes of such systems are hardly understandable by intuition (Booth Sweeny and Sterman 2000; Dörner 1996; Reason 1990; Sterman 1989). Furthermore, analytical solutions are usually not available to "solve" these systems. Thus, simulation serves as a method to observe systems behavior. Simulations require a quantification of all elements and the linkages between the elements of a system. If no mathematical function can be formulated, it can be graphically and ad hoc presented by so-called table functions (Sterman 2000, 551n). Regularly this is the case when fuzzy or hard to quantify variables have a significant effect and therefore need to be incorporated into a system dynamics model. Such variables are included no matter how difficult it is to measure and to quantify them because neglecting them would certainly lead to an erroneous model and simulation. Examples are the influence that organizational factors (like work satisfaction, stress, and motivation) have on worker productivity in a model of skill formation in new production systems (Diawati et al. 1994) or perceived delivery delays that affect product sales in a model of corporate growth and capacity restrictions (Forrester 1968b). The effects of such variables are often described with the help of estimations that are formulated by competent individuals in the form of "educated guesses" (Ford and Sterman 1998a). Estimated parameters and functions indicate starting points for additional empirical research, when they demonstrate an effect on systems behavior in the simulation runs (which can be tested by sensitivity analyses). Furthermore, such initial models can be calibrated using numerical methods (Rahmandad et al. 2015).

Computer Simulation

To better understand dynamic systems, system dynamics provides a graphical syntax in which flow and stock variables are identified and combined to larger structures (Forrester 1968a, ch. 5; Lane 2000). The graphical representation of systems using this syntax proves to be a valuable tool for understanding complex issues; it is usually named stock-flow diagram (for an example, see Fig. 2). There are some commercially available

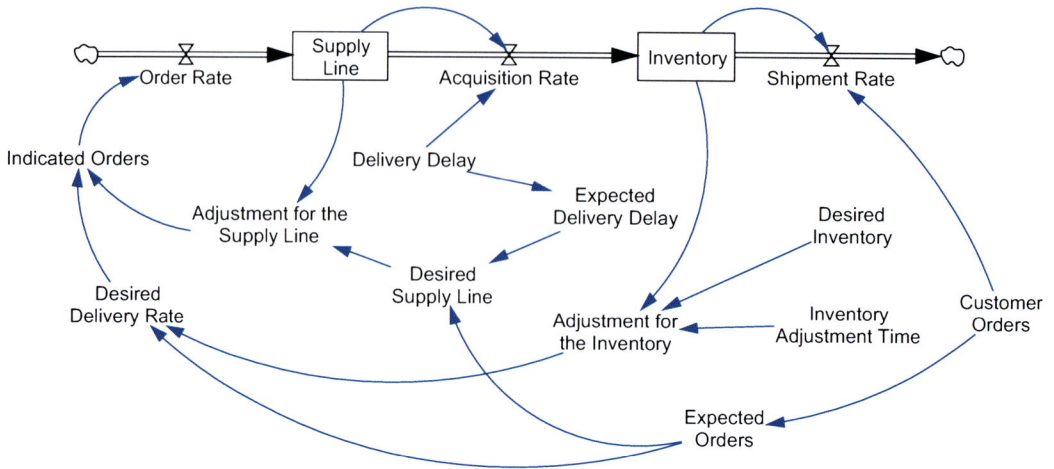

System Dynamics and Operations Management, Fig. 2 Example of a stock-flow diagram; here: material and information flows controlling the (delayed) adjustment of an inventory via a supply line based on the shipment rate

affected by customer orders (Sterman 2000, 688). Variables in boxes are stocks, pipes with valves are flows, single arrows are causal links, and other variables are parameters or auxiliaries

software packages to support the process of modelling and simulation (for instance, Vensim, Powersim, or Stella). By quantification of variables and linkages between variables, a system of differential equations is created that can be simulated by numerical algorithms (Sterman 2000, 903n).

Simulation is a feature of system dynamics that distinguishes it from other forms of systems thinking (and general systems theory). Simulation is the generation of behavior over time for a given model structure (for an example, see Fig. 3). In system dynamics, a model-based analysis is incomplete without simulation (Forrester 1994). A reason for this is the difficulty people have when trying to induce behavior from reasonably complex system structures (Dörner 1996; Frensch and Funke 1995; Sterman 1994). Oscillations, exponential growth, and overshoot and collapse are behavior modes that are difficult for humans to comprehend completely and correctly (Dörner 1980; Sterman 1989). By way of simulation, such effects can be explored and analyzed; dangerous, unethical, impossible, costly, or just very creative decisions can be tested on their effects on a system. Simulations can be started with arbitrary initial settings and as often as useful (Pidd 2004).

Modelling Process

Many authors emphasize the importance of the modelling process for gaining insights into the problem (Lane 1995; Sterman 1988; Vennix 1996). The reason for this emphasis on the process is that system dynamics models are usually descriptive and not prescriptive, in the sense that they do not result in a singular best solution, which has two implications. First, insights from the model can only indirectly be deduced by formulating its equations and running different simulations (as quantitative scenario analyses). Second, while being a modelling endeavor, system dynamics research is substantially empirical: regularly, it does not involve the development of ideal models, but rather the depiction of reality as precisely as possible. Based on a valid picture of what is the case, attempts are made to gain insight into how the real system can be improved.

In the first step of the modelling process, the problem to be modelled is defined, and key variables are identified (see Fig. 4; inspired by Forrester 1994). An important question to answer in this phase concerns the system boundary, i.e., it is determined, which factors should be included and which should not. As a rule of thumb, important feedback loops should not be cut in system dynamics models. In the second phase, a stock-

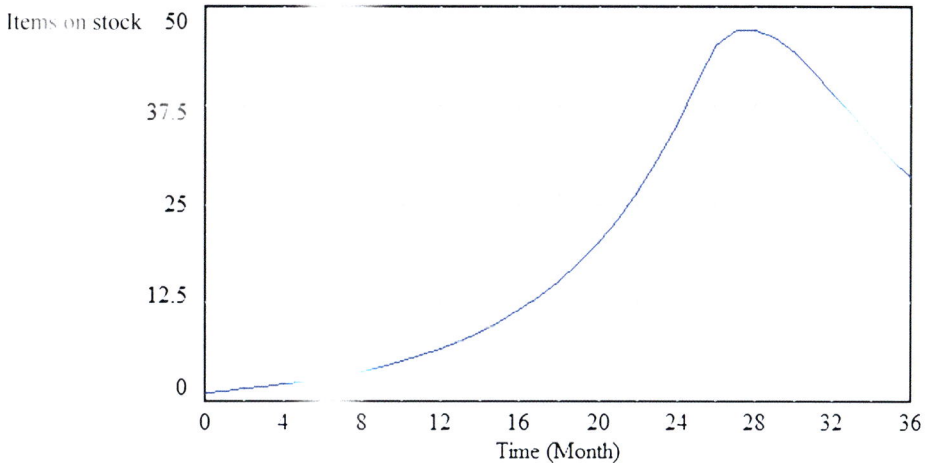

System Dynamics and Operations Management, Fig. 3 Example of simulation output; here: development of inventory level as response to (estimated) increasing demand (leading to an overshoot because of linear extrapolation of past demand)

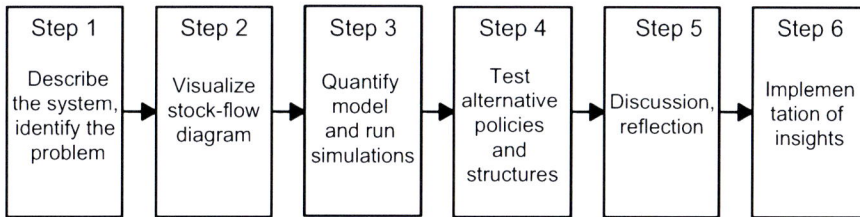

System Dynamics and Operations Management, Fig. 4 Prototypical system dynamics process (iterations not depicted)

flow diagram is built that represents the system structure as identified before. This model is quantified in the next (third) step, i.e., mathematical functions are defined that determine how values of a variable are calculated based on the values of other variables. Furthermore, values of constants and initial parameters are explicated. Based on this quantification process, simulation runs are conducted to generate the behavior over time of the structure represented by the stock-flow diagram and elaborated in the model's equations. In the following step (4), additional policies and structures are tested, either to try out other conceivable givens or to find alternatives for improvement of the system. The fifth step acts as a reflection point: insights from the modelling and simulation activities are discussed and tested for their validity. In the last step, model-based solutions are implemented to improve the real-world system.

Applications of System Dynamics in Operations Management

System dynamics has been used in six major areas in operations management which are listed below. For each category, a sample of prototypical studies is referenced that could serve as starting points for a further review of the usage of system dynamics within operations management.

1. Production and supply chain flow issues (e.g., Akkermans and Dellaert 2005; Akkermans and Vos 2003; Anderson et al. 2000; Forrester 1958; Fowler 1999; Georgiadis and Michaloudis 2012; Gupta and Gupta 1989; Morecroft 1983; Spengler and Schröter 2003; Sterman 1989; Towill 1982, 1996; Wikner et al. 1991)

2. Improvement programs in operations (e.g., Burgess 1998; Godinho Filho and Uzsoy 2014; Jambekar 2000; Keating et al. 1999; Mohammadi et al. 2015; Morrison 2015; Repenning and Sterman 2001; Salge and Milling 2006; Sterman et al. 1997; Thun 2006)
3. Project management (e.g., Abdel-Hamid and Madnick 1991; Bendoly 2014; Ford and Sterman 1998b, 2003; Lyneis and Ford 2007; Parvan et al. 2015; Tan et al. 2010)
4. New product development, innovation, and diffusion (e.g., Anderson and Parker 2002; Bloodgood et al. 2015; Kreng and Wang 2013; Maier 1998; Milling 1996; Repenning 2002; Smets et al. 2013; Wunderlich et al. 2014)
5. Effects of specific operations decisions and technologies (e.g., Chuang and Oliva 2015; De Marco et al. 2012; Ebrahimpour and Fathi 1984; Größler et al. 2015; Zahn et al. 1987, 1998)
6. Specific supply chain conditions, like disruptions, closed-loop supply chains, forecasting failures (e.g., Ardi and Leisten 2016; Bam et al. 2017; Choi et al. 2012; Das and Dutta 2013; Georgiadis and Besiou 2008; Lehr et al. 2013; Peng et al. 2014; Syntetos et al. 2011)

Two of these studies should be described here in more detail in order to demonstrate the usefulness and contribution of system dynamics to theory and practice in operations management. In Lehr et al. (2013), the authors investigate strategies in and consequences of closed-loop supply chains. More precisely, they develop a system dynamics model that allows an original equipment manufacturer of computer hardware to test different value recovery strategies for business-to-business products in a closed-loop supply chain. One issue Lehr and colleagues put attention on is the interdependency between primary and secondary market dynamics (i.e., in the market of recycled parts and goods). For instance, an important aspect that is discussed in their paper is that a shock in the primary market not only leads to instabilities in that specific market but also has major consequences for the secondary market (like a transfer of the bullwhip effect over markets).

Based on a case study from the motor vehicle industry, Morrison (2015) addresses the questions why resource shortages in operational firms persist for an extended time frame and do not disappear given the emphasis on efficiency improvement programs in operations (like lean manufacturing). He develops a system dynamics model to understand chronic resource shortages and the double role of work-arounds that operational personnel employs to succeed despite these shortages. Workarounds are a solution to an operational problem but also camouflage underlying system weaknesses. The model exemplifies how the actions of various groups in an organization (e.g., managers and production workers) interact with each other and with the structural characteristics of the workplace to sustain resource shortages, which are clearly identified as problematic.

Future Directions

A growing field of study in operations management which benefits from system dynamics is behavioral operations management (BOM; Bendoly et al. 2010; Morrison and Oliva 2017). There are two reasons why system dynamics is well-suited for investigating effects of cognition and behavior on operations' tasks. First, system dynamics models and analyses are by design incorporating "soft" variables that are often disregarded by other modelling methods. For instance, system dynamics models regularly make use of variables like "worker motivation," "learning capability," or "company reputation." Second, system dynamics allows to easily build laboratory versions of dynamic tasks which can be used to run experiments, which is the prime way of creating knowledge in behavioral operations. In these tasks, participants of experiments can work in a well-defined and controlled setting in the form of computer simulation games. Exemplary studies in BOM based on or inspired by system dynamics modelling and simulation are Bendoly and Cotteleer (2008), Croson and Donohue (2006), Croson et al. (2014), Fransoo and Wiers (2006), Gino and Pisano (2008), Schweitzer and Cachon (2000), Sterman and Dogan (2015), Strohhecker and Größler (2013), Weinhardt et al. (2015), and Wu and Katok (2006).

Bibliography

Abdel-Hamid TK, Madnick SE (1991) Software project dynamics: an integrated approach. Prentice Hall, Englewood Cliffs

Akkermans H (1993) Participative business modelling to support strategic decision making in operations – a case study. Int J Oper Prod Manag 13(10):34–48

Akkermans H, Bertrand W (1997) On the usability of quantitative modelling in operations strategy decision making. Int J Oper Prod Manag 17(10):953–966

Akkermans H, Dellaert N (2005) The dynamics of supply chains and networks. Syst Dyn Rev 21(3):173–186

Akkermans H, Vos B (2003) Amplification in service supply chains: an exploratory case study from the telecom industry. Prod Oper Manag 12(2):204–223

Anderson EG Jr, Morrice DJ, Lundeen G (2003) The 'physics' of capacity and backlog management in service and custom manufacturing supply chains. Syst Dyn Rev 21(3):217–247

Anderson EG Jr, Parker GG (2002) The effect of learning on the make/buy decision. Prod Oper Manag 11(3):313–329

Anderson EG Jr, Fine CH, Parker GG (2000) Upstream volatility in the supply chain: the machine tool industry as a case study. Prod Oper Manag 9(3):239–261

Ardi R, Leisten R (2016) Assessing the role of informal sector in WEEE management systems: a system dynamics approach. Waste Manag 57:3–16

Bam L, McLaren ZM, Coetzee E, von Leipzig KH (2017) Reducing stock-outs of essential tuberculosis medicines: a system dynamics modelling approach to supply chain management. Health Policy Plan 32:1127–1134 czx057 (forthcoming)

Barlas Y (1996) Formal aspects of model validity and validation in system dynamics. Syst Dyn Rev 12(3):183–210

Bendoly E (2014) System dynamics understanding in projects: information sharing, psychological safety, and performance effects. Prod Oper Manag 23(8):1352–1369

Bendoly E, Cotteleer MJ (2008) Understanding behavioral sources of process variation following enterprise system deployment. J Oper Manag 26(1):23–44

Bendoly E, Hur D (2007) Bipolarity in reaction to operational 'constraints': OM bugs under an OB lens. J Oper Manag 25(1):1–13

Bendoly E, Croson R, Goncalves P, Schultz K (2010) Bodies of knowledge for research in behavioral operations. Prod Oper Manag 19(4):434–452

Bertrand JWM, Fransoo JC (2002) Operations management research methodologies using quantitative modeling. Int J Oper Prod Manag 22(2):241–264

Bloodgood JM, Hornsby JS, Burkemper AC, Sarooghi H (2015) A system dynamics perspective of corporate entrepreneurship. Small Bus Econ 45(2):383–402

Booth Sweeny L, Sterman JD (2000) Bathtub dynamics: initial results of a systems thinking inventory. Syst Dyn Rev 16(4):249–286

Burgess TF (1998) Modelling the impact of reengineering with system dynamics. Int J Oper Prod Manag 18(9/10):950–963

Cachon G, Terwiesch C (2013) Matching supply with demand. 3rd edn. McGraw-Hill, New York

Choi T, Narasimhan R, Kim SW (2012) Postponement strategy for international transfer of products in a global supply chain: a system dynamics examination. J Oper Manag 30(3):167–179

Chuang HHC, Oliva R (2015) Inventory record inaccuracy: causes and labor effects. J Oper Manag 39:63–78

Croson R, Donohue K (2006) Behavioral causes of the bullwhip effect and the observed value of inventory information. Manag Sci 52(3):323–336

Croson R, Donohue K, Katok E, Sterman J (2014) Order stability in supply chains: coordination risk and the role of coordination stock. Prod Oper Manag 23(2):176–196

Das D, Dutta P (2013) A system dynamics framework for integrated reverse supply chain with three way recovery and product exchange policy. Comput Ind Eng 66(4):720–733

De Marco A, Cagliano AC, Nervo ML, Rafele C (2012) Using system dynamics to assess the impact of RFID technology on retail operations. Int J Prod Econ 135(1):333–344

Diawati L, Kawashima H, Hayashi Y (1994) Skill formation and its impact on the adaptation process of new production systems. Syst Dyn Rev 10(1):29–47

Dörner D (1980) On the difficulties people have in dealing with complexity. Simul Games 11(1):87–106

Dörner D (1996) The logic of failure. Metropolitan Books/Henry Holt, New York

Ebrahimpour M, Fathi BM (1984) Dynamic simulation of a Kanban production inventory system. Int J Oper Prod Manag 5(1):5–14

Ferdows K, De Meyer A (1990) Lasting improvements in manufacturing performance – in search of a new theory. J Oper Manag 9(2):168–184

Ford DN, Sterman JD (1998a) Expert knowledge elicitation for improving mental and formal models. Syst Dyn Rev 14(4):309–340

Ford DN, Sterman JD (1998b) Dynamic modeling of product development processes. Syst Dyn Rev 14(1):31–68

Ford DN, Sterman JD (2003) The liar's club: concealing rework in concurrent development. Concurr Eng Res Appl 11(3):211–220

Forrester JW (1958) Industrial dynamics: a major breakthrough for decision makers. Harv Bus Rev 36(4):37–66

Forrester JW (1961) Industrial dynamics. MIT Press, Cambridge

Forrester JW (1968a) Principles of systems. MIT Press, Cambridge

Forrester JW (1968b) Market growth as influenced by capital investment. Ind Manag Rev 9(2):83–105

Forrester JW (1969) Urban dynamics. Productivity, Cambridge

Forrester JW (1971) World dynamics. Wright-Allen, Cambridge

Forrester JW (1976) A new view of business cycles. J Portfolio Manag 3(1):22–32

Forrester JW (1989) The beginning of system dynamics. Working paper D-4165-1, System Dynamics Group at MIT, Cambridge

Forrester JW (1994) System dynamics, systems thinking, and soft OR. Syst Dyn Rev 10(2/3):245–256

Fowler A (1999) Feedback and feedforward as systemic frameworks for operations control. Int J Oper Prod Manag 19(2):182–204

Fransoo JC, Wiers VC (2006) Action variety of planners: cognitive load and requisite variety. J Oper Manag 24(6):813–821

Frensch PA, Funke J (eds) (1995) Complex problem solving – the European perspective. Lawrence Erlbaum, Hillsdale

Georgiadis P, Besiou M (2008) Sustainability in electrical and electronic equipment closed-loop supply chains: a system dynamics approach. J Clean Prod 16(15):1665–1678

Georgiadis PC, Michaloudis (2012) Real-time production planning and control system for job-shop manufacturing: a system dynamics analysis. Eur J Oper Res 216(1):94–104

Gino F, Pisano G (2008) Toward a theory of behavioral operations. Manuf Serv Oper Manag 10(4):676–691

Glaser BG, Strauss AL (1967) The discovery of grounded theory: strategies for qualitative research. Aldine, Chicago

Godinho Filho M, Uzsoy R (2014) Assessing the impact of alternative continuous improvement programmes in a flow shop using system dynamics. Int J Prod Res 52(10):3014–3031

Gonçalves P, Hines J, Sterman JD (2003) The impact of endogenous demand on push-pull production systems. Syst Dyn Rev 21(3):187–216

Größler A, Thun J-H, Milling PM (2008) System dynamics as a structural theory of strategic issues in operations management. Prod Oper Manag 17(3):373–384

Größler A, Bivona E, Li F (2015) Evaluation of asset replacement strategies considering economic cycles: lessons from the machinery rental business. Int J Model Oper Manag 5(1):52–71

Gupta YP, Gupta MC (1989) A system dynamics model for a multi-stage multi-line dual-card JIT-Kanban system. Int J Prod Res 27(2):309–352

Gupta S, Verma R, Victorino L (2006) Empirical research published in production and operations management (1992–2005): trend and future directions. Prod Oper Manag 15(3):432–448

Henderson BD (1984) The application and misapplication of the experience curve. J Bus Strateg 4(3):3–9

Hill T, Nicholson A, Westbrook R (1999) Closing the gap: a polemic on plant-based research in operations management. Int J Oper Prod Manag 19(2):139–156

Jambekar AB (2000) A systems thinking perspective of maintenance, operations and process quality. J Qual Maint Eng 6(2):123–132

Keating EK, Oliva R, Repenning NP, Rockart S, Sterman JD (1999) Overcoming the improvement paradox. Eur Manag J 17(2):120–134

Kim DH, Senge PM (1994) Putting systems thinking into practice. Syst Dyn Rev 10(2/3):277–290

Kreng VB, Wang BJ (2013) An innovation diffusion of successive generations by system dynamics – an empirical study of Nike Golf company. Technol Forecast Soc Chang 80(1):77–87

Lane DC (1995) On a resurgence of management simulations and games. J Oper Res Soc 46(5):604–625

Lane DC (1999) Social theory and system dynamics practice. Eur J Oper Res 113(3):501–527

Lane DC (2000) Diagramming conventions in system dynamics. J Oper Res Soc 51(2):241–245

Lane DC (2007) The power of the bond between cause and effect: Jay Wright Forrester and the field of system dynamics. Syst Dyn Rev 23(2/3):95–118

Lee HL, Padmanabhan V, Whang S (1997) The bullwhip effect in supply chains. Sloan Manag Rev 38(3):93–102

Lehr CB, Thun J-H, Milling PM (2013) From waste to value–a system dynamics model for strategic decision-making in closed-loop supply chains. Int J Prod Res 51(13):4105–4116

Liehr M, Größler A, Klein M, Milling PM (2001) Cycles in the sky: understanding and managing business cycles in the airline market. Syst Dyn Rev 17(4):311–332

Lorenz H-W (1992) Complex dynamics in low-dimensional continuous-time business cycle models: the Silnikov case. Syst Dyn Rev 8(3):233–250

Lyneis JM, Ford DN (2007) System dynamics applied to project management: a survey, assessment, and directions for future research. Syst Dyn Rev 23(2/3):157–189

Maier F (1998) New product diffusion models in innovation management – a system dynamics perspective. Syst Dyn Rev 14(4):285–308

Meadows DL (1970) Dynamics of commodity production cycles. Productivity Press, Cambridge

Meadows DH (1982) Whole earth models and systems. CoEvolution Quarterly, Summer, pp 98–108

Meredith J (1998) Building operations management theory through case and field research. J Oper Manag 16(4):441–454

Milling P (1996) Modeling innovation processes for decision support and management simulation. Syst Dyn Rev 12(3):211–234

Mohammadi H, Ghazanfari M, Nozari H, Shafiezad O (2015) Combining the theory of constraints with system dynamics: a general model (case study of the subsidized milk industry). Int J Manag Sci Eng Manag 10(2):102–108

Morecroft JDW (1983) Managing product lines that share a common capacity base. J Oper Manag 3(2):57–66

Morecroft JDW (2007) Strategic modelling and business dynamics. Wiley, Chichester

Morrison B (2015) The problem with workarounds is that they work: the persistence of resource shortages. J Oper Manag 39:79–91

Morrison JB, Oliva R (2017) Integration of behavioral and operational elements through system dynamics. In: Donohue K, Katok E, Leider S (eds) handbook of behavioral operations. Wiley, New York (forthcoming)

Olaya C (2015) Cows, agency, and the significance of operational thinking. Syst Dyn Rev 31:183–219

Oliva R, Sterman JD (2001) Cutting corners and working overtime: quality erosion in the service industry. Manag Sci 47(7):894–914

Parvan K, Rahmandad H, Haghani A (2015) Inter-phase feedbacks in construction projects. J Oper Manag 39:48–62

Peng M, Peng Y, Chen H (2014) Post-seismic supply chain risk management: a system dynamics disruption analysis approach for inventory and logistics planning. Comput Oper Res 42:14–24

Pidd M (2004) Computer simulation in management science, 5th edn. Wiley, Chichester

Rahmandad H, Oliva R, Osgood ND (eds) (2015) Analytical methods for dynamic modelers. MIT Press, Cambridge

Randers J, Göluke U (2007) Forecasting turing points in shipping freight rates: lessons from 30 years of practical effort. Syst Dyn Rev 23(2/3):253–284

Reason J (1990) Human error. Cambridge University Press, Cambridge

Repenning NP (2002) A simulation-based approach to understanding the dynamics of innovation implementation. Organ Sci 13(2):109–127

Repenning NP, Sterman JD (2001) Nobody ever gets credit for fixing problems that never happened – creating and sustaining process improvement. Calif Manag Rev 43(4):64–88

Richardson GP (1991) Feedback thought in social science and systems theory. Pegasus, Waltham

Richmond B (1993) Systems thinking: critical thinking skills for the 1990s and beyond. Syst Dyn Rev 9(2):113–133

Rungtusanatham MJ, Choi TY, Hollingworth DG, Wu Z, Forza C (2003) Survey research in operations management: historical analyses. J Oper Manag 21(4):475–488

Ryzhenkov AV (2000) Unfolding the eco-wave: why renewal is pivotal. Wiley, Chichester

Salge M, Milling PM (2006) Who is to blame, the operator or the designer? Two stages of human failure in the chernobyl accident. Syst Dyn Rev 22(2):89–112

Schweitzer ME, Cachon GP (2000) Decision bias in the newsvendor problem with a known demand distribution: experimental evidence. Manag Sci 46(3):404–420

Scudder GD, Hill CA (1998) A review and classification of empirical research in operations management. J Oper Manag 16(1):91–101

Slack N, Lewis M (2015) Operations strategy, 4th edn. Pearson, Harlow

Slack N, Chambers S, Johnston R (2010) Operations management, 6th edn. Pearson, Harlow

Smets LP, Oorschot KE, Langerak F (2013) Don't trust trust: a dynamic approach to controlling supplier involvement in new product development. J Prod Innov Manag 30(6):1145–1158

Snabe B, Größler A (2006) System dynamics modelling for strategy implementation – concept and case study. Syst Res Behav Sci 23(4):467–481

Spengler T, Schröter M (2003) Strategic management of spare parts in closed-loop supply chains – a system dynamics approach. Interfaces 33(6):7–17

Sterman JD (1985) A behavioral model of the economic long wave. J Econ Behav Organ 6(1):17–53

Sterman JD (1988) A skeptic's guide to computer modeling. In: Grant L (ed) Foresight and national decisions. University Press of America, Lanham, pp 133–169

Sterman JD (1989) Misperceptions of feedback in dynamic decision making. Organ Behav Hum Decis Process 43(3):301–335

Sterman JD (1994) Learning in and about complex systems. Syst Dyn Rev 10(2/3):291–330

Sterman JD (2000) Business dynamics – system thinking and modeling in a complex world. McGraw-Hill, Boston

Sterman JD, Dogan G (2015) "I'm not hoarding, I'm just stocking up before the hoarders get here.": behavioral causes of phantom ordering in supply chains. J Oper Manag 39:6–22

Sterman JD, Kofman F, Repenning NP (1997) Unanticipated side effects of successful quality programs: exploring the paradox of organizational improvement. Manag Sci 43(4):503–520

Sterman JD, Oliva R, Linderman K, Bendoly E (2015) Editorial: system dynamics perspectives and modeling opportunities for research in operations management. J Oper Manag 39/40:1–5

Strohhecker J, Größler A (2013) Do personal traits influence inventory management performance? – the case of intelligence, personality, interest and knowledge. Int J Prod Econ 142(1):37–50

Swamidass PM (1991) Empirical science: new frontier in operations management research. Acad Manag Rev 16(4):793–814

Syntetos AA, Georgantzas NC, Boylan JE, Dangerfield BC (2011) Judgement and supply chain dynamics. J Oper Res Soc 62(6):1138–1158

Tan B, Anderson EG, Dyer JS, Parker GG (2010) Evaluating system dynamics models of risky projects using decision trees: alternative energy projects as an illustrative example. Syst Dyn Rev 26(1):1–17

Thun J-H (2006) Maintaining preventive maintenance and maintenance prevention: analysing the dynamic implications of total productive maintenance. Syst Dyn Rev 22(2):163–179

Towill DR (1982) Dynamic analysis of an inventory and order based production control system. Int J Prod Res 20(6):671–687

Towill DR (1996) Industrial dynamics modelling of supply chains. Logis Inform Syst 9(4):43–56

Vennix JAM (1996) Group model building – facilitating team learning using system dynamics. Wiley, Chichester

Warren K (2002) Competitive strategy dynamics. Wiley, Chichester

Warren K (2008) Strategic management dynamics. Wiley, Chichester

Weinhardt JM, Hendijani R, Harman JL, Steel P, Gonzalez C (2015) How analytic reasoning style and global thinking relate to understanding stocks and flows. J Oper Manag 39:23–30

Wikner J, Naim M, Towill DR (1991) Smoothing supply chain dynamics. Int J Prod Econ 22(3):231–249

Wu DY, Katok E (2006) Learning, communication, and the bullwhip effect. J Oper Manag 24(6):839–850

Wunderlich P, Größler A, Zimmermann N, Vennix JAM (2014) Managerial influence on the diffusion of innovations within intra-organizational networks. Syst Dyn Rev 30(3):161–185

Yin RK (1994) Case study research: design and methods, 2nd edn. Sage, Thousand Oaks

Zahn EOK, Bunz A, Hopfmann L (1987) System dynamics models: a tool for strategic planning of flexible assembly systems. Syst Dyn Rev 3(2):150–155

Zahn EOK, Dillerup R, Schmid U (1998) Strategic evaluation of flexible assembly systems on the basis of hard and soft decision criteria. Syst Dyn Rev 14(4):263–284

System Dynamics Applied to Project Management: A Survey, Assessment, and Directions for Future Research

David N. Ford[1] and James M. Lyneis[2]
[1]Zachry Department of Civil Engineering, Texas A&M University, College Station, TX, USA
[2]System Design and Management Program, Massachusetts Institute of Technology, Cambridge, MA, USA

Article Outline

Glossary
Context
Structures Underlying Project Dynamics
Common Project Behaviors
Discussion and Conclusions
Bibliography

Glossary

Knock-on effects Secondary and tertiary relationships and impacts of processes and human reactions to project conditions

Project A series of activities or tasks that have a specific objective to be completed within certain requirements; have defined start dates, end dates, and funding limits; and consume resources

Rework cycle A structure in which a fraction of the work performed is discovered to require additional effort for completion

Ripple effects Unintended side effects of project control on rework and productivity that generate policy resistance

Tipping point A structure and conditions that can create qualitatively different behavior modes that move the system toward distinct attractors

Context

Projects abound in industry, public service, and many other endeavors. As a series of activities or tasks that (1) have a specific objective (scope) to be completed within certain specifications (requirements), (2) have defined start and end dates, (3) have funding limits, and (4) consume and/or utilize resources (Project Management Institute 2000), projects have proven challenging to plan and manage. This is largely because project conditions and performance evolve over time as a result of feedback responses, many involving non-linear relationships, and to accumulations of project progress and resources. This has made the application of system dynamics to project management a fertile and productive field of study. This entry surveys the large body of system dynamics work on projects, evaluates its progress, and suggests directions for future development.

Many different types of models have been developed to improve project management. These models include some of the system features and characteristics addressed by system dynamics. For example, basic project models such as the critical path method explicitly model causally linked development activities and phases, and cost control models use forecasted performance gaps (e.g., budget deficits) to allocate funds. More advanced models, such as the computational models developed by Levitt et al. (1999) and others, are quite system dynamics-like, as they include linked development activities as well as feedback. Another body of work models multiple projects, using system dynamics as well as other approaches. Surveying all of these works is beyond the scope of a single article. Therefore, we focus here on models of single projects built using the system dynamics methodology as described in the domain literature. But even models of single

An earlier version of this paper appeared with the same title in System Dynamics Review, Vol. 23, No. 2/3, (Summer/Fall 2007): 157–189

Originally published in
R. A. Meyers (ed.), *Encyclopedia of Complexity and Systems Science*, © Springer Science+Business Media LLC 2019, https://doi.org/10.1007/978-3-642-27737-5_658-1

projects are too numerous to describe structures or applications in detail. Therefore, in this entry we focus on the most important and general model structures in conceptual form and provide references to additional details. Our work is based primarily on the published literature in journals and books, and our experience using system dynamics to model projects. In particular, including our experience provides a significant contribution beyond the extant literature, with reference to the results of some system dynamics project work that has not, and very likely never will be, published and otherwise widely available.

The literature on system dynamics models of projects varies widely in the level of detail provided, especially in model structure descriptions, from complete model disclosure to almost none. Some authors focus on model structure, while others focus on model use and describe model structure only in general terms (e.g., the "Strathclyde" work of Ackermann et al. 1997; Eden et al. 1998, 2000). Our assessments are necessarily limited when model equations or detailed structure information is not available. However, our review reveals a direct and positive relationship between the access provided by authors to model details and the subsequent use of those models by other researchers and practitioners.

The remainder of the entry is structured as follows. Important conceptual model structures are described in a way that relates them to system dynamics principles and in the approximate chronological order of development. This provides a brief history as well as a meta-structure of system dynamics project-model structures. Model structures are followed by some typical project behaviors they produce. The entry then discusses applications, policy lessons, and future research directions organized by traditional areas of project management and finishes with a general assessment of the work to date and suggestions for future development.

Structures Underlying Project Dynamics

The basic structures that system dynamicists have used to model projects can be described in four groups based on the central system dynamics

concept that they integrate into project models. The categorization provides a meta-structure of project model structures and relates those structures to the system dynamics methodology. The four model structure groups are:

1. **Project features**: System dynamics focuses on modeling features found in actual systems. In projects these include development processes, resources, managerial mental models, and decision making. One version of the history of system dynamics project models could describe the incorporation of important project features in models. Modeling important components of actual projects increases the ability to simulate realistic project dynamics and relate directly to the experiences of practicing managers.

2. **A rework cycle**: System dynamics has a set of canonical structures that drive much of the dynamics of specific model types. The inventory-WIP structure in supply lines (Sterman 2000) and the aging structure in Urban Dynamics (Forrester 1969) are examples. As described below, the canonical structure of system dynamics project models is the rework cycle.

3. **Project control**: Modeling, analyzing, and improving the control of dynamic systems is the objective of applying system dynamics in many domains. Since controlling projects is the goal of project management, modeling the controlling feedback loops through which management attempts to close gaps between project performance and targets directly applies one foundation of system dynamics to project management.

4. **Ripple and Knock-on effects**: Policy resistance and unintended consequences are fundamental explanations used by system dynamics for many adverse behaviors. "Ripple Effects" is the name commonly used in projects to describe the primary side effects of well-intentioned project control efforts. Modeling ripple effects in projects captures and leverages the concept of policy resistance. "Knock-on effects" refer to the secondary impacts of project control efforts, i.e., the impacts of ripple effects, often caused by

processes that produce excessive or detrimental concurrence or human factors that amplify the negative effects via channels such as morale. Capturing knock-on effects in project models uses the concept of unintended side effects to explain project behavior and performance.

Project Features

System dynamics models of projects almost always consist of a collection of tasks that are performed in parallel and in series. Therefore, a principal feature of all system dynamics project models is the use of development tasks or work packages as the primary material flowing through a project. Development tasks typically start in a stock of tasks *to be done* and are processed through the project's development process until the stock of tasks *done* reaches the level of project completion. In a model of a specific project, tasks in the development process may represent the entire project or may be disaggregated into more detailed development phases (e.g., into developing specifications or coding software). Another feature of projects represented in system dynamics models is the application of resources to manage the flows in the development process, based on management's perceptions of project conditions.

Roberts (1964, 1974) developed the first published model of a project and introduced the flows of project work in terms of "job units" based on resources applied and productivity. In addition, he introduced several important concepts that represent management's understanding of project conditions: (1) perception gaps – differences between perceived progress and real progress, and between perceived productivity and real productivity; and (2) underestimating scope and effort required. These errors can cause under- or misallocation of resources that ultimately feed back to affect project performance.

Robert's work was followed by a succession of modelers that improved the richness of project models by adding other features found in actual projects, including both development processes and management. Improved representations of development processes included, but are not limited to: distinguishing work done correctly from work done incorrectly (first by Pugh-Roberts Associates (PRA) (Pugh-Roberts Associates is now a part of the PA Consulting Group), Cooper 1980 and Richardson and Pugh 1981); multiple project phases (first by PRA, Cooper 1980); separate effort for quality assurance first by Abdel-Hamid 1984); nonlinear constraints of work availability on progress (first by Homer et al. 1993); development projects as value-adding aging chains (first by Ford and Sterman 1998a); and concurrence constraints limiting how much work can be done in parallel (Ford and Sterman 1998a; Madachy 2002).

Simultaneously, modelers were improving project models by adding features that reflect the human aspects of projects, especially project management features and processes such as the "freezing" and "unfreezing" of designs due to changes and uncertainties (Strathclyde (The Strathclyde models include Ackermann et al. (1997), Eden et al. (1998), (2000). Henceforth this group's work will be referred to as Strathclyde)), releasing completed work to downstream phases (Ford 1995), using contingency funds (Ford 2002) and schedule buffers (Park and Pena-Mora 2004), and resource allocation policies (Joglekar and Ford 2005). These features clearly exploit the power of system dynamics to model human decision-making, such as modeling decisions driven by gaps, delays in human processes, and nonlinear relationships. Most formulations of these features apply traditional system dynamics structures (e.g., supply provided proportional to demand, but delayed) that are described in other system dynamics literature.

The Rework Cycle

The rework cycle is, in our opinion, the most important single feature of system dynamics project models. The rework cycle's recursive nature in which rework generates more rework that generates more rework, etc., creates problematic behaviors that often stretch out over most of a project's duration and are the source of many project management challenges. PRA developed the first rework cycle model, shown conceptually in Fig. 1 (Cooper 1980, 1993) (Versions of the PRA rework cycle with equations are given in Richardson and Pugh (1981) and Lyneis (2003)). In this form the rework cycle includes four stocks

**System Dynamics
Applied to Project
Management: A Survey,
Assessment, and
Directions for Future
Research, Fig. 1** The
rework cycle. (Adapted
from Cooper 1993)

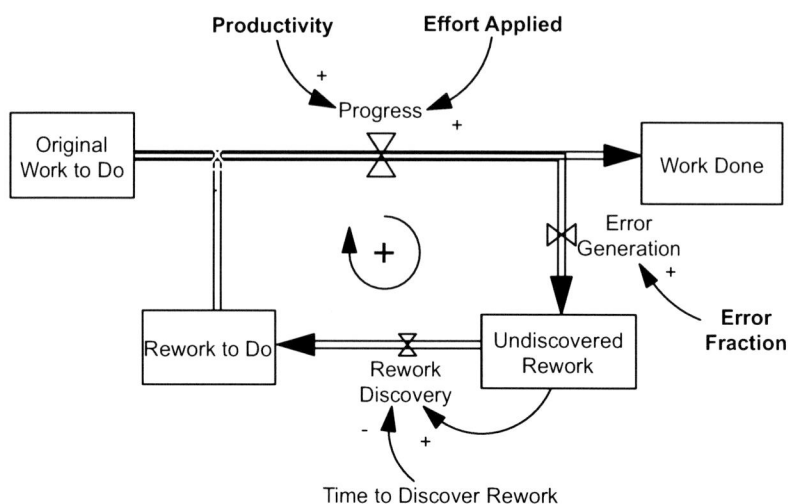

of work. At the start of a project or project stage, all work resides in the stock "Original Work to Do." Progress is made by applying effort that moves work between stocks. A fraction of the work being done at any point in time contains errors ("Error Generation"). Work done correctly (Progress less Error Generation) enters the "Work Done" stock and never needs rework (unless later changes obsolete that work). However, work containing errors enters the "Undiscovered Rework" stock. Errors are not immediately recognized, but are detected as a result of doing downstream work or testing. This "Rework Discovery" may occur months or even years after the rework was created. Once discovered, the backlog of "Rework to Do" demands the application of additional effort. Reworking an item can generate or reveal more rework that must be done. Therefore, some re-worked items flow through the rework cycle one or more subsequent times, generating the rework cycle's recursive effects. The rework cycle forms a reinforcing feedback loop in which a fraction of the completed Rework to Do reenters the rework cycle each iteration. Except when all work is discovered to require rework, the rework cycle "leaks" completed work done correctly, creating progress toward the ultimate completion of the project.

Subsequent modelers have developed other rework cycles, principally Abdel-Hamid (1984) and Ford and Sterman (1998a, 2003b). They retain

the rework cycle's recursive nature, but add other features or use other model structures. For example, Ford and Sterman's aging chain structure moves work through a series of backlogs and improvement activities that initially complete, then test, and then release work with the rework cycle linked to the aging chain at the Quality Assurance backlog. This structure uses a separate quality assurance effort and adds parallel rework cycles in co-flow structures to distinguish between errors that are generated within a phase and those generated by upstream phases which contaminate downstream work. Other authors, such as Park and Pena-Mora (2003), elaborate on the work flows and distinguish between rework to correct flawed work (e.g., removing and replacing poor construction) and externally generated changes.

The importance of the rework cycle is indicated by the fact that all known system dynamics project models subsequent to PRA's original work have included a rework cycle in some form. The details of some recent models vary from the original structures in form (e.g., Van Oorschot et al. (2010) and Akkermans and van Oorschot (2016)), but maintain the basic rework cycle and feedback structures.

One significant extension to the basic dynamic structural theory of projects has been developed in the last decade. Rahmandad and Hu (2010) elaborate on the basic rework cycle in two areas: (1) allowing for multiple defects per task via a co-flow structure and (2) representing different characteristics of the testing process in terms of the action of defects per task discovered (rather than being the average in a typical co-flow structure). Historical project model formulations conceptualize tasks as atomic work packages that are correct or flawed (with a maximum of one error per task at any point in time). As a result, such models are unable to adequately describe (and match the system data) if the tasks and errors reflect large units of work. Therefore, the ability to capture multiple defects per task can be useful when a task, the unit of work in the model, is defined at a high level (e.g., function points or modules). In this case, it is highly likely that there will be multiple defects or errors in a given task. The authors also suggest that the nature of the testing process causes rework discovery on tasks with multiple defects to be biased toward the end of the project, which is an alternative explanation for a "90% syndrome" behavior mode (see further below).

Controlling Feedback

In modeling controlling feedback, system dynamicists have focused on the information processing of project managers. Project performance is typically measured in terms of schedule, cost, quality, and scope. Management actions to control a project's performance are modeled as efforts to close the target-performance gap in one or more performance dimensions. The two basic methods available to practicing project managers have been modeled: move project behavior closer to targets (e.g., work overtime), or move targets toward project behavior (e.g., slip a deadline). Both methods use negative (controlling) feedback loops, with managerial responses typically being proportional to gap size. However, limits often exist on the amounts and speeds of adjustments and both methods impose costs (monetary and other types). Project targets are often set for future dates (e.g., cost when a phase of work or the entire project is completed), and therefore modelers have often included managerial forecasting of target-performance gaps at milestones as the basis for decision-making. This has facilitated

work on the roles of forecasting and bias on the effectiveness of controlling feedback and on project performance. As noted above, system dynamicists have consistently modeled perceived conditions separately from actual conditions, with the former driving project control actions and the latter driving actual progress. In several models, the structures used to model perceived conditions reflect managerial mental models and are not just delayed actual conditions. For example, managers generally include undiscovered rework in work believed to be completed and therefore overestimate progress. This, combined with reporting systems that often estimate productivity based on work believed to be done to date divided by hours spent to date (rather than a delay of current computed productivity), can overestimate progress early in the project and underestimate it later (e.g., Ford 1995; Lyneis et al. 2001). As will be discussed, these generate adverse feedback effects in the form of ripple effects.

Controlling to meet a deadline is common in project management practice and has been a particular focus of many system dynamics models of projects. We therefore use it here to illustrate a model of managerial action for project control. Three common actions can be taken to correct a forecasted missing of a deadline: (1) hire additional workforce (most project models starting with Roberts, 1994), (2) work overtime (PRA, Ford and Sterman, Van Oorschot, et al., and the Strathclyde models), and (3) work faster (PRA, Abdel-Hamid, and Strathclyde models). As indicated in Fig. 2, these form the "Add People," "Work More," and "Work Faster/Slack Off" feedback loops. In these loops, an expected completion delay, as indicated by more time required to finish the work remaining than the time remaining to the project deadline, initiates hiring, overtime (which increases effort via more hours per worker), higher intensity of work (which increases productivity – e.g., output per person-hour), or a combination. In isolation these actions increase progress, reduce the remaining work, and thereby reduce the expected completion delay. Note, however, that if the expected completion delay is zero or negative (i.e., the work remaining seems doable in less than the time remaining),

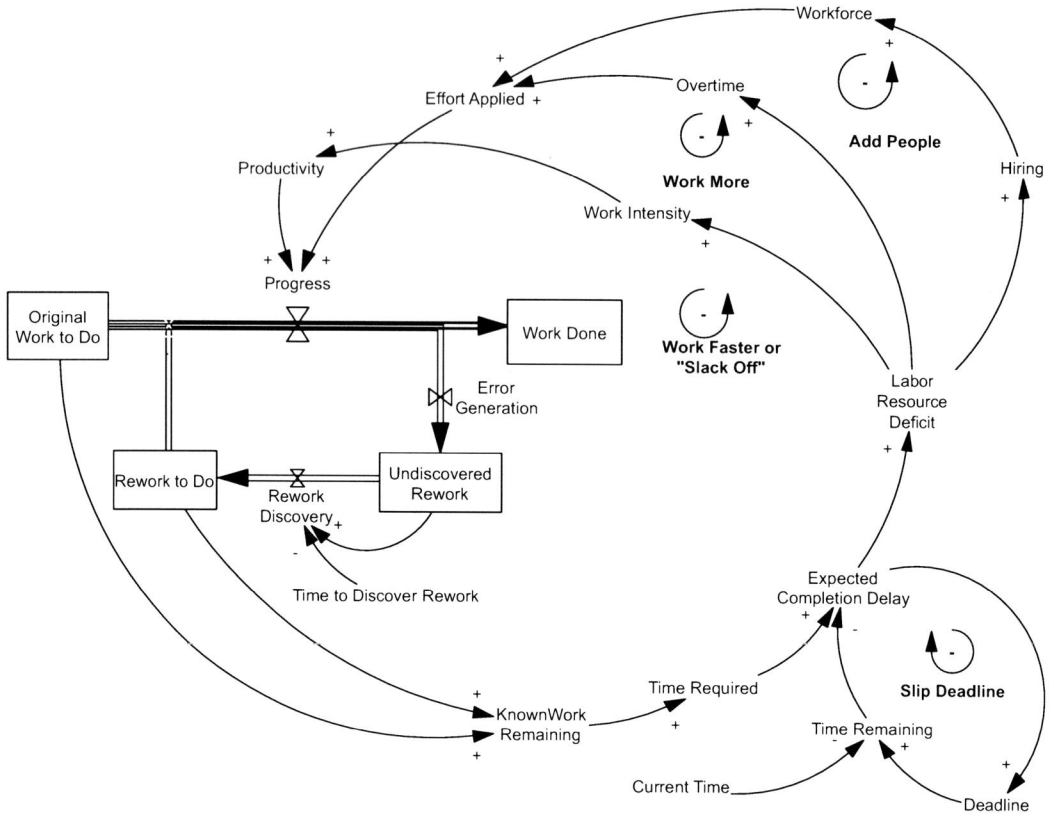

System Dynamics Applied to Project Management: A Survey, Assessment, and Directions for Future Research, Fig. 2 Controlling feedback loops for achieving a target schedule (deadline)

work intensity is often reduced, thereby creating the "Slacking Off" variation on the "Work Faster" loop – work intensity and productivity decrease and work remaining does not fall as fast as originally planned. Another variation on this "Slacking Off" loop, not shown for clarity, is a "gold-plating" loop whereby slack in the project leads designers to add "unnecessary" features and capabilities, thereby eliminating the slack. Another possible action, slipping the deadline, is indicated by the negative loop in the lower right of Fig. 2. Deadline slip is often taken only as a last resort when the adding resource loops fail to completely solve the problem.

At the project level, the relative flexibilities, effectiveness, and costs of moving forecasted performance toward targets and moving targets toward forecasted performance vary across projects, their environments, and their stakeholders. All three characteristics are important to

practitioners. For example, an effective control action that is unaffordable or cannot be changed to the effective degree is not useful to practitioners. Therefore, the traditional system dynamics goal of identifying high leverage (i.e., most effective) parameters may be inadequate to make practical recommendations. At the project control level, similar types of variations exist among using changes in workforce size, overtime, and working faster to manage schedule performance and other actions to manage other performance dimensions. Future work can start with existing model structures to integrate all of the most important project control characteristics into recommendations for practitioners.

Ripple Effects

Unfortunately, actions taken to close a gap between project performance and targets have unintended side effects that generate policy resistance. These ripple effects are the primary impacts of project control on rework and productivity. Figure 3 adds four important ripple effect feedbacks of the three project control actions shown in Fig. 2. These effects typically reduce productivity or quality (by increasing the error fraction and rework). Hiring can dilute experience as workers with less skill and/or less familiarity with the project are brought on, and because they require experienced developers to divert time to training instead of doing development (most models since Roberts that include adding staff to a project). Larger workforces can increase congestion and communication difficulties, which increase errors and decrease productivity (PRA, Abdel-Hamid, and Ford-Sterman models). Overtime leads to fatigue (after a delay)

that also increases errors and decreases productivity (all models that include overtime). Higher work intensity increases errors (PRA, Abdel-Hamid, and Strathclyde models). Reduced productivity and increased rework keeps the amount of work remaining greater than it would have otherwise been, thereby increasing labor resources needed to finish on time. These effects form the Experience Dilution, Too Big to Manage, Burnout, and Haste Makes Waste loops. Consistent with system dynamics theory, they are reinforcing loops which can cause a project to spin out of control. While these ripple effect feedbacks are characteristic of many of the early project models, they were usually not clearly diagrammed or highlighted by authors. Some of these loops appear to be explicitly

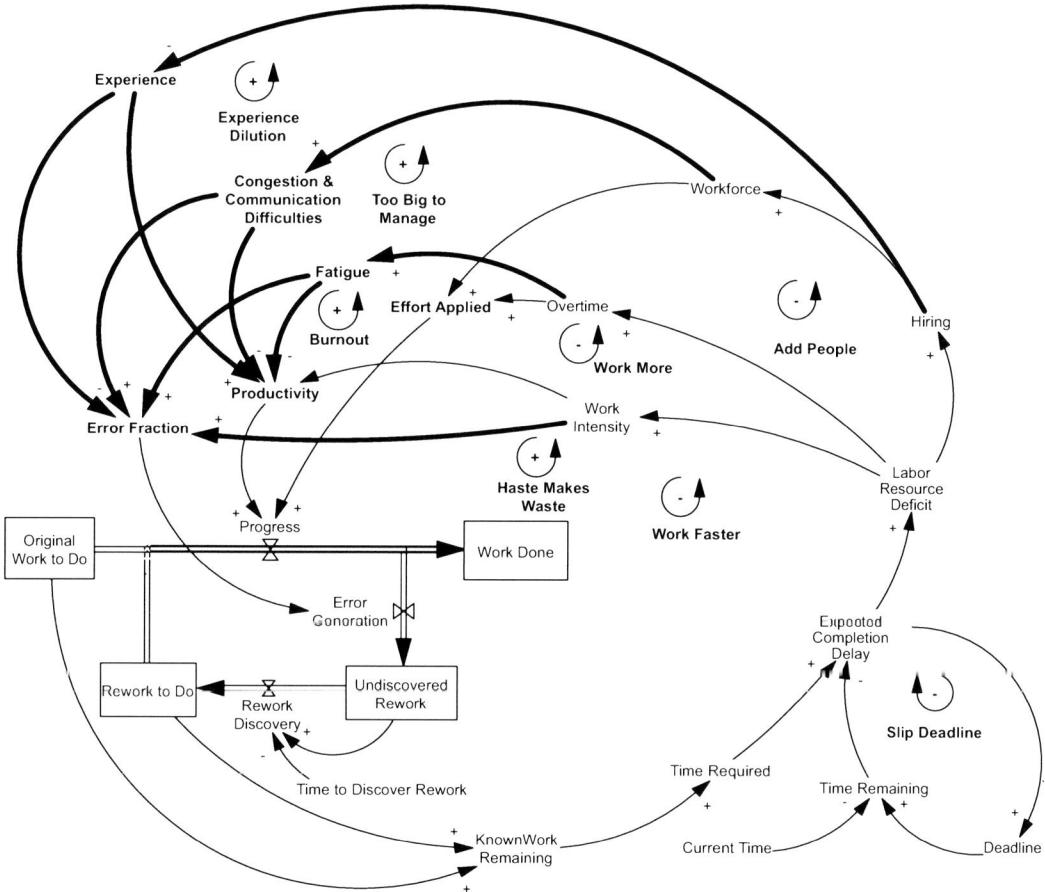

System Dynamics Applied to Project Management: A Survey, Assessment, and Directions for Future Research, Fig. 3 Policy resistance via ripple effects of rework and controlling feedback to improve schedule performance

diagrammed for the first time in Sengupta and Abdel-Hamid (1993) and Cooper (1994).

Knock-on Effects

Ripple effects generate secondary and tertiary feedbacks – some are consequences of physical processes related to work flow through projects that propagate from upstream work to downstream work, both within a phase of work (e.g., design), and between phases of work (e.g., from design to construction), while others are due to "human" reactions to project conditions. Many of these effects are generated by the activation of ripple effects structures described above. Figure 4 builds from Fig. 3 to illustrate that these "knock-on" relationships can generate significant harmful dynamics, including:

- "Haste Creates Out-of-sequence work" – trying to accomplish more tasks in parallel than physical or information constraints allow,

whether by adding resources or exerting schedule pressure, can cause work to be done concurrently, out of the desired sequence, or both. This reduces productivity and increases errors (PRA and Ford -Sterman models).

- "Errors Build Errors" – undiscovered errors in upstream work products (e.g., design packages) that are inherited by downstream project phases (e.g., construction) reduce the quality of downstream work as these undiscovered problems, are built into downstream work products. Coded software is a good example of this contamination effect (PRA, Abdel-Hamid, Ford-Sterman, Strathclyde models, and Akkermans - van Oorschot models).

- "Errors Create More Work" – the process of correcting errors can increase the number of tasks that need to be done in order to fix the problem or can increase the work required because fixing the errors takes more effort than doing the original work. Taylor and Ford

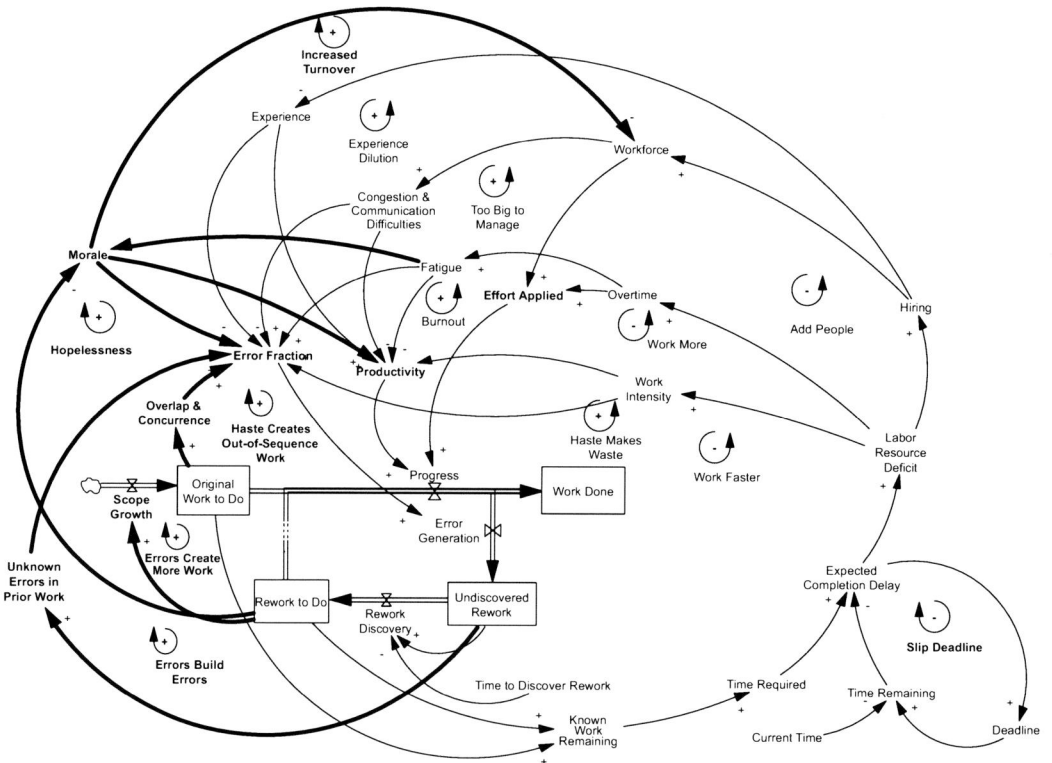

System Dynamics Applied to Project Management: A Survey, Assessment, and Directions for Future Research, Fig. 4 Policy resistance via "knock-on" effects to controlling feedback to improve schedule performance

(2006) demonstrate that this feedback can create "tipping point" dynamics through which fraction complete can stop increasing and begin to decline, often resulting in project cancellation.

- "Hopelessness" – Morale problems can exacerbate the effects – fatigue and rework can create a sense of "hopelessness" that increases errors and reduces productivity, and which also increases turnover (PRA and Strathclyde models).

Finally, while the primary adverse ripple and knock-on feedbacks as typically modeled by system dynamicists are internal to the project (often including suppliers and subcontractors), adverse feedbacks through clients and customers can initiate or amplify internal project dynamics (Rodrigues and Williams 1998; Reichelt 1990; McKenna 2005). Examples of these external actions include:

- Clients often change scope or requirements, activating project control actions, ripple effects, and knock-on effects, thereby degrading projects that were otherwise successful.
- Projects which are under-budgeted can lead to efforts by the contractor to increase the budget via change orders, which divert efforts from other project work.
- Poor schedule performance and slipping of deadlines can reduce client trust in the project team, with the resultant demands for more progress reports; more time spent on progress reporting and interacting with the client reduces productivity, slows progress, and necessitates additional schedule slip through a reinforcing loop.
- Reduced client trust can also lead to reluctance by the client to tolerate further deadline slippage, which increases schedule pressure and aggravates project control problems.
- In the extreme, if project problems lead to litigation (while the project is still on-going), then diversion of management attention to litigation activities can reduce attention to the project itself, and thereby exacerbate project performance.

These "external" feedbacks are sometimes included in project models.

Assessment of System Dynamics Project Model Structures and Research Needs

The structural theory of project dynamics described above is, in our view, relatively complete. Based on our project management experience and modeling, we believe that the development of project models has now captured the majority of the important features of development projects: the characteristics of the rework cycle, controlling feedback loops, and ripple and knock-on effect re-enforcing loops. While specific models differ in their level of detail with regard to the phases of work represented, the complexity of the rework cycle, and the feedback effects that are represented (few, if any, models capture all feedback effects), formulations of these processes have been developed and are documented for others to use.

There are, however, two areas of structural research which we feel warrant additional work:

1. Nearly all the ripple and knock-on effect feedbacks manifest themselves through nonlinear relationships (as represented by graphical functions). There is relatively little discussion of the nature and strength of these relationships, and in particular, how they might differ by phase of work (e.g., design vs. construction) or by type of project (e.g., software vs. hardware), or as a result of changes in process and tools (e.g., CAD systems might reduce the strength of errors on errors feedback, and make error fraction less sensitive to people factors), and how different strengths may alter any policy heuristics.

Ford and Sterman (1998b) provide one approach for estimating relationships, and examples of how they differ by phase of work. Praven et al. (2015) conduct a postmortem study of 30 actual construction projects. They verify and quantify several important feedback connections (design quality on construction quality and productivity, construction progress on design rework discovery). They find that error rates are typically higher in the design phase than in the construction phase, and that it takes longer to discover design errors than construction errors. They then quantify the impact of undiscovered design errors (design quality) on construction error rates and construction productivity and of construction progress on design rework discovery. Jalili and Ford (2016), in a theoretical analysis, quantify the size of the impacts of three reinforcing feedback loops on project duration and thereby project performance. The results show that, even without other structures or exogenous changes, failing to account for even a few common reinforcing loops can cause enormous delays. These studies are an important beginning, but much more work is needed.

2. While nearly all system dynamics project models represent aspects of the above secondary impacts of project controls to achieve project performance targets, the secondary consequences of adjusting targets have not been as well studied. While some modelers have represented slipping schedule as well as adding resources, and sometimes compute a value for the damages of late delivery, they rarely explicitly examine the secondary impacts of such slips on performance of the product in the market.

An example of this type of investigation is Rahmandad and Weiss (2009), which discussed how delivering projects with "bugs" can create demands for maintenance which take staff time away from the current

development effort. If resources are short, this can cause the current project to be shipped with "bugs" and potentially lock the organization into a permanent situation of always shortchanging current projects and shipping low-quality products. While not strictly the market dynamics of a single project, it does highlight an adverse feedback effect of failure to achieve product quality targets before delivery. While representing the market feedbacks from an individual project may not be useful (as opposed to the feedbacks from a portfolio of projects), much more work is needed in this area.

Common Project Behaviors

The most common behavior mode of actual projects cited in both the project and the system dynamics literature describe failures to meet performance targets (e.g., see Lyneis et al. 2001). System dynamicists have used project structures described above to explain these failures and suggest improvements. Figure 5 illustrates typical (but by no means all) possible behaviors for project staffing: *planned* staffing, or workforce, often builds up to a peak, and then gradually declines; *actual* staffing, however, can deviate significantly from the plan. Often the ramp-up of staff is delayed, then staff levels overshoot the planned peak and extend longer (sometimes with a second hump). The actual staffing patterns produce both a schedule and budget overrun. Many of the papers

System Dynamics Applied to Project Management: A Survey, Assessment, and Directions for Future Research, Fig. 5 Some Rework cycle and productivity/quality effects on project staffing dynamics. (Adapted from Lyneis et al. 2001)

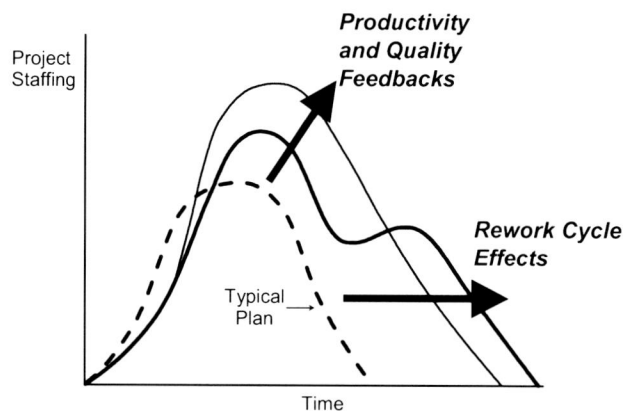

cited give examples of this behavior with data from actual projects.

How do the above feedback structures contribute to this behavior? The ramp-up of staff is often delayed either because of external conditions such as late finish of other projects, but also because of management's underestimates of true project scope, delays in getting resources, and overestimates of productivity. The "hiding" of undiscovered rework in the rework cycle further delays the recognition of staffing needs. As the need for additional staff is recognized and controlling actions taken, ripple effects and knock-on effects on productivity and quality of managerial responses tend to increase effort required and therefore staff needs, pushing project staffing (and labor through overtime) up, above the planned peak and to the right (later in time). Cycling of work through the rework cycle also pushes project completion later in time.

The evolution of the fraction of work completed is also often used to describe progress on actual projects. As illustrated in Fig. 6, typically a period of relatively steady apparent progress is followed by a period of slow and decelerating progress before completion. This behavior results initially because of underestimates of true work scope and the hiding of undiscovered rework, and later as managerial responses initiate ripple and knock-on feedbacks which slow progress by reducing productivity and increasing errors. In the end, progress is constrained by the discovery and cycling of the last bits of rework through the rework cycle (often called the "90% syndrome"). Ford and Sterman (2003a) examine this syndrome; Rahmandad and Hu (2010) provide an alternative explanation involving tasks with multiple defects per task, and a testing process which tends to drag out the discovery of these multiple defects rather than discovering them all at once. Some researchers have documented projects that also experienced a slower initial start-up period that formed an elongated "S" behavior mode for progress (Reichelt and Lyneis 1999; Ford 1995; Ford and Sterman 2003b). This basic behavior mode is sometimes augmented by periods of little or no net progress (i.e., project is "stalled," often because of the recognition of undiscovered rework and actions to execute that work in a timely fashion), or by temporary (Ford 1995) or permanent (Taylor and Ford 2006) declines in net progress (i.e., a "decaying" project) because of added work to execute rework.

The decline in progress during a project evolution noted in Fig. 6 (dashed line on right) highlights the emergent discussion of "tipping point" dynamics in projects. Tipping point structures create conditions in which the system's behavior mode can fundamentally shift to a qualitatively different attractor, in which external forces or managerial actions can push the system across its tipping point conditions and create very unexpected behavior and performance (see Jalili 2017, for an analysis of definitions of tipping points and a definition that distinguishes among structure, conditions, and behavior). A tipping point is

System Dynamics Applied to Project Management: A Survey, Assessment, and Directions for Future Research, Fig. 6 Some rework cycle and productivity/quality effects on Fraction Complete. (Adapted from Lyneis 2003)

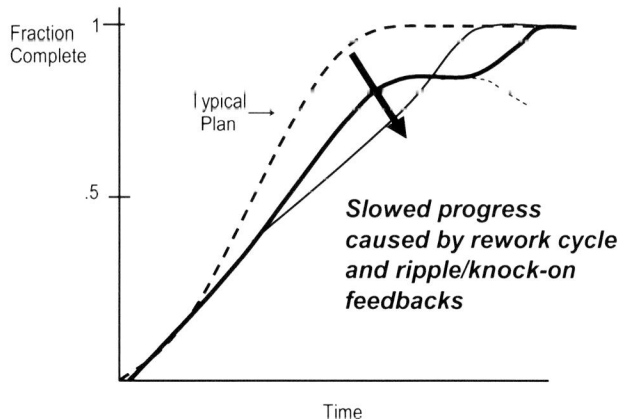

Fraction Complete

Typical Plan

Slowed progress caused by rework cycle and ripple/knock-on feedbacks

Time

most obvious on individual projects in the project's fraction of work complete. The normal and desired behavior mode is monotonically increasing from 0% at the beginning of the project to 100% complete ("Typical Plan" in Fig. 6 above). Although projects typically do get to completion, the path may not be smooth, potentially due to tipping points. Managers experience tipping point dynamics when progress stalls and begins to decline away from project completion. As an example, Taylor and Ford (2006) compared actual and simulated percent complete on the Watts Bar Unit 2 nuclear power plant construction project (Fig. 7) and modeled a tipping point to explain the behavior and how crossing the tipping point can be triggered by either exogenous changes or caused solely by endogenous feedback.

Repenning (2000) first used system dynamics to raise the issue of tipping points in projects, although he addressed the behavior of a series of projects rather than an individual project and did not emphasize the concept of a tipping point structure. Rahmandad and Weiss (2009) modeled a similar tipping point dynamic for software projects based on a new software development phase and client support and maintenance. Both

Repenning and Rahmandad described how tipping point conditions are created by a combination of inadequate resources and a focus on short-term fire-fighting in response, which worsens the resource shortage over time. They discussed strategies for increasing system robustness to these temporary shocks and strategies for recovery and reversing the tipping point behavior in sequences of projects.

Tipping point dynamics also apply to single projects. Taylor and Ford (Taylor and Ford 2006, 2008) and Jalili and Ford (2016) focused on three reinforcing feedback loops in Fig. 4 that can create tipping points on projects: (1) the rework cycle, (2) side effects feedbacks of project control (using work intensity as a proxy for all three control actions and their ripple effects), and (3) the "errors create more work loop" ripple effect loop. Taylor and Ford subjected projects to an external change (in this case a temporary increase in error rate) and analyzed under what conditions the project can recover and under what conditions the project crosses its tipping point into its declining fraction complete behavior mode and fails. Oorschot et al. (2010) also discussed tipping points on single projects. Their

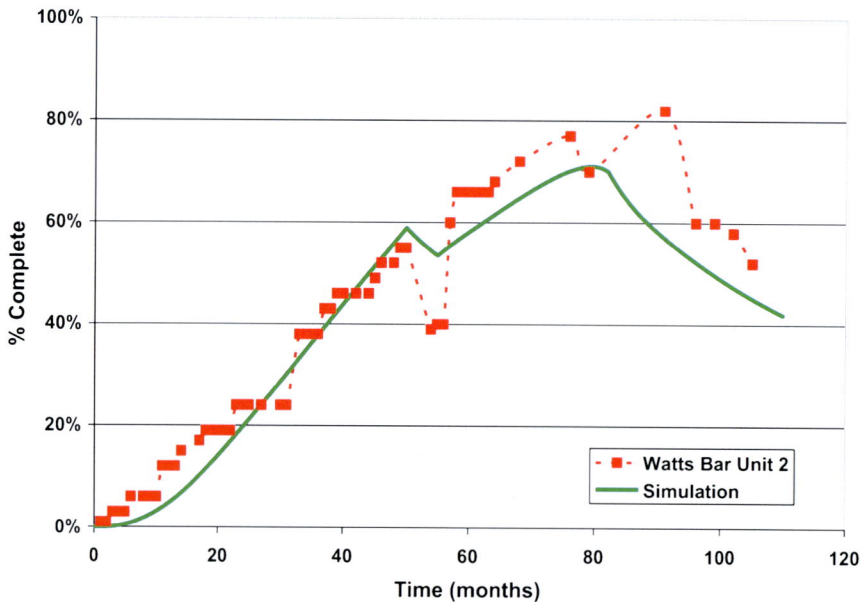

System Dynamics Applied to Project Management: A Survey, Assessment, and Directions for Future Research, Fig. 7 Actual and simulated project tipping point behavior (Taylor and Ford 2006, Fig. 3)

work is similar to Repenning and Rahmandad in that the trigger for crossing the tipping behavior is inadequate resourcing. Specifically, they represented a project as a multiphase effort with explicit stage gates at the end of each stage. They showed that under-resourcing the first phase and passing the project on at its scheduled finish date means that the second phase builds from work that is incomplete and/or contains errors. In this case a tipping point is crossed when the system allows the resource shortage and incompleteness/quality problems to grow from phase to phase. Tipping point dynamics are important to practicing project managers, and this research should be generalized along the lines of work by Jalili (2017).

In the next section, we discuss what system dynamicists have learned regarding how managers can improve these behavior patterns, and what further work should be done to improve the practice of project management.

Applications and Dynamic (Policy) Insights

Summary/Overview

The theory work of several of those cited above has spawned significant follow-on work by numerous other researchers, consultants, and companies. The number of real-world applications is particularly significant. By our count as of 2007, more than 50 companies have used system dynamics for project management on at least one project and some companies on many projects. PRA alone are known to have applied system dynamics to over 100 projects. Together with the efforts of other organizations, therefore, the total number of such applications most likely exceeds 200. Anecdotal evidence suggests that the number of applications to real-world projects continues to grow.

While these applications have been in a wide range of industries (aerospace, automotive, civil construction, energy, and software to name a few), none are known to require major deviations from the basic structures described above. The usefulness of these structures across so many applications makes them as fundamental to system dynamics as "classic" structures such as the supply chain described by Forrester (1961) and the

infection, commodity, and growth structures described in Sterman (2000). However, while the basic structures of projects across industries have many dynamic similarities, projects in different sectors and industries have some unique features. Hardware projects differ in some respects from software projects; consumer goods differ from defense projects; one-of-a-kind, first-of-a-kind projects differ from product development projects. In most cases, researchers and consultants have adapted one of the generic models to a specific application area; in other cases, notably software and civil construction, a stream of research and application has occurred such that more specific versions of the structures underlying project dynamics, and specific policy issues, have been developed for these industries.

Research and application in project dynamics has focused on understanding the drivers of cost and schedule overrun in particular situations and then on developing actions that either avoid or minimize the overruns, or on obtaining compensation for the additional costs. While there is some overlap, the research and applications address one of four general categories of project management: (1) Post-Mortem Assessments for Disputes and Learning; (2) Project Estimating and Risk Assessment; (3) Change Management, Risk Management, and Project Control; and (4) Management Training and Education. We have chosen to discuss the research and applications in this order because it reflects both the historical and the logical application of system dynamics to project management (e.g.,, in order to apply system dynamics to a new project within an organization, it is desirable to first model and do an assessment of a project that has already been completed). We briefly describe the general area of application and the role system dynamics models have played, then illustrate with specific applications that have been published, and conclude with policy insights and directions for future research in each area.

Postmortem Assessments for Disputes and Learning

Many applications of system dynamics involve a postproject assessment of what happened – how did the project deviate from the original plan, and why? The most numerous of these involve disputes between the owner/financer of the project and the contractor/executer of the project. For example, a dispute between Ingalls Shipbuilding (contractor) and the US Navy (owner) is described in Cooper (1980). But postmortem assessments also involve attempts to learn from one project to the next within an organization.

Disputes. Projects involving an owner and a contractor often engender disputes over interpretations of the specific requirements of the contract, changes requested by the owner, or external events such as strikes. In many cases, such disputes involve claims of "delay and disruption" – in our terminology, ripple, and knock-on effects – that might result from these specific problems. In these cases, when trying to understand why project performance differed from the plan, all changes from conditions assumed in the original plan must be specified, regardless of responsibility.

Client responsible changes often include: increasing project scope or altering the original design requirements, taking longer than specified in the contract to review and approve design drawings, to provide information about equipment provided by the client or another contractor, or to provide the actual equipment or test equipment. Contractor responsible changes might include failure to obtain resources in a timely fashion, perhaps as a result of delays in other projects. These changes and delays from the contracted scope of work are often referred to as the "direct impact" of client (or contractor) actions on the project.

These direct impacts often trigger controlling feedbacks and resultant "ripple and knock-on effects" caused by the vicious circle feedbacks described earlier – "delay and disruption" in the jargon of project management disputes. In its application to such disputes, the system dynamics model is used to quantify and explain the impact of these direct changes to the project on its final cost, including the ripple effects. The model can be set up to represent the project as it actually occurred, including the direct impacts, and calibrated to the actual performance of the project. Then client-responsible direct impacts are

"removed" and the model re-simulated to determine what "would-have" happened "but-for" the actions of the client. The difference between the two simulations is the full cost of the client actions, including ripple and knock-on effects. Because of its representation of the vicious circle feedbacks, system dynamics is ideally suited to determine the magnitude of these ripple effects, and explain their origins. It can also apportion costs to the client, to other parties, and to the contractor through simulations removing different groups of direct impacts.

A significant number of the "real-world" applications of project dynamics have been for delay and disruption disputes. As of 2005, PRA had done more than 45 such projects (Stephens et al. 2005) and more since then. All have been settled out of court on favorable terms to the contractor (the usual PRA client), with the typical award averaging 50% more than with traditional dispute resolution approaches, supporting the power of system dynamics to add value to these investigations. PRA's dispute work is described by Cooper (1980), Weil and Etherton (1990) and summarized by Sterman (2000). The Strathclyde Group have also successfully used system dynamics to support six delay and disruption claims ranging in value from US $50 m to $350 m, as cited in Howick and Eden (2001).

Project-to-Project Learning. Another important use of system dynamics modeling has been in post-project evaluation. In many ways, the process of postproject evaluation is similar to a delay and disruption analysis: the model is set up to represent what actually happened on the project, including the direct impact of any changes to the project from the original plan. These changes can include externally caused changes as noted above, as well as internally generated changes such as delays in obtaining staff or other resources, the implementation of new processes or procedures, and changes in management policy. Then the direct impacts of these changes are removed as inputs to the simulation, one at a time, to identify their contribution to any project overrun. In this way, project managers can learn which changes had the greatest impact on the project and thereby identify *risks* that might be planned for on future projects. In addition, to the extent that any new management initiatives were introduced on a project, project managers can test their impact on performance and thus decide if they should be implemented on other projects.

For example, Lyneis et al. (2001) and Cooper et al. (2002) described the use of a model by PRA to assess the lessons learned from a comparison three command and control system projects Hughes Aircraft Company. The effort identified the major external and internal drivers of differences in project costs and thereby identified management initiatives to be adopted on future projects. Abdel-Hamid and Madnick (1991) applied their software project dynamics model to five organizations during model development and five others after model completion (two by the authors and three by others). These assessments were used primarily to determine what happened on the projects and what would-have happened had different estimating methods been used, or other staffing/schedule decisions been taken. Van Oorschot et al. (2013) documented, via a longitudinal process study involving extensive co-incident interviews and data collection, an assessment of what leads to cancelation of projects and how cancellation might have been avoided (understaffing early stages of work and resultant failure to satisfy stage-gate criteria). This paper supports the modeling work in the Oorschot et al. (2010). As noted above, Praven et al. (2015) conducted a postmortem assessment of 15 projects focusing on the impact of design progress and quality on construction progress and construction progress on design rework discovery. They used data from the 15 projects to quantify these effects, and then use that quantification to estimate cost and schedule on 15 other projects held from the initial analysis. They discussed the process of conducting such a postmortem and how the results can be used as a method of estimating cost and schedule for future projects.

Postmortem Assessments – Insights and Future Work. A significant number of the real world system dynamics applications have involved post-mortem assessments. Especially on disputes, the payoff for demonstrating delay and disruption is high, the costs of modeling relatively low, and the data is complete and generally available – all conditions which make the use of system dynamics feasible and high value to real projects. Postmortem assessments for learning have been less numerous, but for project-based organizations can be just as valuable. System dynamics is a scientific method for assessing what went right and what went wrong on a project, and therefore provides the raw materials for many of the other uses of system dynamics discussed below (estimating, risk assessment, control).

Perhaps, then, the greatest need in this arena is for greater documentation and discussion of the process of using models for such assessments and for published success stories and lessons learned that can serve as both models of success and for "marketing" future work. Greater discussion of the process is warranted because there seems to be some divergence in approach. For example, PRA start with calibrating the model to what actually happened on the project and remove direct impacts, while the Strathclyde group generally starts with a calibration to the plan and adds direct impacts. Are both approaches acceptable? When is which approach preferable? There may also be some methodological (and perhaps legal) issues about the proper way to conduct "would-have" analyses, i.e., in what order should the direct impacts be removed to assess delay and disruption. PRA usually remove the direct impacts of client actions one at a time in reverse chronological order, which makes intuitive sense but has not to the authors' knowledge been rigorously studied. Another process issue is how uncertainty in parameter inputs (and model structure) should be considered in arriving at a "number" for the magnitude of any delay and disruption. Graham et al. (2002a, b) discuss the use of Monte Carlo simulations to develop confidence bands around this magnitude. Howick (2005) discussed different audiences for a model used in litigation (e.g., lawyers, other expert modelers, arbiters, and judges) and how each audience's prior experience (or not) with SD models, and other models, colors their expectations for the appropriateness, validation, and use of the SD model. She suggests expectations and needs that must be addressed in order for use of the model to be successful.

Given the continued poor performance of projects, the failure of project organizations to devote significant effort to project-to-project learning seems inexplicable. Documentation of success stories, both for disputes and for learning, can demonstrate the value of using system dynamics for project management. A common progression within a project organization is (1) first use on a dispute, (2) followed by use for estimating and management of a new project, and (3) finally use on additional projects and project-to-project learning. Documentation of success stories and process can facilitate this progression.

Project Estimating, Planning, and Risk Assessment
Strong anecdotal evidence suggests that, in addition to changes to the plan, another common trigger for adverse project dynamics is underestimating work scope or under budgeting for the estimated work scope. While postproject assessments are essential for understanding what happened, their greatest value may be in improving project estimating and risk assessment – how can we develop project budgets and plans that are more realistic and robust?

Project Estimating. Abdel-Hamid and Madnick (1991) and Abdel-Hamid (1993b) discussed the use of system dynamics models in conjunction with more traditional estimation approaches (such as COCOMO for software) to develop project estimates of effort and time requirements. He argued that this can and should be done at three stages during a project: (1) upfront to adjust traditional estimates based on known or expected deviations (risks) from typical projects; (2) during the project in order to determine the degree of any project underestimation earlier than would typically occur; and (3) after the project finishes, to assess what the project should have cost had other decisions been taken, including better initial estimates. This last assessment is critical, as he demonstrated how project estimates can affect the final schedule and cost of a project – projects which are underestimated end up costing more because of the adverse ripple effect dynamics incurred once the underestimate is realized; projects which are overestimated also end up costing more than they otherwise would because of

the tendency to slack off and or "gold-plate" when there is insufficient schedule pressure. He also showed how similar dynamics can lock in project-to-project underperformance of productivity enhancing tools (Abdel-Hamid 1996), and demonstrated this phenomenon in a series of controlled experiments using a gaming version of the model (Sengupta and Abdel-Hamid 1996). Adjusting estimates based on a system dynamics model can help reduce or eliminate these dynamics. Praven et al. (2015) discussed how information from postmortem assessments can then be used to forecast the costs of future projects.

Planning. The cost of a project is affected not only by its scope and schedule, but also by how it is structured and managed. Decisions that are typically taken in the project planning phase of the project include:

- Selecting the process model, for example, waterfall versus iterative versus agile
- Deciding how much to subcontract, make vs. buy
- Defining teams and responsibilities (e.g., integrated product team versus functional)
- Staffing strategy (should the project pay extra for experience, hire generalists or specialists, co-locate of geographically disperse, how much training of new staff)
- Scheduling (how overlap between phases and concurrency within a phase)
- Determining project controls (what to measure, monitor, and actions to take)
- Identifying risks and developing risk management plans

Many of these decisions will affect project dynamics, yet very little work has been published addressing these decisions. Work by Van Oorschot et al. (2010, 2013, 2017) examine the impact of different staffing strategies in conjunction with a stage-gate development process on project performance.

A major exception is SD work on concurrent project development. (In the construction industry concurrent project development is referred to as "fast tracking.") This work started with Ford and

Sterman (2003a), who used concurrence to partially explain the 90% syndrome in which a project appears to be mostly (90%) complete on schedule but rework delays completion to well after the original deadline. (Rahmandad and Hu (2010) provide an alternative explanation for the 90% syndrome based on rework discovery.) They explicitly modeled the contamination of work in downstream phases by the release of flawed work by upstream phases and showed the impacts of varying amounts of concurrence on the three predominant project performance dimensions: time, effort (proxy for cost), and quality. Lin et al. (2008) subsequently found similar results using a similar model structure of mobile phone design projects. Rahmandad and Weiss (2009) included fixed amounts of concurrence in their modeling of software development and showed how resource constraints can also generate concurrence dynamics. As described above, Praven et al. (2015) quantified the strengths of some of these impacts. These works generally conclude that (ceteris paribus) more concurrence increases detrimental impacts on downstream phases and project performance, potentially overwhelming the schedule benefits that the concurrence was designed to capture. For example, Graham (2000) recommends that project managers:

> Avoid the tendency to start downstream work too early and thereby increase unplanned concurrence, reallocate "excess" staff that will be needed later when the rework is discovered, or both. Another consequence of undiscovered rework is that a project is likely to be further behind than typical reporting systems indicate. This can lure managers into earlier downstream phase initiation or staff reductions that generate knock-on effects.

However, other researchers of project concurrence have found benefits of inter-phase feedback. Akkermans and van Oorschot (2016) modeled several benefits of more concurrency. They assess the optimal amount of concurrency for the complex projects in their case studies and conclude that:

> ...major improvements occur when more concurrency is allowed, because in projects of such complexity team learning is critical ...and more concurrency feeds learning in two major ways, even when it also results in rework or throwing

away previous work based on outdated assumptions. First, downstream stages can begin learning sooner, albeit at a slower pace, such that they are ready "to hit the ground running" when the definitive outputs arrive from the preceding stage. Second, learning in the downstream stages results in valuable upstream feedback... (p. 88)

Like SD work on the fundamental drivers of project dynamics, the SD work on concurrence in projects is relatively robust, having been applied to a variety of project types and processes with consistent results. We view these papers as a good start, but much more work is needed to identify and describe tradeoffs and to develop heuristics based on project characteristics so that practicing project managers can apply these lessons to actual projects.

Risk Assessment. All project plans make assumptions about uncontrollable factors that might impact a project. These include, but are not limited to, that resources and/or skills can be obtained as planned, delays in receiving information and/or materials from other projects and vendors will not be excessive, that technical uncertainties will be resolved in a timely manner, and that new tools and organization structures will work as planned. Risk assessment asks the question: "If certain assumptions in the baseline plan are not met, what would be the impact on project performance?" System dynamics has been used for risk assessment in two ways: (1) postproject evaluations determine the magnitude of changes that actually occurred on projects as a guide for what might occur on future projects; and (2) preproject simulations test the consequences of similar risks for the current project. For example, PRA did a postmortem analysis of 11 development projects at a major automotive company. This analysis identified and quantified five continuing sources of risk (causes of overrun) for such development projects: (1) late information and/or changes; (2) resource availability (slow ramp up, lower peak, forced ramp down to meet budget, inadequate skills mix); (3) new processes, missing enablers, or new materials; (4) organization and/or geographic changes; and (5) aggressive program assumptions (compressed timing, inadequate budget, lean allowance for prototypes). They also suggested actions to mitigate these risks. While this work, and similar work by others that the authors are aware of, has proved to be of significant benefit to the company, it has not been published.

Project Estimating, Planning, and Risk Assessment – Insights and Future Work. Project plans sow the seeds for project success or failure. Why are managers biased toward continued underestimates of true costs in the face of continued evidence of project overruns, and how can they be convinced that the first step to avoiding adverse project dynamics is to bid and plan the project correctly?

Bias toward underestimation. In all likelihood, managers' continued underestimating of project budgets partly reflects the inherent difficulties in estimating the scope of work on a complex development and partly management's (and staff's) bias toward optimistic estimates of the effort required. This underestimation, at least on management's part, likely reflects some combination of (1) fear that the project will not be accepted or continued if the budget estimate is too high; (2) desire to put pressure on staff to avoid the "gold-plating" and slacking-off phenomena noted above; (3) failure to adequately budget for the hours and time needed to perform even normal rework, or for the productivity and quality costs of planned concurrence; and (4) the belief that trying to achieve an unachievable plan does not have any adverse consequences (underestimate of ripple effects). Some managers apparently believe that: "Sure, the project is underestimated, but what's the worst that can happen? We'll add resources or schedule and end up with what a reasonable plan would have produced. And maybe we'll be able to pressure the staff to do better and actually reduce our costs." Project planners find it seductively easy to ignore the adverse dynamics created when a project falls behind and actions are taken to bring it back on schedule.

One significant contribution that system dynamics has, and can continue to make, is to convincingly persuade management that trying to achieve an infeasible plan actually makes the performance of the project worse. What factors cause continued underestimation (and budgeting)

of projects in spite of evidence of its adverse consequences? We see three possible explanations. First, practitioners may fail to understand the impacts of project dynamics on performance. This could be because most project planning tools ignore rework and productivity changes in projects, and/or it could be because SD modelers have not communicated the importance of feedback impacts on projects. Publications that use system dynamics to both rigorously model these impacts and explain them in ways that non-SD practitioners understand are part of a solution. For example, Abdel-Hamid (2011) explained the role of single loop project control and its challenges to a project management (not SD) audience. Getting very simple project models that can be used and explained by non-SD people into the hands of project management educators, trainers, and practitioners can also fill this gap. The project management community can use models such as the one used and made available by Jalili and Ford (2016) to learn themselves and then disseminate the lessons that SD has developed about project dynamics. But significant additional efforts to overcome this barrier are needed.

Second, SD modelers have not communicated the *magnitude* of rework and the ripple effect feedbacks on cost and schedule. Some SD project work addresses this gap. Reichelt and Lyneis (1999) provide one such quantification from a survey of 13 projects. They show that the amount of design rework averages 40–60% on complex projects, and quantify the degree of productivity loss from feedback effects at an average of 33%. Cooper and Lee (2008) express the impacts of feedback in a different way – in terms of the cost of "secondary" impacts relative to the direct costs of changes to the project, including adding required scope not included in the original estimate. They note that the secondary impact "can easily be as large as the *direct* costs of the changes, and sometimes 2, 3, or 4 *times* the direct change costs" (p. 11). More recently Jalili and Ford (2016) demonstrate the relative sizes of contributions of three common reinforcing project feedback loops to project performance. Their results highlight the important contribution of project dynamics to the magnitude of cost and schedule overruns on projects and suggest impact sizes similar to those found by Cooper and Lee based on their empirical work.

But it would appear that even pro planners that understand project vulnerability feedback dynamics continue to ignore or underestimate rework and the ripple effects of project control actions. A third explanation for underestimation (and budgeting) of projects in spite of evidence of its adverse consequences is that project planners may understand project dynamics but knowingly underestimate resource requirements (including time to complete) and overestimate performance as part of "overselling" projects to get them approved and resourced. During projects they may also knowingly mislead others to manage policy resistance caused by the entrenched beliefs and practices of senior management. Yaghootkar and Gil (2012) support the latter explanation when they discuss why managers ignore the consequences of feedbacks even when they know they exist:

> [The long-term degradation in performance from short-term fire-fighting actions] is consistent with Repenning's (2001) claim that fire-fighting is self-reinforcing. It is also in agreement with prior studies which suggest that the urgency to resolve short-term problems often motivates organizations to overlook the long-term effects of short-term fixes (Repenning 2001). *Interestingly, our fieldwork reveals practitioners are aware that a schedule-driven management policy can harm the long-term organization's capability to deliver projects efficiently. But the empirical findings also suggest that conventional wisdom has been inadequate to talk top management out of deep-seated practice.* [Emphasis added.] p. 136.

System dynamics project work could expand model boundaries to include the dynamics of getting projects approved, underway, and overcoming policy resistance within management chains of command. Research is needed on why managers continue to underestimate the cost and schedule of projects, rework, and the impact of ripple effect feedbacks on cost and schedule, and then how system dynamicists can better work with practicing project planners and managers to address these challenges.

Improving Project Estimates. Once the reasons for underestimation are identified, the question of how better estimates should be determined remains. As discussed above, postmortem analysis of prior projects can provide a baseline model estimate of the planned project. However, even with these estimates, the true scope and cost of the next project cannot be known with certainty in advance. Therefore, is it better to err on the high or the low side? Under what conditions? What degree of slack or buffers are appropriate, and who on project teams should know about and have control over what buffers? Initial work by Ford (2002) on the management and use of buffers in projects should be expanded.

In addition to better estimating, system dynamics can provide insights into how to design the project to improve cost and schedule, and mitigate adverse project dynamics. What process model works best for the complexity and uniqueness of this project? How much overlap between project phases is appropriate to maximize rework discovery but minimize error propagation? There are many "planning" decisions to be made in designing the project. Work like that on concurrence noted above and described below about risk management in public private partnerships should be continued to explain and resolve the seemingly conflicting results, and expanded to other project design (vs. product design) questions.

Finally, uncertainty is omnipresent in projects, causing risks to meeting objectives. Although, as described, risk has been addressed by system dynamics project modelers, the full power of the strategic perspective possible with system dynamics has not been used to design or analyze risk strategies that apply several risk management tools or integrate with existing risk management theory. Focusing more narrowly on system dynamics, robustness – the ability for good performance under uncertainty – is a holy grail of project management that system dynamics suggests is attainable with good planning for dynamics and the appropriate use of adaptive control. But robustness is difficult to measure and harder to design. This makes robustness a rich opportunity for advancement via system dynamics. How can robustness be measured and used as a project planning and management performance measure? Which project components and policies impact robustness most? What processes and policies improve robustness? Initial efforts such as by Taylor and Ford (2006) should be extended and expanded.

Change Management, Risk Management, and Project Control

Even with improved planning, projects will rarely go exactly as planned. When problems occur, how should management best respond? To what extent can additional budget be obtained ("change management")? How can risks be mitigated ("risk management")? What mix of adding resources (e.g., by hiring, overtime, work intensity), changing the schedule (both final and interim milestones), reducing scope, cutting activities such as QA, and so on will provide the most satisfactory outcome ("project control")? A system dynamics model can provide valuable input into such decisions by taking into consideration feedback in projects, especially the adverse ripple effects of management actions.

Change Management. When customers make changes to projects, the original plan almost always becomes infeasible. Change management entails the pricing and mitigating of proposed changes as they occur on an on-going project (rather than waiting for disputes to occur after the project ends). Cooper and Reichelt (2004) demonstrate that the full cost of changes, including ripple effects, increases nonlinearly with the cumulative size of all changes, and as the changes occur later in the project. Eden et al. (2000) call these "Portfolio Effects," where combinations of changes produce impacts greater than the sum of the individual impacts alone. Many clients and project managers overlook these ripple effects when requesting, pricing, and accepting changes – they typically price the changes at the estimated cost of the directly added scope. As a result, the project overruns the schedule and/or budget, and disputes are likely to arise.

Examples of applications of system dynamics to change management include:

- Williams (1999) describes the use of their model to assess optimal schedule extensions when changes are introduced on a project.

- Howick and Eden (2001) examine the consequences of attempting to compress a project's schedule, at the request of the client, once the project has started, and demonstrate that too often contractors ask for insufficient monies as compensation because they ignore the ripple effects.

- Fluor Corporation proactively uses project models to forecast and mitigate change impacts, including quantifying the changes' effects, diagnosing the causes, and planning and testing mitigating actions to reduce project costs. Fluor reports that their clients welcome their use of these models, appreciating the foresight that helps avoid project cost surprises and minimize capital expenditures. The Fluor work is described in Godlewski et al. (2012):

> On our most complex projects, we have implemented a system dynamics model-based system that has improved our project management, transformed our change management, and brought large quantified business benefits to us and our clients. The model can be rapidly set up and tailored to each major engineering and construction project. We use it to foresee the future cost and schedule impacts of project changes, and most important, to test ways to avoid the impacts. Since 2005, Fluor has used the system on over 100 projects and has trained hundreds of project managers and planners in its ongoing internal use. Quantitative business benefits exceed $800 million to date for Fluor and our clients. It has also transformed the mindset of our managers away from the industry's typical retrospective view, in which disputes could become the channel for resolving cost responsibility, and replaced it with a proactive approach, in which we work with our clients to find, in advance, ways to mitigate impacts and reduce costs—a win-win situation for Fluor and our clients.

This work has won two important awards: the 2008 System Dynamics Applications Award and Finalist for the 2011 Franz Edelman Award for Achievement in Operations Research and the Management Science (INFORMS).

Risk Management. System dynamics project models have also been applied to investigate risk management as an aspect of project management distinct from project control. While system dynamicists have used project models to investigate the effectiveness and use of specific risk management tools or strategies, they also develop insight by focusing on risk management approaches instead of specific policies. Managerial flexibility is an example. Ford and others use system dynamics to operationalize real options theory in projects for risk management. Case studies (Ford and Ceylan 2002; Alessandri et al. 2004; Johnson et al. 2006) and comparisons with other approaches (Cao et al. 2006) establish a basis for the feedback role of managerial real options. System dynamics model structures specify real options decision making and test option valuation theory (Ford and Bhargav 2006). Ford and Sobek (2005) applied this approach to a product development project to more fully describe Toyota's unique product development approach to managing design risk and to partially explain Toyota's industry-leading performance. Adopting a similar approach, Johnson et al. (2006) used system dynamics to model and value flexibility in equipment delivery strategies in a large petrochemical project. Finally, Johnson et al. (2009) in work for the US Navy used a model to understand the drivers of cost reduction on a series of submarines in a design-build-design-build … sequence, and assure that expected "learning curve" improvements actually occur through appropriate management actions.

Managerial mental models toward risk provide a second example. Project managers simultaneously seek project structures and policies that maximize project performance yet perform well when faced with a range of uncertainties that reflect risks. System dynamics research has shown that managers who tailor polices for specific project assumptions can outperform those that manage for a wide range of conditions, *if those project assumptions materialize*. But if conditions deviate from those assumed by management, tailored policies generate much worse performance. Several researchers in different contexts have identified this fundamental tradeoff between project robustness, the ability to perform well across a range of uncertain conditions, and performance under known specific conditions. Repenning (2000) first identified this tradeoff using system dynamics. Ford (2002) found a similar trade-off by modeling practitioner mental

models of budget contingency management. Park and Pena-Mora (2004) propose and test a strategy of schedule buffer allocation that includes overlapping to allow more time for quality assurance in downstream activities. They found that sizing and locating schedule buffers can improve project performance by reducing the impacts of changes (a particular type of risk).

A recent development in the use of system dynamics to investigate project risk management is the application to issues in public private partnerships (PPP), a project delivery approach that is organizationally complex because of the non-traditional multiple diverse roles played by owners, designers, and builders. For example, funding is often provided by owners and developers and the project's design, construction, and operation are performed by a single legal entity. This makes appropriately allocating and rewarding risk paramount for successful PPP projects. Ford et al. (2015) use extreme risk allocations and simple tipping point structures to explain PPP project success and failure. This work is the first known to investigate the interaction of multiple tipping points, an area for future research. Paez-Perez and Sanchez-Silva (2016) used system dynamics with agent-based modeling to investigate the principal-agent problem in PPP. They demonstrate the relationships among PPP, the physical project, the contract design, and the natural environment and the usefulness of their approach to addressing risk sharing and other issues (e.g., moral hazard). Damnjanovic et al. (2016) investigated the impacts of continuous refinancing based on cash flow risk on risk sharing and suggest a risk analysis tool for PPP design. The work in this domain demonstrates the potential for system dynamics to contribute to project management through the modeling of complex project structures and issues beyond the design and construction phase planning and management that has dominated the work historically.

Project Control. Projects rarely go as planned. When problems occur, how should management best respond? Building from the research and applications in the field, and especially Graham (2000), Cooper (1994), and Smith et al. (1993), we summarize the key project control lessons that come from understanding project dynamics into two categories: managing the rework cycle and minimizing ripple and knock-on effects.

The rework cycle is central to many adverse project dynamics. Doing work multiple times almost always delays the project and increases its costs. If the rework cycle is recognized, management can take actions to minimize its consequences. Specifically, system dynamics project models have been used to identify that managers can:

- Make efforts to improve quality/reduce errors, even if those efforts reduce productivity – doing work fewer times, even at lower productivity, is generally beneficial. One approach is to slow down and do work right the first time (i.e., reduce work intensity), even if this might cause some "slacking-off." Another approach uses integrated product teams, which improve quality and rework discovery at the expense of reduced productivity from greater communications overheads. Graham (2000) argues that to be effective, these teams should include customers, and all functions, and that the people on the teams should have the knowledge and authority to make decisions that will improve the end product.

- Recognize the existence of undiscovered rework and avoid its consequences, primarily the "errors create more errors" dynamic. Undiscovered rework can be reduced, for example, by prioritizing rework detection and correction over starting new work. Park and Pena-Mora (2004) investigate how to have staff spend time (re) checking before starting new work to operationalize this strategy (see also Lyneis et al. 2001). Early testing to discover problems rather than testing to pass tests, can also reduce rework cycle consequences.

- Avoid the tendency to start downstream work too early and thereby increase unplanned concurrence, reallocate "excess" staff that will be needed later when the rework is discovered, or both. Another consequence of undiscovered rework is that a project is likely to be further behind than typical reporting systems indicate.

This can lure managers into earlier downstream phase initiation or staff reductions that generate knock-on effects.

- Use a formal model to help implement improved policies. Even if recognized and designed well, the project control actions above are often difficult to implement because implementation initiates a "worse before better" behavior mode. By effectively demonstrating any worse-before-better dynamics and the eventual benefits of implementation, a formal model can give managers the courage to stick with implementation. For example, allocating more resources to QA reduces "perceived progress" while actually increasing "real progress." Such actions may be difficult to stick with unless managers have confidence the "better" will occur after the "worse."

In addition to improving behavior through management of the rework cycle, managers can significantly improve project performance through efforts to manage ripple and knock-on effects – how managers respond when the existence of an infeasible initial plan is discovered, or when changes or other risks materialize (thereby making the plan infeasible), has a significant impact on project dynamics. Two types of project control actions are available:

- Ease performance targets, such as by slipping the completion or milestone deadlines, increasing the budget, reducing the scope, or accepting a higher fraction of flaws in the final product. These actions reduce ripple effects by reducing the need to change project management and progress. Easing targets would seem to be more attractive when it is difficult or expensive to change the performance of the project.

- Increase effective resources, such as by adding staff, working overtime, or increasing work intensity or by using staff more efficiently. These actions can initiate ripple and knock-on effects.

Change Management, Risk Management, and Project Control – Insights and Future Work. While the literature stresses the importance of minimizing the rework cycle and avoiding vicious circle feedbacks, it is unhelpful on specifics. For example, Smith et al. (1993) offered the following policy advice: (1) extra time spent during requirements and design result in a higher quality project at lower cost; (2) increasing personnel on a project is usually counter-productive – implement a project with a fixed number of staff from the beginning; (3) sustained overtime does not increase productivity in the long run; (4) a moderate amount of schedule pressure is optimal; and (5) the use of "experts" can significantly improve project performance. While these seem logical, are they true on all projects, and what specific actions should be taken? What are "sustained," or "moderate" levels that should be avoided or followed? Beyond Graham's (2000) assertion that "managers should avoid use of sustained overtime (longer than 3 months)," there is little quantitative guidance or support for the general advice offered. Advice such as "use overtime for a couple of months while hiring by up to 20 percent new staff, then cut back as new staff gain experience" is not forthcoming. Why?

The lack of published advice may reflect the conflict between seeking recommendations that are both widely applicable and have rigor based on specific projects and issues. But the lack of published advice may also reflect the fact that research in this area is lacking. While researchers and practitioners have examined specific situations, few have attempted to generalize recommendations. This is an important area for future research. What project conditions should managers use as the basis for project control decisions? What should managers do when a project is forecasted to fail to meet performance targets? What combination, order, and duration of easing targets and increasing effective resources bring the project closest to its performance goals and minimize the project's vulnerability to tipping point-induced failure? The best approach to getting the project back on track is not obvious. From a dynamic systems perspective, this is particularly difficult to ascertain because the strengths of the feedback loops differ across projects and are dynamic during projects. What heuristics for managing project dynamics improve performance?

What qualitative and quantitative models help develop, teach, and train about these heuristics? Each project is different, so the literature cannot offer specific advice such as "use x% overtime while hiring no more staff." However, more work is needed on how managers can use the insights from the system dynamics literature or from project-specific models to develop such guidelines for their particular projects. Ford et al. (2007) have initiated one research project along these lines.

Sengupta and Abdel-Hamid (1993) used their model in a gaming format to show how system dynamics models can be used to generalize about project management. They showed that student managers perform best when given cognitive feedback (e.g., information on fraction of workforce experienced, productivity, communication overhead), worse when given feed forward feedback (e.g., heuristics for hiring), and worst when given outcome feedback (e.g., estimated progress, hours spent). That is, students do worst with typical management information, but can improve with heuristics and information targeted at the cause of adverse dynamics. Similar questions apply to other project controls. For example, Lee, Ford, and Joglekar (2007) investigate the interaction of resource allocation delays and different amounts of control imposed by managers. Their results suggest general but counterintuitive project control recommendations, such as to exert less control to decrease project durations. Assuming these and other general project control lessons prove effective, how can project managers be convinced to adopt them?

While at least some work has been done examining actions to bring a project back on schedule, little work has been done on the consequences of closing the performance-target gap by changing project deadlines. More generally, the secondary consequences of adjusting project targets have not been as well studied in system dynamics as the secondary consequences of adjusting efforts. PRA usually model slipping schedule as well as adding resources; however, they rarely examine secondary impacts of such slips on performance of the product in the market. Additional work is needed in understanding the secondary consequences of controls to achieve quality and budget targets and of actions to adjust the targets themselves. However, adequately representing the secondary consequences of target adjustments will expand the boundary of project models to include interactions with the market (impact of delays in reaching the market, project and product cost, and product features on market success) and other parts of the firm (impact of resource usage on other projects).

Multiple performance measures that vary in importance across project participants and time are a hallmark of projects. While schedule targets are often the top priority on projects, cost, delivered quality, and/or scope can also be critical. Relatively little analysis of the dynamics created by attempts to meet these targets, along the lines shown in Fig. 4, has been done. How does management of various performance measures differ? The management of different performance measures is complicated by the interdependence of performance measures (e.g., longer projects can cost more because of "marching army" fixed costs). System dynamics project models can demonstrate traditional performance tradeoffs (e.g., between duration and cost) inherent in project management (e.g., the use of overtime). But to add value system dynamics needs to contribute deeper insights about multiple performance measures. The work to date has demonstrated many ways in which managers can *fail*. How can managers proactively *succeed* when faced with conflicting performance measures? Most system dynamics models of projects assume a single set of performance priorities. But project practice typically includes important differences in performance priorities across the project team. What is "best" often differs among project participants (e.g., owner, designer, builder). For example, to an owner the best solution to a late project may be increased builder's staff. Although that may increase costs and reduce the builder's profit, the owner can retain the planned benefits. In contrast, the best solution from the builder's perspective in the same project may be to slip the completion deadline. Although this may delay and therefore reduce the owner benefits, it can minimize the builder's costs and retain the builder's profit.

How can the competition among project participants with different targets and priorities be modeled and improved? System dynamics models can be developed to address the relative winners and losers within project teams when projects are managed with certain approaches and policies.

Management Training and Education

Projects are often used to teach system dynamics. System dynamics project models have been used with both practicing project managers and students in formal educational settings. The familiarity of projects and the challenges of project management to many people (as suggested by the popularity of the comic strip "Dilbert") make projects common parts of system dynamics courses. These applications typically are limited to a focused case study or building relatively simple models to illustrate the system dynamics method with a few of the structures and dynamics described above. The ability of system dynamics to clearly and richly explain how project structures and behavior interact also makes it effective for teaching project management. Most uses for this purpose in schools are graduate level courses that depend on a fundamental understanding of project management from undergraduate coursework or practice.

A number of the project models have been converted into gaming simulators for use in management training, for example, at BP, Bath, Ford, Hughes/Raytheon, and The World Bank. Trainees manage a simulated project, making typical decisions such as hiring, use of overtime, exerting schedule pressure, allocation of staff to various activities, when to start downstream phases of work, and so on. After such a simulation, the reasons for simulated project performance are analyzed, usually using diagrams such as those developed above, and diagnostic output from the models. Trainees are then often allowed to manage the simulated project again. The Strathclyde group (Howick and Eden 2001) have converted their claim models into a generic project model they have used for management training. Abdel-Hamid has developed a simulator version of his software dynamics model, but seems to have used it primarily to conduct experiments on how managers make decisions (as discussed earlier). In addition, Barlas and Bayraktutar (1992), Repenning (2000), MacInnis (2004), and others have developed gaming simulators for use with students and managers.

Management Training and Education – Insights Future Work. As an aid in teaching system dynamics, project models have and will continue to play a valuable role. However, project models have only begun to fulfill their potential in teaching project management and project control. The experience of one of the authors in teaching project management indicates that two of the largest challenges for students, especially those with little or no industry experience, are to understand the complexity of projects that the many causal relations create, and to appreciate at a visceral level the challenges of project management. Abdel-Hamid (2011) wrote a very good overview paper to introduce students and managers to the concepts of feedback, mental vs. computer models, Brooks Law, setting up a model for a specific project, and using a model to determine the optimal strategy. This is a good but very small start.

System dynamics project models can help effectively address both of these challenges by being the basis for project management flight simulators. Beyond using project models as black box proxies for actual projects, the transparent nature of the models developed for project management training and education can be leveraged to demonstrate how a scientific investigation of the causal feedback structures of projects can improve managerial understanding, policies, and performance. Existing models can be the basis for such teaching tools. However, a good project model is only one part of an effective management flight simulator – supporting diagnostic and educational materials are also needed. Developing a useful system dynamics project model-based teaching tool for widespread project management training and education is a relatively large endeavor, but could make major advances in project management education, potentially when added to established project management training approaches such as the PRINCE2 system in Britain or the Project Management Institute in the United States, and the application of system dynamics. Existing and new models and tools should be developed for widespread use.

Discussion and Conclusions

As evidenced above, both in terms of academic research and numbers of "real world" applications, using system dynamics to understand and improve project management has been a great success. This success includes significant advancement in new system dynamics theory, new and improved model structures, value addition in practice, and growth of the system dynamics field. We summarize what has been accomplished, and what remains to be done, in three categories: (1) theory development, (2) guidance in improving project management and education, and (3) applications.

Theory Development. The primary causes of project dynamics – project features, the rework cycle, project controls, and ripple and knock-on effects – are in our view nearly completely and adequately represented in existing project models, especially as they pertain to controls to achieve schedule performance. While no doubt special structures are needed to represent particular types of projects, these special structures are at the fine-tuning level and do not represent fundamental additions to the drivers of project dynamics. One important area of research may be in identifying how the strengths of the various ripple and knock-on effects may differ by phase of work (e.g., design vs. construction) or by type of project (e.g., software vs. hardware), or as a result of changes in process and tools (e.g., the adoption of CAD systems), and how different strengths may alter any policy heuristics. Ideally, empirical studies would be conducted to identify differences, and simulation analyses to translate those differences into policy guidance. Additional work is needed in understanding the secondary consequences of controls to achieve quality and budget targets and of actions to adjust the targets themselves. However, adequately representing the secondary consequences of target adjustments will expand the boundary of project models to include interactions with the market and other parts of the firm.

Guidance in Improving Project Management and Education. While our understanding of what causes project dynamics is fairly complete, much work remains in translating that theory into improved project management and education. In which directions should system dynamics project modeling develop? What issues and topics are the most fertile for continuing to improve our understanding of project dynamics? Our review of the domain suggests valuable issues and questions in several areas. We offer these, not only as suggestions for future work, but also to initiate and catalyze a discourse among system dynamics project modelers about future directions.

- *Postmortem Assessments*: While some research on the approach to conducting simulation experiments is warranted, as discussed above, perhaps the greatest need is for published success stories and lessons learned that can serve as both models of success and for "marketing" future work.
- *Project Estimating, Planning, and Risk Assessment:* The persistent and large underestimation of projects may provide the best opportunity for system dynamics to improve projects. Why do managers continue to underestimate projects in the face of continued evidence of project overruns? How much do they underestimate ripple effects, and why? What can be done to improve these practices? Case and empirical studies of projects would seem to be the best approach to these questions. If we could get managers to attempt to make more accurate estimates, how should this best be done? How should the project be designed (process model, overlap and concurrence, staffing strategies, etc.) to improve its performance and mitigate dynamics? While we may have prior project experience, and even models, to develop base estimates from, the true scope and cost of the next project cannot be known in advance. Therefore, is it better to err on the high or the low side? Under what conditions? What degree of slack, buffers, and budget contingencies are appropriate, and who on project teams should know about and have control over what buffers? Here, simulation analyses

in concert with real applications would seem to be the best approach.
- *Change Management, Risk Management, and Project Controls:* What project conditions should managers use as the basis for project control decisions? What should managers do when a project is forecasted to fail to meet performance targets? In what amounts, durations, combinations, and orders of application should individual project controls such as hiring or overtime be applied to improve performance as much as is possible? What heuristics for managing project dynamics improve performance? What qualitative and quantitative models help develop, teach, and train about these heuristics? These are all questions that should rigorously be studied using simulation models. Similar questions apply to other project controls. While schedule targets are often the top priority on projects, cost, delivered quality, and/or scope can be critical. Relatively little analysis of the dynamics created by attempts to meet these targets, along the lines shown in Fig. 4, has been done. How does management of various performance measures differ? Project performance measures are interdependent. How do performance measure dependencies impact project management success?
- *Management Training and Education:* While project-management flight simulators exist, and project dynamics ideas are taught in a few courses, the use of system dynamics concepts to help managers understand project behavior and improve management is very limited. Perhaps the biggest need are teaching materials that can be used by both system dynamicists and by others, at a range of depth and duration, to first enhance management awareness and then to improve the skills of those that might use SD models in support of project management.

Applications. Even though the application of system dynamics to project management has been significant relative to other system dynamics work, *system dynamics is used on a relatively small percentage of projects.* In addition, system dynamics modeling is typically applied to individual projects, not to all projects across an organization. Why? And how can we increase its use in a field that clearly needs better management? Based on our survey, three things might improve penetration: (1) publication of more success stories, especially in the project management literature; (2) making system dynamics models easier and less costly to develop; and (3) integration of system dynamics models with more traditional, and widely used, project management tools.

In spite of the hundreds of applications of system dynamics to project management, relatively few papers have been written and even fewer in the journals widely read by practicing project managers. Moreover, few project management courses include system dynamics as a component. Increased publications and training can only lead to more applications, in a virtuous circle. Like other areas of system dynamics, applications to project management are time-intensive and as a result costly. In addition to greater training, development of "packaged" project models, components, and tools to support model setup and refinement (e.g., integration with data bases and calibration software) will help to reduce the effort and cost of applications.

Finally, some have suggested that integration of system dynamics with more traditional tools can help spread its use. System dynamics modeling is more strategic/tactical in nature than more traditional operational project management tools such as work breakdown structures, critical path modeling, and component cost estimating (Rodrigues and Bowers 1996). Almost all projects of any complexity make use of these operational tools. The ability of system dynamics modeling to complement traditional project management tools by adding a strategic and tactical perspective suggests that it can add value by being applied in combination with traditional tools, a common experience in system dynamics project model applications. Rodrigues (2000) and Rodrigues and Bowers (1996) described a methodology to integrate the use of system dynamics within the established project management processes. In Rodrigues (2001), this is further extended to integrate the use of system dynamics modeling

within the PMBOK risk management process, providing a useful framework for managing project risk dynamics. Pfahl and Lebsanft recommended integration of system dynamics models with descriptive process modeling and goal-oriented measurement. Park and Pena-Mora (2003, 2004; Park 2001; Lee and Pena-Mora 2007) propose and apply an integration of dynamic schedule buffering using a system dynamics model with critical path modeling. Williams (2002) devoted a chapter to the topic, discussing how system dynamics can improve traditional models, traditional tools can be used to create system dynamics models, and how system dynamics and traditional models can be used to inform the other. Integrating system dynamics project models and other project management tools might improve application and therefore warrants further research on integration techniques, trial applications, and dissemination of successful approaches. White and Sholtes (2016) extend the CPM concepts to include explicit stocks of work to do and work done, and dynamic productivity for each task, but they sacrifice representation of undiscovered rework and rework, and many process and control feedbacks, to achieve the detail complexity. This seems to be the fundamental tradeoff – in order to represent detail complexity, models sacrifice dynamic complexity. Striking the right balance is critical, and we may need to recognize that physically merging more detailed CPM-like models with more strategic system dynamics models is not possible, impractical, and potentially not desirable. The best solution may be to facilitate the information transfer between models.

In summary, the application of system dynamics to project management has clearly demonstrated the power of the methodology to build theory and improve practice. But more impact is possible. Expansion of models and applications can be facilitated with a broader dissemination of existing and future work. The nature and diversity of projects provide ample opportunities for continued development and application. We welcome a dialog with academics and practitioners to further the work.

Bibliography

Abdel-Hamid TK (1984) The dynamics of software development project management: an integrative system dynamics perspective. PhD thesis, MIT Sloan School of Management, Cambridge, MA

Abdel-Hamid TK (1993a) A multiproject perspective of single-project dynamics. J Syst Softw 22(3):151–165

Abdel-Hamid TK (1993b) Adapting, correcting, and perfecting software estimates: a maintenance metaphor. Computer 26:20–29

Abdel-Hamid TK (1996) The slippery path to productivity improvement. IEEE Softw 13:43–52

Abdel-Hamid TK (2011) Single-loop project controls: reigning paradigms or straitjackets? Proj Manag J 42(1):17–30

Abdel-Hamid TK, Madnick SE (1991) Software project dynamics: an integrated approach. Prentice-Hall, Englewood Cliffs

Ackermann F, Eden C, Williams T (1997) Modelling for litigation: mixing qualitative and quantitative approaches. Interfaces 27(2):48–65

Akkermans H, van Oorschot KE (2016) Pilot error? Managerial decision biases as explanation for disruptions in aircraft development. Proj Manag J 47(2):79–102

Alessandri T, Ford D, Lander D, Leggio K, Taylor M (2004) Managing risk and uncertainty in complex capital projects. Q Rev Econ Financ 44(5):751–767

Barlas Y, Bayraktutar I (1992) An interactive simulation game for software project management (Softsim). In: Proceedings of the 1992 international system dynamics conference, Utrecht

Bayer S, Gann D (2006) Balancing work: bidding strategies and workload dynamics in a project-based professional service organization. Syst Dyn Rev 22(3):185

Cao Q, Ford D, Leggio K (2006) The application of real options to the R&D outsourcing decision. In: Schniederjans M, Schniederjans A, Schniederjans D (eds) Outsourcing management information systems. Idea Group Publishing, London

Cooper KG (1980) Naval ship production: a claim settled and a framework built. Interfaces 10(6):20–36

Cooper KG (1993) The rework cycle (a series of 3 articles): Why projects are mismanaged; how it really works . . . and reworks . . .; benchmarks for the project manager. PMNETwork Magazine, February 1993 for first two articles; Project Management Journal March 1993 for third article

Cooper KG (1994) The $2,000 hour: how managers influence project performance through the rework cycle. Proj Manag J 25(1):11

Cooper K, Lee G (2008) Managing the dynamics of projects and changes at fluor. [reference SD Society Website Link]

Cooper KG, Reichelt KS (2004) Project changes: sources, impacts, mitigation, pricing, litigation, and excellence. In: PWG M, Pinto JK (eds) The Wiley guide to managing projects. Wiley, Hoboken, pp 743–772

Cooper KG, Lyneis JM, Byrant BJ (2002) Learning to learn, from past to future. Int J Proj Manag 20:213–219

Damnjanovic I, Johnson S, Ford DN (2016) Financial stress testing of toll road projects: the effect of feedback loop dynamics. J Struct Financ 21(4):51–64. https://doi.org/10.3905/jsf.2016.2016.1.047

Eden CE, Williams TM, Ackermann FA (1998) Dismantling the learning curve: the role of disruptions on the planning of development projects. Int J Proj Manag 16(3):131–138

Eden CE, Williams TM, Ackermann FA, Howick S (2000) On the nature of disruption and delay (D&D) in major projects. J Oper Res Soc 51(3):291–300

Ford DN (1995) The dynamics of project management: an investigation of the impacts of project process and coordination on performance. PhD thesis, MIT Sloan School of Management, Cambridge, MA

Ford DN (2002) Achieving multiple project objectives through contingency management. ASCE J Constr Eng Manag 128(1):30–39

Ford DN, Bhargav S (2006) Project management quality and the value of flexible strategies. Eng Constr Archit Manag 13(3):275–289

Ford DN, Ceylan K (2002) Using options to manage dynamic uncertainty in acquisition projects. Acquis Rev Q 9(4):243–258

Ford DN, Sobek D (2005) Modeling real options to switch among alternatives in product development. IEEE Trans Eng Manag 52(2):1–11

Ford DN, Sterman JD (1998a) Dynamic modeling of product development processes. Syst Dyn Rev 14(1):31–68

Ford DN, Sterman JD (1998b) Expert knowledge elicitation for improving mental and formal models. Syst Dyn Rev 14(4):309–340

Ford DN, Sterman JD (2003a) Overcoming the 90% syndrome: iteration management in concurrent development projects. Concurr Eng Res Appl 111(3):177–186

Ford DN, Sterman JD (2003b) The liar's club: impacts of concealment in concurrent development projects. Concurr Eng Res Appl 111(3):211–219

Ford DN, Anderson S, Damron A, de Las Casas R, Gokmen N, Kuennen S (2004) Managing constructability reviews to reduce highway project durations. ASCE J Constr Eng Manag 130(1):33–42

Ford DN, Lyneis JM, Taylor T (2007) Project controls to minimize cost and schedule overruns: a model, research agenda, and initial results. In: Proceedings of the 2007 international system dynamics conference, Boston, MA

Ford DN, Damnjanovic I, Johnson S (2015) Public-private partnerships: a study of risk allocation design envelops. In: Johnston EW (ed) Governance in the information era: theory and practice of policy infomatics. Routledge, New York

Forrester JW (1961) Industrial dynamics. MIT Press, Cambridge, MA

Forrester JW (1969) Urban dynamics. MIT Press, Cambridge, MA

Godlewski E, Lee G, Cooper K (2012) System dynamics transforms fluor pr... ...ct and change management. I... ... faces 42(1):17–3...

Graham AK (2000) ... ...d PM101: lessons for ma... large developmen... ...grams. Proj Manag J 31(...

Graham AK, Choi ... Mullen TW (2002a) Us... constrained Monte ...arlo trials to quantify con... ...c in simulation mod...l outcomes. In: Proceedings ... ...c 35th annual Hawaii conference on systems sci... ...es. IEEE, Los Alamitos

Graham AK, Moore J, Choi CY (2002b) How robust are conclusions from a complex calibrated model, really? A project management model benchmark using fit constrained Monte Carlo analysis. In: Proceedings of the 2002 international system dynamics conference, Palermo

Homer JB, Sterman JD, Greenwood B, Perkola M (1993) Delivery time reduction in pulp and paper mill construction projects: a dynamic analysis of alternatives. In: Proceedings of the 1993 international system dynamics conference, Cancun

Howick S (2005) Using system dynamics models with litigation audiences, European Journal of Operational Research 162:239–250

Howick S, Eden C (2001) The impact of disruption and delay when compressing large projects: going for incentives? J Oper Res Soc 52:26–34

Jalili Y (2017) Tipping point indicators in dynamic systems. PhD thesis, Texas A&M University

Jalili Y, Ford DN (2016) Quantifying the impacts of rework, schedule pressure, and ripple effect loops on project schedule performance. Syst Dyn Rev 32(1):82

Joglekar N, Ford DN (2005) Product development resource allocation with foresight. Eur J Oper Res 160(1):72–87

Johnson S, Taylor T, Ford DN (2006) Using system dynamics to extend real options use: insights from the oil & gas industry. In: Proceedings of the 2006 international system dynamics conference, Nijmegen, July 23–27

Johnson DC, Drakeley GM, Plante TN, Dalton WJ, Trost CS (2009) Managing change on complex programs: VIRGINIA class cost reduction. Naval Eng J 121(4):79–94

Lee SH, Peña-Mora F (2007) Understanding and managing iterative error and change cycles in construction. Syst Dyn Rev 23(1):35

Lee Z, Ford DN, Joglekar N (2007) Resource allocation policy design for reduced project duration: a systems modeling approach. Syst Res Behav Sci 24:1–15

Levitt RE, Thomsen J, Christiansen TR, Kunz JC, Jin Y, Nass C (1999) Simulating project work processes and organizations: toward a micro-contingency theory of organizational design. Manag Sci 45(11):1479–1495

Lin J, Chai KH, Wong TS, Brombacher AC (2008) A dynamic model for managing overlapped iterative product development. Eur J Oper Res 185:378–392

Lyneis JM (2003) Course notes for MIT course ESD.36J: system and project management, fall 2003. http://ocw.

mit.edu/OcwWeb/Engineering-Systems-Division/ESD-36JFall-2003/CourseHome/index.htm

Lyneis JM, Cooper KG, Els SA (2001) Strategic management of complex projects: a case study using system dynamics. Syst Dyn Rev 17:237–260

MacInnis DV (2004) Development of a system dynamics based management flight simulator for new product development. MSc thesis, System Design and Management Program, MIT

Madachy RJ (2002) Software process concurrence. In: Proceedings of the 2002 international system dynamics conference, Palermo

McKenna N (2005) Executing major projects through contractors. In: Proceedings of the 2005 international system dynamics conference, Boston

Paez-Perez D, Sanchez-Silva M (2016) A dynamic principal-agent framework for modeling the performance of infrastructure. Eur J Oper Res 254:576–594

Park M (2001) Dynamic planning and control methodology for concurrent construction projects. PhD dissertation, Massachusetts Institute of Technology, Cambridge, MA

Park M, Pena-Mora F (2003) Dynamic change management for construction: introducing the change cycle into model-based project management. Syst Dyn Rev 19(3):213–242

Park M, Pena-Mora F (2004) Reliability buffering for construction projects. ASCE J Constr Eng Manag 130(5):626–637

Pfahl D, Lebsanft K (1999) Integration of system dynamics modelling with descriptive process modelling and goal-oriented measurement. J Syst Softw 46(2/3):135–150

Praven K, Rahmandad H, Haghani A (2015) Inter-phase feedbacks in construction projects. J Oper Manag 39–40:48–62

Project Management Institute (2000) A guide to the project management body of knowledge (PMBOK® Guide). Project Management Institute, Newtown Square

Rahmandad H, Hu K (2010) Modeling the rework cycle: capturing multiple defects per task. Syst Dyn Rev 26(4):291–315

Rahmandad H, Weiss DM (2009) Dynamics of concurrent software development. Syst Dyn Rev 25(3):224

Reichelt KS (1990) Halter marine: a case study of the dangers of litigation. Unpublished Masters thesis, Massachusetts Institute of Technology

Reichelt KS, Lyneis JM (1999) The dynamics of project performance: benchmarking the drivers of cost and schedule overrun. Eur Manag J 17(2):135–150

Repenning NP (2000) A dynamic model of resource allocation in multi-project research and development systems. Syst Dyn Rev 16(3):173–212

Repenning NP (2001) Understanding fire fighting in new product development. Journal of Product Innovation Management 18:285–300

Repenning NP, Sterman JD (2001) Nobody ever gets credit for fixing problems that never happened: creating and sustaining process improvement. Calif Manag Rev 43(4):64–88

Repenning NP, Sterman JD (2002) Capability traps and self-confirming attribution errors in the dynamics of process improvement. Adm Sci Q 47:265–295

Richardson GP, Pugh AL III (1981) Introduction to system dynamics modeling with dynamo. MIT Press, Cambridge, MA

Roberts EB (1964) The dynamics of research and development. Harper and Row, New York

Roberts EB (1974) A simple model of R&D project dynamics. R&D Manag 5(1):1. Reprinted in Roberts EB (ed) (1978) Managerial applications of system dynamics. Productivity Press, Cambridge, MA

Rodrigues A (2000) The application of system dynamics to project management: an integrated methodology (SYDPIM). PhD dissertation thesis. Department of Management Science, University of Strathclyde

Rodrigues AG (2001) Managing and modelling project risk dynamics: a system dynamics-based framework. Presented at the Fourth European project management conference, PMI Europe 2001, London, 6–7 June

Rodrigues AG, Bowers J (1996) System dynamics in project management: a comparative analysis with traditional methods. Syst Dyn Rev 12(2):121–139

Rodrigues AG, Williams TM (1998) System dynamics in project management: assessing the impacts of client behavior on project performance. J Oper Res Soc 49(1):2–15

Sengupta K, Abdel-Hamid TK (1993) Alternative conceptions of feedback in dynamic decision environments: an experimental investigation. Manag Sci 39(4):411–428

Sengupta K, Abdel-Hamid TK (1996) The impact of unreliable information on the management of software projects: a dynamic decision perspective. IEEE Trans Syst Man Cybern 26(2):177–189

Smith BJ, Nguyen N, Vidale RF (1993) Death of a software manager: how to avoid career suicide through dynamic software process modeling. Am Program 6(5):10–17

Stephens CA, Graham AK, Lyneis JM (2005) System dynamics modeling in the legal arena: meeting the challenges of expert witness admissibility. Syst Dyn Rev 21(2):95–122

Sterman JD (2000) Business dynamics: systems thinking and modeling for a complex world. Irwin/McGraw Hill, Chicago

Taylor T, Ford DN (2006) Tipping point dynamics in development projects. Syst Dyn Rev 22(1):51–71

Taylor TRB, Ford DN (2008) Managing tipping point dynamics in complex construction projects. J Constr Eng Manag 134(6):421–431

Van Oorschot K, Sengupta K, Akkermans H, van Wassenhove L (2010) Get fat fast: surviving stage-gate® in NPD. J Prod Innov Manag 27(6):828–839

Van Oorschot KE, Akkermans KS, Wassenhove LNV (2013) Anatomy of a decision trap in complex new product development projects. Acad Manag J 56(1):285–307

Van Oorschot K, Eling K, Langerak F (2017) Measuring the knowns to manage the unknown: how to choose the gate timing strategy in NPD projects. J Prod Innov Manag 35(2):164–183

Weil HB, Etherton RL (1990) System dynamics in dispute resolution. In: Proceedings of the 1990 international system dynamics conference, Utrecht

White JC, Sholtes RM (2016) The Dynamic Progress Method: Using Advanced Simulation to Improve Project planning and Management. Boca Raton, FL: CRC Press

Williams TM (1999) Seeking optimum project duration extensions. J Oper Res Soc 50:460–467

Williams T (2002) Modelling complex projects. Wiley, Chichester

Yaghootkar K, Gil N (2012) The effects of schedule-driven project management in multi-project environments. Int J Proj Manag 30(12):127–140

Delay and Disruption in Complex Projects

Susan Howick[1], Fran Ackermann[2], Colin Eden[1] and Terence Williams[3]
[1]Strathclyde Business School, University of Strathclyde, Glasgow, UK
[2]Curtin Business School, Curtin University, Perth, Western Australia, Australia
[3]University of Hull, Hull, UK

Article Outline

Glossary

Cause map A cause map is similar to a cognitive map; however, it is not composed of an individual's perception but rather the views/statements from a number of participants. It follows the same formalisms as cognitive mapping but does not reflect cognition as it is composite.

Cognitive map A cognitive map is a representation of an individual's perception (cognition) of an issue. It is graphically depicted illustrating concepts/statements connected together with arrows representing causality. They are created using a set of established formalisms.

Complex project A complex project is a project in which the project behaviors and outcomes are difficult to predict and difficult to explain post hoc.

Disruption and delay Disruption and delay (D&D) is primarily the consequence of interactions that feed on themselves as a result of an initial disruption or delay or portfolio of disruptions and delays.

Project A project is a temporary endeavor undertaken to create a unique product or service (Project Management Institute 2000).

Definition of the Subject

There are many examples of complex projects suffering massive time and cost overruns. If a project has suffered such an overrun, there may be a need to understand why it behaved the way it did. Two main reasons for this are: (i) to gain learning for future projects or (ii) because one party of the project wishes to claim compensation from another party and thus is trying to explain what occurred during the project. In the latter case, system dynamics has been used for the last 30 years to help understand why projects behave the way they do. Its success in this arena stems from its ability to model and unravel complex dynamic behavior that can result in project over runs. Starting from the first use of system dynamics in a claim situation in the late 1970s (Cooper 1980), it has directly influenced claim results worth millions of dollars. However, the number of claims which system dynamics has been involved in is still small as it is not perceived by

Originally published in
R. A. Meyers (ed.), *Encyclopedia of Complexity and Systems Science*, © Springer Science+Business Media LLC 2017, https://doi.org/10.1007/978-3-642-27737-5_118-4

project management practitioners as a standard tool for analyzing projects. System dynamics has a lot to offer in understanding complex projects, not only in a postmortem situation, but it could also add value in the pre-project analysis stage and during the operational stage of a project.

Introduction

In this chapter, we discuss the role of system dynamics (SD) modeling in understanding, and planning, a complex project. In particular, we are interested in understanding how and why projects can go awry in a manner that seems surprising and often very difficult to unravel.

When we refer to projects we mean "a temporary endeavor undertaken to create a unique product or service" (Project Management Institute 2000). Projects are a specific undertaking, which implies that they are "one-shot," nonrepetitive, time-limited, and, when complex, frequently bring about revolutionary (rather than evolutionary) improvements, start (to some extent) without precedent, and are risky with respect to customer, product, and project. If physical products are being created in a project, then the product is in some way significantly different to previous occasions of manufacturing (e.g., in its engineering principles, or the expected operating conditions of the product), and it is this feature that means there is a need to take a project orientation.

Complex projects often suffer massive cost overruns. In recent decades, those that have been publicized relate to large public construction projects, for example, airports, bridges, and public buildings. Some examples include Denver's US$5 billion airport that was 200% overspent (Szyliowicz and Goetz 1995), the 800 million Danish Kroner Oresund bridge that was 68% overspent (Flyvberg et al. 2003), and the UK's Scottish Parliament, which was ten times the first budget (Scottish Parliament 2003). The Major Projects Association (1994) talks of a calamitous history of cost overruns of very large projects in the public sector. Flyvberg et al. (2002) describe 258 major transportation infrastructure projects showing 90% of projects overspent. Morris and

Hough (1987) conclude that "the track record of projects is fundamentally poor, particularly for the larger and more difficult ones. ... Projects are often completed late or over budget, do not perform in the way expected, involve severe strain on participating institutions or are canceled prior to their completion after the expenditure of considerable sums of money" (p. 7).

"Complex" projects are ones in which the project behaviors and outcomes are difficult to predict and difficult to explain post hoc. Complex projects, by their nature, comprise multiple interdependencies, and involve nonlinear relationships (which are themselves dynamic). For example, choices to accelerate might involve the use of additional overtime which can affect both learning curves and productivity as a result of fatigue – each of which are non-linear relationships. In addition many of the important features of complex projects are manifested through "soft" relationships – for example managers will recognize deteriorating morale as projects become messy and look a failure, but assessing the impact of morale on levels of mistakes and rate of working has to be a matter of qualitative judgment. These characteristics are amenable particularly to SD modeling which specializes in working with qualitative relationships that are non-linear (Forrester 1961; Ackermann et al. 1997; Lyneis and Ford 2007).

It is therefore surprising that simulation modeling has not been used more extensively to construct post-mortem analyzes of failed projects, and even more surprising because of SD's aptitude for dealing with feedback. Nevertheless the authors have been involved in the analysis of ten projects that have incurred time and cost overruns and PA Consulting Group have claimed to have used SD to explain time and cost overruns for over 30 litigation cases (Lyneis et al. 2001). Although in the mid-1990s, attempts to integrate SD modeling with more typical approaches to project management were emerging, their use has never become established within the project management literature or practice (Ford 1995; Rodrigues and Bowers 1996a, b). In addition, recognition that the trend toward tighter project delivery and accelerated development times meant that parallelism in project tasks was becoming endemic, and the impact of increasing parallelism could result in complex feedback dynamics where vicious cycles exist (Williams et al. 1995a). These vicious cycles are often the consequence of actions taken to enforce feedback control designed to bring a project back on track.

As project managers describe their experiences of projects going wrong they will often talk of these "vicious cycles" occurring, particularly with respect to the way in which customer changes seem to generate much more rework than might be expected, and that the rework itself then generates even more rework. Consider a small part of a manager's description of what he sees going on around him:

> For some time now we've been short of some really important information the customer was supposed to provide us. As a consequence we've been forced to progress the contract by making engineering assumptions, which, I fear, have led to more mistakes being made than usual. This started giving us more rework than we'd planned for. But, of course, rework on some parts of the project has meant reopening work that we thought we'd completed, and that, in turn has reopened even more past work.

Engineering rework has led to the need for production work-arounds and so our labour in both engineering and production have been suffering stop/starts and interruptions – and each time this happens we take time to get back up to speed again This has led to productivity dropping because of unnecessary tasks, let alone productivity losses from the workforce getting fed-up with redoing things over and over again and so just becoming demoralized and so working slower. Inevitably all the rework and consequential productivity losses have put pressure on us to accelerate the project forcing us to have to make more engineering assumptions and do work-arounds.

Figure 1 shows a "cause map" of the arguments presented by this project manager – the words used in the map are those used by the project manager and the *arrows* represent the causality described by the manager. This description is full of vicious cycles (indeed there are 35 vicious cycles discussed – see Fig. 1) all triggered by a shortage of customer furnished information and resulting in the rework cycle (Cooper 1993a, b, c, 19931994) and the need to accelerate in order to keep the

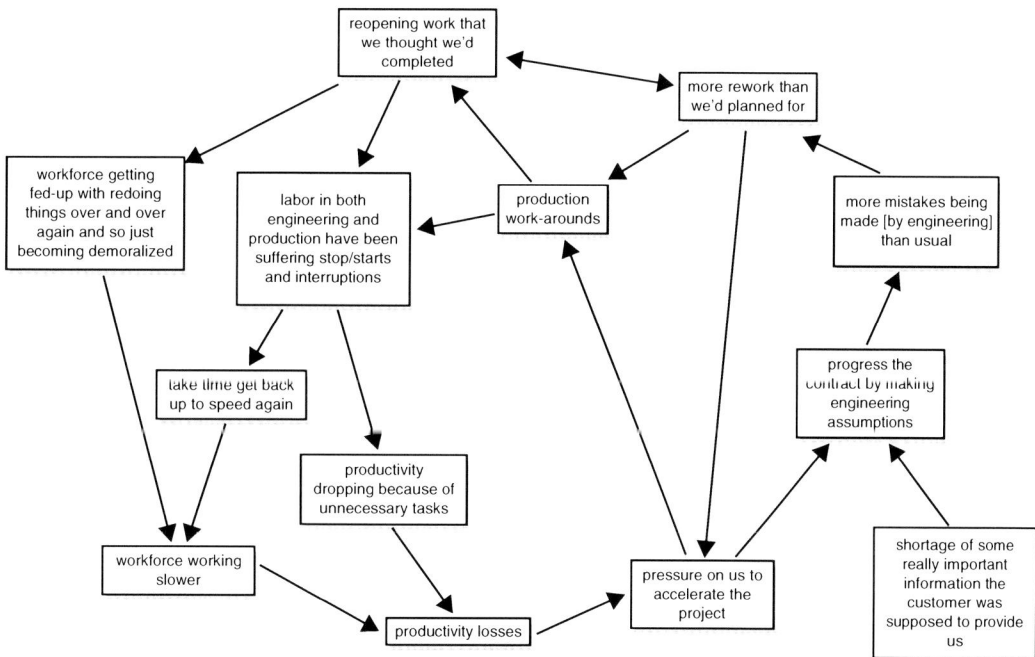

Delay and Disruption in Complex Projects, Fig. 1 Cause map showing the interactions described by a project manager and illustrating the feedback loops resulting from the complex dynamics behavior of a project under duress. The *arrows* represent causality

project on schedule. Using traditional project management models such as Critical Path Method/Network Analysis cannot capture any of the dynamics depicted in Fig. 1, but SD simulation modeling is absolutely appropriate (Eden et al. 2000).

So, why has SD modeling been so little used? Partly it is because in taking apart a failed project the purpose is usually associated with a contractor wishing to make a claim for cost-overruns. In these circumstances, the traditions of successful claims and typical attitudes of courts tend to determine the approach used. A "measured mile" approach is common, where numerical simplicity replaces the need for a proper understanding (Eden et al. 2004).

It was not until the early 1980s that the use of simulation modeling became apparent from publications in the public-domain. The settlement of a shipbuilding claim (Cooper 1980) prompted interest in SD modeling and Nahmias (1980), in the same year, reported on the use of management science modeling for the same purpose. It was not surprising that this modeling for litigation generated interest in modeling where the purpose was oriented to learning about failed projects (indeed the learning can follow from litigation modeling [Williams et al. 2005], although it rarely does).

As Fig. 1 demonstrates, it is not easy to understand fully the complex dynamic behavior of a project under duress. Few would realize that 35 feedback loops are encompassed in the description that led to Fig. 1. Indeed one of the significant features of complex projects is the likelihood of underestimating the complexity due to the dynamics generated by disruptions. Major Projects Association (1994) has reported on the more specific difficulty of understanding feedback behavior and research in the field of managerial judgment reinforces the difficulties of biases unduly influencing judgment (Kahneman et al. 1982).

In the work presented here we presume that there is a customer and a contractor, and there is a bidding process usually involving considerations of liquidated damages for delays and possibly strategic reputational consequences for late delivery. Thus, we expect the project to have a clear beginning and an end when the customer (internal or external) signs off a contract. Finally, we do not explore the whole project business life cycle, but that part where major cost overruns occur: thus, we start our consideration when a bid is to be prepared, consider development and manufacturing or construction, but stop when the product of the project is handed over to the customer.

Thus, in this chapter we shall be concerned specifically with the use of SD to model the consequences of disruptions and delays. Often these disruptions are small changes to the project, for example, design changes (Williams et al. 1995b). The work discussed here is the consequence of 12 years of constructing detailed SD simulation models of failed complex projects. The first significant case was reported in Ackermann et al. (1997) and Bennett et al. (1997). In each case, the prompt for the work was the reasonable prospect of the contractor making a successful claim for damages. In all the cases, the claim was settled out of court and the simulation model played a key role in settling the dispute.

The chapter will firstly consider why modeling disruption and delay (D&D) is so difficult. It will discuss what is meant by the term D&D and the typical consequences of D&D. This will be examined using examples from real projects that have suffered D&D. The contribution of SD modeling to the analysis of D&D and thus to the explanation of project behavior will then be discussed. A process of modeling that has been developed over the last 12 years and one that provides a means of modeling and explaining project behavior will be introduced. This process involves constructing both qualitative cause maps and quantitative system dynamics models. The chapter will conclude by considering potential future developments for the use of SD in modeling complex projects.

Disruption and Delay

(The following contains excerpts from Eden et al. (2000), which provides a full discussion on the nature of D&D).

The idea that small disruptions can cause serious consequences to the life of a major project, resulting in massive time and cost overruns, is well established. The terms "disruption and delay" or "delay and disruption" are also often used to describe what has happened on such projects. However, although justifying the direct impact of disruptions and delays is relatively easy, there has been considerable difficulty in justifying and quantifying the claim for the indirect consequences. Our experience from working on a series of such claims is that some of the difficulty derives from ambiguity about the nature of disruption and delay (D&D). We now consider what we mean by D&D before moving onto considering the types of consequences that can result from the impact of D&D.

What Is a Disruption?

Disruptions are events that prevent the contractor completing the work as planned. Many disruptions to complex projects are planned for at the bid stage because they may be expected to unfold during the project. For example, some level of rework is usually expected, even when everything goes well, because there will always be "normal" errors and mistakes made by both the contractor and the client. The disruption and delay that follows would typically be taken to be a part of a risk factor encompassed in the base estimate, although this can be significantly underestimated (Eden et al. 2005). However, our experience suggests that *there are other types of disruptions that can be significant in their impact and are rarely thought about during original estimating*. When these types of disruptions do occur, their consequences can be underestimated as they are often seen by the contractor as aberrations with an expectation that their consequences can be controlled and managed. The linkage between risk assessment and the risks as potential triggers of D&D is often missed (Ackermann et al. 2007). Interferences with the flow of work in the project is a common disruption. For example, when a larger number of design comments than expected are made by the client, an increased number of drawings need rework. However, it also needs to be recognized that these comments could have been made by the contractor's own engineering staff. In either case, the additional work needed to respond to these comments, increases the contractor's workload and thus requires management to take mitigating actions if they still want to deliver on time. These mitigating actions are usually regarded as routine and capable of easily bringing the contract back to plan, even though they can have complex feedback ramifications.

Probably one of the most common disruptions to a project comes when a customer or contractor causes changes to the product (a Variation or Change Order). For example, the contractor may wish to alter the product after engineering work has commenced and so request a direct change. However, sometimes changes may be made unwittingly. For example, a significant part of cost overruns may arise where there have been what might be called "giveaways". These may occur because the contractor's engineers get excited about a unique and creative solution and rather than sticking to the original design, produce something better but with additional costs. Alternatively, when the contractor and customer have different interpretations of the contract requirements unanticipated changes can occur. For example, suppose the contract asks for a door to open and let out 50 passengers in 2 min, but the customer insists on this being assessed with respect to the unlikely event of dominantly large, slow passengers rather than the contractor's design assumptions of an average person. This is often known as "preferential engineering". In both instances there are contractor and/or customer requested changes that result in the final product being more extensive than originally intended.

The following example, taken from a real project and originally cited in Eden et al. (2005), illustrates the impact of a client induced change to the product:

Project 1: The contract for a "state-of-the-art" train had just been awarded. Using well-established design principles – adopted from similar train systems – the contractor believed that the project was on track. However, within a few months problems were beginning to emerge. The client team was behaving very differently from previous experience and using the contract specification to demand performance levels beyond that envisioned by the estimating team. One example of these performance levels emerged during initial testing, 6 months into the contract, and related to watertightness. It was discovered that the passenger doors were not sufficiently watertight. Under extreme test conditions a small (tiny puddle) amount of water appeared. The customer demanded that there must be no ingress of water, despite acknowledging that passengers experiencing such weather would bring in more water on themselves than the leakage.

The contractor argued that no train had ever met these demands, citing that most manufacturers and operators recognized that a small amount of water would always ingress, and that all operators accepted this. Nevertheless, the customer interpreted the contract such that new methods and materials had to be considered for sealing the openings. The dialog became extremely combative and the contractor was forced to redesign. An option was presented to the customer for their approval, one that would have ramifications for the production process. The customer, after many tests and after the verdict of many external experts in the field, agreed to the solution after several weeks. Not only were many designs revisited and changed, with an impact on other designs, but also the delays in resolution impacted the schedule well beyond any direct consequences that could be tracked by the schedule system (www.Primavera.com) or costs forecasting system.

What Is a Delay?

Delays are any events that will have an impact on the final date for completion of the project. Delays in projects come from a variety of sources. One common source is that of the client-induced delay. Where there are contractual obligations to comment upon documents, make approvals, supply information or supply equipment, and the client is late in these contractually defined duties, then there may be a client-induced delay to the expected delivery date (although in many instances the delay is presumed to be absorbed by slack). But also a delay could be self-inflicted: if the subassembly designed and built did not work, a delay might be expected.

Different types of client-induced delays (approvals, information, etc.) have different effects and implications. Delays in client approval, in particular, are often ambiguous contractually. A time to respond to approvals may not have been properly set, or the expectations of what was required within a set time may be ambiguous (e.g., in one project analyzed by the authors the clients had to respond within n weeks – but this simply meant that they sent back a drawing after n weeks with comments, then after the drawing was modified, they sent back the same drawing after a further n weeks with more comments). Furthermore, excessive comments, or delays in comments, can cause chains of problems, impacting, for example, on the document approval process with subcontractors, or causing overload to the client's document approval process.

If a delay occurs in a project, it is generally considered relatively straightforward to cost. However, ramifications resulting from delays are often not trivial either to understand or to evaluate. Let us consider a delay only in terms of the CPM (Critical Path Method), the standard approach for considering the effects of delays on a project (Wickwire and Smith 1974). The consequences of the delay depend on whether the activities delayed are on the Critical Path. If they *are* on the Critical Path, or the delays are sufficient to cause the activities to become on the critical path, it is conceptually easy to compute the effect as an Extension Of Time (EOT) (Scott 1993). However, even in this case there are complicating issues. For example, what is the effect on other projects being undertaken by the contractor? When this is not the first delay, then to which schedule does the term "critical path" refer? To the original, planned program, which has already been changed or disrupted, or to the "as built," actual schedule? Opinions differ here. It is interesting to note that, "the established procedure in the USA [of using as-built CPM schedules for claims] is almost unheard of in the UK" (Scott 1993).

If the delay is *not* on the Critical Path then, still thinking in CPM terms, there are only indirect co For example, the activities on the Critical Path likely to be resource dependent, and it is rarely to hire and fire at will – so if noncritical activities delayed, the project may need to work on tasks nonoptimal sequence to keep the workforce oc pied; this will usually imply making guesses engineering or production, requiring later rework. less productive work, stop/starts, workforce over-crowding, and so on.

The following example, taken from a real project, illustrates the impact of a delay in client furnished information to the project:

Project 2: A state-of-the-art vessels project had been commissioned that demanded not only the contractor meeting a challenging design but additionally incorporating new sophisticated equipment. This equipment was being developed in another country by a third party. The client had originally guaranteed that the information on the equipment would be provided within the first few months of the contract – time enough for the information to be integrated within the entire design. However, time passed and no detailed specifications were provided by the third party – despite continual requests from the contractor to the client.

As the project had an aggressive time penalty, the contractor was forced to make a number of assumptions in order to keep the design process going. Further difficulties emerged as information from the third party trickled in demanding changes from the emerging design. Finally, manufacturing, which had been geared up according to the schedule, were forced to use whatever designs they could access in order to start building the vessel.

Portfolio Effect of Many Disruptions

It is not just the extent of the disruption or delay but the number of them that may be of relevance. This is particularly the case when a large number of the disruptions and/or delays impact immediately upon one another, thus causing a portfolio of changes. These portfolios of D&D impacts result in effects that would probably not occur if only one or two impacts had occurred. For example, the combination of a large number of impacts might result in overcrowding or having to work in poor weather condition (see example below). In these instances, it is possible to identify each individual item as a contributory cause of extra work and delay but not easy to identify the combined effect.

The following example, taken from another real project, illustrates the impact of a series of disruptions to the project:

Project 3: A large paper mill was to be extended and modernized. The extension was given extra urgency by new antipollution laws imposing a limit on emissions being enacted with a strict deadline.

Although the project had started well, costs seemed to be growing beyond anything that made sense, given the apparent minor nature of the disruptions. Documents issued to the customer for "information only" were changed late in the process. The customer insisted on benchmarking proven systems, involving visits to sites working with experimental installations or installations operating under different conditions in various different countries. In addition, there were many changes of mind about where equipment should be positioned and how certain systems should work. Exacerbating these events was the circumstance of both the customer's and contractor's engineers being colocated, leading to "endless" discussions and meetings slowing the rate of both design and (later) commissioning.

Relations with the customer, who was seen by the contractor to be continually interfering with progress of the project, were steadily deteriorating. In addition, and in order to keep the construction work going, drawings were released to the construction team before being fully agreed upon. This meant that construction was done in a piece-meal fashion, often inefficiently (e.g., scaffolding would be put up for a job, then taken down so other work could proceed, then put up in the same place to do another task for which drawings subsequently had been produced). As the construction timescale got tighter and tighter, many more men were put on the site than was efficient (considerable overcrowding ensued) and so each task took longer than estimated.

As a result the project was behind schedule, and, as it involved a considerable amount of external construction work, was vulnerable to being affected by the weather. In the original project plan (as used for the estimate), the outer shell (walls and roof) was due to be completed by mid-Autumn. However, the project manager now found himself undertaking the initial construction of the walls and roofing in the middle of winter! As chance would have it, it was the coldest winter for decades, which resulted in many days being lost while it was too cold to work. The combination of the particularly vicious winter and many interferences resulted in an unexpectedly huge increase in both labor hours and overall delay. Overtime payments (for design and construction workers) escalated. The final overspend was over 40% more than the original budget.

Consequences of Disruptions and Delays

Disruption and delay (D&D) is primarily the consequence of interactions that feed on themselves as a result of an initial disruption or delay or portfolio of disruptions and delays. If an unexpected variation (or disruption) occurs in a project, then, if no intervention was to take place, a delivery delay would occur. In an attempt to avoid this situation, management may choose to take actions to prevent the delay (and possible penalties). In implementing these actions, side effects can occur, which cause further disruptions. These disruptions then cause further delays to the project. In order to avoid this situation, additional managerial action is required. Thus, an initial disruption has led to a delay, which has led to a disruption, which has led to a further delay. A positive feedback loop has been formed, where both disruption and delay feed back on themselves causing further disruptions and delays. Due to the nature of feedback loops, a powerful vicious cycle has been created, which, if there is no alternative intervention, can escalate with the potential of getting "out of control." It is the dynamic behavior caused by these vicious cycles that can cause severe disruption and consequential delay in a project.

The dynamic behavior of the vicious cycles that are responsible for much of the D&D in a project make the costing of D&D very difficult. It is extremely difficult to separate each of the vicious cycles and evaluate their individual cost. Due to the dynamic behavior of the interactions between vicious cycles, the cost of two individual cycles will escalate when they interact with one another; thus, disruptions have to be costed as part of a portfolio of disruptions.

Returning to Project 2, the vessel case, as can be seen in Fig. 2, the client caused both disruptions (continuous changes of mind) and delays (late permission to use a particular product). Both of these caused the contractor to undertake rework, and struggle with achieving a frozen (fixed) design. These consequences in turn impacted upon staff morale and also developed as noted above dynamic behavior – where rework resulted in more submissions of designs, which led to further comments, some of which were inconsistent and therefore led to further rework. As mentioned in the introduction, the rework cycle (Cooper 1993a, b, c, 1994) can be a major driver of escalating feedback within a complex project.

Managerial Actions and the Consequences of D&D

The acceleration of disrupted projects to avoid overall project delays is common practice by managers who are under pressure from the client and/or their own senior management to deliver on time. However, the belief that this action will always help avoid delays is naive as it does not take into account an appreciation of the future consequences that can be faced. For example, one typical action designed to accelerate a project is to hire new staff. In doing so, some of the difficulties that may follow are:

- New staff take time to become acquainted with both the project and thus their productivity is lower than that of an existing skilled worker.

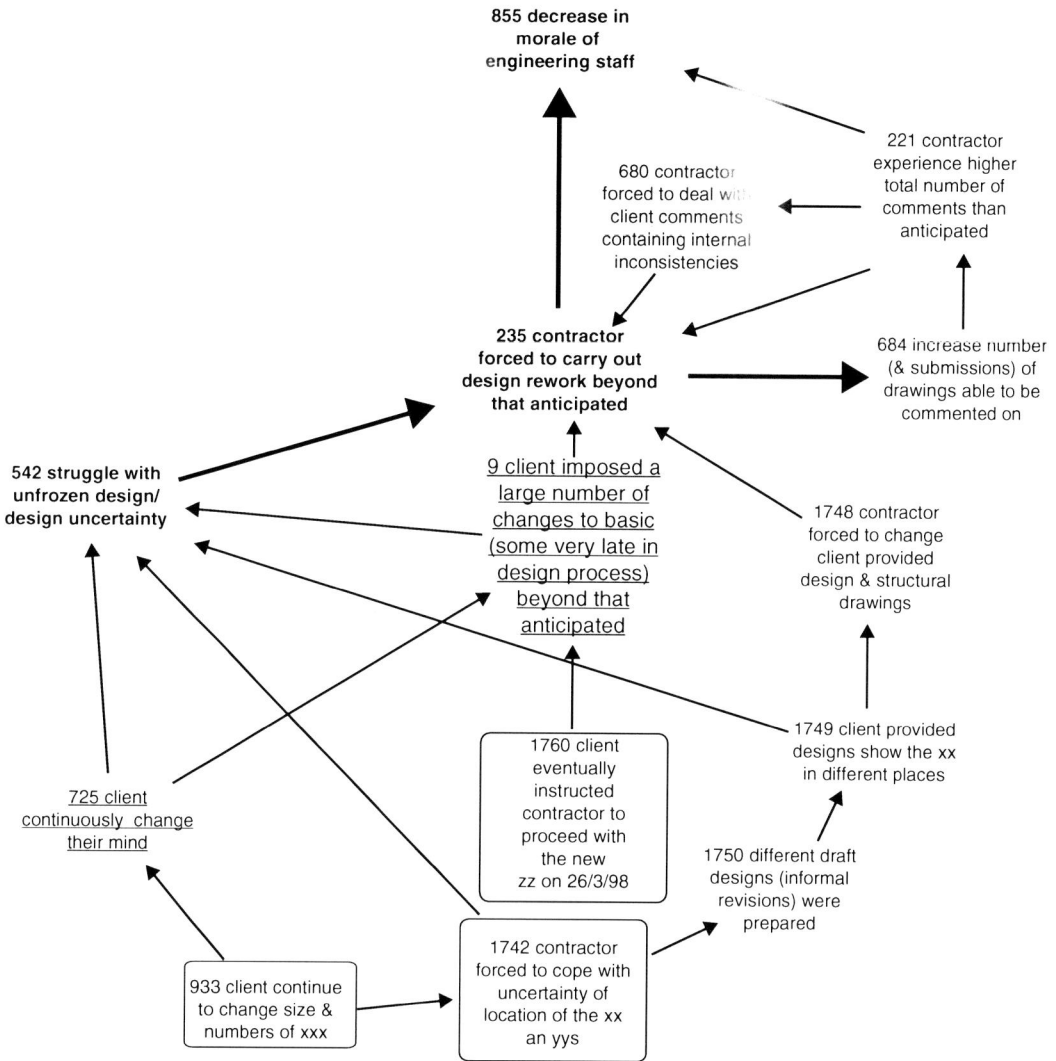

Delay and Disruption in Complex Projects, Fig. 2 Excerpt from a cause map showing some of the consequences of disruption and delay in Project 2. *Boxed* statements are specific illustrations with statements *underlined* representing generic categories (e.g., changes of mind). Statements in *bold text* represent the SD variables with the remainder providing additional context. All links are causal; however, those in bold illustrate sections of a feedback loop. The numbers at the beginning of concept are used as reference numbers in the model

- New staff require training on the project and this will have an impact on the productivity of existing staff.
- Rather than hiring new staff to the organization, staff may be moved from other parts of the organization. This action results in costs to other projects as the other project is short of staff and so may have to hire workers from elsewhere, thereby suffering many of the problems discussed above.

Many of the outcomes of this action and other similar actions can lead to a reduction in expected productivity levels. Low productivity is a further disruption to the project through a lack of expected progress. If management identifies this lack of progress, then further managerial actions may be taken in an attempt to avoid a further delay in delivery. These actions often lead to more disruptions, reinforcing the feedback loop that had been set up by the first actions.

Two other common managerial actions taken to avoid the impact of a disruption on delivery are (i) the use of overtime and (ii) placing pressure on staff in an attempt to increase work rate. Both of these actions can also have detrimental effects on staff productivity once they have reached particular levels. Although these actions are used to increase productivity levels, effects on fatigue and morale can actually lead to a lowering of productivity via a slower rate of work and/or additional work to be completed due to increased levels of rework (Cooper 1994; Howick and Eden 2001). This lowering of productivity causes a delay through lack of expected progress on the project, causing a further delay to delivery. Management may then attempt to avoid this by taking other actions, which in turn cause a disruption, which again reinforces the feedback loop that has been set up.

Analyzing D&D and Project Behavior

The above discussion has shown that whilst D&D is a serious aspect of project management, it is a complicated phenomenon to understand. A single or a series of disruptions or delays can lead to significant impacts on a project, which cannot be easily thought through due to human difficulties in identifying and thinking through feedback loops (Sterman 1989; Diehl and Sterman 1995). This makes the analysis of D&D and the resulting project behavior particularly difficult to explain.

SD modeling has made a significant contribution to increasing our understanding of why projects behave in the way they do and in quantifying effects. There are two situations in which this is valuable: the claim situation, where one side of the party is trying to explain the project's behavior to the other (and, usually, why the actions of the other party has caused the project to behave in the way it has) and the post-project situation, where an organization is trying to learn lessons from the experience of a project. In the case of a claim situation, although it has been shown that SD modeling can meet criteria for admissibility to court (Stephens et al. 2005), there are a number of objectives that SD, or any modeling method, would need to address (Howick 2003). These include the following:

1. Prove causality – show what events triggered the D&D and how the triggers of D&D caused time and cost overruns on the project.
2. Prove the "quantum" – show that the events that caused D&D created a specific time and cost overrun in the project. Therefore, there is a need to replicate over time the hours of work due to D&D that were over-and-above those that were contracted, but were required to carry out the project.
3. Prove responsibility – show that the defendant was responsible for the outcomes of the project. Also to demonstrate the extent to which plaintiffs management of the project was reasonable and the extent that overruns could not have been reasonably avoided.
4. All of the above have to be proved in a way that will be convincing to the several stakeholders in a litigation audience.

Over the last 12 years, the authors have developed a model building process that aims to meet each of these purposes. This process involves constructing qualitative models to aid the process of building the "case" and thus help to prove causality and responsibility (purposes 1 and 3). In addition, quantitative system dynamics models are involved in order to help to prove the quantum (purpose 2). However, most importantly, the process provides a structured, transparent, formalized process from "real world" interviews to resulting output that enables multiple audiences, including multiple nonexperts as well as scientific/expert audiences to appreciate the validity of the models and thus gain confidence in these models and the consulting process in which they are embedded (purpose 4). The process is called the "Cascade Model Building Process." The next section describes the different stages of the model building process and some of the advantages of using the process.

Cascade Model Building Process

(The following contains excerpts from Howick et al. (2007), which contains a full description of the Cascade Model Building process).

**Delay and Disruption in
Complex Projects,
Fig. 3** The Cascade Model
Building Process

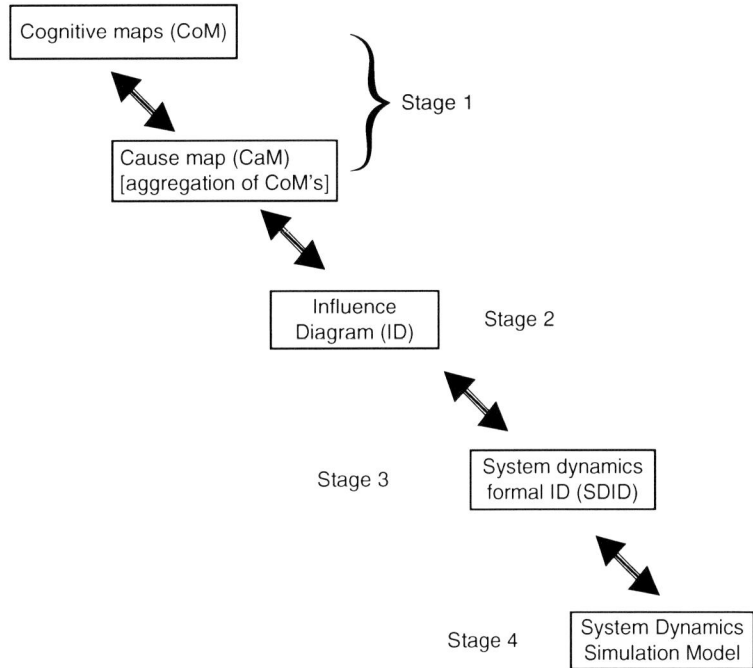

**Delay and Disruption in
Complex Projects,
Fig. 3** The Cascade Model
Building Process

The "Cascade Model Building Process" involves four stages (see Fig. 3), each of which is described below.

Stage 1: Qualitative Cognitive and Cause Map

The qualitative cognitive maps and/or project cause map aim to capture the key events that occurred on the project, for example, a delay as noted above in the vessel example in Project 2. The process of initial elicitation of these events can be achieved in two ways. One option is to interview, and construct cognitive maps (Eden 1988; Ackermann and Eden 2004; Bryson et al. 2004) for each participant's views. Here the aim is to gain a deep and rich understanding that taps the wealth of knowledge of each individual. These maps act as a preface to getting the group together to review and assess the total content represented as a merged cause map (Shaw et al. 2003) in a workshop setting. The second option is to undertake group workshops where participants can contribute directly, anonymously, and simultaneously, to the construction of a cause map. The participants are able to "piggy back" off one another, triggering new memories, challenging views, and developing together a comprehensive overview (Ackermann and Eden 2001). As contributions from one participant are captured and structured to form a causal chain, this process triggers thoughts from others and as a result a comprehensive view begins to unfold. In Project 1, this allowed the relevant design engineers (not just those whose responsibility was the watertight doors, but also those affected who were dealing with car-body structure, ventilation, etc.), methods personnel, and construction managers to surface a comprehensive view of the different events and consequences that emerged

The continual development of the qualitative model, sometimes over a number of group workshops, engenders clarity of thought predominantly through its adherence to the coding formalisms used for cause mapping (Ackermann et al. 2005). Members of the group are able to debate and consider the impact of contributions on one another. Through bringing the different views together it is also possible to check for coherency – do all the views fit together or are there inconsistencies? This is not uncommon as different parts of the organizations (including different discipline groups within a division, e.g., engineering) encounter particular effects. For example, during an engineering project, manufacturing can often find themselves bewildered by engineering processes – why are designs so late? However, the first stage of the cascade process enables the views from engineering, methods, manufacturing, commissioning, etc., to be integrated. Arguments are tightened as a result, inconsistencies identified and resolved and detailed audits (through analysis and features in the modeling software) undertaken to ensure consistency between both modeling team and model audience. In some instances, the documents generated through reports about the organizational situation can be coded into a cause map and merged into the interview and workshop material (Eden and Ackermann 2004).

The cause map developed at this stage is usually large – containing up to 1,000 nodes. Computer supported analysis of the causal map can inform further discussion. For example, it can reveal those aspects of causality that are central to understanding what happened. Events that have multiple consequences for important outcomes can be detected. Feedback loops can be identified and examined. The use of software facilitates the identification of sometimes complex but important feedback loops that follow from the holistic view that arises from the merging of expertise and experience across many disciplines within the organization.

The resulting cause map from stage 1 can be of particular use in proving causality. For example, Fig. 4 represents some of the conversations made regarding the water ingress situation described in the above case. In this figure, consequences such as additional engineering effort and engineering delays can be traced back to events such as client finding water seeping out of door.

Stage 2: Cause Map to Influence Diagram

The causal model produced from stage 1 is typically very extensive. This extensiveness requires that a process of "filtering" or "reducing" the content be undertaken – leading to the development of an Influence Diagram (ID) (the second step of the cascade process). Partly, this is due to the fact that many of the statements captured while enabling a detailed and thorough understanding of the project are not relevant when building the SD model in stage 4 (as a result of the statements being of a commentary-like nature rather than a discrete variable). Another reason is that for the most part SD models comprise fewer variables/auxiliaries to help manage the complexity (necessary for good modeling as well as comprehension).

The steps involved in moving from a cause map to an ID are as follows:

Step 1: Determining the core/endogenous variables of the ID	1. Identification of feedback loops: it is not uncommon to find over 100 of these (many of these may contain a large percentage of common variables) when working on large projects with contributions from all phases of the project
	2. Analysis of feedback loops: once the feedback loops have been detected they are scrutinized to determine (a) whether there are nested feedback "bundles" and (b) whether they traverse more than one stage of the project. Nested feedback loops comprise a number of feedback loops around a particular topic where a large number of the variables/statements are common but with variations in the formulation of the feedback loop. Once detected, those statements that appear in the most number of the nested feedback loops are identified as they provide core variables in the ID model
	Where feedback loops straddle different stages of the process, for example,

(continued)

	from engineering to manufacturing note is taken. Particularly interesting is where a feedback loop appears in one of the later stages of the project, e.g., commissioning which links back to engineering. Here care must be taken to avoid chronological inconsistencies – it is easy to link extra engineering hours into the existing engineering variable; however, by the time commissioning discover problems in engineering, the majority if not all engineering effort has been completed
Step 2: Identifying the triggers/exogenous variables for the ID	The next stage of the analysis is to look for triggers – those statements that form the exogenous variables in the ID. Two forms of analysis provide clues that can subsequently be confirmed by the group:
	1. The first analysis focuses on starting at the end of the chains of argument (the tails) and laddering up (following the chain of argument) until a branch point appears (two or more consequences). Often statements at the bottom of a chain of argument are examples that when explored further lead to a particular behavior, e.g., delay in information, which provides insights into the triggers
	2. The initial set of triggers created by (i) can be confirmed through a second type of analysis – one that takes two different means of examining the model structure for those statements that are central or busy. Once these are

(continued)

	identified, they can be examined in more detail through creating hierarchical sets based upon them and thus "tear drops" of their content. Each of these teardrops is examined as possible triggers
Step 3: Checking the ID	Once the triggers and the feedback loops are identified, care is taken to avoid double counting – where one trigger has multiple consequences some care must be exercised in case the multiple consequences are simple replications of one another

The resulting ID is comparable to a "causal loop diagram" (Lane 2000), which is often used as a precursor to an SD model. From the ID structure it is possible to create "stories" where a particular example triggers an endogenous variable that illustrates the dynamic behavior experienced (Fig. 5).

Stage 3: Influence Diagram to System Dynamics Influence Diagram (SDID)

When an SD model is typically constructed after producing a qualitative model such as an ID (or causal loop diagram), the modeler determines which of the variables in the ID should form the stocks and flows in the SD model, then uses the rest of the ID to determine the main relationships that should be included in the SD model. However, when building the SD model there will be additional variables/constants that will need to be included in order to make it "work" that were not required when capturing the main dynamic relationships in the ID. The SDID is an influence diagram that includes all stocks, flows, and variables that will appear in the SD model and is, therefore, a qualitative version of the SD model. It provides a clear link between the ID and the SD model.

Delay and Disruption in Complex Projects, Fig. 4 Excerpt from a cause map showing some of the conversations regarding the water ingress situation in Project 1. As with Fig. 2, statements that have borders are the illustrations, those with *bold font* represent variables with the remainder detailing context. *Dotted arrows* denote the existence of further material which can be revealed at anytime

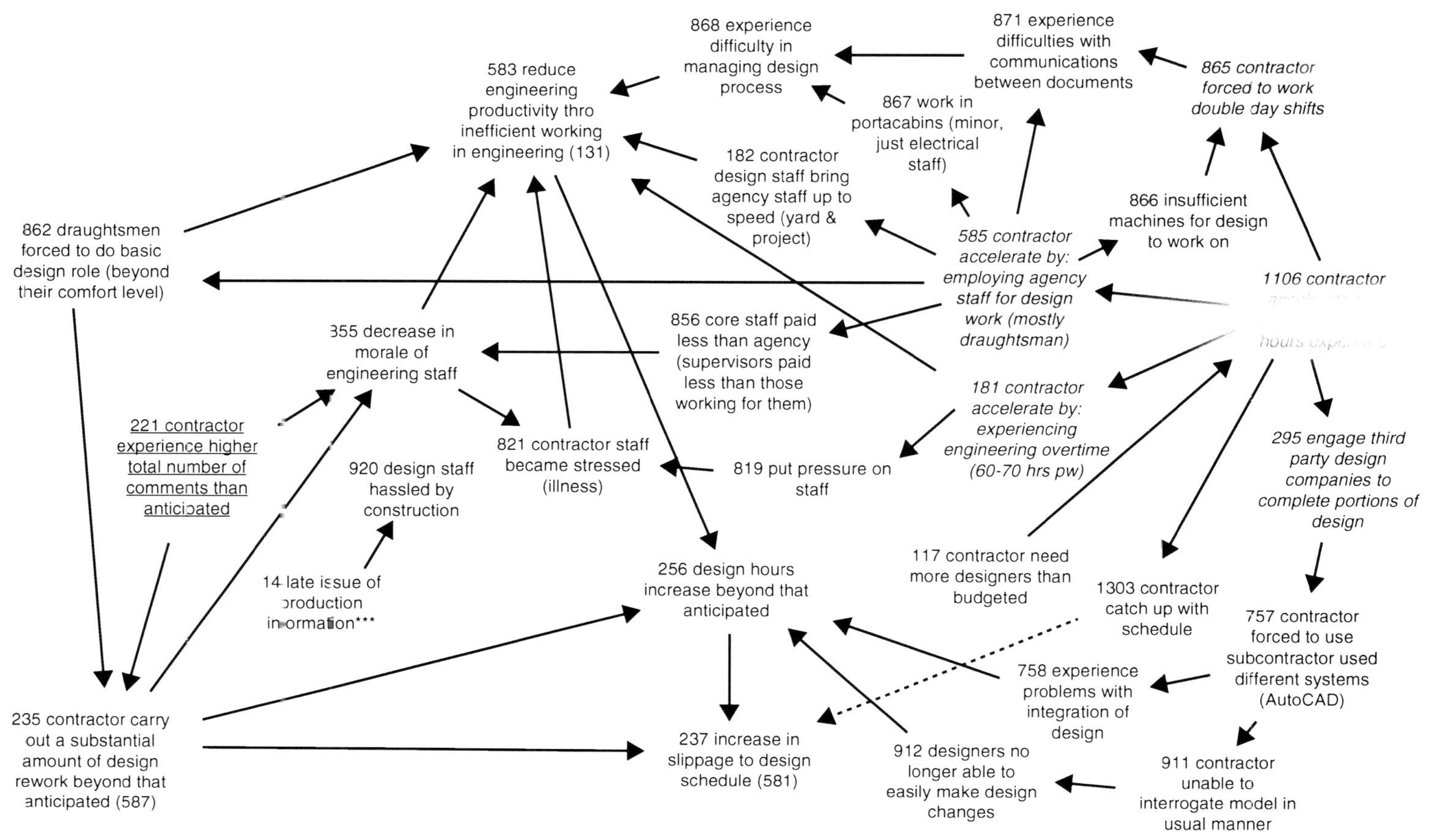

Delay and Disruption in Complex Projects, Fig. 5 A small section of an ID from Project 2 showing mitigating actions (*italics*), triggers (*underline*) and some of the feedback cycles

The SDID is therefore far more detailed than the ID and other qualitative models normally used as a precursor to an SD model.

Methods have been proposed to automate the formulation of an SD model from a qualitative model such as a causal loop diagram (Burns 1977; Burns and Ulgen 1978; Burns et al. 1979) and for understanding the underlying structure of an SD model (Oliva 2004). However, these methods do not allow for the degree of transparency required to enable the range of audiences involved in a claim situation, or indeed as part of an organizational learning experience, to follow the transition from one model to the next. The SDID provides an intermediary step between an ID and an SD model to enhance the transparency of the transition from one model to another for the audiences. This supports an auditable trail from one model to the next.

The approach used to construct the SDID is as follows:

The SDID is initially created in parallel with the SD model. As a modeler considers how to translate an ID into an SD model, the SDID provides an intermediary step. For each variable in the ID, the modeler can do either of the following:

1. Create one variable in the SD & SDID: If the modeler wishes to include the variable as one variable in the SD model, then the variable is simply recorded in both the SDID and the SD model as it appears in the ID.
2. Create multiple variables in the SD & SDID: To enable proper quantification of the variable,

additional variables need to be created in the SD model. These variables are then recorded in both the SD model and SDID with appropriate links in the SDID, which reflect the structure created in the SD model.

The SDID model forces all qualitative ideas to be placed in a format ready for quantification. However, if the ideas are not amenable to quantification or contradict one another, then this step is not possible. As a result of this process, a number of issues typically emerge including the need to add links and statements and the ability to assess the overall profile of the model though examining the impact of particular categories on the overall model structure. This process can also translate back into the causal model or ID model to reflect the increased understanding.

Stage 4: The System Dynamics Simulation Model

The process of quantifying SD model variables can be a challenge, particularly as it is difficult to justify subjective estimates of higher-level concepts such as "productivity" (Ford and Sterman 1998). However, moving up the cascade reveals the causal structure behind such concepts and allows quantification at a level that is appropriate to the data-collection opportunities available. Figure 6, taken from the ID for Project 1, provides an example. The quantitative model will require a variable "productivity" or

Delay and Disruption in Complex Projects,
Fig. 6 Section of an ID from Project 1 showing the factors affecting productivity

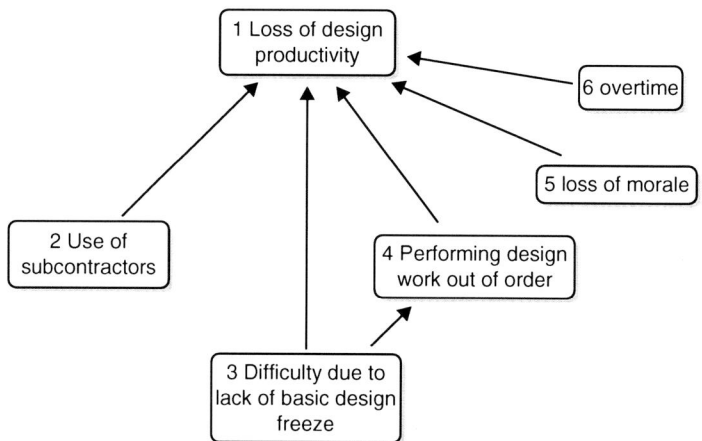

"morale," and the analyst will require estimation of the relationship between it and its exogenous and (particularly) endogenous causal factors. While the higher-level concept is essential to the qualitative model, simply presenting it to the project team for estimation would not facilitate justifiable estimates of these relationships.

Reversing the Cascade

The approach of moving from stage 1 through to stage 4 can increase understanding and stimulate learning for all parties. However, the process of moving back up the cascade can also facilitate understanding between the parties. For example, in Fig. 7 the idea that a company was forced to use subcontractors and thus lost productivity might be a key part of a case for lawyers. The lawyers and the project team might have come at Fig. 7 as part of their construction of the case. Moving back up from the ID to the Cause Map (i.e., Figs. 7 and 8) as part of a facilitated discussion not only helps the parties to come to an agreed definition of the

(often quite ill-defined) terms involved, it also helps the lawyers understand how the project team arrived at the estimate of the degree of the relationship. Having established the relationship, moving through the SDID (ensuring well-defined variables, etc.) to the SD model enables the analysts to test the relationships to see whether any contradictions arise, or if model behaviors are significantly different from actuality, and it enables comparison of the variables with data that might be collected by (say) cost accountants. Where there are differences or contradictions, the ID can be reinspected and if necessary the team presented with the effect of the relationship within the SD model explained using the ID, so that the ID and the supporting cause maps can be reexamined to identify the flaws or gaps in the reasoning. Thus, in this example, as simulation modelers, cost accountants, lawyers, and engineers approach the different levels of abstraction, the cascade process provides a unifying structure within which they can communicate, understand each other, and equate terms in each other's discourse.

Advantages of the Cascade

The Cascade integrates a well-established method, cause mapping, with SD. This integration results in a number of important advantages for modeling to explain project behavior:

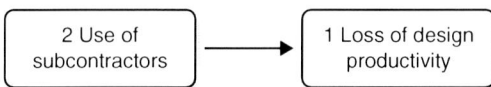

Delay and Disruption in Complex Projects, Fig. 7 Section of an ID from Project 1 indicating the influence of the use of subcontractors on productivity

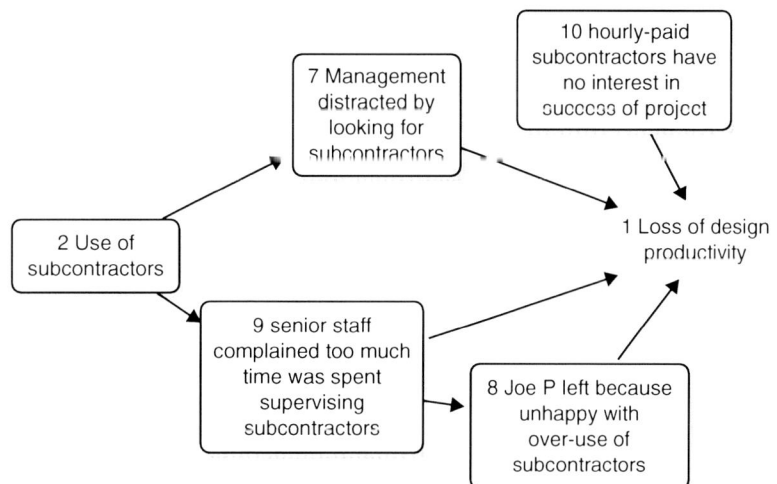

Delay and Disruption in Complex Projects, Fig. 8 Section of a Cause Map from Project 1 explaining the relationship between the use of subcontractors and productivity

Achieving Comprehensiveness

Our experience suggests that one of the principal benefits of using the cascade process derives from the added value gained through developing a rich and elaborated qualitative model that provides the structure (in a formalized manner) for the quantitative modeling. The cascade process immerses users in the richness and subtlety that surround their view of the projects and ensures involvement and ownership of all of the qualitative and quantitative models. The comprehensiveness leads to a better understanding of what occurred, which is important due to the complex nature of D&D, and enables effective conversations to take place across different organizational disciplines.

The process triggers new contributions as memories are stimulated and both new material and new connections are revealed. The resultant models thus act as organizational memories providing useful insights into future project management (both in relation to bids and implementation). These models provide more richness and therefore an increased organizational memory when compared to the traditional methods used in group model building for system dynamics models (e.g., Vennix 1996). However, this outcome is not untypical of other problem structuring methods (Rosenhead and Mingers 2001).

Testing the Veracity of Multiple Perspectives

The cascade's bidirectionality enabled the project team's understandings to be tested both numerically and from the perspective of the coherency of the systemic portrayal of logic. By populating the initial quantitative model with data (Ackermann et al. 1997), rigorous checks of the validity of assertions were possible.

In a claim situation, blame can be the fear of those participating in accounting for history and often restricts contributions (Ackermann et al. 2005). When initiating the cascade process, the use of either interviews or group workshops increases the probability that the modeling team will uncover the rich story rather than partial explanations or as is often the case with highly politicized situations, "sanitized" explanations. By starting with "concrete" events that can be verified, and exploring their multiple consequences, the resultant model provides the means to reveal and explore the different experiences of various stakeholders in the project.

Modeling Transparency

By concentrating the qualitative modeling efforts on the capture and structuring of multiple experiences and viewpoints the cascade process initially uses natural language and rich description as the medium that facilitates generation of views and enables a more transparent record to be attained.

There are often insightful moments as participants viewing the *whole* picture realize that the project is more complex than they thought. This realization results in two advantages. The first is a sense of relief that they did not act incompetently given the circumstances, that is, the consequences of D&D took over – which in turn instills an atmosphere more conducive to openness and comprehensiveness (see Ackermann et al. 2005). The second is learning – understanding the whole, the myriad and interacting consequences, and in particular the dynamic effects that occurred on the project (that often acts in a counterintuitive manner) provides lessons for future projects.

Common Understanding Across Many Audiences

Claim situations involve numerous stakeholders, with varying backgrounds. The cascade process promotes ownership of the models from this mixed audience. For example, lawyers are more convinced by the detailed qualitative argument presented in the cause map (stage 1) and find this part of greatest utility and hence engage with this element of the cascade. However, engineers get more involved in the construction of the quantitative model and evaluating the data encompassed within it.

A large, detailed system dynamics model can be extremely difficult to understand for many of the stakeholders in a claim process (Howick 2005). However, the rich qualitative maps developed as part of the cascade method are presented in terms that are easier for people with no modeling experience to understand. In addition, by moving back up the cascade, the dynamic results that are output by the simulation model are given a grounding in the key events of the project, enabling the audience to be given fuller explanations and reasons for the D&D that occurred on the project.

Using the cascade method, any structure or parameters that are contained in the simulation model can be easily, and quickly, traced back to information gathered as a part of creating the cognitive maps or cause maps. Each contribution in these maps can then normally be traced to an individual witness who could defend that detail in the model. This auditable trail can aid the process of explaining the model and refuting any attacks made on the model.

Clarity

The step-by-step process forces the modeler to be clear in what statements mean. Any illogical or inconsistent statements highlighted require the previous stage to be revisited and meanings clarified, or inconsistencies cleared up. This results in clear, logical models.

Confidence Building

As a part of gaining overall confidence in a model, any audience for the model will wish to have confidence in the structure of the model (e.g., Ackoff and Sasieni 1968; Rivett 1972; Mitchell 1993; Pidd 2003). When assessing confidence levels in a part of the structure of an SD model, the cascade process enables any member of the "client" audience to clearly trace the structure of the SD model directly to the initial natural language views and beliefs provided from individual interviews or group sessions.

Scenarios are also an important test in which the confidence of the project team in the model can be considerably strengthened. Simulation is subject to the demands to reproduce scenarios that are recognizable to the managers capturing a portfolio of meaningful circumstances that occur at the same time, including many qualitative aspects such as morale levels. For example, if a particular time-point during the quantitative simulation is selected, clearly the simulated values of all the variables, and in particular the relative contributions of factors in each relationship, can be output from the model. If we consider Fig. 6, the

simulation might show that at a particular point in a project, loss of productivity is 26% and that the loss due to:

> "Use of subcontractors" is 5%.
> "Difficulty due to lack of basic design freeze" is 9%.
> "Performing design work out of order" is 3%.
> "Loss of morale" is 5%.
> "Overtime" is 4%.

Asking the project team their estimates of loss of productivity at this point in time, and their estimation of the relative contribution of these five factors, will help to validate the model. In most cases, this loss level is best captured by plotting the relative levels of productivity against the time of critical incidents during the life of the project. Discussion around this estimation might reveal unease with the simple model described in Fig. 6, which will enable discussion around the ID and the underlying cause map, either to validate agreed upon, or possibly to modify it and return up the cascade to further refine the model. In this scenario, validation of the cascade process provides a unifying structure within which the various audiences can communicate and understand each other.

The Cascade Model Building Process provides a rigorous approach to explaining why a project has behaved in a certain way. The cascade uses rich, qualitative stories to give a grounding in the key events that drive the behavior of the project. In addition, it provides a quantifiable structure that allows the over time dynamics of the project to be described. The Cascade has therefore contributed significantly in understanding why projects behave in the way they do.

This chapter has focused on the role of SD modeling in explaining the behavior of complex projects. The final two sections will consider the implications of this work and will explore potential future directions for the use of SD modeling of projects.

Implications for Development

So what is the current status of SD modeling of projects? What is the research agenda for studying projects using SD? Below we consider each aspect of the project life cycle in turn, to suggest areas where SD modeling may be applied, and to consider where further work is needed.

The first area is pre-project risk analysis. Risk analysis traditionally looks at risks individually, but looking at the systemicity in risks has clear advantages (Williams et al. 1997). Firstly, the use of cause mapping techniques by an experienced facilitator, aided by software tools, is a powerful means of drawing out knowledge of project risk from an individual manager (or group of managers), enhancing clarity of thought, allowing investigation of the interactions between risks, and enhancing creativity. It is particularly valuable when used with groups, bringing out interactions between the managers and helping to surface cultural differences. And it clearly enables analysis of the systemicity, in particular identification of feedback dynamics, which can help explicate project dynamics in the ways discussed above. The influence of such work has led to the ideas of cause maps, influence diagrams, and SD to be included into risk practice standard advice (the UK "PRAM" Guide, edition 2 (APM Publishing Ltd 2004) – absent from Edition 1). In one key example (Ackermann et al. 2007), the work described above enabled the team to develop a "Risk Filter" in a large multinational project-based organization, for identifying areas of risk exposure on future projects and creating a framework for their investigation. The team reviewed the system after a few years; it had been used by nine divisions, on over 60 major projects, and completed by 450 respondents; and it was used at several stages during the life of a project to aid in the risk assessment and contribute to a project database. The system allowed investigation of the interactions between risks, and so encouraged the management of the causality of relationships between risks, rather than just risks, thus focusing attention on those risks and causality that create the most frightening ramifications on clusters of risks, as a system, rather than single items. This also encouraged conversations about risk mitigation across disciplines within the organization. Clearly, cause mapping is useful in risk analysis,

but there are a number of research questions that follow, for example:

- In looking at possible risk scenarios, what are appropriate methodologies to organize and facilitate heterogeneous groups of managers? And how technically can knowledge of systemicity and scenarios be gathered into one integrated SD model and enhance understanding? (Howick et al. 2006).
- How can SD models of possible scenarios be populated to identify key risks? How does the modeling cascade help in forward-looking analysis?
- There are many attempts to use Monte-Carlo simulation to model projects, without taking the systemic issues into account – leading to models that can be seriously misleading (Williams 2004). SD models can give a much more realistic account of the effect of risks – but how can essentially deterministic SD models as described above be integrated into a stochastic framework to undertake probabilistic risk analyses of projects, which acknowledges the systemicity between the risks and the systemic effects of each risk?
- The use of SD is able to identify structures that give projects a propensity for the catastrophic systemic effects discussed in the Introduction. In particular, the three dimensions of structural complexity, uncertainty, and severe time-limitation in projects can combine together to cause significant positive feedback. However, defining metrics for such dimensions still remains an important open question. While a little work has been undertaken to give operational measures to the first of these (e.g., Williams 1999; Shenhar 2001), and de Meyer et al. (2002) and Shenhar and Dvir (2004) suggest selecting the management strategy based on such parameters, there has been little success so far in quantifying these attributes. The use of the SD models discussed above needs to be developed to a point where a project can be parameterized to give quantitatively its propensity for positive feedback.
- Finally, SD modeling shows that the effects of individual risks can be considerably greater

than intuition would indicate, and the effects of clusters of risks particularly so. How can this be quantified so that risks or groups of risks can be ranked in importance to provide prioritization to managers? Again, Howick et al. (2006) give some initial indications here, but more work is needed.

The use of SD in operational control of projects has been less prevalent (Lyneis et al. (2001) refer to and discuss examples of where it has been used). For a variety of reasons, SD and the traditional project management approach do not match well together. Traditional project-management tools look at the project in its decomposed pieces in a structured way (networks, work breakdown structures, etc.); they look at operational management problems at a detailed level; SD models aggregate into a higher strategic level and look at the underlying structure and feedback. Rodrigues and Williams (1997) describe one attempt at an integrated methodology, but there is scope for research into how work with the SD paradigm can contribute to operational management of projects, and Williams (2002) provides some suggestions for hybrid methods.

There is also a more fundamental reason why SD models do not fit in easily into conventional project management. Current project management practice and discourse is dominated by the "Bodies of Knowledge" or BoKs (Dixon 2000), which professional project management bodies consider to be the core knowledge of managing projects (Project Management Institute 2000; Stevens 2002), presenting sets of normative procedures that appear to be self-evidently correct. However, there are three underlying assumptions to this discourse (Williams 2005)

- Project Management is self-evidently correct: it is rationalist (Lundin 1995) and normative (Packendorff 1995).
- The ontological stance is effectively positivist (Linehan and Kavanagh 2004).
- Project management is particularly concerned with managing scope in individual parts (Koskela and Howell 2002a).

These three assumptions lead to three particular *emphases* in current project management discourse and thus in the BoKs (Williams 2005):

- A heavy emphasis on planning (Packendorff 1995; Koskela and Howell 2002b).
- An implication of a very conventional control model (Hodgson 2004).
- Project management is generally decoupled from the environment (Malgrati and Damiani 2002).

The SD modeling work provided explanations for why some projects severely overrun, which clash with these assumptions of the current dominant project management discourse.

- Unlike the third assumption, the SD models show behavior arising from the complex interactions of the various parts of the project, which would *not* be predicted from an analysis of the individual parts of the project (Lindkvist et al. 1998).
- Against the first assumption, the SD models show project behavior that is complex and nonintuitive, with feedback exacerbated through management response to project perturbations, conventional methods provide unhelpful or even disbeneficial advice and are not necessarily self-evidently correct.
- The second assumption is also challenged. Firstly, the models differ from the BoKs in their emphasis on, or inclusion of, "soft" factors, often important links in the chains of causality. Secondly, they show that the models need to incorporate not only "real" data but management perceptions of data and to capture the socially constructed nature of "reality" in a project.

The SD models tell us why failures occur in projects that exhibit complexity (Williams 1999) – that is, when they combine *structural complexity* (Baccarini 1996) – many parts in complex combinations – and *uncertainty*, in project goals and in the means to achieve those goals (Turner and Cochrane 1993). Goal uncertainty in particular is lacking in the conventional project management discourse (Linehan and Kavanagh 2004; Engwall 2002), and it is when uncertainty affects a structurally complex traditionally managed project that the systemic effects discussed above start to occur. But there is a third factor identified in the SD modeling. Frequently, events arise that compromise the plan at a faster rate than that at which it is practical to replan. When the project is heavily *time-constrained,* the project manager feels forced to take acceleration actions. A structurally complex project when perturbed by external uncertainties can become unstable and difficult to manage, and under time-constraints dictating acceleration actions when management has to make very fast and sometimes very many decisions, the catastrophic overruns described above can occur. Work from different direction seeking to establish characteristics that cause complexity projects come up with similar characteristics (e.g., Shenhar and Dvir 2004). But the SD modeling explains *how* the tightness of the time-constraints strengthens the power of the feedback loops, which means that small problems or uncertainties can cause unexpectedly large effects; it also shows how the type of underspecification identified by Flyvberg et al. (2003) brings what is sometimes called "double jeopardy" – underestimation (when the estimate is elevated to the status of a project control-budget) that leads to acceleration actions that then cause feedback, which causes much greater overspend than the degree of underestimation.

Because of this, the greatest contribution that SD has made – and perhaps can make – is to increase our understanding of why projects behave in the way they do. There are two situations in which this is valuable: the claim situation, where one side of the party is trying to explain the project's behavior to the others (and, usually, why the actions of the other party has caused the project to behave in the way it has) and the post-project situation, where an organization is trying to learn lessons from the experience of a project.

The bulk of the work referred to in this chapter comes in the first of these, the claim situation. However, while these have proved popular among SD modelers, they have not necessarily found universal acceptance among the practicing project-management community. Work is needed, therefore, in a number of directions. These will be discussed in the next section.

Future Directions

We have already discussed the difficulty that various audiences can have in comprehending a large, detailed system dynamics model (Howick 2005), and that gradual explanations that can be given by working down (and back up) the cascade to bring understanding to a heterogeneous group (which might include jurors, lawyers, engineers, and so on) and so link the SD model to key events in the project. While this is clearly effective, more work is needed to investigate the use of the cascade. In particular, ways in which the cascade can be most effective in promoting understanding, in formalizing the methodology and the various techniques mentioned above to make it replicable, as well as how best to use SD here (Howick 2005), outlines nine particular challenges the SD modeler faces in such situations). Having said this, it is still the case that many forums in which claims are made are very set in conventional project-management thinking, and we need to investigate more how the SD methods can be combined with more traditional methods synergistically, so that each supports the other (see, e.g., Williams 2003).

Significant unrealized potential of these methodologies are to be found in the post-project "lessons learned" situation. Research has shown many problems in learning generic lessons that can be extrapolated to other projects, such as getting to the root causes of problems in projects, seeing the underlying systemicity, and understanding the narratives around project events (Williams 2007, which gives an extensive bibliography in the area). Clearly, the modeling cascade, working from the messiness of individual perceptions of the situation to an SD model, can help in these areas. The first part of the process (Fig. 3), working through to the cause map, has been shown to enhance understanding in many cases; for example, Robertson and Williams (2006) describe a case in an insurance firm, and

Williams (2004) gives an example of a project in an electronics firm, where the methodology was used very "quick and dirty" but still gave increased understanding of why (in that case successful) project turned out as it did, with some pointers to lessons learned about the process. However, as well as formalization of this part of the methodology and research into the most effective ways of bringing groups together to form cause maps, more clarity is required as to how far down the cascade to go and the additional benefits that the SD modeling brings. "Stage 4" describes the need to look at quantification at a level that is appropriate to the data-collection opportunities available, and there might perhaps be scope for SD models of parts of the process explaining particular aspects of the outcomes. Attempts to describe the behavior of the whole project at a detailed level may only be suitable for the claims situation; there needs to be research into what is needed in terms of Stages 3 and 4 for gaining lessons from projects (or, if these stages are not carried out, how the benefits such as enhanced clarity and validity using the cause maps, can be gained).

One idea for learning lessons from projects used by the authors, following the idea of simulation "learning labs," was to incorporate learning from a number of projects undertaken by one particular large manufacturer into a simulation learning "game" (Williams et al. 2005). Over a period of 7 years, several hundred Presidents, Vice-Presidents, Directors, and Project Managers from around the company used the simulation tool as a part of a series of senior management seminars, where it promoted discussion around the experience and the effects encountered, and encouraged consideration of potential long-term consequences of decisions, enabling cause and effect relationships and feedback loops to be formed from participants' experiences. More research is required here as to how such learning can be made most effective.

SD modeling has brought a new view to project management, enabling understanding of the behavior of complex projects that was not accessible with other methods. The chapter has described methodology for where SD has been used in this domain.

This last part of the chapter has looked forward to a research agenda into how the SD work needs to be developed to bring greater benefits within the project management community.

Bibliography

Ackermann F, Eden C (2001) Contrasting single user and networked group decision support systems. Group Decis Negot 10(1):47–66

Ackermann F, Eden C (2004) Using causal mapping: individual and group: traditional and new. In: Pidd M (ed) Systems modelling: theory and practice. Wiley, Chichester, pp 127–145

Ackermann F, Eden C, Williams T (1997) Modeling for litigation: mixing qualitative and quantitative approaches. Interfaces 27:48–65

Ackermann F, Eden C, Brown I (2005) Using causal mapping with group support systems to elicit an understanding of failure in complex projects: some implications for organizational research. Group Decis Negot 14(5):355–376

Ackermann F, Eden C, Williams T, Howick S (2007) Systemic risk assessment: a case study. J Oper Res Soc 58(1):39–51

Ackoff RL, Sasieni MW (1968) Fundamentals of operations research. Wiley, New York

Baccarini D (1996) The concept of project complexity – a review. Int J Proj Manag 14:201–204

Bennett PG, Ackermann F, Eden C, Williams TM (1997) Analysing litigation and negotiation: using a combined methodology. In: Mingers J, Gill A (eds) Multimethodology: the theory and practice of combining management science methodologies. Wiley, Chichester, pp 59–88

Bryson JM, Ackermann F, Eden C, Finn C (2004) Visible thinking: unlocking causal mapping for practical business results. Wiley, Chichester

Burns JR (1977) Converting signed digraphs to Forrester schematics and converting Forrester schematics to differential equations. IEEE Trans Syst Manag Cybern SMC 7(10):695–707

Burns JR, Ulgen OM (1978) A sector approach to the formulation of system dynamics models. Int J Syst Sci 9(6):649–680

Burns JR, Ulgen OM, Beights HW (1979) An algorithm for converting signed digraphs to Forrester's schematics. IEEE Trans Syst Manag Cybern SMC 9(3):115–124

Cooper KG (1980) Naval ship production: a claim settled and a framework built. Interfaces 10:20–36

Cooper KG (1993a) The rework cycle: benchmarks for the project manager. Proj Manag J 24:17–21

Cooper KG (1993b) The rework cycle: how it really works and reworks. PMNETwork VII:25–28

Cooper KG (1993c) The rework cycle: why projects are mismanaged. PMNETwork VII:5–7

Cooper KG (1994) The $2,000 hour: how managers influence project performance through the rework cycle. Proj Manag J 25:11–24

De Meyer A, Loch CH, Rich MT (2002) Managing project uncertainty: from variation to chaos. MIT Sloan Manag Rev 43(2):60–67

Diehl E, Sterman JD (1995) Effects of feedback complexity on dynamic decision making. Organ Behav Hum Decis Process 62(2):198–215

Dixon M (ed) (2000) The Association for Project Management (APM) Body of Knowledge (BoK), 4th edn. Association for Project Management, High Wycombe

Eden C (1988) Cognitive mapping: a review. Eur J Oper Res 36:1–13

Eden C, Ackermann F (2004) Cognitive mapping expert views for policy analysis in the public sector. Eur J Oper Res 152:615–630

Eden C, Williams TM, Ackermann F, Howick S (2000) On the nature of disruption and delay. J Oper Res Soc 51:291–300

Eden C, Ackermann F, Williams T (2004) Analysing project cost overruns: comparing the measured mile analysis and system dynamics modelling. Int J Proj Manag 23:135–139

Eden C, Ackermann F, Williams T (2005) The amoebic growth of project costs. Proj Manag J 36(2):15–27

Engwall M (2002) The futile dream of the perfect goal. In: Sahil-Andersson K, Soderholm A (eds) Beyond project management: new perspectives on the temporary-permanent dilemma. Libe Ekonomi, Copenhagen Business School Press, Malmo, pp 261–277

Flyvberg B, Holm MK, Buhl SL (2002) Understanding costs in public works projects: error or lie? J Am Plan Assoc 68:279–295

Flyvberg B, Bruzelius N, Rothengatter W (2003) Megaprojects and risk: an anatomy of ambition. Cambridge University Press, Cambridge

Ford DN (1995) The dynamics of project management: an investigation of the impacts of project process and coordination on performance. Massachusetts Institute of Technology, Boston

Ford D, Sterman J (1998) Expert knowledge elicitation to improve formal and mental models. Syst Dyn Rev 14(4):309–340

Forrester J (1961) Industrial dynamics. Productivity Press, Portland

Hodgson DE (2004) Project work: the legacy of bureaucratic control in the post-bureaucratic organization. Organization 11:81–100

Howick S (2003) Using system dynamics to analyse disruption and delay in complex projects for litigation: can the modelling purposes be met? J Oper Res Soc 54(3):222–229

Howick S (2005) Using system dynamics models with litigation audiences. Eur J Oper Res 162(1):239–250

Howick S, Eden C (2001) The impact of disruption and delay when compressing large projects: going for incentives? J Oper Res Soc 52:26–34

Howick S, Ackermann F, Andersen D (2006) Linking event thinking with structural thinking: methods to improve client value in projects. Syst Dyn Rev 22(2):113–140

Howick S, Eden C, Ackermann F, Williams T (2007) Building confidence in models for multiple audiences: the modelling cascade. Eur J Oper Res 186:1068–1083

Kahneman D, Slovic P, Tversky A (1982) Judgment under uncertainty: heuristics and biases. Cambridge University Press, Cambridge

Koskela L, Howell G (2002a) The theory of project management: explanation to novel methods. In: Proceedings 10th annual conference on lean construction, IGLC-10, August 2002, Gramado

Koskela L, Howell G (2002b) The underlying theory of project management is obsolete. In: Proceedings of PMI (Project Management Institute) research conference 2002, Seattle, pp 293–301

Lane (2000) Diagramming conventions in system dynamics. J Oper Res Soc 51(2):241–245

Lindkvist L, Soderlund J, Tell F (1998) Managing product development projects: on the significance of fountains and deadlines. Organ Stud 19:931–951

Linehan C, Kavanagh D (2004) From project ontologies to communities of virtue. Paper presented at the 2nd international workshop, making projects critical. University of Western England, 13–14th December 2004

Lundin RA (1995) Editorial: temporary organizations and project management. Scand J Manag 11:315–317

Lyneis JM, Ford DN (2007) System dynamics applied to project management: a survey, assessment, and directions for future research. Syst Dyn Rev 23:157–189

Lyneis JM, Cooper KG, Els SA (2001) Strategic management of complex projects: a case study using system dynamics. Syst Dyn Rev 17:237–260

Major Projects Association (1994) Beyond 2000: a source book for major projects. Major Projects Association, Oxford

Malgrati A, Damiani M (2002) Rethinking the new project management framework: new epistemology, new insights. In: Proceedings of PMI (Project Management Institute) research conference 2002, Seattle, pp 371–380

Mitchell G (1993) The practice of operational research. Wiley, Chichester

Morris PWG, Hough GH (1987) The anatomy of major projects. A study of the reality of project management. Wiley, Chichester

Nahmias S (1980) The use of management science to support a multimillion dollar precedent-setting government contact litigation. Interfaces 10:1–11

Oliva R (2004) Model structure analysis through graph theory: partition heuristics and feedback structure decomposition. Syst Dyn Rev 20(4):313–336

Packendorff J (1995) Inquiring into the temporary organization: new directions for project management research. Scand J Manag 11:319–333

Pidd M (2003) Tools for thinking: modelling in management science. Wiley, Chichester

Project Management Institute (2000) A guide to the Project Management Body of Knowledge (PMBOK). Project Management Institute, Newtown Square

APM Publishing Ltd (2004) Project risk analysis and management guide. APM Publishing Ltd, High Wycombe

Rivett P (1972) Principles of model building. Wiley, London

Robertson S, Williams T (2006) Understanding project failure: using cognitive mapping in an insurance project. Proj Manag J 37(4):55–71

Rodrigues A, Bowers J (1996a) The role of system dynamics in project management. Int J Proj Manag 14:213–220

Rodrigues A, Bowers J (1996b) System dynamics in project management: a comparative analysis with traditional methods. Syst Dyn Rev 12:121–139

Rodrigues A, Williams TM (1997) Systems dynamics in software project management: towards the development of a formal integrated framework. Eur J Inf Syst 6:51–66

Rosenhead J, Mingers J (2001) Rational analysis for a problematic world revisited. Wiley, Chichester

Scott S (1993) Dealing with delay claims: a survey. Int J Proj Manag 11(3):143–153

Scottish Parliament (2003) Corporate body issues August update on Holyrood. Parliamentary News Release 049/2003

Shaw D, Ackermann F, Eden C (2003) Approaches to sharing knowledge in group problem structuring. J Oper Res Soc 54:936–948

Shenhar AJ (2001) One size does not fit all projects: exploring classical contingency domains. Manag Sci 47:394–414

Shenhar AJ, Dvir D (2004) How project differ and what to do about it. In: Pinto J, Morris P (eds) Handbook of managing projects. Wiley, New York, pp 1265–1286

Stephens CA, Graham AK, Lyneis JM (2005) System dynamics modelling in the legal arena: meeting the challenges of expert witness admissibility. Syst Dyn Rev 21:95–122.35

Sterman JD (1989) Modelling of managerial behavior: misperceptions of feedback in a dynamic decision making experiment. Manag Sci 35:321–339

Stevens M (2002) Project management pathways. Association for Project Management, High Wycombe

Szyliewicz JS, Goetz AR (1995) Getting realistic about megaproject planning: the case of the new Denver International Airport. Policy Sci 28:347–367

Turner JR, Cochrane RA (1993) Goals-and-methods matrix: coping with projects with ill defined goals and/or methods of achieving them. Int J Proj Manag 11:93–102

Vennix J (1996) Group model building: facilitating team learning using system dynamics. Wiley, Chichester

Wickwire JM, Smith RF (1974) The use of critical path method techniques in contract claims. Public Contract Law J 7(1):1–45

Williams TM (1999) The need for new paradigms for complex projects. Int J Proj Manag 17:269–273

Williams TM (2002) Modelling complex projects. Wiley, Chichester

Williams TM (2003) Assessing extension of time delays on major projects. Int J Proj Manag 21(1):19–26

Williams TM (2004) Learning the hard lessons from projects – easily. Int J Proj Manag 22(4):273–279

Williams TM (2005) Assessing and building on project management theory in the light of badly overrun projects. IEEE Trans Eng Manag 52(4):497–508

Williams TM (2007) Post-project reviews to gain effective lessons learned. Project Management Institute, Newtown Square

Williams TM, Eden C, Ackermann F (1995a) The vicious circles of parallelism. Int J Proj Manag 13:151–155

Williams TM, Eden C, Ackermann F, Tait A (1995b) The effects of design changes and delays on project costs. J Oper Res Soc 46:809–818

Williams TM, Ackermann F, Eden C (1997) Project risk: systemicity, cause mapping and a scenario approach. In: Kahkonen K, Artto KA (eds) Managing risks in projects. E & FN Spon, London, pp 343–352

Williams TM, Ackermann F, Eden C, Howick S (2005) Learning from project failure. In: Love P, Irani Z, Fong P (eds) Knowledge management in project environments. Elsevier, Oxford

System Dynamics Modeling to Inform Defense Strategic Decision-Making

Alan C. McLucas and Sondoss Elsawah
School of Engineering and Information
Technology, Capability Systems Centre,
University of New South Wales, Canberra,
Australian Defence Force Academy, Canberra,
Australia

Article Outline

Introduction: Scope and Definition of the Subject
Conflict
Disclaimer Regarding Classified or
    Sensitive Work
Modeling and Simulation of Force-on-Force
    Engagements
Dynamic Intuition in Military Command and
    Control
Defense Capability and Force Structure Analysis
Defense Capital Asset Acquisition
Helicopter Operations: Crews and Missions
System Dynamics Modeling to Inform Suitability
    of System Reliability and Performance Targets
Future Work
Bibliography

Glossary

Combat analysis Combat analysis involves detailed scientific study of real historical and simulated hypothetical war fighting with a view to determining the impacts on outcomes that derive from command and control; doctrine; weapons and weapons effects; logistics; and the composition, deployment, and engagement of force elements and force element groupings to enable optimal exploitation of defense force equipment and personnel to achieve the best possible war-fighting capability.

Command and control (C2) A set of organizational and technical attributes and processes that employs human, physical, and information resources to solve problems and accomplish missions to achieve the goals of an organization or enterprise (Vassiliou et al. 2015: 1).

Defense analysis Defense analysis is a strategic scenario-planning approach to determining how governments, combined military services, government agencies, and industry might best respond to real and possible threats to national security. Top down, through its various levels, such analysis exploits scientific analysis to assist in designing possible responses and military assessments to determine how to apply limited national defense resources to possible responses according to evolving threat assessments.

Defense preparedness Being a critical element of military capability, preparedness is further refined within the context and combination of *readiness* and *sustainment,* that being the ability to undertake military operations. The effective generation of military capability consists of a range of inputs that are fundamental to success. Readiness denotes a force's ability to be committed to operations within a specified time. Readiness refers to the availability and proficiency/serviceability of personnel, equipment, facilities, and consumables allocated to a force. Sustainment is the provision of personnel, logistic, and other support, including recovery and reconstitution required to maintain and prolong operations or combat until successful accomplishment of the mission (AS DoD (1988) ADFP 4).

Fundamental inputs to capability Fundamental inputs to capability enable the effective generation of defense capabilities. They are personnel; organization; collective training; major systems including significant platforms, fleets of equipment, and operating systems; supplies; facilities;

© Springer Science+Business Media, LLC, part of Springer Nature 2020
B. Dangerfield (ed.), *System Dynamics*,
https://doi.org/10.1007/978-1-4939-8790-0_657

Originally published in
R. A. Meyers (ed.), *Encyclopedia of Complexity and Systems Science,* © Springer Science+Business
Media LLC 2019, https://doi.org/10.1007/978-3-642-27737-5_657-1

support; and command and management (AS DoD (1988) ADFP 4).

Military capability Capability is the power to achieve or influence an effect. Military capability is the ability to achieve a desired effect in a specific operating environment. It is the combination of force structure and preparedness, which enables the nation to exercise military power (AS DoD (1988) ADFP 4).

Operational availability (*Ao*) Operational availability is a measure used in reliability analysis to indicate for a specific system or sub-system its real average availability over a period of time. Calculating *Ao* requires that all sources of downtime, that is, resulting in the system or sub-system being unavailable for operational use, be taken into account. This includes administrative downtime, logistic downtime, time waiting in backlog, etc.

Strategic defense planning Defense analysis provides the framework within which strategic defense planning can occur. At the more detailed levels, this involves planning for a wide range of possible military responses; examining how fundamental inputs to capability combine to create force structures, that is, force elements, force element groupings with appropriate support, and supporting infrastructure to deliver selected, plausible military responses; evaluating force structures in terms of efficacy in prosecuting specific military responses and the likely costs of doing so; and reporting to decision-makers on force structure options and associated costs.

Introduction: Scope and Definition of the Subject

The term *defense* (or UK *defence*) is used broadly to describe roles and activities of military forces, together with enabling and supporting civilian organizations, being capable and prepared to respond in a diverse portfolio of ways to prosecute government directives in the protecting of national interests.

This entry addresses the nature of complex systemic problems that arise in a defense context and describes how qualitative system dynamics and quantitative (computational) system dynamics

modeling and simulation apply as aids to defense analysis, strategic planning, and management.

Defense analysis involves considerations of:

- Ways in which the means might be created, and effects delivered, to overpower adversaries
- Constituting forces in terms of size and weapons effects
- Being prepared to prosecute the necessary military responses
- Being able to sustain the forces needed to prosecute the necessary military responses
- Being able to respond concurrently to a diverse range threats, including new threats
- Being able to achieve these in ways that do not excessively drain or exhaust national resources

Strategic defense planning involves:

- Identification and systematic analysis of changing international, national, and domestic threats
- Development of scenarios describing plausible ways that threats might manifest themselves
- Formulating a wide range of possible military responses to the threat scenarios
- Examining how fundamental inputs to capability combine to create force structures, that is, force elements, force element groupings with appropriate support, and supporting infrastructure to deliver selected, plausible military responses
- Evaluating force structures in terms of efficacy in prosecuting specific military responses and the likely costs of doing so
- Reporting to decision-makers on force structure options and associated costs

The analytical effort spans:

- Dynamic combat analysis, including simulations and gaming of possible war-fighting scenarios, to gauge the relative effectiveness of forces, engagement strategies, and tactics
- Identification of the fundamental inputs to capability, and how to bring them together to create the means of delivering an effect, either on the battlefield or in response to possible domestic or terrorist threats

- Identification of how to achieve specific levels of preparedness of the capability, that is, being able to deliver the necessary military responses within specific timeframes
- Determination of the cost and resource implications of each of these

The fundamental purpose of analytical efforts is to provide evidence-based options for decision-makers to consider in detail. Options are alternative ways of achieving military capability, force structures, and preparedness suited to delivering diverse and demanding roles, meeting continually changing threats, and doing so with limited resources.

If indeed measurable as an outcome, military effectiveness is achieved through the allocation of resources to the military for the purposes of transforming them into effective war-fighting capability. Effectiveness must also take into account diverse factors such as extant *modus operandi* (doctrine), command and control, force structures, competence, and extent to which skills are current. While a country may provide its military with generous budgets and large cadres of manpower, if the military's doctrine is misguided, the training ineffective or not up-to-date, the leadership unschooled, or the organization inappropriate, military capability will suffer RAND (1992).

The problems of measuring military capability are, in many respects, quite similar to the difficulties faced in measuring national power. Certainly, one or two individual measures – the number of personnel under arms, for example, or the number of tanks or missile launchers in a nation's inventory – are unlikely to capture the key factors for assessing military power, just as a single measure does not provide a useful assessment of a country's overall power.

Analytical methodologies must be amenable in systemic and dynamic problem situations, where decisions and actions produce consequences dependent upon previous decisions or actions and delayed consequences. Systemic and dynamic challenges that arise in a defense context include how to:

- Develop highly effective strategies for the recruiting and training of personnel in anticipation of forming them operational units, force elements, and force element groupings in order to have them ready when required for deployment on operations. This includes having only sufficient resources continually ready, but able to be expanded and raised to heightened levels of preparedness, possibly with foreshortened warning periods, in time operational deployment.
- Provide the necessary sustainment, logistics support, and materiel replacements to assure that units, force elements, and force element groupings continue to have capacity and capability to operate.
- Acquire capital material and weapons systems, when these may have long development or procurement lead times and being able to do so without wasteful expenditures or diminished opportunities to respond in a balanced way to meet other resourcing and/or funding priorities.
- Develop war-fighting strategies, in the context of future scenario situations that result in superiority in force-on-force engagements.

Mapping Causal Hypotheses: Qualitative System Dynamics Diagramming Conventions

Directed graphs used in qualitative system dynamics to map out cause-and-effect relationships and problem structures take the forms of either causal loop diagrams (CLDs) (Sterman 2000) or influence diagrams (ID) (Coyle 1996 and Wolstenholme 1990), though there has been much debate (Richardson 1986 and Lane 2000) about the relative merits of each.

The influence diagramming (ID) technique is used in this entry as a matter of preference as it:

- Enables differentiation between types of variables, viz., accumulations, rate-controlling variables, and parameters
- Allows for specificity of the variables to be combined when formulating of decision rules as they apply to rate-controlling variables, in particular
- Leads directly to conversion of a problem's decision rules into algebraic expressions ready to be incorporated directly into the coded form of the model

- Obviates the need for two types of diagram found more commonly in the System Dynamics literature, CLDs, and stock-and-flow (SnF) diagrams

Conventions for SD influence diagramming are included at the end of this entry, Fig. 16. A useful variation to the ID conventions is reserve UPPER CASE for labelling stocks (levels, or accumulations, i.e., mathematical integrations).

Where original work being cited was based on the author(s) using causal loop diagrams (CLD) or stock-and-flow diagrams (SnF), the original forms generally have been retained. Figures containing causal diagrams are labelled according to their use of ID, CLD, or SnF technique.

Qualitative and Quantitative System Dynamics Applied to Defense Problem Analysis

Defense *capability development* (DoD AS 2014), *force structure development*, and *preparedness management* (Paterson 2003; McLucas and Linard 2000) are dynamic and complex, which makes their management inherently confounding. In addition to being characterized by detail complexity, they variously exhibit the attributes of *dynamic complexity* described by Sterman (2000):

(a) They change continuously.
(b) The actors involved interact and strongly influence each other.
(c) Actions taken by an actor, who may be a decision-maker seeking to make a desirable change, cause changes that come back to create often undesired influences at the original point of intervention.
(d) Effects are rarely proportional to the cause, and effects may appear in unexpected forms seemingly unrelated to a specific cause.
(e) Taking one course of action often excludes other courses.
(f) Changes may appear to be spontaneous, produced as unanticipated responses by unexpected or unidentified actors.

(g) Capabilities and decision rules of actors in complex problems change over time and often are difficult to discover.
(h) Cause and effect may not be proximate in either space or time with the consequence that effective strategies are either not obvious or when formulated appear to be counterintuitive.
(i) Seemingly obvious solutions to problems often fail or make a problematical situation worse.
(j) Time delays in physical and/or information feedback channels mean that long-run response and short-run responses can be quite different, thereby creating dynamic instability and confusion, or at least uncertainty for decision-makers, about likely efficacy of alternate interventions.

Information is often limited, leading to the likelihood of erroneous interpretations of meaning. Information may be incomplete or appear erroneous because of the means used to collect it. In a defense context, information may be deliberately obscured, distorted, subverted, or corrupted by adversaries. While how to treat deliberately corrupted information is beyond the scope of this topic, it is important to be able to differentiate between information, which is limited in its availability and that which may have been deliberately corrupted by an adversary in order to mislead.

Ambiguity arises when it is difficult to discriminate between alternate theories or explanations for changes observed in a problematical situation or in specific variables of interest. Before any form of analysis is to be done effectively, it is necessary to be able to select, observe, measure, and correctly interpret the meaning of observations in cause-and-effect terms.

When delays and multiple circular causality pathways are involved, qualitative modeling is unlikely to be sufficient. Quantitative (computational) modeling and simulation may be necessary to test which of a select set of alternate hypotheses represents the most plausible explanation for the observed behaviors. Further, such modeling may be essential to reduce or remove the effects of human bounded rationality and misperceptions of feedback.

Undertaking such quantitative modeling and simulation analysis provides valuable opportunities to put to the test those potentially flawed cognitive maps that analysts and decision-makers alike might otherwise use in interpreting a particular problematical situation. Further, modeling and simulation proceed through making causal structures explicit, thereby challenging inferences about dynamic mechanisms that might otherwise remain untested.

Defense strategists and decision-makers are not immune to unscientific reasoning and errors in judgment. They face additional challenges when presented with unique problems and situations for which their prior experiences may have limited relevance and then have to make critical decisions, which can have irreversible consequences.

Depending on their background and experiences, they are likely to bring their own biases into problem analysis and decision-making settings. Decision-making challenges can be exacerbated by group dynamics, covert wielding of political power, differences in espoused theories, and theories in use, Argyris and Schön (1978), and Argyris (1980, 1982, 1991, 1993, 1994), undeclared and untested assumptions Mason and Mitroff (1973) which detract from making informed and balanced decisions based on analysis of causes and their possible effects.

System dynamics modeling (Forrester 1968) relies on a fundamental principle that in each problematic situation, there is an underlying causal structure that gives rise to the ways changes play out over time. Hence, analysis and development of possible intervention strategies require the hypothesizing, modeling of the causal structure, and testing the response of these structures to changes in selected variables (policy) or combinations of variables (strategy) (Coyle 1996).

Though modeling and simulation interventions do not guarantee consensus, acceptance, or even accommodation of alternate strategies by decision-makers, conducted openly and transparently (Vennix 1996), they create valuable opportunities to:

- Surface and test assumptions decision-makers and analysts might hold about given problem situations.
- Potentially limit the invoking by decision-makers of defensive routines relating to selection and subsequent implementation of strategies.
- Exercise the power of decision-makers' position to override or limit consideration of possible alternate strategies.

Many problems in the subclasses described in this entry are candidates for analysis using qualitative and/or quantitative (computational) system dynamics. Here it is important to recognize that analysis of complex problems is not exclusively the domain of system dynamics modeling and simulation. Modeling and simulation in its diverse forms provide valuable opportunities for the analysis of the ways that the implementation of particular strategies could create alternate outcomes, without the threat of implementation in real world producing irreversible changes. Whether conducted in qualitative or quantitative terms, system dynamics modeling seeks to recognize these challenges and improve outcomes. This entry provides examples of both qualitative and quantitative system dynamics modeling and simulation in support of the analysis of various defense problems.

Conflict

Internal and External Threats to Nations

Nations do not exist in isolation. National economies are affected by the needs to provide national security against domestic internal, external, and international threats. The ability of a nation to sustain itself, maintain security, and pursue defense goals will depend upon the strength of its economy and relationships with other nations be they allies or adversaries.

Wils et al. (1998), in analyzing internal and external threats to nations arising from conflict and giving rise to conflict, present a system dynamics model of violent behavior at the national level. They sought to untangle the sources and consequences of conflict, Fig. 1., through formulating and analyzing a system dynamics model based on the theory of lateral pressure, hypothesizing inter alia arguments such as:

> ... military build-up ... [is related to military expenditure, in that] ... expenditure is taken as a user-defined portion of real gross national product ... when the gross national product increases, absolute military expenditures rise also... The higher expenditures accumulate in greater military force. Enhanced military force, in turn, increases the probability of conflict. Military force is drained through depreciation of obsolete weaponry, plus, during a conflict, a portion of the stock of military force is used up annually thus draining the stock faster. Furthermore, conflict, if it occurs on national territory, reduces GDP.

System dynamics modeling simulations based on their Vensim™ model provided insights into the consequences of violent internal and external conflict for 13 countries in Africa, Europe, and the USA, concluding that particular types of conflict arise and subsequently devising strategies within the model's parametric specifications would not necessarily lead to stopping conflict once it had started.

Conflict: Causal Analysis of War in Angola

The Angolan Civil War started in late 1975 after Angola became independent from Portugal. An intense struggle for power ensued between two former liberation movements, the People's Movement for the Liberation of Angola (MPLA) and the National Union for the Total Independence of Angola (UNITA). War continued until 2002 when MPLA won. An estimated 500,000 people died in the war. Many people suffered horrific injuries from landmines. Much of the country's infrastructure and many buildings were badly damaged. Most Angolans lacked access to basic medical care, and many did not have access to water. Thirty percent of Angolan children would die before the age of 5, with an overall life expectancy of less than 40 years.

Coyle (1998) explained how qualitative system dynamics analysis of the causal drivers of complex dynamic problems such as the continual conflict in Angola enabled creation of a graphic overview which could prove valuable from a diagnostic viewpoint. For all intents and purposes, the Angolan situation was both hopeless and intractable.

Sustainability of the efforts of the various combatants depended upon their ability to control access to Angola's resources. Having a greater share of resources, viz., oil and diamonds, provided an enhancement to a combatant's capability to engage in conflict and to pursue their goals. Should it have been able to muster sufficient military strength, the government might set out to

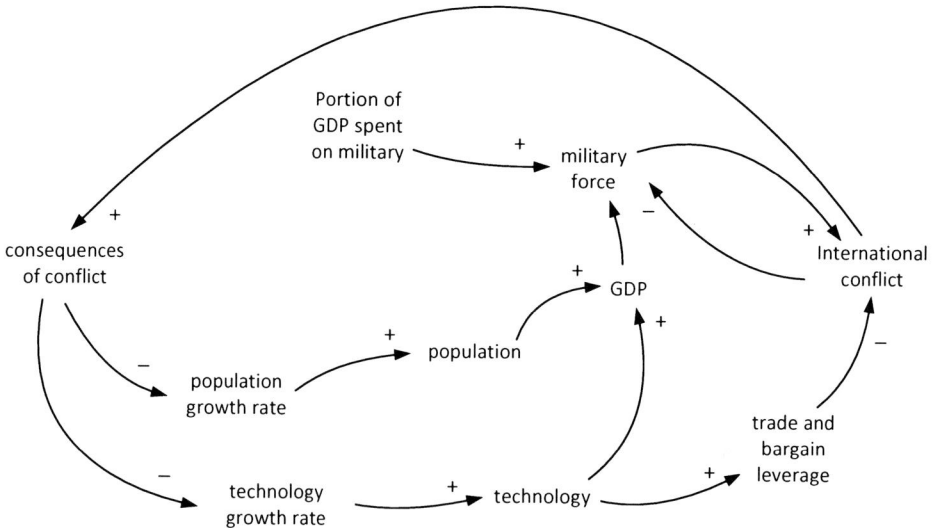

System Dynamics Modeling to Inform Defense Strategic Decision-Making, Fig. 1 CLD conflict and consequences. (After Wils et al. 1998)

control resources, though by itself (and for some time) this would increase the occurrence of conflict. Over time, sustained balance of power in favor of the government would reduce UNITA's control over diamond areas and hence reduce UNITA's diamond sales and ability to obtain arms and build its strength. This would lead to a further increase in the balance of power in favor of the government. Faced with providing for starving population, it would be difficult for the government to find money to build military strength.

Many interconnected cause-and-effect mechanisms were found to be at play. Coyle's analysis identified some 57 variables, 101 instances of 1 variable having some causal influence on another variable, and around 30 feedback loops. Each cause-and-effect relationship between pairs of variables represents the basis of a dynamic hypothesis, and each of the feedback loops tells its own story about chains of causal influences.

Arguably, the value of such depictions of such a complex problematic situation is that they enabled hypothesizing about how the situation could change over time. They also offer the opportunity to identify possible interventions at single points or simultaneously at multiple points, with a view to identifying how these might influence the problematic behaviors. However, dealing with this level of detail complexity brings its own complications and increases to risk of both erroneous inferences about causality and the effectiveness of plausible interventions. The value of such a depiction is that plausible cause-and-effect arguments can be traced in detail a view to discovering where the pressure points or leverage points exist, that is, where a change might be made to produce a paradigm shift (Senge 1990)

The value of such analysis depends on:

- Being able to identify the variables involved
- Having a justifiable basis upon which to develop and explicitly state causal hypotheses about the influence variables have on each other
- Selection of the problem boundary, which gives rise to what to include and what to exclude from the analysis

- Careful choice of the level of aggregation to be portrayed in the causal mapping
- The extent to which each causal relationship can be validated and, hence, the extent to which it can be claimed that the resulting causal mapping is a necessary and sufficient representation of real-world causes and their effects

The massively complex original cause-and-effect representation was distilled into a select number of summary depictions, each taking the form of a high-level influence diagram, noting that at high levels of aggregation, an influence diagram is effectively the same as causal loop diagram. The depiction in Fig. 2 provides opportunity to proffer and analyze scenarios and plausible interventions which might lead to the ending of the war.

The period of some 27 years that this war continued saw changes in the extent to which international and domestic players influenced what was happening in Angola. Had the international influences been captured in a mapping representation at higher level of aggregation, it might have been depicted as a proxy extension of the Cold War with influences by Cuba on behalf of the USSR and Brazil and South Africa on behalf of the USA.

Conflict: Causal Analysis of War in Afghanistan

In 2009, a consulting firm was engaged to work with the US military in analyzing the dynamics and challenges of operations in Afghanistan. Qualitative system dynamics cause-and-effect analysis of the war sought to create an understanding of the complexity of the Afghanistan situation and to identify why one of the longest wars in history was continuing with no end in sight. When a summary PowerPoint™ of the massively complex problem situation was leaked to a reporter, a media storm ensued.

The analysis, depicted in a single slide, was a comprehensive set of hypotheses, which sought to explain why and how, at any point in time, the Afghani population adopted one of the following five camps:

- Actively supporting government and special forces activities
- Sympathizing with government

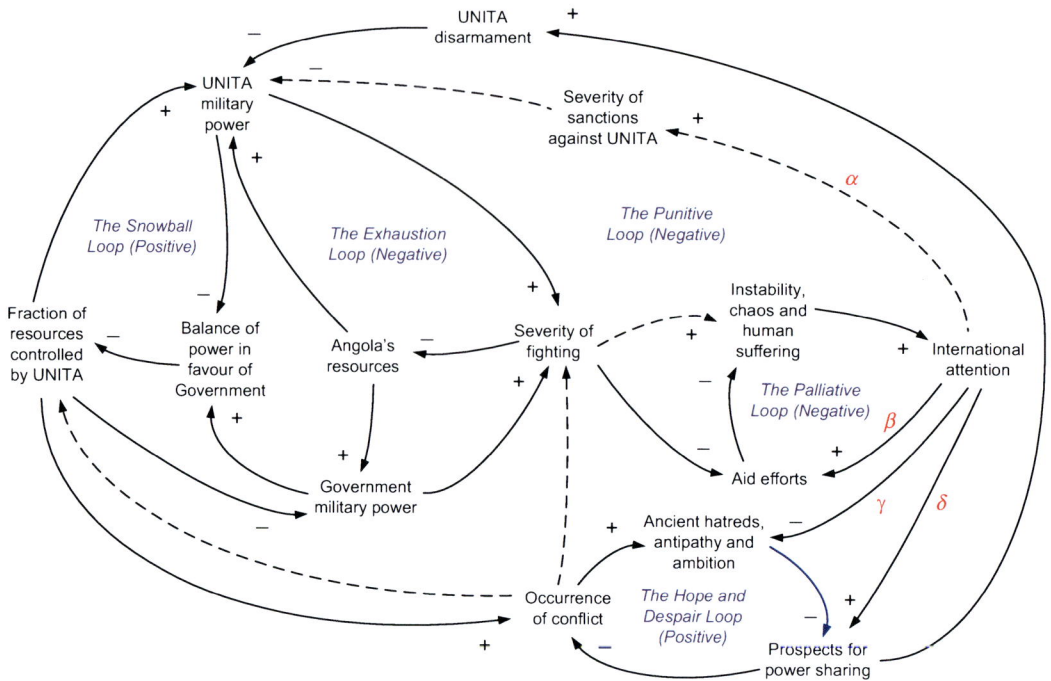

System Dynamics Modeling to Inform Defense Strategic Decision-Making, Fig. 2   High-level influence diagram –
Angolan Civil War. (Coyle 1998)

- Neutral
- Sympathizing with insurgents
- Actively supporting insurgency

Understanding the underlying reasons for the
Afghani population adopting or aligning with a
particular camp would have proven most valuable
in the design of interventions, which might influence allegiances and the adopting of the most
favorable camps.

The massively complicated depiction of the
causal relationships on a single PowerPoint™
slide, and the confusion this created for a first-
time reader, who could not know the rationale
behind any specific cause-and-effect relationship,
arose because there was far too much information
being presented. The diagram sought to depict
how, in combination, perspectives arising in ten
different sectors of the problem create their own
influences on the Afghani population leading
them to adopt one of the five camps listed. Choosing to adopt one camp in preference to others
would depend on potential attractiveness of

government actions and ideology compared to
that of insurgents or to choose to make a change
in their choice, noting that an adoption delay is
involved. Herein would lie opportunities to
change the allegiances of the Afghani people,
possibly leading to long-term peace.

Disclaimer Regarding Classified or Sensitive Work

Reporting the details of original analytical work in
any military-related field is constrained by officially imposed access controls, particularly where
the associated documentation contains specific
details of the composition of forces, military
response options, and actual states of readiness of
force elements or force element groupings. It is
inescapable that much related documentation has
been or remains classified, is contained within
classified sources, or is sensitive. This limits the
publication of articles and papers, which address
the content of the official reports on subjects such

as modeling and simulation of force-on-force engagements and defense capability and preparedness; the authors of those articles and papers necessarily avoid discussing sensitive information.

Where particular analytical techniques might be reported upon, their descriptions become generalized with, for example, selected information being redacted as part of a declassification process. The inevitable consequence is that their value as primary resources for research purposes is diminished. In many cases the original work cannot be part of the publicly accessible literature. In some particularly sensitive instances, it can neither be confirmed nor denied that the work was ever conducted or not, cannot be confirmed or denied.

Modeling and Simulation of Force-on-Force Engagements

In a conventional warfare setting, that is, where escalation of warfare to include the use of nuclear weapons is considered unlikely, the success in battle of one force over another in conventional warfare will depend on a number of reasonably well-known aspects. These will be considered in detail by opposing military commanders in anticipation of possible battle engagements. These aspects include size and organization of the combatant forces, *force structure*, effectiveness of *command and control*, momentum in attack and speed of maneuver in response to opponent's action, ability to acquire targets and to respond by bringing weapons to bear, weapons effectiveness and capacity to sustain rates of fire, etc. For detailed explanation of these aspects, how they interact, and descriptions of various models that have been developed to analyze force-on-force engagements, see Bracken et al. (1995).

Earliest examples of system dynamics models being developed to demonstrate how simulated force-on-force engagements might play out under specific circumstances can be found in the Cold War era. Models of this period were based on hypothetical Warsaw Pact vs NATO engagement scenarios, referred to by NATO as the *AirLand Battle*.

Langsaeter (1984) examined how a Warsaw Pact ORANGE force attack on NATO forces in

Europe might specifically involve Norway, which was on the Western flank of a possible ORANGE force attack.

Norway's BLUE force would seek to defend three vital areas, doing do for as long as possible. The modeling scenario presumed that ORANGE force maneuvering would be constrained by Norway's difficult terrain and that BLUE forces would have opportunities to redeploy in the event that any one vital area was overrun. The key question was: How long could BLUE force expect to hold the ORANGE force attack before being overwhelmed?

SD simulations were based on a mathematical model of attrition rates under the expected conditions, a modified Lanchester Combat Model. Lanchester's laws (Lanchester 1956) suggest that the effectiveness of a force in battle would not simply depend on its size but that substantial advantage would accrue from, in this case the defender's ability to acquire and engage targets presented by an attacker when the attacker was compelled to change formation because of terrain. In such case the advancing enemy would be channeled into areas where the defender could bring fire to bear with greatest effect.

BLUE forces would rely on knowing that an attack was imminent, and when the attack was launched, BLUE forces would engage the ORANGE attacking force from a series of well-prepared defensive positions. BLUE's advantage over the ORANGE attacking force would derive from being able to bring high rates of accurate fire to bear on them, inflicting high casualty rates on the attacking force, albeit for a finite amount of time.

In the more general case, RED-on-BLUE force engagement simulations by NATO would envision a massively overwhelming assault led by Warsaw Pact (notionally RED) armored formations supported by ground attack aircraft and artillery. NATO forces (notionally BLUE) would first fight a defensive battle in depth, repeatedly falling back to previously prepared alternate defensive positions while seeking to inflict maximum damage on the attacking force. Having inflicted the maximum damage and causing the RED force to change formation and lose momentum, the BLUE force would launch a coordinated counter-attack in the air and on land.

Though the exact details of the military responses and plans, and the models used to examine them, were highly classified at the time, simplified versions of some models have become publicly accessible. Such modeling and simulations presented many opportunities to learn about ways conventional warfare could develop and battles ensue.

Wolstenholme and Al-Alusi (1987) describe the armored advance model (Fig. 3). Wolstenholme (1988, 1990) also describes the modeling approach used. The armored advance model was developed for purpose of examining the effects of alternative formation change strategies by an attacking RED force under a variety of fire delivery strategies on the part of the BLUE defending force. The BLUE force strategies for delivering fire were based on criteria involving RED's distance, speed, and momentum (speed of advance and

numbers advancing). The RED strategies considered ranged from changing formation at a fixed distance of advance to changing formation at a variable distance of advance, using a range of different variables on which to base this decision. The performance measure used to assess those strategies was that of the momentum of the RED force on arrival at the BLUE position (force size/arrival time).

With modifications, the model might have been used to demonstrate the possible effects of ammunition shortages and alternate BLUE strategies involving the "dumping" of ammunition stocks. With prepositioning of large stocks of ammunition, it would be possible for BLUE to deliver significantly higher rates of fire and sustain those rates for longer periods, thereby significantly slowing RED forces or causing the premature breaking of battalion formations and thereby further slowing RED's rate of advancing. Similarly, the model was not designed originally to include specific details of

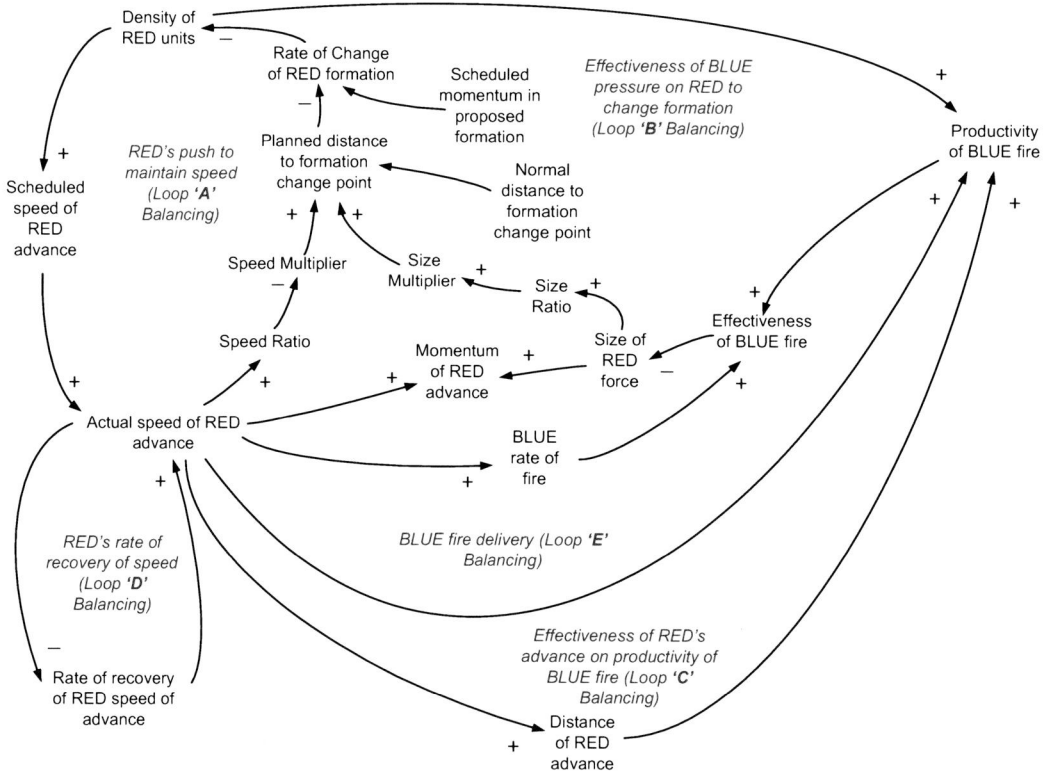

System Dynamics Modeling to Inform Defense Strategic Decision-Making, Fig. 3 The underlying feedback casual structure of the armored advance model (ID). (After Wolstenholme 1990)

battlefield terrain and geographical features, which might have been used by BLUE force to check RED's advance or cause direction of advance momentum to change, that is, a development of Langsaeter's (1984) model.

Nonetheless, the modeling and simulations were immensely valuable. A series of simulations demonstrated that high levels of interaction occurred between BLUE's strategies and those of RED could be expected enact in response. In particular, BLUE could control RED's speed and size by adjusting rates of fire and (undesirably) could allow RED to regain momentum by slowing rates of fire. When RED chose to maintain speed by holding its battalion formation, the total advance time and, hence, exposure to BLUE's fire would be reduced. BLUE's cumulative fire delivered would then be less and hence RED's speed, size, and momentum higher.

Analyses of this type also identified the need, and substantiated the priories, for the development and acquisition of particular types of weapons systems, those that could strongly influence the course of the *AirLand Battle*, should it ever occur. One example of a "force multiplier" effect for NATO forces came about with the development and introduction of the M270 Multiple Launch Rocket System (MLRS), developed jointly by the UK, USA, West Germany, France, and Italy and adopted by various NATO countries circa 1983. The MLRS had the potential to deliver a variety of munitions accurately to a target area equivalent to 20 football fields obliterating all ground forces, before rapidly redeploying and rearming ready to strike again. To achieve this, the MLRS also required intimate support of a highly mobile munitions supply system capable of quickly handling, deploying, redeploying, and reloading massive loads of rockets.

Dynamic Intuition in Military Command and Control

The capacity and ability to formulate strategies, observe changing battlefield situations, and then make timely, effective command decisions are invaluable attributes for military commanders.

The costs of developing these attributes through wartime experience are extreme and the results highly variable. Further, commanders should have attained the desired capacity and ability in anticipation of the possibility of war. Combat Dynamic Intuition (CDI), defined by Bakken (1993), is the cognitive capability possessed by a military commander when conducting operations.

Being able to measure CDI of individual commanders through involvement in a variety of realistic combat situations leads to opportunities to design training that could significantly improve the capability and effectiveness of commanders. System dynamics simulation games designed as microworld experiments in contexts representing "real-world" *dynamically complex* operational challenges offer the possibilities of measuring the performance of commanders and then through purposeful design significantly improving command and control training.

Developing CDI and enhancing learning through practice in SD microworlds rely on an appreciation of Dynamic Decision Theory (DDT), that is:

- People seek confirmation for their theories and as a consequence are often stuck in suboptimal decision strategies.
- Decision-makers do not seek out alternative strategies when they are satisfied with outcomes, thereby avoiding rigorous and time-consuming testing of alternate hypotheses.
- In dynamic environments decision-makers frequently underestimate dynamic processes along with side effects and self-reinforcing dynamics.
- People fail to adjust their decision strategies to account for delays in the system, expecting feedback to arrive before the system can pro vide such information.
- Decision-makers go into dynamic scenarios with inappropriate mental models based on apparent task characteristics, and they apply little effort to challenge the appropriateness of those mental models.

Noting that the history of warfare is replete with logistic and resource management failures resulting in defeat, and far fewer examples where logistic and resource management successes were critical enablers for victory, Bakken et al. developed Minimalist Decision Training (MDT), a simulation-supported microworld where the commander or command group is required not only to develop and enact their plans but to manage the resources needed to facilitate that enactment. The system dynamics associated with logistics management readily demonstrate needs to accumulate and preposition stocks such as ammunition, fuel, water, and rations; develop strategies for resource usage rates; and to accommodate and prepare for the consequences of systemic delays affecting the enacting of any chosen strategy.

The MDT is characterized by simplifying the commander's operating environment, compressing time and space while giving continuous feedback about the unfolding of the conflict consequential to earlier decisions made. The results observed included improvements in performance with the number of practice trials.

Defense Capability and Force Structure Analysis

Building *defense capability* involves assembling fundamental inputs to capability, personnel, organization, collective training, major systems, supplies, facilities, support, and command and management in creating *force structures* and maintaining such at appropriate state of *preparedness*. Limited availability of funds for military capital acquisition programs and constrained training and operating budgets means that level of achieved preparedness at any time will be a compromise.

Defense planners and decision-makers are concerned with preparing for these diverse roles and activities by achieving the following, within resource and budgetary constraints:

(a) Military capability, which includes force structure development, that is, creating force elements and forming these into force element groupings that are capable of delivering specific effects and countering threats to national security

(b) Preparedness, which includes:
  (i) Readiness within specified timeframes to be deployed on military operations
  (ii) Sustainability to continue operating as an effective military force

Planning and decision-making in a defense context are conducted through highest to lowest levels, they being strategic, operational, and tactical, with each level being inextricably linked to the other.

Within current budgets and forecast funding availability, there will be trade-offs among a diverse set of demands, to:

- Invest for reasons of upgrading or replacing capital assets or structuring forces for new or changing roles.
- Train and retrain frequently enough to maintain skills of individuals.
- Engage in collective training exercises sufficiently to remain current in skills needed to operate as teams or crews.
- Be prepared to respond to changing in threat levels, being able to transition as quickly as is needed from a current state of preparedness to being fully prepared and capable of deploying and then sustaining military forces to deliver the required effects.

The complexity of modern force structure design and development and the use of analytical techniques for strategic level evaluation of the likely effectiveness of alternate force structures in a Canadian defense context are explained by Wesolkowski and Eisler (2015) (Fig. 4).

Based on national threat assessments, strategic defense scenarios are formulated, with each being a story line describing how plausible threats might develop to the point that a specific military response would be needed. Military response options are formulated in detail and prioritized according to strategic intelligence assessments of their likelihood. The range of military response options that a nation might expect to be capable of delivering as discrete responses will be constrained by resource availability. Where concurrent action would be

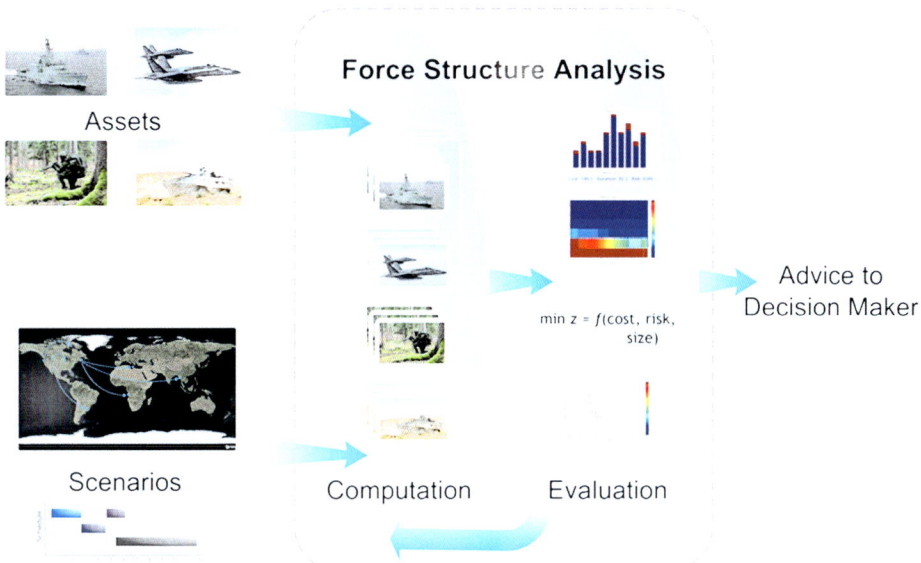

Force Structure Analysis

Assets

Scenarios

Computation   Evaluation

$$\min z = f(\text{cost, risk, size})$$

Advice to
Decision Maker

System Dynamics Modeling to Inform Defense Strategic Decision-Making, Fig. 4 Capability-based models for force structure computation and evaluation.

(Reproduced with permission of Defence Research and Development Canada after Wesolkowski and Eisler (2015))

required, the range of military responses would be further limited.

Practical constraints on the availability of the fundamental inputs to capability, that is, available defense assets forming the basis for creating operational force elements and force element groupings, will drive the selection of what will become the essential set military response options upon which national defense plans could be built. Exact details of which and how many military response options that a nation is capable of delivering, separately and concurrently, will be matters of national secrecy.

Defense Preparedness Modeling

Defense preparedness planning builds on consideration on a broad spectrum of threat scenarios and possible military responses in the context of those scenarios. Being prepared to respond in various ways will involve timeframes that are likely to vary, but all timeframes will be characteristically long.

Threat, estimated in terms of likelihood and extent, must logically lead to selection of the most appropriate military response(s). This has direct implications for resource allocations and

priorities for making changes in preparedness. As threats change these must be matched, indeed, anticipated as effective changes in preparedness levels. It is important that the processes leading to changed level of preparedness be efficient.

Setting the timeframe has two fundamental considerations:

- By when?
- For how long?

The first brings into play the question of how much notice will the military have to change from peacetime capability to being on a war footing. The secondary seeks to define the length of time for which a heightened state of readiness be needed. For example, a force element normally ready to deploy within 5 days ordered to be ready to deploy within 8 h. Knowing how long the heightened level of preparedness will be required will be important for this force element in continuously sustaining the effort to remain prepared and for any others that might have to relieve it, in turn, having raised themselves to the same heightened level of preparedness.

Several hours may be needed to prepare a military aircraft for a specific a mission. Mission planning, crew briefing, arming, and refueling have to be completed. Having the luxury of long warning times will provide great flexibility in mission planning and loading of munitions for the specific of mission. In contrast, very short warning times not only limit mission planning and briefings but may constrain the munition load to being generic. When warning times are excessively short, crews will be in their aircraft with engines running. Even though a mission may be aborted, a maintenance debt continues to accrue, fuel continues to be consumed, and crews become fatigued. Threat levels that demand continuously high degrees of preparedness, inherently having very short notice to mission launch time, are costly and deplete resources at exceedingly high rates.

Consequences of having to maintain high levels of preparedness are ubiquitous and can be insidious. Personnel held at high levels of readiness for extended periods and continuously retrained sustain injuries at high rates. Burnout can increase with degradation of decision quality. Equipment used frequently for training wears out and requires earlier replacement or will have less residual life if actually required for operations.

Response relates to the challenge of being able to generate sufficient and necessary capability to respond to the threat. A very large repertoire of responses has to be developed and practiced. These could include naval blockades, special operations, strategic reconnaissance, and strike, through to large-scale deployments on conventional operations.

Depending on the type and extent of the threat, indigenous independent response or response with allies may be required. Planned operating with allies will require that interoperability between force elements and force element groupings be developed, tested, and rehearsed. This may generate the need for specific changes to resources such as urgent upgrading of equipment and modus operandi to ensure interoperability.

Unless government resource and funding allocations accompany a change in threat, it will be expected that extant military assets and resource allocations will suffice. Further, the minimum resource requirement for one response might preclude use of another because of resource limits. Further, some responses require exceedingly long lead times before it is possible to execute them: if

a country does not retain the response as a continually resourced capability, it will not be available for a long time.

Defense strategic analysis relates to the ways governments describe threats to national security and define response options in the context of wider political conversations. In cold war Europe, it was acceptable, indeed essential, for planning purposes and setting governments' resource agendas for NATO countries, to identify the Soviet Union as a specific threat. The existence of a clear and present threat provides impetus for the scheduling of training, achievement of military competencies and conduct of combined military exercises. In the absence of a set of clear and present threats, planned responses are more likely to be generic, being described in terms of broad capabilities and typical response timeframes. It is critically important that strategic planning be able to cope with both risk and uncertainty: this requires flexibility and responsiveness in addition to being able to determine what activities, resources, and timeframes are needed to create the necessary changes in preparedness arising from changing threats.

Defense preparedness has two constituent parts:

1. Readiness, being ready within specified timeframes to deploy on military operations:

   (a) Recruiting and training individuals in military skills before bringing them together for collective training through which they undergo further skills development to enable them to operate as teams, sub-units, units, force elements, and force element groupings

   (b) Acquiring and having available the individual weapons and equipment with which to equip and arm individuals and prime capital equipment and weapon systems to enable collective training leading to building the capability to form teams, sub-units, units, force elements, and force element groupings for the purposes of conducting military operations

   (c) Having facilities and infrastructure including bases needed for training, to provide personnel and medical support, warehousing, maintenance support to equipment and weapons, etc.

2. Sustainability (AS) or Sustainment (US), being the provision of logistics and personnel services required to maintain and prolong operations until successful mission accomplishment.

Readiness

An important consequence for any nation deliberately seeking to avoid the excessively high costs associated with maintaining readiness on a "knife edge" is that the practical level of readiness achievable will be significantly less than desired. When threats begin to grow, there will be a need to increase levels of readiness as a matter of urgency. Achieving the desired level of readiness is a matter of overcoming inertia. Even when a massive effort is directed to achieving growth in readiness, this growth will be nonlinear. It is unlikely that actual achievements will undershoot the desired targets and the extent of this undershooting will get worse over time. Further, the extent of underachievement would be

exacerbated by the initial gap between current and desired level of readiness.

While increasing the level of readiness is notionally a managerial challenge demanding the rapid harnessing of resources, it is manifestly more complicated. It is confounded by:

• The interconnectedness of diverse activities, many of which have to be undertaken concurrently
• Concurrent activities competing for the same limited resources
• Unavoidable delays, which cannot be foreshortened significantly even by massive injections of resources and effort

Planning in advance of choosing deliberately to increase readiness is essential and indeed critical if budgetary constraints apply.

In peacetime, defense preparedness managers seek to (Fig. 5):

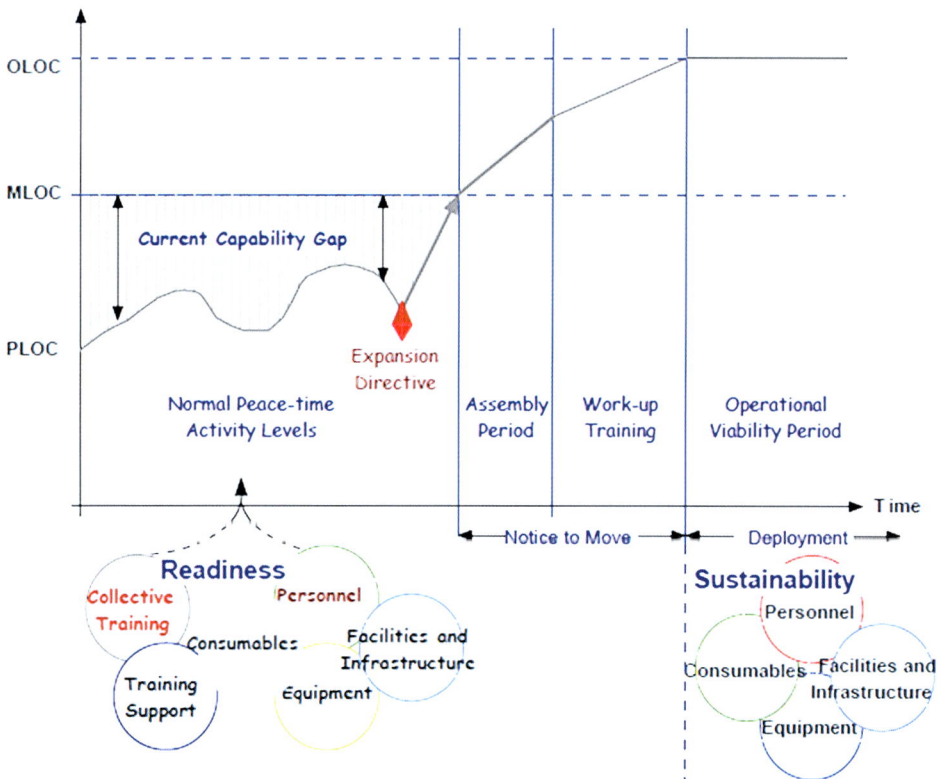

System Dynamics Modeling to Inform Defense Strategic Decision-Making, Fig. 5 General concept of preparedness, readiness, and sustainability. (McLucas (2001: 298), developed after Australian Defence Force Publication No 4)

- Ensure that the minimum present (peacetime) level of capability, that is, PLoC is measurable and continually monitored.
- Ensure that the PLoC is as close as is practicable to a target level of preparedness known as the maintenance level of capability (MLoC) which is theoretically set to enable a force to ramp up to the full operational level of capability (OLoC) quickly and cost-effectively.
- Respond, with appropriate forewarning, in time to deal with changes in threat levels, that is, escalation of known threats and unexpected changes in threats or the nature of threats

Terms equivalent, or similar, to these (Australian Defence Force Publication No. 4) are presumed to be defined and in use by other defense forces.

The context within which the various levels of preparedness are managed is characterized by tension between having capable force structures ready to be deployed as needed and the cost of overinvesting and consequent reduction in government ability to spend on other high-priority needs such as health, education, infrastructure, and social services. Hence, the enormous costs of building defense capability and maintaining high levels of preparedness to enact military responses are in continual conflict with alternative use of government funds and national resources. Preparedness planning also needs to take into account that military forces have concurrent extramural peacetime roles, including responding in emergencies to assist communities recover after natural disasters, conducting search and rescue, assisting other military organizations in international peacekeeping, and so on.

Governments, therefore, must be well equipped to make informed decisions about the priorities for allocating resources to defense activities. The list of actions and associated expenses is long indeed, including acquiring infrastructure and materiel; conducting operations and sustaining deployed forces often for prolonged periods and in remote locations; maintaining, upgrading, and replacing weapon systems; recruiting, training, and forming force elements and force element groups for specific roles; developing of support bases; and enabling defense industries.

In contrast to a naïvely simple count of numbers of armed personnel and their weapons, a nation's defense capability is gauged in terms of a combination of effectiveness of military forces and their preparedness to respond to threats with sufficient anticipation of them arising. Achieving requisite preparedness creates tensions between risk of not being prepared in time, excessive investment in resources, and not having the right mix of force elements formed into force element groupings ready as needed to respond most effectively to particular threats. Military responses include protecting national interests against all adversaries, foreign and domestic, while concurrently providing commanders the capabilities to pursue specific military objectives.

Real World-Preparedness Modeling Challenges

In 1996 the Australian National Audit Office (ANAO) reported critically on the Defence Department's force structure development and preparedness methodology (Minchin et al. 1996). Key criticisms related to a lack of coordination between Army, Navy, and Air Force directed at achieving defined levels of preparedness and inability of the Department to determine the costs associated with changing from one level of readiness to another (higher) level of readiness within a specified timeframe. This led to a detailed investigation by the Department into achieving the required levels of readiness. While the details remain classified, the Department initiated a study, which included extensive system dynamics modeling and simulation by Paterson (2003) and McLucas and Linard (2000).

The problem context presumes that a viable force structure exists at the time a decision is made to increase the level of readiness to a higher level, viz., operational level of capability (OLoC) within a timeframe specified as a number of days. The causal relationships within and spanning three interconnected sectors are depicted in Fig. 6:

This system dynamics influence diagram shows how the primary exogenous driving force, an assessment *Threat*, is changing and directly

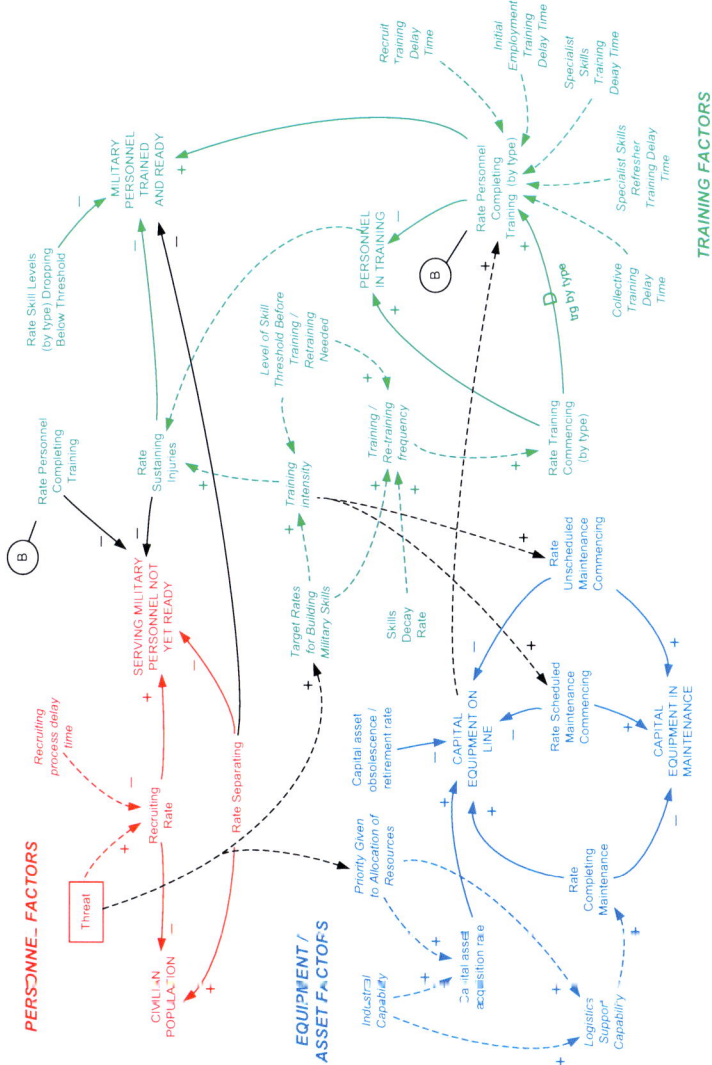

System Dynamics Modeling to Inform Defense Strategic Decision-Making, Fig. 6 Relationships among personnel, training, and equipment factors in preparedness. (Adapted after McLucas 2001 and Paterson 2003)

influences; *Recruiting Rate*, Government *Priority Given to Allocation of Resources*, and *Target Rates for Building Military Skills* of those currently serving and, in turn, those who will be recruited.

Three sets of factors interact strongly, personnel, training, and equipment/assets. These are shown PERSONNEL, EQUIPMENT and TRAINING with arrows and text, respectively, red, green, and blue. Influences and as shown in black. Later these interconnected sectors form the basis of integrated system dynamics model. Commencing with a consideration of personnel factors, personnel will be recruited, while personnel currently serving will undergo individual training, specialist skills training, training to refresh and update specialist skill sets, and collective training. Such training will require access to equipment and assets such as weapons, aircraft, vehicles, and ships. Increased rates of usage of these assets will create the need for increased frequency of maintenance, both scheduled and unscheduled. With training activity rates increased, there will be increases in the need for repair facilities and technicians to restore equipment to serviceability. Not all delays are shown, but there will be numerous delays. Each of the types of training depicted will have its own time to complete, that is, *delay time*. While some sets of activities are concurrent, some will be conducted serially.

As training intensity increases, personnel and equipment casualties will occur.

While skills development and retention among a workforce during times of change are important to every organization (Winch 2001), these are critically important to the military (Paterson

2003; Stothard and Nicholson 2001) because of the risks involved. In cases where specialist skills have to be developed and maintained, such as pilots, these will need to be trained initially, achieve competence in a variety of roles, and be fit and maintain a current journal of competencies to complete a variety of missions.

Figure 7 depicts the initial building of skills or refreshing of skills, leading to the achievement of maximum proficiency, followed by progressive decay of those skills. Of particular concern in both the maintenance of a minimum level of readiness and the rate of retraining is a comprehensive understanding of the ways that skills decay over time. This graph provides the conceptual basis for skills decay over time. The skills decay process would need to be measured carefully, and the graph calibrated before it could be used in any model to calculate the frequency for conducting skills refresher training.

Each of these challenges gives rise to the investigation of alternative interventions for which system dynamics modeling and simulation can be highly effective.

Sustainability

The mechanics of sustainability relate to the design and operation of various supply chains with the aim of delivering items of defense materiel and various classes of goods to theatres of operation according to, or in anticipation of, demand. These supply chains have various rates of production, rates of delivery into depot, depot stocks, rates of forwarding out of depots, shipping delays, rates of delivery to and temporary storage within a theatre

System Dynamics Modeling to Inform Defense Strategic Decision-Making, Fig. 7 Hypothetical growth and decay of skills

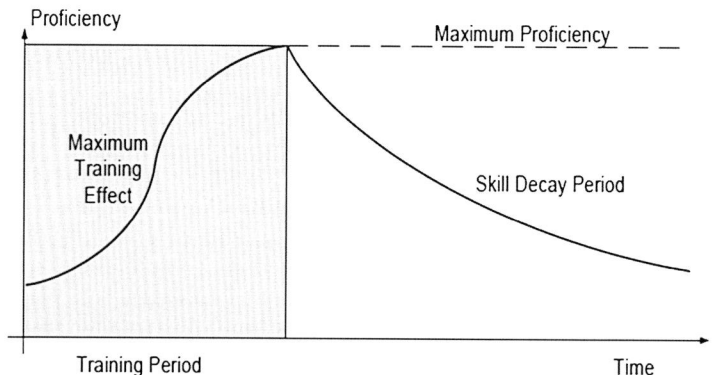

of operations, and delivery to military units. Doctrine and mechanisms for replenishment of various classes of warlike items are generally well defined and are not examined in this entry.

Notable SD modeling and simulation efforts have been directed at managing the manufacture, supply, and delivery of critical items. Critical items are generally capital equipment items, replacement assemblies or repair parts, etc., the availability of which is essential to sustaining military operations at specified rates or tempo. Typical examples of critical items are aircraft, armored vehicles, submarines, aircraft engines, field deployable communications equipment, medical equipment, and repair parts that are required urgently. Critical items are items that are limited in number and are exceptionally scarce and valuable, such that their supply chains have to be intensively managed.

In addressing the challenges of maintaining a submarine fleet to project naval power (Fig. 8), Coyle and Gardiner (1991) use system dynamics influence diagrams to explain that deployment of submarines impacts on other forces and those forces also impact upon the numbers of submarines to be deployed on operations. Like other major capital equipment assets, availability of submarines is a critical issue: the management challenge is to reduce or eliminate the

discrepancy between desired numbers and numbers available.

Considerations of submarines being available at specific points in time cannot be separated from recruiting and training of crews, rotation of crews after repeated deployments, and the time taken to repair, service, and refit submarines.

Defense Capital Asset Acquisition

Project Management

Many billions of dollars are spent every year on acquisition, upgrade, and replacement of defense materiel assets and capital equipment. Universally, projects that involve ab initio research and development of new weapons systems or materiel or the integration of sub-systems into a fully functioning whole typically take much longer than planned with corresponding excessive costs. In many instances, overruns are of the order of two to three times. Many of the underlying reasons for this have been discovered through the modeling and simulation of project activities, especially the need to redo work that was not completed without error in the first instance.

Sterman (2000) describes the Ingalls Shipbuilding case which characterizes the insidious challenges arising from rework. Software

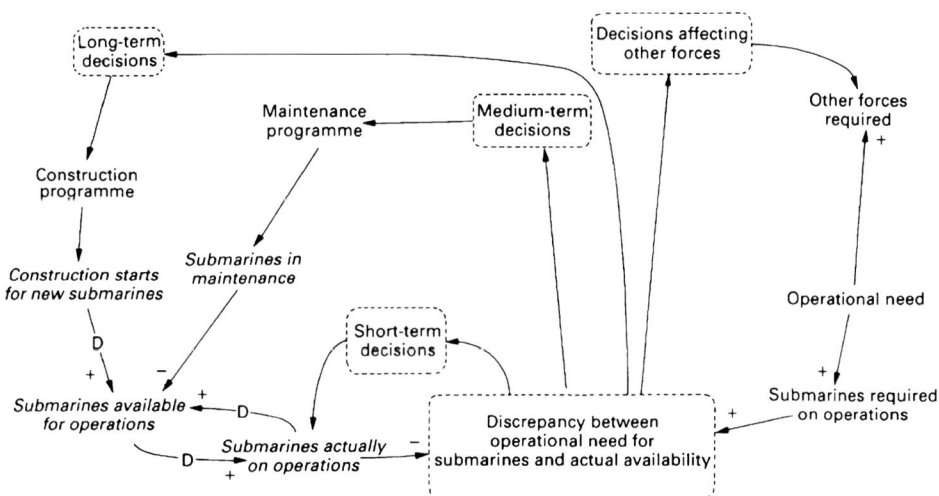

System Dynamics Modeling to Inform Defense Strategic Decision-Making, Fig. 8 Strategic influences on sustainability of a submarine fleet. (Coyle and Gardiner 1991)

development projects (Abdel-Hamid and Madnick 1990) exhibit the same dynamic effects and extending of the time to complete the project. These are examples of research and development and systems integration projects, a special class of projects, very often found in defense contexts. This class of projects is worthy of treatment in its own right. However, system dynamics modeling of the rework cycle has proven valuable in providing greater understanding of the reasons for project cost and schedule overruns and failures of projects involving systems integration. While a comprehensive SD literature, relating to the dynamics of project rework cycles, has developed over nearly two decades, many defense acquisition and systems integration projects continue to suffer serious overruns.

Capital Asset Introduction into Service and Maintenance: Helicopter Fleet Upgrade

Upgrading defense capability involves much more than acquiring new hardware such as weapons or aircraft. This entry demonstrates how system dynamics modeling was used to assist in planning and management of the introduction into service of a new generation multi-role helicopter type. It describes the challenges of managing resources and the complex interrelationships between tasks such as the training of pilots and aircrew, conducting maintenance on the aircraft, and the achievement of defined levels of capability to conduct military operations. While the modeling task focused initially on the management of human resources, it soon became obvious that complex dynamic problems are best addressed using a top-down approach to achieve optimization at the system level rather than attempting to optimize subsystems. The modeling approach exploited trusted, functional modules of system dynamics structure rather than ab initio model construction. How this aided model construction and verification is described by McLucas et al. (2006).

A familiar consequence of aging of equipment fleets is that they become technologically outdated, expensive to maintain and unfit for new or changing roles. This affects military aircraft, weapons, ships and vehicles and their support systems. With several types of helicopters in service, some dating back to the 1960s and 1970s, designed for specific roles that are no longer relevant, there comes a time when an aircraft type is considered as a replacement. Introducing into service a helicopter that is capable of flying many different types of missions, able to operate from a ship in a variety of weather conditions by day and night, carrying troops and stores over long distances, and operating in support of armed helicopters is an attractive proposition. This demands conversion training for pilots who were trained on various older types to be able to fly the new type, including by night under blackout conditions using night vision aids. The new type requires a crew of four rather than two or three as did the older types. To reduce its susceptibility to being degraded in a maritime environment is made of composite materials, which if damaged require vastly different repair techniques. As the new helicopters are delivered, having completed certification and postproduction testing, they begin to be used for the training of pilots as well as flying operational missions. Priority is given to operational missions. Both types of flying serve to build up flying hours, which bring the aircraft closer to needing operational level maintenance (OLM) or deep level maintenance (DLM) (Fig. 9).

Maintenance actions must be scheduled according to the operational and flying training demands, but the numbers of flying hours before maintenance is conducted cannot be exceeded. Initially there are relatively few new aircraft and heavy demands for both types of flying. There is pressure to retire the older types because of limited stocks of repair parts and decreasing reliability. The new aircraft requires new maintenance facilities, ground support equipment, and new procedures with which the maintainers must become familiar. Operational commanders and trainers are interested having at least sufficient aircraft for their tasking and training programs. This model demonstrated the possible ways of achieving the numbers of "available helos" through the management of the various rates such as "accumulating OLM service debt" and applying limited resources to completing OLM servicing tasks, that is, "rate exiting OLM" and similarly for DLM.

This is but one sector of the model. Other sectors of the model addressed:

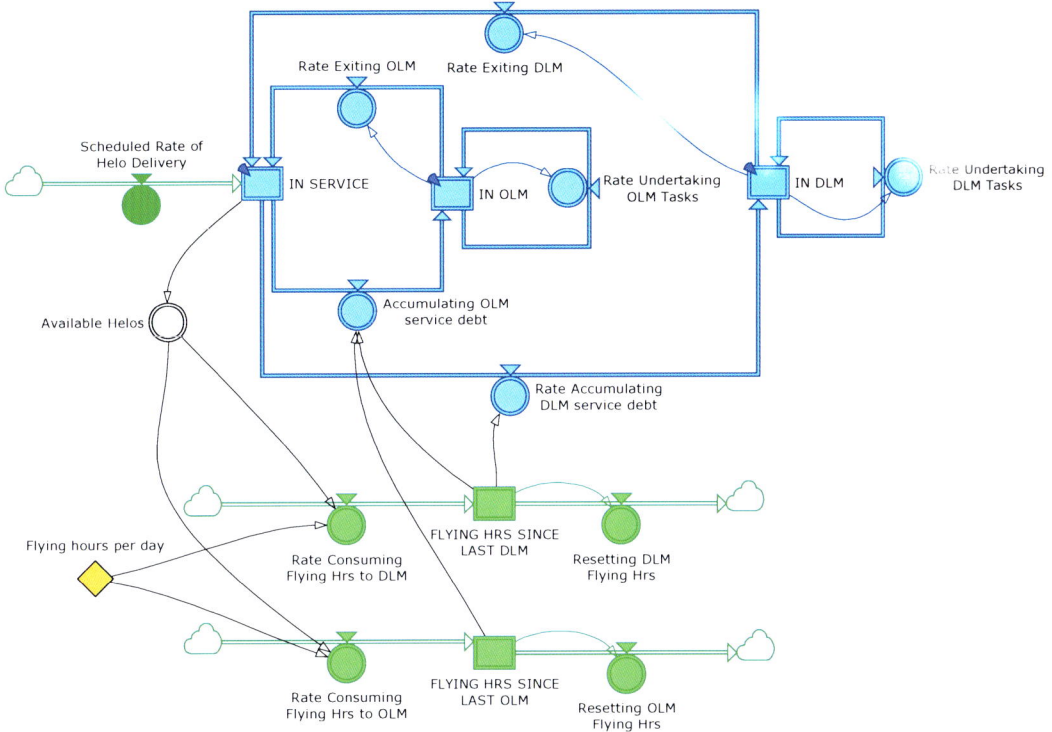

System Dynamics Modeling to Inform Defense Strategic Decision-Making, Fig. 9 Helicopter introduction into service and maintenance model structure (SnF)

- Aircraft acquisition, certification, testing, and final delivery
- Training of flying instructors
- Training of pilots
- Training of aircrew, including loadmasters, aircraft handlers, and weapons technicians
- Preparation for new operational flying roles including day and night, all weather, and amphibious and other specialist tasks
- Allocating of aircraft to training or operational tasks
- Pilot training for transition from various other types to new helicopter type
- Loadmaster training for transition to new helicopter type
- The need to conduct routine operational and deep level maintenance and do so efficiently
- Technical "train-the-trainer" training and specialist technician training
- Commissioning and access to special-to-type maintenance facilities

- Collective training of pilots and loadmasters in forming operational crews
- Availability of qualified flying instructors and qualified loadmaster instructors
- Achievement of defined levels of operational capability

Helicopter Operations: Crews and Missions

Yue (2002) (Fig. 10) analyzed the demands on crews and helicopters created by the need to fly both operational missions and training flights.

System Dynamics Modeling to Inform Suitability of System Reliability and Performance Targets

Defense acquisition projects often have specified targets for operational availability Ao of a system

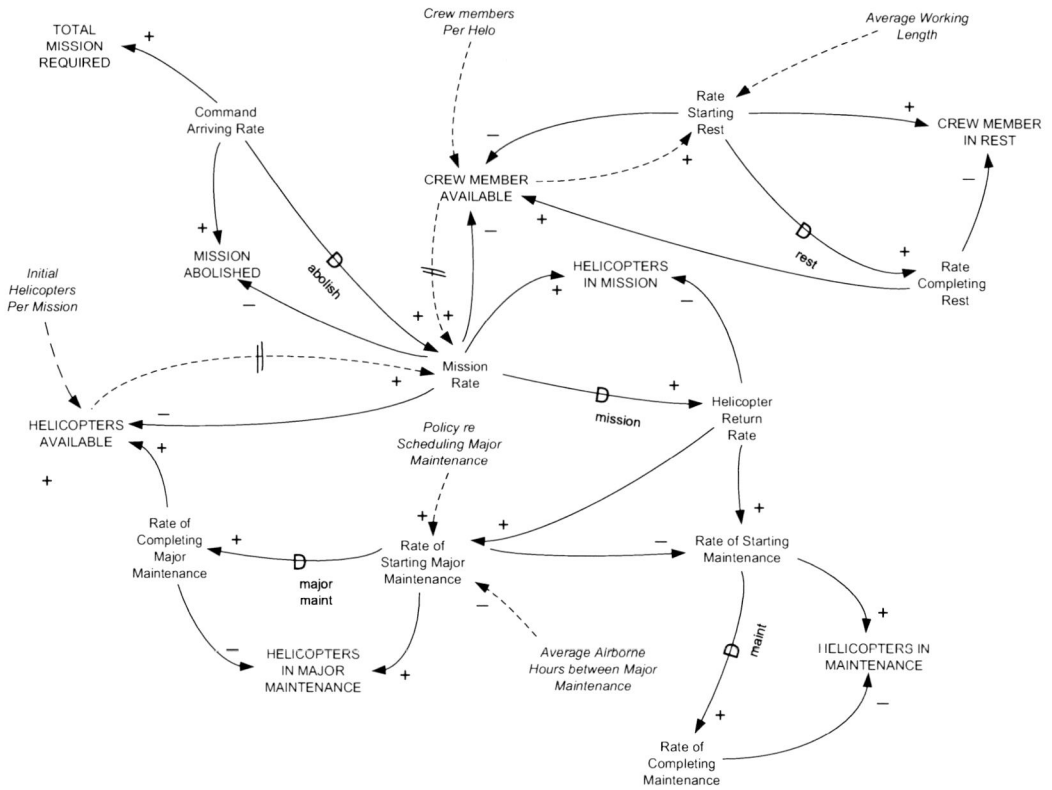

System Dynamics Modeling to Inform Defense Strategic Decision-Making, Fig. 10 A system dynamics model for helicopter operations (ID). (Adapted after Yue 2002)

of capital asset. The actual Ao can only be known well after the system has been in operational use. However, decisions about possible alternate system configurations and sub-system designs have to be made before the system reaches its final design stage. In this setting, it can be highly valuable to create a conceptual design of a system having various sub-systems each of a hypothetical reliability, specified as a Mean Time Between Failures (MTBF). Modeling of such a system provides for the testing of the sensitivity of the overall system to possible variations in reliability of each of the sub-systems. This can also provide valuable insights into how the system might be operated and how susceptible the sub-systems and hence the system overall may be to failures. In turn, this creates opportunities to implement preemptive maintenance or replacement of critical sub-systems and help map out a

reliability growth program, all ahead of the system's introduction into service (McLucas 2011) (Figs. 11 and 12).

Manpower and Career Succession Planning

Manpower is a fundamental input to capability and is much more than a number of individuals. Individuals need to be:

- Trained, that is, having specified individual skills and competence in their trade, mustering, or specialization, such as for an infantry soldier, being able to correctly identify a man-sized standing target at a range of 300 m under specified conditions and within 5 s engage that target with an individual weapon, hitting the target with at least eight out of ten rounds fired in less than 30 s

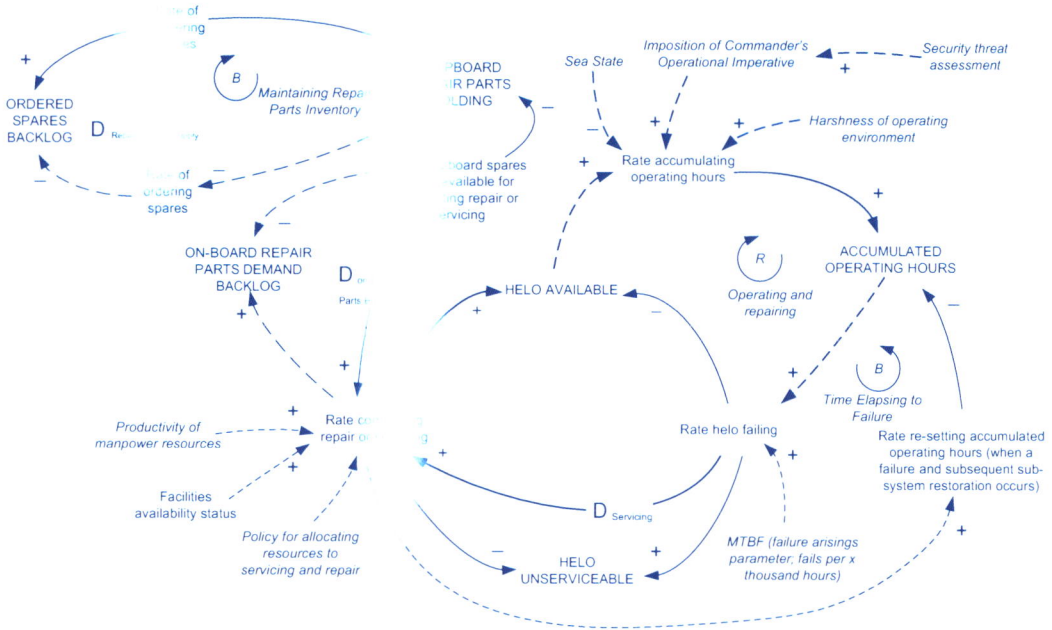

System Dynamics Modeling to Inform Defense Strategic Decision-Making, Fig. 11 Causal influences on shipborne helicopter UAV availability (ID)

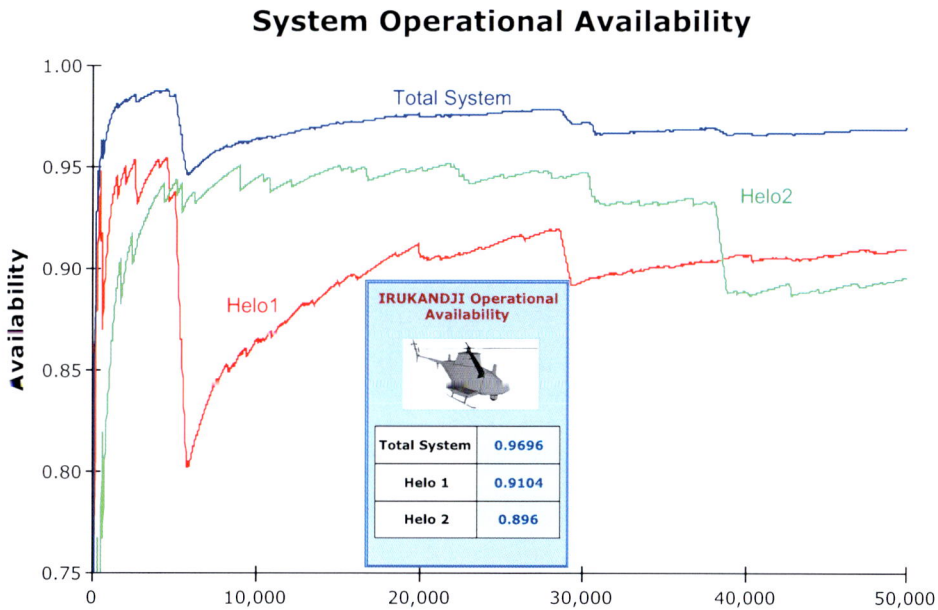

System Dynamics Modeling to Inform Defense Strategic Decision-Making, Fig. 12 Simulated system behavior: expected operational availability with respect to hours of operation

- Current and up-to-date with skills that have been refreshed, as recently as is needed to ensure their ability to complete their tasks, such as, according to trade, mustering or specialization, such as, as a helicopter pilot being able fly a specified mission at night under blackout conditions wearing night-vision goggles
- Organized into teams, units, force elements, and force element groupings; organized, having been trained together, capable of operating in organized ways according to specific and current *modus operandi* or doctrine
- Capable of operating as part of a combat team, operating major weapons systems
- Capable of operating as part of a logistics (medical, medical support, or other) team operating designated logistics (medical, medical support, or other) systems
- Competent in decision-making and commanding sub-units, units, force elements, and force element groupings as appropriate
- Able to respond quickly and appropriately to command and control, etc.

The management of military manpower is a critical part of capability development, force structure, defense preparedness, and sustainability. There is continual dynamic tension between:

- The need to maintain the minimum necessary manning levels for force elements and force element groupings
- Retention of sufficient numbers of personnel with requisite skills and expertise forming a nucleus for force expansion
- Training intensity and frequency of refresher training to provide skills retention and currency and the cost-effectiveness of such training
- Career and employment aspirations of individuals
- Achieving a mix of full-time service personnel and reservists, with opportunities to transfer from part-time to full-time and vice versa to provide for variations in operational tasking and changes in threat levels

Military manpower managers require to set targets for recruiting, completing basic training, completing initial employment training, and promoting within their trade, mustering, or specialization. They operate within constrained sets of rules specified according to length of service; time in current rank; state of fitness and employability according to trade, mustering, or specialization; age; achieving and maintaining a journal of competencies applicable to their particular military role; etc. The various modeling and simulation techniques used for workforce planning are described by Jnitova et al. 2017.

Generic Structure for Manpower and Career Succession Modeling

A simple bathtub metaphor (Fig. 13) provides a useful start point for appreciating the challenging dynamics of manpower management and career succession planning in military organizations.

Manpower managers seek at all times to have continuously predicable and sufficient contents in each bathtub (McLucas 2005). There will be some initial contents in each bathtub (rank level stock). Recruits enter the first bathtub (each bathtub being a stock, level, accumulation, or mathematical integration) as a consequence of having been recruited when the organization seeks to fill gaps and/or meet expected demand for personnel in more senior ranks and experience.

Each bathtub (stock) is:

- Added to under influence of the rate of inflowing which, in turn, operates under various control policies, resource constraints, and constrained maximum numbers in the stock linked to maximum establishment size and payroll constraints (Huntsinger et al. 2012.
- Depleted under the influence of:
  - Rates of promotion for those who have developed the necessary competencies, have spent at least the minimum time in rank, and are ready for promotion of the basis of experience, if there is a vacancy at the next level.
  - Attrition rates, which might include wastage as personnel seek alternate employment, have reached their prescribed length of service or age limitation.

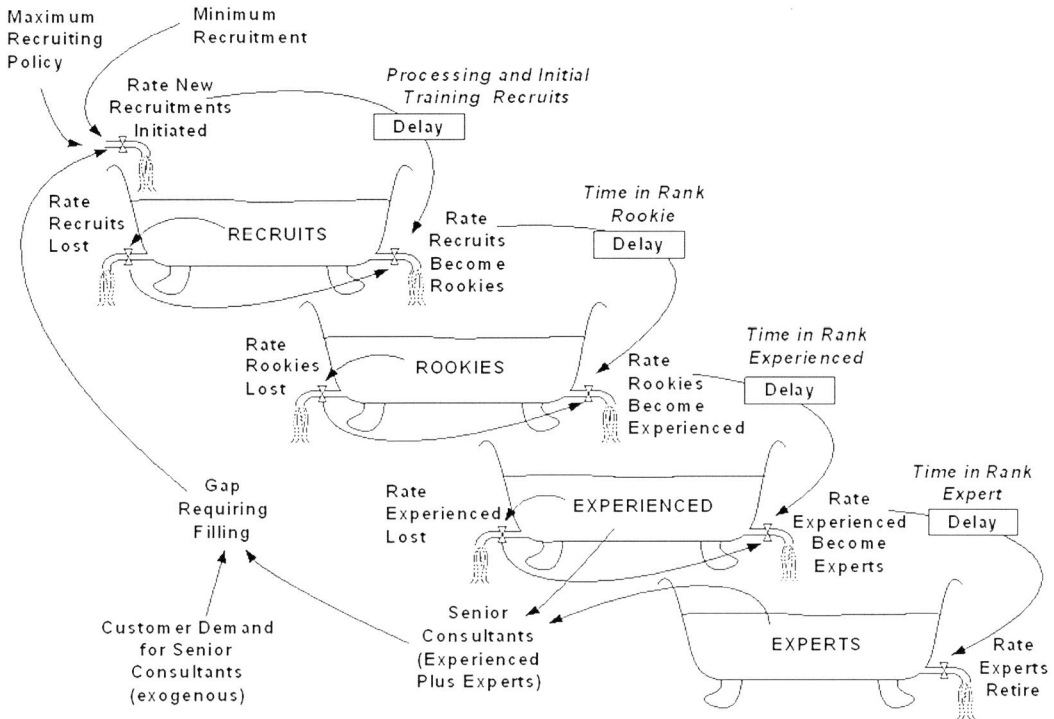

System Dynamics Modeling to Inform Defense Strategic Decision-Making, Fig. 13 Metaphorical bathtub model for career succession

This metaphor helps in identifying which control actions are possibly available to manpower managers. Rates of promotion insofar as competencies are acquired as an individual completes prescribed training provided, or sponsored, by employers are indirectly manageable by manpower managers. Also underlying the rates of promotion are the opportunities for promotion which are created as a consequence of organizational activities, restructuring, growing, or shrinking.

Rates of attrition are largely out of the control of manpower managers as these are largely a consequence of individual choice upon completion of a current period of enlistment or engagement.

This figure depicts a "pull" model, that is, the target recruiting rates are set in response to the demands for senior personnel, which creates a gap to be filled depending on current numbers of senior personnel. A "push" model would see excessive recruiting with a consequent overfilling of the first and subsequent bathtub. In practice, other than at times of war, military manpower models tend to be a

combination "pull" and "push," with greatest influence being given to demand ("pull").

SD manpower planning/career succession models seek to capture discrete events with sets of deterministic rules applying to time in rank, age, length of service, and competence. Rules that might be formulated to represent attrition and some aspects of promotion relating to vacancies becoming available involve an element of chance and/or external forces. For example, defense forces have variously investigated the reasons for service personnel leaving. Many reasons identified are related directly to service conditions, but others depend on job opportunities presented to a spouse or a desire for children to remain at a particular school.

In building system dynamics manpower management and career succession models, researchers (Wang 2007; Armenia et al. 2012; Cavana et al. 2007; Linard et al. 1999; Kearney 1998; Séguin 2015) have sought to discover options for exercising control over valuable manpower resources in diverse contexts, specialist trades, and employment categories.

High-value manpower assets such as pilots, their training, and career progression warrant detailed study. Depending on the actual training pathway, it can take 5–7 years to train a military pilot for operational flying. During this time, many recruits are found to be unsuitable, some become injured or fail to acquire the very high levels of skill demanded by the military. Pilots are trained by experienced pilots who themselves become qualified flying instructors (QFI). Availability of QFI, classrooms, training simulators, and aircraft all have to come together to create the pathway for pilot progression to the point of graduation and ultimate achievement flying proficiency.

Models have been created with the aim of demonstrating to manpower managers the effects of (within trade, specialization, or mustering):

- Gaps that occur between actual and desirable numbers of personnel at specific rank levels
- Setting recruiting targets in response to current gaps or expected gaps between actual and desirable numbers of personnel at specific rank levels
- Slowing rates of promoting
- Expediting and promoting
- Achieving targeted rates of separation at each rank level

The main elements of a career succession model, shown in stock-and-flow notation, for four rank levels in a trade, mustering, or specialization might be depicted as shown in Fig. 14:

This extends the previous multi-bathtub metaphor in a number of significant ways. The double line around specific icons indicate that the relevant stocks, rates, converters, and constants have contents depending on the specific year of the simulation (stocks) or contain specific decision rules that apply only to a cohort in a specific year:

- Individuals enter each rank level and spend a minimum time in rank. Individuals might also transfer directly from another trade or mustering into that rank level with a number of years of seniority.

- The rates of flowing will be governed by rules relating to current or previous years of service (YoS).
- Upon completion of a year of service, an individual will move to the next level of seniority in the rank level, that is, the next cell in the array. If during a particular year an individual reaches the end of his/her current engagement, they may choose to separate.
- If an individual is currently engaged and is not likely to end their engagement in the next period, say 2 years, or has re-engaged in advance, that individual may be eligible for promotion, if:
  – Medically fit
  – Currently competent in their trade or mustering
  – A vacancy exists
  – The individual is recommended suitable for promotion

Each of the four modules has a similar structure. A relatively complex causal structure, not shown, depicts the interrelationships between variables such as numbers holding the senior ranks or a specific rank that will influence the setting of target rates of recruiting, promotion, and separation. Having numbers in a particular rank level that are excessive may lead to offers of voluntary redundancy, whereas encouraging individuals not to separate has proven problematic for manpower managers.

Alternate multi-dimensional model with user-selectable number of rank levels is, Fig. 15:

Defense manpower managers face a multitude of challenges as they seek to have the required number of personnel available when required. Individuals are managed within employment category, trade, or mustering with their skill sets and time to acquire those skill sets even more diverse. From an endogenous perspective, the management challenge involves coping with feedback structures and delays ranging from months to years.

Even during times of relative peace, there can be high rates of deployment and redeployment for those in selected employment categories, trades, or musterings. Repeated deployments and operational rotations are particularly problematic for special forces and submarine crews and may

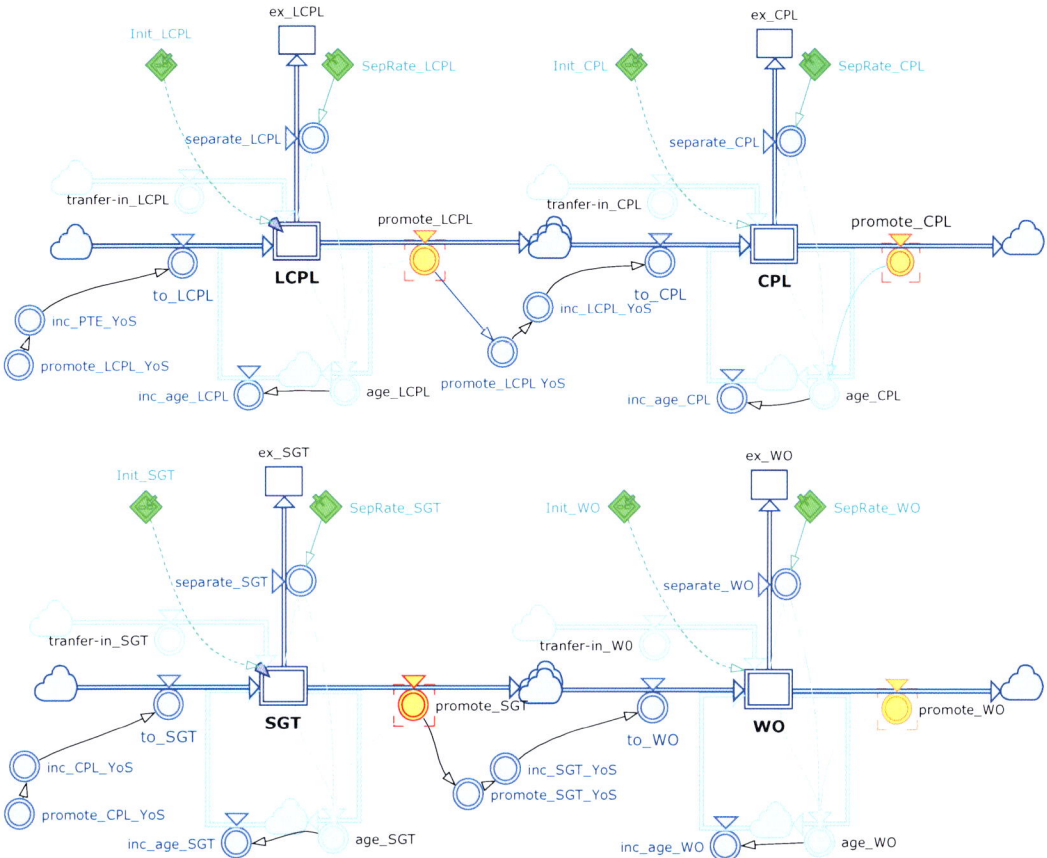

System Dynamics Modeling to Inform Defense Strategic Decision-Making, Fig. 14 A four-tier model for career succession planning of a trade specialization (SnF)

become problematic for individuals in almost any other employment category depending on, at a particular point in time, there being a gap between numbers ready and numbers needed.

Manpower managers need to be planning for long-run dynamics created by the lengths of delays, such as for recruiting, training, acquiring and reacquiring competence, and promoting. They also need to be prepared to invoke short-run action plans in response to step changes or perturbations in demand for personnel to be deployed.

Perhaps the greatest challenge is to identify the underlying causes of highly variable separation rates. Separations occur when an individual:

- As a matter of course chooses not to re-engage
- Despite having previously indicated an intent to re-engage, chooses to separate.

Reasons for separating from the military are rarely based on a single factor, but multiple, interacting factors contribute to the decision. Contributing factors may derive from the nature of service and recent or ongoing service experiences or may be personal and private to an individual.

Service-related causes for separation may include:

- Being overloaded by lengthy and repeated deployments
- Being redeployed with insufficient rest
- Pressure being applied by commanders to improve operational performance in already demanding circumstances
- Quality of the service experience being perceived as declining

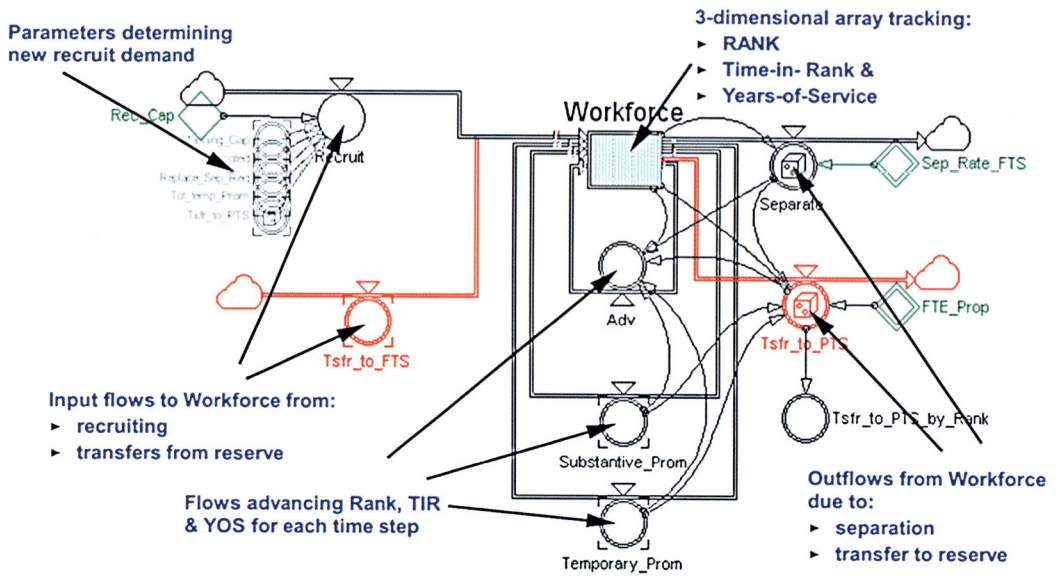

System Dynamics Modeling to Inform Defense Strategic Decision-Making, Fig. 15 A generic multitier model for career succession planning (SnF)

- Exciting service opportunities appearing to favor a select few
- Limited promotion prospects despite having high levels of skills and competence
- Pressure being applied in individuals to seek promotion, with concomitant increase in time to serve
- Previous recruiting initiatives leading there being a current excess in numbers
- Perceived disadvantage of military service compared to advantages of civilian career options
- Colleagues choosing to separate
- Becoming aware of growing pressures to increase retention rates

The details of studies into reasons for retention or separation, as initiated by Service Chiefs or Defense Departments, are sensitive and are generally unavailable.

For manpower managers normal to high separation rates are desirable when there are surplus numbers, but when numbers are less than needed, they struggle to design and implement retention initiatives. In all cases any initiative or, indeed, lack of action will produce a delayed and potentially undesirable response.

The effects of each of these can be dramatic and unpredictable with potentially serious consequences for recruiting targets and previously planned promotion rates.

Often the appropriate targets for rates of recruiting can only be derived through trial and error, that is, multiple simulation runs are conducted with a view of discovering the effects that variations in separation rates have, that is, the range of strategies for setting targeted recruiting rates are informed by multiple simulations.

Rules such as relate to time in rank, completing journal of competencies (demonstrated competence), being recommended for promotion, existence of a vacancy, age, and/or remaining service commitment, are relatively straightforward to formulate. The critical variables are the most difficult to formulate and to manage they being related to separation rates. Personnel will separate from military service for common and obvious reasons in many cases. However, they will often separate for surprising reasons including some that are endogenous, such as impact of spouse's career or change in financial circumstances or timing of a job offer. Models that represent separation rates as a percentage of personnel at a particular rank level are

Types of link in an Influence Diagram

Physical Flow

Information transmission
Control action
Behaviour of nature

Links have + or - signs

Parameter – – – – –▶

Sign may be omitted, but a + or - may also be used

Delays in Physical Flow

──D₁─▶

D denotes delay, subscript identifies a particular delay, sign is always positive

Smoothing of Information

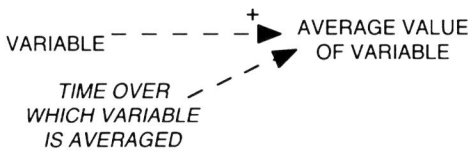

VARIABLE – – – – – ▸ AVERAGE VALUE
OF VARIABLE

TIME OVER
WHICH VARIABLE
IS AVERAGED

External Forces

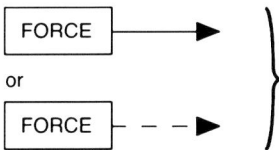

FORCE ────▶

or

FORCE – – ▶

These are forces outside the system over which the system's 'managers' have no control. They may be physical or influences from the behaviour of nature. The link has a + or - sign.

Constrained Flows

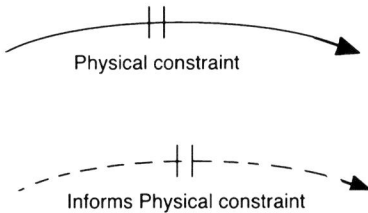

Physical constraint

Physical constrained flows are used to indicate that flow is only possible whilst material remains available to flow in direction of the arrow.

Informs Physical constraint

A dashed line constrained flow is used to remotely control when physical flows can occur, that is, where a physical flow is constrained by activity which originates elsewhere in the system.

Signs for Loops

If the loops has an EVEN number of - signs (0 is an even number) then the loop is POSITIVE
If the loops has an ODD number of - signs then the loop is NEGATIVE

System Dynamics Modeling to Inform Defense Strategic Decision-Making, Fig. 16 System dynamics influence diagramming conventions. (After Coyle 1996)

relatively easy to build. However, these are unlikely to be sufficiently faithful to the real world and so will have limited utility for personnel managers trying to determine how to manage available personnel against demands for personnel with skills in specific skills at specific ranks at specific points in time to meet operational needs or the short notice arising of a particular vacancy.

High-fidelity models will be essential to informing the creation of strategies such as, under particular circumstances:

- Offering retention bonuses for those with particularly valuable skills to continue their service for a further specific period
- Offering premium salaries or salary loadings for valuable skills or for those volunteering for specific types of mission
- Offering voluntary redundancies to encourage increased rates of separation for particular ranks, trades of specialization, length of service, or age
- Offering new career opportunities to those who might volunteer to change their trade or specialization to become part of an alternate career stream
- Extending or removing age-related or time-in-rank limits (Figs. 14 and 15)

Future Work

The review presented in this entry showcases the potential of system dynamics in supporting various planning and management areas in defense. However, there are still areas where more effort is still needed. In this section, we highlight those efforts.

First, there is emerging evidence in the modeling literature on the value of hybrid modeling approaches combining system dynamics, discrete event simulation, and agent-based modeling. Mykoniatis (2015) cited benefits including ability to leverage the strengths of different modeling techniques and represent the phenomenon of interest at different levels of aggregation. This is a promising venue that warrants exploration. For example, agent-based modeling can be integrated with system dynamics to examine the influence of different social dynamics, such as the patterns in recruitments and attrition trends.

Second, there is considerable scope using modern software applications to develop simulations of hypothetical engagements, including high-fidelity representations of terrain derived from real geographical information systems (GIS) data

of potential battlefields. Such applications provide for the detailed analysis of the influences that terrain variations have on advancing troop formations and effectiveness of fire delivered by forces occupying prepared defensive positions. This also provides for the possibility for integrating virtual and live simulations in which forces enact their virtual role as they maneuver across real ground engaging a virtual enemy.

Alternatively, models might be developed into high-fidelity interactive computer simulations in which players acting in the roles of RED or BLUE force commanders, each following designated military doctrine, could engage interactively in the command of their own forces of known structure and preparedness levels. Such developments are considered to be beyond the scope of this entry. Details of alternative approaches to the modeling and simulation of warfare can be found in Bracken et al. (1995).

Bibliography

Primary Literature

Abdel-Hamid T, Madnick SE (1990) Software project dynamics: an integrated approach. Prentice – Hall, Englewood Cliffs

Argyris C (1980) Inner contradictions of rigorous research. Academic Press, New York

Argyris C (1982) How learning and reasoning processes affect organisational change. In: Goodman PS, Associates (eds) Change in organisations: new perspectives on theory, research and practice. Jossey-Bass, San Francisco

Argyris C (1991) Teaching smart people to learn. Harv Bus Rev 69:99–109

Argyris C (1993) Knowledge for action: a guide to overcoming barriers to organisational change. Jossey-Bass, San Francisco

Argyris C (1994) On organisational learning. Blackwell, Cambridge, MA

Argyris C, Schön D (1978) Organisational learning. Addison-Wesley, Reading

Armenia S, Centra A, Cesarotti V et al (2012) Military workforce dynamics and planning in the Italian Airforce. In: 30th international conference of systems dynamics society, St. Gallen, 22–26 July 2012, 1, pp 123–160

AS DoD (1988) Australian Defence Force Publication No 4, ADFP 4. Preparedness, Unpublished

Bakken BE (1993) Learning and transfer of understanding in dynamic decision environments. Dissertation, Massachusetts Institute of Technology

Bakken TB, Gilliam M (2003) Dynamics intuition in military command and control: why it is important and how it should be developed. Cogn Tech Work 5(3):205

Bracken J, Kress Rosenthal RE (1995) Warfare modelling. Military Operations Research Society, Alexandria

Cavana AY, Boyd DM, Taylor AJ (2007) A systems thinking study of retention and recruitment issues for the New Zealand army electronic technician trade group. Syst Res Behav Sci J 24(2):201–216

Coyle RG (1998) The practice of system dynamics: milestones, lessons and ideas from 30 years' experience. Syst Dyn Rev 14(4):343–365

Coyle RG (1996) System dynamics modelling: a practical approach. Chapman and Hall, London

Coyle RG, Gardiner PAA (1991) System dynamics model of submarine operations and maintenance schedules. J Oper Res Soc 42(6):453–462

DoD AS (2014) Defence capability development handbook. Capability Development Group, Department of Defence, Canberra

Forrester JW (1968) Principles of systems. Productivity Press, Cambridge, MA

Huntsinger GC, Burk RC, Trainor TE (2012) The US army projects the effect of merit pay on payroll growth. Interfaces 42(4):395–405

Jnitova V, Elsawah S, Ryan M (2017) Review of simulation models in military workforce planning and management context. J Def Model Simul Appl Methodol Technol 14(4):447–463

Kearney J (1998) Defence applications of system dynamics models. J Battlefield Technol 1:33–34

Lanchester FW (1956) Mathematics in warfare in the world of mathematics, vol 4. In: Newman JR (ed). Simon and Schuster, New York, pp 2138–2157

Lane DC (2000) Diagramming conventions in system dynamics. J Oper Res Soc 51:241–245

Langsaeter T (1984) The land operations model: an application of system dynamics for modeling land combat at the divisional level. In: Huber RK (ed) Systems analysis and modeling in defense: developments, trends, and issues. Permagon Press, Elmsford, pp 23–138

Linard K, Blake M, Paterson D (1999) Optimising workforce structure the system dynamics of employment planning. In: Proceedings of international systems dynamics conference 1999. System Dynamics Society, Canada

Mason RO, Mitroff II (1973) A program for research on management information systems. Manag Sci 19(5):475–487

McLucas AC (2001) An investigation into the integration of qualitative and quantitative techniques for addressing systemic complexity in the context of organisational strategic decision-making. PhD dissertation, University of New South Wales, Canberra

McLucas AC (2005) System dynamics applications. Argos Press, Canberra

McLucas AC (2011) Using system dynamics modelling to aid in establishing realistic availability for complex systems. In the proceedings of systems engineering test and evaluation conference, Adelaide

McLucas AC, Linard KT (2000) System dynamics practice in a non-ideal world: modelling Defence preparedness. In the proceedings of system dynamics 2000, international system dynamics conference, System Dynamics Society, Bergen, Aug 2000

McLucas A, Lyell D, Rose B (2006) Defence capability management: introduction into service of multi-role helicopters. In the proceedings of 24th international conference of the System Dynamics Society, Nijmegen, 23–27 July 2006

Minchin T, Robinson P, Long T (1996) Management of Australian defence force preparedness. In: The Auditor-General preliminary study audit report no 17 1996-96. Australian Government Publishing Service, Canberra

Mykoniatis K (2015) A generic framework for multi-method modeling and simulation of complex systems using discrete event, system dynamics and agent based approaches. PhD thesis, University of Central Florida

Paterson D (2003) An investigation into the application of system dynamics modeling to planning resource allocation for military preparedness in the Australian context. ME research thesis, University of New South Wales, Canberra

Rand Corporation (1992) Planning reserve force mobilization. Rand Corporation, US

Richardson GP (1986) Problems with causal-loop diagrams (1976). Syst Dyn Rev 2(2):158–170

Séguin R (2015) PARSim, a simulation model of the Royal Canadian Air Force (RCAF) pilot occupation: an assessment of the pilot occupation sustainability under high student production and reduced flying rates. Defense Research Development Canada Centre for Operational Research and Analysis, Ottawa

Senge P (1990) The fifth discipline: the art and practice of the learning organisation. Doubleday, New York

Sterman JD (2000) Business dynamics: systems thinking and modeling for a complex world. McGraw-Hill, Irwin

Stothard C, Nicholson R (2001) Skill acquisition and retention in training: DSTO support to the army ammunition study DSTO-CR-0218

Vennix JAM (1996) Group model building: facilitating team learning using system dynamics. Wiley, Chichester

Wang JA (2007) System dynamics simulation model for a four-rank military workforce. Land Operations Division report, DSTO-TR-2037

Wesolkowski S, Eisler C (2015) Capability-based models for force structure computation and evaluation. Defence Research and Development, Canada

Wils A, Kamiya M, Choucri N (1998) Threats to sustainability: simulating conflict within and between nations. Syst Dyn Rev 14(2–3):129–162

Winch GW (2001) Management of the "skills inventory" in times of major change. Syst Dyn Rev 17(2, Summer):151–159

Wolstenholme EF (1990) System enquiry: a system dynamics approach. Wiley, Chichester, pp 153–154

Wolstenholme EF, Al-Alusi A-S (1987) System dynamics and heuristic optimization, Syst Dyn Rev 3(2, Summer):102–115

Yue Y (2002) A system dynamics model for helicopter operations Land Operations Division, Defence Science and Technology Organisation, Department of Defence Australia, in Operations Research/Management Science at Work. Springer, Salisbury SA

Books and Reviews

Adamides ED, Stamboulis YA, Varelis AG (2004) Model-based assessment of military aircraft engine maintenance systems. J Oper Res Soc 55(9):957–967

Australian Army (1997) Blackhawk board of enquiry, documents for public release. Canberra. Unpublished defence report

Bakken T, Gilljam M (2003) Training to improve decision making – system dynamics applied to higher-level military operations. J Battlefield Technol 6(1):33–42

Betts RK (1995) Military readiness; concepts, choices, consequences. Brookings Institution, Washington, DC

Checkland PB (1990) Systems thinking, systems practice. John Wiley and Sons, Chichester

Clark RH (1993) The dynamics of force reduction and reconstitution. Def Anal 9:51–68

Coyle RG (1981) The dynamics of the third world war. J Oper Res Soc 32.555–565

Coyle RG (1983) Who rules the waves? A case study in system description. J Oper Res Soc 34(9):885–898

Coyle RG (1985) A system description of counter-insurgency warfare. Policy Sci 18:55–78

Coyle RG (1987) A model for assessing the work-processing capacity of military command and control systems. Eur J Oper Res 28:27–43

Coyle RGA (1992a) System dynamics model of aircraft carrier survivability. Syst Dyn Rev 8:193–212

Coyle RG (1992b) The optimisation of defence expenditure. Eur J Oper Res 56:304–318

Coyle RG (1996) System dynamics applied to Defense analysis: a literature survey. Def Anal 12(2):141–160

Coyle RG, Millar CJ (1996) A methodology for understanding military complexity: the case of the Rhodesian counter-insurgency campaign. Small Wars Insurgencies 7(3):360–378

Coyle JM, Exelby D, Holt I (1999) System dynamics in defence analysis: some case studies. J Oper Res Soc 50:372–382

Evans JR (2009) Optimising training of ET in FFGs. Internal RAN research for NPCMA Workforce Planning Task

Forrester JW (1961) Industrial dynamics. MIT Press, Cambridge, MA

GAO US (1994) DOD needs to develop a more comprehensive measurement system. United States General Accounting Office, Washington, DC

Garza A, Kumara S, Martin JD et al (2014) System dynamics-based manpower Modeling. In: Proceedings of the industrial and systems engineering research conference (eds Y Guan and H Liao), Norcross, Georgia, USA. ProQuest Central, Ann Arbor, pp 3683–3693

Jajo NK (2015) The trade-off between DES and SD in modeling military manpower. J MSL Online Publication 5, pp 369–376, Epub ahead of print 4 February 2015. https://doi.org/10.5267/j.msl.2015.2.002, Growing Science Ltd., Online Referencing, www.GrowingScience.com/msl. Accessed 23 May 2016

Johnstone M, Le V, Khan B et al (2015) Modeling a helicopter training continuum to support system transformation. In: Interservice/Industry Training, Simulation, and Education Conference (I/ITSEC). National Training and Simulation Association, Arlington, pp 1–11

Kearney, J. (1998) Defence applications of system dynamics models. J Battlefield Technol; 1: 33-34

Linard KT, Blake M, Paterson D (1999) Optimising workforce structure – the system dynamics of employment planning. In: Proceedings of the System Dynamics Society conference (eds RY Cavana, AM Vennix, AJA Rouwette et al), Wellington, 20–23 July 1999

Markham JY (2008) Framing user confidence in a systems dynamics model: the case of workforce planning in a New Zealand Army. PhD thesis, Victoria University of Wellington

McGinnis ML, Kays JL, Slaten P (1994) Computer simulation of US Army Officer professional development. In: Proceedings of the winter simulation conference (eds JD Tew, S Manivannan, DA Sadowski, et al.), Walt Disney World Swan Hotel, Orlando, 11–14 Dec, pp 813–820. Piscataway: IEEE Conference Publications

McLucas AC (2003) Decision making: risk management, systems thinking and situation awareness. Argos Press, Canberra

Mehmood A (2007) Application of system dynamics to human resource management of Canadian naval reserves. College of Engineering, U.A.E., 2007, Online Referencing, http://www.systemdynamics.org/conferences/2007/proceed/papers/MEHMO253.pdf on 15 May 16. Accessed 25 June 2016

Moffat I (1996) The system dynamics of future warfare. Eur J Oper Res 90:609–618

Moore C (1991) Measuring military readiness and sustainability. In: R-3842-DAG. RAND Corporation, US

Reece RL (1990) An analysis of the effect of frequency of task performance on job performance measurement. Research report, Naval Postgraduate School, Monterey

Richardson G (1999) Reflections for the future of system dynamics. J Oper Res Soc 50(4):440–449

Schnieder CAPT (USAF) TA (1990) An assessment of graduated mobilization response. Airforce Institute of Technology, Ohio

Skraba A, Kljajic M, Knaflic A et al (2007) Development of a human resources transition simulation model in Slovenian Armed Forces, 2007. Online Referencing, http://www.systemdynamics.org/conferences/2007/proceed/. Accessed 7 March 2016

Straver M (2013) Development of a naval reserve training and career progression analysis application. Defence Research and Development, Canada. Online Referencing, http://cradpdf.drdcrddc.gc.ca/PDFS/

unc194/p800952_A1b.pdf on 14 May 16. Accessed 24 May 2016

Thomas DA, Kwinn BT, McGinnis M, et al. (1997), The U.S. Army enlisted personnel system: a system dynamics approach, in: Proceedings of the IEEE international conference on computational cybernetics and simulation, 12–15 October, 1263–1267. Piscataway: IEEE Conference Publications

Vassiliou M, Alberts DS, Agre JR (2015) C2 re-envisioned: the future of the enterprise. CRC Press, New York

Wang K-D, Huang Z-J (2000) The system dynamics (SD) is used for researching on Battle simulation. J Syst Sci Syst Eng 9:367–372

Wolstenholme EF (1985) A methodology for qualitative system dynamics. In the proceedings of the system dynamics conference, Denver

Wolstenholme EF (1986) Defense operational analysis using system dynamics. Eur J Oper Res 34:10–18

Wolstenholme EF (1988) Defence operational analysis using system dynamics. Eur J Oper Res 34(1):16–18

Wolstenholme EF (1999) Qualitative vs. quantitative modeling: the evolving balance. J Oper Res Soc 50(4):422–428

Wolstenholme EF, Coyle RG (1983) The development of system dynamics as a rigorous procedure for system description. J Oper Res Soc 35(1):77

System Dynamics Models of Environment, Energy, and Climate Change

Andrew Ford
School of the Environment, Washington State University, Pullman, WA, USA

Article Outline

Glossary

CGCM Coupled general circulation model, a climate model which combines the atmospheric and oceanic systems.

CO$_2$ Carbon dioxide is the predominant greenhouse gas. Anthropogenic CO$_2$ emissions are created largely by the combustion of fossil fuels.

C-ROADS Climate Rapid Overview and Decision Support simulator, a system dynamics model developed by Climate Interactive.

GCM General circulation model, a term commonly used to describe climate models maintained at large research centers.

GHG GHG is a greenhouse gas such as CO$_2$ and methane. These gases contribute to global warming by capturing some of the outgoing infrared radiation before it leaves the atmosphere.

GT Gigaton, a common measure of carbon storage in the global carbon cycle. A GT is a billion metric tons.

IPCC The Intergovernmental Panel on Climate Change was formed in 1988 by the World Meteorological Organization and the United Nations Environmental Program. It reports research on climate change. Their assessments are closely watched because of the requirement for unanimous approval by all participating delegates.

Definition of the Subject

System dynamics is a methodology for studying and managing complex systems which change over time. The method uses computer modeling to focus attention on the information feedback loops that give rise to the dynamic behavior. Computer simulation is particularly useful when it helps us understand the impact of time delays and nonlinearities in the system. A variety of modeling methods can aid the manager of complex systems. Coyle (1977, 2) puts the system dynamics approach in perspective when he describes it as that "branch of control theory which deals with socio-economic systems, and that branch of management science which deals with problems of controllability." The emphasis on controllability can be traced to the early work of Jay Forrester (1961) and his background in control engineering (Forrester 2000). Coyle highlighted controllability again in a highly pragmatic definition:

> *System dynamics is a method of analyzing problems in which time is an important factor and which involve the study of how a system can be defended against, or made to benefit from, the shocks which fall upon it from the out-side world.*

© Springer Science+Business Media, LLC, part of Springer Nature 2020
B. Dangerfield (ed.), *System Dynamics*,
https://doi.org/10.1007/978-1-4939-8790-0_541

Originally published in
R. A. Meyers (ed.), *Encyclopedia of Complexity and Systems Science*, © Springer Science+Business
Media LLC 2018, https://doi.org/10.1007/978-3-642-27737-5_541-4

The emphasis on controllability is important as it directs our attention to understanding and managing the system, not to forecasting the future state of the system. Although point predictions are the objective of some models, system dynamics models are used to improve our understanding of the general patterns of dynamic behavior. System dynamics has been widely used in business, public policy, and energy and environmental policy making.

Introduction

This entry describes some of the many applications to energy and environmental systems, paying particular attention to the problem of global climate change. The applications are similar to applications to other systems described in this encyclopedia. They usually begin with the recognition of a dynamic pattern that represents a problem. System dynamics is based on the premise that we can improve our understanding of the dynamic behavior by the construction and testing of computer simulation models. The models are especially helpful when they illuminate the key feedbacks that give rise to the problematic behavior. The applications selected for this entry illustrate the power of the method in promoting an interdisciplinary understanding of complex problems.

System Dynamics is explained in the core article in this volume, in the early texts by Forrester (1961), Coyle (1977), and Richardson and Pugh (1981) and in more recent texts on strategy by Warren (2002) and Morecroft (2015). The most comprehensive explanation is provided in the text on business dynamics by Sterman (2000). Applications to environmental systems are explained in the text by Ford (2010). The most widely read application to the environment is undoubtedly *The Limits to Growth* (Meadows et al. 1972). Collections of environmental applications appear in a special issue of the *System Dynamics Review* (Ford and Cavana 2004).

The models are normally implemented with visual software such as Stella (http://www.iseesystems.com), Vensim (http://www.vensim.com/) or Powersim (http://www.powersim.com/).

These programs use stock and flow icons to show where the accumulations of the system take place. They also help us see the information feedback in the simulated system. The programs use numerical methods to show the dynamic behavior of the simulated system. The examples selected for this entry made use of Stella and Vensim.

This entry begins with textbook examples of environmental resources in the western USA. The management of water levels at Mono Lake in northern California is the first example. It shows a hydrological model to simulate the decline in lake levels due to water exported out of the basin. The second example involves the declining salmon population in the Tucannon River in eastern Washington. These examples demonstrate the clarity of the approach, and they illustrate the potential for interdisciplinary modeling.

The entry then turns to the topic of climate change and global warming. The focus is on the global carbon cycle and the growing concentration of carbon dioxide (CO_2) in the atmosphere. A wide variety of models have been used to improve our understanding of the climate system and the importance of anthropogenic CO_2 emissions. System dynamics models are presented to show how they can improve our understanding of the climate and climate policy impacts. The models provide a platform for interdisciplinary analysis and for role-playing exercises dealing with international agreements on emission reductions.

System dynamics has also been widely applied in the study of energy problems, especially problems in the electric power industry. The final section describes two applications to electric power. The first involved the financial problems of regulated electric utilities in the USA during the 1970s. It demonstrates the usefulness of the method in promoting an interdisciplinary understanding of the utilities' financial problems. The second study dealt with the CO_2 emissions in the large electricity system in the western USA and Canada. It demonstrated how the power industry could lead the way in reducing CO_2 emissions in the decades following the implementation of a market in carbon allowances.

The Model of Mono Lake

Mono Lake ... an ancient inland sea ... e east side of the S... n Nevada Mountains ... ornia. Microscopic ...gae thrive in its salin ... rs, and the algae support huge populations ... e flies and brine shrimp which can, under ... t conditions, provide a virtually limitless ... supply for migratory and nesting birds.

Starting in 1941, stream flows t... i Mono Lake were diverted into an aqueduct ... xport to Los Angeles. The large export depri... the lake of the historical flows, and the vol...e shrunk over the next four decades. By 1980. the lake's volume was cut approximately in half, and its salinity nearly doubled. Higher salinity levels posed risks to the ecosystem, and environmental scientists feared for the future of the lake ecosystem. Various groups filed suit in the 1970s to limit exports, and the California Supreme Court ruled in 1983 that public trust doctrine mandated a reconsideration of the management of the waters of the Mono Basin. That reconsideration led to a long-term plan to limit exports until the lake's elevation would return to safer levels.

Figure 1 shows a system dynamics model to simulate water flows and storage in the Mono Basin. The goal was to understand the pattern of decline over four decades and to study the responsiveness of the lake to a change in export policy. The model is implemented with the Stella software, and Fig. 1 shows how the model appears when using the software. A single stock variable is used to represent the storage in the basin. The main flow into the lake is the flow from gauged streams that bring runoff from the Sierra to the lake. The aqueduct system diverts a portion of this flow south to Los Angeles, leaving flow past the diversion points to reach the lake. The main outflow is evaporation. It depends on the surface area of the lake and the evaporation rate. The surface area depends in a nonlinear way on the volume of water in the lake. Figure 1 shows that this model follows the standard, system dynamics practice of using familiar names to convey the meaning of the variables in the model. (These particular names match the terms used by water managers and hydrological models of the basin.)

Figure 2 shows the simulated decline in the lake if exports were allowed to continue at high levels for 50 years. The lake would decline from 6374 to around 6342 feet above sea level, a value which is designated as a hypothetical danger level for this simulation. The long, gradual decline

System Dynamics Models of Environment, Energy, and Climate Change, Fig. 1 Stella diagram of the model of Mono Lake

System Dynamics Models of Environment, Energy, and Climate Change, Fig. 2 Simulated decline in Mono Lake elevation if historical export were allowed to continue until the year 2040

System Dynamics Models of Environment, Energy, and Climate Change, Fig. 3 Simulated recovery of Mono Lake elevation if export is set to zero for the second half of the simulation

matches projections by the other hydrological models used in the management plan for the basin. The lake will continue to fall until the declining evaporation leads to a balance of flows in and out of the basin.

Figure 3 shows the simulated responsiveness of the lake to a change in export. The export is cut to zero mid-way through the simulation, and the elevation increases rapidly in the ensuing decade. The simulation reveals an immediate and rapid response, indicating that there is little downward momentum associated with the hydrology of the basin. This responsiveness is highly relevant to the management plan. When the lake falls to a dangerous level, the export could be reduced, and the lake would climb to higher elevations within a few years after the change in policy. This rapid response supports the "wait and see" argument by

those who advocated waiting for full signs of a dangerous salinity before changing export policy. ("Wait and see" may be supported by an analysis of the hydrology of the basin, but it does not necessarily make sense when considering the long delays in the political and managerial process to change water export.) But there is far more than hydrology at work in this system. The waters of Mono Lake support a complex ecosystem which may or may not recover as quickly as the lake elevation. To explore the larger system requires an interdisciplinary model, one that looks at both hydrology and population biology.

Figure 4 shows a model of the population of brine shrimp that live in Mono Lake. The life cycle begins when the adult females deposit cysts in the summer. A stock is assigned to the over wintering cysts. The nauplii and juvenile

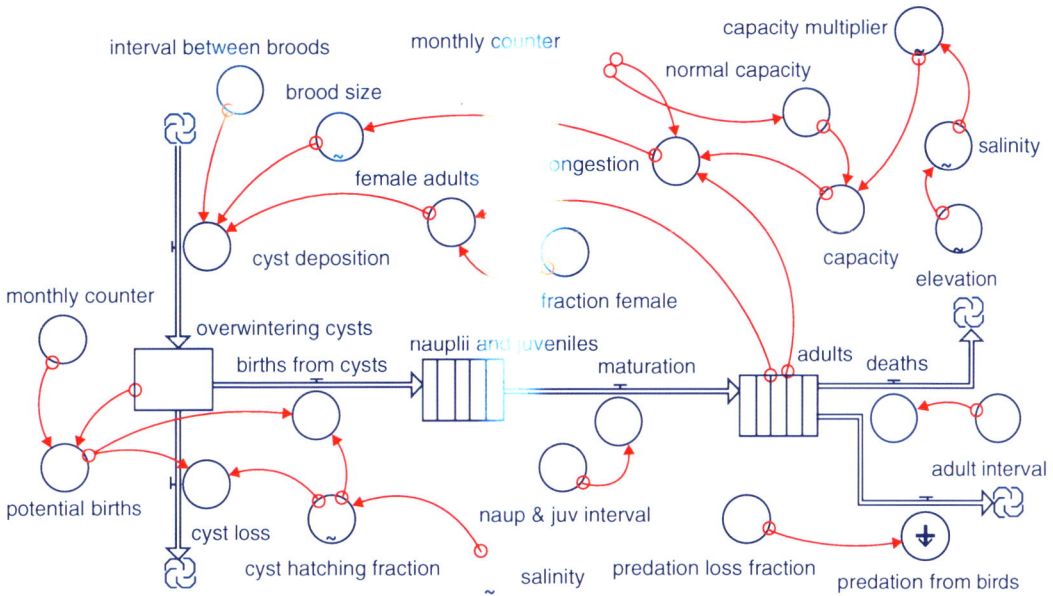

System Dynamics Models of Environment, Energy, and Climate Change, Fig. 4 Stella model of the brine shrimp population of Mono Lake

phases are combined into a second stock, and the maturation leads to a new population of adults in the following summer. The model operates in months and is simulated over a long time interval to show the population response to long-term changes in elevation and in salinity. The model shows the population's response to changes in lake elevation, so one can learn about the delays in the population's response to the changes in lake elevation. Since the shrimp life cycle is 12 months, one would expect the population to rebound rapidly after the increase in elevation and the reduction in salinity. The model confirms that the shrimp population would increase rapidly in the years following the elimination of water export from the basin.

The Mono Lake models are textbook models (Ford 2010). They demonstrate the clarity that the system dynamics approach brings to the modeling of environmental systems. The stock and flow icons help one see the structure of the system, and the long variable names help one appreciate the individual relationships. The simulation results help one understand the downward momentum in the system. In this particular case, there is no significant downward momentum

associated with either the hydrological dynamics or the population dynamics.

The model in Fig. 1 allows for a system dynamics portrayal of the type of calculations commonly performed by hydrologists. Compared to the previous methods in hydrology, system dynamics adds clarity and ease of experimentation. The population model in Fig. 4 is a system dynamics version of the type of modeling commonly performed by population biologists. System dynamics adds clarity and ease of experimentation in this discipline as well.

The main theme of this entry is that system dynamics offers the opportunity for interdisciplinary modeling and exploration. The Mono Lake case illustrates this opportunity with the combination of the hydrological and biological models that allows one to simulate management policies that control export based on the size of the brine shrimp population. The new model is no longer strictly hydrology nor strictly population biology; it is an interdisciplinary combination of both. And by using stock and flow symbols that are easily recognized by experts from many fields of study, the system dynamics enables quick transfer of knowledge. The ability to combine perspectives from different disciplines is one of the most useful aspects of the system dynamics approach to environmental and energy systems. This point is illustrated further with each of the remaining examples.

The Model of the Salmon in the Tucannon River

The next example involves the decline in salmon populations in the Snake and Columbia River system of the Pacific Northwest. By the end of the 1990s, the salmon had disappeared from 40% of their historical breeding ranges despite a public and private investment of more than $1 billion. The annual salmon and steelhead runs had dwindled to less than a quarter of the runs from one hundred years ago. Figure 5 shows a system dynamics model one of the salmon runs, the population of Spring Chinook that spawn in the Tucannon River. The river rises in the Blue Mountains of Oregon and flows 50 miles toward the Snake River in eastern Washington. It is estimated

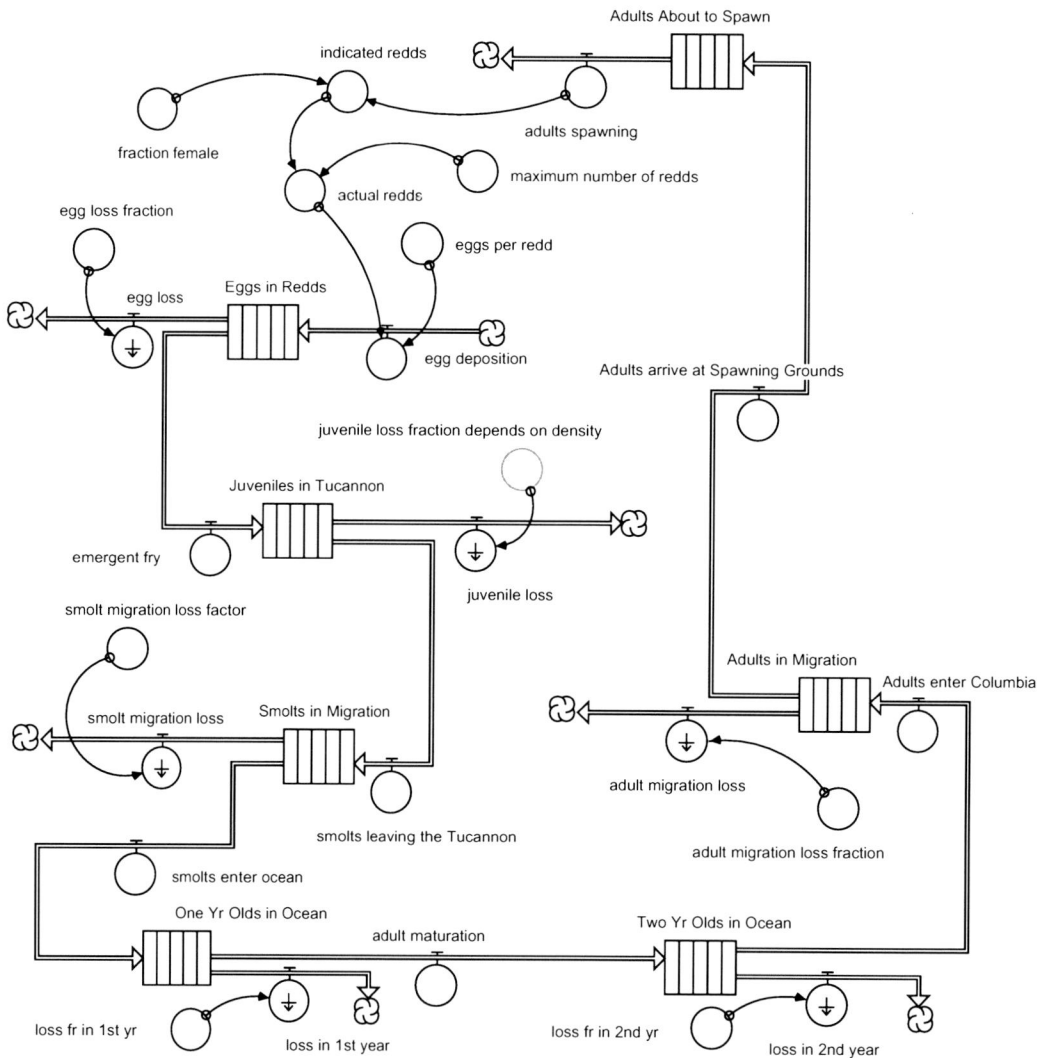

System Dynamics Models of Environment, Energy, and Climate Change, Fig. 5 Stella diagram of the model of the salmon life cycle

that the river originally supported runs of 20 thousand adults. But the number of returning adults has declined substantially due to many changes in the past 60 years. These changes include agricultural development in the Tucannon watershed, hydro-electric development on the Snake and Columbia and harvesting in the ocean.

Each of the stocks in Fig. 5 corresponds to a different phase in the salmon life cycle (see Table 1), with a total life-cycle of 48 months. The parameters represent predevelopment conditions, the conditions prior to agricultural development in the Tucannon watershed and hydro-electric development on the Snake and Columbia. Each of these parameters is fixed regardless of the size of the salmon populations. One of the most important variables is the "juvenile loss fraction depends on density." It can be as low as 50% when there are only a few emergent fry each spring. With higher densities, however, juvenile survival becomes more difficult due to crowding in the cool and safe portions of the river.

Figure 6 shows the model results over a 480 month period with the population parameters in Table 1. The simulation begins with a small number to see if the population will grow to the 20 thousand adults that were thought to have returned to the river in earlier times. The time graph shows a rapid rise to around 20 thousand adults within the first 120 months of the

System Dynamics Models of Environment, Energy, and Climate Change, Table 1 Inputs to simulate the salmon population under predevelopment conditions

Months in each phase		Population parameters	
Adults ready to spawn	1	Fraction female	50%
Eggs in redds	6	Eggs per redd	3900
Juveniles in Tucannon	12	Egg loss fraction	50%
Smolts in migration	1	Smolt migration loss factor	90%
One year olds in ocean	12	Loss factor for first year	35%
Two year olds in ocean	12	Loss factor for second year	10%
Adults in migration	4	Adult migration loss fraction	25%

simulation. The remainder of the simulation tests the population response to variability in environmental conditions, as represented by random variations in the smolt migration loss fraction. (This loss tends to be high in years with low runoff and low in years with high runoff.) Figure 6 confirms that the model simulates the major swings in returning adults due to environmental variability. The runs can vary from a low of ten thousand to a high of thirty thousand.

System dynamics models are especially useful when they help us to understand the key feedbacks in the system. Positive feedback loops are essential to our understanding of rapid, exponential growth; negative feedbacks are essential to our understanding of the controllability of the system. Causal loop diagrams are often used to depict the feedback loops at work in the simulated system. Figure 7 shows an example by emphasizing the most important feedback loops in the salmon model.

Readers will immediately recognize the importance of the outer loop which is highlighted by bold arrows in the diagram. Starting near the top, imagine that there are more spawning adults and more eggs in redds. We would then expect to see more emergent fry, more juveniles, more smolts in migration, more salmon in the ocean, more adults entering the Columbia, and a subsequent increase in the number of spawning adults. This is the positive feedback loop that gives the salmon population the opportunity to grow rapidly under favorable conditions.

An equally important feedback works its way around the inner loop in the diagram. If we begin at the top with more spawners, we would expect to see more eggs, more fry, and a greater juvenile loss fraction as the fry compete for space in the river. With a higher loss fraction, we expect to see fewer juveniles survive to be smolts, fewer smolts in migration, and fewer adults in the ocean. This means we would see fewer returning adults and less egg deposition. This "density dependent feedback" becomes increasingly strong with larger populations, and it turns out to be crucial to the eventual size of the population. Simulating density-dependent feedback is also essential to our understanding of the recovery potential of

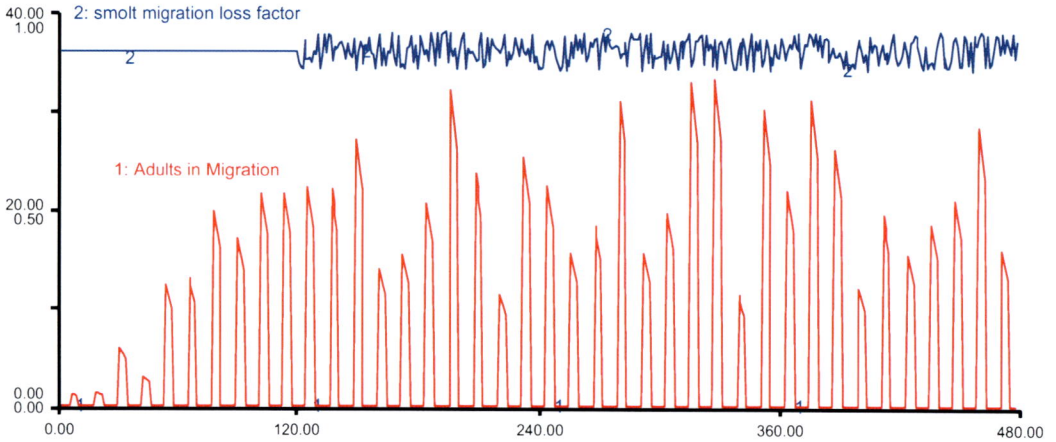

System Dynamics Models of Environment, Energy, and Climate Change, Fig. 6 Test of the salmon model with random variations in the smolt migration losses

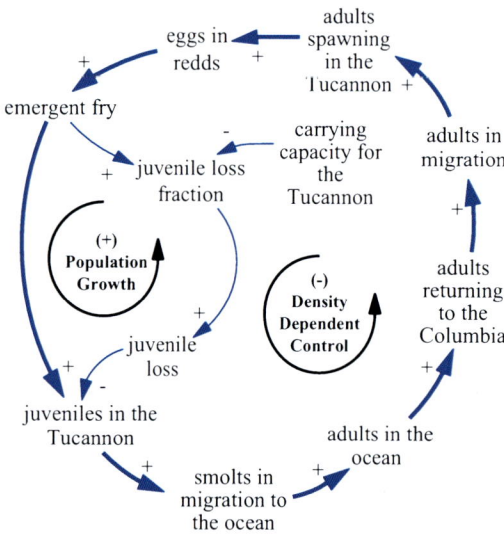

System Dynamics Models of Environment, Energy, and Climate Change, Fig. 7 Key feedback loops in the salmon model

the salmon population. Suppose, for example, that the salmon experience high losses during the adult migration. This will mean that fewer adults reach the spawning grounds. There will be less egg deposition and fewer emergent fry in the following spring. The new cohort of juveniles will then experience more favorable conditions, and a larger fraction will survive the juvenile stage and migrate to the ocean. The density-dependent feedback is crucial to the population's ability to

withstand shocks from external conditions. (The shocks could take the form of changes in ocean mortalities, changes in harvesting, and changes in the migration mortalities. These shocks are external to the boundary of this model, so one is reminded of Coyle's definition of system dynamics. That is, the model helps us understand how the salmon population could withstand the shocks which fall upon it from the out-side world.)

Figure 8 shows a version of the model to encourage student experimentation with harvesting policies. The information fields instruct the students to work in groups of three with one student playing the role of "the harvest manager." The harvest manager's goal is to achieve a large, sustainable harvest through control of the harvest fraction. The other students are given control of the parameters that describe conditions on the Snake and Columbia and in the Tucannon watershed. These students are encouraged to make major and unpredictable changes to test the instincts of the harvest manager.

Models designed for highly interactive simulations of this kind are sometimes called "management flight simulators" because they serve the same function as actual flight simulators. With a pilot simulator, the trainee takes the controls of an electro-mechanical model and tests his instincts for managing the simulated airplane under difficult conditions. The Tucannon Harvesting Model provides a similar opportunity for environmental

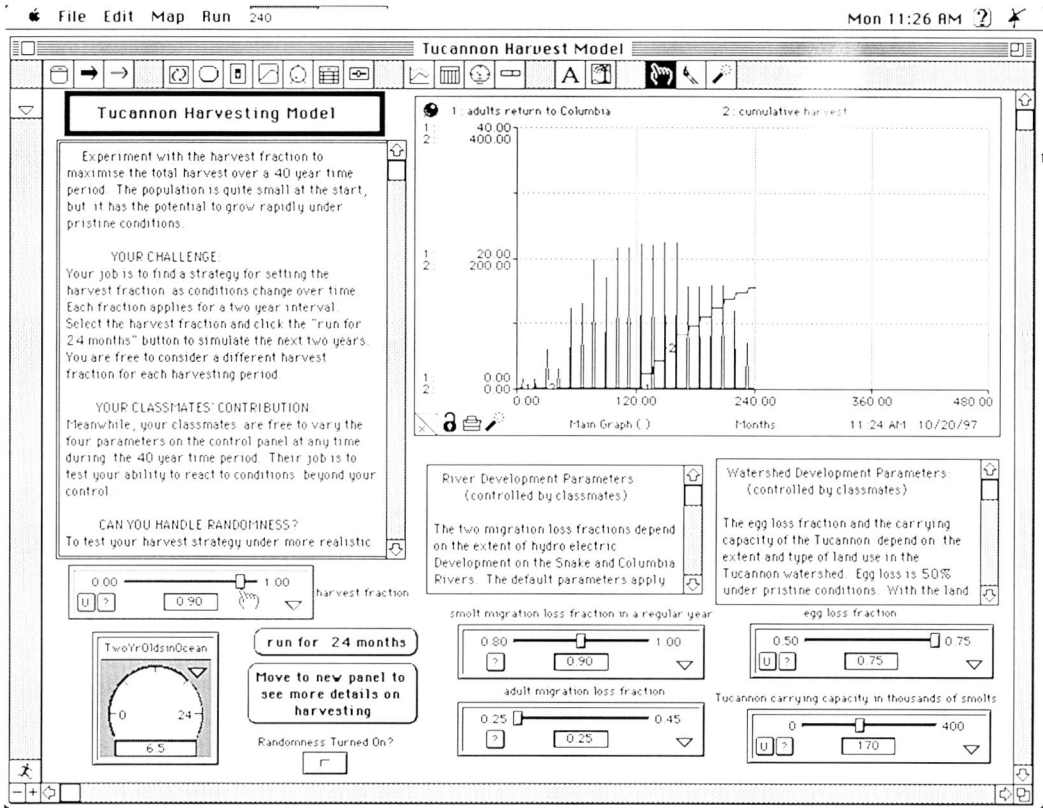

System Dynamics Models of Environment, Energy, and Climate Change, Fig. 8 Salmon harvesting model to encourage student experimentation

students. They can learn the challenge of managing open access fisheries that are vulnerable to over harvesting and the tragedy of the commons (Hardin 1968). In this particular exercise, students learn that they can achieve a sustainable harvest under a wide variety of difficult and unpredictable conditions. The key to sustainability is harvest manager's freedom to change the harvest fraction in response to recent trends in number of returning adults. This is an important finding for fishery management because it reveals that the population dynamics are not the main obstacle to sustainability. Rather, unsustainable harvesting is more likely to occur when the managers find it difficult to change the harvest fraction in response to recent trends. This is the fundamental challenge of an open-access fishery.

The salmon model is a system dynamics version of the type of modeling commonly performed by population biologists. System dynamics adds clarity and ease of experimentation compared to these models. It also provides a launching point for model expansions that can go beyond population biology. Figure 9 shows an example. This is a student expansion to change the carrying capacity from a user input to a variable that responds to the user's river restoration strategy. The student was trained in geomorphology and was an expert on restoring degraded rivers in the west. The Tucannon began the simulation with 25 miles of river in degraded condition and the remaining 25 miles in a mature, fully restored river with a much higher carrying capacity. The new model permits one to experiment with the timing of river restoration spending and to learn the impact on the management of the salmon fishery.

The student's model provides another example of interdisciplinary modeling that aids our

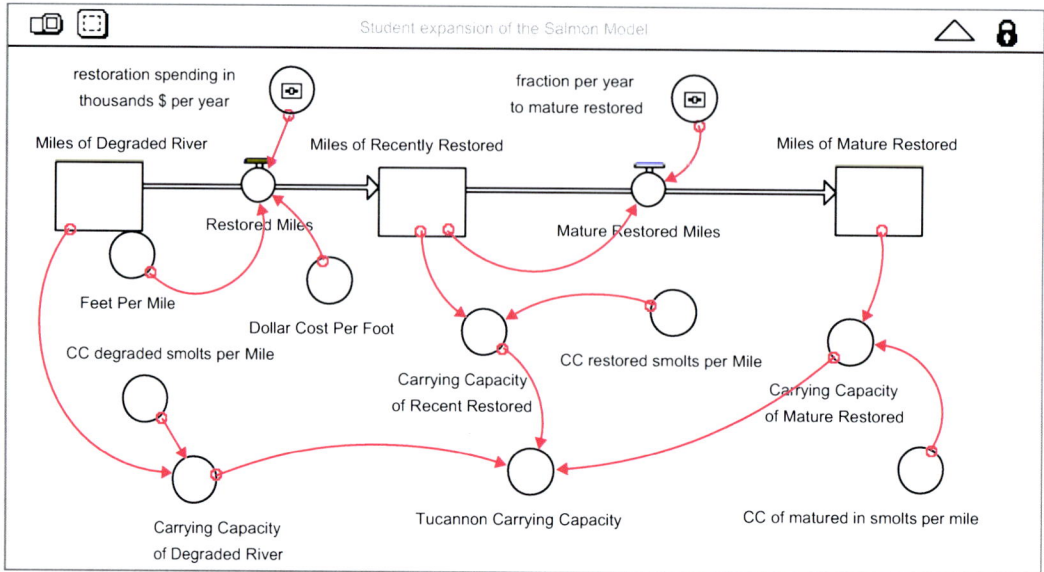

System Dynamics Models of Environment, Energy, and Climate Change, Fig. 9 Student addition to simulate river restoration

understanding of environmental systems. In this particular case, the modeling of river restoration is normally the domain of the fluvial geomorphologist. The model of the salmon population is the domain of the population biologist. Their work is often conducted separately, and their models are seldom connected. This is unfortunate as the experts working in their separate domains miss out on the insights that arise when two perspectives are combined within a single model. In the student's case, surprising insights emerged when the combined model was used to study the economic value of the harvesting that could be sustained in the decades following the restoration of the river. To the student's surprise, the new harvesting could "pay back" the entire cost of the river restoration in less than a decade.

Models of Climate Change

Scientists use a variety of models to keep track of the greenhouse gasses and their impact on the climate. Some of the models combine simulations of the atmosphere, soils, biomass, and ocean response to anthropogenic emissions. The more developed models include CO_2, methane, nitrous oxides, and other greenhouse gas (GHG) emissions and their changing concentrations in the atmosphere. Claussen (2002) classifies climate models as simple, intermediate, and comprehensive. The simple models are sometimes called "box models" since they represent the storage in the system by highly aggregated stocks. The parameters are usually selected to match the results from more complicated models. The simple models can be simulated faster on the computer, and the results are easier to interpret. This makes them valuable for sensitivity studies and in scenario analysis.

The comprehensive models are maintained by large research centers, such as the Hadley Center in the UK. The term "comprehensive" refers to the goal of capturing all the important processes and simulating them in a highly detailed manner. The models are sometimes called GCMs (General Circulation Models). They can be used to describe circulation in the atmosphere or the ocean. Some simulate both the ocean and atmospheric circulation in a simultaneous, interacting fashion. They are said to be coupled general circulation models (CGCMs) and are considered to be the "most comprehensive" of the models available [4]. They are particularly useful when a high spatial resolution is required. However, a disadvantage of the CGCMs is that only a limited number of multidecadal experiments can be performed even when using the most powerful computers.

Intermediate models help scientists bridge the gap between the simple and the comprehensive models. Claussen (2002) describes 11 models of intermediate complexity. These models aim to "preserve the geographic integrity of the Earth system" while still providing the opportunity for multiple simulations to "explore the parameter space with some completeness. Thus, they are more suitable for assessing uncertainty."

Figure 10 characterizes the different categories of models based on their relative emphasis on the number of processes (right axis), detailed treatment of the each process (left axis), and the extent of integration among the different processes (vertical axis). Regardless of the methodology, climate modeling teams must make some judgments on where to concentrate their attention. No model can achieve maximum performance along all three dimensions. (Fig. 10 uses the dashed lines to draw our attention to the impossible task of doing everything within a single model.)

The comprehensive models strive to simulate as many processes as possible with a high degree of detail. This approach provides greater realism, but the models often fail to simulate the key feedback loops that link that atmospheric system with the terrestrial and oceanic systems. (An example is the feedback between CO_2 emissions, temperatures, and the decomposition of soil carbon. If higher temperatures lead to accelerated decomposition, the soils could change from a net sink to a net source of carbon.) The simple models sacrifice detail and the number of processes in order to focus on the feedback effects between the processes. Using Claussen's terminology, one would say that such models aim for a high degree of "integration." However, the increased integration is achieved by limiting the number of processes and the degree of detail in representing each of the processes. System dynamics has been used in a few applications to climate change. These applications fit in the category of simple models whose goal is to provide a highly integrated representation of the system.

System Dynamics Models of the Carbon Cycle

Figures 11 and 12 depict the global carbon cycle. Figure 11 shows the carbon flows in a visual manner. Figure 12 uses the Vensim stock and flow icons to summarize carbon storage and flux in the current system. The storage is measured in

System Dynamics Models of Environment, Energy, and Climate Change,
Fig. 10 Classification of climate models

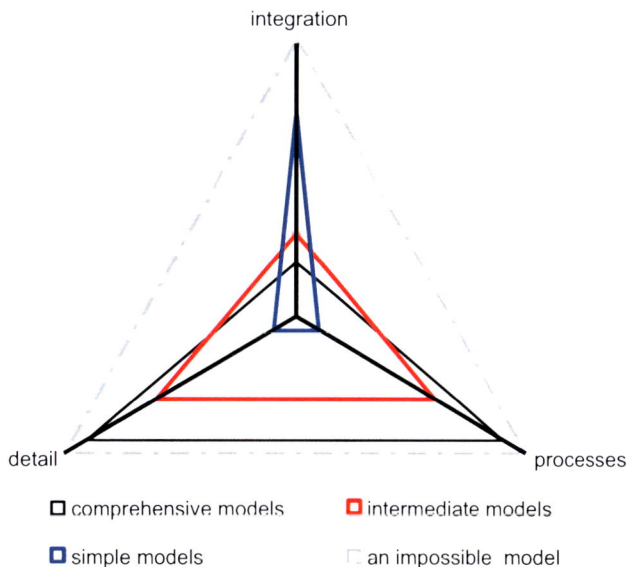

System Dynamics Models of Environment, Energy, and Climate Change, Fig. 11 The global carbon cycle. (Source: United Nations Environmental Program (UNEP) http://www.unep.org/)

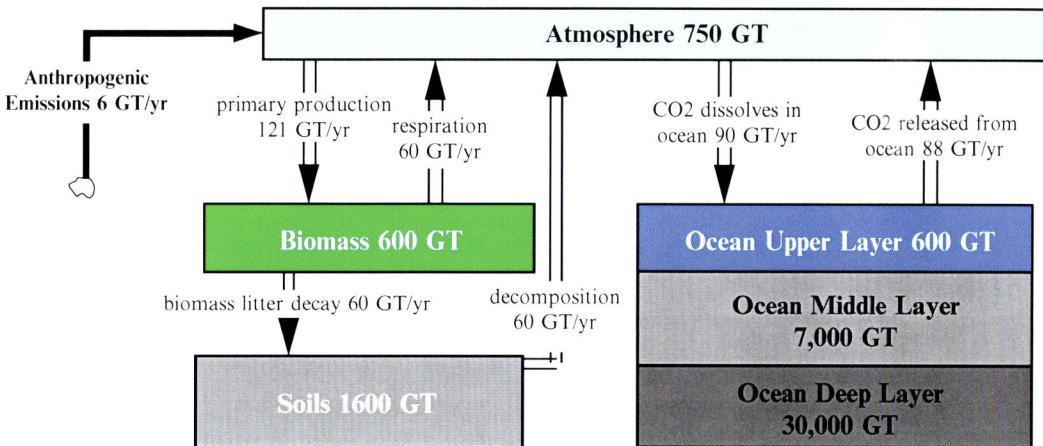

System Dynamics Models of Environment, Energy, and Climate Change, Fig. 12 Diagram of the stocks and flows in the carbon cycle

GT, gigatons of carbon, (where carbon is the C in CO_2). The flows are in GT/year of carbon with values rounded off for clarity.

The left side of Fig. 12 shows the flows to the terrestrial system. The primary production removes 121 GT/year from the atmosphere. This outflow exceeds the return flows by 1 GT/year. This imbalance suggests that around 1 GT of carbon is added to the stocks of biomass and soil each year. So the carbon stored in the terrestrial system would grow over time (perhaps due to extensive reforestation of previously cleared land). The right side of Fig. 12 shows the flows from the atmosphere to the ocean. The CO_2 dissolved in the ocean each year exceeds the annual release back to the atmosphere by 2 GT. The total net-flow out of the atmosphere is 3 GT/year which means that natural processes are acting to negate approximately half of the current anthropogenic load.

As the use of fossil fuels grows over time, the anthropogenic load will increase. But scientists do not think that natural processes can continue to negate 50% of an ever increasing anthropogenic load. On the terrestrial side of the system, there are limits on the net flow associated with reforestation of previously cleared land. And there are limits to the carbon sequestration in plants and soils due to nitrogen constraints. On the ocean side of the system, the current absorption of 2 GT/year is already sufficiently high to disrupt the chemistry of the ocean's upper layer. Higher CO_2 can reduce the concentration of carbonate, the ocean's main buffering agent, thus affecting the ocean's ability to absorb CO_2 over long time periods.

Almost of the intermediate and comprehensive climate models may be used to estimate CO_2 accumulation in the atmosphere in the future. For this entry, it is useful to draw on the mean estimate published in *Climatic Change* (Webster 2003). He used the climate model developed at the Massachusetts Institute of Technology, one of the 11 models of "intermediate complexity" in the review by (Claussen 2002). The model began the simulation in the year 2000 with an atmospheric CO_2 concentration of 350 parts per million (ppm). (This concentration corresponds to around 750 GT of carbon in the atmosphere.) The mean projection assumed that anthropogenic emissions would grow to around 19 GT/year by 2100. The mean projection of atmospheric CO_2 was around 700 ppm by 2100. The amount of CO_2 in the atmosphere would be twice as high at the end of the century.

Figure 13 shows the simplest possible model to explain the doubling of atmospheric CO_2. The stock accumulates the effect of three flows, each of which is specified by the user. Anthropogenic emissions are set to match Webster's assumption. They grow to 19 GT/year by the end of the

**System Dynamics
Models of Environment,
Energy, and Climate
Change, Fig. 13** Simple
model to understand
accumulation of CO_2 in the
atmosphere

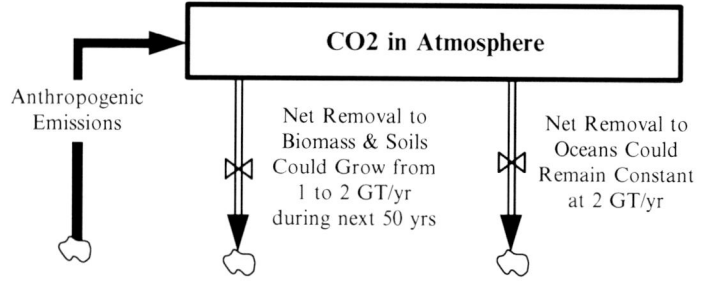

System Dynamics Models of Environment, Energy, and Climate Change, Fig. 13 Simple model to understand accumulation of CO_2 in the atmosphere

century. Net removal to oceans is assumed to remain constant at 2 GT/year for the reasons given previously.

Net removal to biomass and soils is then subject to experimentation to allow this simple model to match Webster's results. A close match is provided if the net removal increases from 1 to 2 GT/year during the first half of the century and then remains at 2 GT/year for the next 50 years. With these assumptions, the CO_2 in the atmosphere would double from 750 to 1,500 GT during the century. This means that the atmospheric concentration would double from 350 to 700 ppm, the same result by (Webster 2003).

The model in Fig. 13 is no more than an accumulator. This is the simplest of possible models to add insight on the dynamics of CO_2 accumulation in the atmosphere. It includes a single stock and only three flows, with all of the flows specified by the user. There are no feedback relationships which are normally at the core of system dynamics models. This extreme simplification is intended to make the point that simple models may provide perspective on the dynamics of a system. In this case, a simple accumulator can teach one about the sluggish response of atmospheric CO_2 in the wake of reductions in the anthropogenic emissions. As an example, suppose carbon policies were to succeed in cutting global emissions dramatically in the year 2050. By this year, emissions would have reached 10 GT/year, so the supposed policy would reduce emissions to 5 GT/year. What might then happen to CO_2 concentrations in the atmosphere for the remainder of the century? Experiments with highly educated adults (Sterman and Sweeney 2007) suggest that some subjects would answer this question with "pattern matching" reasoning. For example, if

emissions are cut in half, it might make sense that CO_2 concentrations would be cut in half as well. But pattern matching leads one astray since the accumulation of CO_2 in the atmosphere responds to the total effect of the flows in Fig. 13. Were anthropogenic emissions to be reduced to 5 GT/year and net removals were to remain at 4 GT/year, the CO_2 concentration would continue to grow, and atmospheric CO_2 would reach 470 ppm by the end of the century.

The model in Fig. 13 demonstrates that useful insights can emerge from a very simple model. Figure 14 turns our attention to a system dynamics model with multiple stocks and several feedback loops. The diagram shows the carbon cycle portion of the climate model by Fiddaman (2002). His purpose was to create a simple model of climate change, with the results checked against corresponding results from GCMs and CGCMs. The model was then expanded to simulate the climate system within a larger system that includes growth in human population, growth in the economy, and changes in the production of energy. The model was organized conceptually as nine interacting sectors with a high degree of coupling between the energy, economic, and the climate sectors.

Fiddaman focused on policy making, particularly the best way to put a price on carbon. In the current debate, this question comes down to a choice between a carbon tax and a carbon market. His simulations add support to those who argue that the carbon tax is the preferred method of putting a price on carbon. The simulations also provide another example of the usefulness of system dynamics models that cross disciplinary boundaries. The economy, energy system, and climate system were simulated within a single, tightly

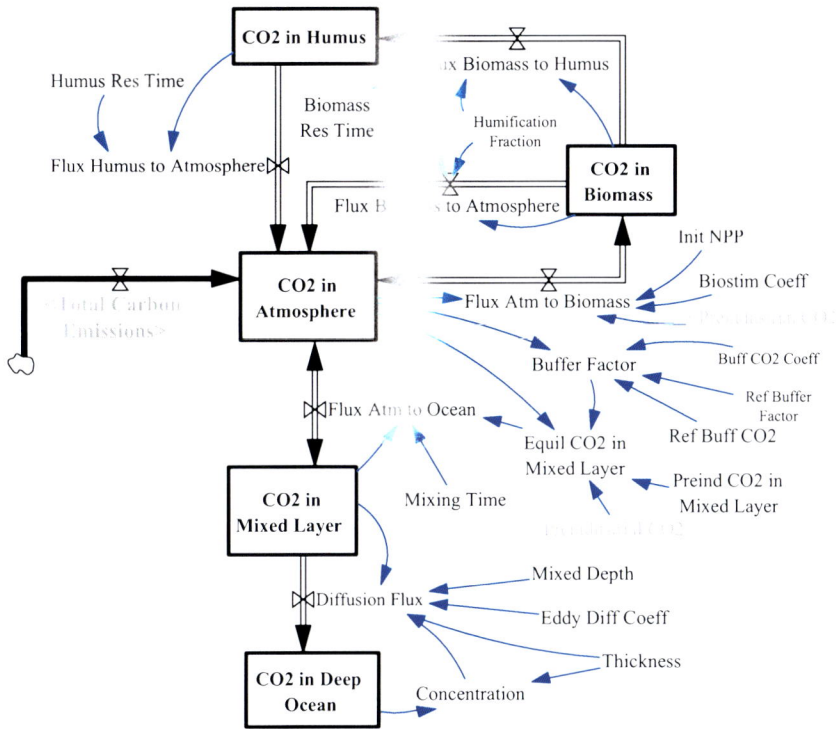

System Dynamics Models of Environment, Energy, and Climate Change, Fig. 14 Representation of the carbon cycle in the model by Fiddaman (2002)

coupled model, providing another example of the power of system dynamics to promote interdisciplinary exploration of complex problems.

Climate Modeling for Decision Support

Fiddaman's model was later adapted to serve as the core carbon cycle, climate sector of a model developed be Climate Interactive (2017). The new model was designed for fast, clear simulations showing the carbon cycle, other GHGs, radiative forcing, global mean surface temperature and change in sea level. The model simulates emissions reductions from different countries, with the interface allowing for one, three, six, or 15 different blocs of countries. The model responds to the users' inputs with rapid simulations giving an overview of climate impacts. The model is known as C-ROADS which stands for Climate – Rapid Overview And Decision Support simulator (Climate Interactive 2017).

C-ROADS is well known for its use in a role playing exercise for participants to experience the challenges of international climate change negotiations. The participants typically number 8–50, but up to 500 participants can be accommodated. A facilitator plays the role of a United Nations official to lead the group. Participants adopt the role of delegates, perhaps from a specific nation or from a block of nations. Their goal is to arrive at a global agreement on emission reductions that limit temperature changes to less than 2 °C over preindustrial temperature.

The role-playing simulations typically last for 2–3 h, plenty of time for participants to face up to the simulated changes in climate and to experience the tension of negotiations among nations. The negotiations exercises have been put to good use in high schools, in universities, and in executive education programs. The negotiations exercise was also featured at a 2015 White House event on climate change education. Thousands of participants have benefited from the role playing exercises, including Nobel-prize winning scientists, oil company executives, staff from the US Forest Service, a former US Secretary of State, and policy makers from the European Union.

The global climate system is one of the many examples of energy and environmental systems that are highly interconnected. Limiting CO_2 emissions in the future will challenge nations to develop cleaner energy systems.

Lessons from the Regulated Power Industry in the 1970s

Environmental and energy systems are often intertwined, and system dynamics has been put to good use in simulating the interactions between energy and environmental systems. Examples of energy applications are given by Bunn and Larsen (1977) and by Ford (2010). A key word frequency count in 2004 revealed nearly 400 energy entries in the System Dynamics bibliography [2]. Many of these applications deal with the electric power industry. A cleaner power system is crucial to limiting future CO_2 emissions, so I have selected two examples from the power industry to illustrate the usefulness of the approach. The first example involves the regulatory and financial challenges of the investor owned electric utilities in the United States.

The 1970s was a difficult decade for the regulated power companies in the United States. The price of oil and gas was increasing rapidly, and the power companies were frequently calling on their regulators to increase retail rates to cover the growing cost of fuel. The demand for electricity had been growing rapidly during previous decades, often at 7%/year. At this rate, the demand doubled every decade, and the power companies faced the challenge of doubling the amount of generating capacity to ensure that demand would be satisfied. Power companies dealt with this challenge in previous decades by building ever larger power plants (whose unit construction costs declined due to economies of scale). But the economies of scale were exhausted by the 1970s, and the power companies found themselves with less internal funds and poor financial indicators.

Utilities worried that the construction of new power plants would not keep pace with demand, and the newspapers warned of curtailments and blackouts.

Figure 15 puts the financial problems in perspective by showing the forecasting, planning, and construction processes. The side by side charts allows one to compare the difficult conditions of the 1970s with conditions in previous decades. Figure 15a shows the situation in the 1950s and 1960s. Construction lead times were around 5 years, so forecasts would extend 5 years into the future. Given the costs at the time, the power company would need to finance $3 billion in construction. This was a substantial but manageable task for a company with $10 billion in assets.

Figure 15b shows the dramatic change in the 1970s. Construction lead times had grown to around 10 years, and construction costs had increased as well. The power company faced the challenge of financing $10 billion in construction with an asset base of $10 billion. The utility executives turned to the regulators for help. They asked for higher electricity rates in order to increase annual revenues and improve their ability to attract external financing. The regulators responded with substantial rate increases, but they began to wonder whether further rate increases would pose a problem with consumer demand. If consumers were to lower electricity consumption, the utility would have less sales and less revenues. The executives might then be forced to request another round of rate increases. Regulators wondered if they were setting loose a "death spiral" of ever increasing rates, declining sales, and inadequate financing.

Figure 16 puts the problem in perspective by showing the consumer response to higher electricity rates alongside of the other key feedback loops in the system. Higher electricity rates do pose the problem which came to be called "the death spiral." But the death spiral does not act in isolation. Figure 16 reminds us that higher rates lead to lower consumption and to a subsequent reduction in the demand forecast and in construction. After delays for the new power plants to come on line, the power companies experiences a reduction in its "rate base" and the "allowed revenues." When

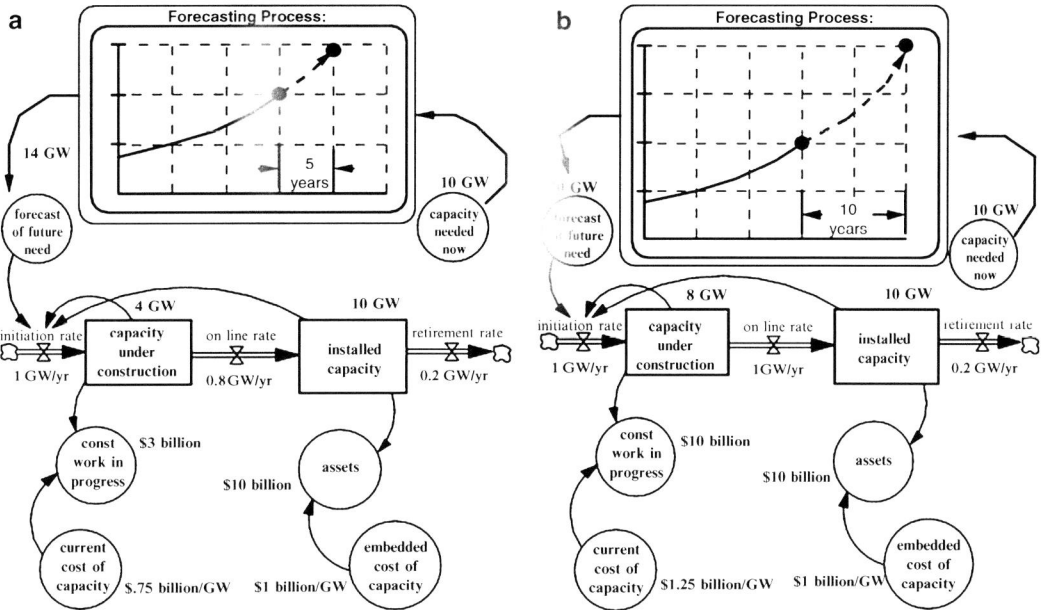

System Dynamics Models of Environment, Energy, and Climate Change, Fig. 15 (**a**) The electric utility's financial challenge during the 1950s and 1960s. (**b**) The electric utility's financial challenge during the 1970s

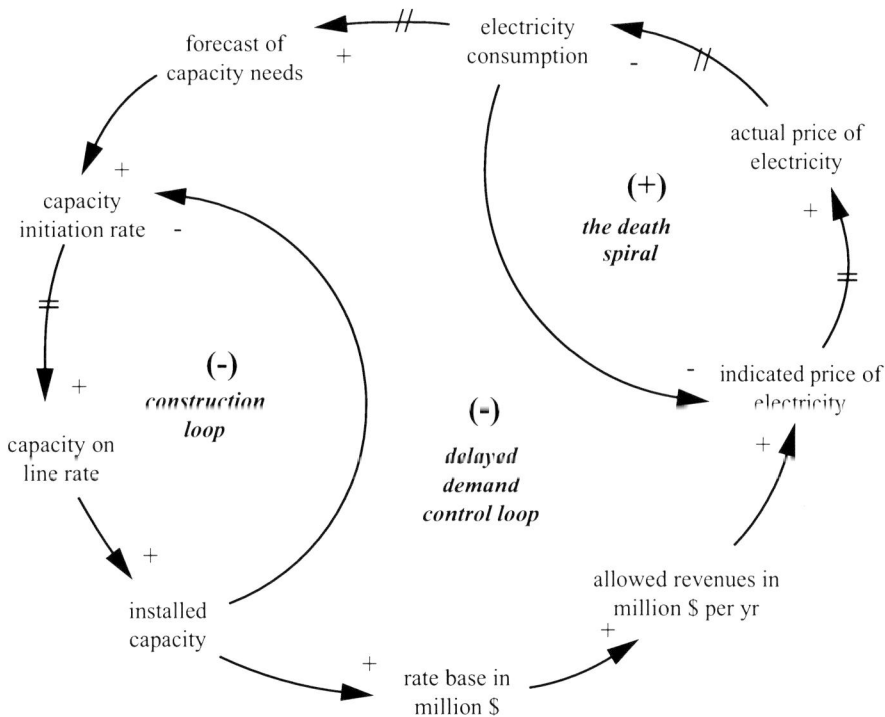

System Dynamics Models of Environment, Energy, and Climate Change, Fig. 16 Key feedbacks and delays faced by power companies in the 1970s

the causal relationships are traced around the outer loop, one sees a negative feedback loop that could act to stabilize the situation. The problem, however, is that the delays around the outer loop are substantially longer than the delay for the death spiral.

The utility companies' financial challenge was the subject of several system dynamics studies in the 1970s and 1980s. The studies revealed that the downward spiral could pose difficult problems, especially if consumers reacted quickly while utilities were stuck with long-lead time, capital intensive power plants under construction. The studies showed that utility executives needed to do more than rely on regulators to grant rate increases; they needed to take steps on their own to soften the impact of the death spiral. The best strategy was to shift the investments to technologies with shorter lead times. (As an example, a power company in coal region would do better to switch from large to smaller coal plants because of the small plants' shorter lead time.) The studies also revealed that the company's financial situation would improve markedly with slower growth in demand. By the late 1970s and early 1980s, many power companies began to provide direct financial incentives to their customers to slow the growth in demand. System dynamics studies showed that the company-sponsored efficiency programs would be beneficial to the both the customers (lower electric bills) and to the power companies (improved financial performance).

An essential feature of the utility modeling was the inclusion of power operations alongside of consumer behavior, company forecasting, power plant construction, regulatory decision making, and company financing. This interdisciplinary approach is common within the system dynamics community because practitioners believe that insights will emerge from simulating the key feedback loops. This belief leads one to follow the cause and effect connections around the key loops regardless of the disciplinary boundaries that are crossed along the way. This approach contrasts strongly with the customary modeling framework of large power companies who were not familiar with system dynamics. Their approach was to assign models to different departments (i.e., operations, accounting and forecasting) and string the models together to provide a view of the entire corporation over the long-term planning interval.

Figure 17 shows what can happen when models within separate departments are strung together. A large corporation might use 30 models, but this diagram makes the point by describing

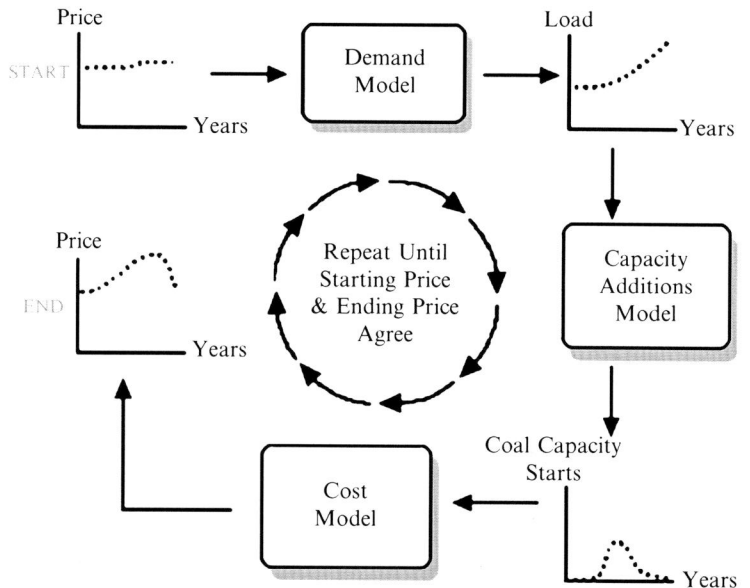

System Dynamics Models of Environment, Energy, and Climate Change, Fig. 17 The iterative approach often used by large power companies in the 1970s

three models. The analysis would begin with an assumption on future electricity prices over the 20-year interval. These are needed to prepare a forecast of the growth in electricity load. The forecast is then given to the planning department which may run a variety of models to select the number power plants to construct in the future. The construction results are then handed to the accounting and rate making departments to prepare a forecast of electricity prices. When the company finally completes the many calculations, the prices that emerge may not agree with the prices that were assumed at the start. The company must then choose whether to ignore the contradiction or to repeat the entire process with a new estimate of the prices at the top of the diagram. This was not an easy choice. Ignoring the price discrepancy was problematic because it was equivalent to ignoring the "death spiral," one of the foremost problems of the 1970s. Repeating the analysis was also problematic. The new round of calculations would be time consuming, and there was no guarantee that consistent results would be obtained at the end of the next iteration.

The power companies' dilemma from the 1970s is described here to make an important point about the usefulness of system dynamics. System dynamics modeling is ideally suited for the analysis of dynamic problems that require a feedback perspective. The method allows one to "close the loop," as long as one is willing to cross the necessarily disciplinary boundaries. In contrast, other modeling methods are likely to be extremely time consuming or fall short in simulating the key feedbacks that tie the system together.

Simulating the Power Industry Response to a Carbon Market

The world is getting warmer, both in the atmosphere and in the oceans. The clearest and most emphatic description of global warming was issued by the intergovernmental panel on climate change (IPCC) in February of 2007. Their summary for policymakers [11, p 4] reported that the "Warming of the climate system is unequivocal,

as is now evident from observations of increases in global average air and ocean temperatures, widespread melting of snow and ice and rising global mean sea level." The IPCC concluded that "most of the observed increase is very likely due to the observed increase in anthropogenic greenhouse gas concentrations." As a consequence of the IPCC and other warnings, policymakers around the world are calling for massive reductions in CO_2 and other greenhouse gas (GHG) emissions to reduce the risks of global warming.

Figure 18 summarizes some of the targets for emission reductions that have been adopted or proposed around the world. In many cases, the targets are specified relative to a country's emissions in the year 1990. So, for ease of comparison, the chart uses 100 to denote emissions in the year 1990. Emissions have been growing at around 1.4%/year. The upward curve shows the future emissions if this trend continues: emissions would reach 200 by 2040 and 400 by 2090. The chart shows the great differences in the stringency of the targets. Some call for holding emissions constant; others call for dramatic reductions over time. Some targets apply to the next two decades; many extend to the year 2050; and some extend to the year 2100. However, when compared to the upward trend, all targets require major reductions relative to business as usual.

The targets from the Kyoto treaty are probably the best known of the goals in Fig. 18. The treaty became effective in February of 2005 and called for the Annex I countries to reduce emissions, on average, by 5% below 1990 emissions by the year 2008 and to maintain this limit through 2012. The extension of the Kyoto protocol beyond 2012 is the subject of ongoing discussions. The solid line from 2010 to 2050 represents the "stabilization path" used in the climate modeling by Webster (2003). The limit on emissions was imposed in modeling calculations designed to stabilize atmospheric CO_2 at 550 ppmv or lower. The scenario assumed that the Kyoto emissions caps are adopted by all countries by 2010. The policy assumed that the caps would be extended and then further lowered by 5% every 15 years. By

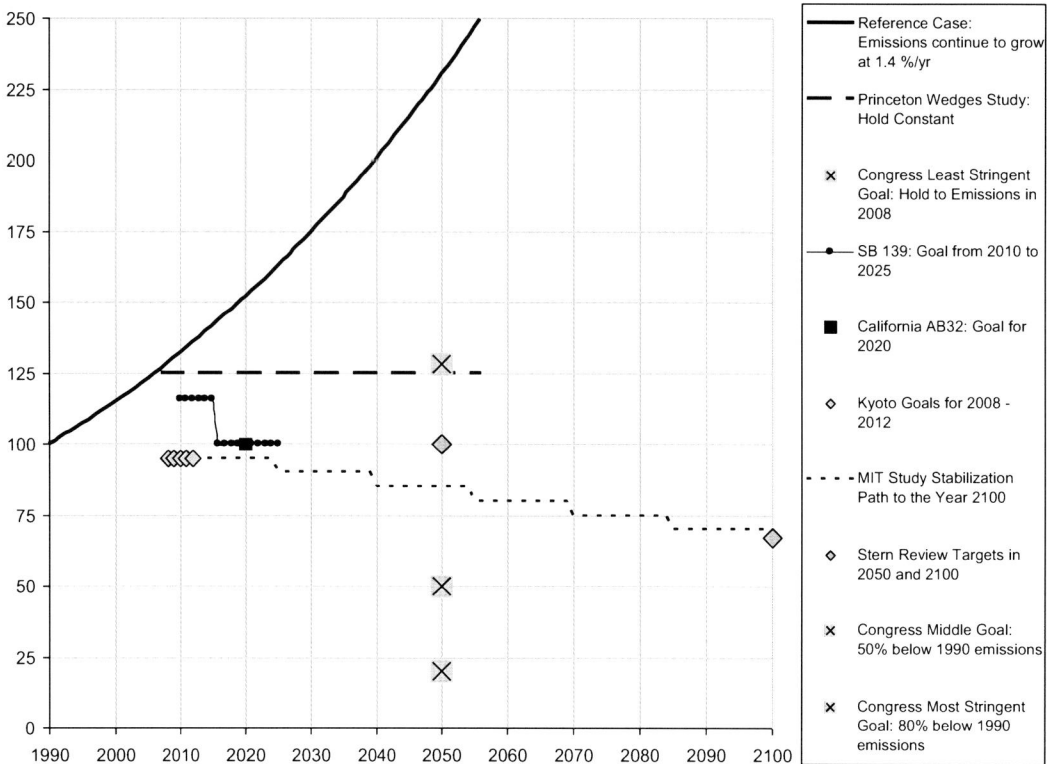

System Dynamics Models of Environment, Energy, and Climate Change, Fig. 18 Comparison of goals for emissions (100 on the vertical axis represent emissions in the year 1990)

the end of the century, the emissions would be 35% below the value in 1990.

This entry concentrates on Senate Bill 139, The Climate Stewardship Act of 2003. Figure 19 shows the S139 targets over the interval from 2010 to 2025. The bill called for an initial cap on emissions from 2010 to 2016. The cap would be reduced to a more challenging level in 2016, when the goal was to limit emissions to no more than the emissions from 1990. S139 was introduced by Senators McCain and Lieberman in January of 2003. It did not pass, but it was the subject of several studies including a highly detailed study by the Energy Information Administration (EIA 2003). The EIA used a wide variety of models to search for the carbon market prices that would induce industries to lower emissions to come into compliance with the cap. The carbon prices were estimated at $22 per metric ton of CO_2 when the market was to open in 2010. They were projected to grow to $60 by the year 2025.

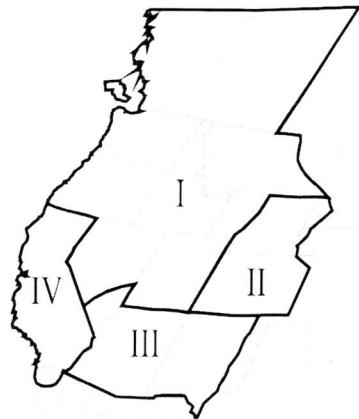

System Dynamics Models of Environment, Energy, and Climate Change, Fig. 19 Map of the western electricity system

The EIA study showed that the electric power sector would lead the way in reducing emissions. By the year 2025, power sector emissions would be reduced 75% below the reference case. This reduction was far beyond the reductions to be achieved by other sectors of the economy. This dramatic response was possible given the large use of coal in power generation and the power industry's wide range of choices for cleaner generation.

A system dynamics study of S139 was conducted at Washington State University (WSU) to learn if S139 could lead to similar reductions in the west. Electricity generation in the western system is provided in a large, interconnected power system shown in Fig. 19. This region has considerably more hydro resources, and it makes less use of coal-fired generation than the nation as a whole. The goal was to learn if dramatic reductions in CO_2 emissions could be possible in the west and to learn if they could be achieved with commercially available technologies.

The opening view of the WSU model is shown in Fig. 20. The model deals with generation, transmission, and distribution to end use customers, with price feedback on the demand for electricity. The model is much larger than the textbook models described earlier in this entry. Fifty

views are required to show the all the diagrams and the simulation results, and the opening view serves as a central hub to connect with the other views.

The opening view uses Vensim's comment icons to draw attention to the CO_2 emissions in the model. The emissions arise mainly from coal-fired power plants, as shown in Fig. 21. A smaller, but still significant fraction of the emissions is caused by burning natural gas in combined cycle power plants. Total emissions vary with the seasons of the year, with the peak normally appearing in the summer when almost all of the fossil-fueled plants are needed to satisfy peak demand. The base case shows annual emissions growing by over 75% by the year 2025.

A major challenge for the system dynamics model is representing power flows across a transmission grid. Finding the flows on each transmission line and the prices in each area is difficult with the standard tools of system dynamics. It simply does not make sense to represent the power flows with a combination of stocks, flows, and feedback processes to explain the flows. It makes more sense to calculate the flows and prices

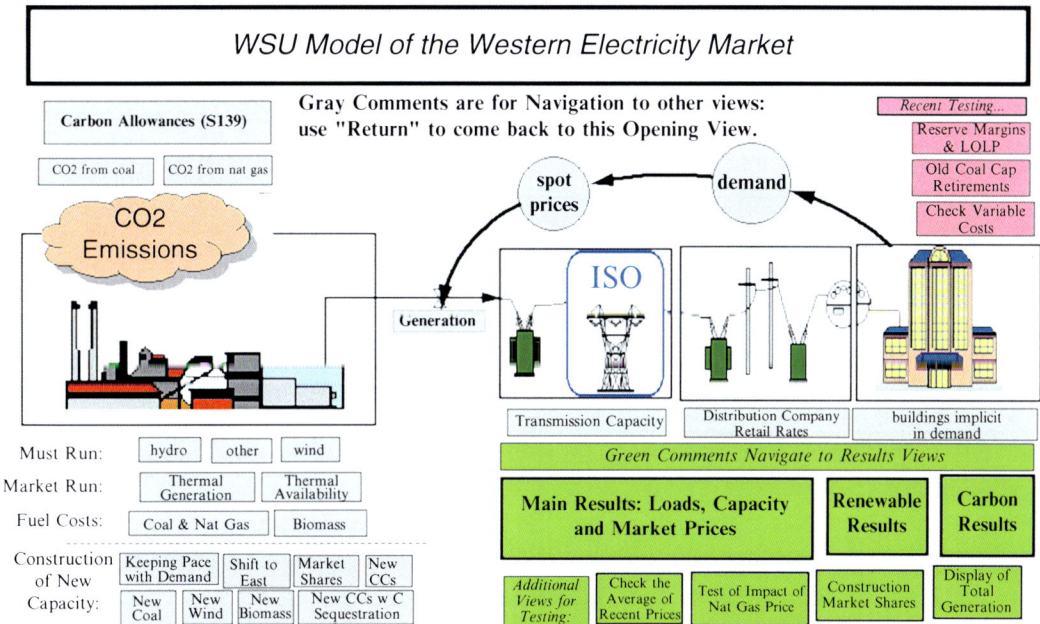

System Dynamics Models of Environment, Energy, and Climate Change, Fig. 20 Opening view of the model of the western electricity system

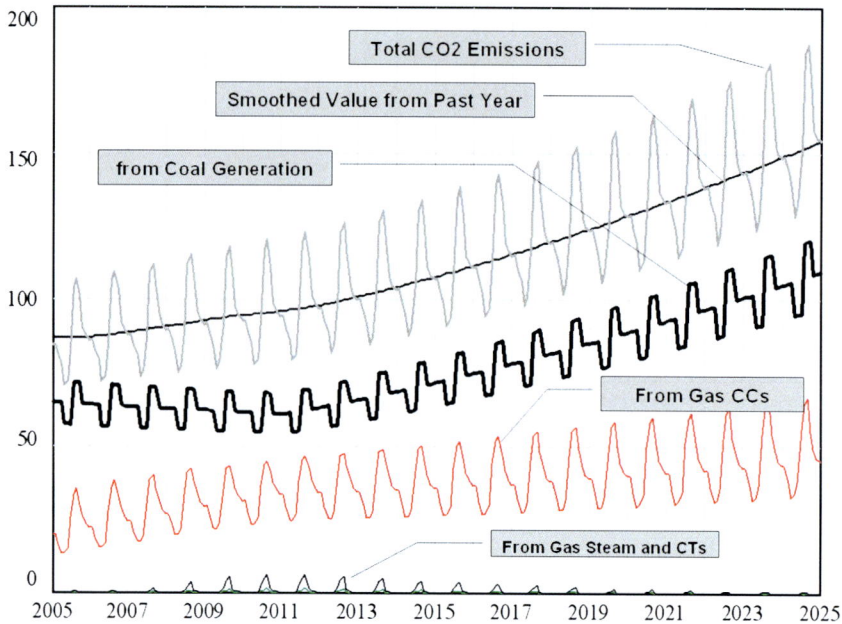

System Dynamics Models of Environment, Energy, and Climate Change, Fig. 21 Annual emissions in a base case simulation (annual emissions are in million metric tons of carbon)

using traditional power systems methods, as explained by Dimitrovski et al. (2007). The power flows were estimated using an algebraic approach which power engineers label as a reduced version of a direct-current optimal power flow calculation. The solution to the algebraic constraints was developed with the Matlab software and then transferred to user-defined functions to operate within the Vensim software. The Vensim simulations were set to run over 20 years with time in months. (A typical simulation required 240 months with changes during a typical day handled by carrying along separate calculations for each of 24 h in a typical day.) These are extensive calculations compared to many system dynamics models, so there was concern that we would lose the rapid simulation speed that helps to promote interactive exploration and model testing. The important methodological accomplishment of this project was the inclusion of network and hourly results within a long-term model without losing the rapid simulation response that encourages users to experiment with the model.

One of the model experiments called for a new simulation with carbon prices set to follow the $20 to $60 trajectory projected by the EIA for S139. These prices were specified as a user input, and the model responded with a change in both short-term operations and long-term investments. The important result was a 75% reduction in CO_2 emissions by the end of the simulation. This dramatic reduction corresponds almost exactly to the EIA estimate of CO_2 reduction for the power industry in the entire USA.

Figure 22 helps one understand how CO_2 emissions could be reduced by such a large amount. These diagrams show the operation of generating units across the western USA and Canada for a typical day in the summer of the final year of the simulation. Figure 22a shows the reference case; Fig. 22b shows the case with S139. The side by side comparison helps one visualize the change in system operation. A comparison of the peak loads shows that the demand for electricity would be reduced. The reduction is 9%, which is due entirely to the consumers' reaction to higher retail electric prices over time.

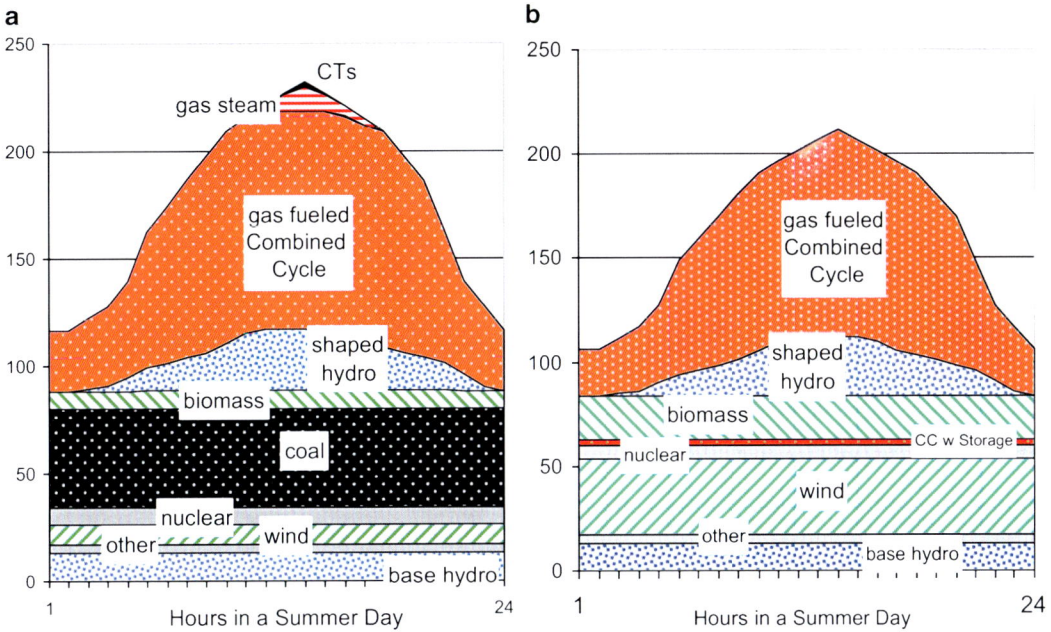

System Dynamics Models of Environment, Energy, and Climate Change, Fig. 22 (**a**) Projected generation for a peak summer day in 2024 in the reference case. (**b**) Projected generation for a peak summer day in 2024 in the S139 case

Figure 22b shows large contributions from wind and biomass generation. Wind generation is carbon free, and biomass generation is judged to be carbon neutral, so these generating units make an important contribution by the end of the simulation. Both of these generating technologies were competitive with today's fuel prices and tax credits. The model includes combined cycle gas generation equipped with carbon capture and storage, a technology that was not commercially available. The model assumes that advances in carbon sequestration over the next two decades would allow this technology to capture a small share of investment near the end of the simulation. By the year 2025, the combined cycle plants with sequestration equipment would provide 2% of the generation.

The most important observation from Fig. 22 is the complete elimination of coal-fired generation in the S139 case. Coal-fired units are shown to operate in a base load mode in the reference simulation. They provide around 28% of the annual generation, but they account for around two-thirds of the CO_2 emissions in the western system. The

carbon prices from S139 make investment in new coal-fired capacity unprofitable at the very start of the simulated market in 2010. As the carbon prices increase, utilities to cut back on coal-fired generation and compensate with increased generation from gas-fired CC capacity. In the simulations reported here, this fuel switching would push the coal units into the difficult position of operating fewer and fewer hours in a day. Eventually this short duration operation is no longer feasible, and coal generation is eliminated completely by the end of the simulation.

The WSU study of the western electric system was selected as the concluding example because of its novel treatment of network flows inside a system dynamics model (Dimitrovski et al. 2007). The model is also interesting for its treatment of daily price changes within a long-term model. (Such changes are important to the simulation of revenues in the wholesale market.) From a policy perspective, the study confirms previous modeling of the pivotal role of the electric power industry in responding to carbon markets. The study indicated that the western electricity system could achieve dramatic reductions in CO_2 emissions within 15 years after the opening of a carbon market, and it could do so with technologies that are commercially available today (Ford 2008).

Conditions for Effective Interdisciplinary Modeling

All of the applications demonstrate the usefulness of system dynamics in promoting interdisciplinary modeling. The entry concludes with comments on the level of effort and the conditions needed for effective, interdisciplinary modeling.

The examples in this entry differ substantially in the level of effort required, from several weeks for the classroom examples by Ford (2010) to several years for the energy studies. The textbook examples involved student expansions of models of Mono Lake and the Tucannon Salmon. The expansions were completed by undergraduate students in projects lasting 2 or 3 weeks. The key was the students' previous education (classes from many different departments) and their receptiveness to an interdisciplinary approach.

Fiddaman's model of the climate and energy system (Fiddaman 2002) was a more ambitious exercise, requiring several years of effort as part of his doctoral research. Bringing multiyear interdisciplinary modeling projects to a successful conclusion requires one to invest the time to master several disciplines and to maintain a belief that there are potential insights at the end of the effort. The follow-on development of C-ROADS was a far more ambitious effort requiring the combined efforts of the Climate Interactive team.

The electric power industry examples were ambitious projects that required several years of effort. The modeling of the western electricity system was a 4-year project with support from the National Science Foundation. The long research period was crucial for it allowed the researchers from power systems engineering, system dynamics, and environmental science to take the time to learn from one another.

The modeling of the electric company problems in the 1970s was also spread over several years of effort. The success of this modeling was aided by utility planners, managers, and modelers who were looking for a systems view of their agency and its problems. They saw system dynamics as a way to tie existing ideas together within an integrated portrayal of their system. Their existing ideas were implemented in models maintained by separate functional areas (i.e., forecasting, accounting, operations). The existing models often provided a foundation for the system dynamics models (i.e., in the same way that the comprehensive climate models in Fig. 11 provide support for the development of the more integrated models). The key to effective, interdisciplinary modeling within such large organizations is support from a client with a strong interest in learning and with managerial responsibility for the larger system.

Future Directions

This entry concludes with future directions for system dynamics applications to climate change. People often talk of mitigation and adaptation. Mitigation refers to the challenge of lowering greenhouse gas emissions to avoid dangerous anthropogenic interference with the climate system. Adaptation refers to the challenge of living in a changing world.

Mitigation: The challenge of lowering CO_2 and other GHG emissions is the fundamental challenge of the coming century. Dynamic simulations such as C-ROADS have already proved useful as nations participate in international negotiations on emission reductions. Achieving the reductions in the USA will be aided by the use of carbon taxes or carbon markets. System dynamics can aid in learning about market design. It is important that we learn how to make these markets work well. And if they do not work well, it is important to speed the transition to a carbon tax policy with better prospects for success. System dynamics can aid in learning about markets, especially if it is coupled with simulating gaming to allow market participants and regulators to "experience" and better understand market dynamics.

Adaptation: The world will continue to warm, and sea levels will continue to rise. These trends will dominate the first half of this century even with major reductions in CO_2 emissions. These and other climate changes will bring a wide variety of problems for management of water resources, public health planning, control of invasive species, preservation of endangered species, control of wildfire, and coastal zone management, just to name a few. Our understanding of the adaptation challenges can be improved through system dynamics modeling. The prospects for insight are best if the models provide an interdisciplinary perspective on adapting to a changing world.

Bibliography

Primary Literature

Bunn D, Larsen E (1977) Systems modelling for energy policy. Wiley, Chichester

Claussen M et al (2002) Earth system models of intermediate complexity: closing the gap in the spectrum of climate system models. Climate Dyn 18:579–586

Climate Interactive (2017) https://www.climateinteractive.org/

Coyle G (1977) Management system dynamics. Wiley, Chichester

Dimitrovski A, Ford A, Tomsovic K (2007) An interdisciplinary approach to long term modeling for power system expansion. Int J Crit Infrastruct 3(1–2):235–264

EIA (2003) United States Department of Energy, Energy Information Administration, Analysis of S139, the Climate Stewardship Act of 2003

Fiddaman T (2002) Exploring policy options with a behavioral climate-economy model. Syst Dyn Rev 18(2):243–264

Ford A (2008) Simulation scenarios for rapid reduction in carbon dioxide emissions in the western electricity system. Energy Policy 36:443–455

Ford A (2010) Modeling the environment, 2nd edn. Island Press, Washington, DC

Ford A, Cavana R (eds) (2004) Special issue of the Syst Dyn Rev

Forrester J (1961) Industrial dynamics. Pegasus Communications, Waltham, MA

Forrester J (2000) From the ranch to system dynamics: an autobiography, in management laureates. JAI Press

Hardin G (1968) The tragedy of the commons. Science 162:1243–1248

IPCC (1997) An introduction to simple climate models used in the IPCC second assessment report. ISBN92-9169-101-1

IPCC (2007) Climate change 2007: the physical science basis, summary for policymakers. www.ipcc.ch/

Kump L (2002) Reducing uncertainty about carbon dioxide as a climate driver. Nature 419:188–190

Meadows DH, Meadows DL, Randers J, Behrens W (1972) The limits to growth. Universe Books, New York

Morecroft J (2015) Strategic modelling and business dynamics. Wiley, Chichester

Richardson J, Pugh A (1981) Introduction to system dynamics modeling with dynamo. Pegasus Communications, Waltham MA

Sterman J (2000) Business dynamics. McGraw-Hill/Irwin, Boston MA

Sterman J (ed) (2002) Special issue of the Syst Dyn Rev

Sterman J, Sweeney L (2007) Understanding public complacency about climate change. Clim Chang 80(3–4):213–238

Warren K (2002) Competitive strategy dynamics. Wiley, Chichester

Webster M et al (2003) Uncertainty analysis of climate change and policy response. Climate Change 61:295–320

Books and Reviews

Houghton J (2004) Global warming: the complete briefing, 3rd edn. Cambridge University Press, Cambridge

Ford A (2010) Modeling the environment, 2nd edn. Island Press, Washington, DC

Forrester J (2000) From the ranch to system dynamics: an autobiography, in management laureates. JAI Press

Meadows DH, Meadows DL, Randers J, Behrens W (1972) The limits to growth. Universe Books, New York

Morecroft J (2015) Strategic modelling and business dynamics. Wiley

Richardson J, Pugh A (1981) Introduction to system dynamics modeling with dynamo. Pegasus Communications, Waltham

Sterman J (2000) Business dynamics. Irwin McGraw-Hill

Warren K (2002) Competitive strategy dynamics. Wiley, Chichester

System Dynamics and Its Contribution to Economics and Economic Modeling

Michael J. Radzicki
Department of Social Science and Policy Studies,
Worcester Polytechnic Institute, Worcester,
MA, USA

Article Outline

Glossary

Feedback Feedback is the transmission and return of information about the amount of "stuff" (information or material) that has accumulated in a system's stocks. Information travels from a stock back to its flow(s), either directly or indirectly, creating a feedback loop. When the return of this information reinforces the behavior of the stocks residing within the loop, the loop is said to be positive. Positive loops are responsible for the growth (or decline) of systems over time. Negative feedback loops represent goal seeking behavior in complex systems. When a negative loop detects a gap between the amount of "stuff" (information or material) in a system's stock and the desired amount of "stuff," it initiates corrective action. If this corrective action is not significantly delayed by one or more additional stocks, the amount of "stuff" in the stock will smoothly adjust to its goal. If the corrective action is delayed by one or more additional stocks, however, the amount of "stuff" in the stock can oscillate by overshooting and undershooting its goal.

Flow Flows of information or material enter and exit a system's stocks and, in so doing, create a system's dynamics. The net flow into or out of a stock determines the stock's rate of change. When human decision-making is represented in a system dynamics model, it appears in the system's flow equations. Mathematically, a system's flow equations are ordinary differential equations, and their format determines whether a system is linear or nonlinear.

Stock Stocks, which are sometimes referred to as "levels" or "states," accumulate (i.e., add up) the information or material that flows into and out of them. Stocks are thus responsible for decoupling flows, creating delays, preserving system memory, and altering the time shape of flows.

Definition of the Subject and Its Importance

System dynamics is a computer modeling technique that has its intellectual origins in control engineering, management science, and digital computing. It was originally created as a tool to help managers better understand and control corporate systems. Today it is applied to problems in a wide variety of academic disciplines, including economics, and is perhaps best thought of as a tool for understanding how a complex system's structure causes its behavior and for redesigning the

© Springer Science+Business Media, LLC, part of Springer Nature 2020   401
B. Dangerfield (ed.), *System Dynamics*,
https://doi.org/10.1007/978-1-4939-8790-0_539

Originally published in
R. A. Meyers (ed.), *Encyclopedia of Complexity and Systems Science*, © Springer Science+Business
Media LLC 2019, https://doi.org/10.1007/978-3-642-27737-5_539-2

system's structure so that it will exhibit improved (and robust) behavior. Of note is that system dynamics models often generate behavior that is both counterintuitive and at odds with traditional economic theory. Historically, this has caused many system dynamics models to be evaluated critically, especially by some economists. However, today economists from several schools of economic thought are beginning to use system dynamics, as they have found it useful for translating their nontraditional ideas into formal models.

Introduction

System dynamics is a computer simulation modeling technique that is used to analyze complex nonlinear dynamic feedback systems for the purposes of generating insight and designing policies that will improve system performance. It was originally created by Jay W. Forrester (1956) of the Massachusetts Institute of Technology as a tool for building computer simulation models of problematic behavior within corporations. The models were used to design and test policies aimed at altering a corporation's structure so that its behavior would improve and become more robust. Today, system dynamics is applied to a large variety of problems in a multitude of academic disciplines, including economics.

System dynamics models are created by identifying and linking the relevant pieces of a system's structure and simulating the behavior generated by that structure. Through an iterative process of structure identification and mapping, followed by simulation, a model emerges that can explain (mimic) a system's problematic behavior and serve as a vehicle for policy design and testing.

From a system dynamics perspective, a system's structure consists of stocks, flows, and feedback loops. Stocks can be thought of as bathtubs that accumulate/de-cumulate a system's flows over time. Flows can be thought of as pipe and faucet assemblies that fill or drain the stocks. Mathematically, the process of flows accumulating/de-cumulating in stocks is called integration.

From a system dynamics point of view, the integration process creates all dynamic behavior be it in a physical system, a biological system, or a socioeconomic system. Examples of stocks and flows in economic systems include a stock of inventory and its inflow of production and its outflow of sales, a stock of the book value of a firm's plant and equipment and its inflow of investment spending and its outflow of depreciation, and a stock of employed labor and its inflow of hiring and its outflow of labor separations.

Feedback is the transmission and return of information about the amount of "stuff" (information or material) that has accumulated in a system's stocks. Information travels from a stock back to its flow(s) either directly or indirectly, and this movement of information causes the system's rates of flow to increase, decrease, stop, or remain constant. Every feedback loop has to contain at least one stock so that a simultaneous equation situation can be avoided, and a model's behavior can be revealed recursively. Loops with a single stock are termed minor, while loops containing more than one stock are termed major.

Two types of feedback loops exist in system dynamics modeling: positive loops and negative loops. Generally speaking, positive loops generate self-reinforcing behavior and are responsible for the growth in, or decline of, the amount of "stuff" residing in the loop's stocks. Any relationship that can be termed a virtuous or vicious circle is thus a positive feedback loop process. Examples of positive loops in economic systems include path-dependent processes, increasing returns, speculative bubbles, learning by doing, and many of the relationships found in macroeconomic growth theory. Forrester (1980), Radzicki and Sterman (1994), Moxnes (1992), Sterman (2000, Chap. 10), Ryzhenkov (2007), and Weber (2007) describe system dynamics models of economic systems that possess dominant positive feedback processes.

Negative feedback loops represent self-regulating, goal-seeking, processes and are responsible for both stabilizing the amount of "stuff" residing in the loop's stocks and causing the amount of "stuff" residing in the loop's stocks to oscillate.

When a negative loop detects a gap between the amount of "stuff" in its stock and the desired amount, it initiates corrective action aimed at closing the gap. If this is accomplished without a significant time delay (i.e., it's a minor loop), the amount of "stuff" in the stock will adjust smoothly to the desired amount. On the other hand, if there are significant time lags in the corrective actions of a negative loop (i.e., if it's a major loop), the amount of "stuff" in the stock can repeatedly overshoot or undershoot its goal and oscillate. Examples of negative feedback processes in economic systems include equilibrating mechanisms ("autopilots") such as simple supply and demand relationships, stock adjustment models for inventory control, any purposeful behavior, and many of the relationships found in macroeconomic business cycle theory. Meadows (1970), Mass (1975), Low (1980), Forrester (1980), and Sterman (1985) provide examples of system dynamics models that generate cyclical behavior at the macroeconomic and microeconomic levels.

From a system dynamics point of view, positive and negative feedback loops fight for control of a system's behavior. The loops that are dominant at any given time determine a system's time path and, if the system is nonlinear, the dominance of the loops can change over time as the system's stocks fill and drain. From this perspective, the dynamic behavior of any economy – that is, the interactions between the trend and the cycle in an economy over time – can be explained as a fight for dominance between the economy's most significant positive and negative feedback loop processes.

In system dynamics modeling, stocks and flows are often conceptualized as having limits. That is, stocks and flows are often seen as being unable to exceed or fall below certain maximum and minimum levels. Indeed, an economic model that can generate, say, either an infinite and/or a negative workforce would be seen as severely flawed by a system dynamicist. As such, when building a model system, dynamicists search for factors that may limit the amount of material or information that the model's flows can process and/or stocks can accumulate. Actual socioeconomic systems possess many limiting factors including physical limits (e.g., the number of widgets a machine can produce per unit of time), cognitive limits (e.g., the amount of information an economic agent can remember and act upon), and financial limits (e.g., the maximum balance allowed on a credit card). When limiting factors are included in a system dynamics model, the system's approach to these factors must be described. Generally speaking, this is accomplished with nonlinear relationships. Figure 1 presents a simple system dynamics model that contains examples of all of the components of system structure described above.

Types of Dynamic Simulation

From a system dynamics point of view, solving a dynamic system – any dynamic system – means determining how much "stuff" (material or information) has accumulated in each of a system's stocks at every point in time. This can be accomplished in one of two ways – analytically or via simulation. Linear dynamic models can be solved either way. Nonlinear models, except for a few special cases, can only be solved via simulation.

Simulated solutions to dynamic systems can be attained from either a continuous (analog) computer or a discrete (digital) computer. Understanding the basic ideas behind the two approaches is necessary for understanding how economic modeling can be undertaken with system dynamics.

In the real world, of course, time unfolds continuously, like a river. Yet, devising a way to mimic this process on a machine is somewhat tricky. On an analog computer, the continuous flow of economic variables into and out of stocks over time is mimicked by the continuous flow of some physical substance such as electricity or water. A wonderful example of the latter case is the MONIAC or Phillips (1950) Machine, which simulates an economy, according to a mix of Keynesian and classical principles, with flows of colored water moving through pipes and accumulating in tanks. Ng and Wright (2007) provide a wonderful description of the history and details of the Phillips Machine.

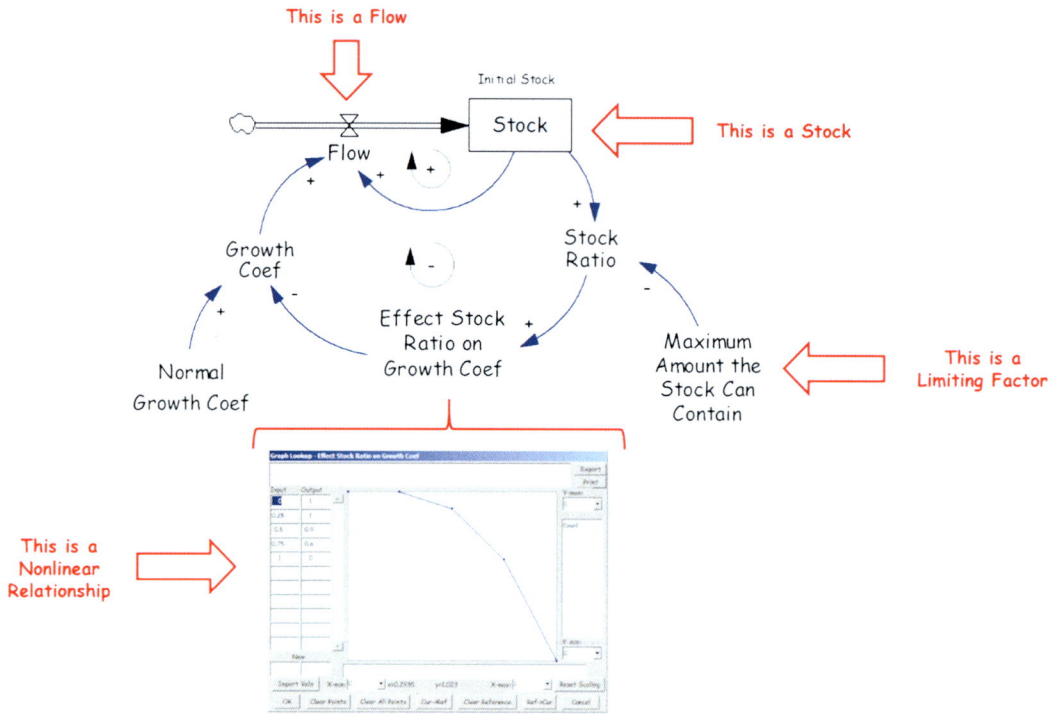

System Dynamics and Its Contribution to Economics and Economic Modeling, Fig. 1 Simple system dynamics model containing examples of all components of system structure

On a digital computer, the continuous flow of economic variables into and out of stocks over time is approximated by specifying the initial amount of material or information in the system's stocks, breaking simulated time into small increments, inching simulated time forward by one of these small increments, calculating the amount of material or information that flowed into and out of the system's stocks during this small interval, and then repeating. The solution to the system will always be approximate because the increment of time cannot be made infinitesimally small, and thus simulated time cannot be made perfectly continuous. In fact, on a digital computer, a trade-off exists between round-off error and integration error. If the increment of time is made too large, the approximate solution can be poor due to integration error. If the increment of time is made too small, the approximate solution can be poor due to round-off error.

In system dynamics modeling, the "true" behavior of the underlying system is conceptualized to unfold over continuous time. As such, mathematically, a system dynamics model is an ordinary differential equation model. To approximate the solution to a continuous time ordinary differential equation model on a digital (discrete) computer, however, a difference equation approach is used. Unlike traditional difference equation modeling in economics, in which the increment of time is typically chosen to match economic data (e.g., a quarter or a year), the increment of time in system dynamics modeling is chosen to yield a solution that is accurate enough for the problem at hand yet avoids the problems associated with significant round-off and integration error. Moreover, the use of a difference equation approach in this context involves asking the computer to step through simulated time according to a numerical integration algorithm, the most common of which is Euler's method.

Since many well-known dynamic economic models have been created with *traditional* difference equations and since solving a system dynamics model on a digital computer involves asking the machine to take discrete steps forward in time, an obvious question to ask is whether or not system dynamics can be used to literally replicate the *traditional* difference equation models. Although doing this deviates from the original ideas embodied in the system dynamics paradigm, it can be done if a modeler feels that analyzing a traditional difference equation model from a system dynamics point of view will yield some additional insight.

Translating Existing Economic Models into a System Dynamics Format

There are three principle ways that system dynamics is used for economic modeling. The first involves translating an existing economic model into a system dynamics format, while the second involves creating an economic model from scratch by following the rules and guidelines of the system dynamics paradigm. Forrester (1961), Richardson and Pugh (1981), Radzicki (1997), and Sterman (2000) provide extensive details about these rules and guidelines. The former approach is valuable because it enables well-known economic models to be represented in a common format, which makes comparing and contrasting their assumptions, concepts, structures, behaviors, etc., fairly easy. The latter approach is valuable because it usually yields models that are more realistic and that produce results that are "counterintuitive" (Forrester 1975) and thus thought-provoking.

The third way that system dynamics can be used for economic modeling is a "hybrid" approach in which a well-known economic model is translated into a system dynamics format, critiqued, and then improved by modifying it so that it more closely adheres to the principles of system dynamics modeling. This approach attempts to blend the advantages of the first two approaches, although it is more closely related to the former.

Generally speaking, existing economic models that can be translated into a system dynamics format can be divided into four categories: written, static (mathematical), difference equation, and ordinary differential equation. Existing economic models that have been created in an ordinary differential equation format can be translated into their equivalent system dynamics representation in a fairly straightforward manner. For example, Fig. 2 presents the Robert Solow's (1956)

continuous time neoclassical economic growth model in a system dynamics format.

Delay Structures and Translating Difference Equation Models into a System Dynamics Format

Translating traditional difference equation models into a system dynamics format is a bit trickier. This requires an understanding of the dynamics of delay (lag) structures.

A fundamental principle of system dynamics modeling is that stocks create delays or lags. In order to represent a delay that exhibits a particular time shape or behavior, the *order* of the delay must be determined. A delay's order is the number of independent stocks its inflow must pass through before reaching its outflow. Hamilton (1980) presents an excellent overview of techniques for formally estimating the order of a delay structure.

In system dynamics modeling, there are essentially four basic delay structures for both information and material flows: first order, third order, nth order, and infinite order. Figure 3 presents the response of each of these delay structures to a step input (with a delay time of 5 months) that knocks the system out of equilibrium. The first order delay immediately reacts to the shock, and its outflow gradually approaches the new, higher, inflow rate. The third order delay begins to react after about a month and exhibits an s-shaped path to the new, higher, inflow rate. The nth (12th) order delay begins to react after about 3 months and exhibits a more egregious s-shaped path to the new, higher, inflow rate. Finally, the infinite order delay does nothing for 5 months and then jumps up in a "reverse z" pattern (the most egregious s-shaped pattern possible) to match the new, higher, inflow rate. Of course, the time shape of the infinite order delay is equivalent to that exhibited by a traditional difference equation and can be used to create a system dynamics representation of these models. As an example, Fig. 4 presents Wynne Godley and Marc Lavoie's (2012, Chap. 3) stock-flow consistent difference equation model in a system dynamics format.

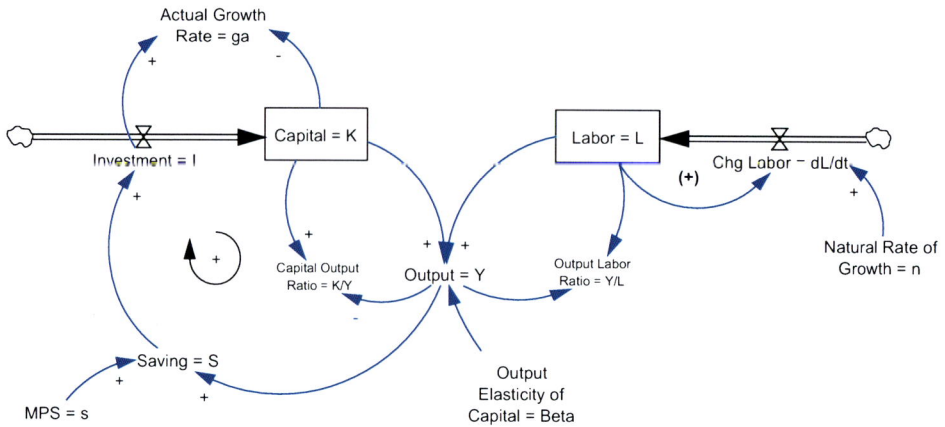

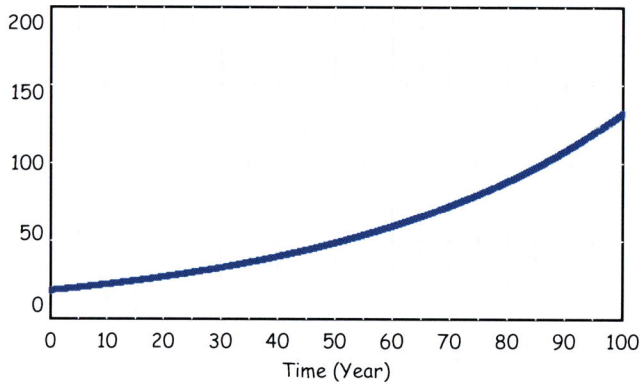

System Dynamics and Its Contribution to Economics and Economic Modeling, Fig. 2 System dynamics representation of Robert Solow's ordinary differential equation growth model

Translating existing static and written economic models and theories into a system dynamics format is a more formidable task. Written models and theories are often dynamic yet are described without mathematics. Static models and theories are often presented with mathematics but lack equations that describe the dynamics of any adjustment processes they may undergo. As such, system dynamicists must devise equations that capture the dynamics being described by the written word or that reveal the adjustment processes that take place when a static system moves from one equilibrium point to another.

An interesting example of a system dynamics model that was created from a written description is Barry Richmond's (1985) model of some of

Adam Smith's ideas presented in the *Wealth of Nations*. This model was created principally from Robert Heilbronner's (1980) written description of Smith's ideas. A classic example of a static model that has been translated into a system dynamics format is a simple two sector Keynesian cross model, as is shown in Fig. 5.

Improving Existing Economic Models with System Dynamics

The simple two sector Keynesian cross model presented in Fig. 5 is an example of a well-known economic model that can be further improved after it has been translated into a system

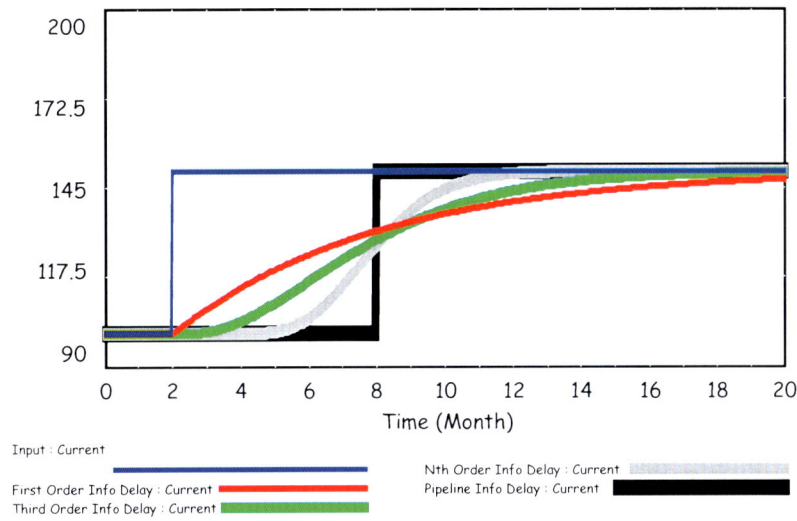

System Dynamics and Its Contribution to Economics and Economic Modeling, Fig. 3 Responses to a step input of a 1st, 3rd, 12th, and infinite order information delay structures

dynamics format. More specifically, in the model presented in Fig. 5, the flow of investment spending does not accumulate anywhere. This violates good system dynamics modeling practice and can be fixed. Figure 6 presents the improved version of the Keynesian cross model, which now more closely adheres to the system dynamics paradigm. This model is essentially a multiplier-accelerator model that exhibits oscillations rather than a smooth transition from one equilibrium to another.

Other well-known examples of classic economics models that have been improved after they have been translated into a system dynamics format and made to conform more closely with good system dynamics modeling practice include the cobweb model (Meadows 1970), Sir John Hicks' multiplier accelerator model (Low 1980), the IS-LM/AD-AS model (Forrester 1982; Wheat 2007a), Dale Jorgenson's investment model (Senge 1980; William Nordhaus' 1992a), DICE climate change model (Fiddaman 1997, 2002), and basic microeconomic supply and demand mechanisms (Mashayekhi et al. 2006). Low's improvement of Hicks' model is particularly interesting because it results in a model that closely resembles Bill Phillips' (1954) multiplier-accelerator model. Senge and Fiddaman's contributions are also very

interesting because they demonstrate how the original economic models are special cases of their more general system dynamics formulations.

Creating Economic Dynamics Models from Scratch

Although translating well-known economic models into a system dynamics format can arguably make them easier to understand and use, system dynamicists believe that the "proper" way to model an economic system that is experiencing a problem is to do so from scratch while following good system dynamics modeling principles. Unlike orthodox economists who generally follow a deductive, logical positivist approach to modeling, system dynamicists follow an inductive pattern modeling or case study process. More specifically, a system dynamicist approaches an economic problem like a detective who is iteratively piecing together an explanation at a crime scene. All types of data that are deemed relevant to the problem are considered including numerical, written, and mental information. The system dynamicist is guided in the pattern modeling process by the perceived facts of the case, as

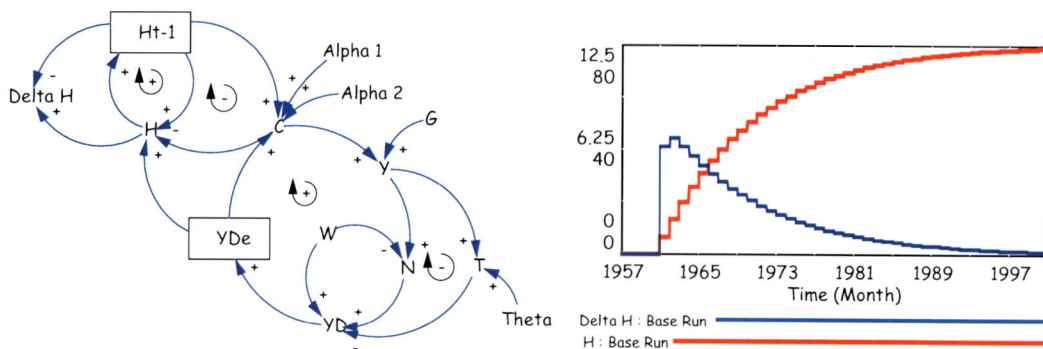

$$C_s = C_d$$

$$G_s = G_d \qquad\qquad G = 20$$

$$T_s = T_d$$

$$N_s = N_d$$

$$YD = W * N_s - T_s \qquad W = 1$$

$$T_d = \theta * W * N_s \qquad \theta < 1 \qquad \theta = .2$$

$$C_d = \alpha_1 * YD^e + \alpha_2 * H_{h-1} \qquad 0 < \alpha_2 < \alpha_1 < 1$$

$$\alpha_1 = .6 \qquad \alpha_2 = .4$$

$$\Delta H_s = H_s - H_{s-1} = G_d - T_d$$

$$\Delta H_h = H_h - H_{h-1} = YD - C_d$$

$$Y = C_s + G_s$$

$$N_d = Y/W$$

$$\Delta H_d = H_d - H_{h-1} = YD^e - C_d$$

$$YD^e = YD_{-1}$$

$$\Delta H_h = \Delta H_s$$

System Dynamics and Its Contribution to Economics and Economic Modeling, Fig. 4 System dynamics representation of Wynne Godley and Marc Lavoie's stock-flow consistent difference equation model

well as by real typologies (termed "generic structures" in system dynamics) and principles of systems. Real typologies are commonalities that have been found to exist in different pattern models, and principles of systems are commonalities that have been found to exist in different real typologies. Paich (1985) discusses generic structures at length, and Forester (1968) lays out a set of principles of systems. Wilber and Harrison (1978) describe the pattern modeling process as used in economics.

Examples of real typologies in economics include Forrester's (1969) *Urban Dynamics* model, which can reproduce the behavior of

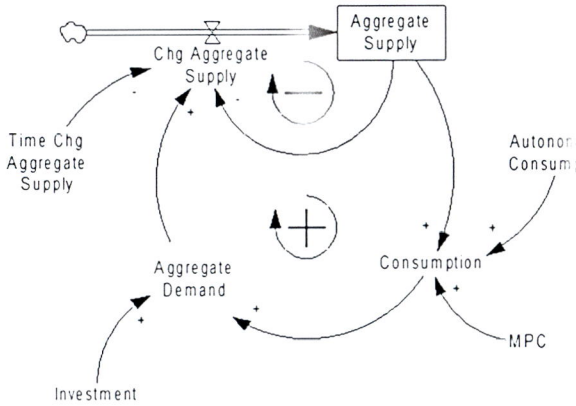

$$Y = C + I$$

$$C = 100 + (.9 * Y)$$

$$I = 200$$

$$Y^e = 3000$$

$$I' = 250$$

$$Y^{e'} = 3500$$

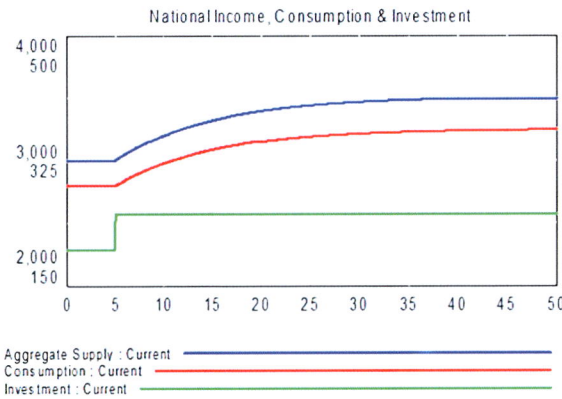

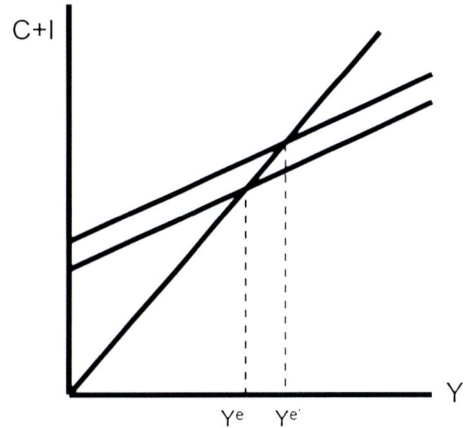

System Dynamics and Its Contribution to Economics and Economic Modeling, Fig. 5 Simple two sector Keynesian cross model in a system dynamics format

many different cities when properly parameterized for those cities, and Homer's (1987) model of the diffusion of new medical technologies into the market place, which can explain the behavior of a wide variety of medical technologies when properly parameterized for those technologies. Examples of fundamental principles of systems include the principle of accumulation, which states that the dynamic behavior of any system is due to flows accumulating in stocks, and the notion of stocks and flows being components of feedback loops. The parallels for these principles in economics can be found in modern post-Keynesian economics, in which modelers try to build "stock-flow consistent models," and in institutional economics, in which the principle of "circular and cumulative causation" is deemed to be a fundamental cause of economic dynamics (Myrdal 1968; Berger 2008). Radzicki (1988, 1990, 2003, 2007) lays out the case for the parallels that exist between methodological concepts in system dynamics and methodological concepts in various schools of economic thought.

The economic models that have been historically created from scratch by following the system dynamics paradigm have tended to be fairly large in scale. Forrester's (1980) national economic model is a classic example, as are the macroeconomic models created by Forrester (1973), Sterman (1981), the Millennium Institute (2007), Radzicki (2007), Wheat (2007a, 2017), and Yamaguchi (2017). With the exception of

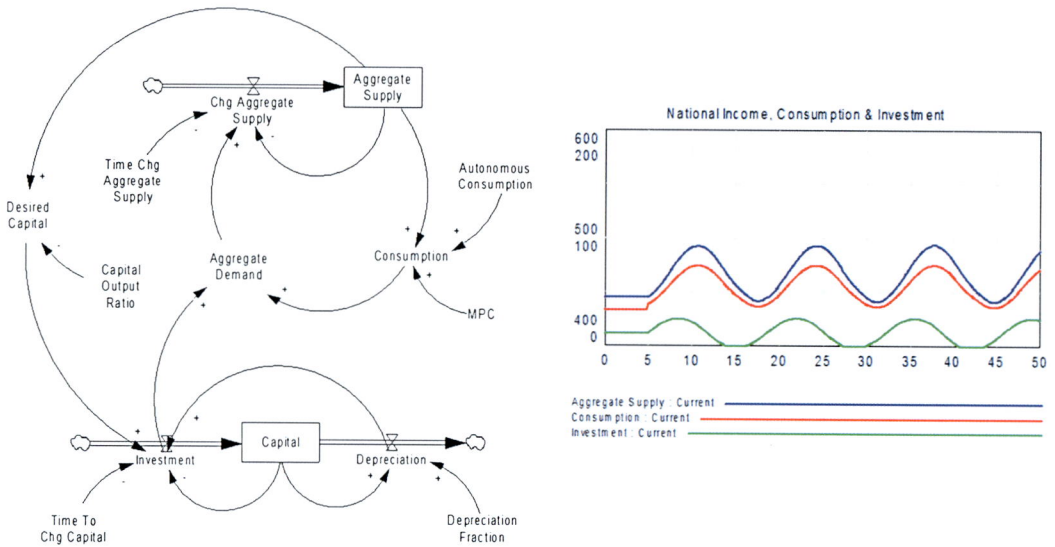

System Dynamics and Its Contribution to Economics and Economic Modeling, Fig. 6 Improved simple two sector Keynesian cross model

Radzicki (2007) and Wheat (2017), whose models are based on ideas from post-Keynesian and institutional economics, these models, by and large, embody orthodox economic relationships.

Model Validity

When a system dynamics model of an economic system that is experiencing a problem is built from scratch, the modeling process is typically quite different from that which is undertaken in traditional economics. As such, the question is raised as to whether or not an original system dynamics model is in any sense "valid."

System dynamicists follow a "pattern modeling" approach (Radzicki 1990) and do not believe that models should be judged in a binary fashion as either "valid" or "invalid." Rather, they argue that confidence in models can be generated along multiple dimensions. More specifically, system dynamicists such as Peterson (1975), Forrester and Senge (1980), and Barlas (1989) have developed a comprehensive series of tests that can be applied to a model's structure and behavior, and they argue that the more tests a model can pass, the more confidence a model builder or user should place in its results. Even more

fundamentally, however, Forrester (1985) has argued that the real value generated through the use of system dynamics comes not from any particular model but from the modeling *process* itself. In other words, it is through the iterative *process* of model conceptualization, creation, simulation, and revision that true learning and insight are generated, and *not* through interaction with the final model.

Another issue that lies under the umbrella of model validity involves fitting models to time series data so that parameters can be estimated and confidence in model results can be raised. In orthodox economics, of course, econometric modeling is almost universally employed when doing empirical research. Orthodox economic theory dictates the structure of the econometric model, and powerful statistical techniques are used to tease out parameter values from numerical data.

System dynamicists, on the other hand, have traditionally argued that it is not necessary to tightly fit models to time series data for the purposes of parameter estimation and confidence building. This is because:

1. The battery of tests that are used to build confidence in system dynamics models go well beyond basic econometric analysis.

2. The particular (measured) time path that an actual economic system happened to take is merely one of an infinite number of paths that it could have taken and is a result of the particular stream of random shocks that happened to be historically processed by its structure. As such, it is more important for a model to mimic the basic character of the data, rather than fit it point-by-point (Forrester 2003).

3. Utilizing the pattern modeling/case study approach enables the modeler to obtain parameter values via observation below the level of aggregation in the model, rather than via statistical analysis. As an example, Graham (1980) suggests that observations on the gradual deterioration of particular buildings, knowledge of the age of a particular building when it is demolished, or knowledge of what a developer must consider when deciding to buy and demolish a building could be used to estimate the parameters for the rate at which houses are demolished in an urban model.

4. The result of a system dynamics modeling intervention is typically a set of policies that improve system performance and increase system robustness. Such policies are usually feedback-based rules (i.e., changes to institutional structure) that do not require the accurate point prediction of system variables.

Although the arguments against the need to fit models to time series data are well-known in system dynamics, many system dynamicists feel that it is still a worthwhile activity because it adds credibility to a modeling study. Moreover, in modern times, advances in software technology have made this process relatively easy and inexpensive. Although several techniques for estimating the parameters of a system dynamics model from numerical data have been devised, perhaps the most interesting is David Peterson's (1975, 1980) Full Information Maximum Likelihood with Optimal Filtering (FIMLOF). FIMLOF is a sophisticated technique for estimating the parameters of a system dynamics model while simultaneously fitting its output to numerical data. Its intellectual origins can be traced to control engineering and the work of Fred Schwepe. David

Peterson pioneered a method for adapting FIMLOF for use in system dynamics modeling.

Figure 7 presents a run from a system dynamics version of the well-known Harrod growth model, to which an adaptive expectations structure has been added, after it has been fit via FIMLOF to real GDP (quarterly), labor supply (quarterly), and investment (monthly) data for the US economy for the years 1946–2016. The fit is outstanding, and the estimated parameter values are consistent with those from more traditional econometric studies. See Radzicki (2020, forthcoming) for a detailed description of the model and its parameter estimates.

Controversies

Since system dynamics modeling is undertaken in a way that is significantly different from traditional economic modeling, it should come as no surprise that many economists have been extremely critical of some system dynamics models of economic systems. For example, Forrester's (1969) *Urban Dynamics* and (1971) *World Dynamics* models have come under severe attack by economists, as has (to a lesser degree) his national economic model. On the other hand, the first paper in the field of system dynamics is Forrester (1956), which is essentially a critique of traditional economic modeling.

Greenberger et al. (1976) present a nice overview of the controversies surrounding the *Urban Dynamics* and *World Dynamics* models. Forrester and his colleagues' replies to criticisms of the *Urban Dynamics* model are contained in Mass (1974) and Schroeder et al. (1975).

One of the harshest critics of the *World Dynamics* (WORLD2) model has been Nordhaus (1972). Nordhaus (1992b) has also been very critical of the well-known follow-up study to *World Dynamics* known as *The Limits to Growth* Meadows et al. (1972). Meadows et al. (1992, 2002) contain updates to the original *Limits to Growth* (WORLD3) model, as well as replies to the world modeling critics.

Forrester (1980) presents a nice overview of his national economic model, and the critiques by

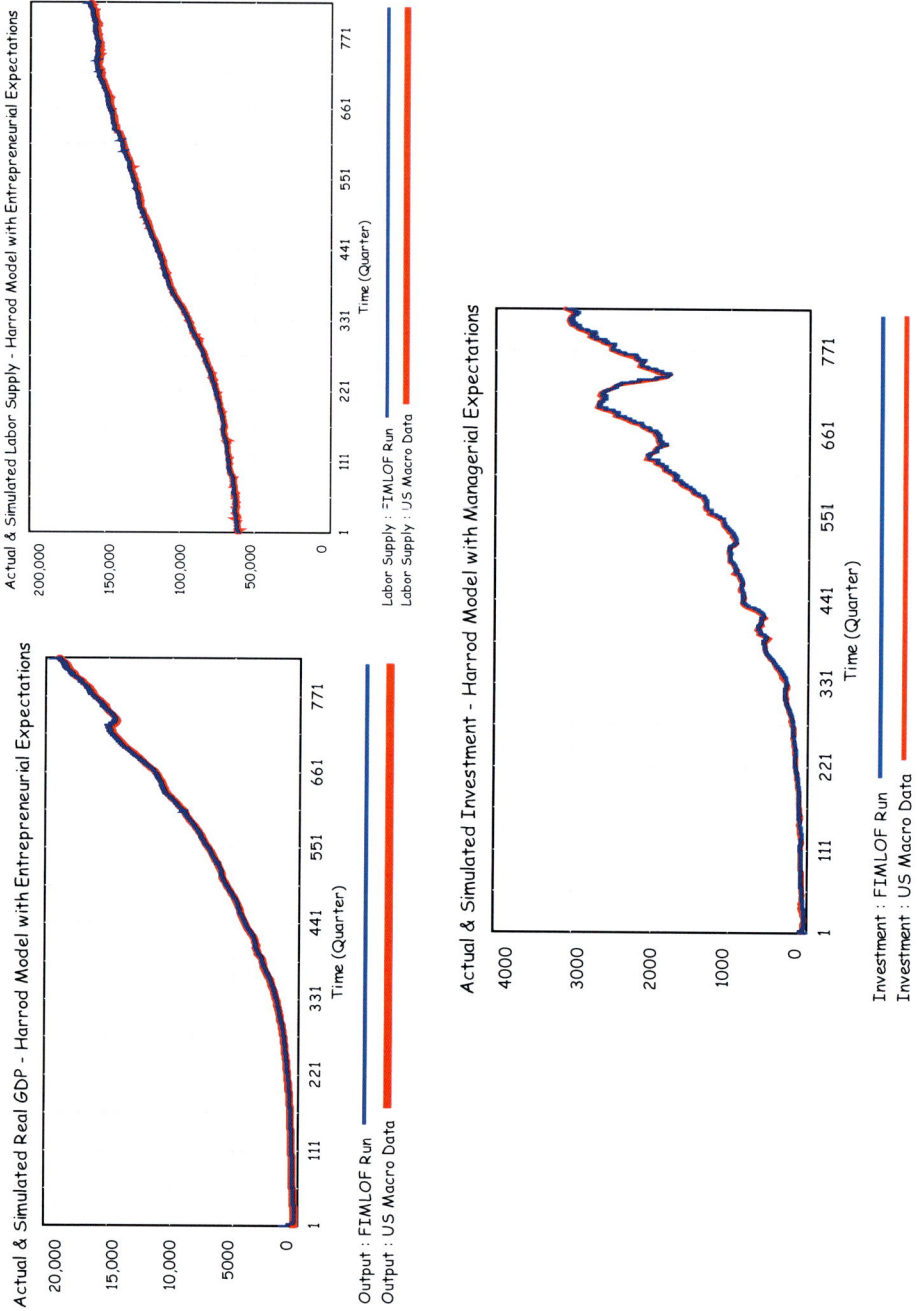

System Dynamics and Its Contribution to Economics and Economic Modeling, Fig. 7   Fit of the Harrod growth model to US macroeconomic data for the years 1929–2004

Stolwijk (1980) and Zellner (1980) are typical of the attitude of professional economists toward macroeconomic modeling that is undertaken by following the traditional system dynamics paradigm. The criticism of Forrester's national economic model by the economics profession has probably been less severe, relative to the criticisms of the *Urban Dynamics* and world models, because most of its details are still largely unpublished at the time of this writing.

Future Directions

Historically, system dynamicists who have engaged in economic modeling have almost never been trained as professional economists. As such, they have had the advantage of being able to think about economic problems differently from those who have been trained along traditional lines but have also suffered the cost of being seen as "amateurs" or "boy economists" (Radzicki 1990) by members of the economics profession. The good news is that there are currently several schools of economic thought, populated by professional economists, in which system dynamics fits quite harmoniously. These include post-Keynesian economics (Keen 1995), institutional economics (Atkinson 2004), ecological economics (Bergh and Nijkamp 1991), and behavioral economics (Sterman 1985). Historically, the economists in these schools have rejected many of the tenets of traditional economics, including most of its formal modeling methods, yet have failed to embrace alternative modeling techniques because they were all seen as inadequate for representing the concepts they felt were important. However in the modern era, with computers having become ubiquitous and simulation having become in some sense routine, system dynamics is increasingly being accepted as an appropriate tool for use in these schools of economic thought. The future of economics and system dynamics will most probably be defined by the economists who work within these schools of thought, as well as by their students. The synthesis of machine learning algorithms and big data sets with system dynamics-based economics models,

as well as the diffusion of system dynamics models of economic systems through their translation into user-friendly interactive "learning environments," will most likely also be of great importance (See Mashayekhi et al. 2006; Wheat 2007b).

Bibliography

Primary Literature

Atkinson G (2004) Common ground for institutional economics and system dynamics modeling. Syst Dyn Rev 20(4):275–286

Barlas Y (1989) Multiple tests for validation of system dynamics type of simulation models. Eur J Oper Res 42(1):59–87

Berger S (2008) Circular cumulative causation (CCC) á la Myrdal and Kapp – political institutionalism for minimizing social costs. J Econ Issues 42(2):1–9

Bergh J, Nijkamp P (1991) Aggregate dynamic economic-ecological models for sustainable development. Environ Plan A 23:1409–1428

Fiddaman T (1997) Feedback complexity in integrated climate-economy models. Ph.D. dissertation, Sloan School of Management, Massachusetts Institute of Technology. Available from http://www.systemdynamics.org/

Fiddaman T (2002) Exploring policy options with a behavioral climate-economy model. Syst Dyn Rev 18(2):243–267

Forrester J (1956) Dynamic models of economic systems and industrial organizations. System dynamics group memo D-0. Massachusetts Institute of Technology. Available from http://www.systemdynamics.org/

Forrester J (1961) Industrial dynamics. Pegasus Communications, Inc., Waltham

Forrester J (1968) Principles of systems. MIT Press, Cambridge

Forrester J (1969) Urban Dynamics. Pegasus Communications, Inc., Waltham

Forrester J (1971) World dynamics. Pegasus Communications, Inc., Waltham

Forrester N (1973) The life cycle of economic development. Wright-Allen Press, Cambridge

Forrester J (1975) Counterintuitive behavior of social systems. In: Forrester J (ed) Collected papers of Jay W. Forrester. Pegasus communications, Inc., Waltham, pp 211–244

Forrester J (1980) Information sources for modeling the National Economy. J Am Stat Assoc 75(371):555–567

Forrester N (1982) A dynamic synthesis of basic macroeconomic theory: implications for stabilization policy analysis, Ph.D. dissertation, Alfred P. Sloan school of management, Massachusetts Institute of Technology. Available from http://www.systemdynamics.org/

Forrester J (1985) 'The' model versus a modeling 'Process'. Syst Dyn Rev 1(1 and 2):133–134

Forrester J (2003) Economic theory for the new millennium. In: Eberlein R, Diker V, Langer R, Rowe J (eds) Proceedings of the twenty-first annual conference of the system dynamics society. Available from http://www.systemdynamics.org/

Forrester J, Senge P (1980) Tests for building confidence in system dynamics models. In: Legasto AA Jr, Forrester JW, Lyneis JM (eds) TIMS studies in the management sciences: system dynamics, vol 14. North Holland Publishing Company, Amsterdam, pp 209–228

Forrester J, Low G, Mass N (1974) The debate on world dynamics: a response to Nordhaus. Policy Sci 5:169–190

Godley W, Lavoie M (2012) Monetary economics: an integrated approach to credit, money, income, production and wealth. Palgrave Macmillan, New York

Graham A (1980) Parameter estimation in system dynamics modeling. In: Randers J (ed) Elements of the system dynamics method. Pegasus Communications, Inc., Waltham, pp 143–161

Greenberger M, Crenson M, Crissey B (1976) Models in the policy process: public decision making in the computer era. Russell Sage Foundation, New York

Hamilton M (1980) Estimating lengths and orders of delays in system dynamics models. In: Randers J (ed) Elements of the system dynamics method. Pegasus Communications, Inc., Waltham, pp 162–183

Heilbroner R (1980) The worldly philosophers, 5th edn. Simon & Schuster, New York

Hicks J (1950) A contribution to the theory of the trade cycle. Oxford University Press, London

Homer J (1987) Homer J. A diffusion model with application to evolving medical technologies. Technol Forecast Soc Chang 31(3):197–218

Keen S (1995) Finance and economic breakdown: modeling Minsky's financial instability hypothesis. J Post Keynesian Econ 17:607–635

Kreutzer W, Garrett M (2000) Business dynamics: systems thinking and modeling for a complex world. Irwin-McGraw-Hill, New York

Low G (1980) The multiplier-accelerator model of business cycles interpreted from a system dynamics perspective. In: Randers J (ed) Elements of the system dynamics method. Pegasus Communications, Inc., Waltham, pp 76–94

Mashayekhi A, Vakili K, Foroughi H, Hadavandi M (2006) Supply demand world: an interactive learning environment for teaching microeconomics. In: Grosler A, Rouwette A, Langer R, Rowe J, Yanni J (eds) Proceedings of the of the twenty-fourth international conference of the system dynamics Society. Available at http://www.systemdynamics.org/

Mass N (1974) Readings in urban dynamics: volume I. Wright-Allen Press, Cambridge

Mass N (1975) Economic cycles: an analysis of the underlying causes. MIT Press, Cambridge

Meadows D (1970) Dynamics of commodity production cycles. MIT Press, Cambridge

Meadows D, Meadows D, Randers J, Behrens W III (1972) The limits to growth: a report for the Club of Rome's project on the predicament of mankind. Universe Books, New York

Meadows D, Meadows D, Randers J (1992) Beyond the limits: confronting global collapse, envisioning a sustainable future. Chelsea Green Publishing, White River Junction

Meadows D, Meadows D, Randers J (2002) Limits to growth: the 30-year update. Chelsea Green Publishing, White River Junction

Millennium Institute (2007) Introduction and purpose of threshold 21. http://www.millennium-institute.org/resources/elibrary/papers/T21Overview.pdf

Moxnes E (1992) Positive feedback economics and the competition between 'hard' and 'soft' energy supplies. J Sci Ind Res 51:257–265

Ng T, Wright M (2007) Introducing the MONIAC: an early and innovative economic model, Reserve Bank of New Zealand. Bulletin 70(4):46–52

Nordhaus W (1972) World dynamics: measurement without data. Econ J 83(332):1156–1183

Nordhaus W (1992a) The "DICE" model: background and structure of a dynamic integrated climate-economy model of the economics of global warming. Cowles Foundation for Research in Economics at Yale University, Discussion Paper No. 1009

Nordhaus W (1992b) Lethal model 2: the limits to growth revisited. Brook Pap Econ Act 2:1–59

Paich M (1985) Generic structures. Syst Dyn Rev 1(1 and 2):126–132

Paich M (1994) Managing the global commons. MIT Press, Cambridge

Peterson D (1975) Hypothesis, estimation, and validation of dynamic social models – energy demand modeling. Ph.D. dissertation, Department of Electrical Engineering, Massachusetts Institute of Technology. Available from http://www.systemdynamics.org/

Peterson D (1980) Statistical tools for system dynamics. In: Randers J (ed) Elements of the system dynamics method. Pegasus Communications, Inc., Waltham, pp 226–245

Phillips W (1950) Mechanical models in economic dynamics. Economica (67):283–305

Phillips W (1954) Stabilization policy in a closed economy. Econ J 64(254):290–323

Radzicki M (1988) Institutional dynamics: an extension of the institutionalist approach to socioeconomic analysis. J Econ Issues 22(3):633–665

Radzicki M (1990) Methodologia Oeconomiae et Systematis Dynamis. Syst Dyn Rev 6(2):123–147

Radzicki M (1997) Introduction to system dynamics. Free web-based system dynamics tutorial. Available at https://www.systemdynamics.org/tips

Radzicki M (2003) Mr. Hamilton, Mr. Forrester and a foundation for evolutionary economics. J Econ Issues 37(1):133–173

Radzicki M (2007) Institutional economics, post Keynesian economics, and system dynamics: three strands of a heterodox economics braid. In: Harvey J, Garnett R (eds) The future of heterodox economics. University of Michigan Press, Ann Arbor

Radzicki M (2020, forthcoming) Behavioral expectation formation and the knife edge: another look at Harrod. In: Cavana R, Dangerfield B, Pavlov O, Wheat D (eds) Feedback economics. Springer

Radzicki M, Sterman J (1994) Evolutionary economics and system dynamics. In: Englund R (ed) Evolutionary concepts in contemporary economics. University of Michigan Press, Ann Arbor, pp 61–89

Richardson G, Pugh A (1981) Introduction to system dynamics modeling with DYNAMO. Pegasus Communications, Inc., Waltham

Richmond B (1985) Conversing with a classic thinker: an illustration from economics. Users guide to STELLA. Chapter 7. High Performance Systems, Inc., Lyme, pp 75–94

Ryzhenkov A (2007) Controlling employment, profitability and proved non-renewable reserves in a theoretical model of the U.S. economy. In: Sterman J, Oliva R, Langer R, Rowe J, Yanni J (eds) Proceedings of the of the twenty-fifth international conference of the system dynamics society. Available at http://www.system dynamics.org/

Schroeder W, Sweeney R, Alfeld L (1975) Readings in urban dynamics, vol II. Wright-Allen Press, Cambridge

Senge P (1980) A system dynamics approach to investment function formulation and testing. Socio Econ Plan Sci 14:269–280

Solow R (1956) A contribution to the theory of economic growth. Q J Econ 70:65–94

Sterman J (1981) The energy transition and the economy: a system dynamics approach, Ph.D. dissertation, Alfred P. Sloan School of Management, Massachusetts Institute of Technology. Available at http://www.system dynamics.org/

Sterman J (1985) A behavioral model of the economic long wave. J Econ Behav Organ 6:17–53

Sterman J (1988) A Skeptic's guide to computer models. In: L Grant (ed) Foresight and national decisions. University Press of America, Lanham, pp 133–169; Revised and reprinted In: Barney G, Kreutzer W, Garrett M (1991) Managing a nation. Westview Press, Boulder, pp 209–2e30

Sterman (2000) Business dynamics: Systems thinking and modeling for a complex world. The McGraw-Hill Companies, Boston

Stolwijk J (1980) Comment on 'information sources for modeling the National Economy' by Jay W. Forrester. J Am Stat Assoc 75(371):569–572

Weber L (2007) Understanding recent developments in growth theory. In: Sterman J, Oliva R, Langer R, Rowe J, Yanni J (eds) Proceedings of the twenty-fifth international conference of the system dynamics society. Available at http://www.systemdynamics.org/

Wheat D (2007a) The feedback method: a system dynamics approach to teaching macroeconomics. Ph.D. dissertation. University of Bergen, Bergen. https://bora.uib.no/handle/1956/2239

Wheat D (2007b) The feedback method of teaching macroeconomics: is it effective? Syst Dyn Rev 23(4):391–413

Wheat D (2017) Teaching endogenous money with systems thinking and simulation tools. Int J Pluralism Econ Educ 8(3):219–243

Wilber C, Harrison R (1978) The methodological basis of institutional economics: pattern model, storytelling, and holism. J Econ Issues 12(1):61–89

Yamaguchi K (2017) Money and macroeconomic dynamics: accounting system dynamics approach. Japan Futures Research Center, Awaji Island

Zellner A (1980) Comment on 'information sources for modeling the National Economy' by Jay W. Forrester. J Am Stat Assoc 75(371):567–569

Books and Reviews

Alfeld L, Graham A (1976) Introduction to urban dynamics. Wright-Allen Press, Cambridge

Lyneis J (1980) Corporate planning and policy design: a system dynamics approach. PA consulting, Cambridge

Meadows D, Meadows D (1973) Toward global equilibrium: collected papers. Wright-Allen Press, Cambridge

Meadows D, Robinson J (1985) The electronic Oracle: computer models and social decisions. Wiley, New York

Meadows D, Behrens W III, Meadows D, Naill R, Randers J, Zahn E (1974) Dynamics of growth in a finite world. Wright-Allen Press, Cambridge

Myrdal G (1968) Asian Drama: an inquiry into the poverty of nations. Pantheon, New York

Richardson G (1999) Feedback thought in social science and systems theory. Pegasus Communications, Inc., Waltham

System Dynamics and Organizational Learning

Kambiz Maani
Massey Business School, Massey University,
Albany, Auckland, New Zealand

Article Outline

Glossary

Causal loops Causal loops (model) are visual maps that connect a group of variables with known or hypothesized cause and effect relationships. A causal loop can be open or closed. Causal loops can be used for complex problem solving/decision-making, consensus building, conflict resolution, priority setting, and group learning.

Delay Cause and effect relationships are often not close in time or space. The lapse time between a cause and its effect is called a systems delay or simply delay. Because some delays in physical, natural, and social systems are rather long, they mask the underlying or earlier causes when effects become evident. This provides confusion and unintended consequences, especially in social systems, such as economics, education, immigration, judicial systems.

Feedback In a cause and effect chain (system), feedback is a signal from the effect/s to cause/s as to its/their influence on downstream effect/s. Feedback can be information, decision, or action. For example, if X causes or changes Y, Y in turn could influence or change X directly or through other intervening variables. This creates a *closed* "causal loop" with either a positive or amplifying feedback (reinforcing – R) or a negative feedback with damping, counteracting, or (balancing – B) effect.

Flow Flow or rate represents change or movement in a stock, such as buying assets, building inventories, adding capacity, losing reputation or morale. Flow is measured as "per unit of time" like hiring rate (employees hired per year, production rate (units made per day), or rainfall (inches of rain per month).

Leverage Leverage refers to decisions and actions for change and intervention which have the highest likelihood of lasting and sustainable outcomes. Leverage decisions are best reached by open discussion after the group develops a deep understanding of system dynamics through a causal loop or stock and flow modeling process.

Microworld Microworlds are simulation models of real systems such as a firm, a hospital, a market, or a production system. They provide a "virtual" world where decision-makers can test and experiment their policies and strategies in a laboratory environment before implementation. Microworlds are constructed using system dynamics software with user-friendly interfaces.

Reference Mode Reference mode is the actual/observed pattern of a key variable of interest to decision-makers or policy analysts. It represents the actual behavior of a variable over time which is used to compare with the simulated pattern of the same variable generated by a simulation model to validate the accuracy of the model.

Simulation A computer tool and methodology for modeling complex situations and challenging problems where mathematical tools fail to operate.

Stock In system dynamics stock is a concept representing accumulation and the state of a variable, such as assets, inventory, capacity,

© Springer Science+Business Media, LLC, part of Springer Nature 2020
B. Dangerfield (ed.), *System Dynamics*,
https://doi.org/10.1007/978-1-4939-8790-0_543

Originally published in
R. A. Meyers (ed.), *Encyclopedia of Complexity and Systems Science*, © Springer Science+Business
Media LLC 2018, https://doi.org/10.1007/978-3-642-27737-5_543-2

reputation, morale. Stock can be measured at any point of time. In mathematical terms, stock is the sum over time (integral) of one or more flows.

Systems thinking Systems thinking is a paradigm for viewing reality based on the primacy of the whole and relationships. Systems thinking is one of the key capabilities (disciplines) for organizational learning which consists of a series of conceptual and modeling tools such as behavior over time, causal loop diagrams, and systems archetypes. These tools reveal cause and effect dynamics over time and assist understanding of complex, nonlinear, and counterintuitive behaviors in all systems – physical, natural, and social.

Definition of the Subject

System dynamics (SD) is "a methodology for studying and managing complex feedback systems... While the word system has been applied to all sorts of situations, feedback is the differentiating descriptor here. Feedback refers to the situation of X affecting Y and Y in turn affecting X perhaps through a chain of causes and effects... Only the study of the whole system as a feedback system will lead to correct results" (System Dynamics Society website 2017).

Sterman (2000, p. 4) defines system dynamics as "a method to enhance learning in complex systems." "System dynamics is fundamentally interdisciplinary... It is grounded in the theory of nonlinear dynamics and feedback control developed in mathematics, physics, and engineering. Because we apply these tools to the behavior of human as well as physical and technical systems, system dynamics draws on cognitive and social psychology, economics, and other social sciences."

Wolstenholme (1997) offers the following description for system dynamics and its scope:

A rigorous way to help thinking, visualizing, sharing, and communication of the future evolution of complex organizations and issues over time; for the purpose of solving problems and creating more robust designs, which minimize the likelihood of unpleasant surprises and unintended consequences; by creating operational maps and simulation models which externalize mental models and capture the interrelationships of physical and behavioral processes, organizational boundaries, policies, information feedback and time delays; and by using these architectures to test the holistic outcomes of alternative plans and ideas; within a framework which respects and fosters the needs and values of awareness, openness, responsibility and equality of individuals and teams.

History of System Dynamics

(Source: US Department of Energy 2007)

System dynamics was created during the mid-1950s by Professor Jay W. Forrester of the Massachusetts Institute of Technology. His early work at MIT was in the Servomechanism Laboratory, conducting pioneering research in feedback control mechanisms for military equipment. Forrester's work for the Laboratory took him to the aircraft carrier Lexington during World War II to repair the ship's radar system. The Lexington was torpedoed while Forrester was on board, but he survived.

After the war, Forrester turned his attention to the creation of an aircraft flight simulator for the US Navy. The design of the simulator led to the creation of one of the earliest digital computers. In 1947, the MIT Digital Computer Laboratory was founded and placed under the directorship of Professor Forrester. The Laboratory's first task was the creation of WHIRLWIND I, MIT's first general-purpose digital computer. As part of this project, Forrester invented and patented random-access magnetic computer memory (RAM), which became the industry standard for computer memory for approximately 20 years.

Another outcome of the WHIRLWIND project was the insights that it developed for the challenges faced by corporate managers. Forrester's experiences as a manager led him to conclude that the biggest challenges were not in technical or engineering but in management. This is because, he reasoned, social systems are much harder to understand and control than are physical systems. This insight, which led to the creation of system dynamics, was triggered, to a large degree, by his consulting work at General Electric during the mid-1950s. At that time, the managers at GE were experiencing wide fluctuations instability in employee numbers, which not be explained by business cycles. Using hand-drawn simulations of GE's employee dynamics, Forrester was able to show that the instability in GE employment was due to the internal structure of the firm (e.g., hiring and layoffs) and not due to external forces such as the business cycle. These simulations were the beginning of the field of system dynamics.

During the late 1950s and early 1960s, Forrester and a team of graduate students converted the nascent field of system dynamics from the hand-simulation stage to computer modeling, which led to the creation of DYNAMO (DYNAmic MOdels) by Phyllis Fox and Alexander Pugh in 1959. DYNAMO became the industry standard for system dynamics software for over 30 years.

In 1961, Forrester published the first, and still classic, book in the field titled *Industrial Dynamics* (Forrester 1961). Another seminal book in the field is *Urban Dynamics* (Forrester 1969), which represents the first major noncorporate application of system dynamics. The book was and still remains controversial, because it illustrates why many well-known urban policies are either ineffective or make urban problems worse. For example, it illustrates that counterintuitive policies – those that at first appear to be wrong – often yield significant favorable results. As an example, the book demonstrates how a policy of building low-income housing ends up being a poverty trap that helps to decay a city, while a policy of integrating low-income housing into normal neighborhoods creates jobs and a rising standard of living for city's population.

The applications of SD modeling have grown extensively since its inception and today encompass work in every domain and sector of society, including (System Dynamics Society website 2017):

- Climate policy
- Corporate planning and policy design
- Public management and policy

- Biological and medical modeling
- Energy and the environment
- Theory development in the natural and social sciences
- Dynamic decision-making
- Complex nonlinear dynamics

The next two sections provide an overview of system dynamics and then its related fields of systems thinking and organizational learning.

Systems Thinking and Modeling Methodology

(Source: Maani and Cavana 2007b)

System dynamics modeling is one of the five phases of systems thinking and modeling intervention methodology (Cavana and Maani 2004; Maani and Cavana 2007b). These distinct but related phases are as follows:

1. Problem structuring
2. Causal loop modeling
3. System dynamics modeling
4. Scenario planning and modeling
5. Implementation and organizational learning (learning lab)

These phases follow a process, each involving a number of steps, as outlined in Table 1. This process does not require all phases to be undertaken, nor does each phase require all the steps listed. Which phases and steps are included in a particular project or intervention depends on the issues or problems that have generated the systems enquiry and the degree of effort that the organization is prepared to commit to.

This phase follows the causal modeling phase. Although it is possible to go into this phase directly after problem structuring, performing the causal modeling phase first will enhance the conceptual rigor and learning power of the systems approach. The completeness and wider insights of systems thinking is generally absent from other simulation modeling approaches, where causal modeling does not play a part. The following

System Dynamics and Organizational Learning, Table 1 The five phase process of systems thinking and modeling (Source: Cavana and Maani 2004)

Phases	Steps	
1	Problem structuring	Identify problems or issues of concern to management, collect preliminary information and data
2	Causal loop modeling	Identify main variables, prepare behavior over time graphs (reference mode), develop causal loop diagram (influence diagram), analyze loop behavior over time, and identify loop types, identify system archetypes, identify key leverage points, develop intervention strategies
3	System dynamic modeling	Develop a systems map or rich picture, define variable types and construct stock-flow diagrams, collect detailed information and data, develop a simulation model, simulate steady-state/ stability conditions, reproduce reference mode behavior (base case), validate the model, perform sensitivity analysis, design and analyze policies, develop and test strategies
4	Scenario planning and modeling	Plan general scope of scenarios, identify key drivers of change and keynote uncertainties, construct forced and learning scenarios, simulate scenarios with the model, evaluate robustness of the policies and strategies
5	Implementation and organizational learning	Prepare a report and presentation to management team, communicate results and insights of proposed intervention to stakeholders, develop a microworld and

(*continued*)

System Dynamics and Organizational Learning, Table 1 (continued)

Phases	Steps	
		learning lab based on the simulation model, use learning lab to examine mental models and facilitate

steps are generally followed in the system dynamics modeling phase.

1. Develop a high-level map or systems diagram showing the main sectors of a potential simulation model, or a "rich picture" of the main variables and issues involved in the system of interest.
2. Define variable types (e.g., stocks, flows, converters) and construct stock-flow diagrams for different sectors of the model.
3. Collect detailed, relevant data including media reports, historical and statistical records, policy documents, previous studies, and stakeholder interviews.
4. Construct a computer simulation model based on the causal loop diagrams or stock-flow diagrams. Identify the initial values for the stocks (levels), parameter values for the relationships, and the structural relationships between the variables using constants, graphical relationships, and mathematical functions where appropriate. This stage involves using specialized computer packages like Stella, Vensim, Powersim, AnyLogic, and more.
5. Simulate the model over time. Select the initial value for the beginning of the simulation run, specify the unit of time for the simulation (e.g., hour, day, week, month, year). Select the simulation interval (DT) (e.g., 0.25, 0.5, 1.0) and the time horizon for the simulation run (i.e., the length of the simulation). Simulate model stability by generating steady-state conditions.
6. Produce graphical and tabular output for the base case of the model. This can be produced using any of the computer packages mentioned above. Compare model behavior with

historical trends or hypothesized reference modes (behavior over time charts).

7. Verify model equations, parameters, and boundaries, and validate the model's behavior over time. Carefully inspect the graphical and tabular output generated by the model.

8. Perform sensitivity tests to gauge the sensitivity of model parameters and initial values. Identify areas of greatest improvement (key leverage points) in the system.

9. Design and test policies with the model to address the issues of concern to management and to look for system improvement.

10. Develop and test strategies (i.e., combinations of functional policies, for example, operations, marketing, finance, human resources).

Organizational Learning

(This section is adapted from Maani and Cavana (2007b).)

Organizational learning is the ability of organizations to enhance their collective capacity to learn and to act, harmoniously. According to Senge (1990a) "Real learning gets to the heart of what it means to be human. Through learning we re-create ourselves. Through learning we become able to do something we never were able to do. Through learning we re-perceive the world and our relationship to it. Through learning we extend our capacity to create, to be part of the generative process of life. There is within each of us a deep hunger for this type of learning." Organizational learning extends this learning to the organization and its members.

Peter Senge popularized the concept through his seminal book: *The Fifth Discipline* (Senge 1990a). He describes a learning organization as one "which is continually expanding its ability to create its future." He identifies five core capabilities (disciplines) of the learning organization that are derived from three "higher orientations": creative orientation, generative conversation, and systems perspective. "The reality each of us sees and understands depend on what we believe is there. By learning the principles of the five

disciplines, teams begin to understand how they can think and inquire that reality, so that they can collaborate in discussions and in working together create the results that matter [to them]."

As Fig. 1 shows, learning organization capabilities are dynamically interrelated, and collectively they lead to organizational learning. Senge maintains that *creative orientation* is the source of a genuine desire to excel. It is the source of an intrinsic motivation and drive to achieve. It relinquishes personal gains in favor of the common good. *Generative conversation* refers to a deep and meaningful dialog to create unity of thought and action. *Systems perspective* is the ability to see things holistically by understanding the interconnectedness of the parts. The foregoing elements give rise to the five core capabilities of learning organizations, namely, personal mastery, shared vision, mental models, team learning and dialog, and systems thinking. These five disciplines are described below. Figure 1 shows the core capabilities and their relationships.

Personal Mastery

Senge (1990a) describes personal mastery as the cornerstone and "spiritual" foundation of the learning organization. It is born out of a creative orientation and systemic perspective. Personal mastery instills a genuine desire to do well and to serve a noble purpose. People exhibiting high levels of personal mastery focus "on the desired result itself, not the process or the means they assume necessary to achieve that result." These people can "successfully focus on their ultimate intrinsic desires, not on secondary goals. This is a cornerstone of Personal Mastery." Personal mastery also requires a commitment to truth, which means to continually challenge "theories of why things are the way they are." Without committing to the truth, people all too quickly revert to old communication routines which can distort reality and prevent them from knowing where they really stand.

Shared Vision

It is commonly assumed that in contemporary organizations senior management can develop a vision which employees will follow with genuine

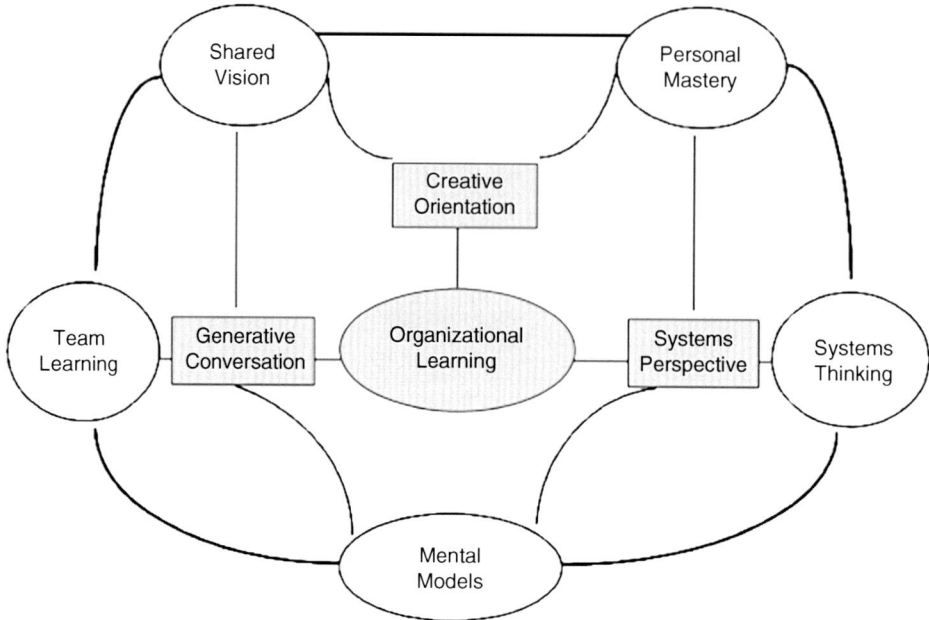

System Dynamics and Organizational Learning, Fig. 1 The core capabilities of a learning organization (Source: Maani and Benton 1999)

commitment. This is a fallacy. Simply promoting a "vision statement" could result in a sense of apathy, complacency, and resentment. Instead, there needs to be a genuine endeavor to understand what people will commit to. The overriding vision of the group must build on the personal visions of its members. Shared vision should align diverse views and feelings into a unified focus.

This is emphasized by Arie de Geus (1997) when he describes what makes a truly extraordinary organization. "The feeling of belonging to an organization and identifying with its achievements is often dismissed as soft. But case histories repeatedly show that a sense of community is essential for long term survival." For example, when Apple Corporation challenged IBM, it was in its "adolescent" years, characterized by creativity, confidence, and even defiance. This is similar to the spirit in Team New Zealand when it competed against the bigger-budget syndicates! Within these organizations there is a real passion for the outcome, a common vision for success (Maani and Benton 1999).

Creating a shared vision is the most fundamental job of a leader (Nadler and Tushman 1990). By creating a vision, the leader provides a vehicle for people to develop commitment, a common goal around which people can rally, and a way for people to *feel* successful. The leader must appeal to people's emotions if they are to be energized toward achieving the goal. Emotional acceptance of, and belief in, a vision is far more powerful in energizing team members than is intellectual recognition that the vision is simply a "good idea." One of the most powerful ways of communicating a vision is through a leader's personal example and actions, demonstrating behavior that symbolizes and furthers that vision.

Mental Model and Leadership

Mental models reflect beliefs, assumptions, and feelings that shape one's world views and actions. They are formed through family, education, professional and social learning based, on the most part, on cultural and social norms. Mental models, however, can be altered and aligned.

Organizations are often constrained by deep-seated belief systems, resulting in preconceived ideas on how things ought to perform. Goodstein and Burke (1991, p. 10), pioneers in the field of social psychology of organizations, observed that "the first step in any change process is to unfreeze the present patterns of behavior as a way of managing resistance to change." The leader has a pivotal role in dismantling negative mental models and shaping new ones.

In order to get people to engage in open discussions of issues that affect the organization, a leader must appeal to their emotions and must get beyond the superficial level of communication. In the 1970s, Shell Oil undertook major changes in its leadership approach and communications style. According to a manager at Shell, "When I tried to talk personally about an issue rather than say 'here's the answer', it was powerful. It caused me to engage in dialog with others that resulted in mutual learning on all sides" (Cohen and Tichy 1997, p. 71).

The leader is a "designer," and part of that role is designing the governing ideas of purpose and core values by which people will live (Senge 1990a, b). In this role, the leader must propose and model the manner in which the group has to operate internally. This provides ample opportunities for leaders to examine their deeply held assumptions about the task, the means to accomplish it, the uniqueness of the people, and the kinds of relationship that should be fostered among the people. Only after people have observed and *experienced* the organizational values in practice would these values become the basis for prolonged group behavior. These values should be manifested first and should be most visible in the leader's own behavior.

Leadership, especially in knowledge-based organizations, must be distributed and shared to a far greater extent than it was in the past. For example, in the Chicago Bulls basketball team, Michael Jordan changed his role: it became not only that of an individually brilliant player but *also* that of a leader whose job was to raise the level of play of other team members. After this transition, the Bulls began their record run of championship seasons (Cohen and Tichy 1997).

Team Learning and Dialog

The word "dialog" comes from the Greek words *dia* and *logos*. It implies that when people engage in dialog, the meaning *moves through* them – thus, it enables them to "see through words" (Isaacs 1993). Dialog is an essential requirement for organizational learning. It results from generative conversation, shared vision, and transparent mental models. Dialog creates a deep sense of listening and suspending one's own views. Feedback is an integral aspect of dialog.

Communication routines in organizations are generally anti-learning and promote mediocrity. They include "defensive routines" (Argyris 1992) – statement that can stifle dialog and innovative thinking. Exposing and unlearning such routines, and understanding the powerful detrimental impact they have on learning, are serious challenges many organizations face if they are to create effective learning environments.

Many leaders are charismatic and are highly eloquent when it comes to presenting their ideas; that's often why they get to the top of the organization. However, many appear to lack the ability to extract the very best from employees in a non-threatening manner. Without this ability, leaders may miss many good ideas or might act on many bad ones.

In a group context, encouragement from the leader and mutual encouragement among group members are essential. Furthermore, personal differences must be put aside in order for effective dialog to ensue.

How Organizations Learn

(Source: http://en.wikipedia.org/wiki/Organizational_learning)

Argyris and Schon were the early pioneers of organizational learning. In their seminal book (Argyris and Schon 1978), they propose single-loop and double-loop learning, related to Gregory Bateson's concepts of first- and second-order learning. In single-loop learning, individuals, groups, or organizations review and modify their actions according to the difference between expected and actual outcomes. In double-loop learning, they question the values, assumptions, and policies that led to the actions in the first place and whether they are able to alter those. In this scenario, second-order or double-loop learning has taken place. In other words, double loop learning is the learning about single-loop learning.

Since Argyris and Schon's seminal work, other researchers have advanced the field of organizational learning. Some of notable contributions on OL are summarized below:

- March and Olsen's work (1975) links individual and organizational learning based on the premise that individual beliefs lead to individual action. Individual action, in turn, may lead to an organizational action and the response may induce improved individual beliefs – the cycle then repeats over and over. In this model, learning occurs as enhanced beliefs produce better actions.
- Kim (1993), in his article "The Link Between Individual and Organizational Learning," integrates Argyris, March and Olsen, and another model by Kofman into a single comprehensive model. He further analyzes all the possible breakdowns in the information flows in the model, leading to failures in organizational learning. For instance, what happens if an individual action is rejected by the organization for political or other reasons and therefore no organizational action takes place?
- Nonaka and Takeuchi (1995) developed a four-stage spiral model of organizational learning. They started by differentiating Polanyi's concept of "tacit knowledge" from "explicit knowledge" and describe a process of alternating between the two. Tacit knowledge is personal, context-specific, subjective knowledge, whereas explicit knowledge is codified, systematic, formal, and easy to communicate. The tacit knowledge of key personnel within the organization can be made explicit, codified in manuals, and incorporated into new products and processes. They called this process "externalization." The reverse process (from explicit to implicit) is called "internalization" because it involves employees internalizing an organization's formal rules, procedures, and other forms of explicit knowledge. They also use the term "socialization" to denote the

sharing of tacit knowledge and the term "combination" to denote the dissemination of codified knowledge. According to this model, knowledge creation and organizational learning take a path of socialization, externalization, combination, internalization, socialization, externalization, combination, etc. in an infinite spiral.

- Flood, in his book (Flood 1999), aims to "rethink" Senge's *The Fifth Discipline* through systems theory. He discusses Senge's notion of organizational learning from and the origins of the theory by Argyris and Schon and integrates them with related concepts from key theorists such as Bertalanffy, Churchman, Beer, Checkland, and Ackoff. Conceptualizing organizational learning in terms of structure, process, meaning, ideology, and knowledge, Flood provides insights into *The Fifth Discipline* within the context of the philosophy of science and the way in which systems scientists were influenced by twentieth-century advances in challenging the assumptions of classical science.
- Nick Bontis et al. (2002) empirically tested a model of organizational learning that encompassed both stocks and flows of knowledge across three levels: individual, team, and organization. Results showed a negative and statistically significant relationship between the misalignment of stocks (knowledge) and flows (learning) and organizational performance.
- Imants (2003) provides a theoretical framework for organizational learning in schools by distinguishing learning communities from communities of practice in the context of teachers' professional communities. He analyzes and highlights the underlying paradoxes for organizational learning in schools and proposes two mechanisms critical for professional development and effective organizational learning: (1) steering information about teaching and learning and (2) encouraging interaction among teachers and workers.
- Common (2004) discusses the concept of organizational learning in a political environment to improve public policy-making. He details

the initial uncontroversial reception of organizational learning in the public sector and the development of the concept with the learning organization. Noting research in UK local government, he addresses definitional problems in applying the concept to public policy and suggests the following barriers for organizational learning in the public sector: (1) overemphasis of the individual, (2) resistance to change and politics, (3) social learning is self-limiting, i.e., individualism, and (4) political "blame culture." The concepts of *policy learning* and *policy transfer* are then defined with detail on the conditions for realizing organizational learning in the public sector.

- Maani (2017) extends the principles of organizational learning to multistakeholder, multiagency decision-making scenarios. Through real-world case studies, he demonstrates how the concepts and the process of systems thinking and modeling by novice users can induce group and organizational learning, leading to greater harmony, shared vision, and commitment to action. His "learning lab" model and associated processes can be applied to a wide range of situations from decision-making within small teams to large-scale multiagency "wicked" problems such as poverty and disease alleviation, security, and climate change.

The Learning Laboratory

(Source: Maani and Cavana 2007b, Chap. 6)

A *learning laboratory* is a setting as well as a process in which a group can learn together. The purpose of the learning lab is to enable decision-makers to test their assumptions and to experiment and "foresee" the consequences of their actions, policies, and strategies. This often results in finding inconsistencies and the discovery of *unintended* consequences of actions and decisions, *before* they are implemented. System dynamics situation-specific models known as Microworlds or Management Flight Simulators (MFS) are the "engine" behind the learning lab. "Just as an airline uses flight simulators to help pilots learn, system dynamics is, partly, a method for developing management flight simulators, often computer simulation models, to help us

learn about dynamic complexity, understand the sources of policy resistance, and design more effective policies" (System Dynamics Society website 2017, p. 4).

A learning lab is distinct from the so-called management games. In management games, the players are required to compete, i.e., design the "best" strategy and "beat" other players or teams. The competitive nature of management games often encourages aggressive and individualistic behavior with scant regard for group learning and gaining deep insights. The learning lab, in contrast, aims to enhance *learning*: to test individual and group mental models and to provide deeper understanding and insights into why systems behave the way they do. This will help the participants to test their theories and discover inconsistencies and "blind spots" in policies and strategies *before* they are implemented.

A significant benefit of the learning lab stems from the process in which participants examine, reveal, and test their mental models and those of their organization. The learning lab can also help the participants

- To align strategic thinking with operational decisions
- To connect short-term and long-term measures
- To facilitate integration within and outside the organization
- To undertake experimentation and learning
- To balance competition with collaboration

Group Model Building

Group model building (GMB) is a powerful tool for dealing with complex or "messy" (Ackoff 1999) problems.

Messy problems are situations where there exist large differences of opinion about the nature of the problem or even on the question of whether there is a problem. These situations, especially in multistakeholder scenarios, make it difficult for the group to make decisions let alone reaching consensus.

GMB offers an opportunity to align and share piecemeal mental models and create the possibility of assimilating and integrating partial mental models into a holistic system description (Vennix

1995, 1996). GMB and SD can help uncover misperception that may occur due to the fact that the definition of a problem may be a socially constructed phenomenon that has not been put to test (Maani 2002, p. 84).

In general, the *process* of group model building can be an effective conduit for collective learning. In semi- or ill-structured decision situations systems thinking and, in particular, system dynamics modeling can be used to enhance organizational learning through rapid experimentation and feedback and the ability to test participants' assumptions and mental models (Vennix 1996). As discussed earlier, dealing effectively with mental models is one of the core competencies for organizational learning.

Managerial Practice Field

Team and teamwork are parts of the lexicon of numerous organizations today. Company after company has reorganized their work around a variety of team concepts. From factories to hospitals, *titles* like "manager" and "supervisor" have been replaced by *roles* such as "facilitator" and "team leader." Despite this level of attention to team and teamwork the expected benefits have been marginal at best.

However, if we examine *other* teams, such as sporting teams, orchestras, or ballet companies, closely, they all share one key characteristic. That is they *practice* a lot more than they *perform*. Practice involves allowing time and space to experiment with new ways, try different approaches, and most importantly, make mistakes without the fear of failure. As such, making mistakes is indispensable to learning. One cannot learn from doing things right all the time! Yet a great deal of organizational energy and attention is devoted to the prevention and masking of mistakes.

So, where is the *practice field* for management teams? The fact is that the notion and reality of practice field is curiously absent from organizations and managers work. In other words, there is no time or space in organizations for management teams to "practice." That is, to experiment, make mistakes and learn together from their mistakes. This is most noticeable during disruptive activities

such restructuring, downsizing, when pressure, confusion, and lack of time abound. One could say that time pressure coupled with short-term thinking is the greatest impediments to managerial and organizational learning. In the late 1990s, a full-page *New York Times* advert placed by IBM read "*Innovative Thinking! We don't even have time for bad thinking.*" In contrast, Lord Rutherford, the Noble Laureate in physics, credited for splitting the atom, said about his lab's budget "*we don't have money, so we need to think.*"

The pace of work in today's organization is so unrelenting that there is virtually no opportunity for managers to slow down, experiment, reflect, and learn. The consequences of the lack of experimentation, reflection, and learning while unmeasured are grave. Jay Forrester, the father of system dynamics (Forrester 1994), asserts most organizations only achieve about 5% of their potential.

The learning lab concept discussed earlier provides a practice field for organizations and decision-makers. It allows learning to become an integral part of managerial routine work and helps facilitate learning to become institutionalized within the culture of the organization (Kim 1993).

Aligning Mental Models Through the Learning Lab

Mental models are formed throughout one's life. Family, school, culture, religion, profession, and social norms play important roles in this formation. Therefore, recognizing and modifying one's mental model is not a small matter. The most effective way to check one's mental models is to *experience* alternative realities at first hand and see their implications with a new "lens" (Brown 1991).

There are rarely any opportunities in the course of a manager's daily work to engage in lengthy cycles of experimentation and reflection. Learning in a "laboratory" setting is a viable and powerful alternative. Fortunately, advanced computers and sophisticated system dynamics software have enabled the creation of learning labs where managers/decision-makers can experiment, test their theories, and learn rapidly. Thus, learning labs can play a significant role in clarifying and changing mental models. Learning labs deal with mental models at three levels (Senge and Sterman 1991), as described below.

- *Mapping* mental models. This step begins at the conceptualization phase. Here, the learning lab participants articulate and clarify their assumptions, views, opinions, and biases regarding the issue at hand.
- *Challenging* mental models. The participants identify and discuss inconsistencies and contradictions in their assumptions. This step will begin at the conceptualization phase and will continue to the experimentation phase.
- *Improving* mental models. Having conducted experimentation and testing, the participants reflect on the outcomes. This may cause them to alter, adjust, improve, and harmonize their mental models.

The laboratory setting provides a neutral and "safe" space for the participants to create a shared understanding of complex and endemic issues. The following characteristics of the learning lab provide a powerful catalyst for alignment of divergent mental models in the organization.

- The laboratory environment is neutral and nonthreatening. The emphasis is on learning and theory building (what we *don't* know), not on winning or display of knowledge.
- Lack of hierarchy. Managers and staff are equal in this environment. The traditional hierarchy is minimized in the laboratory setting.
- The response time is fast. Hence, the feedback cycle is short, which leads to rapid learning.
- There is no cost or "loss of face" as a consequence of failure. Hence, it is safe and liberating to make mistakes. In fact, mistakes are a conduit for learning, as is the case with flight simulators.
- Participants can see the consequences of their actions first hand. No one attempts to convince or teach anyone else or force his or her preconceived views on others. People best learn by themselves and through group interactions.

Implications for Management

The practice field and the learning lab concepts offer fresh and challenging implications for managers and their role. They suggest that a leader/manager should think as a *scientist*, be open to and welcome hard questions, experiment with new ideas and test new questions or "hypotheses," and be prepared to be proven *wrong*. In this regard, systems thinking provides a powerful tool set and skills for managers not just for "solving" problems but for communication, team building, and organizational learning. This means that an effective leader needs to be the *"designer"* of the ship and not its captain (Rothfeder 2003). Once a leader has designed a new structure, strategy, policy, and procedures, then she/he should create a "practice field" to allow the employees to test its assumptions and experiment with plausible scenarios that may unfold. The outcome would be a shared understanding leading to alignment of thoughts and actions. This is the essence of organizational learning.

Future Directions

Agent-Based Modeling (ABM)

Agent-based modeling (ABM) is a modeling technology which draws its theories and techniques from complexity science (Rothfeder 2003). While system dynamics and agent-based modeling (ABM) use different modeling philosophies and approaches, they can be used complementarily and synergistically.

System dynamics focuses on modeling structures (i.e., relationships, policies, strategies) that underlie behavior of systems. This may be viewed as a weakness of system dynamics approach in that behavior is assumed to be solely a function of structure (model relationships defined a priori). In contrast, an agent-based model consists of a system of agents and their relationships. ABM models systems/organizations as aggregates of semiautonomous decision-making elements – purposeful individuals called *agents*. Each agent individually assesses its situation and "makes decisions" based upon value hierarchies representing goals, preferences, and standards for behavior. Thus, high-level macro-behavior is not modeled separately but *emerges* from the micro-decisions of individual agents. In other words, in agent-based modeling, *"emergent"* behavior is expected as a result of agents' interactions. This is a key difference between the two approaches.

While system dynamics acknowledges the critical role of individual and organizational mental models (e.g., motivations, values, norms, biases), it does not explicitly model them. Furthermore, SD utilizes factual data or "cold knowledge" and does not take into account decision-makers "mood" and emotions. In contrast, ABM attempts to capture "warm knowledge," representing emotional and human context of decision-making (Maani et al. 2003).

Recent advances in video game technology allow the development of multiagent, artificial "society" simulators with capabilities for modeling physiology, stress, and emotion in decision-making (Silverman et al. 2002).

This new approach enables superior understanding of the complexity in organizations and their relevant business environments. This in turn provides an opportunity for new sophistications in game-play that enhances decision-making. Experience with agent-based modeling shows that even a simple agent-based model can exhibit complex behavior patterns and provide valuable information about the dynamics of the real-world system that emulates them. The applications of ABM are varied and rapidly growing, including city planning, healthcare policy, economics development, stock market prediction, and crowd movements/stampede prevention to name a few.

Despite these differences, SD and ABM can be used in a complementary fashion. Both ABM and SD are powerful tools for transforming information into knowledge and understanding leading to individual and group learning. However, the transition from knowledge to understanding may not be immediate or transparent. This requires a deep shift in mental models through experimentation and group learning (Maani 2017).

Systems Thinking and Sustainability

Systems thinking has a natural affinity with sustainability modeling and management.

Sustainability issues are complex; cut across several disciplines; involve multiple stakeholders; and require long-term integrated approaches. Hence, the systems paradigm and tools have direct and powerful applications in sustainability issues and management. In fact, the applications of system dynamics in sustainability go back to the early 1970s with seminal publications including "World Dynamics" (Forrester 1971), "The Limits to Growth" (Meadows et al. 1972), and "Beyond the Limits" (Meadows et al. 1992). Since then, the attention to the environment and associated global politics has evolved dramatically causing wide awareness and debates on the causes and consequences of the environmental challenge (Randers 2000). As an example, today, concern over CO_2 emissions has led to carbon trading become an international currency and market place.

Sustainability has brought a global challenge for governments, businesses and industries, scientists, farmers, and world's citizens alike. This is an unprecedented test for the citizens of our home planet – mankind – to find fresh, collective, united, and systemic approaches. Systems thinking and system dynamics can make real and valuable contributions to addressing this challenge.

Bibliography

Primary Literature

Ackoff RA (1999) Re-creating the corporation – a design of organizations for the 21st century. Oxford University Press, Oxford

Argyris C (1992) The next challenge for TQM: overcoming organisational defences. J Qual Particip 15:26–29

Argyris C, Schon D (1978) Organizational learning: a theory of action perspective. Addison-Wesley, Reading

Bontis N, Crossan M, Hulland J (2002) Managing an organizational learning system by aligning stocks and flows. J Manag Stud 39(4):437–469

Brown JS (1991) Research that reinvents the corporation. Harv Bus Rev 68:102–111

Cavana R, Maani K (2004) A methodological framework for integrating systems thinking and system dynamics. In: System dynamics society proceedings, Oxford

Cohen E, Tichy N (1997) How leaders develop leaders. Train Dev, May, p 58

Common R (2004) Organisational learning in a political environment: improving policy-making in UK government. Policy Stud 25(1):35–49

De Geus A (1997) The living company. Harv Bus Rev 75(2):51–59

Flood RL (1999) Rethinking the fifth discipline: learning within the unknowable. Routledge, London

Forrester JW (1961) Industrial dynamics. Productivity Press, Cambridge

Forrester JW (1969) Urban dynamics. Productivity Press, Cambridge

Forrester JW (1971) World dynamics. Wright-Allen Press, Cambridge, MA. (Subsequently re-published by Productivity Press, and Pegasus Communications)

Forrester JW (1994) Building a foundation for tomorrow's organizations. In: Systems thinking in action video collection, vol 1. Pegasus Communications, Cambridge

Goodstein L, Burke W (1991) Creating successful organisation change. Organ Dyn 19(4):5–17

Imants J (2003) Two basic mechanisms for organizational learning in schools. Eur J Teach Educ 26(3):293–311

Isaacs W (1993) Taking flight: dialogue, collective thinking and organisational learning. Organ Dyn 22(2):24–39

Kim DH (1993) The link between individual and organizational learning. Sloan Manag Rev 35(1):37–50

Maani K (2002) Consensus building through systems thinking – the case of policy and planning in healthcare. Aust J Inform Syst 9(2):84–93

Maani K (2017) Multi-stakeholder decision making for complex problems: a systems thinking approach with cases. World Scientific, Singapore

Maani K, Benton C (1999) Rapid team learning. Lessons from team New Zealand's America's cup campaign. Organ Dyn 27(4):6

Maani K, Cavana R (2007a) Systems methodology. Syst Think 18(8):2–7

Maani K, Cavana R (2007b) Systems thinking, system dynamics – managing change and complexity, 2nd edn. Prentice Hall, Pearson Education, Auckland

Maani K, Pourdehnad J, Sedehi H (2003) Integrating system dynamics and intelligent agent-based modelling – theory and case study. Euro INFORMS, Istanbul

March JG, Olsen JP (1975) The uncertainty of the past; organizational ambiguous learning. Eur J Polit Res 3:147–171

Meadows DH, Meadows DL, Randers J, Behren W (1972) The limits to growth. Universe Press, New York

Meadows DH, Meadows DL, Randers J (1992) Beyond the limits. Chelsey Green, Post Mills

Nadler DA, Tushman ML (1990) Beyond the charismatic leader: leadership and organisational change. Calif Manag Rev 32(2):77–97

Nonaka I, Takeuchi H (1995) The knowledge creating company. Oxford University Press, New York

Randers J (2000) From limits to growth to sustainable development or SD (sustainable development) in a SD (system dynamics) perspective. Syst Dyn Rev 16(3):213–224

Rothfeder J (2003) Expert voices: Icosystem's Eric Bonabeau. CIO Insights

Senge P (1990a) The fifth discipline: the art and practice of the learning organisation. Currency, New York, Boston

Senge P (1990b) The leader's new work: building learning organizations. Sloan Manag Rev 32:7–23

Senge P (1994) Building learning organizations. J Qual Particip, p 379

Senge P, Sterman JD (1991) Systems thinking and organizational learning: acting locally and thinking globally in the organization of the future. In: Kochan T, Useem M (eds) Transforming organizations. Oxford University Press, Oxford

Silverman BG et al (2002) Using human models to improve the realism of synthetic agents. Cogn Sci Q 3

Sterman JD (2000) Business dynamics, systems thinking and modeling for a complex world. McGraw-Hill, Irwin

System Dynamics Society website (2017) http://www.systemdynamics.org/

US Department of Energy (2007) Introduction to system dynamics, A systems approach to understanding complex policy issues. US Department of Energy. http://www.systemdynamics.org/DL-IntroSysDyn/inside.htm

Vennix JAM (1995) Building consensus in strategic decision-making: system dynamics as a support system. Group Decis Negot 4(4):335–355

Vennix JAM (1996) Chapter 5. In: Group model-building: facilitating team learning using system dynamics. Wiley, Chichester

Wolstenholme E (1997) System dynamics in the elevator (SD1163), e-mail communication, 24 Oct 1997. system-dynamics@world.std.com

Books and Reviews

Bloodgood JM, Hornsby JS, Burkemper AC, Sarooghi H (2015) A system dynamics perspective of corporate entrepreneurship. Small Bus Econ 45(2):383–402

Carlsson B (ed) (2012) Industrial dynamics: technological, organizational, and structural changes in industries and firms, vol 10. Springer Science & Business Media, Boston/London

Edmondson A, Moingeon B (1998) From organizational learning to the learning organization. Manag Learn 29(1):5–20

Gephart MA, Marsick VJ (2016a) Leveraging system dynamics for strategic learning. In: Strategic organizational learning. Springer, Berlin/Heidelberg, pp 139–161

Gephart MA, Marsick VJ (2016b) Strategic organizational learning. Springer, Berlin/Heidelberg

Grösser SN, Zeier R (2012) Systemic management for intelligent organizations: concepts, model-based approach, and applications. Springer, Berlin

Hall RI (1976) A system pathology of an organization: the rise and fall of the old Saturday evening post. Adm Sci Q 21(2):185–211

Jiang H (2012) System dynamics application in organizational learning evaluation, 62nd IIE annual conference and expo 2012, Orlando

Kim D (1989) Learning laboratories: designing a reflective learning environment. In: Milling PM, Zahn EOK (eds) Computer-based management of complex systems: international system dynamics conference. Springer, Berlin

Kim DH (1990) Toward learning organizations: integrating total quality control and systems thinking. (Working paper no. D-4036). System Dynamics Group, Sloan School of Management, MIT, Cambridge

Lane DC, Rouwette EA, Vennix JA (2006) System dynamics in organizational consultation: modelling for intervening in organizations. Syst Res Behav Sci 23(4):443–449

Martins H (2017) Strategic organizational learning–using system dynamics for innovation and sustained performance. Learn Organ 24(3):198

Maula M (2006) Organizations as learning systems: 'living composition' as an enabling infrastructure, vol 4. Emerald Group Publishing, Burlington

Morecroft JDW (1988) System dynamics and microworlds for policymakers. Eur J Oper Res 35(3):301–320

Morecroft JDW, Sterman JD (eds) (1992) Modelling for learning. Eur J Operat Res Special Issue 59(1)

Morecroft JD, Sterman JD (2000) Modeling for learning organizations. Productivity Press, New York

Rajbhoj A, Saxena K (2016) Early experience with system dynamics modeling for organizational decision making. In: ModSym+ SAAAS@ ISEC, 2nd AMS Modelling Symposium, Goa, India, pp 4–9

Richmond B (1990) Systems thinking: a critical set of critical thinking skills for the 90's and beyond. In: Andersen DF, Richardson GP, Sterman JD (eds) International system dynamics conference, Massachusetts

Rodrigues LLR (2008) Managing the learning dynamics within the projects based organization- A novel approach through modelling and simulation. In: Proceedings of the 2nd Asia international conference on modelling and simulation, AMS 2008, Kuala, Lumpur

Rouwette EA, Vennix JA (2006) System dynamics and organizational interventions. Syst Res Behav Sci 23(4):451–466

Schwaninger M (2008) Intelligent organizations: powerful models for systemic management. Springer Science & Business Media, Berlin

Senge PM (1990a) Catalyzing systems thinking within organizations. In: Masarik F (ed) Advances in organization development. Abex, Norwood

Senge PM (1990b) The fifth discipline: the art and practice of the learning organization. Doubleday Currency, New York

Senge PM (1992) Systems thinking and organizational learning: acting locally and thinking globally in the organization of the future. Eur J Oper Res 59(1):137

Snabe B (2007) The usage of system dynamics in organizational interventions: a participative modeling approach supporting change management efforts. Springer Science & Business Media. DUV, Wiesbaden

Spector JM (2006) How can organizational learning be modeled and measured. Eval Program Plann 29(1):63

Stabler SG, Ewaldt JW (1998) Simulation modeling and analysis of complex learning processes in organizations. Account Manag Inf Technol 8(4):255–263

Stacey R (2011) Strategic management and organisational dynamics: the challenge of complexity, 6th edn. F T Press, London

Stacey R, Mowles C (2016) Strategic management and organisational dynamics : the challenge of complexity to ways of thinking about organisations, 7th edn. Pearson Education, London

Stata R, Almond P (1989) Organizational learning: the key to management innovation. The training and development sourcebook, 2:31–42

Tucker JS (2005) Dynamic systems and organizational decision-making processes in nonprofit. J Appl Behav Sci 41(4):482

Xu XG (2012) An innovative approach of the mechanism of organizational synergetic learning in emerging economies. Probl Perspect Manag 10(3)

Yang MM, Young S, Li SJ, Huang YY (2017) Using system dynamics to investigate how belief systems influence the process of organizational change. Syst Res Behav Sci 34(1):94–108

Zaini RM (2017) Organizational dissent dynamics: a conceptual framework. Manag Commun Q 31(2):258

Zhang SS (2008) Organizational innovation based on systems thinking: learning organization. In: International conference on wireless communications, networking and mobile computing, Wicom, Dalian

System Dynamics and Workforce Planning

Siôn Cave[1] and Graham Willis[2]
[1]Decision Analysis Services Limited, Basingstoke, UK
[2]Robust Futures Limited, Winchester, UK

Article Outline

Glossary

Feedback loop A closed loop of causality that acts to counterbalance or reinforce prior change in a system state.

Flow A rate of change variable affecting a stock, such as trainees flowing into a workforce or attrition of trainees flowing out.

Full-Time Equivalents (FTE) A measure of the workforce size, taking into account the number of hours worked per person.

Headcount (HC) A measure of the workforce size. The number of people in employment, irrespective of their hours worked.

Stock An accumulation or state variable, such as the size of the workforce.

System Dynamics A modelling approach that enables complex systems to be better understood and their behavior over time to be projected using computer simulation.

Workforce Demand The workforce required by an organization, whether in terms of workforce size (headcount and FTE), skills, composition, etc.

Workforce Planning The practice of determining the demand that will be placed on the workforce of an enterprise at some time in the future, in terms of required effort, and hence the number, skills, and proportion of people required. Following this, determining how those demands will be met through developing workforce or human resources (HR) plans.

Workforce Supply The workforce delivered by the organization to meet the workface demand.

Introduction

This paper describes the complexities associated with strategic workforce planning and how system dynamics has been applied in support of the workforce planning process.

Workforce planning is the practice of determining the demand that will be placed on the workforce of an enterprise at some time in the future, in terms of required effort, and determining how this demand can best be met through developing workforce plans. For the purposes of this paper, an enterprise could be an organization in the public, private, or third sector or a national workforce capability.

System dynamics is a modelling approach that enables complex systems to be better understood and their behavior over time to be projected using computer simulation. The method has been applied numerous times across different sectors to support workforce planning.

This entry has been written with two audiences in mind:

1. People with an interest in workforce planning who seek more information about the system

Originally published in
R. A. Meyers (ed.), *Encyclopedia of Complexity and Systems Science*, © Springer Science+Business
Media LLC 2019, https://doi.org/10.1007/978-3-642-27737-5_659-1

dynamics approach with respect to workforce planning and details about how it has been applied within the workforce planning process

2. System dynamics practitioners with the aim of better understanding the complexities associated with workforce planning and where it has been supported by system dynamics

This paper is organized as follows. The next section defines what we mean by workforce planning and explains the complexities that need to be overcome. This is followed by a description of a general strategic workforce planning process and where systems approaches can be used to better understand and manage the associated complexity and uncertainty. A brief description of the system dynamics approach is then provided with respect to supporting strategic workforce planning.

A review of the many applications of system dynamics to workforce planning is then presented. This includes a sector-based review of applications from the literature and a detailed case study where system dynamics was used to support national workforce planning in the English health and social care system.

Finally, future directions for applying system dynamics for workforce planning are discussed.

The Complexities of Strategic Workforce Planning

What Is Workforce Planning?

Strategic workforce planning is concerned with aligning the workforce of an organization with its strategic goals and priorities. It has two parts. First, determining the demand that will be placed on the workforce at some time in the future, in terms of required effort, and hence the number, skills, and proportion of people required. Second, determining how these demands will be met through workforce planning. The planning should be a continuous process, closely linked to the wider business planning process.

There are many different definitions of workforce planning. Taylor (2005, p.78) provides the well-known definition of right people, with the right skills, in the right places at the right time.

Taylor outlines the stages in a planning cycle, starting with forecasting future demand, then internal and external supply, and then formulating responses to the forecasts. This suggests that workforce planning is an exact science and that the future can be accurately forecast. This may not be possible or indeed sensible in times of uncertainty.

In contrast, the HR Society (2013) defines workforce planning as being about aligning the workforce to business requirements and discusses the importance of using scenario planning to test the impact of alternative futures.

RAND (Emmerichs et al. 2004) has developed an approach that any organization can use for workforce planning focused on four thematic areas:

(a) The critical workforce characteristics needed in the future to deliver the strategic goals and priorities
(b) The characteristics of today's workforce
(c) The characteristics that the future workforce will have if current plans unfold as expected
(d) The changes in plans needed to bring the projected future workforce closer to the desired workforce

An important distinction is made between workforce characteristics (e.g., measurable workforce parameters such as numbers, age, and skills) and how these characteristics are distributed between different jobs, roles, locations, or other groups within the organization.

The RAND approach introduces the idea of where an organization's workforce will be if current plans unfold as expected, often called the "base case," which might be very different from the actual future requirements. This is the gap that workforce planning must close or reduce. However, the uncertainty of the future means that different possibilities need to be considered. This should be done within the strategic workforce planning process.

It is important to note that the pressures faced by an enterprise will vary depending on the nature of their business and whether they are in the private or public sectors. Private sector enterprises typically face competition and develop their business strategy and plans accordingly. Public sector enterprises, for example, the health and education sectors, often face stricter regulation and legislation, and professional staff may be required to meet a defined and assessed standards of competence. Furthermore, they must operate and deliver services as required by government policies, targets, and fiscal constraints.

Understanding Supply and Demand

Whatever the nature of the workforce, workforce planning is about trying to achieve a balance – or at least reduce the imbalance – between workforce supply and workforce demand. This is the same whether it is numbers of people, their skills and competences, or other workforce characteristics that are needed to achieve the strategic goals of an organization.

While supply and demand may appear to be simple concepts, measuring and modelling them are often quite difficult. Both depend on having good quality data.

Workforce demand is simply the workforce required by the business. Workforce supply is the quantity available. In simple economic terms (Tulchinsky and Varavikova 2014, pp. 578–581), demand depends on the quantity wanted to produce "goods" – which in turn depends on the "price" of the goods. So, if the price is too high, demand will fall. For most enterprises, these terms can easily be translated and applied.

In the public sector, supply and demand are not the same as in a free market, and they are not independent of each other. Regulations and economics may restrict the amount of supply. Demand may be limited by the cost and the ease of access to services. If services are largely free or made easier to access, then extra demand may be induced.

To add to the complexity, demand is not the same as the actual use (utilization) of services. The need for healthcare as seen by a prospective consumer of services may not be the same as seen by a medical professional. Demand is where the consumer perceives a need and decides to act on it. Utilization is the intersection of this demand with supply. A critical question in healthcare is thus how to measure need and who determines what is necessary.

Once supply and demand have been defined, workforce planning depends on being able to quantify the amounts today and to project these measures into the future.

Projecting supply requires understanding of workforce inflows and outflows and delays in the system. For example, it takes time to recruit and hire staff and in some sectors to educate and train staff. Not all staff pass interviews, complete their training, or even join the workforce if they are eligible. Within the workforce, staff become sick, take career breaks from which some may not return, leave for other jobs, and retire.

Projecting workforce demand requires a clear understanding of what is meant by demand – as has been previously discussed. The difficulty here is that it requires an understanding of how the future might evolve and whether the workforce of the future will behave in the same way as today, for example, their working patterns and productivity. There is also the matter of organizational structure and whether the workforce will be performing the different roles in different ways to today.

Finally, there is the process of allocating the organization's workforce to the demand, assigning and sharing tasks between different workforce groups.

Achieving Balance

One of the frequently stated requirements for effective workforce planning is to achieve a balance or match between workforce supply and demand now and into the future. Perfect balance (supply exactly meeting demand) is near impossible to achieve over time due to:

- **Imperfect information** – supply and demand projections may not be accurate, so that the planned vs. actual numbers are very different.
- **Delays** – there may be a delay in noticing a gap between supply and demand. Even if a gap is noticed immediately, supply will often lag due to the time to recruit and/or train workers.
- **Linkage between supply and demand** – if there is an excess of supply, for example, more staff in customer-facing roles, the resulting reduction in service time may cause

demand to increase above what might otherwise have been expected.

Zurn et al. (2004) discuss the concept of imbalance and that effective workforce planning should try to reduce the imbalance between supply and demand. In looking at the health workforce, they describe a skill imbalance (a shortage or a surplus) as being the result of continuous fluctuations between labor market supplies and demand for occupational skills, so there is a disequilibrium. A key question then is how long an imbalance will last, and whether it is temporary or permanent.

A dynamic imbalance will generally resolve over time in a competitive labor market. A static imbalance occurs when supply does not change, or changes only slowly, so that market equilibrium is not achieved. This can be because of the long time to train new workers, which is a problem in some sectors. For example, it can take over 15 years to train a hospital doctor. Imperfect information or slow changes to wages can also lead to static imbalance.

The paper (ibid) also differentiates between qualitative and quantitative imbalance. For example, where labor is scarce, enterprises may not find candidates with the right skills but nonetheless must still recruit. Using vacancy rate as an indicator will not show a problem as vacancies are being filled. Nevertheless, from the employer's perspective, there is a skills shortage. Sousa et al. (2013) provide a useful framework for analysis of labor market dynamics.

While the future cannot be forecast with any degree of certainty, the workforce of tomorrow will surely need very different skills to that of today, such as creativity and flexibility, as well as other emerging but unknown skills.

Planning and Uncertainty

There is always a level of uncertainty associated with forecasting how the future will unfold, which means it is extremely difficult to create the "perfect" workforce plan that will deliver the desired results whatever happens. Plans need to be flexible enough to cope with this uncertainty, and trade-offs may need to be made.

One approach is to consider how workforce strategies and plans might work across different – and challenging – futures. This allows decisions to be made about which would work best against this uncertainty. A robust decision is not one that necessarily has the best performance in a single future but performs best across all the different future states. It does require thinking about the plausibility of the different futures made trade-offs. A term sometimes used instead of "robust" is "least regret," i.e., the decision that we would regret least whatever the future unfolds. For further information on the use of scenarios and robust decision making, see Lempert et al. (2003).

The Workforce Planning Process

In today's uncertain world, it is necessary to develop a workforce plan that takes account of the wider business system and environment. This naturally leads to thinking about the "system" that the workforce organization is operating in, the scope and boundaries of this system, the system components, and how they interact. In addition, the consideration of uncertainty and different futures is important for developing robust workforce plans. The different steps generally adopted in a workforce planning process are described below. This brings together the main stages reported by other authors (ibid), together with ideas from systems and futures thinking. The stages are as follows:

1. Define the problem.
2. Capture the current workforce state.
3. Understand the system.
4. Define the required future workforce.
5. Define the expected future workforce.
6. Identify challenging futures.
7. Stress test workforce plans.
8. Decide, implement, and revise.

As illustrated in Fig. 1, the stages are not necessarily followed in a strict sequence, and there is iteration between them. The following sections describe each stage in more detail.

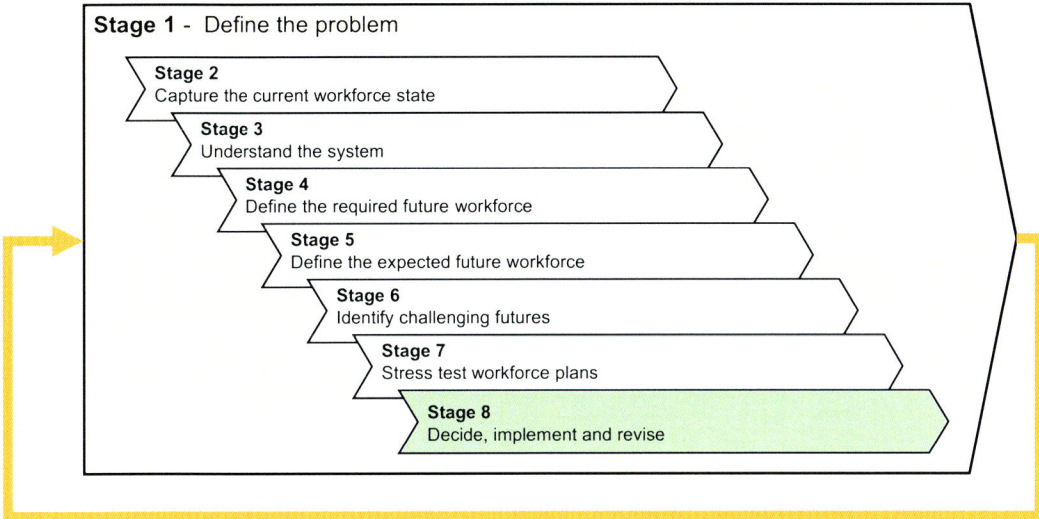

System Dynamics and Workforce Planning, Fig. 1 A generic approach to strategic workforce planning

Stage 1: Define the Problem

The starting point is to define the purpose of the strategic workforce planning activity. This requires a clear statement of the goals, the timescale for the planning exercise, how far to look to the future, and any specific problems that need to be addressed. It is helpful to describe not only what is in scope but what is specifically excluded and any constraints to the planning such as implementation costs. It is important that the planning process is collaborative and that the right stakeholders are involved. The stakeholders include those whose collective judgment will inform the process; those who will implement the workforce plans; those who have the capacity to influence the outcomes, either supporting or opposing; and those who will evaluate the success. It is important to consider stakeholders who are often marginalized or forgotten. The plans may need to include the distribution of the workforce across multiple locations or cover mobile working. The frequency for updating and reviewing the plans should also be decided. For example, staffing rotas may need to be revised monthly, but a whole organization workforce plan may need to be reviewed annually or at longer intervals.

Defining the problem helps to provide early identification of key issues, data, and people. It is important that the rationale for planning is clear,

so that time and effort is not wasted in trying to solve the wrong problem in the wrong way with the wrong people.

Stage 2: Capture the Current Workforce State

The starting point for all planning is to understand the situation today, and how it has been reached, in terms of historic plans and goals. Trends over time may be important to see if key metrics are getting better or worse.

The workforce characteristics of interest depend on the nature of the workforce planning exercise but can include the following:

- **Composition** – Key metrics for the composition of the workforce include the headcount, full-time equivalents, skills and competences, age distribution and average age, gender, ethnicity, tasks and roles, and their distribution across the organization, including geography.
- **Employment and benefits** – This includes the employment type (full- or part-time, temporary, contract, etc.), salary, and other benefits.
- **Productivity and performance** – Productivity metrics concern service outputs and quality measures, for example, production, sales or service measures, and performance standards. Engagement and culture are important for workforce

productivity and consistency, including employee feedback and job satisfaction.

- **Supply and demand** – The availability of workforce supply, which may be from internal or external sources, versus the current demand for the workforce. This may be for specific workforce jobs, roles, or skills.

There are wider workforce planning dimensions which may need to be considered for planning including the labor market, rate of unemployment, job vacancies, and customer satisfaction.

Stage 3: Understand the System

All organizations operate within a wider business system, which include partners, suppliers, customers, and competitors. Strategic workforce planning must work within this complex environment.

Most system where workforce planning could be applied are complex, involving people, technologies, and resources, with many activities, feedback processes, and time delays. They may behave in unexpected ways. To avoid problems, it is necessary to understand the forces and factors that influence the system and their cause and effect relationships. As understanding is gained, it may be necessary to revise the definition of the problem (Stage 1) or collect more information (Stage 2).

Systems methods may be applied to build understanding. System maps can be used to capture the relationships between factors, feedbacks, and time delays. Understanding these factors provide insight into system behavior and what plans are likely to be most effective. It is helpful to involve stakeholders in building system maps. This helps people to visualize the system under investigation, challenge their assumptions, and build understanding. These maps serve as a "common language" between people of different backgrounds.

System mapping may be done using causal loop diagrams to visualize the relationships between factors (Meadows 2008) or stock and flow diagrams (Sterman 2000) that provide a different visual representation. Analysis of system maps can identify the forces and factors that are driving future change.

It is helpful when thinking about the system to build a list of potential planning options. At this stage the aim is to capture as many as possible and not to exclude unconventional approaches. Brainstorming sessions can help to generate a range of ideas.

Stage 4: Define the Required Future Workforce

Given the current state of the workforce and the wider workforce system, thought needs to be given to the workforce that will be required in the future to meet the goals and objectives. This necessitates consideration of the strategic direction for the organization. The characteristics of the future workforce may vary greatly from that of today, not simply in changing numbers and proportions, but characteristics that do not currently exist such as new skills and roles. External influences on work and the workforce will also need to be addressed. Shifting demographic patterns may change the age, gender, or ethnicity of the population and the available workforce. Technological change will almost certainly continue, but the impact on many businesses may be highly uncertain. The global economy, oil, and other resources may all have an impact.

Thinking about the future, workforce composition is difficult but critical for strategic workforce planning. This definition should be revised as knowledge is gained in subsequent stages.

Stage 5: Define the Expected Future Workforce

The expected future workforce should not be confused with a forecast for the future. It is the workforce that is most likely if current and future policies and plans unfold as expected, and social, economic, political and technological trends continue. Changes occur that are reasonably certain to happen, and there are no sudden or discontinuous changes or surprises. If a plan is expected to work or fail, then this should be included. The expected future workforce is not a forecast or a prediction.

There are two parts to defining the expected future workforce: firstly, considering what the future is likely to look like, in terms of the organization, the business environment, and the workforce requirements, and secondly, deciding the future workforce characteristics needed to meet these requirements.

Stage 6: Identify Challenging Futures

The process of identifying challenging but plausible futures surfaces issues, promotes debate on creative solutions to the problem, and provides insight into which workforce plans and actions are likely to be the most effective.

There are many paths that the future might take, but they cannot be forecast with any certainty. By imagining different futures, stakeholders are challenged to think in creative ways and question their current viewpoint, so that knowledge is built on the problem and how best to resolve it.

A common method to think about the future is to engage with stakeholders and help them to create challenging but plausible narrative scenarios by considering the key forces and factors of change, as identified in the previous stage. This is a common approach to Scenario Generation (Wright and Cairns 2011). Each scenario represents an extreme future that is consistent but is highly unlikely to happen. But they can be used, in combination with modelling and simulation of the future workforce, to stress test workforce plans, starting from the list of possible options developed in Stage 2.

Stage 7: Stress Test Workforce Plans

There are three parts to stress testing. The first is the development of a simulation model built following sound software engineering principles. System dynamics is a good method for developing models that can represent the underlying cause and effect dynamics and will be described in greater detail later along with how it has been applied in support of workforce planning. System dynamics is a modelling approach that enables complex systems to be better understood and their behavior over time to be projected using computer simulation.

The second part is to quantify critical parameters needed in the model that vary according to how the future may unfold. For example, the economic situation may impact workforce supply and demand,

and working time may change as staff seek a better work-life balance. These will be different in each scenario. Although these parameters cannot be forecast, methods such as expert elicitation can be used to quantify them, together with their uncertainty and the reason behind this (Lamba et al. 2015).

The third and final part is to use the model to simulate these different futures and quantify the impact of planning options toward achieving the desired goal.

Stage 8: Decide, Implement, and Revise

The final stage is to decide which workforce options to implement to achieve the required future workforce. This will depend not only on the performance of the different options from Stage 7 but other considerations such as cost, risk and organizational culture, and alignment to existing business practices.

The essential technique to use here is option appraisal, whereby options are reviewed by analyzing their costs and effect (HM Treasury 2013) against each one of the future scenarios – in other words analyzing the outputs from the model. Option appraisal includes non-cost metrics such as robustness, uncertainty, and acceptability.

Once the workforce plan is implemented, the outcomes must be assessed to see if the desired effects are being achieved. If not, it may need to be revised which may mean revisiting some or all the previous stages. This reflection and reassessment can be considered as a part of an enterprises' ongoing consideration of strategic workforce planning.

System Dynamics to Support Workforce Planning

As discussed previously, workforce planning is complex, demanding an understanding of the wider system that an organization operates in the internal and external forces and factors driving change, and the different futures that may arise. Specific systemic properties of a workforce system that should be considered include:

- The delays inherent in developing a workforce, for example, times to train or develop skills and skills fade
- The attractiveness of the workforce to new recruits
- Loss rates from training pipelines and career stages
- The complex endogenous and exogenous interactions that can drive demand
- Interactions between workforce supply and workforce demand

System dynamics has been successfully applied to addressing these issues in strategic workforce planning. It can be used to gain a better systemic understanding and in turn to quantify the outcomes and impacts associated with different workforce planning options. For example, the Centre for Workforce Intelligence made use of system dynamics to support robust workforce planning of the English health and social care system (Willis et al. 2018).

System dynamics was first developed in the 1960s by Jay Forrester (1961), where he explicitly mentioned the applicability of the approach to workforce planning. Forrester discussed the dynamic

effects that result due to an organization's workforce policies impacting upon the size of the workforce and workforce utilization. Furthermore, Forrester (1969) applied system dynamics directly to the consideration of a workforce in an urban area in terms of those employed and those underemployed.

Several important texts describing the system dynamics approach have been published since Forrester's early books. For example, see Sterman (2000), Warren (2007), and Morecroft (2015). Richardson (2015) describes the basic elements of system dynamics within this encyclopedia.

The system dynamics approach is iterative in nature and when applied effectively includes extensive stakeholder involvement. Figure 2, adapted from Wolstenholme (2009), illustrates the stages that are often in a system dynamics project. There are commonalities between Wolstenholme's overall approach and the generic workforce planning approach illustrated in Fig. 1. For example, the system dynamics approach can be of use during "Understand the system" (Stage 3), "Define the expected future workforce" (Stage 5), and "Stress test workforce plans" (Stage 7).

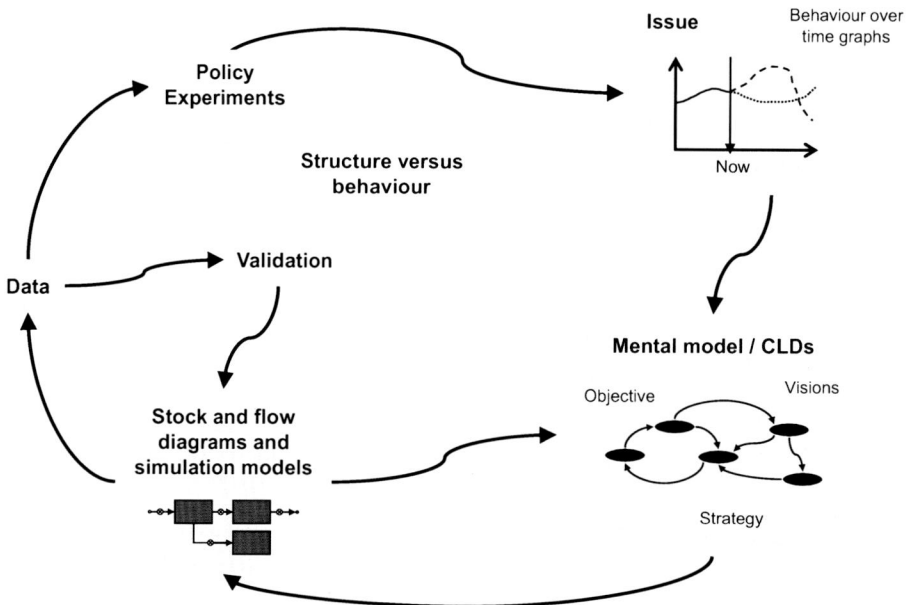

System Dynamics and Workforce Planning, Fig. 2 An overall approach to system dynamics. (Adapted from Wolstenholme 2009)

The system dynamics approach can be considered to have two key parts. The first is mapping the system to better understand it. The second is using computer simulation to calculate system behavior over time. Both parts have shown great utility to workforce planning and are described in more detail below with respect to workforce planning and in particular to workforce supply. We have focused on workforce supply as the issues are similar across different workforces and sectors. The demand for workforce and the potential interactions between demand and supply are highly dependent on the workforce that is being analyzed.

Mapping the System to Understand Behavior

The first stage of a system dynamics-based project involves mapping the cause and effect relationships that drive system behavior. System dynamics uses specific diagramming notation such as stock and flow diagrams or causal loop diagrams to map the system.

A causal loop diagram is used to capture major feedback mechanisms. The diagram includes variables and arrows (causal links) linking these variables together.

A stock and flow diagram captures the main stocks in the system and the flows that act to increase and decrease the size of the stocks. Figure 3 illustrates the differences between the diagrams based on the same simplified workforce supply system.

These diagrams should be created with the stakeholders in the system. The completed diagrams represent a shared understanding of the system, which can be used in many ways, for example, to investigate potential points where interventions could be made. Meadows (1999) discusses where best to intervene in a system.

A method of developing qualitative system dynamics models with stakeholders (facilitated modelling) is known as group model building (GMB). GMB actively engages stakeholders in the process of qualitative modelling to achieve a shared understanding of the problems in the system and their solutions (Rouwette and Vennix 2006). A good resource for facilitated model building is Scriptapedia (2017) which is a freely distributed collection of workshop scripts. Dill et al. (2015) give an example of applying workshop scripts within a workforce planning context.

Simulating the System to Quantify Behavior

Once an agreed diagrammatic representation of the system has been created, specialist software can be used to quantify the relationships. The completed simulation model can be used to test system interventions in a risk-free environment. Simulation model provides a means to calculate change over time, depending on the data sources, the underlying assumptions, and the proposed interventions. The models can be developed to produce outputs using desired performance measures and can be validated against historic performance.

Many authors have produced guidance on producing robust system dynamics models; for example, see Sterman (2000), Keating (1999), Randers (1980), and Cave (2014).

Bounding the System Dynamics Model

A critical challenge in any system dynamics study is how much of the system to model. The model boundary will be highly dependent on the requirements of the modelling. One of the advantages of the system dynamics approach is that a broad system view can be adopted. Figure 4 (adapted from Reilly (1996) and updated based on information in the previous sections of this entry) illustrates how broad the model boundary can be when considering workforce planning.

Generally, qualitative system dynamics modelling will consider a much wider boundary than quantitative modelling, which can be limited by data availability.

System Dynamics Tools

There are a variety of tools available for developing system dynamics models. Each can be used to map the system and then produce quantified simulations. There is no consensus on the tool most appropriate for workforce-based models. For example, authors report using among other software tools:

Causal Loop Diagram (CLD)

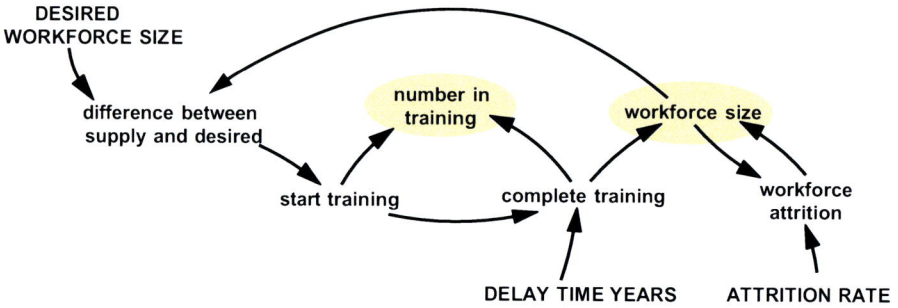

Stock and Flow Diagram (SFD)

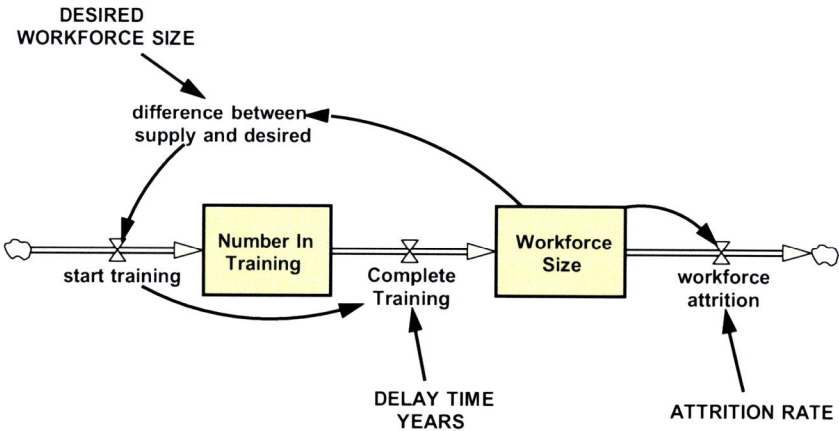

The *number in training* is increased by *start training* and reduced by *complete training*. Trained people then move onto the *workforce size* stock, where they leave as a result of the *attrition rate.* The rate people start training is based on the *desired workforce size* and the *workforce size.*

System Dynamics and Workforce Planning, Fig. 3 Different views on the same system using system dynamics diagramming nomenclature

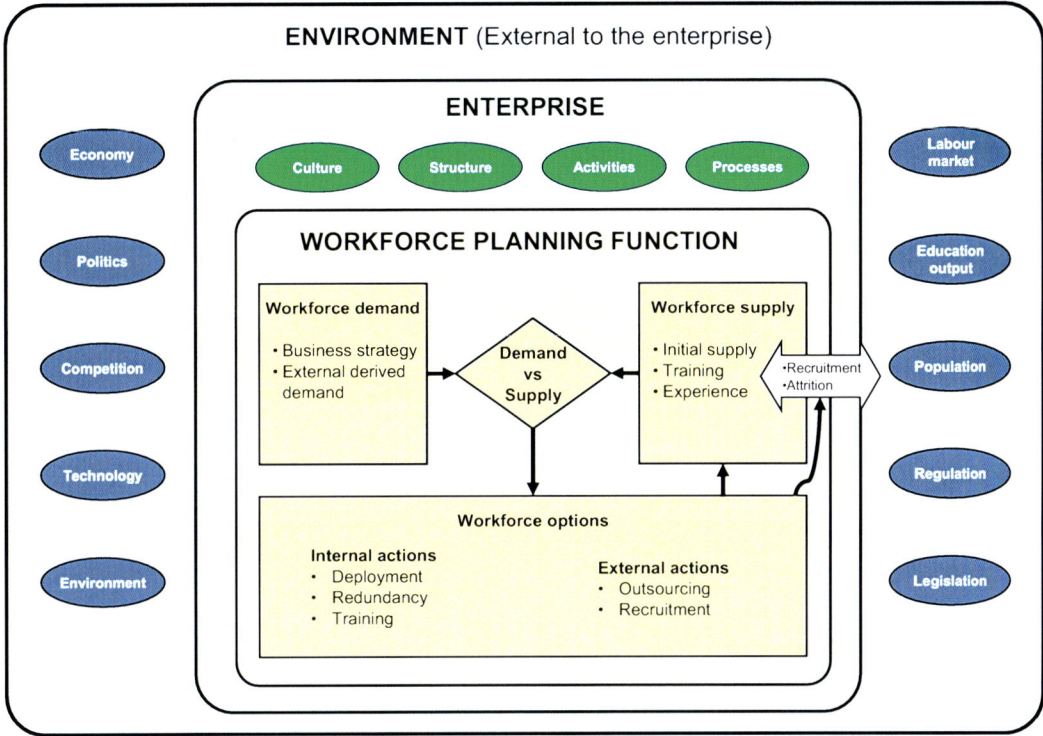

System Dynamics and Workforce Planning, Fig. 4 Workforce planning modelling boundaries

- **Vensim**: Ezzatabadi et al. (2017), Cave et al. (2016), Größler and Zock (2010)
- **Powersim**: McLucas and Lewis (2008), Cave et al. (2011), Barber and López-Valcárcel (2010), Senese et al. (2015)
- **iThink/Stella**: Collofello et al. (1998), Ishikawa et al. (2013), Brailsford and De Silva (2015)
- **AnyLogic**: Marin et al. (2006)

The authors of the references in this entry gave little, if any, justification for the tool that they used. Some authors did describe the benefits of using system dynamics tools over spreadsheets, which are more commonly used for workforce planning (e.g., Masnick 2009a and Birch et al. 2005). It is our opinion, based on knowledge of these tools, that they all have similar levels of functionality, and so a workforce planner could legitimately consider any of them.

Applications of System Dynamics to Workforce Planning

The previous sections have described the complexities associated with strategic workforce planning and the applicability of the system dynamics approach. The following sections give examples where system dynamics has been applied to workforce planning.

The first section provides a number of applications of system dynamics across a range of different sectors, countries, and types of workforce. The review describes the models that were developed and the effect of the system dynamics analysis on organizational workforce policy. The papers' referenced in these reviews are all available in the public domain.

This is followed by a detailed case study of the use of system dynamics to inform national workforce policy for the English health and social care system. The study describes why system dynamics was particularly appropriate in this instance, how system dynamics was integrated into an overarching workforce planning approach, and the overall impact of the analysis.

Sector-Based Review of Applications of System Dynamics to Workforce Planning

This section focuses on those papers that describe where system dynamics has been used to inform workforce policy. In general terms, this is where system dynamics-based models have been used to test different workforce policies and have in turn either been used to make recommendations on the appropriate course of action or provide the workforce planners with a range of potential scenarios to plan against. Included within this review are papers that describe system dynamics models, which, although they may not have been used directly to define workforce policy, have been created with the aim of providing an exploratory learning environment, so that the workforce planners and policy makers better understand their workforce system.

This review has identified a significant number of papers across a range of sectors. However, the literature is dominated by examples from the healthcare sector. This is not to say that system dynamics is not being extensively applied in other sectors, but there is not as much evidence within the public domain. Even so, system dynamics is still not widely adopted even in the healthcare sector. For example, a review of workforce models carried out as part of a European Union-funded research project (Malgieri et al. 2015) found that out of seven countries reviewed, system dynamics was only being used to inform policy in two countries.

The scope of the models described in the reviewed papers varies significantly. The models range from qualitative models used to help managers better understand workforce issues, single-stock quantitative models to understand issues associated with an aging workforce, to highly complex whole-system models covering supply and demand. Several of the papers describe the bullwhip effect, with oscillating shortfall and surpluses over time, which has been used to justify a more detailed analysis using system dynamics.

In addition, the extent of endogeneity varies significantly between the models. For example, some of the models contain explicit linkages between workforce demand and supply, whereas others treat the management of supply against exogenous demand as a management policy decision to be tested.

It is interesting to note that there is much commonality between the structures used to model workforce supply. For example, there will typically be stocks representing the training pipeline. In contrast, the calculation of workforce demand varies significantly depending on the specific workforce activity that drives demand.

The papers described in this review are categorized by workforce sector to help workforce planners identify relevant approaches.

Health and Social Care

As described in the case study, the health and social care workforce has attributes that make modelling their supply complicated, for example, the long training pipelines for medical specialists and the wide range of specialist skills that they apply. On the demand side, there are concerns that workforce planners especially need to consider (Office of Health Economics 2015):

- **Demographic shift:** Aging and growing populations meaning more work needs to be done.
- **Changing technology:** Which means that more can be done (although sometimes technology creates more work as new services become available).
- **Increasing expectations:** The population is demanding more care.

The health and social care models need to take these demand factors into account and often include epidemiological based models for the conditions associated with the workforce of interest. Some of the key papers are summarized below.

Birch et al. (2005) is a comprehensive report describing a regional workforce planning study for four Canadian provinces (Nova Scotia, Newfoundland and Labrador, Prince Edward Island, and New Brunswick) commissioned by the Atlantic Health Human Resources Association. The goal of study was to carry out a detailed analysis of the regional requirements for health professionals in these areas along with the educational/training programs. A specific objective of the project was to develop a system dynamics-based model that would allow the region to simulate gaps in the supply of, and need for, the workforce groups and to then test the effectiveness of policy initiatives in dealing with gaps prior to the full implementation of the policy intervention. A diverse range of workforce groups were investigated, including audiologists, family physicians, health records professionals, licensed practical nurses, medical laboratory technologists, medical radiation technologists, physiotherapists, registered nurses, respiratory therapists, and speech-language pathologists. The system dynamics model was developed in Vensim and produced 40-year projections. Population demography, need, and workforce productivity were included in the demand side of the model. The training pipeline and an age-segmented workforce were included in the supply side. A series of system-based and workforce specific recommendations were made because of the work.

Barber and López-Valcárcel (2010) describe system dynamics modelling used to represent the supply and demand/need for 43 medical specialties in Spain. The purpose of the model was to simulate the consequences of different policies aimed at improving the capacity of the Spanish health system. The Delphi method was used to inform projected demand for specialists per capita. The model identified potential specialist shortfalls which were projected to get worse and has been used to inform medical school intake and specialist training numbers.

Ishikawa et al. (2013) considered the total supply of physicians and the obstetrician and gynecologist specialists in Japan. The work was carried out because of shortages of physicians in Japan and the desire of the authors for workforce

planners to consider future dynamic changes in physician numbers. The system dynamics model focused on the workforce supply, with the training pipeline starting at medical school, and simulated from 2008 to 2030. Workforce demand was based on a desired number of physicians per capita. The model projected growth in the workforce in the baseline case over the simulation period; however, there was a projected shortfall until 2026. The authors proposed a variety of methods based on model insights for bridging the gap.

Dill et al. (2015) also investigate physician demand and supply, in this instance with an emphasis of the modelling for a more local area. The authors describe their experiences in applying the approach at a pilot site in the United States city of Cleveland (Ohio), with especial emphasis on their experiences with group model building. Although the quantitative model described in the paper was still under development, the authors provided key structures and initial results and stated their plan to continue the development and roll out the methodology and model to other locations.

Ansah et al. (2015) describe a projection model developed to represent the demand for and supply of ophthalmologists in Singapore. The work was funded by the Singapore Ministry of Health's National Medical Research. The model had a time frame from 2010 to 2040. The model had three core components which considered the prevalence of different eye diseases by age and ethnicity, the demand module which included time spent in care and an ophthalmologist supply module which represented the training pipeline. The model was used to analyze a set of plausible scenarios and found that in all cases Singapore's aging population would lead to a significant increase in public sector eye care demand and therefore an increased requirement for ophthalmologists. The model provided policy makers with an overview of the levers available to manage this issue. Ansah et al. (2017) then built on this work through comparing alternative system dynamics representations for modelling ophthalmologist demand (workforce-to-population, needs-based, utilization-based, and an integrated approach). This included testing the sensitivity of the projections.

In an extended abstract, Ezzatabadi et al. (2017) describe the work carried out to model the projected gap between demand and supply of neurosurgeons in Iran out to 2020. The demand model was stated to be based upon epidemiological, demographic, and utilization parameters. The supply model included surgeon attrition, migration, and retirements. The work identified potential regional shortfalls.

Taba et al. (2015) carried out an analysis of the Australian radiologist workforce with the intent of assessing whether new policies of increasing the number of radiologists were a sustainable long-term solution for their demand and supply imbalances. Their model compared the demand and supply of Australian radiologists over 40 years between 2010 and 2050 and suggested that current policies may not meet the projected supply gap.

Vanderby et al. (2014) describe a model developed to simulate a single-specialty medical workforce for Canada with the intent of providing a tool to assist future resource planning. The authors present a sample analysis for cardiac surgeons over a 22-year period from 2008. As might be expected, demand is driven by demographics, and the supply side includes workforce aging effects. Interestingly, the model includes feedback loops that capture the effects workforce shortages have on productivity and unemployed graduates have on program enrollment. The authors stated that the model was an effective tool for communicating the workforce system and possible future scenarios to both surgeons and students while also providing insights that are valuable to those making decisions regarding training programs.

Masnick (2009a, b) presents a model that allows the simulation of the training pipeline for a workforce with multiple entry and multiple exit points and then applies the model to the eye care workforce (e.g., eye care nurses and optometrists) in Thailand over a 20-year period. The model was used to support the development of a flexible competencies-based multiple entry and exit training system that matches and adapts training to the prevailing population and service needs and

demands. This scheme was commented on favorably by local experts.

System dynamics has also been used to assess the impact that specific exogenously defined policies could have on a workforce. For example, Ratnarajah (2004) examined the possible impact of a new European Working Time Directive on the UK's doctor workforce. The intention of this policy was to limit the number of hours worked each week to 48. The model examined the impact not only on workforce size (e.g., through attrition and in-service training) but also softer attributes such as moral and work-life balance. The model focused on supply-side dynamics, projected out 15 years, and identified the potential for a decline in patient care should the implications of the policy not be addressed.

Masnick and McDonnell (2010) present an endogenously focused qualitative model of healthcare dynamics. The purpose of the model is to enable workforce planners to better understand their real-world operating parameters. The structural map of a health system was intended to show the linkages between the tasks performed by a health workforce and the different types of personnel that could supply those tasks. The maps were based on three linked components: the population to be served, the clinical workforce to serve it, and the workload generated by both the population and the clinical workforce. Although qualitative in nature, the model does provide many of the building blocks for creating a generic whole system quantitative model focused on healthcare workforce planning, which the authors intend to develop.

Kephart et al. (2014) applied system dynamics to a long-term Canadian strategy to ensure an adequate future supply of appropriately skilled nurses over a 10-year period across the Canadian provinces. Canada has historically had cyclical patterns of nursing shortages and oversupply. The analysis focused on three regulated nursing professions (registered nurses, licensed/registered nurses, and registered psychiatric nurses). The model was focused on the nursing supply, with an emphasis on workforce age distributions, recruitment, and retention. Workforce age was a significant issue, with anticipated high levels of future retirements. The authors acknowledged that future modelling could incorporate demand-side structures, covering aspects such as future population, health, and other work environment matters. However, this was not required for this study. The modelling provided valuable insights into the reasons for the then crises and gave recommendations into how they could be resolved in the short and long term.

Senese et al. (2015) combined system dynamics with mixed-integer programming to support decision-making around the number of residency grants to be financed by the national Italian government and the number and mix of supplementary grants to be funded by the regional budget. The system dynamics model projected the evolution of the supply of medical 43 types of specialist out to 2030. The supply model began with entry to specialty training and included doctors working within the public and private sector. Supply was compared against demand scenarios that incorporate demography, service utilization and hospital beds. The mixed-integer programming (MIP) then assigns medical specialization grants for each year of the projection calculated from the system dynamics model.

National policy decisions regarding the Sri Lankan dental workforce were informed with the system dynamics model described in Brailsford and De Silva (2015). The model represented both dentist workforce supply and the demand for dental care services. The supply side of the model represented career progression from initial entry to dental school through different career paths to retirement and was implemented using system dynamics software. The demand was calculated in a separate spreadsheet and was based on population projections and other activity measures and was an input into the supply model. The model had a 15-year time horizon and was used by the Sri Lankan government to plan better provision of state-funded dental care and the future university intake of dental students. Subsequently, De Silva (2017) carried out a similar analysis of doctors in Sri Lanka projecting demand and supply from 2017 to 2032. Although no immediate need for additional doctor training was identified, it recommended carrying out a system dynamics-based analysis of demand and supply every

5 years to determine the future number of doctors to be trained.

An earlier example of modelling the dental workforce is described by Bronkhorst (1995). This model was created to enable analysis of the Dutch dental system by the Dutch government. The model is comprehensive and covers both workforce demand (based on population, epidemiology, economic, and sociological components) and workforce supply (which included dental hygienists). Workforce supply included the training pipeline and the employment status of the dental workforces. The model is documented in detail and was validated against historic data and subsequently used for scenario analysis. Surprisingly, this model was constructed in Pascal, rather than specific system dynamics software.

The Centre for Workforce Intelligence (CfWI) adopted a system dynamics-based workforce planning approach for the English health and social care system and carried out strategic studies for workforces such as doctors, dentists, speech and language therapists, etc. The system dynamics models considered workforce demand and supply in detail, including workforce aging and gender. The models typically had a time horizon of 10 years. This work is described in greater detail in the following case study (Willis et al. 2013, 2018; Cave et al. 2016).

Interestingly, although there have been numerous applications of system dynamics for workforce planning in the healthcare sector, there has been comparatively little within social care. Social care workforce provision is generally more complex as compared to healthcare workforces. For example, the great majority of social care workforces undergo a much shorter training period. Onggo (2012) demonstrates how system dynamics could be used to inform workforce planning in the adult social care system based on data from England. Onggo's model had a time horizon of 20 years from 2010. The author considered the workforce to be composed of migrant and local care workers, and the total demand for care was driven by previous demand projections. Workforce supply was influenced by job satisfaction and job attractiveness, which were in turn affected by issues such as working conditions and wage attractiveness. The model was framed as a learning tool, rather than for developing projections for policy analysis. However, the author identified some key feedback loops that would need to be incorporated into a full analysis of the social care workforce.

Another study which included social care was carried out by Health Workforce New Zealand (2011), which considered whole system issues out to 2026. This study was part of a project to review the needs of older people and how services are provided now, propose how services could be met in the future, and consider how this could be achieved. A qualitative model which covered the workforce, population health, and service delivery was used to identify and describe different initiatives. A quantitative model was then used to create plausible projections to estimate future demand and workforce requirements for the different initiatives. The study recommendations included more emphasis on prevention, greater support (e.g., training) for caregivers, and greater workforce integration between health and social care.

Finally, the CfWI used system dynamics models to assess the future demand for workforce skills out to 2035. Their modelling represented the whole health and social care workforce system, including the contribution of informal carers and volunteers. The model was based on the skills required from the workforces rather than simply workforce numbers to enable a more direct comparison between workforce groups (CfWI 2015).

Defense

Military workforces follow a highly structured training and career progression with a pyramid-like military rank chain. This is not dissimilar to the well-defined stages of medical workforce training. As such, supply-side modelling can use similar structures and architectures. However, there are some key differences between medical and defense workforces. For example, there tends to be a single route into the workforce at the lowest rank, and workforce shortfalls cannot be filled with external resources through direct entry. System dynamics has been applied to workforce problems by several defense

organizations including the Italian Armed Forces, the United States Department of Defense, and the Norwegian Ministry of Defence, and Australian Department of Defence.

Armenia et al. (2012) describe how workforce planning is carried out by the Italian Armed Forces with specific reference to the aircrews' service branch (pilots and navigators). A workforce planning model was used to assess the impact of management decisions. Armenia's model considered three components: recruitment planning (hiring policy); training (focused on the initial 2-year training program); and the workforce (rank progression). Workforce demand is an external variable defined in terms of the target workforce, which is based on Italian legislation. The model was validated against 13 years of historic data, and policy analysis was carried out over the period from 2009 to 2020. The modelling explored a series of scenarios to suggest possible strategies to achieve the required workforce size and emphasized how a whole system model would be able to assist in strategy development for the planned reductions across the whole of the Italian armed services.

Although the publications do not contain operational data, Wang (2007) describes a workforce model developed by the Australian Department of Defence. The model itself was supply-side focused and represented four ranks and associated delays and flows between the ranks. The utility of the model was described as particularly useful for policy analysis (e.g., to avoid potential bullwhip/oscillation effects) and recommended expanding the use of system dynamics as a planning tool for the military.

Linard (2002) describes several structures that can be used to represent defense workforce career progression. This includes methods for tracking multiple workforce attributes, for example, within aging chain stock and flow structures, multidimensional arrays on stocks (e.g., a 3-dimensional array structure tracking staff numbers by rank, by time-in-rank, by years of service), co-flows, and using decimal numbers as flags.

The experiences of applying system dynamics to the Norwegian Armed Forces are described by Bakken et al. (2005). In this example, a specific issue of the potential impact of lifelong employment on the workforce composition was explored. The analysis led to a new strategic policy which changed guaranteed lifelong employment. The modelling enabled this policy to be convincingly presented and its implementation success secured, which the authors claim has in turn led to greater force agility.

Information Technology

Information technology, and in particular software development, has long been the subject of system dynamics analysis. For example, in the book by Abdel-Hamid and Madnick (1991), the authors investigated the dynamics of projects requiring industrial-scale software development, with attention to workforce issues such as the number and quality of people working on a project.

More recently, McLucas and Lewis (2008) describe how system dynamics modelling was used to assist a large government organization. The role of the organization was to deliver information and communications technology business services and new capability projects. The modelling was required to assist a better understanding of the required numbers of people with requisite skills and knowledge, now and into the future. The authors describe how the system dynamics approach created learning throughout the development process. For example, system mapping helped to identify leverage points and pressure points in the system. In addition, the quantitative model was developed iteratively, leading to refinements in the model scope and the potential interventions. These insights led to greater confidence in the proposed strategic recommendations.

Collofello et al. (1998) describe a system dynamics model based on the software development process which was constructed to enable different workforce retention policies to be explored in terms of cost and completion time. The specific policies explored included replacing staff as they leave, initial overstaffing, and doing nothing. The model was calibrated using expert judgment and real project data and was used as a learning laboratory to enable the implications of staffing responses to attrition. Similarly, Cave et al. (2011) describe a model that enabled Microsoft to assess the impact on software development schedules based upon team composition and allocation of the team to different tasks. This model was also used as a learning tool, in this case by Microsoft engineers in Seattle, India, and China.

Although not based on any organization or project, An et al. (2007) describe a system dynamics model which represents the acquisition and development of a company's skill pool. The demand side is "project based" with different requirements for skill. The supply side is represented with an available skill pool which can be utilized to work on project demands. The model was framed as an exploratory learning model.

The demand for the information technology workforces is generally defined by a project profile, indicating tasks that need to be carried out by certain dates.

National Infrastructure Workforces

Sing et al. (2016) describe a model to represent the national demand for and supply of the workforce for the national construction and civil engineering industry. The model includes feedback between demand and supply, for example, workforce shortages leading to a reduction in the commencement of construction projects. The model was validated against data for Hong Kong, and the paper explored two potential scenarios, including changes in government policy and public investment.

Finally, a model investigating the supply of the English nuclear sector workforce is being used to help support national policy development (Cave et al. 2017; NSSG 2018). The model was developed in response to a Nuclear Strategy Skills Group requirement to describe the supply of skills to the nuclear industry, in a way that complements an already developed demand-side picture. It allows scenarios to be designed that in turn inform policy decisions on the level and timing of training and recruitment to meet the UK nuclear program. The model, developed in Vensim, represents up to 25 different high-level resource codes at the 8 different role levels.

Other Sectors

NASA Kennedy Space Center has explored applying system dynamics to explore workforce dynamics (Marin et al. 2006). They investigated internal issues such as changing technical programs and external issues such as the availability of suitably skilled labor and the local economic climate. NASA's intent was to combine system dynamics with agent-based modelling to provide support to strategic workforce planning.

Größler and Zock (2010) describe a study carried out in 2008 where a system dynamics aging chain model was used to improve the recruiting and training process for a large unidentified German service provider in the field of logistics. One of the major services the firm provides requires the availability and timely provision of highly skilled operator staff. The company's managers described their long-term planning scheme as suboptimal, primarily because they perceived the staff situation to be characterized by transient but prolonged periods of staff shortages, followed by periods of staff surpluses. The company decided to complement its regular planning process, which consisted of forecasting future workforce demand through spreadsheet analyses, with a modelling approach based on system dynamics. The key findings were that the aging chain of service operators within the company was affected by a variety of delays in, for example, recruiting, training, and promoting employees and that the structure of the planning process generated cyclic phases of workforce surplus and shortage. The authors reported that the company revised its workforce hiring policies in light of the insights generated from the simulation.

Although not used directly for workforce planning, the work carried out by Nanda et al. (2005) demonstrate that system dynamics can be used for projecting education output, which is one of the key drivers for the availability of a suitably qualified labor market. In this instance the authors focused on the Indian agricultural sector and tested variations against current agricultural education policy and trends. Agriculture is a critical component of the Indian economy. There is limited information about the model in the paper, but it does consider the whole workforce, including

unemployment levels. The model considered workforce supply and demand and was validated against data for 1991 to 2000. The model was used to project scenarios from 2001 to 2020.

Kunc (2008) developed a workforce model of a professional services firm in the financial industry. The purpose was to examine the issues associated with maintaining the workforce balance in terms of staff seniority (i.e., between junior, mid-level, and senior staff) without incurring overtime. The model, which had a 24-month projection period, was developed to help the management team define their workforce budget and understand the implications of their workforce decisions. In a similar vein, Sveiby et al. (2002) discuss the various structures that can be used to represent a workforces' "value adding capacity." The work was undertaken for the Australian Federal Public Service by the UNSW Centre for Business and used a professional services organization to illustrate the relevant model structures.

Although no specific organization is addressed, the dynamic structures considered by Mutingi (2012) for the workforce issues associated with the effective planning and management of workforce for new product development (NPD) projects may be of interest. The author develops a model that can be used to investigate workforce issues associated with typical stages of new product development (design, prototyping, and production). The model itself can be considered as a learning tool around the associated dynamic workforce issues. The model would require parameterization and calibration if used with a specific goal in mind.

Case Study: Applying System Dynamics to National Workforce Planning

The case study below provides a detailed description of how the system dynamics approach has been used to inform national policy for health and social care workforce planning in England across a variety of workforce groups and across the whole system. This case is based on material published in Cave et al. (2016) and Willis et al. (2013).

Context

The English health and social care system is large, complex, and serves the needs of a population of 54 million people (Office for National Statistics 2014). The total number of jobs within the health and social care system in 2013 was 3.4 million (Office for National Statistics 2014). Of this, the National Health Service (NHS) in England employs 1.4 million staff (Health and Social Care Information Centre 2014). The health and social care system has a critical impact on the population of England.

As described previously, making improvements to a complex system is difficult. Changes can have unforeseen consequences, take time to happen, and may occur at a location in the system distant from the original intervention. It is inherently difficult to predict how a complex system will change over time, and there may be resistance to change due to the way the system is organized. A system will often involve many people, each with different perspectives on the system, and different ambitions and fears. The English health and social care system exhibits all these properties, and innovative methods are required to improve how it functions.

The Centre for Workforce Intelligence (CfWI) was formed in July 2010, and through to March 2016 was a key contributor to the planning of future workforce requirements for health and social care in England. The Department of Health (DOH), as well as Health Education England (HEE) and Public Health England (PHE), commissioned the CfWI to inform national and local workforce planning and policy decisions. The CfWI led more than 30 major studies over this period, which impacted the millions of people working in health, public health, and social care. System dynamics was adopted by the CfWI as a key modelling methodology and was fully integrated into their approach to strategic workforce planning.

Overall Approach to Workforce Planning

The CfWI developed a framework for carrying out strategic workforce analysis, which was called the Robust Workforce Planning (RWP) framework. The purpose of this framework was to support making robust workforce decisions within a complex system, taking uncertainty into account so that future workforce problems could be evaluated and well-informed choices made. The result was an advanced approach that allowed policy interventions to be tested across a range of plausible and challenging futures. System dynamics modelling was fully integrated into the framework and in turn informed its development.

The framework has four stages:

- **Horizon Scanning** to understand the system and what drives future behavior
- **Scenario Generation** to explore the future and produce challenging scenarios
- **Workforce Modelling** to simulate these futures and quantify what they look like
- **Policy Analysis** to make robust decisions about which solutions are the least vulnerable to uncertainty

Each study would be bounded by a **focal question**, which is a clear statement of the project purpose and scope. This overall approach is consistent with the generic workforce planning approach illustrated in Fig. 1. The focal question is addressed in Stage 1. Horizon Scanning is addressed by Stage 2 (Capture the current workforce state), Stage 3 (Understand the system), and Stage 4 (Define the required future workforce). Scenario Generation is covered by Stage 5 (Define the expected future workforce) and Stage 6 (Identify challenging futures). Workforce Modelling is Stage 7 (Stress test workforce plans). Finally, Policy Analysis is Stage 8 (Decide, implement, and revise).

The CfWI method for developing system dynamics-based workforce models had four stages: model scoping, model construction, model documentation, and model testing (Cave 2014). A key document was the model specification created during the model scoping stage. This helped gain the confidence of stakeholders and permit concerns or questions to be raised by them or the developer before the build began. It would often open doors to previously unknown datasets or information.

As with the stages of the overall framework, the stages of the model development approach overlap and may iterate. The approach is illustrated in Fig. 5, with reference to the RWP framework stages.

The diagram illustrates how the Workforce Modelling stage is influenced by, and influences, all stages of the RWP. For example, the focal question helps to bound the model, the Horizon Scanning stage helps to scope what needs to be represented, the Scenario Generation stage develops the scenarios to be modelled, and the Policy Analysis Stage is where the model is applied. The development approach is consistent with guidance provided by the UK Government on producing quality analysis for government (HM Treasury 2015).

The System Dynamics Models

The system dynamics models developed in support of the CfWI's studies tended to be more focused on the supply side, with each stage of the training pipeline and career represented explicitly, along with possible inflows and outflows from each of the stocks. For example, Fig. 6 gives the high-level stages from the MDSI model used to determine the required intake to medical school to meet future demand.

Each of the workforce stocks were segmented by age and gender. This was to enable issues such as the variation of attrition with age and different working practices by gender to be addressed.

The demand side of the models was generally based on changes in demographics (e.g., changing population numbers and age) and the key activities and conditions that drive demand. Potential changes in need, productivity, and service delivery were also integrated into the demand calculations.

The supply and demand sides of the models would be developed based on discussion with relevant stakeholder groups. In addition, model behavior was explored with the stakeholder groups. The models underwent strict testing protocols prior to use for policy analysis.

The models would be used to project demand, supply, and the potential gap between them under different circumstances, for example:

- The expected future workforce (Stage 5 in the planning process; see Fig. 1).
- A set of plausible and consistent but challenging scenarios (Stage 6; see Fig. 1).
- The impact that different policy interventions have under each of the scenarios (Stage 7; see Fig. 1).

An important part of the analytical process was to have a clear representation of the impact of uncertainty on the model outputs. This was achieved in two ways:

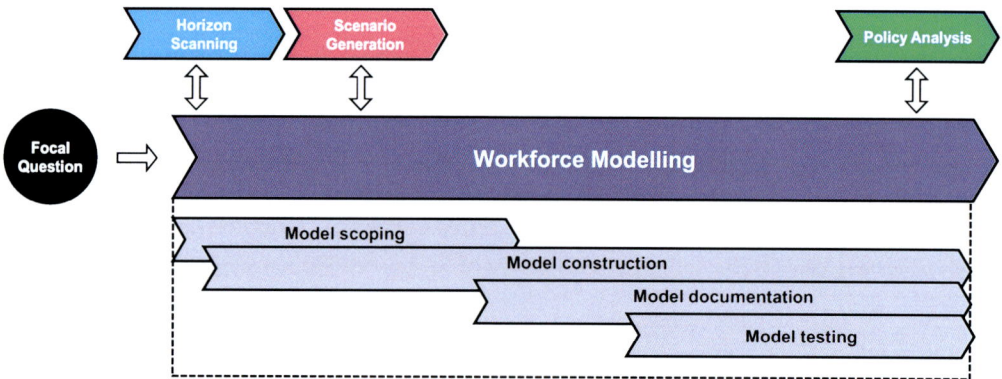

System Dynamics and Workforce Planning, Fig. 5 Integration of system dynamics model development within robust workforce planning framework (Adapted from Cave 2014)

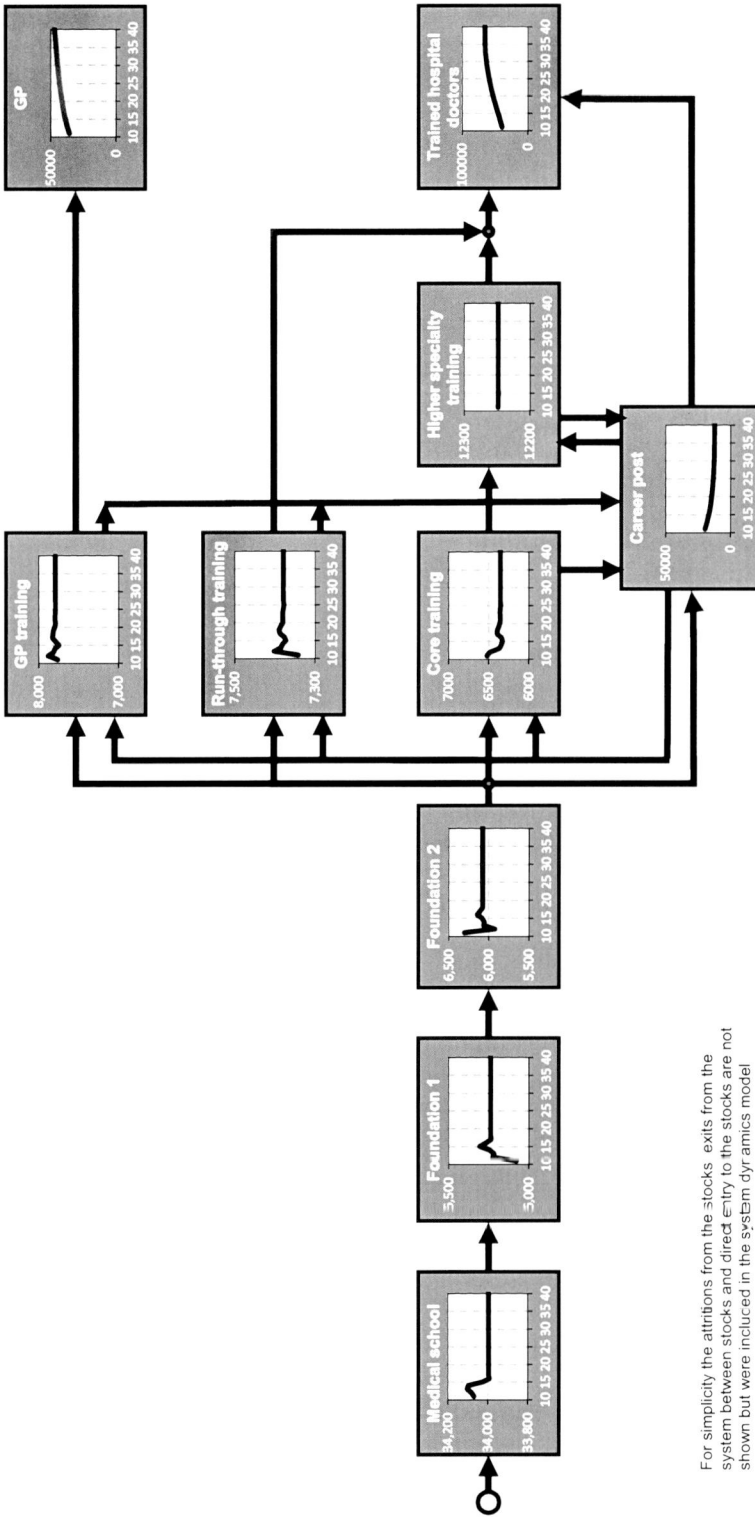

For simplicity the attritions from the stocks exits from the system between stocks and direct entry to the stocks are not shown but were included in the system dynamics model

System Dynamics and Workforce Planning, Fig. 6 Example high-level stock and flow diagram describing the key stocks in the English medical workforce (Willis et al. 2013)

1. Key uncertain input variables were elicited from expert groups through workshops, as probability distributions. The simulations were then executed in "Monte Carlo" mode so the variation of the inputs on the model outputs could be determined. In Monte Carlo analysis, the model is run many times to sample from the probability distribution.
2. Adjustments were made to each of the input data items separately to identify the variables that showed the strongest influence on the key outputs.

The models were developed in Vensim DSS, with a Microsoft Excel data repository to hold the input data and the results of simulation runs. This architecture enabled the large amounts of data generated to be effectively managed.

Impact

Figure 7 gives an approximate timeline for selected system dynamics-based workforce studies carried out by the CfWI. It does not represent all the studies since a number were not system dynamics-based. The timeline represents the duration for the complete project, i.e., all stages of the RWP framework, rather than just the period where system dynamics modelling took place.

In each of these workforce studies, the commissioners were confident to use the results of the modelling to support their decision-making process, in part due to the transparent process in developing the models and their involvement throughout all stages.

The impacts of the system dynamics modelling included:

- Workforce commissions for many of the professions and specialties in the English health and social care system were influenced by CfWI modelling. The CfWI's work has been cited by HEE in their commissioning and investment plans, and empirical influence is present in many of the changes to trainee numbers (Health Education England 2014, 2015, 2016).
- Stakeholders such as Royal Colleges became more supportive and trusting of the CfWI's

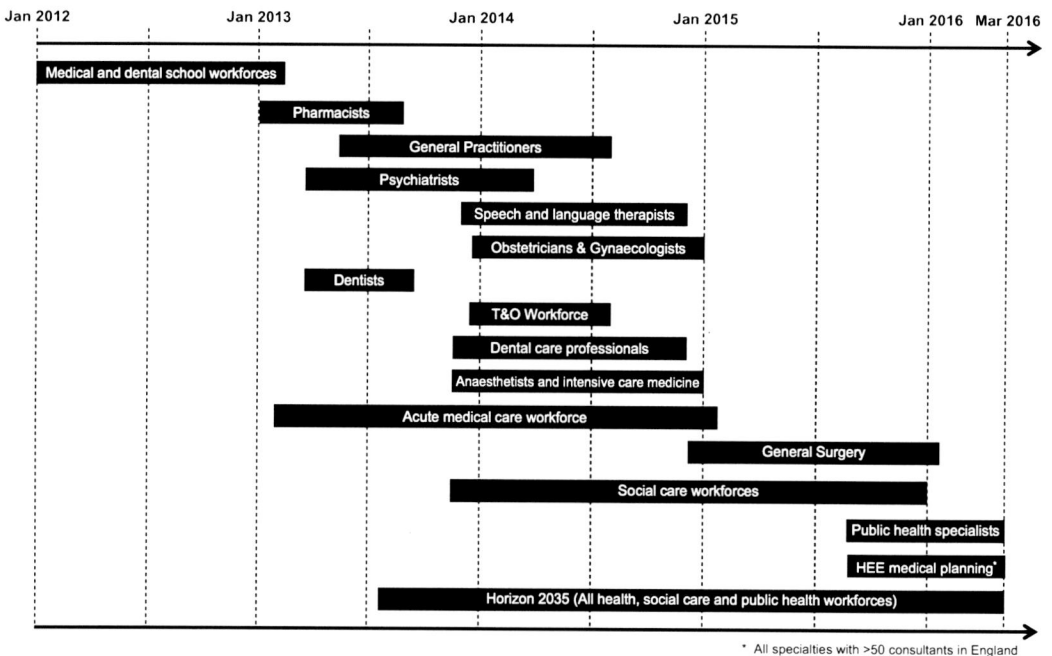

System Dynamics and Workforce Planning, Fig. 7 Approximate timeline for the application of system dynamics for workforce studies at the CfWI (Adapted from Cave et al. (2016))

work. The Royal Colleges are responsible for development of and training in one or more medical specialties. This improved level of trust was attributed to the more transparent approaches adopted by the CfWI, partly by using graphical models.

- The value of modelling has been highlighted by the CfWI embedding it in a framework that helps decision-makers. The realization that the models are only as good as the data they use has been understood. During the lifetime of the CfWI, there has been more standardization of workforce data (HEE are developing data and standards) and expansion in the scope of data collection, notably for the primary care workforce. This shows a confidence in evidence-based workforce planning.
- The profile of workforce planning and the standardization of models allowed more project time to be given to addressing the core problems. Previously, custom Microsoft Excel models had to be created for each project.
- The published work of the CfWI has an international dimension, garnering interest from European member states and other international health economies. For example, the CfWI worked with the DH to share workforce planning capabilities with the European Union Joint Action on Health Workforce Planning and Forecasting network.

Future Directions for System Dynamics in Workforce Planning

There are several areas where the application of system dynamics could develop in the future with respect to workforce planning. The first, and perhaps the most significant, is simply in terms of the uptake of the method. This paper has identified many benefits in using system dynamics for supporting strategic workforce planning. However, evidence in the public domain suggests that its use is not as widespread as might be expected. The area where there has been wider adoption is the health and social care system, but even here its applications are limited (Malgieri et al. 2015). Reasons for low adoption of system dynamics

and other systems methods have been identified by Atkinson et al. (2015) to include a lack of locally available expertise in system dynamics, unfamiliarity among policy makers of the benefits it offers and the difficulty of establishing confidence of the technique among stakeholders. This could be resolved through more effective communications around what system dynamics is and the benefits it brings.

Beyond greater uptake, there are some technical areas where future development of the approach could deliver greater benefits. For example:

- **Generic system dynamics models:** As noted in this paper, workforce planners tend to think in terms of supply-side and demand-side models. Demand-side models tend to be quite specific to the workforce being considered, driven by the specific workforce activities. Supply-side models, although structured by the naming conventions of the workforce being studied, tend to contain quite similar structures, for example, training pipelines, aging stocks, and skill development co-flows. As such there is scope for developing generic supply-side models that could be easily configured to meet the requirements of the workforce planner. This could build directly upon the various structures described within the literature and authors who have described the use of model molecules/components (Hines 2005; Cave et al. 2016).
- **Integration into enterprise data architecture:** For a model to be used by a workforce planner to make strategic decisions, it needs to use the most up-to-date data, be readily accessible, and user-friendly. There is scope for integrating system dynamics-based workforce planning models within an organizations' information architecture to make the decision-making process as efficient as possible.
- **Web-based exploratory models:** This paper has identified several projects where the purpose of the modelling was to aid workforce planners to understand the dynamic consequences of workforce planning decisions through learning environments. There is

scope to provide more general system dynamics-based learning environments, for example, to highlight the impact of systemic delays and feedback processes.

- **Explicit uncertainty:** There is greater scope for a more explicit representation of uncertainty within workforce models. The papers tend to examine a limited set of alternative policy options and/or alternative futures. This consideration could be expanded, for example, through the methods described by Lempert et al. (2003) and Kwakkel and Pruyt (2013) and greater application of the Monte Carlo functionality embedded within the system dynamics tools.
- **Simulation method integration:** There is a growing literature base describing how different simulation methods can be integrated (Morgan et al. 2016, 2017; Brailsford et al. 2010). Supply side models need to be initialized with the starting stock of employees. Good operational data would enable population of this data at an entity level. Therefore, there is great scope here for creating a hybrid agent based/discrete event simulation/system dynamics model combining the benefits of each approach.

Bibliography

Primary Literature

Abdel-Hamid T, Madnick SE (1991) Software project dynamics: an integrated approach. Prentice-Hall, Englewood Cliffs

An L, Jeng JJ, Lee YM, Ren C (2007) Effective workforce lifecycle management via system dynamics modeling and simulation. Proceeding of the 2007 Winter Simulation Conference, Washington, DC Published IEEE, Piscataway, NJ, 2187–2195

Ansah JP, De Korne 1 D, Bayer S, Pan C, Jayabaskar T, Matchar DB, Lew N, Phua A, Koh V, Lamoureux E, Quek D (2015) Future requirements for and supply of ophthalmologists for an aging population in Singapore. Hum Resour Health 13(86):1–13

Ansah JP, Koh V, De Korne D, Bayer S, Pan C, Thiyagarajan J, Matchar DB, Quek D (2017) Comparing health workforce forecasting approaches for healthcare planning: the case for ophthalmologists. Int J Healthcare 3(1):84–96

Armenia S, Centra A, Cesarotti V, De Angelis A, Retrosi C (2012) Military workforce dynamics and planning in

the Italian airforce. In: Proceedings of the 30th international conference of the system dynamics society, St. Gallen, Switzerland, 22–26 July 2012

Atkinson JM, Wells R, Page A, Dominello A, Haines M, Wilson A (2015) Applications of system dynamics modelling to support health policy. Public Health Res Pract 25(3):e2531531

Bakken BE, Østby PR, Røksund A (2005) Transforming a military personnel policy –learning from a model supported intervention. In: Proceedings of the 23rd international conference of the system dynamics society, Boston, 17–21 July 2005

Barber P, López-Valcárcel BG (2010) Forecasting the need for medical specialists in Spain: application of a system dynamics model. Hum Resour Health 8:24

Birch S, Kephart G, O'Brien-Pallas L, Tomblin Murphy G (2005) Atlantic health human resources planning study. Med-Emerg Inc, Mississauga

Brailsford S, De Silva D (2015) How many dentists does Sri Lanka need? Modelling to inform policy decisions. J Oper Res Soc 66:1566–1577

Brailsford SC, Desai SM, Viana J (2010) Towards the holy grail: combining system dynamics and discrete event simulation in healthcare. In: Proceedings of the 2010 winter simulation conference (WSC), Baltimore, 5–8 Dec 2010

Bronkhorst EM (1995) Modelling the Dutch dental health care system – a comprehensive system dynamics approach. Dissertation, Radboud University, Nijmegen

Cave S (2014) Developing robust system-dynamics-based workforce models: a best-practice approach. CfWI technical paper series no. 0008. Centre for Workforce Intelligence. http://webarchive.nationalarchives.gov.uk/20161007101116/http://www.cfwi.org.uk/publications/developing-robust-system-dynamics-based-workforce-models-a-best-practice-guide. Accessed 05 Apr 2017

Cave S, Gliniecki M, Johnson S, Nemesszeghy G (2011) Application of system dynamics modelling in support of Microsoft's automation strategy. In: Proceedings of the 29th international conference of the system dynamics society, Washington, DC, 24–28 July 2011

Cave S, Willis G, Woodward A (2016) A retrospective of system dynamics based workforce modelling at the Centre for Workforce Intelligence. In: Proceedings of the 34th international conference of the system dynamics society, Delft, 17–21 July 2016

Cave S, Bennett S, Pleasant R, Woodham E (2017) Assessing the future workforce supply for the UK Nuclear sector. Unpublished report. Decision Analysis Service Ltd, Basingstoke

CfWI (2015) Horizon 2035 – future demand for skill: initial results. Centre for Workforce Intelligence. http://webarchive.nationalarchives.gov.uk/20161007101116/http://www.cfwi.org.uk/publications/horizon-2035-future-demand-for-skills-initial-results. Accessed 05 Apr 2017

Collofello J, Houston D, Rus I, Chauhan A, Sycamore DM, Smith-Daniels D (1998) A system dynamics software process simulator for staffing policies decision support. Proc Hawaii Int Conf Syst Sci 6(3):103–111

De Silva D (2017) How many doctors should we train for Sri Lanka? System dynamics modelling of training needs. Ceylon Med J 62:233–237

Dill M, Hirsch G, Yunker E (2015) Local area physician workforce planning model pilot. In: Proceedings of the 33rd international conference of the system dynamics society, Cambridge, 19–23 July 2015

Emmerichs RM, Marcum CY, Robbert AA (2004) An executive perspective on workforce planning. Santa Monica: RAND Corporation, http://www.rand.org/pubs/monograph_reports/MR1684z2.html. Accessed 05 Apr 2017

European Food Safety Authority (2014) Guidance on expert knowledge elicitation in food and feed safety risk assessment. EFSA J 12(6):3734

Ezzatabadi MR, Zadeh SA, Rafiei S (2017) Forecasting the shortage of neurosurgeons in Iran using a system dynamics model approach. BMJ Open 7:A1–A78

Forrester JW (1961) Industrial dynamics. The MIT Press, Cambridge

Forrester JW (1969) Urban dynamics. The MIT Press, Cambridge

Größler A, Zock A (2010) Supporting long-term workforce planning with a dynamic aging chain model: a case study from the service industry. Hum Resour Manag 49(5):829–848

Health and Social Care Information Centre (2014) NHS hospital and community health service (HCHS) workforce statistics in England, Summary of Staff in the NHS 2003–2013. http://www.hscic.gov.uk/catalogue/PUB13724.

Health Education England (2014) Workforce plan for England: Proposed education and training commissions for 2014/15. https://www.hee.nhs.uk/sites/default/files/documents/Workforce-plan%202014-15.pdf. Accessed 05 Apr 2017

Health Education England (2015) Education and training plan 2016–2017: commissions and training numbers. https://www.hee.nhs.uk/sites/default/files/documents/HEWCommissioningTrainingPlanSept2015.pdf. Accessed 05 Apr 2017

Health Education England (2016) HEE commissioning and investment plan for England 2016/17. https://www.hee.nhs.uk/printpdf/our-work/planning-commissioning/workforce-planning/commissioning-investment-plan-england-2016-17. Accessed 05 Apr 2017

Health Workforce New Zealand (2011) Workforce for the care of older people. 3rd Draft. Health Workforce New Zealand. https://www.health.govt.nz/system/files/documents/pages/care-of-older-people-phase1-report.pdf. Accessed 04 Apr 2017

Hines J (2005) Molecules of structure: building blocks for system dynamics models, V2.02. http://www.systemswiki.org/images/a/a8/Molecule.pdf. Accessed 05 Apr 2017

HR Society (2013) What is workforce planning. http://hrsociety.co.uk/resources/knowledge_resources/Workforce-planning-chapter-one_draft_02.pdf. Accessed 03 Apr 2017

Ishikawa T, Hisateru Ohba H, Yokooka Y, Nakamura K, Ogasawara K (2013) Forecasting the absolute and relative shortage of physicians in Japan using a system dynamics model approach. Hum Resour Health 11:41

Keating EK (1999) Issues to consider while developing a system dynamics model. http://blog.metasd.com/wp-content/uploads/2010/03/SDModelCritique.pdf. Accessed 04 Apr 2017

Kephart G, Maaten S, O'Brien-Pallas L, Tomblin Murphy G, Milburn B (2014) Building the future: an integrated strategy for nursing human resources in canada – simulation analysis report. The Nursing Sector Study Corporation, Ottawa

Kunc M (2008) Achieving a balanced organizational structure in professional services firms: some lessons from a modeling project. Syst Dyn Rev 24(2):119–143

Kwakkel JH, Pruyt E (2013) Exploratory modelling and analysis, an approach for model-based foresight under deep uncertainty. Technol Forecast Soc Chang 80(3):419–431

Lamba S, Cave S, Willis G (2015) Elicitation methods: updated approaches to elicitation. CfWI technical paper series no. 0014. Centre for Workforce Intelligence. http://webarchive.nationalarchives.gov.uk/20161007101116/http://www.cfwi.org.uk/publications/elicitation-methods-updated-approaches-to-elicitation. Accessed 05 Apr 2017

Lempert RJ, Popper SW, Bankes SC (2003) Shaping the next one hundred years: new methods for quantitative, long-term policy analysis. RAND, Santa Monica

Linard K (2002) System dynamics modelling: HR planning & maintenance of corporate knowledge. Australian Defence Force Academy, University of New South Wales

Malgieri M, Michelutti P, Van Hoegaerden M (eds) (2015) Handbook on health workforce planning methodologies across EU countries. Release 1. Deliverable D052. Ministry of Health of the Slovak Republic, Bratislava

Masnick K (2009a) Narrowing the gap between eye care needs and service provision: the service-training nexus. Hum Resour Health 7:35

Masnick K (2009b) Narrowing the gap between eye care needs and service provision: a model to dynamically regulate the flow of personnel through a multiple entry and exit training programme. Hum Resour Health 7:42

Masnick M, McDonnell G (2010) A model linking clinical workforce skill mix planning to health and health care dynamics. Hum Resour Health 8:11

McLucas A, Lewis E (2008) A multi-methodology approach to addressing ICT skill shortages in a government organisation: integration of system dynamics and risk management. In: Proceedings of the 26th international system dynamics society conference, Athens, 20–24 July 2008

Meadows DH (1999) Leverage points: places to intervene in a system. Sustainability Institute. http://donellameadows.org/wp-content/userfiles/Leverage_Points.pdf. Accessed 05 May 2019

Meadows DH (2008) Thinking in systems: a primer. Earthscan, London

Morecroft J (2015) Strategic modelling and business dynamics: a feedback systems approach, 2nd edn. Wiley, Cichester

Morgan J, Belton V, Howick S (2016) Lessons from mixing OR methods in practice: using DES and SD to explore a radiotherapy treatment planning process. Health Syst 5(3):166–177

Morgan J, Belton V, Howick S (2017) A toolkit of designs for mixing discrete event simulation and system dynamics. Eur J Oper Res 257(3):907–918

Mutingi M (2012) Dynamic simulation for effective workforce management in new product development. Manag Sci Lett 2(7):2571–2580

Nanda SK, Rama Rao D, Vizayakumar K (2005) Human resource development for agricultural sector in India: a dynamic analysis. In: Proceedings of the 23rd international conference of the system dynamics society, Boston, 17–21 July 2005

NSSG (Nuclear Skills Strategy Group) (2018) Skills planning to drive sector productivity. Strateg Plan Update Winter 2018. https://www.nssguk.com/media/1472/nssg-strategic-plan-update-2018.pdf. Accessed 2 Apr 2019

Office for National Statistics (2014) Annual mid-year population estimates, 2013. http://ons.gov.uk/ons/dcp171778_367167.pdf. Accessed 22 Mar 2016

Office of Health Economics (2015) Seminar briefing 17: improving the fiscal and political sustainability of health systems through integrated population needs-based planning

Onggo S (2012) Adult social care workforce analysis in England: a system dynamics approach. Int J Syst Dyn Appl 1(4):1–20

Randers J (1980) Guidelines for model conceptualization. In: Randers J (ed) Elements of the system dynamics method. Productivity Press, Portland

Ratnarajah M (2004) How might the European Union working time directive, designed to limit doctors' hours, contribute to junior doctor attrition from the British National Health Service and can desirable outcomes be achieved within these constraints? Executive MBA management report: London Business School. Unpublished – Cited in Morecroft, 2015, chapter 9, medical workforce dynamics and patient care

Reilly P (1996) Human resource planning: an introduction. IES Rep 312:33

Richardson GP (2015) System dynamics, the basic elements of, encyclopaedia of complexity and systems science. Springer Science+Business Media, New York

Rouwette E, Vennix JAM (2006) System dynamics and organizational interventions. Syst Res Behav Sci 23(4):451–466

Scriptapedia (2017) https://en.wikibooks.org/wiki/Scriptapedia. Accessed 05 Apr 2017

Senese F, Tubertini P, Mazzocchetti A, Lodi A, Ruozi C, Grilli R (2015) Forecasting future needs and optimal allocation of medical residency positions: the Emilia-Romagna region case study. Hum Resour Health 13:7

Sing MCP, Love PED, Edwards DJ, Liu J (2016) Dynamic modeling of workforce planning for infrastructure projects. J Manag Eng 32(6):1–12

Sousa A, Scheffler RM, Nyoni J, Boerma T (2013) A comprehensive health labour market framework for universal health coverage. Bull World Health Organ 91(11):892–894

Sterman JD (2000) Business dynamics. McGraw-Hill Higher Education, New York, NY, USA

Sveiby KE, Linard K, Dvorsky L (2002) Building a knowledge-based strategy: a system dynamics model for allocating value adding capacity. http://www.sveiby.com/articles/sdmodelkstrategy.pdf. Accessed 04 Apr 2017

Taba ST, Atkinson SR, Lewis S, Chung KS, Hossain L (2015) A systems life cycle approach to managing the radiology profession: an Australian perspective. Aust Health Rev 39(2):228–239

Taylor S (2005) People resourcing, 3rd edn. Chartered Institute of Personnel and Development, London

HM Treasury (2013) The Green Book: appraisal and evaluation in central government. https://www.gov.uk/government/publications/the-green-book-appraisal-and-evaluation-in-central-governent. Accessed 05 Apr 2017

HM Treasury (2015) The Aqua Book: guidance on producing quality analysis for government. https://www.gov.uk/government/uploads/system/uploads/attachment_data/file/416478/aqua_book_final_web.pdf. Accessed 05 Apr 2017

Tulchinsky TH, Varavikova EA (2014) The new public health, 3rd edn. Elsevier Academic Press, San Diego, CA

Van Hoegaerden Marin M, Zhu Y, Meade PT, Sargent M, Warren J (2006) System dynamics and agent-based simulations for workforce climate. In: Proceedings of the 2006 winter simulation conference (WSC), Monterey, 3–6 Dec 2006

Vanderby SA, Carter MW, Latham T, Feindel C (2014) Modelling the future of the Canadian cardiac surgery workforce using system dynamics. J Oper Res Soc 65(9):1325–1335

Wang J (2007) A system dynamics simulation model for a four-rank military workforce. Land operations division. DSTO-TR-1688. DSTO Defence Science and Technology Organisation. Department of Defence. Australian government

Warren K (2007) Strategic management dynamics. Wiley, Chichester

Willis, GJ, Woodward, A, Cave S (2013) Robust workforce planning for the English medical workforce. In: Proceedings of the 31st International conference of the system dynamics society, Cambridge, 21–25 July 2013

Willis GJ, Cave S, Kunc M (2018) Strategic workforce planning in healthcare: a multi-methodology approach. Eur J Operl Res vol 267(1):250–263

Wolstenholme E (2009) Health care in the United Kingdom and Europe, system dynamics applications to Encyclopaedia of complexity science

Wright G, Cairns G (2011) Scenario thinking: practical approaches to the future. Palgrave, Lon...

Zurn P, Dal Poz MR, Stilwell B, Adams ... (.04) Imbalance in the health workforce. Hum Res... Health 2:13

Books and Reviews

Brailsford S, Churilov L, Dangerfield BC (eds) (2014) Discrete-event simulation and system dynamics for management decision making. Wiley, Chichester

Coyle RG (1996) System dynamics modelling: a practical approach. Chapman and Hall, London

Kunc M (2015) Modelling supply, demand and need: a literature review. CfWI technical paper series no. 0013. Centre for Workforce Intelligence. http://webarchive. nationalarchives.gov.uk/20161007101116/http://www.

cfwi.org ... /publications/technical-paper-modelling-supply-d... ...-and-need-a-literature-review. Accessed 05 Apr

Pidd M (2... ...ools for thinking: modelling in management s... 3rd edn. Wiley, Chichester

Schwenke... ...ulf T (eds) (2013) Scenario-based strategic pla...ng: developing strategies in an uncertain world. Springer Gabler, Wiesbaden

Tomblin Murphy G, Birch S, MacKenzie BS, Elliot Rose A (2016) A synthesis of recent analyses of human resources for health requirements and labour market dynamics in high-income countries. Hum Resour Health 14:59

Vester F (2012) The art of interconnected thinking, 2nd edn. MCB Publishing, München

System Dynamics Analysis of the Diffusion of Innovations

Peter M. Milling[1] and Frank H. Maier[2]
[1]Industrieseminar der Universität Mannheim, Mannheim University, Mannheim, Germany
[2]Geschäftsführender Gesellschafter, Simcon GmbH, Bensheim, Germany

Article Outline

Glossary

Adopters The cumulated number of persons who have bought a product over time.

Diffusion The spread of a new product, process, or concept in the market. The process of bringing innovation into wide use.

Imitator An imitator buys a new product because he observed or communicated with customers who have already bought the product. The buying decision of imitators is influenced by the adoption of other customers.

Innovation The process of bringing new technology into use.

Innovator A customer with general interest in innovations making his buying decision independent of others.

Installed base Installed base is defined as the amount of users in a network system.

Invention The process of bringing new technology into being.

Network effects A product is characterized by a network effect, if the utility of that product is a function of the installed base. The utility increases with the installed base.

Definition of the Subject

The chapter describes how system dynamics-based models can contribute to the understanding and improved management of the diffusion of innovations . It emphasizes the importance of an integrated feedback-oriented view of the different stages of innovation processes. The aim is to generate insight in the complexity and the dynamics of innovation processes. Based on the classical Bass model of innovation diffusion, the system dynamics perspective is introduced. In a systematic approach several structures to model the complexity and dynamics of managerial decision-making in the context of the diffusion of innovation are described and analyzed. Aspects covered consider market structure, network externalities, dynamic pricing, manufacturing related decisions, and the link between research and development and the diffusion of a new product in the market place. The chapter concludes with managerial implications.

Introduction

Continuous activities to renew a company's range of products are crucial for the survival in a competitive environment. However, to improve the competitive position or the competitive advantage, ongoing innovation activity through the development, test, and introduction of new products is necessary. At least since the 1970s, it could be observed that new and technically more complex and sophisticated products have to be developed in a shorter time span. Resources have to be allocated to research and development (R&D) projects that are expected to be economically successful. New products have to be introduced to global

Originally published in
R. A. Meyers (ed.), *Encyclopedia of Complexity and Systems Science*, © Springer Science+Business
Media LLC 2017, https://doi.org/10.1007/978-3-642-27737-5_124-2

markets with severe competition. Decisions about the adequate time to market and appropriate pricing, advertising, and quality strategies have to be made. The complexity and difficulties to manage innovation activities partly derive from the comprehensiveness of the innovation processes. To be competitive, companies have to be successful in all stages of the innovation process, i.e., the process of invention, innovation, and diffusion. This becomes obvious when new product failure rates and innovation costs are analyzed. Figure 1 illustrates the cascading process of innovation activity and the related innovation costs.

For one successful new product in the market place, 64 promising ideas must be channeled through the process of invention and innovation. The cost at each stage of the invention and innovation process increases from a $1000 to $5 million per attempt. Not only is failure more expensive in later stages – which requires an effective management to reduce the failure rates – successful new products have to earn all necessary resources for the whole process. This requires the following: (1) to manage R&D projects and processes effectively and efficiently – including thorough and educated assessment of the economic potential of a new product – to reduce failure rates in later stages and (2) to increase management attention in the final stages since failures in late stages of the process are much more expensive.

Models of innovation diffusion can support the complex and highly dynamic tasks. The chapter will briefly examine how system dynamics-based analysis of innovation diffusion can contribute to the understanding of the structures and forces driving the processes of innovation and diffusion. It will show how system dynamics models can support the decision-making and how they can

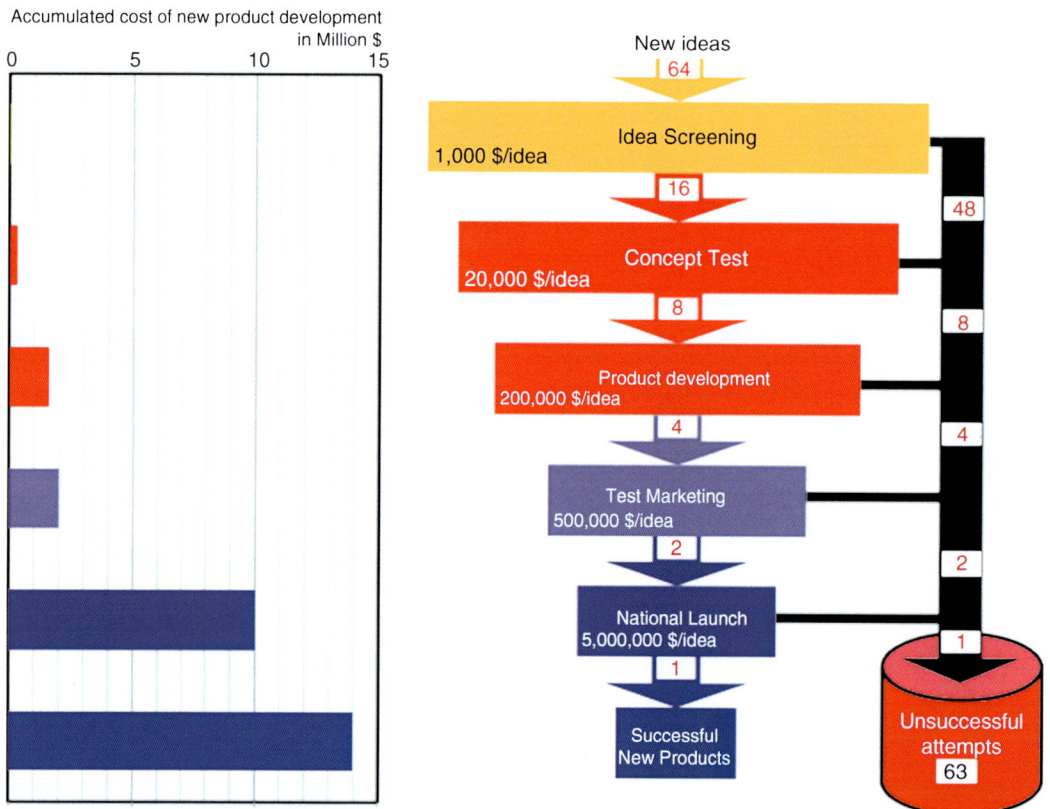

System Dynamics Analysis of the Diffusion of Innovations, Fig. 1 Outcome of activities along the process of invention and innovation

help to reduce failures in the later stages of innovation activities.

Principle Structures to Model the Diffusion of Innovations

Traditional Innovation Diffusion Models from a System Dynamics Perspective

In literature discusses plenty of models about the diffusion of innovations. Many models are based on Frank M. Bass' model of innovation diffusion. In this model, product purchases result from two distinct forms of buying behavior, i.e., innovative purchases and imitative purchases. According to the original Bass model, innovative purchases of a period can be calculated as a fraction of the remaining market potential $(N - X_{t-1})$ with N being the market potential and $X_{t-1} = \sum_{\tau=0}^{t-1} S_\tau$ representing the accumulation of all past purchases of the product S_τ until period $t - 1$.

According to this, innovative purchases S_t^{inno} can be calculated as

$$S_t^{\text{inno}} = \alpha \cdot \left(N - \sum_{\tau=0}^{t-1} S_\tau \right) \qquad (1)$$

where α represents the coefficient of innovation. In the original model, this coefficient is a constant essentially representing the fraction of innovators of the remaining market potential at any point of time. Imitative purchases,

however, are influenced by the number of purchases in the past. Potential adopters of an innovation make their purchasing decision depending on the spread of the product in the market place. The more customers have adopted the product in the past, the higher is the social pressure to purchase the product as well. Imitative demand of a period S_t^{imit} hence can be calculated as

$$S_t^{\text{imit}} = \beta \cdot \frac{\sum_{\tau=0}^{t-1} S_\tau}{N} \cdot \left(N - \sum_{\tau=0}^{t-1} S_\tau \right) \qquad (2)$$

with β representing the coefficient of imitation – a probability that a purchase takes place by someone who observed the use of a product. Together, the total purchases in a period S_t^{total} equal $S_t^{\text{inno}} + S_t^{\text{imit}}$ and hence are calculated as

$$
\begin{aligned}
S_t^{\text{total}} = S_t^{\text{inno}} + S_t^{\text{imit}} = & \; \alpha \cdot \left(N - \sum_{\tau=0}^{t-1} S_\tau \right) \\
& + \beta \cdot \frac{\sum_{\tau=0}^{t-1} S_\tau}{N} \cdot \left(N - \sum_{\tau=0}^{t-1} S_\tau \right).
\end{aligned} \qquad (3)
$$

Innovative and imitative purchases together create the typical product life cycle behavior of the diffusion of an innovation in the market place as shown in Fig. 2.

The model above is a simple mathematical representation of the product life cycle concept, a key

System Dynamics Analysis of the Diffusion of Innovations, Fig. 2 Product life cycle behavior generated by the Bass model

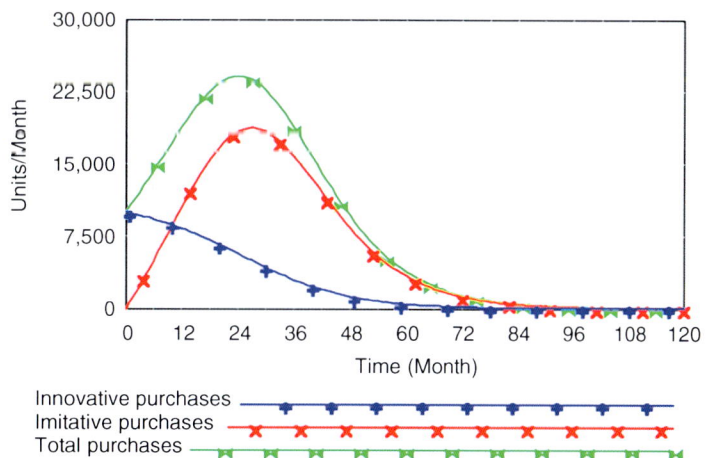

framework in business management. It describes the time pattern a product follows through subsequent stages of introduction, growth, maturity, and decline. Because of its mathematical simplicity and its ability to represent the diffusion of an innovation, the Bass model has been used for parameter estimation and therefore serves as a base for projections of future sales. Although the concept is a powerful heuristic, many models generating this typical behavior do not consider e.g., actual economic environment, competition, capital investment, cost and price effects. Innovation diffusion models, which do not comprise the relevant decision variables, exhibit a significant lack of policy content. They do not explain how structure conditions behavior. They cannot indicate how actions of a firm can promote but also impede innovation diffusion. For an improved understanding of innovation dynamics generated by feedback structures that include managerial decision variables or economic conditions, the system dynamics approach is highly suitable.

Equations (1), (2), and (3) can easily be transformed into the system dynamics terminology. $(N - X_{t-1})$ represents the stock of the remaining market potential at any point in time and X_{t-1} represents the accumulation of all product purchases over time. The sales of a period S_t^{total} are the flows connecting these two stocks as shown in the Fig. 3.

The coefficients α and β represent the probability of a purchase taking place; they are constants in the original Bass model and independent of any decisions or changes over time. For this reason, the model has been criticized and subsequent models have been developed that make the coefficients depending on variables like price or advertising budget. Most of the extensions,

however, include no feedback between the diffusion process and these decision variables. This is a severe shortcoming since in the market place, diffusion processes are strongly influenced by feedback. What in classical innovation diffusions models typically is referred to as word-of-mouth processes is nothing else than a reinforcing feedback process. Adopters of an innovation – represented by the cumulated sales $X_{(t-1)}$ – communicate with potential customers $(N - X_{t-1})$ and – by providing information about the product – influence their behavior. However, feedback in innovation diffusion goes beyond the pure word-of-mouth processes. It also involves the decision processes of a company and the outcome generated by the purchasing decision of the customers like the sales volume generated.

Figure 4 describes as a causal loop diagram the diversity of potential influences of corporate decision variables (marked with hexagons) on demand of the products by making the probability of a purchase – the coefficients α and β – depending on decision variables. It also shows how corporate decisions are interconnected through several feedback structures and influence the diffusion of a new product in the market place. Although being far from a comprehensive structure of potential feedback, the figure gives an impression of the complex dynamic nature of innovation diffusion processes.

Decision variables like pricing or advertising directly influence the purchase probability of a potential customer. The higher the advertising budgets and the lower the price, the higher will be demand for the products of a company. Furthermore, there are indirect and/or delayed effects on the speed of the spread of a new product in the

System Dynamics Analysis of the Diffusion of Innovations, Fig. 3 Stock-flow view of the Bass model

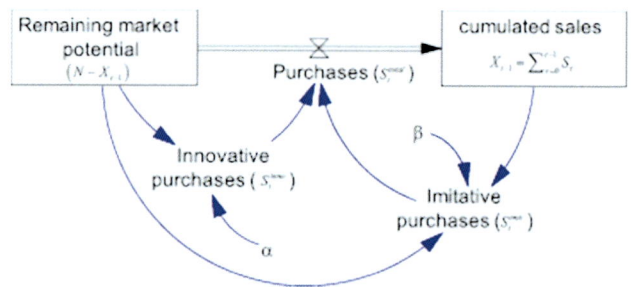

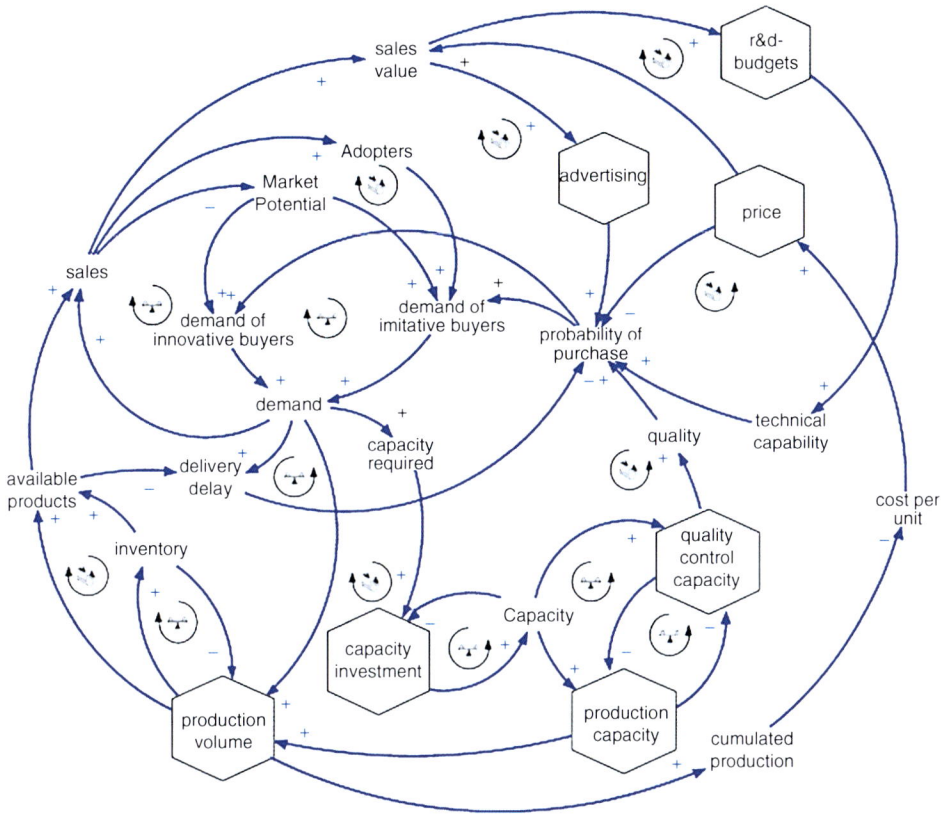

System Dynamics Analysis of the Diffusion of Innovations, Fig. 4 Feedback structures driving innovation processes

market. Actual sales of a product may be limited by insufficient production and inventory levels which increases delivery delays (perceived or actual) and therefore reduce demand. Growing demand, however, motivates the company to expand its capacity and to increase the volume of production. This leads to higher cumulated production and through experience curve effects to decreasing costs per unit, lower prices, and further increased demand. Other influences might reflect that a certain percentage of total available production capacity has to be allocated to ensure the quality of the output – either by final inspection or during the production process. Quality control then will improve product quality, which directly affects demand.

Models developed in this manner can serve as simulators to analyze the consequences of strategies and to improve understanding. They can show e.g.,

how pricing and investment strategies depend on each other and quantify the impact of intensified quality control on production and sales. They are suitable tools to investigate the effects resulting from the impact of a particular management problem on the dynamic complexity of innovation diffusion. Creating an understanding of the processes and interactions is the main purpose of system dynamics-based innovation diffusion models. Subsequently, a base structure of a system dynamics-based model will be described.

Base Structure of a System Dynamics-Based Model of Innovation Diffusion

First, a model will be discussed that maps the diffusion of an innovation in a monopolistic situation or can serve as an industry level model. Secondly, competition between potential and existing companies is introduced. Thirdly, substitution between successive product generations is considered. Each step adds complexity to the model. This approach allows for a better understanding of the forces driving the spread of a new product in the market.

In the following, the coarse structure of a model generating the life cycle in the market of a new product is presented and analyzed in its dynamic implications in section "Representing Managerial Decision Making in Innovation Diffusion Models." Figure 5 gives an aggregated view the main model structure. It also introduces – in contrast to the mathematical terms known from the Bass model, variable names, which are informative and consistent with the use in system dynamics models.

The diffusion of a new product is generated by the behavior of the before mentioned two different types of buyers: innovators and imitators . If the potential customers (PC) – i.e., the remaining market potential of a product – decide to purchase, either as innovators or as imitators, they become adopters *(ADOP)*. The variables *PC* and *ADOP* and their associated transfer rates are the basic variables of the core diffusion process. The untapped market *(UM)* covers latent demand that can be activated by appropriate actions and leads to an increase in the number of potential customers and therefore increases the remaining market potential. Besides the growth resulting from the influx from the untapped market, a decline in market volume can be caused by the loss of

System Dynamics Analysis of the Diffusion of Innovations, Fig. 5 Coarse structure of the innovation diffusion model

potential customers to competitors. This lost demand *(LD)* turned to competing products that are more attractive, e.g., products of a higher level of technological sophistication, quality or lower price.

The differentiation into the two buying categories "innovators" and "imitators" refers to the Bass model of innovation diffusion as described in section "Traditional Innovation Diffusion Models from a System Dynamics Perspective." The distinction is made because these two types of buyers react differently to prices charged, product quality offered, advertisements or the market penetration already achieved. The term "innovator" refers to customers who make their purchasing decision without being influenced by buyers who already purchased the product, the adopters. In the beginning of an innovation diffusion process, innovators take up the new product because they are interested in innovations. The number of innovators is a function of the potential customers.

Mathematically, the purchasing decision of innovators D^{Inno} is defined by a coefficient of innovation α times the number of potential customers PC.

$$D^{\text{inno}}_{(t)} = \alpha_{(t)} \cdot PC_{(t)} \qquad (4)$$

with:

$D^{\text{inno}}_{(t)}$ Demand from innovators
$\alpha_{(t)}$ Coefficient of innovation
$PC_{(t)}$ Potential customers.

The purchasing decision of "imitators" is calculated differently. Imitators buy a new product because they observe or communicate with customers who have already adopted the product. They imitate the observed buying behavior. Innovators initiate new product growth, but the diffusion gains momentum from the word-of-mouth process between potential customers and the increasing level of adopters. The driving force behind the imitation process is communication – either personal communication between an adopter and someone who still does not own the product or just observation of someone already owning and

using the product. Although, the Bass model describes how the imitators' purchases can be calculated – as shown in Eq. (2) – the equation can also be derived from a combinatorial analysis of the number of possible contacts between the adopters and the potential customers. If N is the total number of people in a population consisting of potential customers PC and adopters $ADOP$, the amount of possible combinations C^k_N is

$$C^k_N = \binom{N}{k} = \frac{N!}{k!(N-k)!}. \qquad (5)$$

Here we are only interested in paired combinations ($k = 2$) between the elements in N

$$
\begin{aligned}
C^2_N = \binom{N}{2} &= \frac{N!}{2!(N-2)!} \\
&= \frac{N(N-1)}{2!} = \frac{1}{2}\left(N^2 - N\right).
\end{aligned}
\qquad (6)
$$

Since N represents the sum of elements in PC and in $ADOP$, ($N = PC + ADOP$), the number of combinations between potential customers and adopters is

$$
\begin{aligned}
&= \frac{1}{2}\left[(PC+ADOP)^2 - (PC+ADOP)\right] \\
&= \frac{1}{2}\left[PC^2 + 2\cdot PC\cdot ADOP + ADOP^2 - PC - ADOP\right]
\end{aligned}
\qquad (7)
$$

and after regrouping and collecting terms, we get

$$
= \frac{1}{2}\left(\underbrace{2\cdot PC\cdot ADOP}_{\text{Communication between } PC \text{ and } ADOP} \right.
$$

$$
\left. + \underbrace{PC^2 - PC}_{\text{Communication within } PC} + \underbrace{ADOP^2 - ADOP}_{\text{Communication within } ADOP} \right).
\qquad (8)
$$

Internal communications, both within PC and $ADOP$, generate no incentive to purchase the new product and are neglected; the process of creating imitative buying decisions in Eq. (9) is, therefore, reduced to the first term in Eq. (8), the information exchange between potential customers and adopters.

$$D_{(t)}^{\text{imit}} = \beta^* \cdot PC_{(t)} \cdot ADOP_{(t)} \qquad (9)$$

with:

$D_{(t)}^{\text{imit}}$	Demand from imitators
$\beta_{(t)}^*$	Coefficient of imitation $= \dfrac{\beta_{(t)}}{N}$
$ADOP_{(t)}$	Adopters
N	Initial market potential.

The coefficient of imitation $\beta_{(t)}^*$ represents the original coefficient of innovation β from the Bass model divided by the initial market potential N. β can be interpreted as the probability that the possible contacts between members in PC and $ADOP$ have been established, relevant information has been exchanged, and a purchasing decision is made.

The sum of the demand of innovators and imitators in each period, $D_{(t)}$, establishes the basic equation for the spread of a new product in the market. Together with the state variables of potential customers and adopters the flows of buyers (innovators and imitators) constitute the core model of innovation diffusion, which generates the typical s-shaped pattern of an adoption process over time.

$$
\begin{aligned}
D_{(t)} &= D_{(t)}^{\text{inno}} + D_{(t)}^{\text{imit}} \\
&= \alpha_{(t)} \cdot PC_{(t)} + \beta_{(t)}^* \cdot PC_{(t)} \cdot ADOP_{(t)}.
\end{aligned}
\qquad (10)
$$

Although Eqs. (3) and (10) are based on different interpretations and explanations, they are structurally identical since PC equals $(N - X_{t-1})$ and $ADOP$ equals $X_{t-1} = \sum_{\tau=0}^{t-1} S_\tau$. The only difference is that the coefficients of innovation and imitation, in the context of the model based on (10) are now a variable – rather than a constant – depending on corporate decision variables like price or quality. Furthermore, corporate decisions are not just set as predefined time paths; they are endogenously calculated and depend on the

outcome of the diffusions process itself. Model simulations of this extended innovation diffusion model will be discussed in section "Representing Managerial Decision Making in Innovation Diffusion Models."

Extending the Base Structure to Include Competition

In the model described above, competition is not modeled explicitly. The model only assumes a potential loss in demand, if price, quality or ability to deliver are not within the customers' expectations. The internal corporate structures of competition are not explicitly represented. To generate diffusion patterns that are influenced by corporate decisions and the resulting dynamic interactions of the different competitors in a market, a more sophisticated way to incorporate competition is needed. Therefore, a subscript i ($i = 1, 2,\ldots, k$) representing a particular company is introduced as a convenient and efficient way to model the different competitors. In a competitive innovation diffusion model the calculation of innovative and imitative demand of a company has to be modified. Equation (4) that determines the innovative demand in a monopolistic situation becomes Eq. (11) – in the following discussion, the time subscript (t) is omitted for simplicity. The coefficient of innovation α has to be divided by the number of competitors N to ensure that each company will have the same share of innovative demand as long as there is no differentiation among the competitors' products through, e.g., through pricing or advertising. The subscript i in the coefficient of innovation is necessary because it considers that the decisions of an individual company regarding product differentiation influences its proportion of innovative buyers. A third modification is necessary, because the number of competitors may vary over time. Therefore, the term φ_i represents a factor to model different dates of market entry. It takes the value 1 if a company i is present at the market, otherwise it is 0. Hence, the demand of company i is 0, as long as it is not present at the market and $\sum_{i=1}^{k} \varphi_i$ represents the actual number of competitors. The variable potential customers PC has no subscript because all companies in the market compete for a common group of potential customers, whereas innovative demand has to be calculated for each company.

$$D_i^{inno} = \frac{\alpha_i}{NC} \cdot PC \cdot \varphi_i \qquad (11)$$

with:

α_i coefficient of innovation for company i

NC number of active competitors $= \sum_{i=1}^{k} \varphi_i$

φ_i factor of market presence company i

i subscript representing the companies $i = (1, 2, \ldots, k)$.

The buying decisions of imitators are influenced by observation of, or communication with the adopters *(ADOP)*. In a competitive environment two alternative approaches can be used to calculate imitative demand. These different approaches are a result of different interpretations of the object of the communication processes. In the first interpretation, the 'product related communication', the adopters of a particular company's product communicate information about the product they have purchased e.g., an electronic device like a MP3 player of a particular company. In this case, the calculation of imitative demand has to consider the number of potential contacts between the potential customers PC and the adopters of the products of company i *(ADOP$_i$)* as shown in Eq. (12).

$$D_i^{imit} = \frac{\beta_i}{N} \cdot ADOP_i \cdot PC \cdot \varphi_i \qquad (12)$$

with:

β_i coefficient of imitation for company i.

The second interpretation about the object of communication is the 'product form-related communication'. Here, the adopters communicate information about a product form, for example, DVD players in general and not about an MP3 player of a particular company. The equation to calculate imitative demand for the model of product form related communication is shown in Eq. (13). The sum of adopters for each company i $\left(\sum_{i=1}^{k} ADOP_i\right)$ represents the total number of adopters in the market. The product of the total adopters and the potential customers then represents the total number of potential contacts in the market. Imitative demand of a company i depends on the share of total adopters $\frac{ADOP_i}{\sum_{i=1}^{k} ADOP_i}$ this company holds.

$$D_i^{imit} = \frac{\beta_i}{N} \cdot \frac{ADOP_i}{\sum_{i=1}^{k} ADOP} \cdot PC \cdot \sum_{i=1}^{k} ADOP_i \cdot \phi_i. \qquad (13)$$

If the term that represents a company's share of the total adopters of a market $\frac{ADOP_i}{\sum_{i=1}^{k} ADOP_i}$ is raised to the power of γ as in Eq. (14), weaker ($0 < \gamma < 1$) or stronger ($\gamma > 1$) influences of a company's share of total adopters on demand can be represented explicitly. For $\gamma = 1$, Eq. (14) is identical to Eq. (13).

$$D_i^{imit} = \frac{\beta_i}{N} \cdot \left(\frac{ADOP_i}{\sum_{i=1}^{k} ADOP}\right)^{\gamma} \cdot PC \cdot \sum_{i=1}^{k} ADOP_i \cdot \phi_i \qquad (14)$$

with:

γ factor representing customers' resistance to "Me − too" − pressure.

Figure 6 shows the effects of a company's share of the total adopters for different γ. For a given share of total adopters this means: the higher γ, the lower is the value of the term

$$\left(\frac{ADOP_i}{\sum_{i=1}^{k} ADOP_i}\right)^{\gamma}$$

and the stronger is the importance of a high share of total adopters. The parameter γ can be interpreted as a measure of the importance of customer loyalty or as resistance to "me-too" pressure.

Figure 7 illustrates the coarse structure of an oligopolistic innovation diffusion model as described by Eqs. (11) and (14). The hexahedron

**System Dynamics
Analysis of the Diffusion
of Innovations,**
Fig. 6 Effects of a
company's share of
adopters for different γ

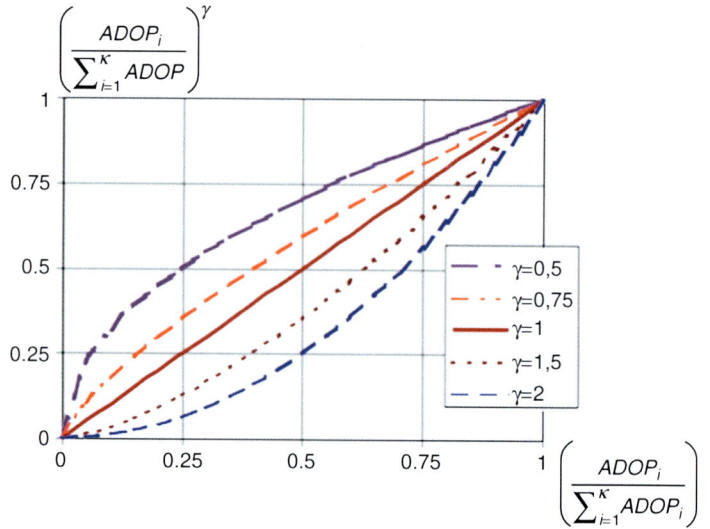

System Dynamics Analysis of the Diffusion of Innovations, Fig. 6 Effects of a company's share of adopters for different γ

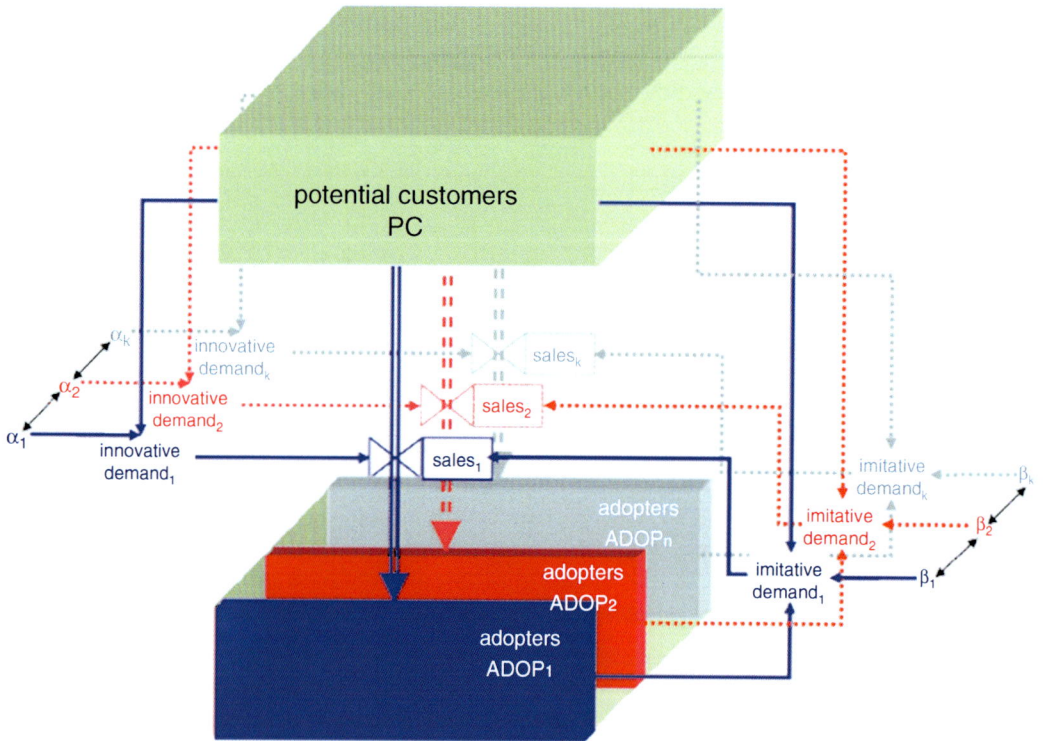

System Dynamics Analysis of the Diffusion of Innovations, Fig. 7 Coarse structure of an oligopolistic innovation diffusion model

at the top represents the stock of potential customers *PC* for the whole market. The blocks with the different shading represent for each company *i* the level of adopters, i.e., the cumulated sales of the company. The total number of adopters of the product form corresponds to the addition of these blocks.

Since the sales are calculated separately for each company *i* there are n outflows from the stock of potential customers to the adopters. Again, sales comprise innovative and imitative demand, which are influenced by the coefficient of innovation α_i and imitation β_i. Both coefficients are influenced by managerial decisions of each company i like pricing, advertising, quality, market entry timing, etc. and measure the relative influence of the decisions compared to the competitor's decisions. Therefore, the values α_i and β_i not only depend on the decisions of company i, they also depend on the competitor's decisions. Both variables are crucial for the speed and the maximum volume of demand for the products of a company i.

Figure 8 shows the results of simulations based on Eq. (11) for innovative demand and Eq. (14) for imitative demand with the effects of a market entry delay of the second company – the influences of other decision variables are switched off. Several model simulations have been made assuming a market entry delay of company 2 between 0 and 12 months.

The plots in Fig. 8 show the development of market share and sales of the second company over time. Since there is no further product differentiation, both competitors have the same market share when they enter the market at the same time. With each month delay of the second company the market share that can be achieved at the end of the simulation decreases. A three months delay reduces the finally achieved market share to 40%; a 12-month delay even causes a decrease in market share down to approximately 25%. Accordingly, the maximum sales volume decreases significantly with each month delay in market entry time.

System Dynamics Analysis of the Diffusion of Innovations,
Fig. 8 Follower's market share and sales for different market entry times

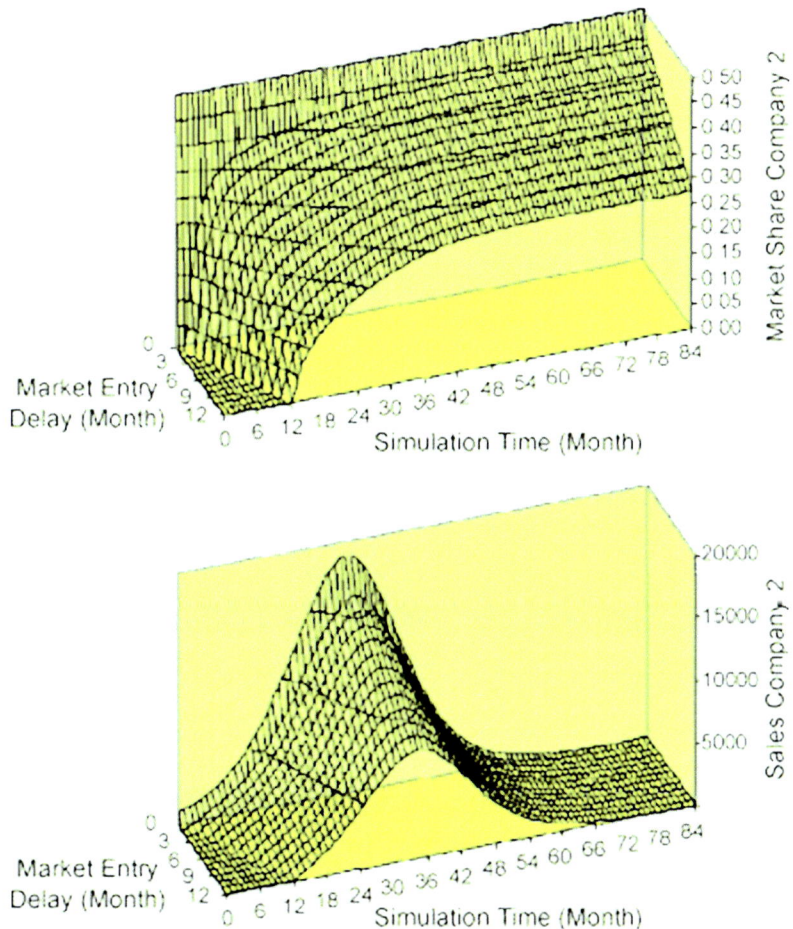

Representing Network Externalities

In the following, we will investigate the diffusion of a specific type of goods in order to show the importance of understanding the diffusion of goods with network effects (based on (Thun et al. 2000)). The trend towards an information society has stressed the relevance of goods satisfying information and communication needs. Many products of this market segment such as electronic mail contain attributes that necessitate a specific examination, since the diffusion of goods showing network effects differs from that of conventional products. The main difference between conventional products and products with network effects is that the utility of the latter cannot be regarded as a constant value. With regard to these products, utility is an endogenous variable which results in a specific diffusion behavior. Two effects are responsible for this particular behavior: the bandwagon effect and the penguin effect . A refined system dynamics model supports a better understanding of this special diffusion process.

The fact that the utility is not constant can be reasoned by a concept commonly referred to as "network effect." A product is characterized by a network effect, if the utility of that product is a function of the installed base, which is defined as the amount of users in a network system. The utility increases with the installed base. This leads to a virtual interdependency of the users inside the network. The interdependency is based on the bandwagon effect and the penguin effect. Starting from the fundamental diffusion model of Bass the characteristics of network effects are integrated into the model in order to simulate the diffusion behavior.

Many definitions for network effects (sometimes also called "positive demand externalities") can be found in the literature. Basically, it can be stated that network externalities exist if the utility of a product for a customer depends on the number of other customers who have also bought and use this product. These network externalities can be indirect or direct, whereby we concentrate on the latter. A typical example for a product with direct network effects is the telephone. The utility that a telephone system can generate increases with the amount of feasible communication connections. Other examples are e-mail, fax, instant messaging, etc. each of which satisfies communication needs. But none of these products can generate any utility for a user on its own. In order to create utility the existence of other users adopting this product is required. Accordingly, the product's utility changes dynamically with the number of users, i.e., the installed base.

The installed base B_t determines the utility of products influenced by network externalities. In terms of direct network externalities the utility is based on B_t exclusively since utility can only be generated by interconnections within the underlying network solely. Accordingly, the utility of a product with direct network externalities is a function of the number of feasible connections I_t. The number of connections is determined by the number of users on the one hand and the technological restriction of the network on the other hand, whereby the latter one represents the number of users being able to communicate via the network simultaneously (for instance, classical telephone system $r = 2$, telephone conferencing $r \leq 2$). Thus, U_t can be calculated by the formula:

$$U_t = U_t(I_t) = \sum_{k=2}^{n} \binom{B_t}{r} = \sum_{k=2}^{n} \frac{B_t!}{r!(B_t - r)!},$$

whereby $r, B_t > 0$.

(15)

Since the achievable utility of a product with direct network externalities depends exclusively on the network size the adoption process depends on the decision of potential users influencing the diffusion process significantly. This leads to two different effects: Firstly, the utility for each user grows exponentially with an increasing amount of actual users according to the formula. This implies that the more people are attracted the more are part of the network leading to an exponential growth of the diffusion process which is referred to as the "bandwagon effect": The higher the number of people the more are decoyed as well resulting in a reinforcing process. Although this effect occurs with conventional products as well, in case of products with direct network externalities, it is much stronger since the exponentially growing utility has a greater impact on the diffusion process.

Secondly, utility must be created by the utilization of the users first in order to establish a new communication network which determines the diffusion process significantly since products influenced by direct network externalities cannot generate an original utility by itself. Accordingly, early adopters of a network are confronted with the risk that they cannot derive sufficient benefit from the network so that they must rely on potential users to follow entering the network in the future. Therefore, the adoption decision of a potential user depends on the future decision of other potential users. All in all, this leads to a hesitating behavior of all potential adopters resulting in an imaginary network barrier , which is based on the risk of backing the wrong horse, which is also known as the "penguin effect."

Finally, another important aspect must be considered when analyzing the diffusion process of products with network externalities. In terms of conventional products the decision to buy a product is the final element of the decision process. Contrary to that, concerning products with network externalities the adoption process is not finished with the decision to enter a network since the subsequent utilization of the product is important for the diffusion process. If the expected utility of the communication product cannot be achieved users may stop using this product leading to a smaller installed base and a lower network utility which is important for other users that may stop their utilization as well and for the adoption process of potential users.

In the following, the basic structure of the underlying model will be described. Analogously to the model presented in the preceding paragraphs there exists a group of potential users (in this model, we only focus on the core diffusion process without considering competitors or latent demand in order to keep the complexity of the model low.) If these potential users decide to adopt the communication product, they become part of the installed base B. The adoption process is illustrated by the adoption rate AR, which is primarily influenced by the variable *word of mouth*. In order to consider the average utility per user – as it is necessary for analyzing products with network externalities –

the imitation coefficient β has been endogenized contrary to the classical Bass model. Therefore, the variable β is influenced by the "word-of-mouth" effect which depends on the average utility per user. If actual utility is bigger than the desired utility all individuals in contact with users adopt and buy the product. If it is smaller, however, only a fraction adopts. The size of this fraction depends on the distance between actual and desired utility.

Figure 9 depicts two simulation runs showing the system behavior of the diffusion process of conventional products and products influenced by direct network externalities. Graph UTIL1 represents the *adoption rate* , i.e., the amount of buyers of a conventional product per period. The graph UTIL1 shows the behavior of the variable *installed base B* which is the accumulation of *adoption rate* (note that the graphs have a different scale). The graphs RISK1 show the system behavior for products influenced by direct network externalities, i.e., the *adoption rate AR* and the corresponding *installed base B*.

A comparison of both simulation runs shows that diffusion needs longer to take off in terms of products influenced by direct network externalities, but showing a steeper proceeding of *Installed Base B* in later periods. This behavior can be verified comparing the adoption rates of the two runs: although adoption starts later with an endogenously generated adoption fraction, it nevertheless has a higher amplitude. This behavior can be interpreted as the penguin effect and the bandwagon effect.

Finally, it has to be taken into account that some users of the installed base might quit to use the product since they are disappointed from its utility. Accordingly, it is an important issue to find ways in order to raise the patience of users to stay within the network. That gives potential users the chance to follow into the network which will increase the utility for the user as well.

From the simulation analysis the following conclusions can be drawn. The importance of the installed base for a success diffusion process is shown. Without a sufficient amount of users it is not possible to generate a utility on a satisfying level which prevents potential users to enter the network or even making users leave the network.

System Dynamics Analysis of the Diffusion of Innovations, Fig. 9 Comparison of diffusion behavior

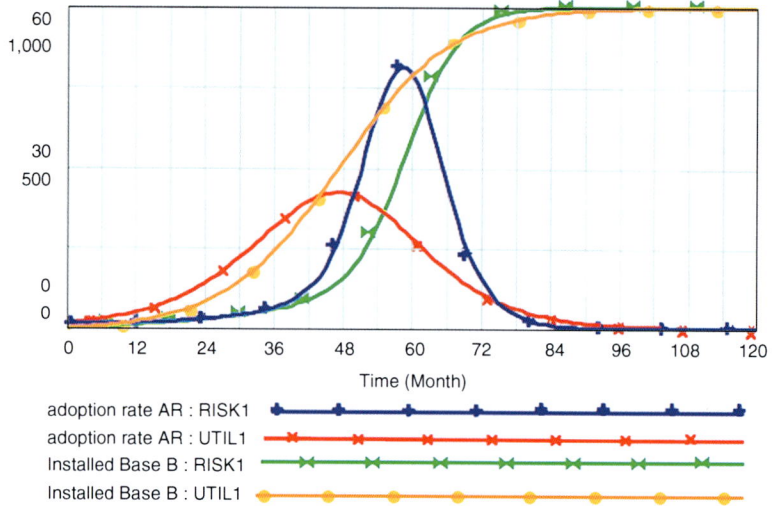

adoption rate AR : RISK1
adoption rate AR : UTIL1
Installed Base B : RISK1
Installed Base B : UTIL1

Accordingly, ways must be found to increase the utility that a network creates for a user in order to reach the critical mass. This can be done in several ways of which some will be discussed briefly. One possible way is to increase the installed base by compatibility to other networks. Furthermore, the risk to back the wrong horse can be mitigated by product pre-announcements in order to lower the imaginary network barrier by making potential users familiar with the product. Another possibility is to increase the group of relevant users, i.e., to enlarge the average group size within the network, since not all users are equally important for a potential user. Furthermore, the technological potential can be improved by introducing multilateral interconnections between the members of a network.

Representing Managerial Decision Making in Innovation Diffusion Models

Subsequently the basic structures of innovation diffusion processes described above will be extended and simulated to demonstrate the impact of managerial decision-making on the diffusion of innovations. The model used for the simulations serves as a simulator to determine how individual strategies can accelerate or hamper market penetration and profit performance. The models are not

designed to predict the basic market success or failure of innovations. Although, they are rather comprehensive, several assumptions apply here as for all models. E.g., in all model runs, the basic market acceptance of the innovation is assumed. Furthermore, the simulations assume for the moment that no competition exists.

Dynamic Pricing Without Direct Competition

In a first step the basic model from section "Base Structure of a System Dynamics-Based Model of Innovation Diffusion" is extended to generate dynamic cost behavior as suggested in Fig. 5. Standard costs are the basis for the calculation of prices – an important decision variable. Experience curve effects are modeled based on cumulated production in order to map the long-term behavior of standard cost. The actual costs of a product in a certain period are derived from the standard cost modified for variations resulting from capacity utilization.

The concept of experience curve effects suggests a direct relationship between cumulated production $X_{(t)}$ and average standard cost per unit $c^s_{(t)}$, adjusted for inflation; where c^s defines standard unit cost at the planned level of production. Every doubling of $X_{(t)}$ is associated with a cost reduction in real terms by a constant percentage according to:

$$c^s_{(t)} = c^n \left(\frac{X_{(t)}}{n}\right)^{-\delta} \qquad (16)$$

where c^n stands for the cost of unit n ($n \subseteq X$) and δ represents a constant depending on the experience rate. For many businesses experience rates of 10% to 20% have been observed and ample empirical evidence for this relationship is available.

The costs of a product in each period of time $C_{(t)}$ are a function of cumulated production $X_{(t)}$ and capacity utilization determined by the production volume of a period $x_{(t)}$ as defined in Eq. (17). Figure 10 shows the behavior of the dynamic cost function

$$C_{(t)} = \Phi\left(X_{(t)}, x_{(t)}\right) . \qquad (17)$$

Furthermore, the model comprises elements of (i) market development, (ii) product pricing and its impact on the profits from producing and selling the products, i.e., the operating results, and (iii) resource allocation, e.g., capital investment, production volume, and quality control. Pricing and quality affects the coefficients of innovation α and imitation β from Eq. (10). Figure 11 shows the run of a model version including market development.

The time behavior of production, demand, and operating results duplicate usual characteristics of the life cycle of a successful innovation. After the

product launch, additional customers can be gained from an untapped market as diffusion and thereby product awareness proceeds and prices decline. The maximum of demand from imitators – the quantitatively most important fraction of demand – is reached when the amount of possible communications between potential customers and adopters reaches its maximum. The decreasing level of potential customers and the depletion of the untapped market cause the decline towards the end of the simulation. The behavior also shows that demand rises much faster than the company can increase its production capacity. The behavior of Fig. 11 will serve as reference mode for further analysis.

Pricing strategies and decisions are additional important elements, which require an extension of the model. The problem of the "right price" for a new product is essential but still unsolved in the area of innovation management. Difficulties to recommend the optimal pricing policy derive in particular from the dynamics in demand interrelations, cost development, potential competition, and the risk of substitution through more advanced products. Regardless of this complex framework, several attempts in management science try to derive and to apply optimal pricing policies. However, they are faced with difficulties, both mathematical and practical. Their results are too complicated to support actual pricing decisions. Therefore simulation studies found more frequently their way into management science.

System Dynamics Analysis of the Diffusion of Innovations, Fig. 10 Dynamic cost function

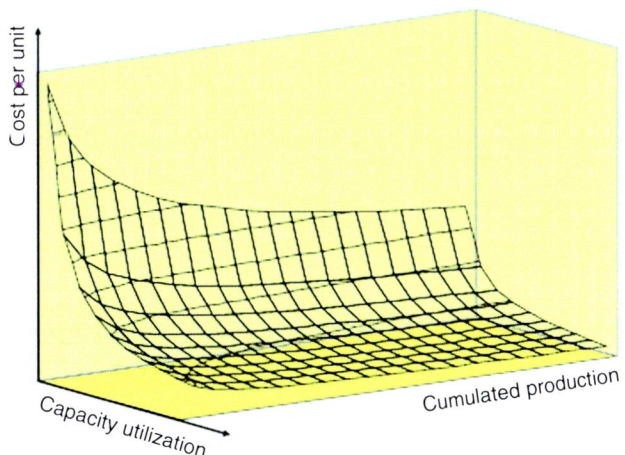

System Dynamics Analysis of the Diffusion of Innovations,
Fig. 11 Reference mode of the basic innovation diffusion model

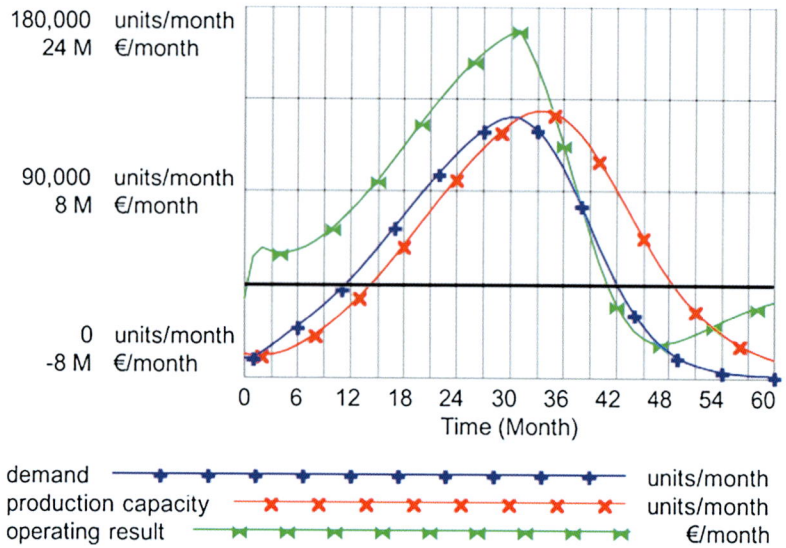

System Dynamics Analysis of the Diffusion of Innovations, Fig. 11 Reference mode of the basic innovation diffusion model

The extended model includes four predefined pricing policies to investigate their impact on market development on operating results:

Myopic profit maximization assuming perfect information about cost and demand. The optimal price p^{opt} is derived from elasticity of demand $\varepsilon_{(t)}$ and per unit standard cost $c^{\text{s}}_{(t)}$ considering the impact of short term capacity utilization:

$$p^{\text{opt}}_{(t)} = c^{\text{s}}_{(t)} \cdot \frac{\varepsilon_t}{\varepsilon_t - 1}. \qquad (18)$$

Skimming price strategy aims at serving innovative customers with high reservation prices and then subsequently reduces prices. The model applies a simple decision rule modifying $p^{\text{opt}}_{(t)}$ through an exponential function that raises the price during the first periods after market introduction:

$$p^{\text{skim}}_{(t)} = p^{\text{opt}}_{(t)} \cdot \left(1 + a \cdot e^{\frac{-t}{T}}\right). \qquad (19)$$

Full cost coverage, i.e., standard cost per unit plus a profit margin π to assure prices above cost level even during the early stages of the life cycle:

$$p^{\text{fcc}}_{(t)} = c^{\text{s}}_{(t)} \cdot \pi. \qquad (20)$$

Penetration pricing aims at rapidly reaching high production volumes to benefit from the experience curve and to increase the number of adopters. It uses a similar policy as for the skimming price, but instead of a surcharge it decreases prices early after market introduction:

$$p^{\text{pen}}_{(t)} = c^{\text{s}}_{(t)} \cdot \pi \cdot \left(1 - a \cdot e^{\frac{-t}{T}}\right) \qquad (21)$$

The simulation runs shown in Fig. 12 give an overview of the development of profits, cumulated profits, and sales for the four pricing strategies discussed above. The model assumes the following: (1) there is an inflow from the untapped market, which depends on the dynamic development of prices; (2) there is no risk of competition; (3) repeat purchases do not occur. Taking profits into account, Fig. 12 indicates that – over the time horizon observed -, the classic pricing rule of profit optimization leads to superior results from a financial point of view. However, if judged by the market development, the strategy of penetration prices is the appropriate strategy. This strategy allows rapid penetration of the market by setting relatively low prices, especially in the early stages of the life cycle. The combined price and diffusion effects stimulate demand and reduce the risk of losing potential customers to upcoming substitution products.

Figure 12 also indicates a disadvantage of the penetration strategy. Since the market is already

System Dynamics Analysis of the Diffusion of Innovations,
Fig. 12 Comparison of the outcome of pricing strategies

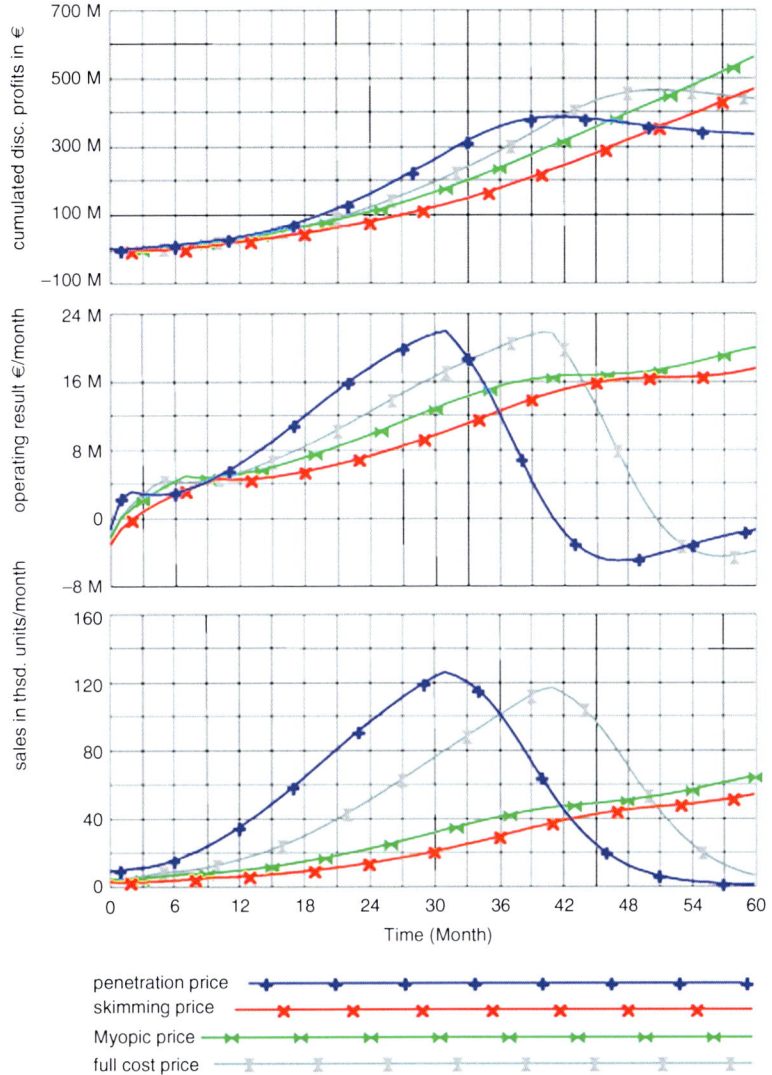

System Dynamics Analysis of the Diffusion of Innovations, Fig. 12 Comparison of the outcome of pricing strategies

completely satisfied after period 54, there is only little time to develop and introduce a new product in the market successfully. The slower market growth of the skimming and optimum price strategy leaves more time for the development of a new product, but the attractive profit situation and the slow development also increase the risk that competitors might enter the market. In a dynamic demand situation where prices influence market growth, where substitution or competition can occur, and where delivery delays eventually accelerate the decision of potential buyers to turn to other products, a strategy of rapid market penetration seems to be the most promising one. It will,

therefore, be the basis for the following simulation runs investigating manufacturing's role in innovation management.

Linking Manufacturing-Related Decision Variables

The role of manufacturing is important for the successful management of innovations. Manufacturing has to provide sufficient capacity to produce the goods sold. The investments to adjust capacity influence a company's ability to meet demand and deliver on time. It is assumed that the necessary financial resources for the investments are available. The aggregated capacity provided by the company includes both, machinery equipment and production personnel. Since the manufacturing function also has to ensure the quality of the output through dedicating a portion of its total available capacity to quality control, the capacity resources can be used to either manufacture the products or to assure the desired level of quality. Capacity allocation to improve quality takes away capacity for production. This additional feedback structure – as indicated in Fig. 5 - maps the allocation of resources for quality control to the achieved ability to meet product demand. If manufacturing capacity does not meet demand, a temporary reduction of capacity for quality assurance seems a plausible strategy. Quality control resources than are allocated to manufacturing rather than testing whether quality standards are met. In this scenario, it would be expected that total cost remain unchanged and the additional manufacturing capacity gained through the reallocation can be used to provide more prod ucts to the customers, increase sales, and improve the overall results.

Figure 13 shows the simulation assuming the same scenario as in the base mode together with penetration prices and reduced quality resources if demand exceeds production capacity. It also shows a quality index plotted as an additional variable. Quality is defined to be 1, if the actual quality capacity equals a standard value of quality resources necessary. It is assumed that 10% of

total production capacity is necessary to assure 100% quality. For values above the 10%-level, quality is better; for values below, it is poorer. The simulation indicates that the policy of reduced quality resources successfully decreases the discrepancy between demand and production as seen in the reference mode of Fig. 11. This results from the increased proportion of capacity used for production and an additional effect caused by lower product quality, which then decreases demand. Although the maximum sales are nearly the same in the simulation of reduced quality control strategy, the peak demand occurs around 5 months later. Instead of gaining higher sales only the shape of the life cycle changed. However, operating results had improved, in particular the sharp decline of profits in the base mode of the simulation could be slowed down and losses could be avoided. The reduced quality control strategy caused a slower capacity build up and therefore, when product sales declined capacity adjustment was easier to achieve. From the financial point of view the strategies of penetration prices and reduced quality control fit quiet well.

The results are different if a strategy of quality reduction is used in combination with a strategy of skimming pricing. Figure 14 compares the outcome of cumulated discounted profits for the strategy of reduced quality and penetration prices or skimming prices with the development of the

System Dynamics Analysis of the Diffusion of Innovations,
Fig. 13 Reduced quality control

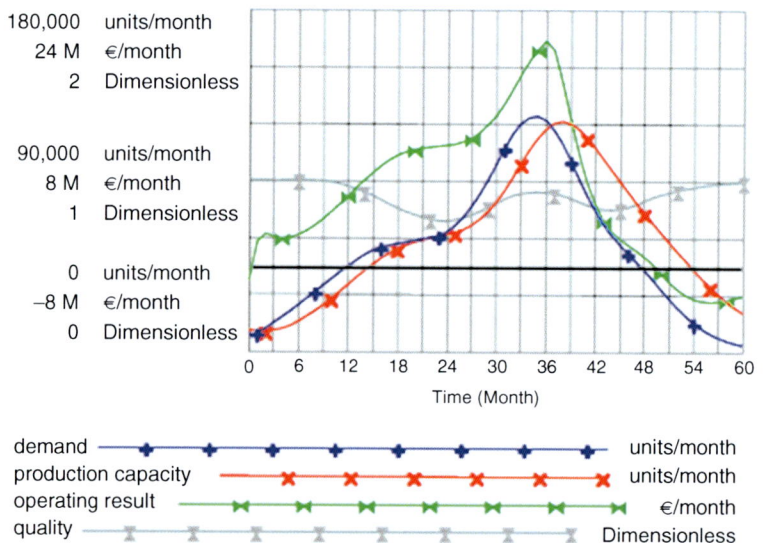

180,000	units/month
24 M	€/month
2	Dimensionless

90,000	units/month
8 M	€/month
1	Dimensionless

0	units/month
–8 M	€/month
0	Dimensionless

Time (Month)

demand	units/month
production capacity	units/month
operating result	€/month
quality	Dimensionless

System Dynamics Analysis of the Diffusion of Innovations,
Fig. 14 Cumulated discounted profits – penetration versus skimming pricing in combination with quality control strategies

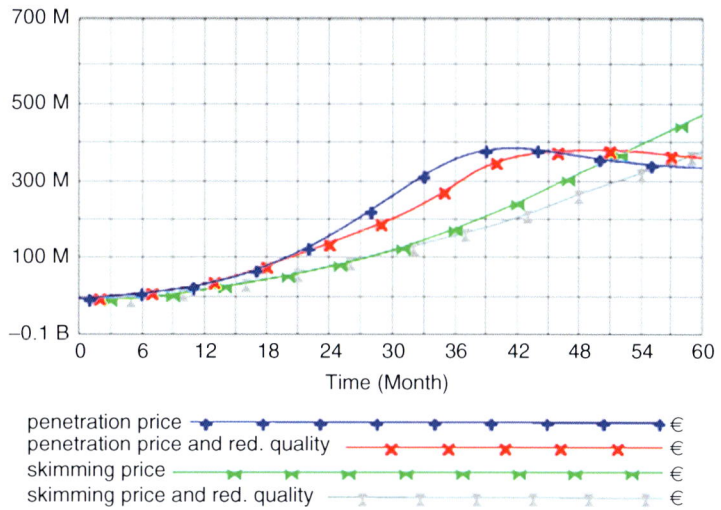

reference mode - the simulations without quality adjustment. The behavior indicates that in the case of skimming prices, quality reductions slow down the development of the market and cumulated profits significantly.

The simulation results raise the question whether emphasizing quality when demand is higher than capacity would be a more appropriate way to react. As the upper part of Fig. 15 points out, the strategy of emphasized quality leads to an accelerated product life cycle in the case of the penetration pricing strategy. Tremendous capacity build-up is necessary after the introduction of the new product. As demand declines, a plenty of capacity is idle, causing significant losses during the downswing of the product life cycle.

Emphasizing quality turns out to be more effective in the case of skimming prices. The additional demand gained from quality improvements also accelerates the product life cycle, but at a much slower rate and leads to improved cumulated profits. Emphasizing quality in combination with skimming or optimum prices leads to improved cumulated profits, compared to both, the simulation without quality reaction and the quality reduction run.

The simulations show the importance of a detailed judgment of strategic choices. Strategies must be consistent with each other and with the real world structures mapped by the model. The simulations above assume a situation without

existing or potential competition. In such an environment there is no interest paid for fast market penetration. Hence, a penetration pricing strategy is the most unfavorable alternative. However, this changes if structural elements are added to the model that incorporate competition - even as in the simple structure from Fig. 5, which considers the loss of demand to a competitor. Lost demand therefore is represented as a process equivalent to the imitative demand from Eq. (9). The calculation of lost demand starts in period 15 through an initial switching of a potential customer to the competitor. This switch starts a process that drives demand for the competitors' products and is influenced through the quality the company offers. If the company provides poor quality , more potential customers and market potential from the untapped market will directly move to the competitor. The accumulation of lost demand corresponds to the number of adopters the competitors gained over time. Simulations with these additional structures give some additional insights (Fig. 16).

Penetration pricing leads again to the fastest market development. In the competitive surrounding, however, emphasizing quality accelerates the market development and leads to better performance than quality reductions. This is in contrast to the simulations without competition shown in Fig. 13 to Fig. 15. Skimming prices in combination with reduced quality control shows the poorest

System Dynamics Analysis of the Diffusion of Innovations, Fig. 15 Emphasized quality in all innovation stages

demand (penetration price) — units/month
production capacity (penetration price) — units/month
operating result (penetration price) — €/month
quality (penetration price) — Dimensionless

demand (skimming price) — units/month
production capacity (skimming price) — units/month
operating result (skimming price) — €/month
quality (skimming price) — Dimensionless

financial and market performance. A strategy of reduced quality control causes in the competitive environment the demand to increase at a slower rate than in the base run, where no quality adjustments were made when demand exceeded capacity. In both cases, the skimming and the penetration price scenario, quality reductions lead to the poorest performance.

Linking R&D and New Product Development

The models discussed above are able to generate under different conditions the typical diffusion patterns of new products in the market place. However, these models do not consider the stage of new product development. New products have to be developed before they can be introduced into the market. A costly, lengthy, and risky period of R&D has to be passed successfully. The diverging trends of shortening product life cycles and increasing R&D costs show the importance of an integrated view of all innovation stages. In the remainder, a comprehensive model comprising both, the process of R&D and an oligopolistic innovation

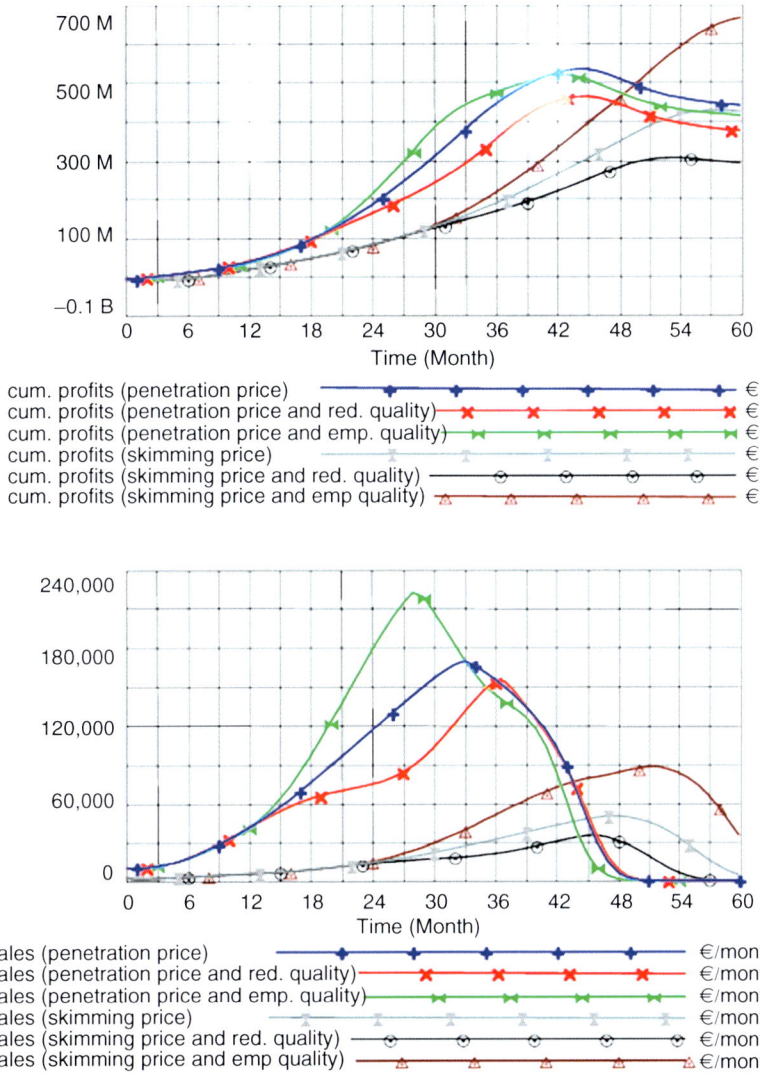

System Dynamics Analysis of the Diffusion of Innovations, Fig. 16 Behavior of the base model including simple competitive structures

diffusion with subsequent product generations is used to investigate the interrelations between the stages of innovation processes. The integration of both modules is shown in Fig. 17.

The volume and the intensity of the research and development activities feed the R&D-process. The number of research personnel determines the volume. Since R&D personnel requires resources like laboratory equipment, material for experiments etc., the intensity of R&D depends on the budget available for each person working in the R&D sector. This information is calculated in

a more comprehensive model in the sector of R&D planning, which also includes policies about resource allocation within the research and development stages, i.e., mainly the question of how much to spend on which new product development project.

Depending on the volume and the intensity of R&D, the technological knowledge of each product generation for each company evolves over time. The module of the R&D-process feeds back the current state of the technological knowledge for each company and product generation.

System Dynamics Analysis of the Diffusion of Innovations,
Fig. 17 Linking R&D-processes with corporate and market structures

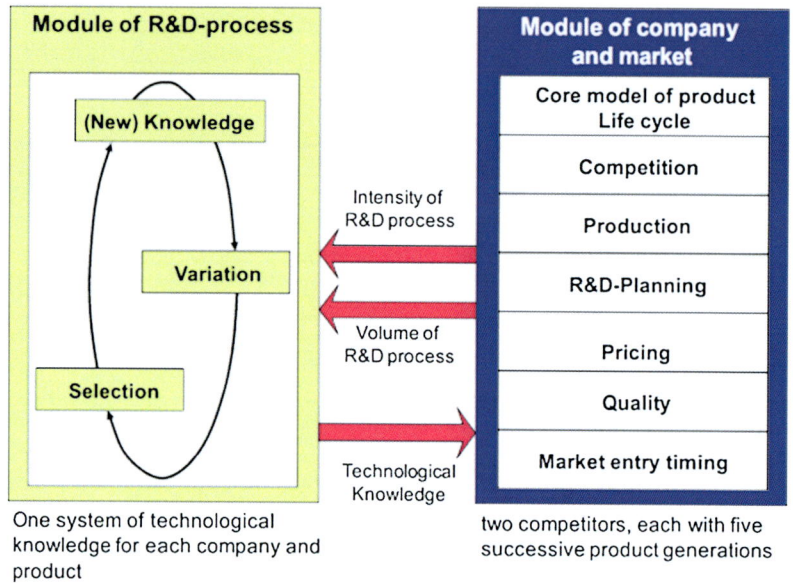

One system of technological knowledge for each company and product

two competitors, each with five successive product generations

The basic assumptions of the model are as follows. The model maps the structures of two competitors. Both competitors can introduce up to five successive product generations. The initial values of the model ensure that all competitors start from the same point. All firms have already introduced the first product generation and share the market equally. The resources generated by the first product are used to develop subsequent product generations. In the base run each company follows the same set of strategies. Therefore, except for minor differences resulting from the stochastic nature of the R&D-process, they show the same behavior over time. Figure 18 provides a simulation run of the model with all modules and sectors coupled.

The curves show for a single company the development of the sales of the products and the total sales. They emphasize the importance of a steady flow of new and improved products. Without on-time replacement of older products, the total sales of the products will flatten or deteriorate like in the simulation around periods 44, 92, and 116. The model also generates the typical s-shaped curves of technological development (lower part of Fig. 18). Each product generation has a higher technological potential and the knowledge developed for the preceding product generations partly

can be used by the successive product generations. For this reason the subsequent product generations start at a level different from zero.

In a dynamic environment such as the computer industry, where investments in R&D and manufacturing equipment are high, the product life cycles are short, and time-to-market as well as time-to-volume are essential variables, it is important to understand the dynamic consequences of decisions and strategies in the different areas. Figure 19 describes some of the important feedback loops linking the process of invention to the processes of innovation and diffusion.

Central element in the figure is the calculation of the sales of a company according to Eqs. (11) and (14). The coefficients of innovation and imitation are influenced by the multiplier of relative competitive advantage, which depends on the relative technical capability and the price advantage of a company. The technical capability of the products is influenced by the strength of its R&D-processes and the total amount of R&D expenditures.

Empirical studies in Germany have shown that measures like sales volume, profits or R&D budgets of earlier periods are quite common as a basis for R&D budgeting. However, using historic sales volume as a basis to determine R&D budgets invokes the positive feedback loop "competing

**System Dynamics
Analysis of the Diffusion
of Innovations,**
Fig. 18 Exemplary
behavior of the integrated
innovation model

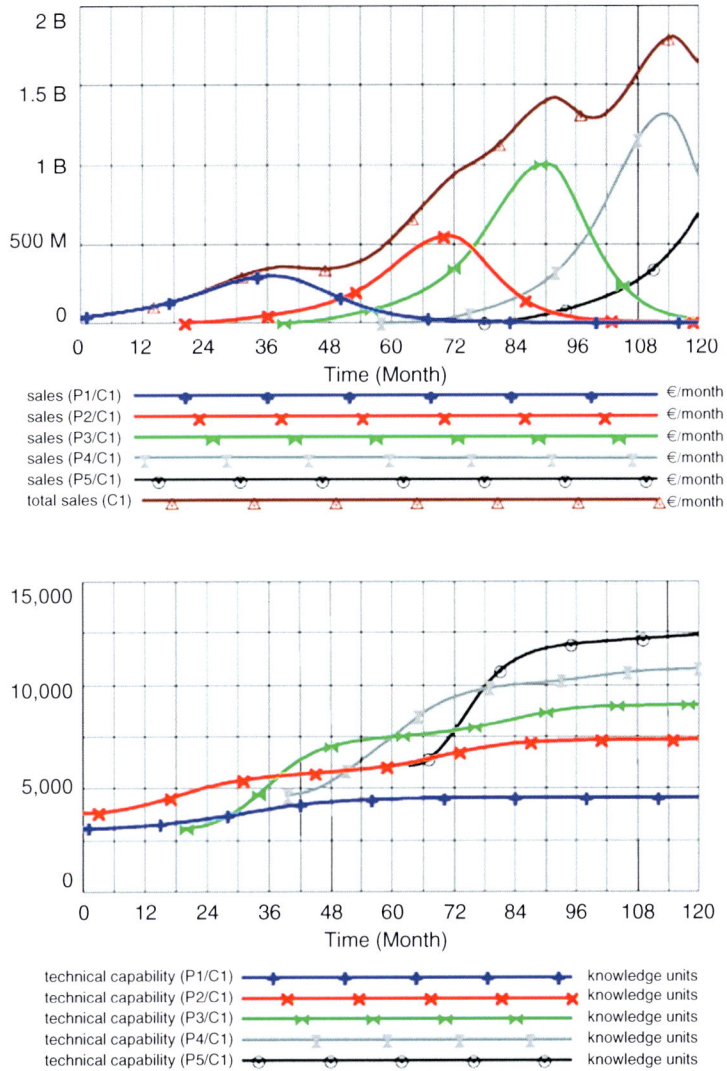

**System Dynamics
Analysis of the Diffusion
of Innovations,**
Fig. 18 Exemplary
behavior of the integrated
innovation model

by technical capability." With an increasing number of products sold and growing value of sales the budget and the number of personnel for R&D grow. This leads to an improved competitive position, if the technical capabilities of a product increases. The higher the sales volume, the better is the resulting competitive position. This produces increasing coefficients of innovation and imitation and leads to higher sales. This budgeting strategy is implemented in the model for the next simulation runs.

The second loop "price competition" links pricing strategies to sales volume. The actual price of a product is influenced by three factors. The first

factor, standard costs, is endogenous. As cumulated production increases, the experience gained from manufacturing causes declining standard costs. The second and third elements influencing the calculation of prices are exogenous: parameters which define the pricing strategy and demand elasticity. Caused by increasing cumulated production, standard costs fall over the life cycle and prices are also declining. Lower prices affect the relative price and improve the effect of price on the coefficients of innovation and imitation, which leads to increased sales and higher cumulated production.

The loop "pricing limits" reduces the effects of the reinforcing loops described above to some

System Dynamics Analysis of the Diffusion of Innovations, Fig. 19 Feedback structure influencing the diffusion process

extent. The standard cost and price reductions induce - ceteris paribus - a decrease in the sales volume and set off all the consequences on the R&D-process, the technical know-how, the market entry time and sales shown in the first feedback loop - but in the opposite direction. Additionally, since standard cost cannot be reduced endlessly this feedback loop will show a goal seeking behavior.

With equivalent initial situations and the same set of strategies, both companies behave in an identical way for all product generations. If one company has a competitive advantage, the reinforcing feedback loops suggest that this company will achieve a dominating position. In the simulation shown in Fig. 20, both competitors have the same competitive position for the first product generation. But the first company will be able to enter the market 2 months earlier than the competitor, because the initial outcome of the R&D process is slightly better than the second company's second product generation. Both competitors follow a strategy of skimming prices and demand elasticity has the value −2.

The initial gain in the outcome of the R&D-process initiates a process of sustained and continuing competitive advantage for the first company. It will improve continuously, since the positive feedback loop "competing by technical capability" dominates. The first company's advantage in the market introduction leads to an increasing readiness for market entry. It is able to launch the third product generation 3 months earlier than the follower and will introduce the fourth product generation in period 112. The follower is not able to introduce its fourth generation during the time horizon of the simulation, i.e., the pioneers advantage has extended to more than 8 months. The first company's competitive advantage is a result of the slightly higher initialization of the knowledge system and the dominance of the positive feedback loops, which causes shortened time-to-market and higher sales volume over all successive product life cycles. Additionally, the technical capabilities of both competitors' product generations show the same reinforcing effect . The difference between the technical capability of both competitors increases in favor of

System Dynamics Analysis of the Diffusion of Innovations,
Fig. 20 Reinforcing effects of initial competitive advantage

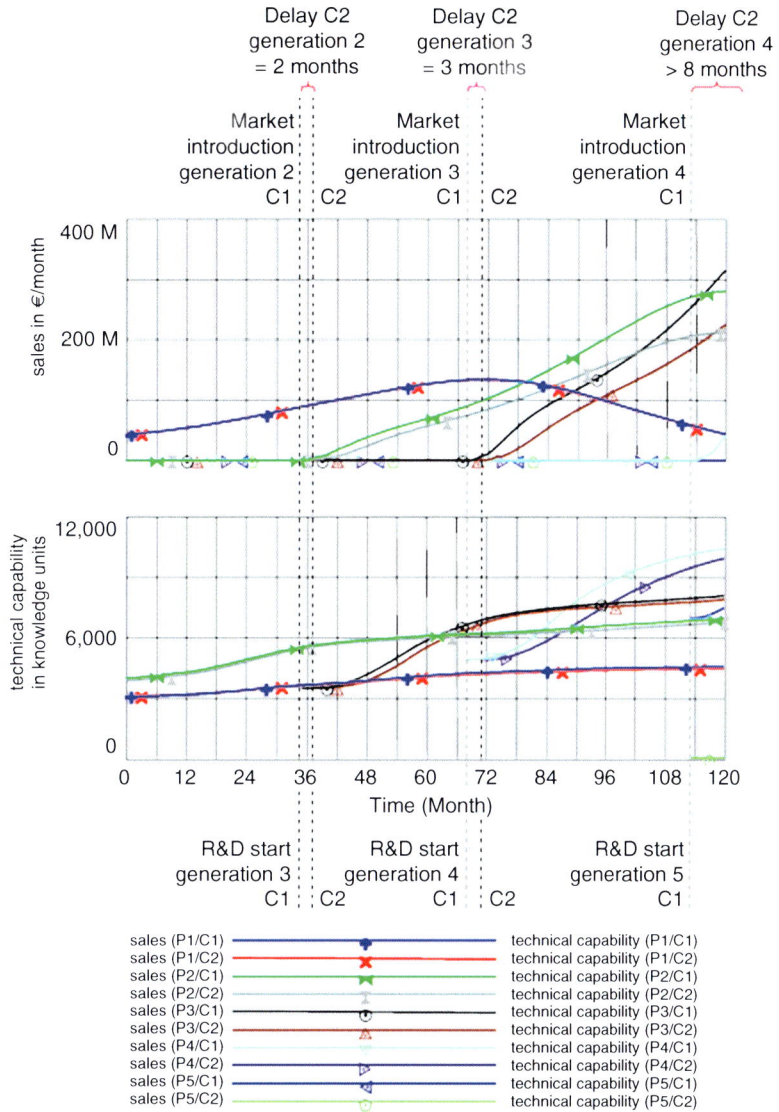

company 1 until they approach the boundaries of the technology.

Although literature discusses a variety of models to find optimal pricing strategies, these models usually only consider the market stage of a new product and neglect the interactions with the development stage of a new product. Pricing decisions not only drive the diffusion of an innovation, but they also have a strong impact on the resources available for research and development. Since the comprehensive innovation model links the stages of developing and introducing a new

product, the following simulations will show the impact of pricing strategies on performance in a competitive environment. In the analysis shown the first company uses the strategy of skimming price for all product generations. The second company alternatively uses a skimming price strategy in the first model run, myopic profit maximization strategy in the second run, and the strategy of penetration prices in the third run. The initial conditions are identical, except the price strategy settings. Sales volume, market position, and cumulated discounted profits are used to judge

the advantages of the alternative pricing strategies. Market entry time is used as a measure of
time-to-market.

The logic behind the skimming price strategy
is to sell new products with high profit margins in
the beginning of a life cycle to receive high returns
on investment , achieve short pay off periods, and
high resources for the R&D- process. However, in
a dynamic competitive setting the strategy of
myopic profit maximization and penetration
prices achieve better results (Fig. 21). Company

1 which uses a skimming price strategy achieves
the lowest sales volume. Myopic profit maximization prices and penetration prices of the second
competitor causes the sales to increase stronger
through the combined price and diffusion effect.

The results are confirmed if the variable market
position - an aggregate of the market share a
company has for its different products - is taken
into account. For values greater than 1 the market
position is better than the one of the competitor.
Using the penetration strategy, company 2 can

**System Dynamics
Analysis of the Diffusion
of Innovations,
Fig. 21** Sales volume and
market position for different
pricing strategies

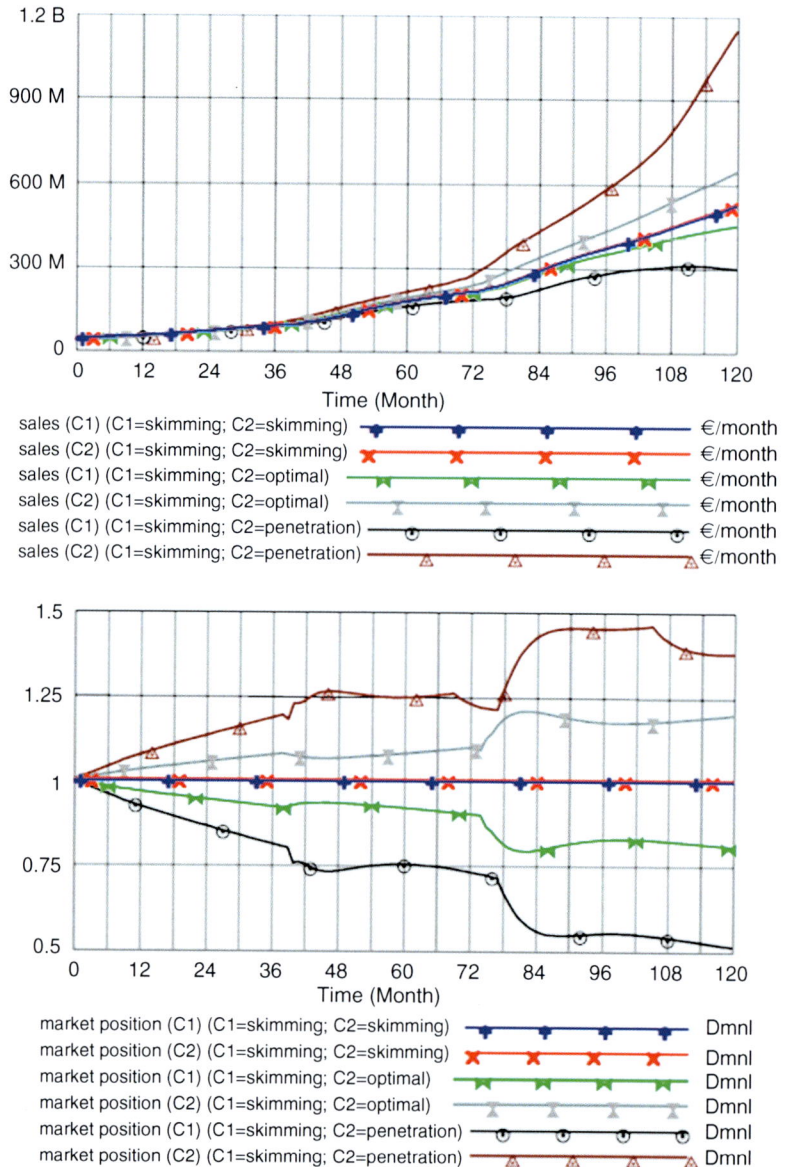

improve its market share, achieve higher sales volume and therefore has more resources available for R&D. This enables it to launch new products earlier than company 1. As shown in Table 1, the advantage of time-to-market increases from product generation to product generation.

The improvement in time-to-market for the first company's second product generation results from the slightly higher sales volume compared to the use of skimming pricing strategies for both competitors. The second company achieves the strongest improvements in time-to-market if it uses a penetration pricing strategy.

In terms of cumulative profits (Fig. 22) one would expect that skimming prices should generate the highest cumulated profits, however, this is not true. Penetration prices generate the highest results followed by skimming prices. The strategy of myopic profit maximization shows the least favorable outcome.

The simulations so far assumed a price response function with a constant price elasticity ε of -2. Since price elasticity influences both, the demand for a product as well as the price level (cf. Fig. 19), the influence of price elasticities have to be investigated before recommendations can be made. Assuming that company 1 uses a strategy of skimming prices and the second competitor follows a strategy of penetration pricing, Fig. 23 shows the time path of cumulated discounted profits and market position for ε between -3.2 and -1.2.

Due to the different profit margins – resulting from myopic profit maximization being the basis for price calculation – the use of the absolute value of the cumulated profits is not appropriate. Therefore, the second company's share of the total cumulated profits is used for evaluation purposes. The measure is calculated as

$$\left(\frac{\text{cum.profits}_2}{\sum_{i=1}^{2} \text{cum.profits}_i} \right).$$

The first graph in Fig. 23 shows that the initial disadvantage of the second company rises with increasing demand elasticity. However, its chance of gaining an advantage increases as well. In the case of lower demand elasticities ($\varepsilon > -1.7$) firm 2 cannot make up the initial disadvantage during the whole simulation. For demand elasticities ($\varepsilon > -1.4$) the cumulated profits ratio even deteriorates. Considering the market position the picture is similar. For demand elasticities $\varepsilon > -1.6$ the penetrations strategy leads to a loss in the market position in the long run. The improvements resulting from the introduction of the successive product generations are only temporary.

Managerial Implications

The simulations above lead to the insight that general recommendations for strategies are not feasible in such complex and dynamic environments. The specific structures like competitive situation, demand elasticity, or strategies followed by the competitors have to be taken into account. Recommendations only can be given in the context of the specific situation. Furthermore, the evaluation of strategies depends on the objectives of a company. If a firm wants to enhance its sales volume or the market share, the strategy of penetration pricing is the superior one. Viewing

System Dynamics Analysis of the Diffusion of Innovations, Table 1 Consequences of pricing strategies on market entry time

Pricing strategy C2[a]	Product generation 2			Product generation 3			Product generation 4		
	C1	C2	Delay C1 to C2	C1	C2	Delay C1 to C2	C1	C2	Delay C1 to C2
Skimming prices	38	38	0	71	71	0	n.i.	n.i.	-
Profit maximization	36	35	1	71	69	2	n.i.	118	> 2
Penetration prices	37	35	2	74	66	8	n.i.	102	> 8

[a]C1 uses skimming prices in all simulations
[b]n.i. = product was not introduced

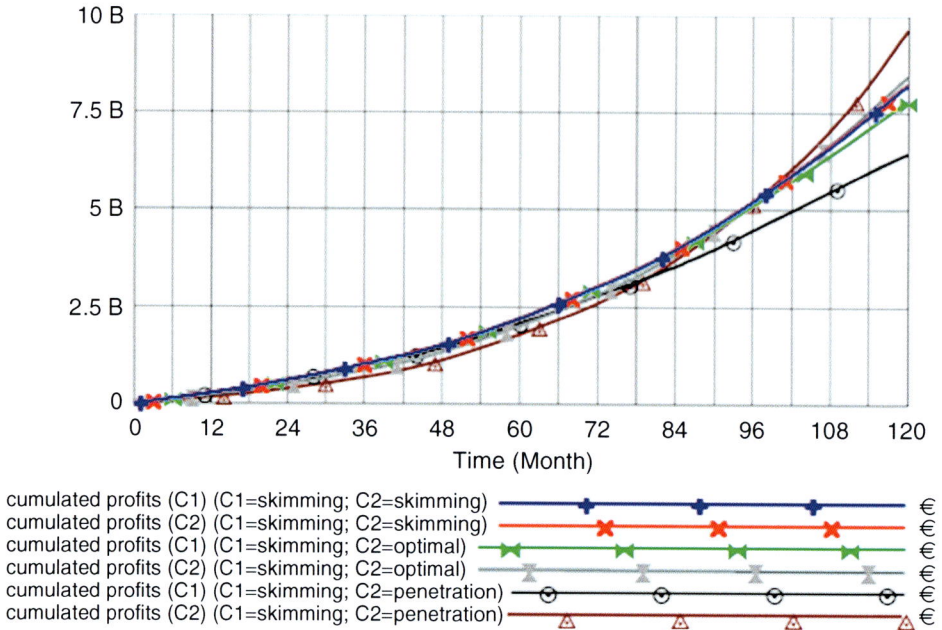

System Dynamics Analysis of the Diffusion of Innovations, Fig. 22 Time path of cumulated profits

cumulative profits and the readiness for market entry as prime objectives, the strategy of skimming prices is the best. However, these recommendations hold only for high demand elasticities. Furthermore, the model does not consider price reactions of competitors. The evaluation of improved strategic behavior would become even more difficult. The outcome and the choice of a particular strategy depend on many factors that influence the diffusion process. The dynamics and the complexity of the structures make it almost unfeasible to find optimal solutions. Improvements of the system behavior gained through a better understanding, even if they are incremental, are steps into the right direction.

Future Directions

The series of models presented here are designed in a modular fashion. They offer the flexibility to be adapted to different types of innovations, to different structures, initial conditions and situations. The models provide the opportunity to investigate courses of action in the setting of a management laboratory. They allow one to investigate different strategies and to learn in a virtual reality. They emphasize the process of learning in developing a strategy rather than the final result. To support learning processes, the models could be combined with an easy-to-use interface and serve as a management flight simulator which allows one to gain experience and understanding from playing.

Although the models cover a variety of different aspects in the management of innovations, they still can be extended. Besides more detailed mapping of corporate structures behind managerial decision processes the structures representing the diffusion process can be extended in various ways. Although some research already discusses the problems of mapping the substitution among successive product generations, this area deserves further attention. In particular in high-tech industries with short product life cycles the interrelations between successive product generations strongly influence the overall success of a company. Furthermore, the diffusion structures could be extended to include cross-buying and up-buying behavior of customers and by that link models of innovation diffusion to the field of customer equity marketing.

System Dynamics Analysis of the Diffusion of Innovations,
Fig. 23 Impact of demand elasticity on performance measures

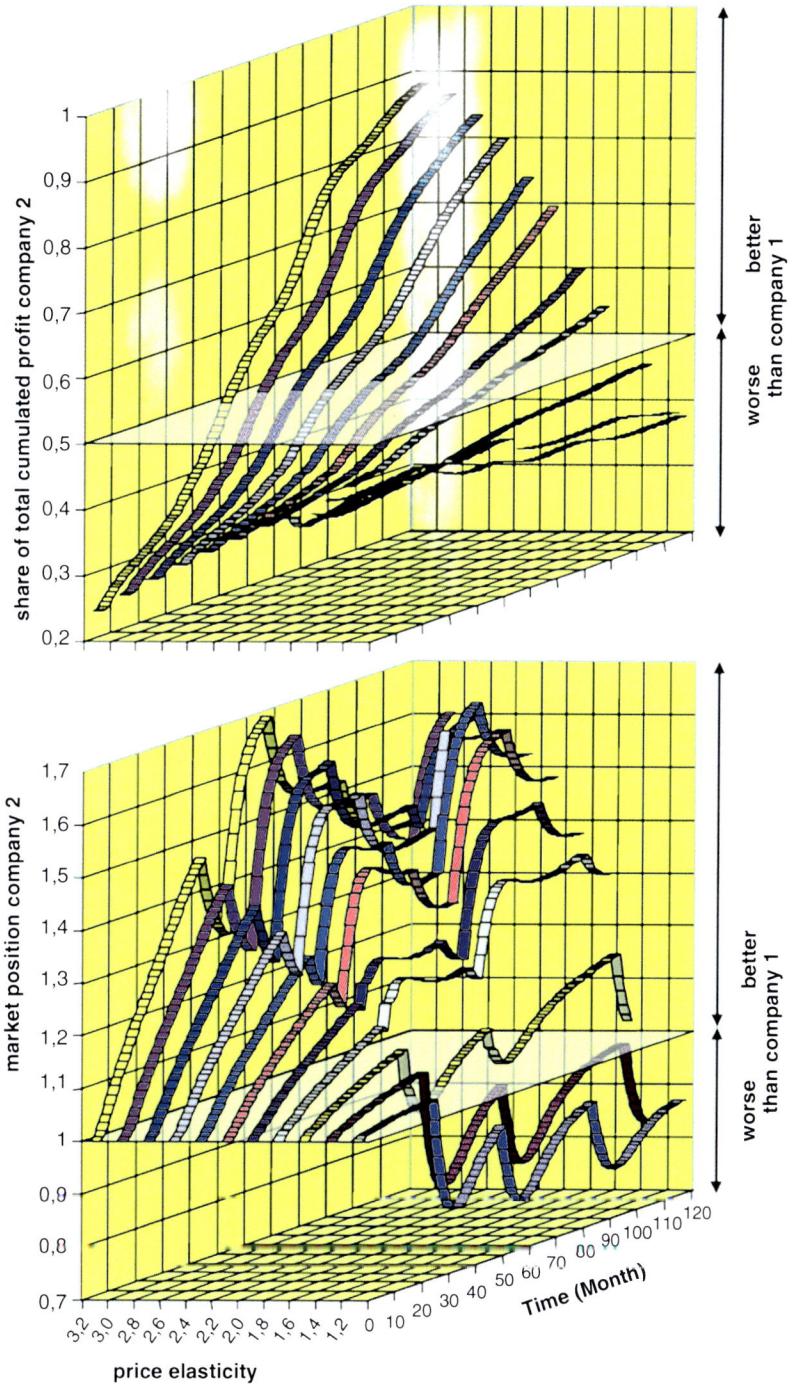

Bibliography

Primary Literature

Abernathy JW, Utterback JM (1988) Patterns of industrial innovation. In: Burgelman RA, Maidique MA (eds) Strategic management of technology and innovation. Irwin, Homewood, pp 141–148

Bailey NTJ (1957) The mathematical theory of epidemics. Griffin, London

Bass FM (1969) A new product growth model for consumer durables. Manag Sci 15:215–227

Bass FM (1980) The relationship between diffusion rates, experience curves, and demand elasticities for consumer durable technological innovations. J Bus 53:50–67

Bental B, Spiegel M (1995) Network competition, product quality, and market coverage in the presence of network externalities. J Ind Econ 43(2):197–208

Boston Consulting Group (1972) Perspectives on experience. Boston Consulting Group Inc., Boston

Bower JL, Hout TM (1988) Fast-cycle capability for competitive power. Harv Bus Rev 66(6):110–118

Brockhoff K (1987) Budgetierungsstrategien fiiur Forschung und Entwicklung. Z Betriebswirt 75:846–869

Brynjolfsson E, Kemerer CF (1996) Network externalities in microcomputer software: an econometric analysis of the spreadsheet market. Manag Sci 42(12):1627–1647

Bye P, Chanaron J (1995) Technology trajectories and strategies. Int J Technol Manag 10(1):45–66

Church J, Gandal N (1993) Complementary network externalities and technological adoption. Int J Ind Organ 11:239–260

Clarke DG, Dolan RJ (1984) A simulation analysis of alternative pricing strategies for dynamic environments. J Bus 57:179–200

Dumaine B (1989) How managers can succeed through speed. Fortune 13:30–35

Easingwood C, Mahajan V, Muller E (1983) A non-uniform influence innovation diffusion model of new product acceptance. Mark Sci 2:273–295

Farrell J, Saloner G (1986) Installed base and compatibility: innovation, product preannouncements, and predation. Am Econ Rev 76(5):940–955

Fisher JC, Pry RH (1971) A simple substitution model of technological change. Technol Forecast Soc Chang 3:75–88

Ford DN, Sterman JD (1998) Dynamic modeling of product development processes. Syst Dyn Rev 14:31–68

Forrester JW (1961) Industrial dynamics. MIT Press, Cambridge

Forrester JW (1968) Industrial dynamics – after the first decade. Manag Sci 14:389–415

Forrester JW (1981) Innovation and economic change. Futures 13(4):323–331

Georgescu-Roegen N (1971) The entropy law and the economic process. Harvard University Press, Cambridge

Graham AK, Senge PM (1980) A long wave hypothesis of innovation. Technol Forecast Soc Chang 17:283–311

Homer JB (1983) A dynamic model for analyzing the emergence of New Medical Technologies. Ph.D. Thesis. MIT Sloan School of Management

Homer JB (1987) A diffusion model with application to evolving medical technologies. Technol Forecast Soc Chang 31(3):197–218

Jeuland AP, Dolan RJ (1982) An aspect of new product planning: dynamic pricing. In: Zoltners AA (ed) TIMS

studies in the management sciences 18. North Holland, Amsterdam, pp 1–21

Katz ML, Shapiro C (1985) Network externalities, competition, and compatibility. Am Econ Rev 75(3):424–440

Kern W, Schröder HH (1977) Forschung und Entwicklung in der Unternehmung. Rowohlt Taschenbuch Verlag, Reinbek

Kotler P (1994) Marketing management. Prentice-Hall, Englewood Cliffs

Leibenstein H (1950) Bandwagon, Snob, and Veblen effects in the theory of consumers' demand. Q J Econ 64(2):183–207

Linstone HA, Sahal D (eds) (1976) Technological substitution. Forecast Techniques Appl. Elsevier, New York

Mahajan V, Muller E (1979) Innovation diffusion and new Product growth models in marketing. J Mark 43:55–68

Maier FH (1992) R&D strategies and the diffusion of innovations. In: Vennix JAM (ed) Proceedings of the 1992 international system dynamics conference. System Dynamics Society, Utrecht, pp 395–404

Maier FH (1995a) Die Integration wissens- und modellbasierter Konzepte zur Entscheidungsunterstützung im Innovationsmanagement. Duncker & Humboldt, Berlin

Maier FH (1995b) Innovation diffusion models for decision support in strategic management. In: Shimada T, Saeed K (eds) System dynamics, vol II. System Dynamics Society, Tokyo, pp 656–665

Maier FH (1996) Substitution among successive product generations – an almost neglected problem in innovation diffusion models. In: Richardson GP, Sterman JD (eds) System dynamics. System Dynamics Society, Boston, pp 345–348

Maier FH (1998) New product diffusion models in innovation management – a system dynamics perspective. Syst Dyn Rev 14:285–308

Mansfield EJ, Rapoport J, Schnee S, Wagner S, Hamburger M (1981) Research and innovation in the modern corporation: conclusions. In: Rothberg RR (ed) Corporate strategy and product innovation. Norton, New York, pp 416–427

Meieran ES (1996) Kostensenkung in der Chip-Fertigung. Siemens Z Special FuE Frühjahr:6–10

Milling P (1986a) Diffusionstheorie und Innovationsmanagement. In: Zahn E (ed) Technologie- und Innovationsmanagement. Duncker & Humboldt, Berlin, pp 49–70

Milling PM (1986b) Decision support for marketing new products. In: Aracil J, Machuca JAD, Karsky M (eds) System dynamics: on the move. The System Dynamics Society, Sevilla, pp 787–793

Milling PM (1987) Manufacturing's role in innovation diffusion and technological innovation. In: Proceedings of the 1987 international conference of the system dynamics society. The System Dynamics Society, Shanghai, pp 372–382

Milling PM (1989) Production policies for high technology firms. In: Murray-Smith D, Stephenson J, Zobel RN (eds) Proceedings of the 3rd European simulation congress. Society for Computer Simulation International, Edinburgh, pp 233–238

Milling PM (1991a) An integrative view of R&D and innovation processes. In: Mosekilde E (ed) Modelling and simulation. Simulation Councils, San Diego, pp 509–514

Milling PM (1991b) Quality management in a dynamic environment. In: Geyer F (ed) The cybernetics of complex systems – self-organization, evolution, and social change. InterSystems Publications, Salinas, pp 125–136

Milling P (1996) Modeling innovation processes for decision support and management simulation. Syst Dyn Rev 12(3):221–234

Milling PM, Maier FH (1993a) Dynamic consequences of pricing strategies for research and development and the diffusion of innovations. In: Zepeda E, Machuca JAD (eds) The role of strategic modeling in international competitiveness - system dynamics. The System Dynamics Society, Cancun, pp 358–367

Milling PM, Maier FH (1993b) The impact of pricing strategies on innovation diffusion and R&D performance. Syst Dyn Int J Policy Model 6:27–35

Milling P, Maier F (1996) Invention, innovation und diffusion. Duncker & Humboldt, Berlin

Milling P, Maier F (2004) R&D, technological innovations and diffusion. In: Barlas Y (ed) System dynamics: systemic feedback modeling for policy analysis. In: Encyclopedia of life support systems (EOLSS), Developed under the Auspices of the UNESCO. EOLSS-Publishers, Oxford, p 39. http://www.eolss.net

Norton JA, Bass FM (1987) A diffusion theory model of adoption and substitution for successive generations of high-technology products. Manag Sci 33:1069–1086

Norton JA, Bass FM (1992) Evolution of technological generations: the law of capture. Sloan Manag Rev 33:66–77

Paich M, Sterman JD (1993) Boom, bust, and failure to learn in experimental markets. Manag Sci 39:1439–1458

Pearl R (1924) Studies in human biology. Williams and Wilkins, Baltimore

Roberts EB (1964) The dynamics of research and development. Harper and Row Publishers, New York

Roberts EB (1978a) Research and development policy making. In: Roberts EB (ed) Managerial applications of system dynamics. Productivity Press, Cambridge, MA, pp 283–292

Roberts EB (1978b) A simple model of R&D project dynamics. In: Roberts EB (ed) Managerial applications of system dynamics. Productivity Press, Cambridge, MA, pp 293–314

Robinson B, Lakhani C (1975) Dynamic price models for new product planning. Manag Sci 21:1113–1122

Rogers EM (1983) Diffusion of innovation, 3rd edn. The Free Press, New York

Schmalen H (1989) Das Bass-Modell zur Diffusionsforschung – Darstellung, Kritik und Modifikation. Schmalenbachs Z Betriebswirtsch Forsch 41:210–225

Schumpeter JA (1961) Konjunkturzyklen – Eine theoretisch, historische und statistische Analyse des kapitalistischen Prozesses, Erster Band. Vandenhoeck & Ruprecht, Göttingen

Senge PM (1994) Microworlds and learning laboratories. In: Senge PM et al (eds) The fifth discipline fieldbook. Doubleday, New York, pp 529–531

Steele LW (1989) Managing technology. McGraw-Hill, New York

Sterman JD (1989) Modeling managerial behavior: misperceptions of feedback in a dynamic decision making environment. Manag Sci 35:321–339

Sterman JD (1992) Teaching takes off – flight simulators for management education. In: OR/MS Today. Lionheart, Marietta, pp 40–44

Sterman JD (1994) Learning in and about complex systems. Syst Dyn Rev 10:291–330

Sterman JD (2000) Business dynamics. Irwin McGraw-Hill, Boston

Thun JH, Größler A, Milling P (2000) The diffusion of goods considering network externalities – a system dynamics-based approach. In: Davidsen P, Ford DN, Mashayekhi AN (eds) Sustainability in the third millennium. Systems Dynamics Society, Albany, pp 204.1–204.14

Weil HB, Bergan TB, Roberts EB (1978) The dynamics of R&D strategy. In: Roberts EB (ed) Managerial applications of system dynamics. Productivity Press, Cambridge, MA, pp 325–340

Xie J, Sirbu M (1995) Price competition and compatibility in the presence of positive demand externalities. Manag Sci 41(5):909–926. Books and Reviews

Dynamics of Income Distribution in a Market Economy: Possibilities for Poverty Alleviation

Khalid Saeed
Worcester Polytechnic Institute, Worcester, USA

Article Outline

Glossary

Absentee owners Parties not present on land and capital resources owned by them.

Artisan owners Parties using own labor together with land and capital resources owned by them to produce goods and services.

Behavioral relations Causal factors influencing a decision.

Capital intensive A process or industry that requires large sums of financial resources to produce a particular good.

Capital Machinery equipment, cash, and material inputs employed for the production of goods and services.

Capitalist sector A subeconomy in which all resources are privately owned and their allocation to production and renting activities is exclusively carried out through a price system.

Capitalist system An economic system in which all resources are in theory privately owned and their allocation to production and renting activities is exclusively carried out through a price system.

Commercial Pertaining to buying and selling with intent to make profit.

Controlling feedback A circular information path that counters change.

Corporate Pertaining to a profit maximizing firm.

Economic dualism Side-by-side existence of multiple subeconomies.

Economic sector A collection of production units with common characteristics.

Entrepreneurship Ability to take risk to start a new business.

Feedback loops Circular information paths created when decisions change information that affects future decisions.

Financial market A mechanism that allows people to easily buy and sell commodities, financial instruments, and other fungible items of value at low transaction costs and at prices that reflect efficient markets.

Household income Income accrued to a household from wages, profits, and rents received by all its members.

Institutionalist economic models Models attributing performance of economies to institutional relationships and advocating selective government intervention to change the behavior that creates dysfunctions.

Iron law of wages David Ricardo's most well-known argument about wages "naturally"

© Springer Science+Business Media, LLC, part of Springer Nature 2020
B. Dangerfield (ed.), *System Dynamics*,
https://doi.org/10.1007/978-1-4939-8790-0_142

Originally published in
R. A. Meyers (ed.), *Encyclopedia of Complexity and Systems Science*, © Springer Science+Business
Media LLC 2017, https://doi.org/10.1007/978-3-642-27737-5_142-2

tending towards a minimum level corresponding to the subsistence needs of the workers.

Keynesian A belief that the total spending in the economy is influenced by a host of economic decisions – both public and private.

Labor intensive A process or industry with significant labor costs.

Labor productivity Output per worker or worker-hour.

Labor Economically active persons in an economy.

Marginal factor cost The incremental costs incurred by employing one additional unit of input.

Marginal revenue product The additional income generated by using one more unit of input.

Market economy An economy which relies primarily on interactions between buyers and sellers to allocate resources.

Marxist economic theory A theory highlighting exploitive mechanisms in an economic system and advocating central governance.

Marxist system A centrally run economic system emphasizing in theory the Marxist axiom "from each according to ability to each according to need."

Model An abstract representation of relationships in a real system.

Neoclassical economic theory A theory highlighting constructive market forces in an economic system and advocating consumer sovereignty and a price system as invisible sources of governance.

Non-linear A system whose behavior can't be expressed as a sum of the behaviors of its parts.

Opportunity cost Real value of resources used in the most desirable alternative, or the amount of one commodity foregone when more of another is consumed.

Ordinary differential equation A relation that contains functions of only one independent variable and one or more of its derivatives with respect to that variable.

Output elasticity Change in output caused by addition of one unit of a production factor.

Perfect market A hypothetical economic system that has a large number of buyers and sellers – all price takers trading a homogeneous product – with complete information on the prices being asked and offered in other parts of the market and with perfect freedom of entry to and exit from the market.

Political economy Interaction of political and economic institutions and the political environment.

Production factor A resource input such as land, labor, or capital contributing to production of output.

Productivity The amount of output created (in terms of goods produced or services rendered) per unit input used.

Purchasing power parity The value of a fixed basket of goods and services based on the ratio of a countries' price levels relative to a country of reference.

Revisionist economic models Models recognizing both constructive and exploitive forces and advocating government intervention against exploitation.

Sector A collection of production units with common characteristics.

Self-employment Work for a self-owned production unit without a defined wage.

Subeconomy A collection of production units and households with common characteristics.

System dynamics A methodology for studying and managing complex feedback systems, such as one finds in business and other social systems.

Theories of value How people positively and negatively value things and concepts, the reasons they use in making their evaluations, and the scope of applications of legitimate evaluations across the social world.

Unearned income Income received as rents.

Wage employment Work for a defined wage.

Definition of the Subject

Poverty is perhaps the most widely written about subject in economic development, although there is little agreement over its causes and how to alleviate it. The undisputed facts about poverty are that it is pervasive, growing, and that the gap between the rich and the poor is widening.

It is widely believed that the governments – irrespective of their ideological inclinations – have the responsibility to intervene to help the poor. Poverty alleviation is also the key mandate of International Bank for Reconstruction and Development (World Bank) and the many civil society organizations. World Bank places poverty line at purchasing power parity of $1 per day, which has improved a bit in terms of percentage below over the past three decades, except in Africa, but remains large in terms of head count. This threshold is however unrealistic since it targets absolutely basket cases. A poverty line at purchasing power parity of $3 per day, which is close to average purchasing power per capita in the poor countries, shows that both poverty head count and gap between rich and poor have been expanding across board. World Bank web site at http://iresearch.worldbank.org/PovcalNet/jsp/index.jsp allows making such computations for selected countries, regions, years, and poverty lines.

Neoclassical economic theory does not explicitly address the process of income distribution among households, although it often views income distribution as shares of profits and wages. In most economic surveys and censuses, however, income distribution is invariably measured in terms of shares of various percentages of the households. The fact that more than 80% of the income is claimed by fewer than 20% of the households who also own most of the capital resources in almost all countries of the world, the theory and the measurement have some common ground. Neoclassical theory has, however, shed little light on the process of concentration of wealth and how can this dysfunction be alleviated.

System dynamics, although rarely used for the design of public policy for addressing poverty, allows us to construct and experiment with models of social systems to understand their internal trends and test policy combinations for changing them. In this entry, I have used system dynamics modeling to understand the process of concentration of wealth and re-evaluate the on-going poverty alleviation effort.

The model, which subsumes resource allocation, production and entitlements, explains the many manifestations of income distribution in a market economy. It generates multiple patterns of asset ownership, wage and employment assumed in neo-classical, Marxist and revisionist perspectives on economic growth while it allows ownership to change through the normal course of buying and selling transactions based on rational though, information-bound criteria. Privately owned resources can be employed through hiring wage-labor, rented out or used for self-employment. In addition to the labor market conditions, the wage rate depends also on the opportunity cost of accepting wage employment as workers may be either self-employed or wage-employed. Since this opportunity cost varies with the capital resources owned by the workers, which may support self-employment, the wage rate is strongly affected by the distribution of ownership. Thus, ownership can become concentrated or widely distributed depending on legal and social norms governing transactions in the economy, which the model replicates. Extended experimentation with this model serves as a basis to identify critical policy instruments that make best use of the system potential for resource constrained growth and poverty alleviation through widening participation in the market and improving income distribution.

Introduction

The opening up of the major centrally planned economies of the world has brought to the fore problems concerning the psychological deprivation, inefficiencies of resource allocation and production, and the lack of dynamism experienced in the working of central planning in a socialist system. The accompanying enthusiasm for free market in a capitalist system has, however, hidden many of the dysfunctional aspects of this alternative. It should be recognized that both systems emerged from time-specific and geography-specific empirical evidence. Since their underlying models treat as given specific economic patterns, the institutional roles and the legal norms associated with each system have inherent weaknesses, which create dysfunctions when implemented in different environmental contexts (Robinson 1955; Streeten 1975). Thus, neither model may furnish an adequate basis for the design of policies for sustainable economic development and poverty alleviation. A search is, therefore, necessary for an organizational framework that might explain the internal trends inherent in each model as special modes of a complex system subsuming the variety of behavioral patterns recognized by specific models before an effective policy for change can be conceived (Saeed 1992).

Using as an experimental apparatus a formal model of the decision structure affecting wage determination, saving and investment behavior, and the disbursement of income, presented earlier in (Saeed 1994a), this entry seeks to identify the fundamental economic relations for creating a dynamic and sustainable market system that may also increase opportunities for the poor, whose market entry is often limited by their financial ability and social position (Sen 1999), to participate in the economy and be entitled to the value it creates. System dynamics provides the technical framework to integrate the various behavioral relations in the system (Forrester 1979; Sterman 2000a).

Notwithstanding the many objections to the abstract models of orthodox economics, which are difficult to identify in the real world (Leontief 1977; Robinson 1979), the model of this entry draws on neo-classical economics to construct a basic structure for growth and market clearing. This structure is, however, progressively modified by relaxing its simplifying assumptions about aggregation of sub-economies, wage determination, ownership, income disbursement, saving and investment behavior, financial markets, and technological differentiation between sub-economies to realize the many growth and income distribution patterns addressed in a variety of economic growth models.

The modified model I finally create represents a real world imperfect market in which expectations formed under bounded rational conditions govern the decisions of the economic actors (Simon 1982), as recognized in the pioneering works of Kaldor (1969), Kalecki (1965), Weintraub (1956), and Robinson (1978). The model also subsumes the concept of economic dualism first recognized by Boeke (1947) and developed further by Lewis (1958), Sen (1966), Bardhan (1973) and others to represent the multiple subeconomies that co-exist especially in the developing countries. Such a model is more identifiable with respect to the real world as compared with the time and geography specific concepts propounded by the various, often controversial, theories of economic growth.

Simulation experiments with the model explore entry points into the economic system for creating an egalitarian wage and income distribution pattern through indirect policy instruments. Also explored are the functions of entrepreneurship and innovation and the mechanisms that may increase the energy of those processes toward facilitating economic growth and alleviating poverty.

The Alternative Economic Models and Their Limitations

The economic models used as the bases for designing development policies over the past several decades have ascended largely from time-specific and geography-specific experiences rather than from a careful study of the variety of behavioral patterns occurring over various time periods and across several geographic locations. Among these, the socialist and the capitalist models are most at odds. They differ in their assumptions about ownership and income distribution patterns, the basis for wage determination, the influence of technology on income growth and the functions of entrepreneurship and innovation (Higgins 1959; Saeed 2005).

The neo-classical economic theory, which is the basis for the capitalist model, is silent on the ownership of capital resources, often assuming it in default to be widely distributed (Barro 1997). Thus, the labor-wage rate may bear little relationship to the income of households, who are also recipients of profits. It is assumed that private control of productive resources is a means for market entry, which creates unlimited potential for economic growth, although private investment is not often seen to be subject to self-finance due to the assumption that financial markets are perfect. The neo-classical economic theory also postulates that short-run labor-wage rates depend on labor market conditions, while in the long run, they are determined by the marginal revenue product of labor. Neo-classical models of economic growth, however, often make the simplifying assumption that equilibrium continues to prevail in both factor and product markets over the course of growth. Thus, only minor fluctuations may occur in wages, profits and prices in the short run, and these can be ignored.

The belief in the existence of such equilibrium is further strengthened by the Keynesian argument for the ineffectiveness of the market mechanisms due to the dependence of prices on long-term wage contracts and production plans which may not respond easily to short-run changes of the market. As a result of the belief in this theory of wage determination, technological choices that increase labor productivity are expected to have a positive effect on wage rates and household income, because they increase the marginal revenue product of labor. Entrepreneurship is viewed as important for new entry into economic activity, which is open to all, and innovation is supposed to benefit society through increased productivity. With these assumptions, the capitalist system advocates minimal government intervention in the economy. This model is widely presented in the many texts on economic development. Pioneering texts include Hirshliefer (1976) and Kindelberger and Herrick (1977).

Marxist economic theory, which underpins the socialist model, assumes on the other hand that ownership of capital resources is concentrated in a minority excluding the workers and that the majority of households receive no part of the profits. Thus, wage payments have a strong effect on household income. The Marxist theory views private ownership as a source of exploitation and postulates labor-wage rates determined by the consumption necessary for a worker to support production in a grossly labor surplus economy following Ricardo's iron law of wages (Marx 1891; Ricardo 1817). The labor-wage rate is, thus, based on the real value of the commodities needed for a worker to subsist, which is more or less fixed, irrespective of the contribution of labor to the production process. In such conditions, technological choices that increase labor productivity may indeed only serve to increase the share of the surplus of product per unit of labor appropriated by the capitalists. In this model, entrepreneurship is viewed as an asocial activity and innovation seen to originate from the need to boost falling returns on capital. Attributing the development of these conditions to market failure, the socialist system assigns control of the economy to the government.

There also exist a number of revisionist models of political economy attempting to explain the nature of interdependence of the multiple sub-economies observed to co-exist in many developing countries in violation of the theoretical premises of the neo-classical model according to which all production factors must eventually move to the most efficient sector. These models often attribute the development of disparities between the various sub-economies to exploitative mechanisms that tend to maintain an upper hand of the stronger influence groups. The revisionist analyses have largely led to making moral appeals for the government policy to target the poor and the disadvantaged in its development effort, which is a stated mandate of the International Bank for Reconstruction and Development (World Bank). Targeting the poor has also been advocated widely by numerous development economists over the past half century. They include such prominent economists as Myrdal (1957), Lipton (1977), Galbraith (1979), and Sen (1999).

Indeed, each economic system can often be endorsed with the help of selected historical evidence, and this has been fully exploited to fuel the traditional debate between the neo-classical and Marxist economic schools. Interesting artifacts of this debate include the normative theories of value suggested by each system to justify the various wage systems, which have little practical significance for development policy (Robinson 1969; Sraffa 1960). This is unfortunate, since contradictions of evidence should clearly indicate the existence of fundamental organizational arrangements in the economic system, which are capable of creating the multiple behavior patterns on which the various economic models are based. Once identified, such arrangements may also serve as entry points for the design of evolutionary changes in an existing pattern. To quote Professor Joan Robinson:

> Each point of view bears the stamp of the period when it was conceived. Marx formed his ideas in the grim poverty of the forties. Marshal saw capitalism blossoming in peace and prosperities in the sixties. Keynes had to find an explanation for the morbid condition of 'poverty in the midst of plenty' in the period between the wars. But each has significance for other times, for in so far as each theory is valid, it throws light upon essential characteristics of the system which have always been present in it and still have to be reckoned with. (Robinson 1955)

Following sections of this entry experiment with a system dynamics model of an economic system, widely found in the developing countries and presented earlier in Saeed (1994a), to understand the variety of economic patterns experienced over time and geography under different legal and social norms. Furthermore, exploratory experimentation with this model helps to outline the basic principles of a market system that can sustain growth, create equitable distribution of benefits and facilitate innovation and productivity improvement, all widely deemed necessary for poverty alleviation.

A System Dynamics Model of Resource Allocation, Production and Entitlements

A system dynamics model subsuming the broad decision rules that underlie resource allocation, production, and income disbursement processes of a developing country economic system was proposed in Saeed (1988) and further experimented with in Saeed (1994a). In this model, capital, labor, and land (which may also be assumed as a proxy for natural resources) are used as production factors. Model structure provides for the functioning of two modes of production, commercial, in which resources are employed on the basis of their profitability and which is managed by the capitalist sector of the

economy; and self-employed, in which workers not employed in the commercial mode make a living. These two modes of production have been referred to variously in the literature, for example as oligopolistic and peripheral firms (Gordon 1972), formal and informal sectors (Lewis 1958), and modern and traditional sub-economies (Fie and Ranis 1966).

It has been assumed in the model that all workers, whether self-employed using their own or rented capital resources or employed as wage-workers by the capitalist sector, are members of a homogeneous socio-economic group with a common interest, which is to maximize consumption. This group is also the sole supplier of labor in the economy since the small number of working capitalists is ignored. On the other hand, the capitalist sector is assumed to maximize profit while it is also the sole wage-employer in the economy (Applebaum 1979; Averitt 1968; Sen 1966).

It is assumed that private ownership is protected by law, but land and capital assets can be freely bought, sold and rented by their owners. Each buying and selling transaction between the two sectors must be accompanied by a corresponding transfer of the cash value of the assets determined by the going market prices. The model also permits the appearance of technological differences between the capitalist and the self-employed sectors, when more than one technologies embodied in the type of capital used (named traditional and modern in the model) are available and the two sectors cannot employ the preferred technology with equal ease (Lipton 1977; Riech et al. 1973), or when the self-employed sector is burdened by excess workers not employed by the commercial sector while it lacks the financial capacity to use its preferred technology.

Figure 1 shows how workers and capital might potentially be retained and employed by the two sectors in the model. Rectangles represent stocks, valve symbols flows and circles intermediate computations following the diagramming convention of system dynamics modeling. The size of each sector is not specified and is determined endogenously by the model, depending on assumptions about the socio-technical environment in which the system functions.

Dynamics of Income Distribution in a Market Economy: Possibilities for Poverty Alleviation, Fig. 1 Potential worker and capital distribution between capitalist and self employed sectors

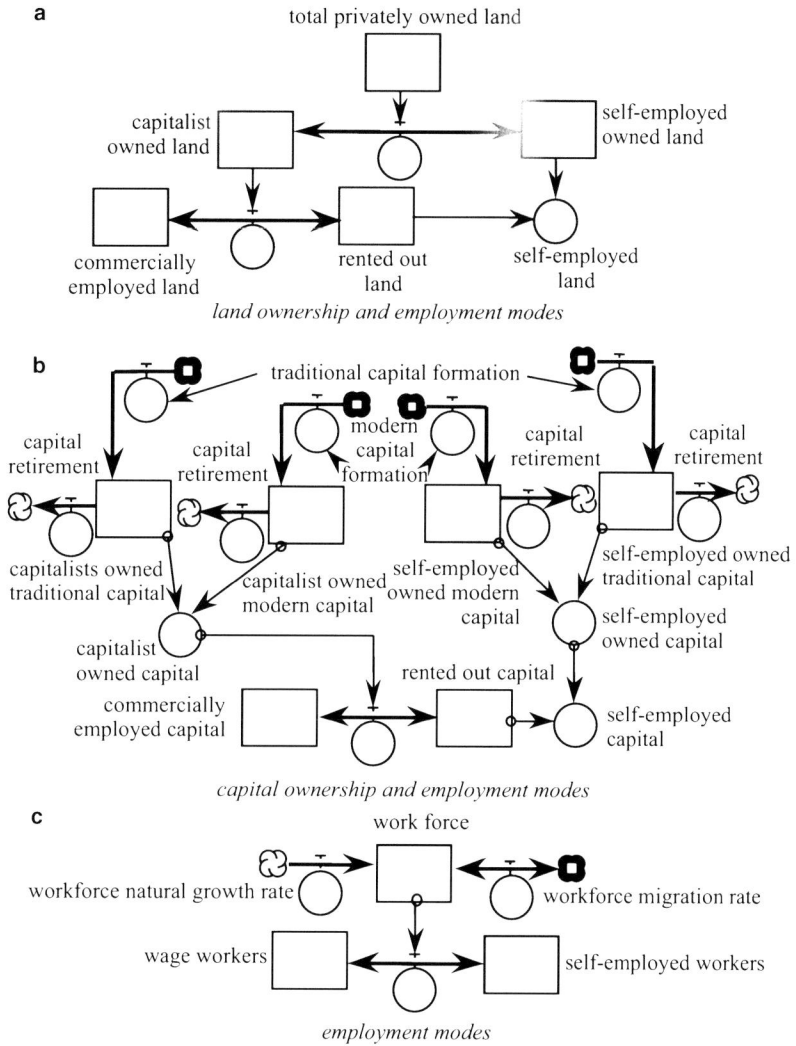

land ownership and employment modes

capital ownership and employment modes

employment modes

The changes in the quantities of the production factors owned or employed by each sector are governed by the decisions of the producers and the consumers of output and by the suppliers of the production factors acting rationally according to their respective motivations within the bounds of the roles defined for them by the system (Simon 1982). The value of production is shared by households on the basis of the quantity of the production factors they contribute and the factor prices they can bargain for (Eichner and Kregel 1975). Income share of the workers, less any investment needed to maintain self-employment, divided by the total

workforce, determines average consumption per worker, which represents the opportunity cost of accepting wage-employment and this is the basis for negotiating a wage (Sen 1966; Sraffa 1960).

Investment and saving rates in the two sectors are decoupled through a balance of internal savings. The financial markets are segmented by sectors and the investment decisions of a sector are not independent of its liquidity position, given by the unspent balance of its savings. Thus, investment decisions depend on profitability criteria, but are constrained by the balance of internal savings of each sector (McKinnon 1973; Minsky 1975).

Figure 2 shows the mechanisms of income disbursement, saving and internal finance incorporated into the model.

The saving propensity of all households is assumed not to be uniform. Since capitalist households receive incomes that are much above subsistence, their saving propensity is stable. On the other hand, the saving propensity of the worker households depends on their need to save to support investment for self-employment and on how their absolute level of income compares with their inflexible consumption (Kaldor 1966; Kalecki 1971; Marglin 1984).

The broad mathematical and behavioral relationships incorporated into the model are given in the Appendix "A.1." Technical documentation

and a machine-readable listing of the model written in DYNAMO code are available from the author on request.

Replicating Income Distribution Patterns Implicit in Models of Alternative Economic Systems

The model is simulated under different assumptions about wages, rents, financial markets and technology and its behavior analyzed in relation to the various theoretical and empirical premises its information relationships represent.

As an arbitrary initial condition, production factors are equally divided between the two sectors and equilibrium in both product and factor

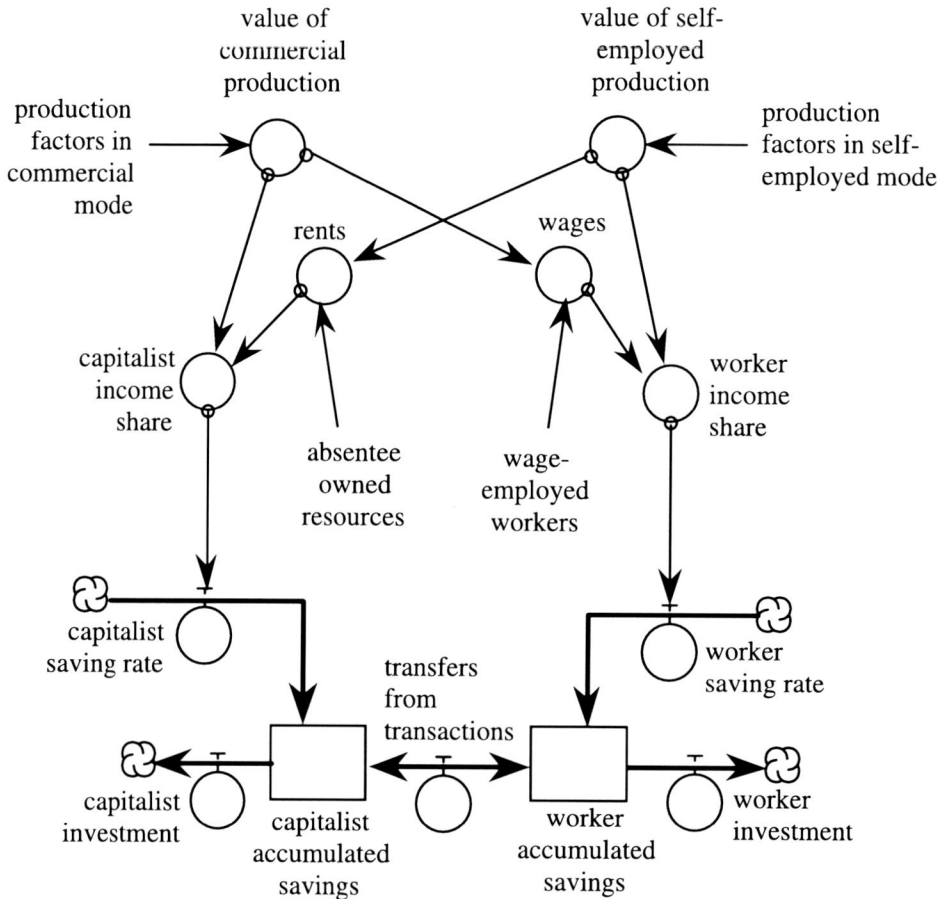

Dynamics of Income Distribution in a Market Economy: Possibilities for Poverty Alleviation, Fig. 2 Income disbursement process

markets is assumed to exist under the conditions of a perfect economic system as described in neo-classical economics. Thus, the marginal revenue products of land and capital are initially assumed to be equal to their respective marginal factor costs determined by an exogenously specified interest rate which represents the general pattern of preferences of the community for current as against future consumption (Hirshliefer 1976). The marginal revenue product of workers is initially set equal to wage rate. The market is initially assumed to be clear and there is no surplus of supply or demand.

Replicating the Theoretical Neo-classical System

This experiment is aimed at understanding internal trends of a system representing the neo-classical economic theory. To transform the model to represent this system, it is assumed that the production factors employed by each sector are owned by it and no renting practice exists (Barro 1997). The wage rate is assumed to be determined by the marginal revenue product of workers and the availability of labor instead of the opportunity cost to the workers of supplying wage-labor. Financial markets are assumed to be perfect and investment decisions of the two sectors are uncoupled from their respective liquidity positions. It is also assumed that the technology of production is the same in the two sectors and, in terms of the model, only traditional capital is available to both of them. The only difference between the two sectors is that the capitalist sector can vary all production factors, including labor to come to an efficient mix, while the self-employed sector may absorb all labor not hired by the capitalist sector, while it can freely adjust other production factors to achieve an efficient mix.

The model thus modified stays in equilibrium when simulated as postulated in neo-classical economic theory. When this equilibrium is disturbed arbitrarily by transferring a fraction of the workers from the capitalist to the self-employed sector, the model tends to restore its equilibrium in a manner

also similar to that described by the neo-classical economic theory. This is shown in Fig. 3.

The transfer raises the marginal revenue product of workers in the capitalist sector, which immediately proceeds to increase its workforce. The transfer also raises the intensity of workers in the self-employed sector as a result of which the marginal revenue products of land and capital in that sector rise. Hence, it proceeds to acquire more land and capital. These activities continue until the marginal revenue products of the factors and their proportions are the same in the two sectors. Note that while the factor proportions and marginal revenue products of the factors are restored by the model to their original values, the absolute amounts of the various factors are different when new equilibrium is reached. There is, however, no difference in endowments per worker between the capitalist and the self-employed sectors.

Since factor payments are determined purely on the basis of contribution to the production process while the quantities of production factors allocated to each sector depend on economic efficiency, the wages and factor allocations seem to be determined fairly and efficiently, as if by an invisible hand. Ownership in such a situation can either be communal or very widely distributed among households since otherwise the wage bargaining process will not lead to fair wages. Renting of production factors among households is irrelevant since transfer to parties who can efficiently employ them is automatic.

Before anything is said about the empirical validity of the simplifying assumptions made in this model, the historical context of these assumptions must be examined carefully. The simplified model is based on Adam Smith's description of an industrial economy observed at the start of the industrial revolution. This economy was run by artisan-turned capitalists and there were many of these capitalists competing with one another, although, none had the financial muscle to outbid the others except through his/her ability to employ resources efficiently (Skinner 1974).

As far as labor wage rate was concerned, although there were instances of exploitation of workers at a later stage of the industrial revolution, the artisan workers could obtain a wage that

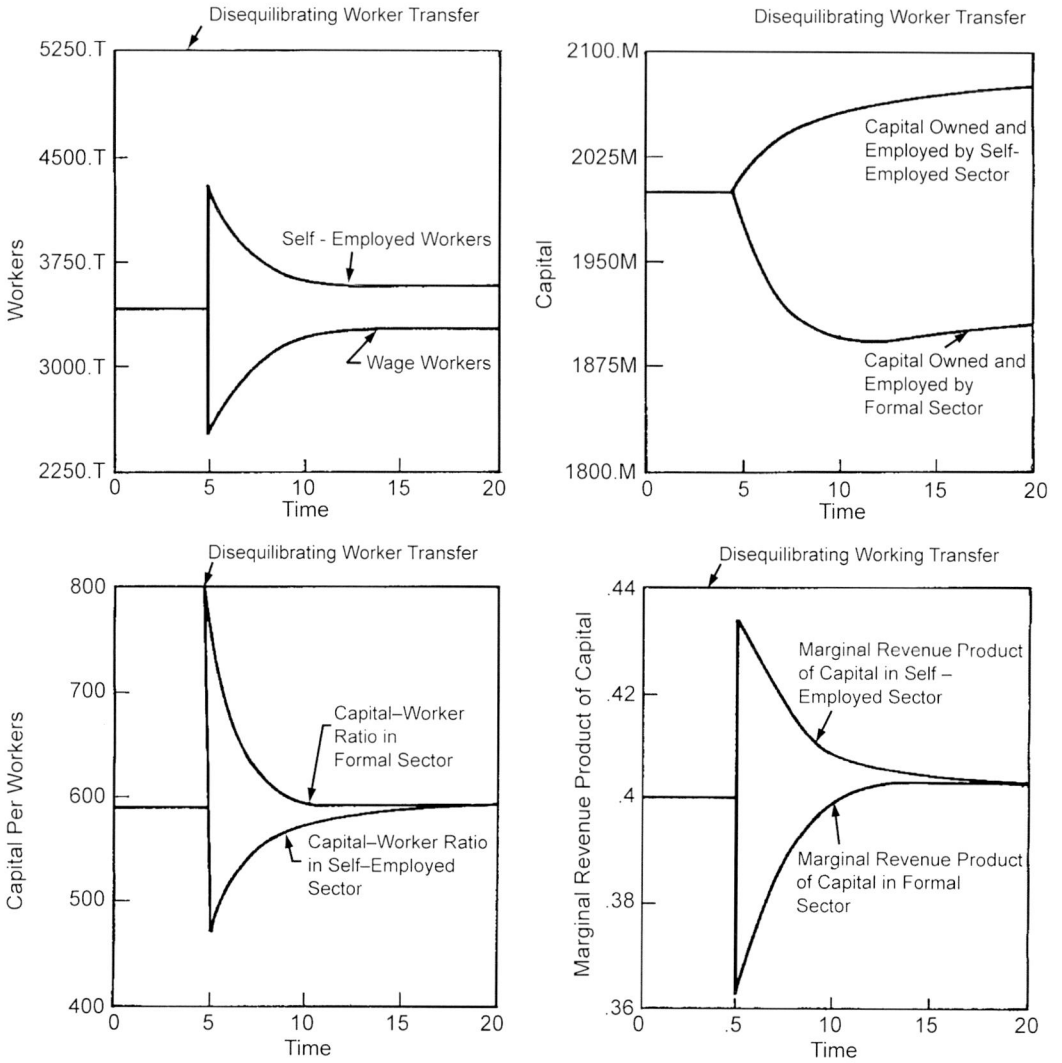

Dynamics of Income Distribution in a Market Economy: Possibilities for Poverty Alleviation, Fig. 3 Recovery from dis-equilibrium in a neo-classical system

was equal to their contribution of labor to the production process, as otherwise they could easily be self-employed since the economy was still quite labor intensive and the tools needed for self-employment may not have cost very much. Also, since ownership of the tools of a trade may have been quite widespread while the contribution of capital resources to the production process was quite small as compared to that of labor, a major part of the income might have accrued to the working households. In such circumstances, the simplifying assumptions of the neo-classical model may appear quite reasonable.

The neo-classical model became irrelevant, however, as the system made progress in the presence of a social organizational framework that legally protected ownership of all types and freely allowed the renting of assets, thus making possible an absentee mode of owning productive resources while technological changes also made the contribution of capital resources to the production process more significant.

Creating Worker Capitalism

It is not only methodologically expedient but also pedagogically interesting to explore what owner-ship and wage patterns might have emerged if labor-wages were determined through bargaining mechanisms incorporated into the model instead of fair payment equal to the marginal revenue product of workers, while all other assumptions

of a perfect market of the experiment of the last section were maintained.

Figure 4 shows a simulation of the model in which wage rate is determined by the average consumption expenditure per worker (as given in Eqs. (1 and 2) of the model described in the Appendix "A.1") while renting of production fac-tors and financial fragmentation of the households are still not allowed. This change in assumptions

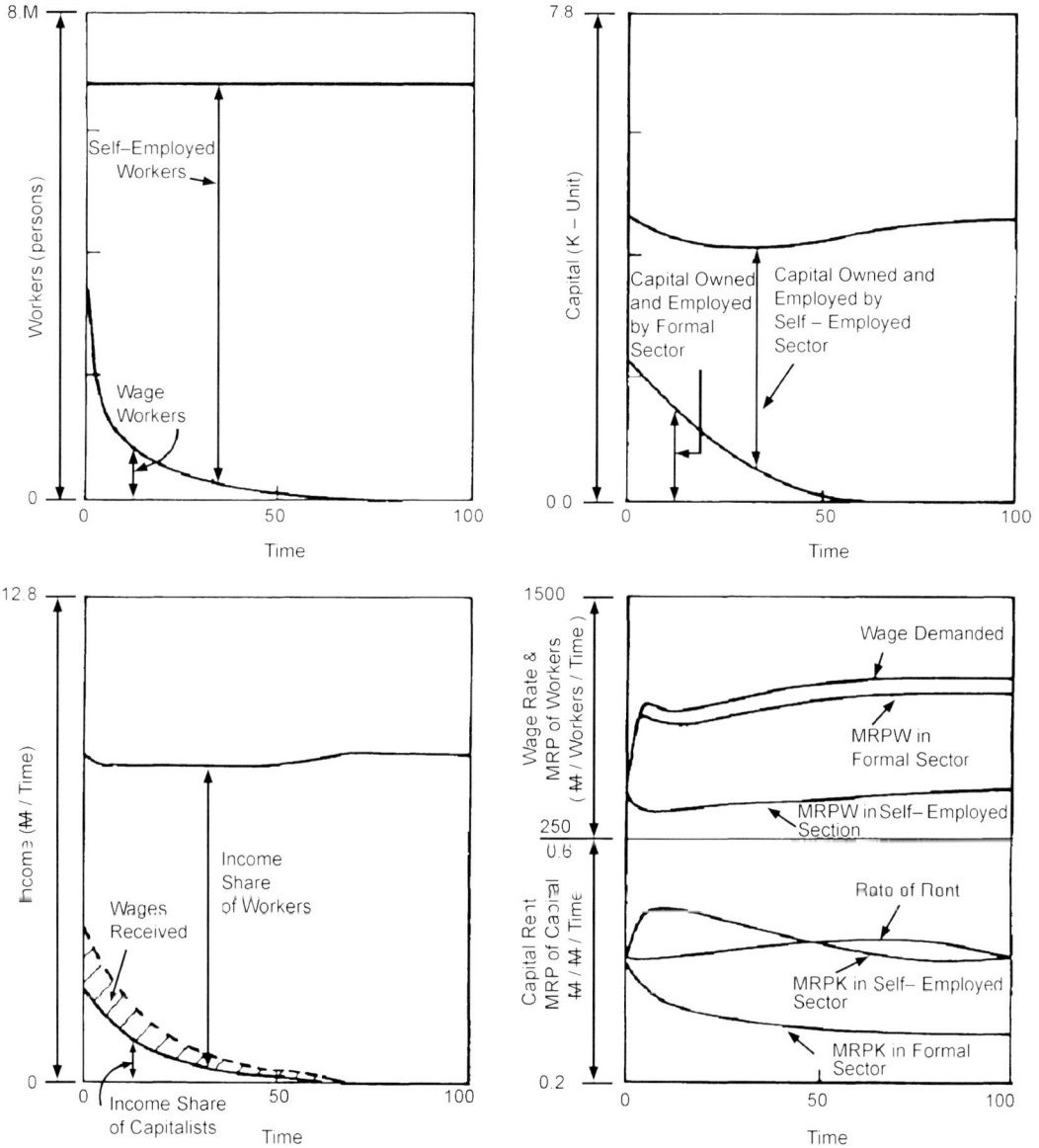

Dynamics of Income Distribution in a Market Economy: Possibilities for Poverty Alleviation, Fig. 4 The development of worker capitalism when wages depend on bargaining position of workers

disturbs the initial market equilibrium in the model thus activating its internal tendency to seek a new equilibrium. No exogenous disequilibrating changes are needed to generate the dynamic behavior in this simulation and in those discussed hereafter.

As a result of this change, the compensation demanded for working in the capitalist sector becomes much higher than the marginal revenue product of the workers. Thus, wage-workers are laid off and accommodated in the self-employed sector. Consequently, the marginal revenue product of land and capital in the self-employed sector increases and its bids for these resources rise. On the other hand, the decrease in the workforce of the capitalist sector increases its land and capital intensities and hence lowers their marginal revenue products. The falling productivity of these resources increases the opportunity cost of holding them. Since renting is not allowed, the capitalist sector is persuaded to sell the resources to the self-employed who can easily buy them since investment in the model is not subject to internal self-finance.

As the self-employed sector increases its land and capital holdings, its production rises. When increases in the production of this sector exceed the wage income lost due to decreasing wage disbursements from the capitalist sector, the net revenue of the workers, and hence their average consumption, rises. The wage rate is thus pushed up further, which necessitates further reductions in wage-workers. These processes spiral into a gradual transfer of all resources to the self-employed sector.

The marginal revenue products of land and labor in the two sectors tend to equilibrate at different values, but the capitalist sector exists only in theory because towards the end of the simulation almost all the resources are owned and managed by the self-employed. Since no part of the income is obtained by absentee owners, and working households may own and manage resources according to the quantity of labor they can supply, the income distribution may appear to be truly egalitarian.

Even though the above simulation is hypothetical, the wage and income distribution pattern shown by it may be experienced when the separation of resources from the households employing them is socially or legally ruled out or the state allocates capital resources and land according to the quantity and quality of labor supplied by a household. Instances of peasant economies having such characteristics have been recorded in history in tribal cultures and, in a somewhat advanced form, in medieval India (Mukhia 1981). Interestingly, such implicit assumptions are also subsumed in the illusive perfect market the neoclassical economic theory is based on.

Appearance of Absentee Ownership

When ownership of resources is legally protected, whether they are productively employed or owned in absentia, many renting and leasing arrangements may appear which may allow a household to own resources without having to employ them for production (Roulet 1976). This is borne out in the simulation of Fig. 5, in which resources are divided by the capitalist sector between commercial production and renting activities depending on the rates of return in each. Rents depend on long-term averages of the marginal revenue products of the respective factors and on the demand for renting as compared with the supply of rentable assets. In the new equilibrium reached by the model, the commercial mode of production and wage-employment gradually disappear but a substantial part of the resources continues to be owned by the capitalist sector, which rents these out to the self-employed sector.

Such a pattern develops because of the combined effect of wage and tenure assumptions incorporated into the model. When workers are laid off by the capitalist sector in response to a high wage rate, the marginal revenue products of land and capital for commercially employing these resources in this sector fall. However, as the laid-off workers are accommodated in the self-employed sector, the marginal revenue products of land and capital, and hence their demand in this sector, rise. Therefore, rents are bid up and the capitalist sector is able to get enough return from renting land and capital to justify maintaining its investment in these.

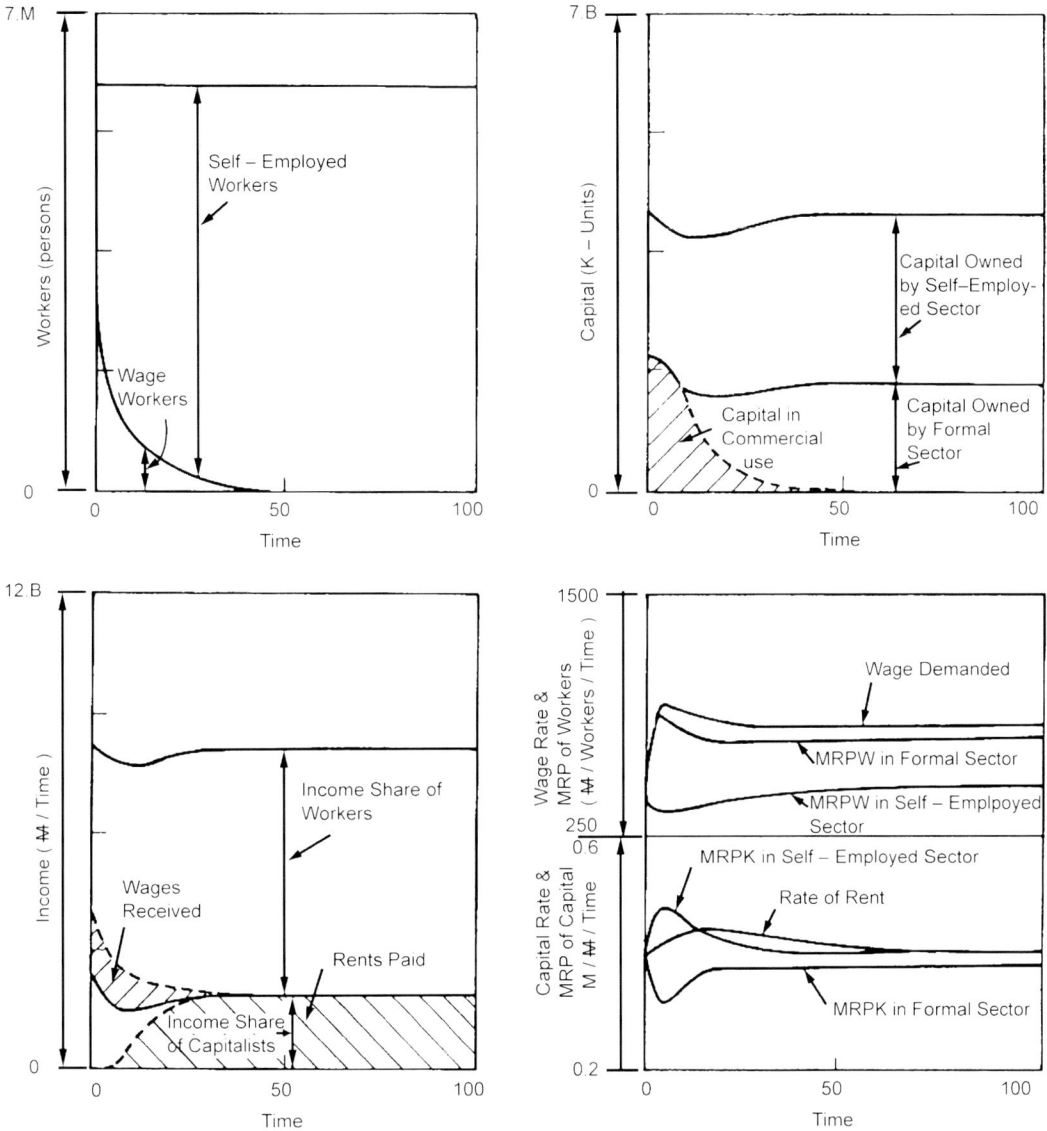

Dynamics of Income Distribution in a Market Economy: Possibilities for Poverty Alleviation, Fig. 5 The appearance of absentee ownership when renting is also allowed

Again, the marginal revenue products of the production factors in the commercial mode of production are only hypothetical as that mode is not practiced towards the end of the simulation. The renting mechanism allows the self-employed sector to adjust its factor proportions quickly when it is faced with the accommodation of a large number of workers. When the economy reaches equilibrium, the marginal rates of return of the production factors in the self-employed

sector are the same as those at the beginning of the simulation. But, the wage demanded equilibrates at a level lower than that for the exclusively self-employed economy described in the simulation of Fig. 4, because a part of the income of the economy is now being obtained by the absentee owners of the capitalist sector in the form of rent.

Note that, although the total income of the economy falls a little during the transition, it rises back to the original level towards the end equilibrium since the technology is uniform, irrespective of the mode of production. Also note that the end equilibrium distribution of income depends on initial distribution of factors when modifying assumptions are introduced, and on the volume of transfers occurring over the course of transition. Thus, an unlimited number of income and ownership distribution patterns would be possible depending on initial conditions and the parameters of the model representing the speeds of adjustment of its variables. The common characteristics of these patterns, however, are the presence of absentee ownership, the absence of a commercial mode of production, and a shadow wage that is less than an exclusively self-employed system.

Separation of Ownership from Workers and the Creation of a Marxist System

The ownership of resources becomes separated from the workers and concentrated in the capitalist sector in the model, irrespective of the initial conditions of resource distribution, when the assumption about the existence of a perfect financial market is also relaxed.

Figure 6 shows the ownership and wage pattern which develops when acquisition of resources by the capitalist and self-employed sectors is made dependent, in addition to their profitability, on the ability to self-finance their purchase. Recall also that the ability to self-finance depends on the unspent balance of savings, and the saving rate of the self-employed sector is sensitive both to the utility of saving in this sector to support investment for self-employment and to the rent burden of this sector compared with the factor contribution to its income from land and capital. The saving rate of the capitalist sector is assumed to be constant.

Such a pattern develops because of an internal goal of the system to employ resources in the most efficient way while the ownership of these resources can only be with the households who have adequate financial ability, which is also not independent of ownership.

Creation of a Dualist System

A dualist system is characterized by the side-by-side existence of both commercial and self-employed modes of production. The former appears to be economically efficient and is often also capital-intensive. The latter is seen to be economically inefficient and is also invariably labor-intensive. The side-by-side existence of these two modes of production in many developing countries has often puzzled observers, since according to the neo-classical economic theory, any inefficient production mode must be displaced by the efficient one.

A stable commercially run capital-intensive production sector existing together with a self-employed labor-intensive sector develops in the model if a technological differentiation is created between the capitalist and self-employed sectors. This is shown in the simulation in Fig. 7, in which an exogenous supply of modern capital is made available after end equilibrium of the simulation in Fig. 6 is reached.

Capital differentiation between the two sectors appears since the scale of the self-employed producers does not allow them to adopt modern technologies requiring indivisible capital inputs. The capitalist sector starts meeting its additional and replacement capital needs by acquiring a mixture of modern and traditional capital while the self-employed sector can use only traditional capital. However, the capital demand of the capitalist sector is met by modern capital as much as the fixed supply permits. The balance of its demand is met by acquiring traditional capital.

The output elasticity of modern capital is assumed to be higher than that of the traditional capital while the use of the former also allows an autonomous increase in output. The output elasticity of land is assumed to remain constant. The assumption of uniform returns to scale is maintained. Thus, the output elasticity of workers decreases when modern capital is introduced. These assumptions serve to represent the high productivity and labor-saving characteristics of the modern capital.

As its capital becomes gradually more modern and potentially more productive, the capitalist sector is able to employ its productive resources

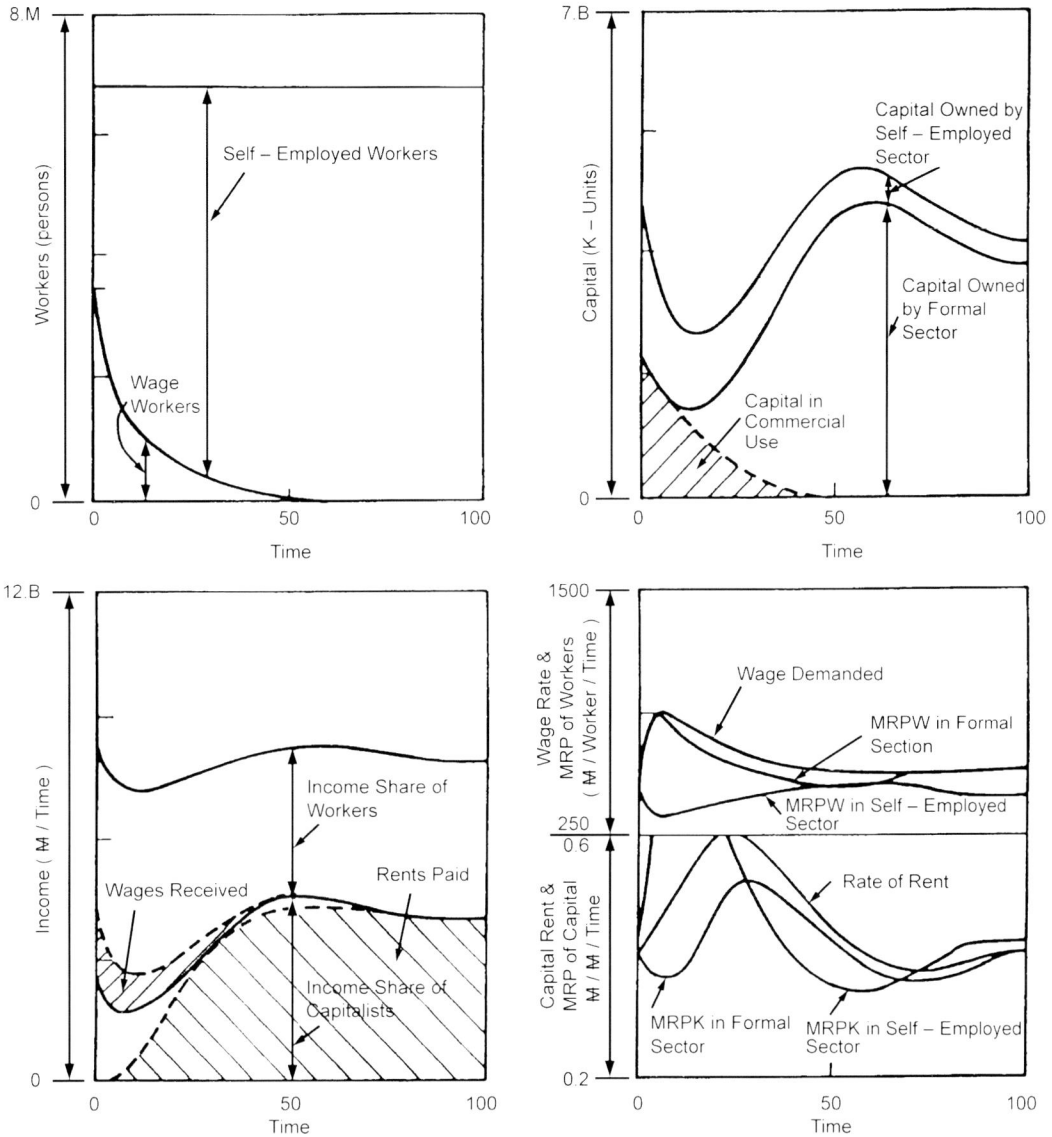

Dynamics of Income Distribution in a Market Economy: Possibilities for Poverty Alleviation, Fig. 6 Separation of ownership from workers as postulated by Marx system when investment must also be internally financed

with advantage in the commercial mode of production, instead of renting these out, and to employ wage-workers at the going wage rate. The increased productivity and income derived from this make it both economically and financially viable for the capitalist sector to invest more. Thus, its share of resources, when a new equilibrium is reached, is further increased.

Since the output elasticity of workers falls with the increase in the fraction of modern capital, the

marginal revenue product of workers in the commercial mode may not rise much with the increase in its output. At the same time, since resources are being transferred away by the capitalist sector from renting to commercial employment, the labor intensity and the demand for renting rises in the self-employed sector. Hence rents are bid up and it again becomes profitable for the capitalist sector to allocate resources to renting. The amount of resources rented out, however, will depend on

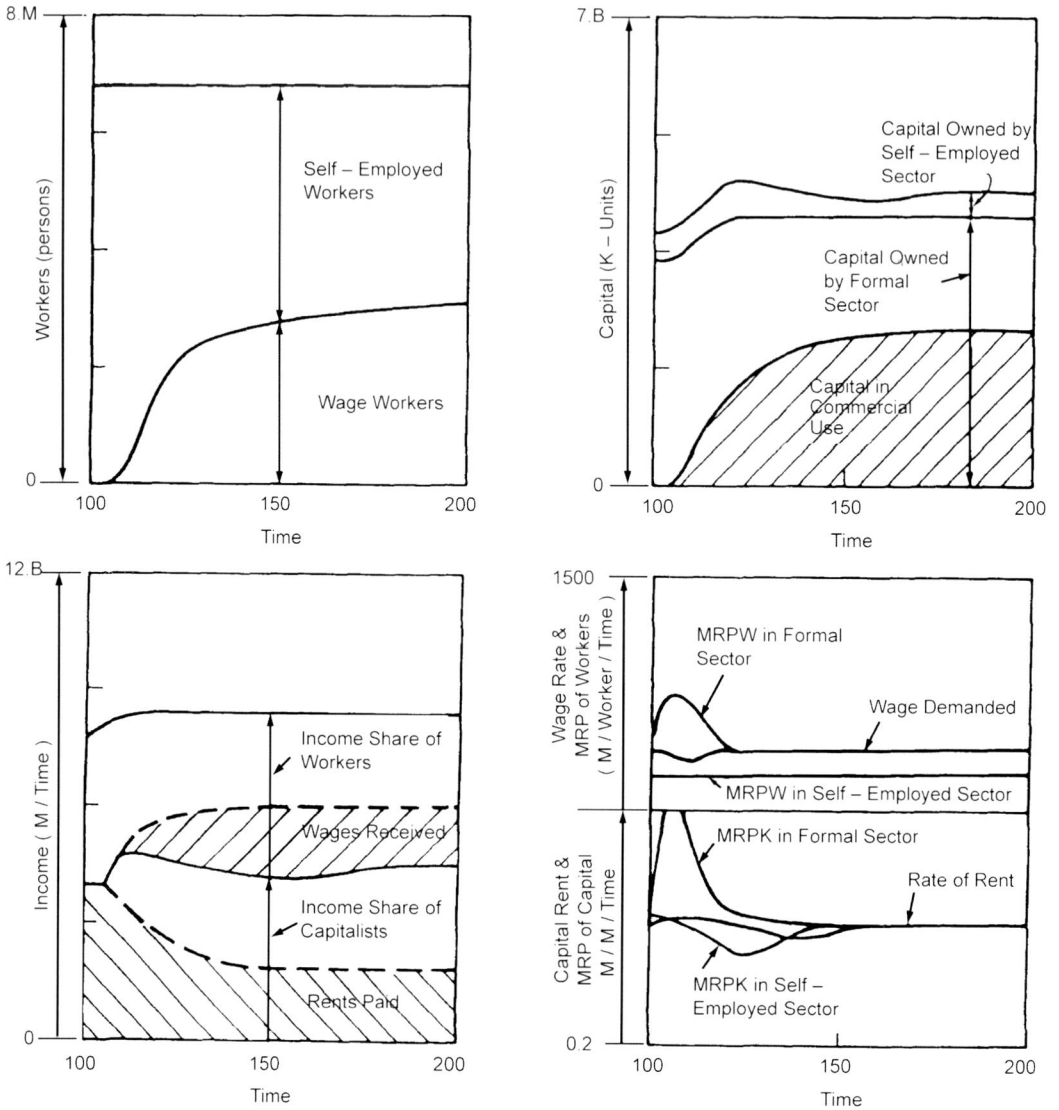

Dynamics of Income Distribution in a Market Economy: Possibilities for Poverty Alleviation, Fig. 7 Creation of dualist system when technological differentiation develops between the capitalist and self-employed sectors

the degree of technological differentiation that may be created between the two sectors.

The wage rate reaches equilibrium at a lower level and the rents at higher levels than without technological differentiation. Rents, however, equal marginal revenue products of land and capital, which rise in the capitalist sector because of employing superior technology and in the self-employed sector due to increased labor intensity.

Interestingly, dualist patterns appeared in the developing countries, both in the agricultural and industrial sectors, only after modern capital inputs became available in limited quantities. Labor-intensive peasant agriculture and small-scale industry and services carried out by the self-employed came to exist side-by-side with the commercially run farms and large-scale industry employing wage labor and modern technologies. However, worker income, both in wage-employment and self-employment, remained low (Griffin and Khan 1978).

Feedback Loops Underlying Wage and Income Patterns

The internal goal of a dynamic system represented by a set of non-linear ordinary differential equations is created by the circular information paths or feedback loops which are formed by the causal relations between variables implicit in the model structure. These causal relations exist in the state space independently of time (unless time also represents a state of the system). The existence of such feedback loops is widely recognized in engineering and they are often graphically represented in the so-called block and signal flow diagrams (Graham 1977; Richardson 1991; Takahashi et al. 1970).

While many feedback loops may be implicit in the differential equations describing the structure of a system, only a few of these would actively control the system behavior at any time. The non-linearities existing in the relationships between the state variables determine which of the feedback loops would actively control the system

behavior. A change may occur in the internal goals of a system if its existing controlling feedback loops become inactive while simultaneously other feedback loops present in its structure become active. Such a shift in the controlling feedback loops of a system is sometimes called a structural change in the social sciences and it can result both from the dynamic changes occurring over time in the states of the system and from policy intervention. The realization of a specific wage and income distribution pattern depends not on assumptions about initial conditions but on legal and social norms concerning ownership, renting, financing of investment and the state of technology, determining which feedback loops would be dominant (Forrester 1987; Richardson 1991).

Figure 8 describes the feedback loops, formed by the causal relations implicit in the model structure that appear to polarize income distribution by separating asset ownership from working households and creating a low wage rate, as shown in Fig. 6. An arrow connecting two variables indicates the direction of the causality while a positive or a negative sign shows the slope of the function relating cause to effect. For clarity, only key variables located along each feedback path are shown.

Dynamics of Income Distribution in a Market Economy: Possibilities for Poverty Alleviation, Fig. 8 Feedback loops creating dysfunctional income distribution trends in the capitalis system

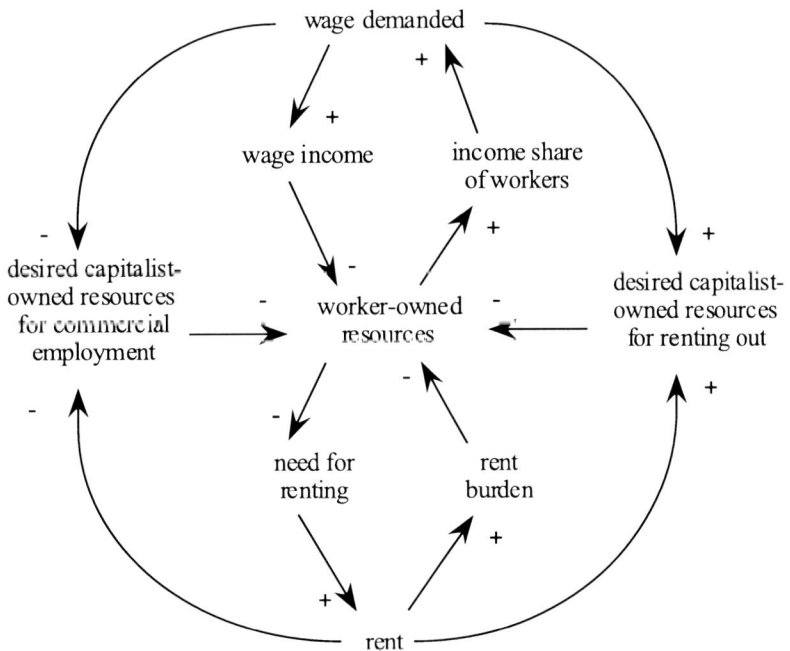

When productive resources can potentially be engaged in commercial or self-employed modes by owners and renters, any autonomous increase in the wage rate would not only decrease the desired capitalist owned resources for commercial employment, it would also concomitantly decrease the utility of investing in resources for self-employment. Thus, while the ownership of resources freed from commercial employment is not transferred to the self-employed sector, the surplus labor released by the commercial sector has to be absorbed in self-employment. As a result, worker income is depressed while the demand for renting rises. Thus, it not only continues to be profitable for the capitalist sector to hold its investments in land and capital, it also gives this sector a financial edge over the self-employed sector, whose savings continue to decline as its rent burden rises. These actions spiral into an expansion of ownership of resources by the capitalist sector even though the commercial mode of production is eliminated due to the high cost of wage labor. This also precipitates a very low wage rate when equilibrium is reached since a low claim to income of the economy creates low opportunity costs for the self-employed workers for accepting wage-employment.

Ironically, the fine distinction between the corporate, artisan and absentee types of ownership is not recognized in the political systems based on the competing neoclassical and Marxist economic paradigms. The former protects all types of ownership; the latter prohibits all. None creates a feasible environment in which a functional form of ownership may help to capture the entrepreneurial energy of the enterprise.

Possibilities for Poverty Alleviation

A functional economic system must incorporate the mechanisms to mobilize the forces of self-interest and entrepreneurship inherent in private ownership of the resources. Yet, it must avoid the conflicts inherent in the inequalities of income and resource ownership that led to the creation of the alternative socialist paradigm, which is devoid of such forces. According to the preceding analysis, the fundamental mechanism which creates the possibility of concentration of resource ownership is the equal protection accorded to the artisan and absentee forms of ownership by the prevailing legal norms. The financial fragmentation of households and the differences in their saving patterns further facilitate the expansion of absentee ownership. Technological differences between the capitalist and self-employed sectors not only make possible the side-by-side existence of the two modes of production, they also exacerbate the dichotomy between ownership of resources and workership. Apparently, the policy agenda for changing resource ownership and income distribution patterns should strive to limit renting and should additionally prevent the development of financial fragmentation and technological differentiation between the commercial and self-employed production modes if the objective is to minimize the conflicts related to income distribution.

Assisting the Poor

Programs to provide technological, organizational, and financial assistance to the poor have been implemented extensively in the developing countries over the past few decades although they have changed neither income distribution nor wage rate as reflected in many learned writings over these decades as well as the data published by UN and World Bank. This occurred because the increased productivity of the self-employed mode first pushed up wage rate, making renting-out resources more attractive for the capitalist sector than commercial production. However, the consequent decrease in wage payments and increase in rent payments pushed down the income share of the workers, which again suppressed the wage rate. Any efforts to facilitate the small-scale sector to increase its productivity through technological development also failed to affect income distribution since the mechanism of renting allowed the gains of the improved productivity to accrue to the absentee owners of the resources (Saeed and Prankprakma 1997). This experience is verified by the simulation of Fig. 9, which incorporates the policies to improve

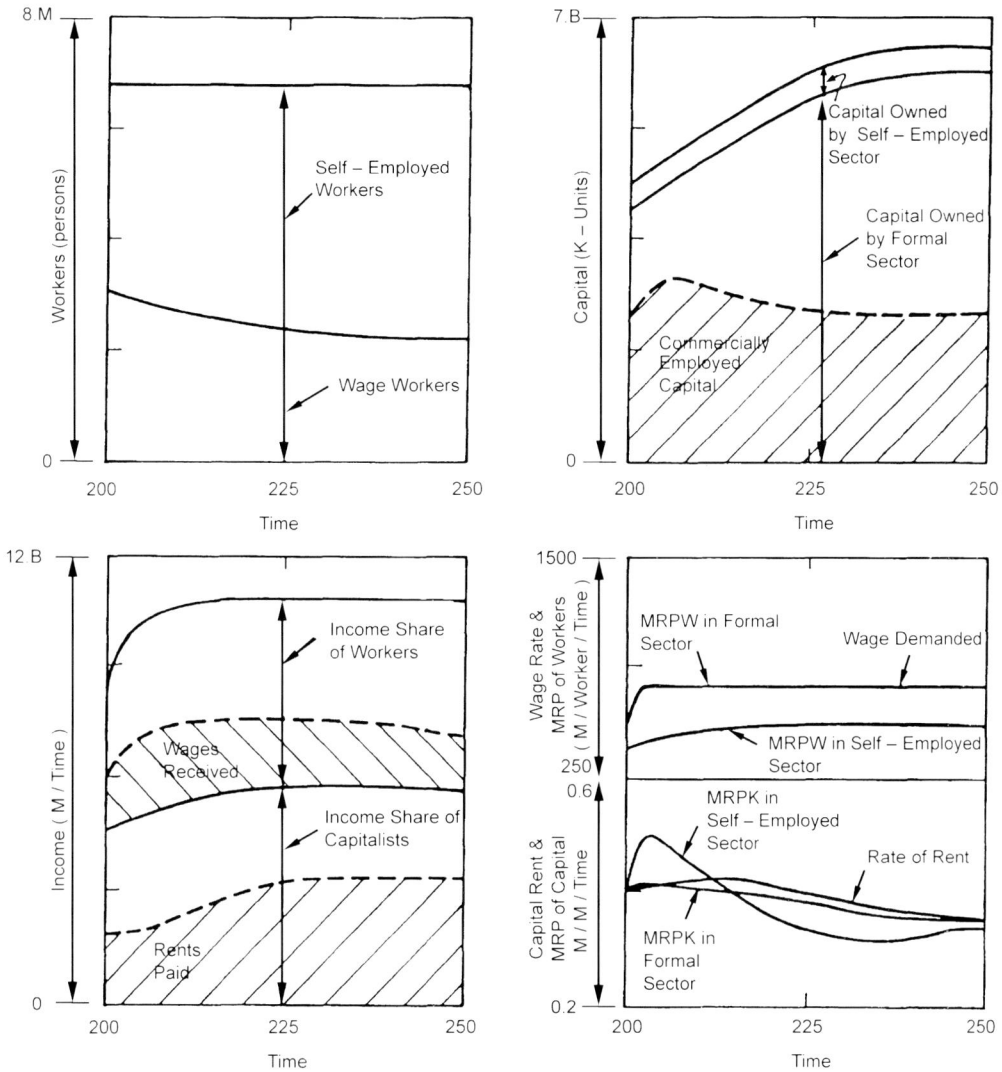

Dynamics of Income Distribution in a Market Economy: Possibilities for Poverty Alleviation, Fig. 9 Perpetuation of low wage and unequal income distribution resulting from widely used economic development policies

productivity, creating financial institutions and assisting the self-employed to adopt modern technologies. These policies only increase the size of the self-employed sector without increasing worker income, due to the possibility of separation of the mode of production from the ownership of resources. This indicates that influencing the decision to retain resources in absentee mode for renting out should be the key element of a policy framework to improve income distribution that should alleviate poverty.

Influencing Income Distribution

The cost of owning capital resources in absentee form can be increased by imposing a tax on rent income. The results of implementing this policy, together with the policies of Fig. 9 are shown in Fig. 10. In the face of a tax on rent income, resources which cannot be employed efficiently under the commercial system are offered for sale to the self-employed instead of being leased out to them. Purchase of these resources by the self-employed raises the entitlement of the workers

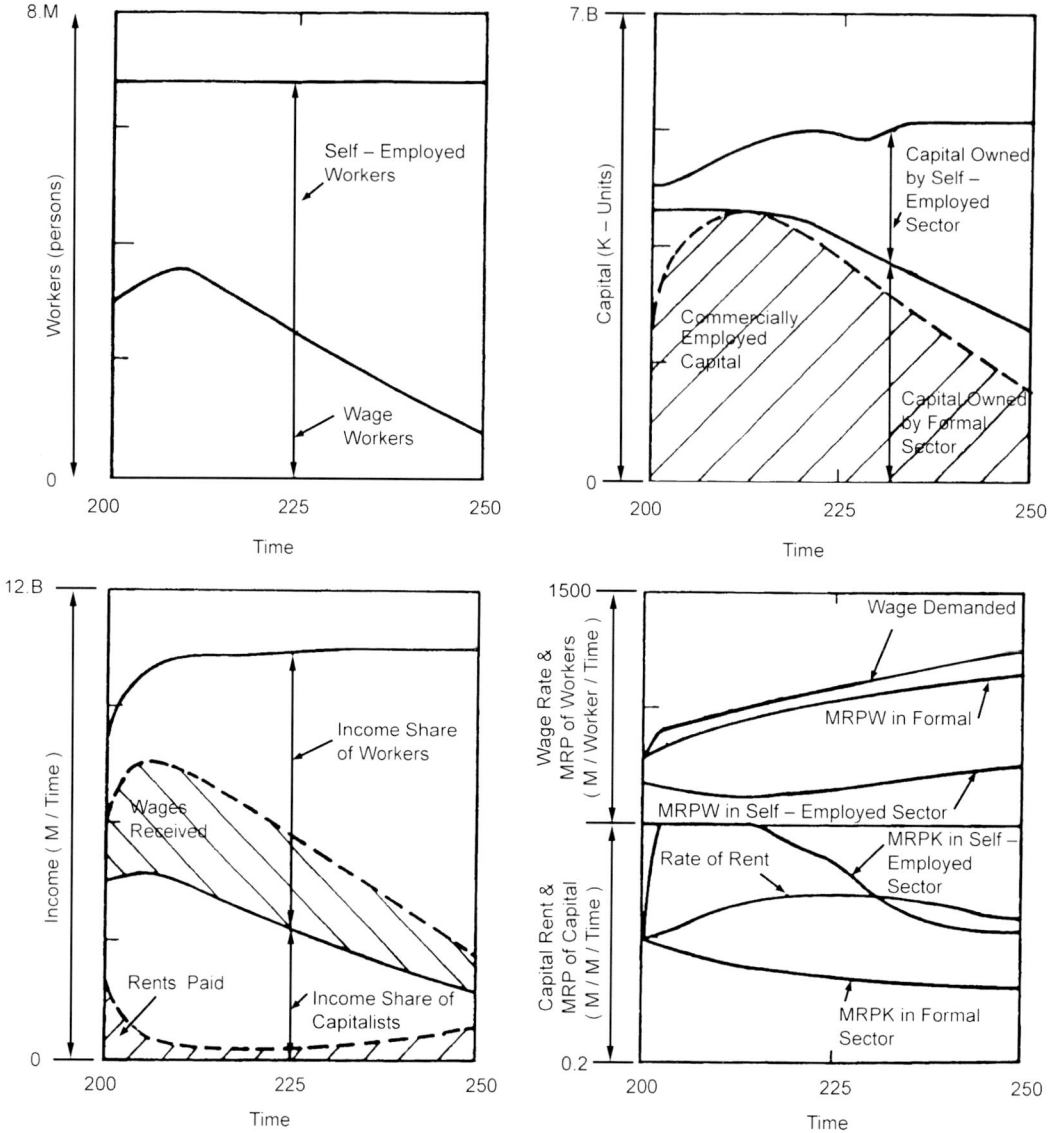

Dynamics of Income Distribution in a Market Economy: Possibilities for Poverty Alleviation, Fig. 10   Changes in wage and income distribution resulting from adding taxation of rent income to the policy package

to the income of the economy, which increases the opportunity cost of supplying wage-labor to the commercial sector. This raises wage rate, which makes the commercial mode of production even more uneconomical, unless it is able to apply a superior technology. Such changes spiral in the long run into a transfer of a substantial amount of resources to the self-employed sector. Concomitant efforts to decrease the financial

fragmentation of households and the technological differentiation between the two modes of production, along with improving productivity, further accelerate these changes.

Facilitation of Innovation and Productivity Improvement

Macroeconomic analyses concerning the industrialized countries show that technological innovation is one of the most important sources of growth (Denison 1974; Solow 1988). Studies conducted at the organizational level in the industrialized countries also show that innovations creating technological progress originate largely from small entrepreneurs or from large companies structured in a way to encourage small independent working groups (Quinn 1985; Roberts 1991). Thus, entrepreneurial activity is often credited with raising productivity through creation of technical and business-related innovations (Baumol 1988). The high and rising cost of labor in the developed countries, possibly also forces the wage-employers into finding innovative ways of maintaining high labor productivity and continuously striving to improve it.

On the other hand, economic growth has been dominated in the developing countries by relatively large and often highly centralized and vertically integrated companies. Technologies of production have mostly been replanted from the industrialized countries and indigenous innovation and technological development have had a poor track record (APO 1985). These technologies often do not perform as well as at their respective sources, but due to the availability of cheap labor, their inefficient performance still yields comfortable profits; hence little effort is made to improve productivity. There also exist serious limitations on the number of small players as large cross-section of the households in the developing countries lack the resources to effectively participate in any form of entrepreneurial activity (Saeed 1994a; Sen 1999). Innovation being a probabilistic process, limited participation drastically limits its scope.

There exists a promising institution in most developing countries, however, which has great potential as a focal point of entrepreneurial activity, which has remained dormant for lack of empowerment. This institution is the small family enterprise in the self-employed sector, which may take the form of a shop-house or an artisan manufacturing firm in the urban sector or a peasant farm in the rural sector. It allows participation from all members of the family while also providing the informal small-group organization

considered conducive to innovation in many studies. Its members are highly motivated to work hard and assume the risk of enterprise because of their commitment to support the extended family. This enterprise is somewhat similar to the small manufacturing units that created the industrial revolution in England in the early nineteenth century. It has also been observed that the small family enterprise tends to maximize consumption; hence its income significantly affects demand, which creates new marketing opportunities (APO 1985; Bardhan 1973). Unfortunately, this enterprise has been systematically suppressed and discriminated against in favor of the large-scale capitalist sector. Even its output remains largely unaccounted for in the national accounting systems of most countries (Eisner 1989; Hicks 1940).

The small family enterprise, variously described as the informal, labor-intensive, traditional, peasant, peripheral and sometimes inefficient sector in the developing countries has been stifled in the first instance by a set of social and legal norms through which the wealth has become concentrated in an absentee ownership mode. Working households are mostly poor and own few assets (Griffin and Khan 1978). The prosperity of these households will not only provide the much-needed financial resources for entrepreneurial activity, their capacity to spend will also create many marketing opportunities for the potential entrepreneur. Thus, influencing income distribution, through the policy framework proposed in the last section, ranks first on the agenda also for developing entrepreneurship and encouraging innovation. Once a significant cross-section of the populace becomes a potential participant in the economic activity, the development of infrastructure and facilitation to manage risk will also appear to be effective instruments to support entrepreneurship and innovation (Saeed 1991). The rise in the wage rates due to the possibility of alternative self-employment opportunities would, at the same time, force the large commercial enterprise to invest in technological innovation for productivity improvement, which should further improve the efficacy of the overall system.

Conclusion

Both neoclassical and Marxist models of economic growth seem to make restricting assumptions about ownership and mechanisms of wage determination, which are linked with specific time- and geography-related historical evidence. These restricting assumptions give internal consistency and a semblance of sustainability to each model, although they remove both from reality. A failure in the free-market system based on the no-classical model occurs when the invisible hand concentrates ownership of resources in a small minority, suppressing wage rate and creating social conflict due to income inequalities. On the other hand, a failure in the socialist system based on the Marxist model occurs, when the visible hand empowered to act in the public interest stifles entrepreneurial energy while also ignoring public interest in favor of its power interests (Rydenfelt 1983; Saeed 1990).

A behavioral model underlying wage and income distribution has been proposed in this entry, in which the opportunity cost of supplying a unit of labor to the capitalist sector is used as a basis for negotiating a wage. Neither this opportunity cost nor the ownership pattern are taken as given, while the dynamic interaction between the two creates a tendency in the system to generate numerous wage and income distribution patterns, subsuming those postulated in the neo-classical and Marxist theories of economics. The realization of a specific wage and income distribution pattern depends on legal and social norms concerning ownership, renting, the financing of investment and the state of technology.

Private ownership seems to have three forms, commercial, artisan and absentee. Predominance of artisan ownership creates an egalitarian wage and income distribution pattern while a healthy competition between the commercial and artisan firms may release considerable entrepreneurial energy. These functional forms can grow only if the renting of resources can be discouraged. On the other hand, absentee ownership creates a low wage rate and an unequal income distribution, while the growth of this form of ownership is facilitated through the renting mechanism.

Potentially, all three ownership forms can exist in an economic system. The problem, therefore, is not to favor or condemn private ownership *per se* as the alternative theories of economics have often advocated, but to understand the reasons behind the development of a particular ownership pattern and identify human motivational factors that would change an existing pattern into a desired one.

The most important reform needed at government level to alleviate poverty is the discouragement of the absentee ownership of capital assets, which would create a wider distribution of wealth. Widespread artisan ownership resulting from this would increase participation in entrepreneurial activity, which would allow adequate performance from the human actors in the system. Such reforms may however not be possible in the authoritarian systems of government pervasive in the developing countries since they must often limit civil rights and public freedoms to sustain power. Hence, the creation of a democratic political system may be a pre-condition to any interventions aimed at poverty alleviation. This, I have discussed elsewhere (Saeed 1990; Saeed 1994a; Saeed 2002a).

Future Directions

While the market system has often been blamed by the proponents of central planning for leading to concentration of wealth and income among few, it in fact offers a powerful means for redistributing income if the process of concentration is carefully understood and an intervention designed on the basis of this understanding. In fact, all economic systems can be reformed to alleviate the dysfunctional tendencies they are believed to have, provided the circular relationships creating such dysfunctions can be understood which should be the first objective of policy design for economic development.

Contrary to this position, economic development has often viewed developmental problems as pre-existing conditions, which must be changed through external intervention. Poverty, Food shortage, poor social services and human resources development infrastructure, technological backwardness, low productivity, resource depletion, environmental degradation and poor governance are cases in point. In all such cases, the starting point for a policy search is the acceptance of a snapshot of the existing conditions. A developmental policy is then constructed as a well-intended measure that should improve existing conditions. Experience shows, however, that policies implemented with such a perspective not only give unreliable performance, they also create unintended consequences. This happens because the causes leading to the existing conditions and their future projections are not adequately understood. The well-intentioned policies addressing problem symptoms only create ad hoc changes, which are often overcome by the system's reactions.

Table 1 collects three key developmental problems, poverty, food shortage and social unrest, and the broad policies implemented over the past several decades to address them. These problems have, however, continued to persist or even become worse.

The policy response for overcoming poverty was to foster economic growth so aggregate income could be increased; that for creating food security was intensive agriculture so more food could be produced; and for containing social unrest, the broad prescription is to strengthen internal security and defense infrastructure so public could be protected from social unrest. The unintended consequences of these policies are many, but in most instances, they include a continuation or worsening of the existing problems.

Thus, poverty and income differentials between rich and poor have in fact shown a steady rise, which is also accompanied by unprecedented debt burdens and extensive depletion of natural resources and degradation of environment. Food shortages have continued but are now accompanied also by land degradation, depletion of water aquifers, the threat of large-scale crop failure due to a reduction in crop diversity and a tremendous growth in population. Social unrest has often intensified together with appearance of organized insurgence burgeoning expenditures on internal security and defense, which has stifled development of social services and human resources and have created authoritarian governments with little commitment to public welfare.

The unintended consequences are often more complex than the initial problems and have lately drawn concerns at the global level, but whether an outside hand at the global level would alleviate them is questionable. This is evident from the failure to formulate and enforce global public policy in spite of active participation by national governments, global agencies like the UN, the World Bank, the World Trade Organization, and advocacy networks sometimes referred to as the civil society. This failure can largely be attributed to the lack of a clear understanding of the roles of the actors who precipitated those problems and whose motivations must be influenced to turn the tide.

Dynamics of Income Distribution in a Market Economy: Possibilities for Poverty Alleviation, Table 1 Developmental problems, policies implemented to address them and unintended consequences experienced

Initially perceived problems	Policies implemented	Unintended consequences
Poverty	Economic growth capital formation sectoral development technology transfer external trade	Low productivity indebtedness natural resources depletion environmental degradation continuing/increased poverty
Food shortage	Intensive agriculture land development irrigation fertilizer application use of new seeds	Land degradation depletion of water aquifers vulnerability to crop failure population growth continuing/increased vulnerability to food shortage
Social unrest	Spending on internal security and defense infrastructure limiting civil rights	Poor social services poor economic infrastructure authoritarian governance insurgence continuing/increased social unrest

Thus, development planning must adopt a problem solving approach in a mathematical sense if it is to achieve sustainable solutions. In this approach, a problem must be defined as an internal behavioral tendency and not as a snap shot of existing conditions. It may represent a set of patterns, a series of trends or a set of existing conditions that appear either to characterize a system or to be resilient to policy intervention. In other words, an end condition by itself must not be seen as a problem definition. The complex pattern of change implicit in the time paths preceding this end condition would, on the other hand, represent a problem. The solution to a recognized problem should be a solution in a mathematical sense, which is analogous to creating an understanding of the underlying causes of a delineated pattern. A development policy should then be perceived as a change in the decision rules that would change a problematic pattern to an acceptable one. Such a problem solving approach can be implemented with advantage using system dynamics modeling process that entails building and experimenting with computer models of problems, provided of course a succinct problem definition has first been created.

Appendix

Model Description

Wage rate WR is assumed to adjust over period $WRAT$ towards indicated wage rate IWR.

$$\mathrm{d}/\mathrm{d}t[WR] = (IWR - WR)/WRAT \quad (1)$$

IWR depends on the wage-bargaining position of the workers, which is determined by their opportunity cost of accepting wage-employment. It is assumed that the opportunity cost of transferring a self-employed worker to wage-work is zero when wage offered is equal to the current consumption expenditure per worker averaged over the whole workforce.

$$IWR = [(R_s{}^*(1 - SP_s) + (AS_s/LAS))/TW], \quad (2)$$

where R_s, SP_s and AS_s are, respectively, income share, saving propensity and accumulated unspent savings of the self-employed sector. LAS and TW are, respectively, life of accumulated unspent savings and total workforce. Subscripts s and f designate, respectively, self-employed and capitalist sectors.

Ownership of land and capital as well as contribution to labor are the bases for claim to income while absentee ownership is possible through leasing arrangements. Thus, R_s is computed by adding together the value of output produced by the self-employed sector VQ_s and the wage payments received by the wage-workers W_f, and subtracting from the sum the rent payments made to the absentee owners. R_f is given by adding together the value of output produced by the capitalist sector VQ_f and the rent payments it receives from the self-employed sector, and subtracting from the sum the wage-payments it makes.

$$R_s = VQ_s + WR * W_f - LR * RL - KR * RK, \quad (3)$$

$$R_f = VQ_f - WR * W_f + LR * RL + KR * RK, \quad (4)$$

where LR, RL, KR, and RK, are, respectively, land rent, rented land, capital rent, and rented capital.

KR and LR depend, respectively, on the long-term averages of the marginal revenue products of capital and land ($AMRPK$ and $AMRPL$) in the economy, and the demand for renting capital and land (RKD and RLD) as compared with the supply of rentable assets (RK and RL). The demand for renting, in turn, depends on the lack of ownership of adequate resources for productively employing the workers in the self-employed sector.

$$KR = AMRPK * f_1[RKD/RK]; \quad f_1' > 0 \quad (5)$$

$$RKD = DKE_s - KO_s. \quad (6)$$

Where DKE_s is desired capital to be employed in the self-employed sector and KO_s is capital owned by it. Land rent LR and demand for renting land RLD are determined similarly.

The saving propensity of all households in not uniform. Since capitalist households associated with the capitalist sector receive incomes which are much above subsistence, their saving propensity is stable. On the other hand, the saving propensity of the worker households depends on their need to save for supporting investment for self-employment and on how their absolute level of income compares with their inflexible consumption. Thus, SP_s in the model is determined by the utility of investment in the self-employed sector arising from a comparison of worker productivity in the sector with the wage rate in the capitalist sector, and the rent burden of this sector compared with the factor contribution to its income from land and capital.

$$SP_s = \mu * f_2[MRPW_s/WR] * f_3[(LR * RL + KR * RK)/(VQ_s - MRPW_s * W_s)], \tag{7}$$

$$SP_f = \mu, \tag{8}$$

where $f_2' > 0$, $f_3' < 0$, μ is a constant, and MRPW is marginal revenue product of workers.

AS represent the balance of unspent savings, which determine the availability of liquid cash resources for purchase of assets. AS are consumed over their life LAS whether or not any investment expenditure occurs.

$$\mathrm{d}/\mathrm{d}t[AS_i] = R_i * SP_i - AS_i/LAS - LA_i * PL - \sum_j KA_i^j * GPL \qquad i = \mathrm{s},\mathrm{f}; \quad j = \mathrm{m},\mathrm{t}, \tag{9}$$

where LA, PL, KA, and GPL are, respectively, land acquisitions, price of land, capital acquisitions, and general price level. Subscript i refers to any of the two sectors, self-employed (s) and capitalist (f), and superscript j to the type of capital, modern (m) or traditional (t).

W_f is assumed to adjust towards indicated workers IW_f given by desired workers DW_f and total workforce TW. TW is assumed to be fixed, although, relaxing this assumption does not alter the conclusions of this entry. All workers who are not wage-employed must be accommodated in self-employment. Thus W_s represents the remaining workers in the economy.

$$\mathrm{d}/\mathrm{d}t[W_f] = (IW_f - W_f)/WAT \tag{10}$$

$$IW_f = TW * f_4(DW_f/TW) \tag{11}$$

$$W_s = TW - W_f \tag{12}$$

where $1 \geq f_4 \geq 0$, and $f_4' > 0$. WAT is worker adjustment time.

The desired workers in each sector DW_i is determined by equating wage rate with the marginal revenue product of workers. A modified Cobb-Douglas type production function is used.

$$DW_i = E_i^w * VQ_i/WR, \tag{13}$$

where E_i^w is the elasticity of production of workers in a sector.

Land and capital owned by the capitalist sector (LO_f and KO_f) are allocated to commercial production (KE_f and LE_f) and renting (RK and RL) activities depending on the desired levels of these factors in each activity. Thus,

$$RK = (DRK/(DRK + DKE_f)) * KO_f \tag{14}$$

$$RL = (DRL/(DRL + DLE_f)) * LO_f \tag{15}$$

$$KE_f = KO_f - RK \tag{16}$$

$$LE_f = LO_f - RL \tag{17}$$

Capital and land employed by the self-employed sector consist of these production factors owned by them and those rented from capitalist sector.

$$KE_s = Ko_s + RK \tag{18}$$

$$LE_s = Lo_s + RL. \tag{19}$$

Desired capital and land to be employed in any sector (DKE_j and DLE_j) are determined on the basis of economic criteria.

$$\mathrm{d}/\mathrm{d}t(DKE_i)/KE_i = f_6[MRPK_i/MFCK] \tag{20}$$

$$\mathrm{d}/\mathrm{d}t(DLE_i)/LE_i = f_5[MRPL_i/MFCL], \tag{21}$$

where f_5' and $f_6' > 0$. $MRPL_i$ and $MRPK_i$ are respectively marginal revenue products of land and capital in a sector, and $MFCL$ AND $MFCK$ are respectively marginal factor costs of land and capital.

$$MRPL_i = \left(E_i^l * VQ_i/LE_i\right) \tag{22}$$

$$MRPK_i = \left(E_i^k * VQ_i/KE_i\right) \tag{23}$$

$$MFCL = PL * IR \tag{24}$$

$$MFCK = IR + (1/LK) * GPL, \tag{25}$$

where E_i^l and E_i^k are, respectively, elasticities of production of land and capital in a sector. PL is price of land, IR is exogenously defined interest rate, LK is life of capital and GPL is general price level.

Changes in the quantities of capital and land desired to be rented out (DRK and DRL) depend on their respective rents KR and LR compared with their marginal factor costs $MFCK$ and $MFCL$.

$$\mathrm{d}/\mathrm{d}t[DRK]/RK = f_7[KR/MFCK]; \quad f_7' > 0 \tag{26}$$

$$\mathrm{d}/\mathrm{d}t[DRL]/RL = f_8[LR/MFCL]; \quad f_8' > 0. \tag{27}$$

The value of output produced by each sector is given by the product of the quantity it produces Q_i and the general price level GPL.

$$VQ_i = Q_i * GPL \tag{28}$$

$$Q_i = A_i * K_i^{E^{ki}} * L_i^{E^{li}} * W_i^{E^{wi}}, \tag{29}$$

where K_i, L_i, and W_i represent capital, land and workers employed by a sector. A_i represent technology constants, which increase with the use of modern capital.

$$A_j = \mathring{A} * f_9\left[K_i^m/\left(K_i^t + K_i^m\right)\right], \tag{30}$$

where $f_9' > 0$ and $\mathring{A}$ is a scaling factor based on initial conditions of inputs and output of the production process.

Ownership is legally protected and the financial market is fragmented by households. Thus, purchase of any productive assets must be self-financed by each sector through cash payments. Land ownership LO_i of each sector changes through acquisitions LA_i from each other. Each sector bids for the available land on the basis of economic criteria, its current holdings, and the sector's liquidity.

$$LA_i = \mathrm{d}/\mathrm{d}t[LO_i] \tag{31}$$

$$LO_i = \left(DLO_i/\sum_i DLO_i\right) * TL, \tag{32}$$

where DLO_i is desired land ownership in a sector and TL is total land which is fixed,

$$DLO_i = LO_i * f_6[MRPL_i/MFCL] * f_{11}[CA_i], \tag{33}$$

where $f_{11}'[CA_i]$ is >0, and CA_i is cash adequacy of a sector.

Cash adequacy of a sector CA_i is given by the ratio of its accumulated unspent savings to the desired savings. The latter is computed by multiplying cash needed to finance investment and the traditional rate of consumption of savings in the sector by cash coverage CC.

$$CA_i = AS_i/(((AS_i/LAS) + (LA_i * PL)$$
$$+ \left(\sum_j KA_{ij} * GPL\right)\right) * CC). \quad (34)$$

Capital ownership in a sector $KO_i = KO_i^t + KO_i^m$ changes through acquisitions KA_i^j and decay. Although there is a preference for modern capital, its acquisition KA_i^m depends on the ability to accommodate the technology represented by it. Inventory availability of each type of capital KIA^j also limits its purchases.

$$d/dt[KO_i] = \sum_j KA_i^j - KO_i/LK, \quad (35)$$

$$KA_i^j = DKA_i^j * KIA^j, \quad (36)$$

$$DKA_i^m = (KO_i/LK) * f_5[MRPK_i/MFCK] * f_{11}[CA_i] * TCF_i, \quad (37)$$

$$DKA_i^t = (KO_i/LK) * f_5[MRPK_i/MFCK] * f_{11}[CA_i] * (1 - TCF_i), \quad (38)$$

where DKA_i are desired capital acquisitions, $f_{11}' \geq 0$, and LK is life of capital. TCF_i represent exogenously defined technological capability. $0 < TCF_i < 1$.

$$KIA^j = f_{12}\left[KI^j/\left(\sum_i DKA_i^j\right) * KIC\right], \quad (39)$$

where $0 \leq f_{12} \leq 1$, $f_{12}' > 0$, and KIC is capital inventory coverage

$$d/dt[KI^i] = KQ^i - \sum_i KA_i^j, \quad (40)$$

where KQ^j represent supply of capital. KQ^m is imported, while KQ^t is created within the economy by allocating a part of the capacity to its production.

$$KQ^t = \sum_i Q_i \cdot \left(\sum_i DKA_i^t/TD\right). \quad (41)$$

The price of land L is assumed to adjust towards indicated price of land IPL which is given by the economy-wide average of the marginal revenue product of land $AMRPL$, interest rate IR and the desired land ownership in each sector DLO_i

$$d/dt[PL] = (IPL - PL)/LPAT, \quad (42)$$

$$IPL = (AMRPL/IR) * f_{13}\left[\sum_i DLO_i/TL\right];$$

where $f_{13}' > 0$. (43)

General price level GPL is determined by supply and demand considerations.

$$d/dt[GPL] = GPLN * f_{14}\left[TD/\sum_i Q_i\right] \quad (44)$$

where $f_{14}' > 0$. GPLN is normal value of GPL and TD is total demand for goods and services to be produced within the economy. TD is given by adding up non-food consumption C_i, traditional capital acquisition KA_i^t and production of traditional capital for inventory, food demand FD and government spending G which is equal to taxes, if any, collected.

$$TD = \sum C_j + \sum_i DKA_i^t$$
$$+ \left(\left(KIC * \sum_i DKA_i^t - KI^j\right)/IAT\right)$$
$$+ FD + G, \quad (45)$$

Dynamics of Income Distribution in a Market Economy: Possibilities for Poverty Alleviation, Fig. 11 Behavioral relationships f_1 through f_8

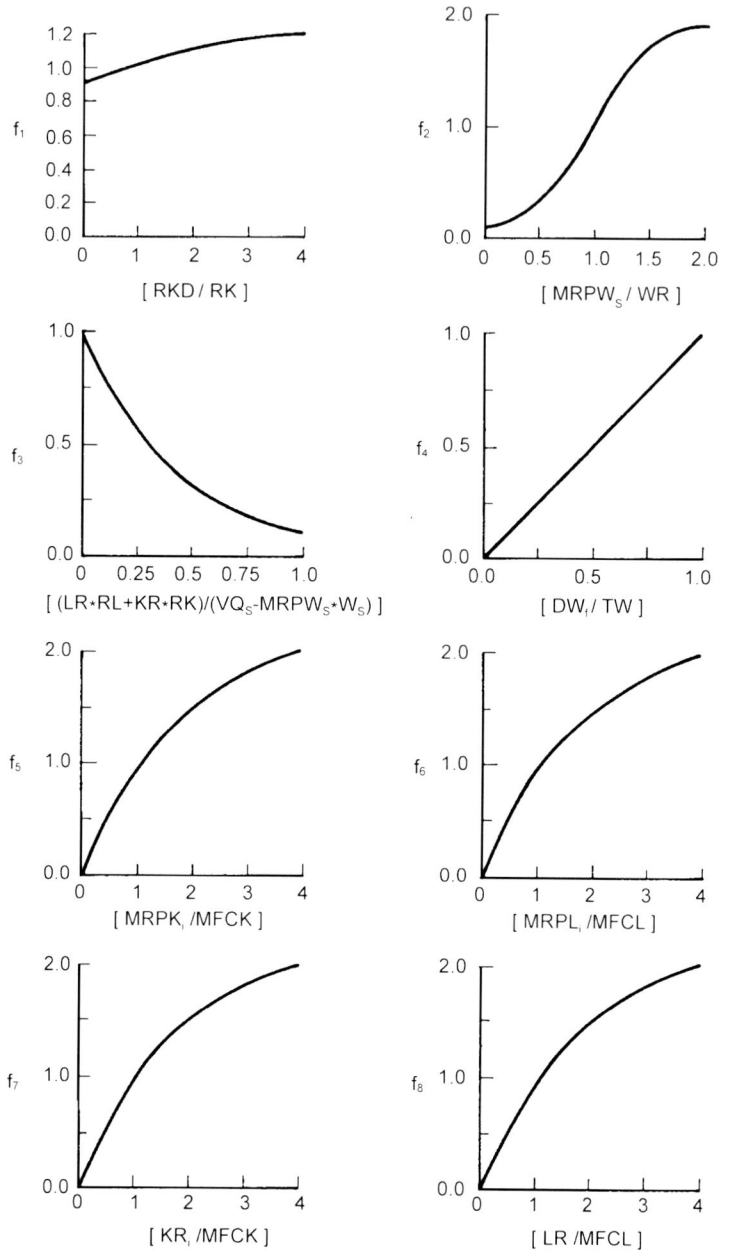

$$d/dt(C_i) = \frac{[(((R_i * (1 - SP_i) + AS_i/LAS)/GPL) * FNFC_i) - C_i]}{CAT} \qquad (46)$$

where *IAT* is inventory adjustment time, *FNFC_i* fraction non-food consumption, and *CAT* is consumption adjustment time. Food demand *FD* is given by multiplying population *P* with normal per

Dynamics of Income Distribution in a Market Economy: Possibilities for Poverty Alleviation, Fig. 12 Behavioral relationships f_9 through f_{16}

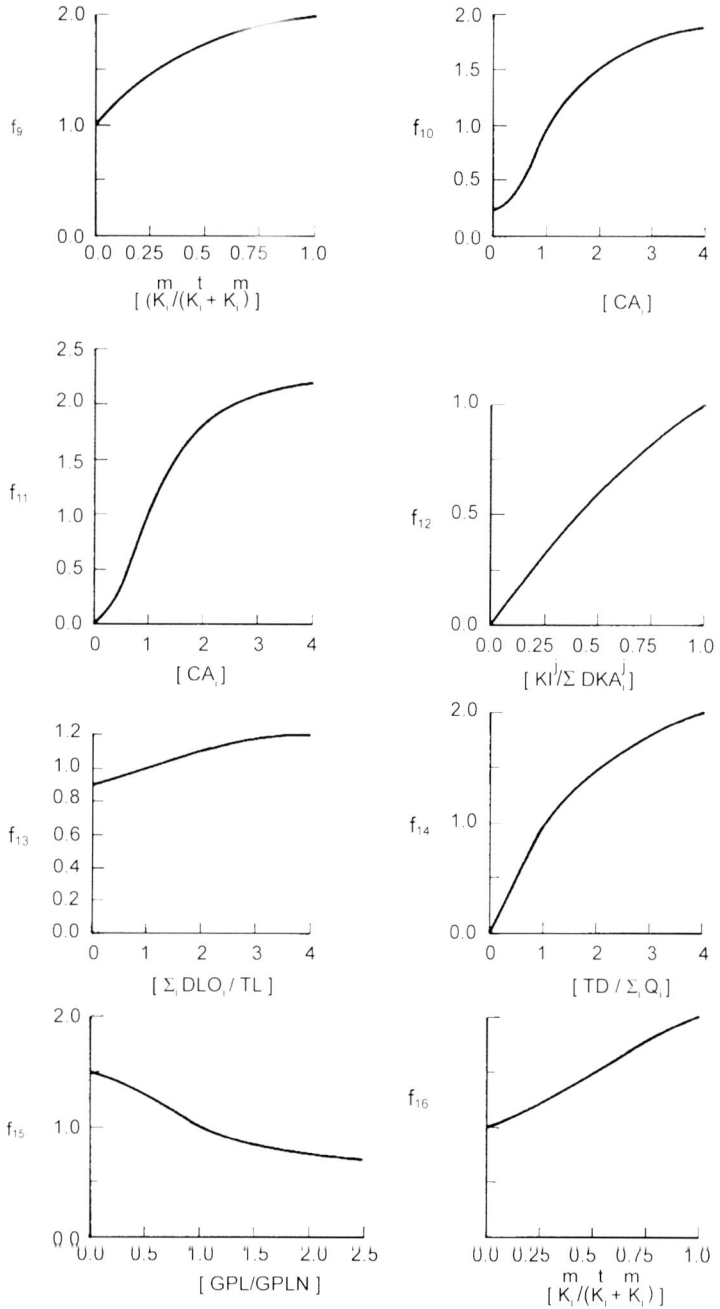

capita food demand $NFPCD$ and a function f_{15} representing a weak influence of price.

$$FD = P * NFPCD * f_{15}[GPL/GPLN], \quad (47)$$

where $f'_{15} < 0$ and P bears a fixed proportion with total workforce TW.

The elasticity of production of land E_i^l is assumed to be constant as is suggested by empirical evidence concerning agricultural economies [Strout 1978, Heady and Dillon 1961]. Elasticity of production of capital E_i^k depends on the technology of production, which is determined by the proportions of traditional and modern capital employed. Since constant returns to scale are assumed, E_i^w is given by (47).

$$E_i^k = f_{16}\left[K_i^m / \left(K_i^t + K_i^m\right)\right]; \quad f'_{16} > 0, \quad (48)$$

$$E_i^w = 1 - E_i^k - E_i^l. \quad (49)$$

Behavioral Relationships

Sixteen behavioral relationships $[f_1 \cdots f_{16}]$ have been incorporated into the model. The slope characteristics of these relationships have already been described in above equations. The graphical forms of the functions representing these relationships are shown in Figs. 11, and 12 placed below. General considerations for specifying such relationships are discussed in (Sterman 2000a).

Bibliography

Primary Literature

APO (1985) Improving productivity through macro-micro linkage. Survey and symposium report. Asian Productivity Organization, Tokyo

Applebaum E (1979) The labor market. In: Eichner A (ed) A guide to post-Keynesian economics. ME Sharpe, White Plains

Averitt RT (1968) The dual economy: the dynamics of american industry structure. Norton, New York

Bardhan PK (1973) A model of growth in a dual agrarian economy. In: Bhagwati G, Eckus R (eds) Development and planning: essays in honor of Paul Rosenstein-Roden. George Allen and Unwin Ltd, New York

Barro RJ (1997) Macroeconomics, 5th edn. MIT Press, Cambridge

Baumol WJ (1988) Is entrepreneurship always productive? J Dev Plan 18:85–94

Boeke JE (1947) Dualist economics, Oriental economics. Institute of Pacific Relations, New York

Cornwall J (1978) Growth and stability in a mature economy. Wiley, London

Denison EF (1974) Accounting for United States economic growth 1929–1969. Brookings Institution, Washington, DC

Eichner A, Kregel J (1975) An essay on post-Keynesian theory: a new paradigm in economics. J Econ Lit 13(4):1293–1314

Eisner R (1989) Divergences of measurement theory and some implications for economic policy. Am Econ Rev 79(1):1–13

Fie JC, Ranis G (1966) Agrarianism, dualism, and economic development. In: Adelman I, Thorbecke E (eds) The theory and design of economic development. Johns Hopkins Press, Baltimore

Forrester JW (1979) Macrobehavior from microstructure. In: Karmany NM, Day R (eds) Economic issues of the eighties. Johns Hopkins University Press, Baltimore

Forrester JW (1987) Lessons from system dynamics modelling. Syst Dyn Rev 3(2):136–149

Galbraith JK (1979) The nature of mass poverty. Harvard University Press, Cambridge

Gordon DM (1972) Economic theories of poverty and underemployment. DC Heath, Lexington

Graham AK (1977) Principles of the relationships between structure and behavior of dynamic systems. PhD thesis. MIT, Cambridge

Griffin K, Ghose AK (1979) Growth and impoverishment in rural areas of Asia. World Dev 7(4/5):361–384

Griffin K, Khan AR (1978) Poverty in the third world: ugly facts and fancy models. World Dev 6(3):295–304

Hicks J (1940) The valuation of the social income. Economica 7(May):163–172

Higgins B (1959) Economic development. Norton, New York

Hirshliefer J (1976) Price theory and applications. Prentice Hall, Englewood Cliffs

Kaldor N (1966) Marginal productivity and the macroeconomic theories of distribution. Rev Econ Stud 33:309–319

Kaldor N (1969) Alternative theories of distribution. In: Stiglitz J, Ozawa H (eds) Readings in modern theories of economic growth. MIT Press, Cambridge

Kalecki M (1965) Theory of economic dynamics, rev edn. Allen and Unwin, London

Kalecki M (1971) Selected essays on dynamics of capitalist economy. Cambridge Univ Press, London

Kindelberger C, Herrick B (1977) Economic development, 3rd edn. McGraw Hill, New York

Leontief W (1977) Theoretical assumptions and non-observable facts. In: Leontief W (ed) Essays in economics, vol II. ME Sharpe, White Plains

Lewis WA (1958) Economic development with unlimited supply of labor. In: Agarwala I, Singh SP (eds) The economics of underdevelopment. Oxford University Press, London

Lipton M (1977) Why poor people stay poor. Harvard University Press, Cambridge

Marglin SA (1984) Growth, distribution and prices. Harvard Univ Press, Cambridge

Marx K (1891) Capital. International Publishers, New York. (Reprinted)

McKinnon RI (1973) Money and capital in economic development. The Brookings Institution, New York

Minsky H (1975) John Maynard Keynes. Columbia University Press, New York

Mukhia H (1981) Was there feudalism in indian history. J Peasant Stud 8(3):273–310

Myrdal G (1957) Economic theory and under-developed regions. Gerald Duckworth Ltd, London

Pack SJ (1985) Reconstructing Marxian economics. Praeger, New York

Quinn JB (1985) Managing innovation: controlled chaos. Harvard Bus Rev 85(3):73–84

Ricardo D (1817) Principles of political economy and taxation, Reprint 1926. Everyman, London

Richardson GP (1991) Feedback thought in social science and systems theory. University of Pennsylvania Press, Philadelphia

Riech M, Gordon D, Edwards R (1973) A theory of labor market segmentation. Am Econ Rev 63:359–365

Roberts EB (1991) Entrepreneurs in high technology, lessons from MIT and beyond. Oxford University Press, New York

Robinson J (1955) Marx, Marshal and Keynes: three views of capitalism. Occasional paper no. 9. Delhi School of Economics, Delhi

Robinson J (1969) The theory of value reconsidered. Aust Econ Pap 8:13–19

Robinson J (1978) Contributions to modern economics. Basil Blackwell, Oxford

Robinson J (1979) Aspects of development and underdevelopment. Cambridge University Press, London

Roulet HM (1976) The Historical context of Pakistan's rural agriculture. In: Stevens RD et al (eds) Rural development in Bangladesh and Pakistan. Hawaii University Press, Honolulu

Rydenfelt S (1983) A pattern for failure. Socialist economies in crisis. Harcourt Brace Jovanavich, New York

Saeed K (1988) Wage determination, income distribution and the design of change. Behav Sci 33(3):161–186

Saeed K (1990) Government support for economic agendas in developing countries. World Dev 18(6):758–801

Saeed K (1991) Entrepreneurship and innovation in developing countries: basic stimulants, organizational factors and hygienes. In: Proceedings of academy of international business conference, National University of Singapore, Singapore

Saeed K (1992) Slicing a complex problem for system dynamics modelling. Syst Dyn Rev 8(3):251–261

Saeed K (1994a) Development planning and policy design: a system dynamics approach. Foreword by Meadows DL. Ashgate/Avebury Books, Aldershot

Saeed K (2002a) A pervasive duality in economic systems: implications for development planning. In: Systems dynamics: systemic feedback modeling for policy analysis, Encyclopedia of life support systems (EOLSS). EOLSS Publishers, Oxford

Saeed K (2005) Limits to growth in classical economics. In: 23rd international conference of system dynamics society, Boston

Saeed K, Prankprakma P (1997) Technological development in a dual economy: alternative policy levers for economic development. World Dev 25(5):695–712

Sen AK (1966) Peasants and dualism with or without surplus labor. J Polit Econ 74(5):425–450

Sen AK (1999) Development as freedom. Oxford University Press, Oxford

Simon HA (1982) Models of bounded rationality. MIT Press, Cambridge

Skinner A (ed) (1974) Adam Smith: the wealth of nations. Pelican Books, Baltimore

Solow R (1988) Growth theory and after. Am Econ Rev 78(3):307–317

Sraffa P (1960) Production of commodities by means of commodities. Cambridge University Press, Cambridge

Sterman J (2000b) Business dynamics. McGraw Hill, Irwin

Streeten P (1975) The limits of development studies, 32nd Montague Burton lecture on international relations. Leads University Press, Leeds

Takahashi Y et al (1970) Control and dynamic systems. Addison-Wesley, Reading

Weintraub S (1956) A macro-economic approach to theory of wages. Am Econ Rev 46(Dec):835–856

Books and Reviews

Atkinson G (2004) Common ground for institutional economics and system dynamics modeling. Syst Dyn Rev 20(4):275–286

Ford A (1999) Modeling the environment. Island Press, Washington DC

Forrester N (1982a) A dynamic synthesis of basic macroeconomic theory: implications for stabilization policy analysis. PhD dissertation. MIT

Forrester N (1982b) The life cycle of economic development. Pegasus Communications, Waltham

Forrester JW (1989) The system dynamics national model: macrobehavior from microstructure. In: Milling PM, Zahn EOK (eds) Computer-based management of complex systems: international system dynamics conference. Springer, Berlin

Hines J (1987) Essays in behavioral economic modeling. PhD dissertation. MIT, Cambridge

Mass NJ (1976) Economic cycles, an analysis of the underlying causes. Pegasus Communications, Waltham

Radzicki M (2003) Mr. Hamilton, Mr. Forrester and a foundation for evolutionary economics. J Econ Issues 37(1):133–173

Randers J (1980) Elements of system dynamics method. Pegasus Communications, Waltham

Richardson GP (1996) Modeling for management: simulation in support of systems thinking. In: Richardson GP (ed) The international library of management. Dartmouth Publishing Company, Aldershot

Saeed K (1980) Rural development and income distribution, the case of Pakistan. PhD dissertation. MIT, Cambridge

Saeed K (1994b) Development planning and policy design: a system dynamics approach. Ashgate/Avebury Books, Aldershot

Saeed K (1998) Towards sustainable development, Essays on system analysis of national policy, 2nd edn. Ashgate Publishing Company, Aldershot

Saeed K (2002b) System dynamics: a learning and problem solving approach to development policy. Glob Bus Econ Rev 4(1):81–105

Saeed K (2003a) Articulating developmental problems for policy intervention: a system dynamics modeling approach. Simul Gaming 34(3):409–436

Saeed K (2003b) Land use and food security – the green revolution and beyond. In: Najam A (ed) Environment, development and human security, perspectives from South Asia. University Press of America, Lanham

Saeed K (2004) Designing an environmental mitigation banking institution for linking the size of economic activity to environmental capacity. J Econ Issues 38(4):909–937

Sterman JD (2000a) Business dynamics. systems thinking and modeling for a complex world. Irwin McGraw Hill, Boston

Wolstenholme E (1990) System enquiry, a system dynamics approach. Wiley, Chichester

CPI Antony Rowe
Eastbourne, UK
March 25, 2020